PLANAR PROCESSING PRIMER

PLANAR PROCESSING PRIMER

GEORGE E. ANNER

Professor Emeritus of Electrical Engineering
University of Illinois
Champaign-Urbana

VNR Van Nostrand Reinhold
New York

Library of Congress Catalog Number 90-34534
ISBN 0-442-20657-7

Printed in the United States of America

Van Nostrand Reinhold
115 Fifth Avenue
New York, New York 10003

Van Nostrand Reinhold International Company Limited
11 New Fetter Lane
London EC4P 4EE, England

Van Nostrand Reinhold
102 Todds Street
South Melbourne, Victoria 3205, Australia

Nelson Canada
1120 Birchmount Road
Scarborough, Ontario M1K 5G4, Canada

16 15 14 13 12 11 10 9 8 7 6 5 4 3 2 1

Library of Congress Cataloging-in-Publication Data

Anner, George E.
Planar processing primer/George E. Anner
p. cm.
Includes bibliographical references.
ISBN 0-442-20657-7
1. Compound semiconductors—Design and construction—Data processing. 2. Silicon compilers. I. Title.
TK7871 99 C65A55 1990
621.381'62—dc20 90-34534
CIP

Preface

Planar Processing Primer is based on lecture notes for a silicon planar processing lecture/lab course offered at the University of Illinois-UC for over fifteen years. Directed primarily to electrical engineering upperclassmen and graduate students, the material also has been used successfully by graduate students in physics and ceramic and metallurgical engineering. It is suitable for self-study by engineers trained in other disciplines who are beginning work in the semiconductor fields, and it can make circuit design engineers aware of the processing limitations under which they must work.

The text describes and explains, at an introductory level, the principal processing steps used to convert raw silicon into a semiconductor device or integrated circuit. First-order models are used for theoretical treatments (e.g., of diffusion and ion implantation), with reference made to more advanced treatments, to computer programs such as SUPREM that include higher order effects, and to interactions among sequential processes. In Chapters 8, 9, and 10, the application of silicon processes to compound semiconductors is discussed briefly.

Over the past several years, the size of transistors has decreased markedly, allowing more transistors per chip unit area, and chip size has increased. Extrapolating these trends forecasts microprocessors by the year 2000 that have 50 million transistors, with 0.1-micrometer active gate lengths, on a one-inch-square chip.[1] Still, the basic processing techniques remain the same; these are the focus of this text, although many updates are included in detail to show how decrease in transistor size has become possible.

---

[1] P. P. Gelsinger et al., "Microprocessors circa 2000," *IEEE Spectrum*, **26**, (10), 43–47, Oct. 1989.

References have been chosen from journals readily available in both university and semiconductor company libraries. Commercial jargon is defined and used so the student will be aware of it.

To give the student a feel for typical orders of magnitude, many illustrative examples are provided. Design curves, data, and sample computer programs for the examples and problems are collected together in Appendix F for convenience.

I am indebted to Prof. Henry Guckel of the University of Wisconsin-Madison for a review of the entire manuscript, and to several others who have made helpful suggestions, among them, Nick Holonyak, Jr., Guenter Bruckmann, Doug Carlyle, John S. Hughes, John Leap, and R. M. Starnes. Ms. Marjorie Spencer and Ms. Elise Oranges have smoothed over many difficulties and have been unending sources of help. My wife, Cynthia, is very forbearing and supportive. I am grateful for her help, which has made this book possible.

Urbana, IL George E. Anner

Contents

Chapter 1

Planar Processing and Basic Devices

Semiconductor devices and integrated circuits (ICs) consist of semiconductors, conductors, and insulators or dielectrics. Their manufacture utilizes a number of processes to arrange and modify layers of these materials to form many electronic devices. In ICs these devices are interconnected to form complete circuits capable of performing one or more desired functions. In recent years, however, the sizes of the devices have decreased, and their number on an IC chip and their functions have increased, by so much that the trend now is to replace the term *integrated circuit* by *microcircuit*.

Introduction

The three types of materials mentioned above are confined to several thin layers just below and above the upper surface of the silicon substrate in which they are fabricated. The layer thickness is typically 10 μm or less, while the substrate will be at least 250 μm thick,[1] so the devices lie essentially in a plane at the upper surface; hence they are considered to be *planar* devices. This also applies to *discrete devices* (devices that are packaged singly).

The processing steps used to make these devices and their interconnections are performed almost entirely on the uppermost surface and comprise what was commonly called *silicon planar processing.* A few steps are required on the wafer bottom side to form an electrical contact there.

Recently, the term *planar processing* has been extended to include intentional steps to maintain surface planarity. As more layers of interconnects and dielectrics are added atop the upper surface, the uppermost or last-deposited layer is nonuniform, departing from planarity (see Figure 10–3). Because of the

[1] By way of calibration, the average human hair diameter is about 3 mil = 3×10^{-3} in. $\approx 75\ \mu$m (Appendix G). Thus 10 μm (micrometers) is roughly 15% of a human hair diameter.

limited depth of focus of the projection lens (Appendix E), this condition adversely affects the sharpness of the image projected onto the surface during photolithography (Section 1.8.5). Hence, with many layers now being used in microcircuit manufacture, a concerted effort is made to maintain planarity as layer upon layer is added. An example of this is given in Section 10.7, where a layer of polyimide is used to smooth out a layer surface by filling in the irregularities. Maintaining planarity also helps in getting good step coverage where a thin conductor layer lies on an irregular surface.

These processes with some modifications are carried over to compound semiconductors; they are basic to essentially all of the semiconductor industry.

In this book we shall be concerned with the basic processes used in fabricating silicon planar devices [1]. To see the need for and the relations among the processes, we first consider the structures of several common silicon planar devices, which will be illustrated by two sketches: one showing an x-ray top view as seen from above the substrate upper surface, and another showing the cross section normal to the upper-surface plane. In these figures the semiconductor regions may be designated as p- or n-type, possibly with typical doping concentrations given numerically. In some cases the cross section will be augmented by another sketch, the *doping profile*, which shows how the doping concentration varies with distance into the wafer from the upper surface (x-direction). A study of these device sketches will show what fabricating processes are needed. The processes are covered in the chapters that follow. Since this is a primer, we study the basics; references are cited for details.

1.1 Basic Devices

Over the years, improved design and the size reduction of semiconductor devices and ICs make it difficult to see where in the silicon one device ends and another begins. Devices have been merged, and they have been stacked to overlap or to lie one above another. To begin with a study of these structures makes for unnecessary complications, so we shall consider seven basic planar devices, initially assuming that they all will be fabricated in an n-type substrate. These devices are shown in basic form in Figure 1–1(a): (A) MOSC or metal-oxide-semiconductor capacitor, (B) Schottky diode, (C) p–n diode, (D) resistor, (E) MOST or metal-oxide-semiconductor transistor (p-channel), (F) BJT or bipolar junction transistor (n–p–n), and (G) ohmic contact to the substrate.

Figure 1–1(a) shows all seven devices in a single substrate, a necessary condition for ICs. Vertical distances have been exaggerated greatly for clarity. Doped regions that extend to the same depth below the wafer surface are *codoped*—i.e., they are doped simultaneously and hence have the same doping levels and profiles. The contact to the substrate bottom surface is *ohmic*: current can flow equally well in both directions across the junction, at least

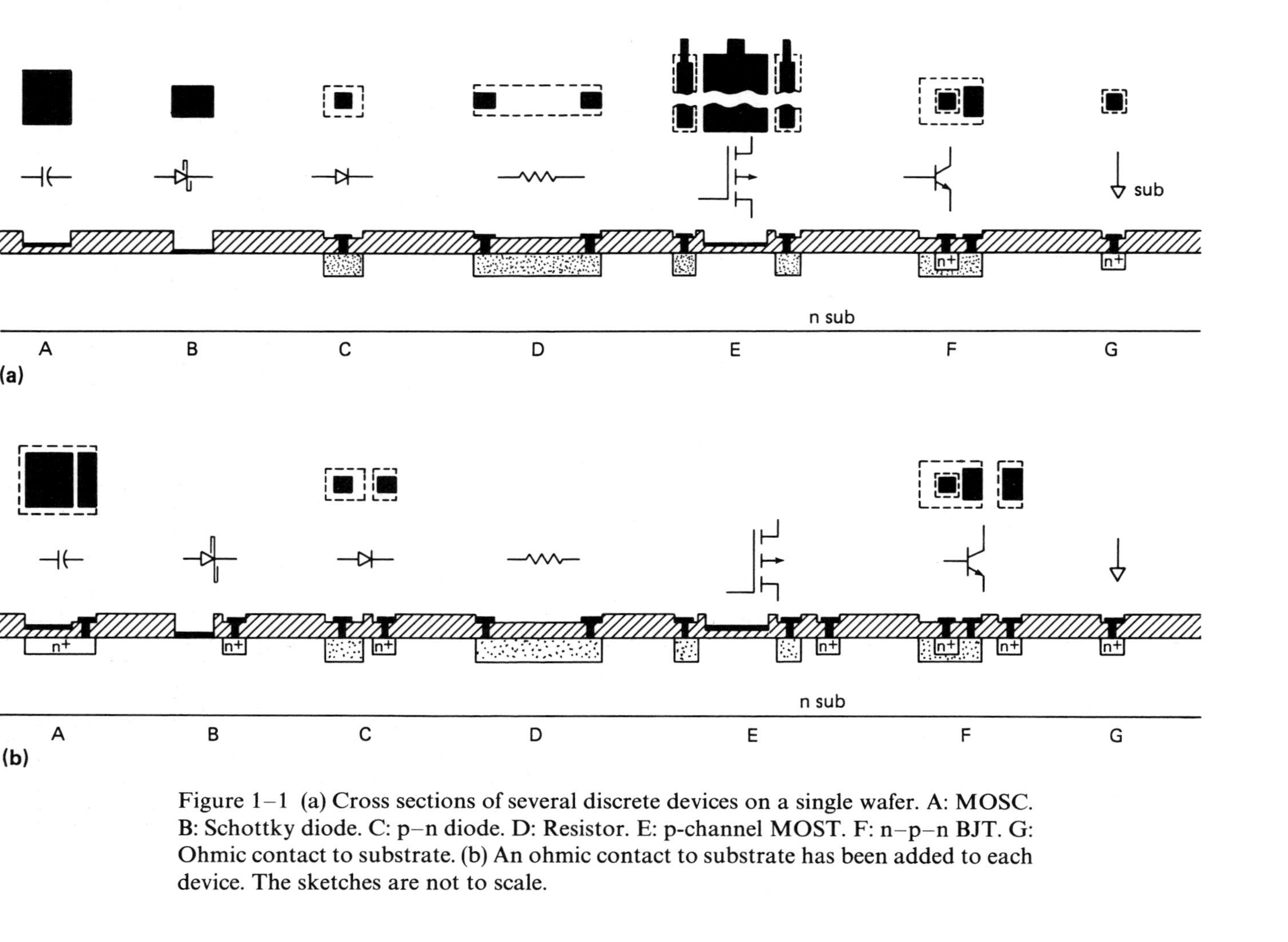

Figure 1–1 (a) Cross sections of several discrete devices on a single wafer. A: MOSC. B: Schottky diode. C: p–n diode. D: Resistor. E: p-channel MOST. F: n–p–n BJT. G: Ohmic contact to substrate. (b) An ohmic contact to substrate has been added to each device. The sketches are not to scale.

for the usual small applied voltages. The general materials key for the diagram, and also for much of the book, is

1. Unhatched areas are silicon. The type is indicated by a p or an n, or a p region may be dotted and n clear.
2. Heavy dark regions are metal—usually aluminum unless stated otherwise.
3. Regions hatched from lower left to upper right are silicon dioxide (SiO_2).
4. Other regions, such as photoresist (PR) or silicon nitride (Si_3N_4) will be clearly identified.

The MOSC and the Schottky diode, shown at (A) and (B), have only one doped region, the substrate, and hence are the easiest to fabricate. It is important to note that (B) will function as a Schottky unit (i.e., it will rectify) only if the doping concentration of the n-type silicon in the region where it is contacted by the aluminum is less than 10^{17} cm^{-3}. A much lower value, say 10^{14} to 10^{15} cm^{-3} will insure the desired rectifying behavior [2].

Devices (C) through (E) in Figure 1–1(a) have both p- and n-type regions to form p–n junctions; thus they require an additional doping process during fabrication to create the p-regions. Diffusion or ion implantation (I^2) are processes that can provide doping in specified areas. The additional doping process makes these devices slightly more complicated to fabricate, but we shall see later that if the four devices (C) through (F) are all produced in a single substrate, the four p-regions may be doped simultaneously. This fact greatly lowers the manufacturing costs of semiconductor devices. *We process entire wafers; not individual devices.*

The diode, resistor, and MOST present no additional problems. The BJT at (F) in Figure 1–1(a), however, has three doped regions and hence requires an additional emitter n^+ diffusion (or ion implant) within the p-type base region. The collector contact for this version of the device is provided by the ohmic contact on the wafer bottom.

Locating the collector contact on the bottom has some interesting implications. First, all the BJTs in the substrate wafer would have a common collector terminal, so they can function only as emitter followers; they cannot be interconnected to function in common emitter mode for IC applications. Therefore, they will be sawed apart and mounted separately as discrete devices.

Second, to keep the resistance between the internal collector and bottom contact small, the final substrate thickness should be small. With the low doping typical of collectors, collector resistivity is high, so the resistor length should be minimized by thinning the wafer. A lower limit must be preserved so the wafer will not break during handling on the production line.

Wafers entering the fabrication facility are often 25 or more mils thick to minimize this loss to breakage. Rear surface grinding or *back-lapping* near the end of production removes bottom material to make the wafer thinner. Back-lapping also removes unwanted diffused layers on the bottom surface that were acquired during processing and aids in making a proper ohmic contact on the wafer bottom surface.[2]

A top-surface ohmic contact is shown at (G) in Figure 1–1(a). Physically it resembles the Schottky diode in having Al in contact with nSi, but the Si has a much higher doping level than the substrate, and electrically the diode is bilateral rather than rectifying. The behavior difference is determined by the doping level and proper processing of the contact. These facts are important when Si contacts Al. At (G) the high doping level is obtained by diffusing or implanting an n^+ layer into the lightly doped n-type substrate. The same explanation applies to the Al/n^+Si emitter in the BJT shown at (F). The base p-region also has been diffused and its profile shows a high p concentration at the surface, so the contact is ohmic.

The thicknesses and concentration levels of the several layers in the Si are all quantities that may be controlled by the process engineer when he sets up the processing schedules. The quantities are chosen to meet device specifications determined by the design engineer.

The top views, or horizontal layouts, of the basic devices also are shown in Figure 1–1. The shapes are quite arbitrary. For example, the square shape of the BJT in Figure 1–1(a) could be a rectangle, as in Figure 1–1(b), a circle, or even a star. Choice of the actual shape lies in the province of the device or design engineer. Two of the factors that influence the choice are rated collector current and β falloff at high values of collector current. Realizing this shape in fabricating the devices is accomplished by using proper photomasks and photolithography, which are discussed in Section 1.8.

We next consider another aspect of fabrication: how the several devices are isolated electrically from each other for use in ICs, even though they are in the same substrate.

1.2 Device Isolation

The lack of isolation due to a common connection among the several units of Figure 1–1 becomes quite clear in Figure 1–2, where the physical devices have been replaced by their usual circuit symbols. Notice that a second n–p–n BJT

---

[2] In commercial practice gold, rather than aluminum, is used as the bottom contact metal. Aluminum is chosen here to illustrate the effect of doping on the behavior of the contact (i.e., whether it is ohmic or rectifying). Also Al is used quite often in teaching laboratories to reduce costs. Contacts of Al/Si are *sintered* (heat-treated) at $\sim 525°C$. The Au/Si contact is discussed in Section 1.8.8.

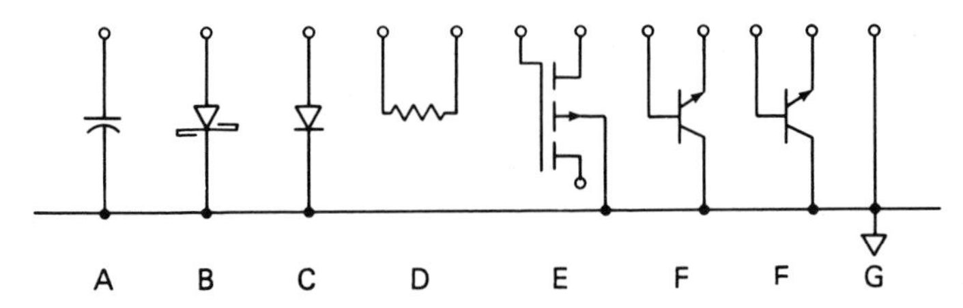

Figure 1–2. Circuit diagram symbols for the devices of Figure 1–1. All the devices, but the resistor, share the substrate as a common lead. An extra BJT has been added at F′.

has been added at (F′) to focus attention on the transistor problem. Since the substrate is the collector for both BJTs, these cannot be used with complete freedom of interconnection in an IC; therefore, isolation is required.

The diffused p-region resistor at (D) in both Figures 1–1 and 1–2 presents a useful concept—the resistor and the substrate form a p–n junction. If both ends of the resistor are always kept negative wrt (with respect to) the substrate, the junction is reverse biased and no direct current can flow between the p- and n-regions. Thus, in a d-c sense the resistor is isolated from the other elements. It is easy to argue that if two or more p-type resistors, each with its own contacts, are diffused into a common substrate, they also will be isolated from each other.

The resistors teach us a method for isolating the other basic devices, namely by using reverse biased p–n junctions. This method is referred to as *junction isolation* or simply JI, and is brought about as shown in Figure 1–3, where each device is in an n-type *tub* surrounded by pSi. The structure can be described as follows. The substrate proper is p-type with an acceptor concentration in the range of $N_a \approx 10^{17} - 10^{18}\ \mathrm{cm}^{-3}$. An *epitaxial layer* of n-type silicon with a donor density of $N_d \approx 10^{15} - 10^{16}\ \mathrm{cm}^{-3}$ has been grown on the upper surface. The term *epitaxial* or simply *epi* implies that the layer is identical to the substrate in crystalline structure and orientation, so crystallographically it is an extension of the substrate. The type and level of doping are different, however. The growth of epi layers is covered in Chapter 9.

The tubs are formed by introducing p-type dopant by diffusion in regions between the desired tub regions to convert the epi to p^+-type. This *isolation diffusion* is deep enough to extend from the upper wafer surface down to the underlying p substrate so that every n tub is completely surrounded on four sides and bottom by a p-type isolation region. Then if all the tubs are reverse biased relative to the substrate, they are d-c isolated from the substrate and from each other by junction isolation. Note that any two tubs are separated by back-to-back p–n junctions, so independent of applied biases, no dc will

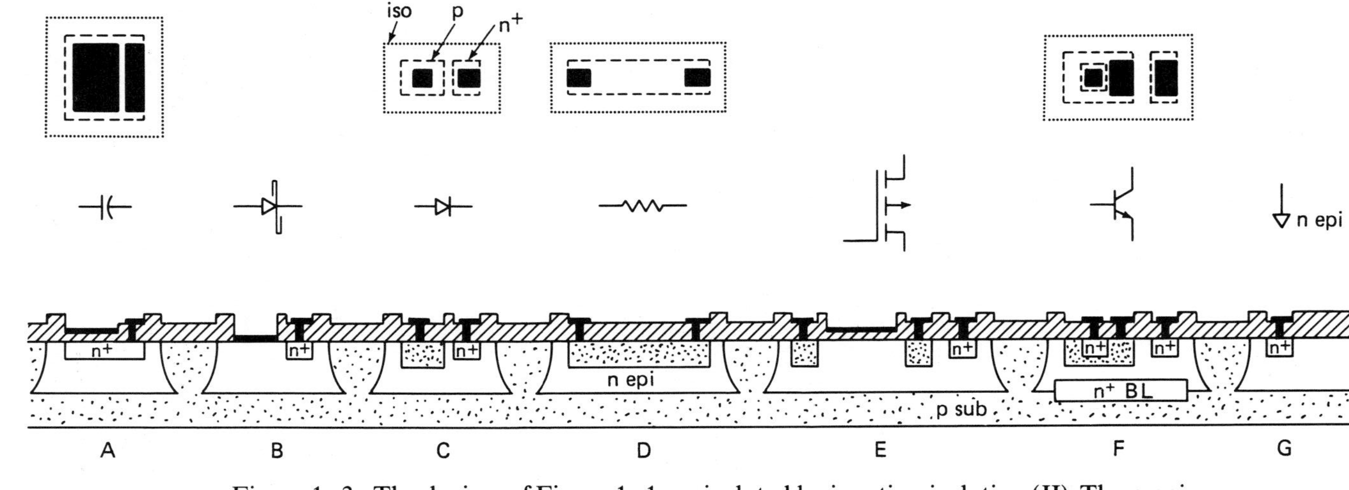

Figure 1–3. The devices of Figure 1–1 are isolated by junction isolation (JI). The n-epi layer on the p-substrate is formed into isolated tubs by deep p^+ isolation regions diffused through the epi layer. The sketches are not to scale.

flow between them; either both junctions will be biased at zero, or one will be reverse biased, thereby providing the no-current-flow isolation condition. When the IC external connections are made, the p substrate will be connected to the most negative voltage. This insures that no n tub can go into forward bias relative to the substrate and lose junction isolation. Another type of isolation, *dielectric isolation* (DI), is discussed in Section 5.9.1, but JI is more common.

Figure 1–3 shows that within an individual n epi tub the devices are fabricated much as they were in the single n substrate in Figure 1–1, but with a few exceptions. First, the n tubs cannot be accessed on the bottom; therefore, top-surface ohmic contacts are provided for access to the transistor collectors and other elements. It is implied by the relative depths of the layers in Figure 1–3 that the n^+ ohmic contact diffused regions are codiffused with the BJT emitters and do not require a separate doping process. Also, in an IC, all interconnections among the various devices are made on the upper surface. This condition also requires epi connections on top. Secondly, certain of the devices in Figure 1–3 have regions doped to n^+ levels that extend into both the epi tubs and the substrate. These regions lie buried under devices and hence are called *buried layers* (BLs). In general, they serve to prevent latchup to the substrate, and in the case of BJTs provide a lowering of the series collector resistance without seriously lowering the value of common-emitter breakdown voltage BV_{CEO}. Buried layers may be diffused into the substrate before the epi layer is deposited or ion implanted *through* the epi, as discussed in Chapter 8. Isolated devices require more processing steps than discrete devices but make ICs possible.

1.3 Interconnects

The interconnections, *interconnects*, or simply *connects* among the several isolated devices on an IC chip serve the same function as the wiring in a conventional circuit. In the early days of ICs, when the number of devices was small, all interconnects could lie in a single layer above the substrate. As integrated circuit complexity increased, more interconnect layers were required—to allow *crossovers*, where one "wire" passes over another without electrical contact. Three types of crossovers are illustrated in Figure 1–4.

At (A), use is made of an existing diffused (or implanted) resistor as one path. A conductor passing over this and insulated by the intervening SiO_2 layer provides the second path. There is some a-c coupling, however, due to the capacitance between the conductor and the p-type resistor body. This capacitance is small. If the SiO_2 layer is 1000 Å thick, which is less than usual except for MOS devices, it will be in the order of 0.2 pF per mil^2 of area. Thicker oxides will result in even smaller *specific capacitance* (capacitance per unit area) values.

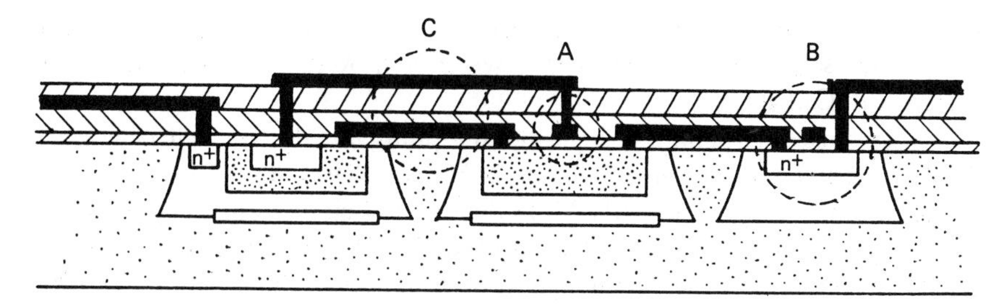

Figure 1–4. Crossovers. (A) A metal connect runs over a diffused resistor, isolated by an oxide layer. (B) A "via" where a metal connect runs over an n^+ diffused region, isolated by an oxide layer. (C) Two metal (or polySi) connects lie at different levels separated by insulating layers.

If no resistor lies where a crossover is required, a very low value resistor may be codiffused or implanted with the BJT n^+ emitters or MOST sources and drains at the desired spot. Again, the second path crosses over this resistor, isolated by an oxide layer as shown at (B) in Figure 1–4. This configuration often is called a *via* since it resembles a viaduct or overpass.

The third crossover type is shown at (C) in Figure 1–4. It is a multilayer structure made up of layers of oxide, conductor, another oxide, another insulating layer of SiO_2 (or silicon nitride, Si_3N_4), another conductor, and so on. To allow connection between different conducting layers, holes, or *windows*, properly located in the insulating layers are required. These windows are produced by *photolithography*, as discussed in Section 1.8.

The first insulating oxide layer in contact with the silicon can be grown by the process of *thermal oxidation* (Section 1.5.1), but the higher levels of oxide or nitride must be deposited by *chemical vapor deposition*, CVD (Section 1.5.2).[3] Thermal oxidation and CVD are covered in greater detail in Chapters 5 and 9, respectively.

Aluminum, either pure or alloyed with small amounts of Si and/or copper (Cu), is the metal most commonly used for contacts and interconnects, because it is easy to deposit by evaporation in a vacuum (Section 1.8.6), it adheres well to SiO_2, and it is ductile. Its ductility is very important because the thermal expansion coefficients of Al and Si do not match well, but the softness of the metal minimizes metal peel-off when large temperature changes are encountered, whether in fabrication, storage, or operation.

Aluminum has a low melting point (MP) of only 660°C, but Al in contact with Si forms an *eutectic* (Section 4.8) that melts at 577°C. Therefore, any processing after Al deposition must stay below this temperature. Those pro-

---

[3] If the insulating organic polymer *polyimide* is used, it may be spun on like photoresist, as described in Section 1.8.3.

cesses that require higher temperatures must precede Al deposition, and temperatures must *never* reach 1410°C, which is the MP of Si. Processing must not melt the substrate!

Consider two *refractory*, or high-melting-point, metals that have been used for ICs. Molybdenum (Mo) melts at 2625°C and has served for top-layer interconnects and gates in early MOS devices and circuits. It is deposited by *sputtering* (Chapter 12), a nonthermal process. Tungsten (W) melts at about 3400°C and matches the thermal expansion of SiO_2 well. It is deposited by sputtering or CVD. Tungsten was used in multiple layer structures as early as 1981. In one example a comparatively thin (4000-Å) layer was sputtered, while a much thicker layer of 18,000 Å was deposited by CVD. Chemical vapor deposition often proceeds much faster than either sputtering or vacuum evaporation, and hence is an attractive method for depositing thick layers. Titanium (Ti) is another refractory metal that has been used for ICs.

Highly doped polycrystalline silicon (*polySi* or *poly*) is also used for interconnects and MOST gates. It has higher resistivity than metals, but with adequate doping is quite satisfactory for many applications. Poly usually is deposited by low-pressure chemical vapor deposition, or LPCVD (Chapter 9).

The metal or polySi connect lines can be very small indeed, being less than 1 μm wide in modern *submicron* technology. Because of this small dimension, huge lengths of connector runs on a multi-interconnect-layer IC chip are possible. For example, an International Business Machines (IBM) chip 4.6 mm × 4.6 mm in size and having three interconnect layers had 2.2 m of interdevice connections!

Connect lines of such small width cannot be deposited directly. Normally, the interconnect material is deposited over the entire insulator surface and subsequently delineated into the desired interconnect paths of proper width by etching away unwanted regions. This is accomplished by photolithography, as described in Section 1.8 and Chaper 11.

1.4 Basic Processes

The previous sections show that the fabrication of devices and ICs requires a number of processes. These now will be overviewed briefly to show how they fit together to make silicon planar fabrication possible. They are considered in detail in later chapters.

1.5 Insulating Layer Fabrication

Figures 1–3 and 1–4 show that SiO_2 layers are needed. If they are to be on silicon, they may be produced by thermally oxidizing the underlying Si layer. Where silicon to combine with O_2 to form the new layer is lacking, chemical vapor deposition (CVD) is used. In this process both components of the oxide, namely Si and O_2, are furnished from external sources. Silicon nitride (Si_3N_4) invariably is deposited by CVD. Silicon dioxide and nitride also may be sputtered.

1.5.1 Thermal Oxidation of Silicon

Thermal oxidation is covered in Chapter 5 but we consider a short overview here. It is a natural oxidation process that takes place when silicon is exposed to oxygen:

$$Si\downarrow + O_2\uparrow \rightarrow SiO_2\downarrow \qquad (1.5\text{-}1)$$

The reaction is temperature dependent, with the oxide growth rate increasing very rapidly with temperature. In silicon technology, oxidation is carried out in a furnace with the Si wafers held in the 900 to 1300°C range. Typically, if the process is maintained for 70 min at 1200°C, roughly 2000 Å ($=0.2\ \mu m$) of oxide will be produced. Faster growth rates can be obtained if water vapor (H_2O) is added to the O_2.

Figure 1–5 shows in schematic form a setup used for both oxidation and diffusion. Three Si wafers are shown in the furnace tube to illustrate the idea of *batch processing*, whereby a process is performed simultaneously on a batch of 50 or more wafers to reduce manufacturing costs. The *carrier* gas indicated in the figure would be an inert gas, such as nitrogen (N_2) or argon (Ar), that does not enter into the chemical reaction. Mixed with this carrier is the X-component shown in the figure. This would be oxygen and/or water vapor for oxidation.

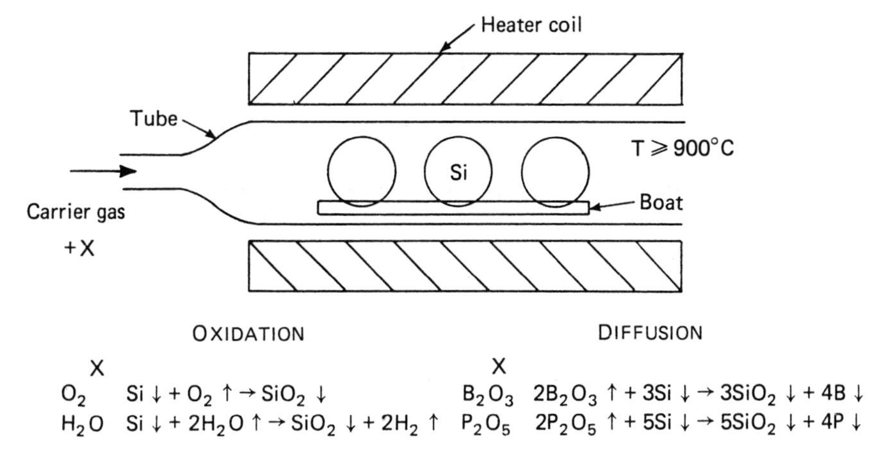

Figure 1–5. Schematic diagram of an oxidation/diffusion furnace. Reactions are indicated for both processes.

The silicon term on the left-hand side (LHS) of Eq. (1.5-1) is furnished by the substrate being oxidized; hence oxidation reduces the Si thickness. This thickness change is seldom shown on device cross sections because it is small wrt the wafer thickness and complicates the diagram. The volume ratio of oxide grown to silicon consumed is known as the *Pilling-Bedworth ratio* and has the numerical value of 2.22 for well-annealed wafers.

Oxide thickness determines the color of light reflected from oxidized silicon. Appendix F.5.8 gives color and thickness data for the sample viewed under white light. The color may be used for estimating oxide thickness, and color differences due to differences in oxide layer thicknesses are important in mask aligning (Section 1.8.5).

1.5.2 Chemical Vapor Deposition

In chemical vapor deposition (CVD) both components of the deposited material (e.g., Si and O_2 for SiO_2, and Si with N_2 for Si_3N_4) must be furnished from external sources. The Si and N_2 gas-phase compounds silane (NH_4) and ammonia (NH_3) are used along with O_2. These are made to react at the substrate surface to give the new compound insulator. For example, the reaction for SiO_2 is

$$SiH_4\uparrow + O_2\uparrow \rightarrow SiO_2\downarrow + 2\,H_2\uparrow \tag{1.5-2}$$

The reaction byproduct, hydrogen, flows out of the CVD reactor as a gaseous effluent and is burned or *flamed off*. This is done to prevent hydrogen from accumulating in the exhaust system, where it could create a hazard.

The CVD reaction for SiO_2 runs at less than 500°C, much lower than for thermal oxidation of silicon, and usually is not carried out in a furnace as shown in Figure 1–5. The reason is that CVD tends to deposit the reaction product on the hottest surfaces in the system. This would be the furnace walls, since they are closest to the furnace heating coils. Therefore, the substrates are made hotter than the walls by using radiant or inductive heating methods, as shown in Figure 9–2(a) and (c). Details are given in Chapter 9.

Silicon nitride (Si_3N_4) also is deposited by LPCVD, the usual reaction being of the oxidation–reduction type:

$$3\,SiH_4\uparrow + 4\,NH_3\uparrow \rightarrow Si_3N_4\downarrow + 12\,H_2\uparrow \tag{1.5-3}$$

The typical reaction temperature is around 850°C.

1.6 Silicon Layer Deposition

Junction isolation requires that an epitaxial layer of n-type Si be deposited on the upper surface of the pSi substrate. This is done by a CVD process that uses *pyrolysis* (chemical decomposition caused by heat) of silane:

$$SiH_4\uparrow \rightarrow Si\downarrow + 2\,H_2\uparrow \qquad (1.6\text{-}1)$$

To make a doped layer, a dopant source gas, such as PH_3 (phosphine) for n-type or B_2H_6 (diborane) for p-type, is fed to the reactor with the silane. The dopant enters into the overall chemical reaction. For example, for an n-type layer

$$x\,SiH_4\uparrow + 2\,PH_3\uparrow \rightarrow (x\,Si + 2\,P)\downarrow + (2x + 3)H_2\uparrow \qquad (1.6\text{-}2)$$

The Si and P components on the RHS of the equation give the doped solid layer. If, as in Eq. (1.6-2), we assume that the Si/P ratios are the same in the deposited solid layer and the gas, it is apparent that the doping level of the Si layer may be controlled through the factor x on the silane supplied to the reactor. This is done by adjusting the relative flow rates of the silane and phosphine. Further details are given in Chapter 9 and Appendix D.

If the CVD process is controlled properly at a high enough temperature, with deposition taking place onto the surface of a *single-crystal* substrate of proper orientation, the deposited layer will be *epitaxial*. As stated earlier, this means that the crystalline structure and orientation of the substrate is propagated into the growing layer so that they are crystallographically the same.

Doped polySi layers also are shown in Figure 1–4, but notice that these lie on layers of either SiO_2 or Si_3N_4 rather than on single-crystal silicon. Even though they are laid down by the same CVD reaction of Eq. (1.6-2), they cannot be epitaxial—there is no single-crystal substrate structure to initiate the epitaxial growth. As a result, the deposited layer will be polySi or, very rarely, amorphous silicon (aSi).

1.7 Selectively Doping Silicon

Figures 1–1 and 1–3 both show selectively doped regions in the silicon substrate or epi layer. The term *selective doping* implies that the type of doping (i.e., p- or n-type), and the regions of the silicon that have been so doped, are under control. Reference to the top views in the figures will make this more clear. For example, when the resistor is made in n-type material, a p-region must be formed, and confined to the proper region in the n material. Two ideas are involved here: doping and lateral confinement of the region being doped. We consider these ideas separately.

Doping of layers *in situ* requires that dopant atoms, either donors or acceptors, be introduced from an external source into the existing Si layer. Furthermore, if the dopant atoms are to affect the electrical properties of the Si properly (i.e., if they are to become *electronically active*) they must eventually locate on crystal lattice sites, substituting for some of the Si atoms. For this reason we need dopants that are *substitutional* in nature. Boron and phosphorus are typical substitutional dopants in Si; gold is not (Section 6.1.1).

To force external atoms into an existing crystal lattice requires energy. Two doping processes are commonly used in planar technology. In *diffusion* the supplied energy is thermal; in *ion implantation* (I^2) it is kinetic, but a subsequent heat treatment (*anneal*) is needed for the dopant atoms to land on substitutional sites.

1.7.1 Diffusion

Because of random thermal agitation, atoms can move about in a host material. If they are not distributed uniformly, they move from regions of high to regions of low concentration. This is the natural process of diffusion. The process speeds up at high temperatures; for diffusion in silicon the temperature range is usually 950 to 1250°C.

If dopant atoms are introduced into a substrate (or host material) from an external source, and if their concentration is greatest at the surface, they will diffuse into the host and cause doping. Diffusion is terminated by lowering the temperature to room value, although diffusion in Si is insignificant at $T \leq 800°C$. Details of the process are covered in Chapter 6.

Figure 1–5 schematically shows the physical setup used for the *predeposition* or *predep* diffusion, in which dopants are introduced into the substrate wafer. While the setup is identical to that for thermal oxidation, different gases are used. Nitrogen or argon still may serve as the inert carrier, but gas-phase dopant sources such as B_2O_3 (boron trioxide) or P_2O_5 (phosphorus pentoxide) may be used as the X-component for boron or phosphorus doping, respectively. (Note that gas-phase B and P cannot incorporate into Si; hence the oxides are used.) A number of ways for introducing these gases into the furnace tube are discussed in Chapter 6.

The reactions for the two cases are shown in the figure. Note that B_2O_3 and Si react at high temperature to form SiO_2 in which solid boron (B) is incorporated. This combination is a *borosilicate glass* that forms on the wafer surface, and is the source for the B atoms that diffuse into the Si.

On the other hand, if P_2O_5 is used, a *phosphosilicate glass* deposits on the wafer surface and P atoms diffuse into the silicon wafer from this P-doped glassy layer, or *glaze*. In either event, what remains of the glassy layer after predep is removed by an etching process known as *deglazing*. This insures that additional dopant cannot be introduced into the silicon in subsequent high-temperature operations. Subsequent processing steps usually require that the glaze be replaced by a new layer of oxide generated by thermal oxidation.

The penetration of the indiffusing dopant atoms into the silicon substrate depends upon the dopant species (e.g., arsenic, boron, or phosphorus) the temperature T at which the diffusion is carried out, the time or duration of diffusion t, and the boundary conditions that are imposed on the diffusing species. (It also depends on the substrate material, but we take that to be silicon.) Say, for example, as in predep, that the dopant concentration at the surface is held constant, that the diffusion process is terminated by lowering the temperature well before the dopant has diffused all the way through the wafer, and that during diffusion the temperature is held constant. Chapter 6 shows that with this set of conditions the dopant concentration profile in the substrate may be modeled, to a first approximation, by the complementary error function, erfc, and the dopant concentration as a function of distance into the wafer may be expressed as

$$N(x,t) = N_0 \,\mathrm{erfc}(x/\sqrt{4Dt}) \qquad (1.7\text{-}1)$$

where

N = concentration of the diffusing species (cm^{-3})

N_0 = concentration at $x = 0$ (i.e., the surface concentration)

x = normal distance into the wafer measured from the surface (cm)

D = *diffusivity* of the diffusing species in the substrate material (nominally Si) at the diffusion temperature (cm^2/s)

T = diffusion temperature (K)

The diffusion temperature is implicit in the equation since D is proportional to $\exp(-b/T)$. The complementary error function is discussed in Chapter 6.

By way of example, Figure 6–6(a) shows the normalized doping profiles, namely $N(x)/N_0$ versus x, for four different diffusion durations. Notice that (1) the surface concentration remains constant as we specified, (2) the atoms move farther into the wafer with increasing time, and (3) the area under the curve increases with time, indicating that more dopant atoms have diffused into the wafer.

The second aspect of selective doping, the lateral confinement of dopants to only preselected regions in the substrate, is covered in Section 1.8.

1.7.2 Ion Implantation

Ion implantation is the second method of introducing doping atoms into a substrate or other Si layer. It is a nonthermal process and may be carried out at room temperature. The dopant atoms first are ionized by stripping off one or two of their orbital electrons. Then they are focused by an electric and/or magnetic field into a beam about 0.5 to 1 cm in diameter, are accelerated by a high electric field, and strike the wafer surface with high velocity. With typical kinetic energies of up to 150 keV ($=2.4 \times 10^{-14}$ J) or more, the ions have sufficient energy to penetrate into the wafer, where they finally are stopped by collisions with lattice atoms and subatomic particles. Since collisions take place randomly, the ions penetrate different distances with some form of statistical variation. They do not necessarily come to rest on lattice sites.

In typical situations the implant energy is so great that damage to the crystal structure results. Often the layer near the surface where the ions come to rest will be rendered essentially amorphous, with no long-range periodic crystal structure. This *radiation damage* can be repaired by *thermal annealing*, which involves heating the substrate to a temperature of 500°C or above for about thirty minutes. The vacancy and dopant diffusivities at this temperature are large enough for both species to move in the damaged substrate. This movement allows the once-damaged layer to regrow itself epitaxially, with the underlying, undamaged substrate acting as the seed crystal for regrowth. Also, the dopant atoms diffuse to lattice sites. Thus, post-implant annealing serves two functions: removal of radiation damage and electronic activation of the dopants.

After anneal, the distribution of the dopant atoms may be modeled to first order by a Gaussian curve whose peak is displaced inward from the surface by a distance R_P, the *projected range*. The equation for the concentration profile is:

$$N(x) = \frac{Q}{\sqrt{2\pi}\Delta R_P} \exp \frac{1}{2}\left(\frac{x - R_P}{\Delta R_P}\right)^2 \tag{1.7-2}$$

where

N = dopant concentration (cm^{-3})

Q = the *dose* = number of ions implanted through 1 cm^2 of wafer surface during the duration of implant t (cm^{-2}).[4]

$\Delta R_P = \sigma_P$ = standard deviation of the Gaussian = the *straggle* (cm)

R_P = projected range = distance of profile peak from the surface (cm)

x = distance from the surface into the wafer (cm)

The implant time or duration t does not appear explicitly in Eq. (1.7-2). It is implicit in Q, however, which depends on t. If the beam current is held

[4] Extreme care must be used in reading the literature, since definitions of the dose vary among authors. The choice here, with dimensions cm^{-2}, is favored in the semiconductor industry.

constant during implant and t doubles, Q and $N(x)$ will double. Figure 8–9(a) shows a typical plot of the equation. For a given dopant species in a given substrate material, both the effective range and the straggle are determined by the ion beam energy and are independent of time and dose.

It might seem that the wafer would become positively charged due to the charged ions, and eventually would start repelling beam ions. Actually this does not happen, because the ions are neutralized when they come to rest. This comes about in the following manner. Ions moving *in vacuo* comprise a flow of electrical current. Current continuity requires the same current to flow everywhere in the closed loop made up of ion source, beam, wafer, power supply, and connecting wires. Thus moving beam ions are balanced by electrons flowing in the return circuit. Positive ions and electrons flow in opposite directions in the wafer. A pair comes to rest simultaneously, combines, and the ion becomes a neutral atom again.

Typically, the beam diameter is much smaller than the wafer, so some scanning motion must take place between the beam and the wafer surface. Since the beam ions have electric charge, they may be deflected by electric fields normal to the direction of beam motion. This and other scanning methods are discussed in Chapter 8.

1.7.3 Junction Formation

The introduction of dopants into an already doped *host* (wafer, substrate) of Si by either predep diffusion or implant may produce a p–n junction. For this to happen, the host and incoming dopants must be of opposite types, one donor and one acceptor. Consider an example where a host is doped with a donor at constant concentration N_d (d for donor). Assume that this distribution is not affected by the predep diffusion that introduces acceptor atoms into the host with constant surface concentration N_{0a} (0 for $x = 0$ and a for acceptor). Assume that $N_{0a} > N_d$. Even at room temperature, both types of dopants will have charge—acceptors being negative, since they have "accepted" an electron, and donors being positive because they have "donated" an electron. We shall sign the dopant concentrations with the sign of their charged state. The *total, net,* or *compensated* concentration will be the *algebraic sum* of the two dopant concentrations. Thus,

$$N(x,t) = N_d - N_a(x,t) \tag{1.7-3}$$

where $N_a(x,t)$ is the indiffusing acceptor profile magnitude given by Eq. (1.7-1). Substituting, with appropriate subscript change, we get:

$$N(x,t) = N_d - N_{0a}\,\mathrm{erfc}\left(\frac{x}{\sqrt{4D_a t}}\right) \tag{1.7-4}$$

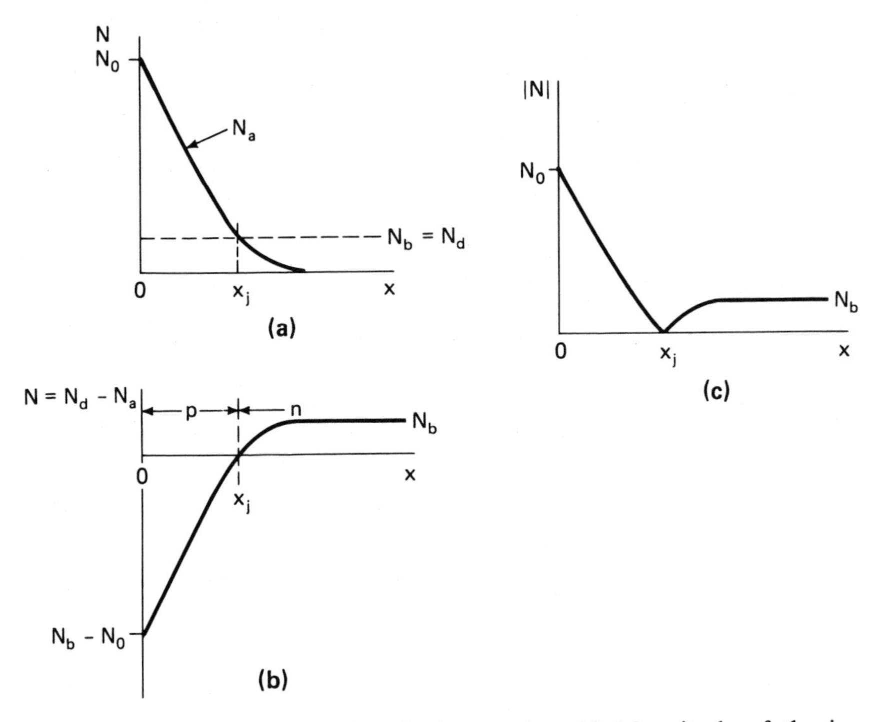

Figure 1–6. Junction formation during predep. (a) Magnitude of the in-diffused N_a, and background $N_b = N_d$, dopant profiles are shown. (b) Plot of the net concentration $N = N_d - N_a$. (c) The profile of $|N|$.

The individual profiles are shown in Figure 1–6(a). The net profile or algebraic sum is shown at (b) in the same figure, and (c) shows the *magnitude* of $N(x, t)$.

To the left of x_j, $N_a(x, t) > N_d$; hence acceptors predominate and the Si is p-type. Donors predominate to the right of x_j and the Si is n-type. Thus, a p–n junction has been formed at x_j, the *junction depth*, where the two concentrations have equal magnitudes.

Observe that $N(x_j, t) = 0$. This is the criterion for finding a junction location; hence Eq. (1.7-4) may be solved for this condition to find the junction depth for the assumed conditions

$$x_j = \sqrt{4D_a t}\,\text{erfc}^{-1}\left(\frac{N_d}{N_{0a}}\right) \tag{1.7-5}$$

where erfc^{-1} is the *inverse complementary error function* (Section 6.3.1).

Junctions also may be formed under proper conditions of ion implantation. Solutions are considered in Chapter 8.

1.8 Pattern Delimitation; Lithography

Recall that selective doping involves two separate processes. Doping by diffusion or ion implantation already has been considered briefly. The second process is lateral confinement of the doping atoms so that only preselected regions of the host will receive them. For example, the top view (F) in Figure 1–3 shows that donors introduced to create the emitter must be confined within the existing base region. Since dopant atoms are introduced into the host from outside, confinement can be accomplished by interposing between source and wafer a *mask* that has holes or *windows* in appropriate locations for the dopant atoms to pass through. The mask windows delimit or delineate the areas to be doped. Pattern delimiting is accomplished by a series of processes that are referred to as *lithography.*

Small window size, say 1 μm × 1 μm or less, rules out use of a thin metal sheet with holes for windows, called a *shadow mask.* But given that SiO_2 is a satisfactory mask for both diffusion and some implantations, it may be used as the mask. But how are such small windows produced in the oxide and at the desired places, and how are the windows properly aligned relative to previously doped regions? To give a frame of reference for answering these questions, we shall first outline the steps of lithography.

1. A *photomask* is prepared photographically, with the required doping pattern reproduced as an optical image in the emulsion on a glass plate.
2. The Si wafer is oxidized.
3. The oxidized wafer is coated with a material, called *photoresist* (PR), that is sensitive to ultraviolet (UV) radiation. (Where UV is used, lithography becomes *photo*lithography.)
4. The PR-coated, oxidized wafer surface is exposed to the photomask by UV in a manner analogous to exposing photographic film.
5. The PR is developed. This results in the optical pattern of the photomask being transferred as open windows in the PR layer. (Sometimes in I^2 the windowed PR is used as a mask and the next step is omitted.)
6. The wafer is exposed to a *selective etch* that cuts windows in the SiO_2 without affecting the PR and underlying silicon. Window positions coincide in the oxide and PR.
7. The PR, having served its purpose, is removed by a *stripping* process.

The exposed, windowed oxide layer then is ready to serve as a mask for diffusion or I^2. The dopant atoms can pass only through the windows and so delineate the doped regions in the Si wafer. Note that the horizontal geometry and locations of the doped areas ultimately are determined by the pattern on the original photomask.

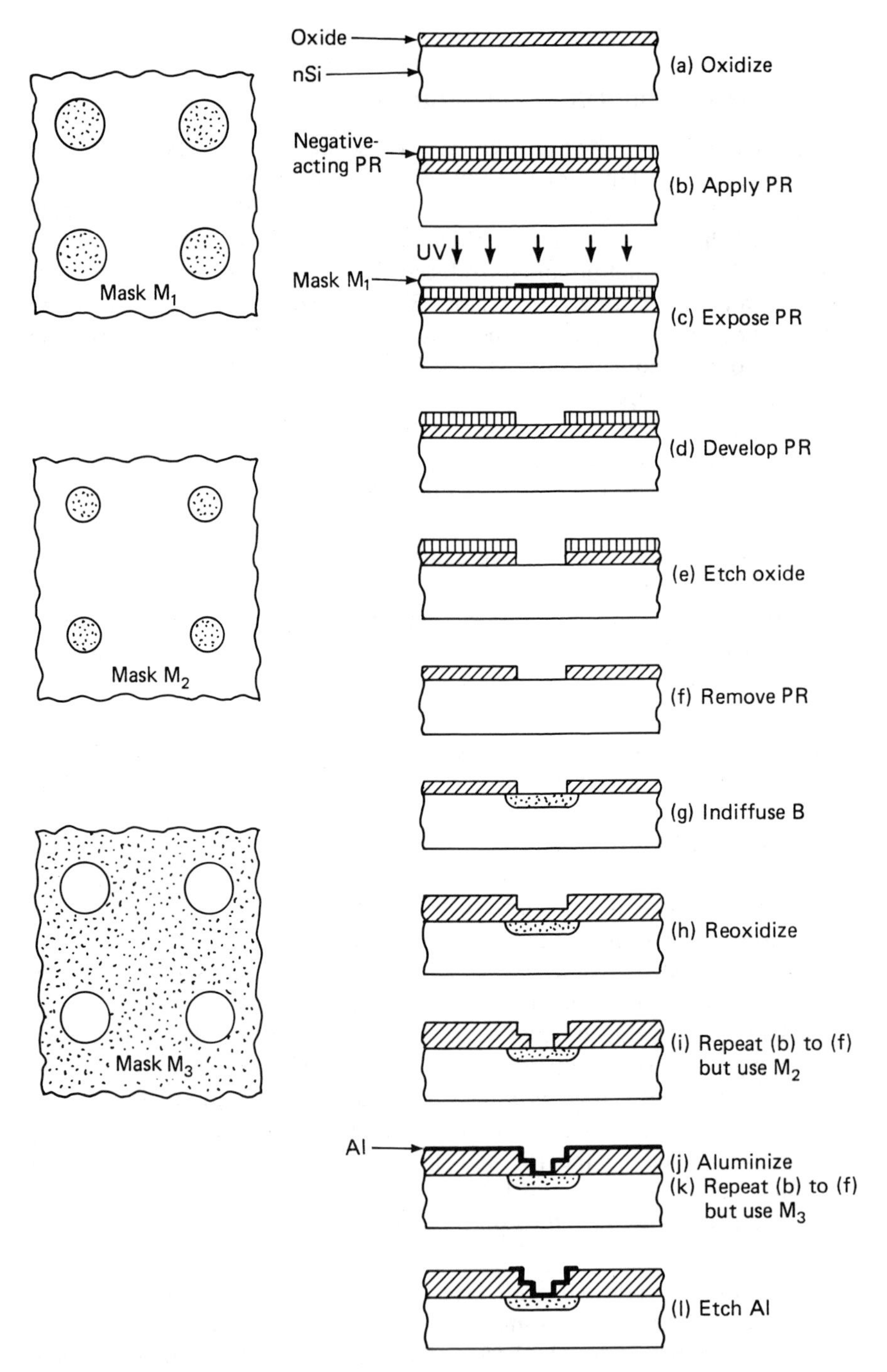

Figure 1–7. Planar diode fabrication.

The foregoing steps are illustrated schematically in Figure 1–7(a) to (f) and should be studied carefully. All the process steps shown there, which result in the windowed SiO_2 mask, comprise photolithography.

Doping and metallization steps plus repetition of the photolithographic steps can be cycled to produce devices and ICs, if proper masks and means of *mask alignment* are used. The latter involves equipment that ensures that each mask is precisely oriented wrt existing features. To illustrate, the entire sequence (a) through (l) in Figure 1–7 will fabricate a planar diode. The cross sections in the figure correspond to one circular pattern on each mask. Also, the dark mask regions cause etched-away areas on the device.

It is easy to extrapolate the diode process to make other devices. For example, if the sequence (a) through (h) is repeated before (i) but with another proper mask, say M_2', to diffuse an n-emitter within the p region that will serve as a base, a BJT will result.

A large number of processing steps is needed to make even a simple diode. The four circular patterns on each partial mask in the figure actually represent *m* patterns on the entire mask. Thus the entire sequence of processing steps would give *m* planar diodes. These subsequently would be cut apart or *diced* into *m* separate diode dice or chips. (The word "die" is the singular of "dice.")

This is another example of batch processing. As early as 1960 it was not uncommon to process more than 10,000 $(=m)$ discrete BJTs on a 2-in. wafer, a typical wafer size then. In modern practice wafer diameter is up to 8 in., and chip size to nearly 1/2 in. × 1/2 in., so *m* may be smaller, and up to 30 masking operations are used. If processing produced only one chip, IC costs would be prohibitive. Batch processing lies at the heart of low prices for complex semiconductor products. Remember, we process wafers, not chips or devices.

We now consider some of these photolithographic steps in more detail.

1.8.1 Oxide Masking

It is a physical fact that a sufficiently thick layer of SiO_2 on the silicon surface can serve as a mask or *stop* against the indiffusion, or sometimes the implant, of the common dopants used in silicon technology—namely boron, phosphorus, and arsenic. This idea is illustrated schematically in Figure 1–8, where the dopant can reach the silicon only through the oxide windows. The required mask thickness for diffusion depends on the dopant species, the mask material, and the diffusion temperature and duration. The determination of its value for a given set of diffusion parameters is discussed in Section 6.6. Empirical values for diffusion that are most commonly used in industry are given by the oxide masking curves of Appendices F.6.6 and F.6.7. A typical set of values is: a minimum thickness of 1000 Å (0.1 μm) of SiO_2 is required to mask against the indiffusion of boron for 70 min at 1050°C.

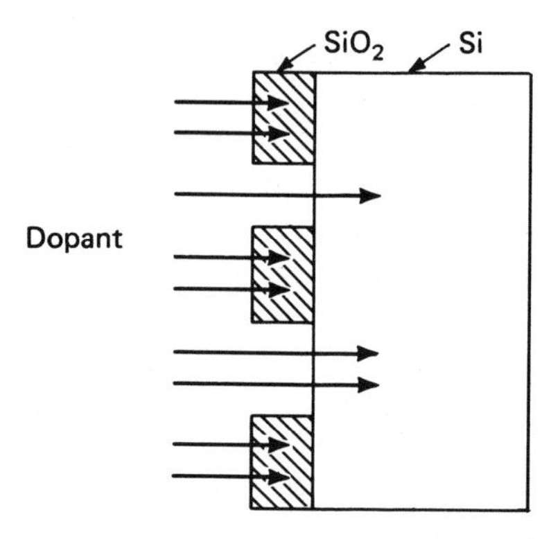

Figure 1–8. Oxide masking. The oxide prevents the dopant from reaching the underlying silicon.

Means of determining the oxide thickness to mask against ion implantation are discussed in Section 8.11. The required masking thickness depends on the dopant species, the mask material, and the ion beam energy. For very high energy implants the masking properties of SiO_2 are augmented by an overlayer of a heavy metal such as gold, platinum, tungsten, or tantalum. Silicon nitride and photoresist also may be used as implant masks.

1.8.2 Selective Etching

There is a need in lithography for chemicals that etch different materials selectively (i.e., at different rates). For example, for mask windows the etch must attack SiO_2 but not Si and PR; to delineate connects in an Al layer it must attack Al but not PR, SiO_2 or Si_3N_4. Fortunately, selective etches are available in both liquid and gas form for these and other material combinations.

Hydrofluoric acid (HF) is the most common liquid selective etch for SiO_2. However, HF is extremely hazardous on contact with human skin and attacks glass vessels. It also etches *isotropically*—its etch rate is the same in all directions. This means that as HF etches the oxide vertically through a PR window, it also etches laterally, undercutting the PR. This is illustrated in Figure 1–9(a). The net effect is that the SiO_2 window will be larger than the defining PR window. This can be catastrophic in fabricating fine-line-geometry ICs. For example, if the undercuts between two adjacent PR windows exceed the spacing between adjacent window edges, the windows merge into one.

There are techniques that permit *anisotropic* etching of SiO_2 (i.e., in one direction). Some are based on *reactive ion etching* (RIE), which uses ionized, gaseous etching source materials (Section 10.18). Silicon also may be etched anisotropically by RIE. Furthermore, since Si appears most often in single-

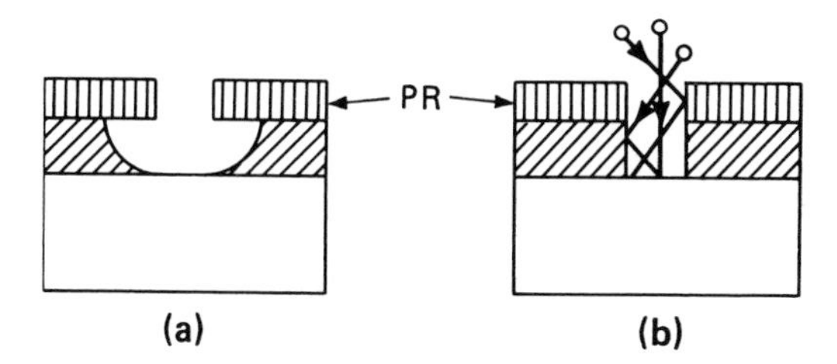

Figure 1–9. Etch profiles for two types of etching. (a) Isotropic. (b) Anisotropic. A gas is assumed as the etchant.

crystal form, liquid etches that are sensitive to crystallographic planes may be used. One such liquid etch is hydrazine (NH_2NH_2). Details on these processes are given in Chapter 10.

Recall from Section 1.3 that interconnects are defined by etching a metal layer that covers an insulating layer over the entire wafer surface. Therefore, selective metal etches are needed. When metal layers are very thin compared to the ultimate interconnect width, anisotropy is not so important. Hot phosphoric acid (H_3PO_4) will selectively etch Al without attacking PR or underlying insulators. Chlorine-based gases also may be used to selectively etch metals (Chapter 10).

1.8.3 Photoresists

The photolithographic process depends on *photoresists* (PR). These materials, based on polymers or resins, are liquids that are photosensitized to ultraviolet (UV) radiation. A resist is applied to the wafer and spun at about 3000 r/min; centrifugal force reduces it to a layer about 0.5 to 1 μm thick. It is then dried and heat-hardened (*prebaked*).

When the wafer is exposed through a photomask to UV and then developed, changes occur in the exposed regions and the photomask pattern is transferred as holes or windows in the PR. There are two types of PR; their behavior is shown in Figure 1–10. With *positive* resist the clear (transparent) regions of the photomask translate into windows in the exposed and developed PR. The effect is reversed with *negative* resist, where dark (opaque) mask regions create windows in the processed PR. A handy mnemonic for the negative type is **d**ark **d**evelops away.

This difference in behavior is based on chemistry. Negative resists use organic polymers, with solvents as developers. Exposed regions *crosslink*, forming long chain molecules that resist development, but these swell in the organic developer and so tend to lose resolution. Positive resists use resins, with water-based inorganic hydroxides as developers. Regions exposed to UV

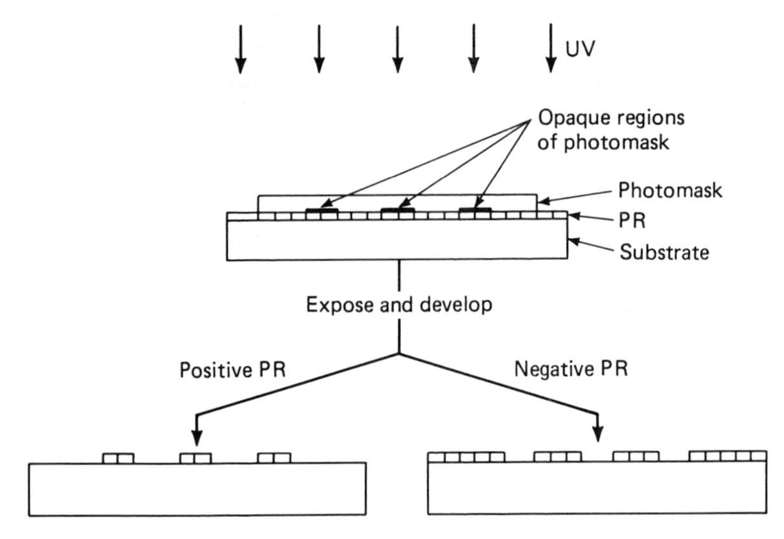

Figure 1–10. Positive and negative photoresist action.

become more soluble in developer and dissolve away. The unexposed regions do not swell, so resolution is inherently better. Positive resists have lower UV sensitivity and require longer exposure times. Since wafers must be exposed one at a time, exposure cannot be a batch process. The exposure *throughput* (wafers processed per hour) is therefore much lower than for negative resists. In both types a post-development heat treatment (*postbake*) hardens the remaining PR. The aqueous chemistry of positive PR is more environment-compatible. Still, for a majority of ICs (where line widths exceed 2.5 μm), industry uses the higher throughput and additional years of experience with negative resists; positive resists tend to be reserved for the more critical, fine-line ICs.

Resists formulated for use on various materials, such as metals and SiO_2, are available to the industry and usually are formulated with *adhesion promoters* to aid in making them stick well to the underlying material. There are adherence problems with SiO_2, however, particularly with positive PR. Adherence may be enhanced by using a primer such as hexamethyldisilizane (HMDS), which is spun on before the PR. Subsequent PR processing is not affected.

After serving its purpose, the resist is removed or *stripped* from the wafer. The stripping agent must be selective to avoid damage to the underlying layer. Liquid strippers for negative resists are based on sulfuric acid (H_2SO_4) or are proprietary mixtures, while hydroxides work with positive resists. Dry stripping can be accomplished through the use of oxygen and other gas plasmas (Section 10.15.3).

1.8.4 Photomasks

A photomask ultimately determines the size and location of all the features in one layer of a device or IC. All its pattern dimensions must be extremely accurate. In one system, the original *artwork* for a single layer of a chip is prepared on a dimensionally stable plastic sheet at, say, 2500 times the final chip size. This is reduced photographically by some 250 times onto an emulsion-covered glass plate, the *reticle*. Another optical reduction from the plate to final size is made in a *step-and-repeat* machine that simultaneously, by means of successive exposures, repeats the pattern in a *diaper* array (i.e., in a two-dimensional matrix). Part of this array, needed for batch processing, is illustrated by the magnified portion of a mask in Figure 1–11. This emulsion-on-glass mask is the master from which *working masks* are copied by contact, photographic printing. The great reduction in size from artwork to working mask reduces any dimensional discrepancies in the artwork and makes for more accurate masks.

Working masks of the emulsion type are short-lived because they are damaged easily. *Hard masks*, where chromium or iron oxide replaces the emulsion on a glass plate, last much longer. They are prepared by etching

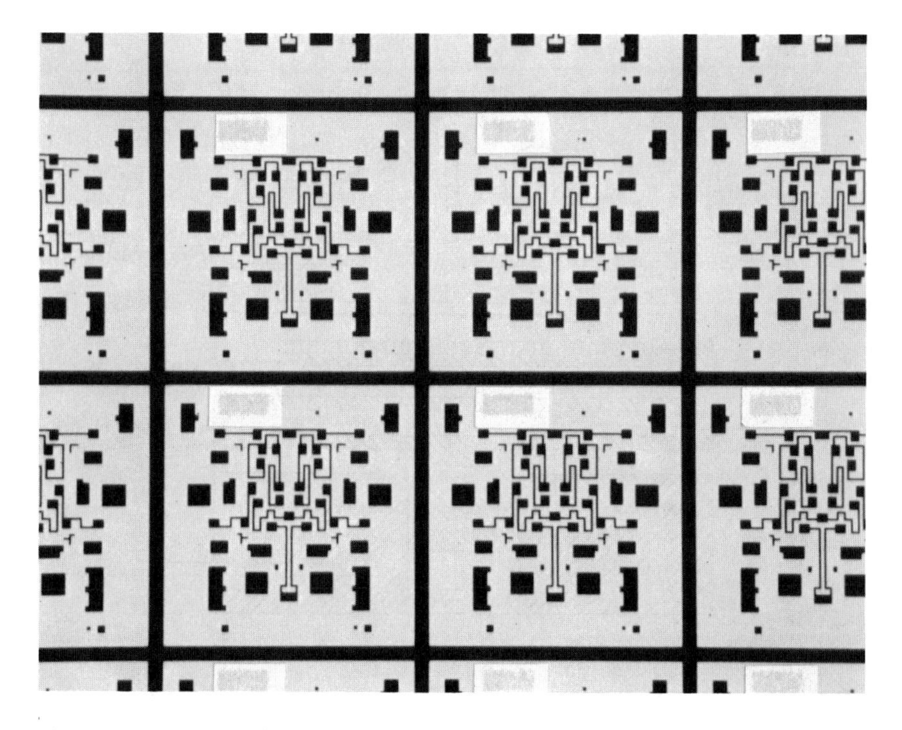

Figure 1–11. Part of an IC photomask. The same pattern is replicated in a two-dimensional array.

through the thin layer of chromium or oxide by photolithography. These layers are thinner than emulsion and so can give finer detail. Furthermore, thin iron oxide is transparent to visible light but opaque to UV, a useful property in visual mask alignment by a human operator.

1.8.5 Mask Alignment

During fabrication, each photomask must be oriented properly relative to features that are already present on the substrate. This insures, for example, that emitters fall properly within bases, interconnects lie properly over contact windows, and so on.

A crucial question is how the alignment operator can locate a diffused base region that is buried under the oxide. Recall that the etching/reoxidation sequence that accompanies base diffusion results in a thinner oxide over the base than over the surrounding region, so the base region is colored differently and can be "seen."

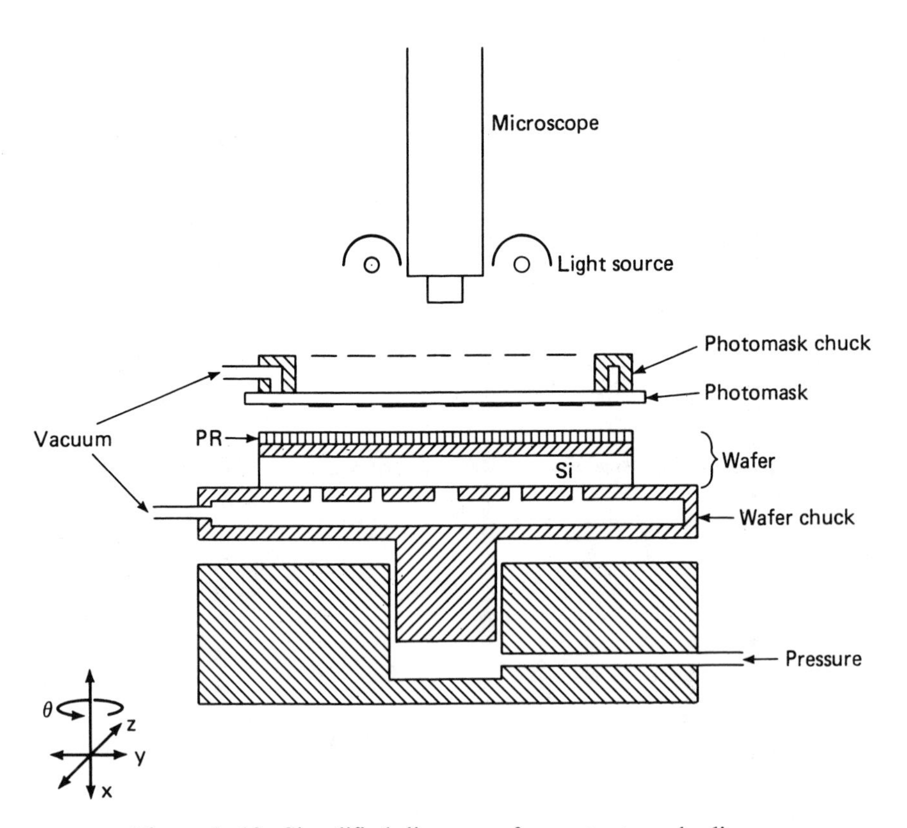

Figure 1–12. Simplified diagram of a contact mask aligner.

The alignment of wafer to mask is performed with a *mask aligner*. Figure 1–12 shows a schematic cross section of a generic, manually operated, *contact* type where mask and wafer are in physical contact during UV exposure. Noncontact aligners are discussed in Chapter 11.

The PR-coated wafer is held on the lower vacuum chuck, which is free to move in the y- and z-directions (parallel to the photomask edges) and vertically. The wafer also may be rotated about the vertical x-axis in the θ-direction. The three motions in the wafer plane are controlled independently by screws, micromanipulators, or *joysticks*.

The photomask is held, pattern side down, by the upper vacuum chuck throughout the entire alignment procedure. The microscope and light source are used for visual checking of the wafer-to-mask alignment. The light color is critical; it must be visible and yet not expose the UV-sensitive photoresist on the wafer.

In use, the wafer and photomask are loaded on the proper chucks, roughly in alignment. The lower piston assembly is raised by air pressure until the wafer is within 1 to 5 mils of the photomask. The operator then uses the microscope to view the relative orientation between mask and wafer patterns while adjusting the y, z and θ controls.[5]

The wafer is raised again by air pressure to close the gap and make the wafer come in physical contact with the mask. The operator checks the alignment visually. If it is unsatisfactory, wafer and mask are separated again and alignment repeated.

When alignment is deemed satisfactory, the microscope assembly is swung aside to make way for a UV lamp that will expose the PR. Exposure time is set manually—typically, for a few seconds—and is controlled automatically. The lower chuck is then dropped to its lowest position, its vacuum supply is turned off, and the exposed wafer is removed.

Mask alignment and other operations involving unexposed photoresist are performed in a *gold room* where the room illumination is from gold-colored or gold-filtered fluorescent lamps. This gives UV-free illumination, to avoid inadvertent PR exposure.

It is the fact that wafer and mask are clamped together during alignment check and UV exposure that causes damage to the mask. Two other aligner types, the *off-contact* and *projection* types, always keep wafer and mask separated so contact damage is eliminated. The projection aligner has another advantage: the image projected onto the wafer may be smaller than the photomask image, which translates into better accuracy. A 4 : 1 reduction is

[5] Aligner microscopes are binocular, and most provide a *split field* option whereby each eye views a different part of the mask/wafer combination. Use of this option can reduce considerably the alignment time. Automatic alignment is used extensively now to eliminate human error and to speed production (Chapter 11).

typical. Further details are discussed in Chapter 11. Projection aligners are now the most commonly used type.

1.8.6 Metallization

Interconnects are formed by etching the desired pattern in a uniform layer of metal, usually aluminum, that has been deposited over the entire wafer surface. One means of depositing this metal layer is *vacuum evaporation* (*vacuum deposition* or *physical vapor deposition*, PVD). No chemistry is involved in PVD in that atoms of the material are transferred from source to substrate without necessarily going through a chemical change as is the case in CVD.

In vacuum evaporation, Figure 1–13, the material to be deposited, say Al, is placed on a *deposition source*. The source and substrate (or wafer) are placed several centimeters apart in a *vacuum chamber* or *bell jar* that is evacuated to a pressure of 5×10^{-5} torr (1 torr = 1 mm Hg = 133 Pa) or less—i.e., to at least seven orders of magnitude below atmospheric pressure (760 torr). Sufficient heat is applied to the Al in the source to cause it to melt and then evaporate. The Al atoms in the vapor have considerable kinetic energy and so move away from the hot source. Since the ambient pressure and corresponding gas concentration are so low, the Al atoms suffer few collisions, if any, and so travel in straight lines. Some are intercepted by the substrate, where they condense out as solid Al to form the film. Also, at such low pressure there is little oxygen or other gas present to react with the Al when it is in vapor phase; hence, the condensed layer will be the same chemically as it was in the source. We saw in Section 1.3 that alloys of Al plus a few percent of Si

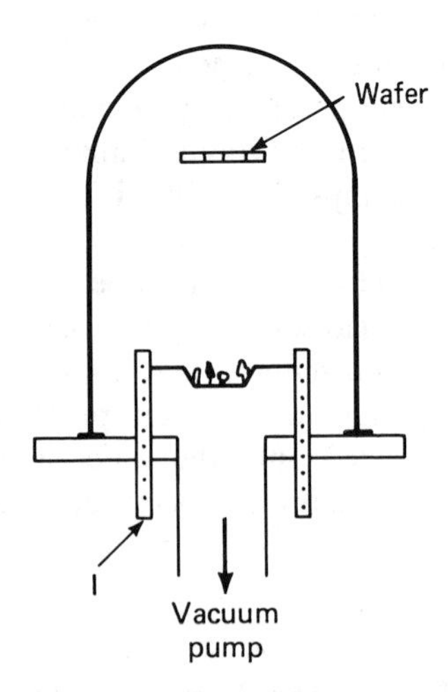

Figure 1–13. Simplified diagram of the vacuum evaporation setup.

and/or Cu are used for deposition on Si rather than pure Al. These alloys evaporate well.

There are three methods of supplying heat to the evaporation source. The first method uses direct *resistance heating* of the source that holds the Al. The source proper is a coil, a dimpled strip, or a boat-shaped strip of refractory (high-melting-point) material that holds the Al. A large current, say 100 A, but dependent upon the particular source in use, is passed through the source to get I^2R heating. For a typical voltage drop across a coil source of 5 V, 500 W would be supplied. Not all of this power goes to the vaporization process; much of it is lost by conduction and radiation from the source. This I^2R heating method is assumed in Figure 1–13.

Tungsten sources usually contain sodium (Na), which can evaporate along with the Al. This raises problems, especially when depositing MOST gates. If this Na migrates into the gate oxide after it condenses on the gate, it can cause threshold voltage shifts, a condition that must be avoided. Two methods of supplying heat to the Al directly, rather than to the source, are available. These eliminate the sodium-contaminated-source problem.

In the first of these methods, called *e-beam evaporation*, a dense beam of high-energy electrons is directed onto the aluminum. Heating results when the electrons give up their kinetic energy on hitting the Al. Typical beam powers range up to a few kW. The electron beam can be focused so that heating is confined to a small region of the metal, giving better heat utilization. The source *crucible* is kept at low temperature, often with water cooling, so it will not emit any of its components.

Another method of supplying heat directly to the aluminum involves induction heating. A nonmetallic crucible holds the Al and is surrounded by a wire heating coil to which radio-frequency (r-f) energy is supplied. Voltages are induced in the Al, large currents flow because of the low resistance, and direct I^2R heating of the Al takes place. A high-resistance material is chosen for the crucible so that it is heated only slightly by the r-f energy. Again water may be used to keep the crucible cool, and if the coil of wire is replaced by a similar coil of metal tubing, the cooling water may circulate through it, also.

Pumpdown of a vacuum system is a slow process, usually taking up to 20 min or more. Thus, it is not commercially feasible to perform PVD on only one wafer at a time. Some form of batch processing is needed so that several wafers can be coated simultaneously for each pumpdown of the vacuum system. The evaporated flux density from typical evaporation sources is directional, so special fixturing is needed to cover all the wafers uniformly. A number of different fixtures have been designed to do this, all of which involve moving the wafers wrt the evaporation source. One type uses a *planetary* system, in which the wafers are fixed to spherical segments. These segments all rotate as a unit about the vertical axis of the source. In addition, each segment rotates about an axis that is inclined to the vertical. In this manner, all wafers are given a nearly equal "view" of the source during the deposition interval. Such fixtures have been designed to hold up to tens of wafers 100 mm in diameter. Smaller numbers of larger wafers can be accommodated, especially if a larger bell jar is provided.

Tungsten and molybdenum also are used for interconnects. These refractory metals vaporize at much higher temperatures than aluminum and so are not amenable to vacuum evaporation. They may, however, be sputtered. Sputtering is a nonthermal process based on momentum transfer from ions to the source material and is discussed in Chapter 12. As we have seen earlier, tungsten also may be deposited by chemical vapor deposition.

1.8.7 Testing

Several tests are made throughout the fabrication procedure, to check diffusion or implant, oxidation, metallization, and junction depths. If processing troubles arise, diagnostic tests can help in locating difficulties. Some of the electrical tests are described in Chapter 3.

Packages for complete devices are expensive, so bad dice should be rejected before packaging. To this end a *100% probe* is made on all dice that are on the wafer. Contact is made to the metal bonding pads of each die in succession by means of external test probes that are connected to appropriate test equipment.

The test probes are held in fixed positions to match the bonding pads on the dice. The wafer is mounted on a vacuum chuck that moves the wafer vertically (the x-direction) until contact is made with the probes. The appropriate electrical tests are made. Then the chuck is lowered and stepped one die width, and raised again for a test on the next die. This procedure continues until all the dice on the wafer have been tested.

Hand probing is slow and tedious, so automatic wafer stepping is used with a *prober* that is computer controlled to step in the y- and z-directions die by die. Bad devices are marked automatically with dark-colored magnetic ink for easy identification. After the dice are separated into chips they may be removed with a magnet. Usually the test results for ICs are stored in an associated computer.

1.8.8 Post-Wafer-Fab Processes

The processes discussed thus far are the principal ones of silicon planar processing. They provide means for material to be added to (diffusion, implantation, CVD, PVD) or subtracted from (etching) the upper or front wafer surface. Another category of steps is required to prepare the wafer for ultimate packaging of its *good* individual chips. In brief, these post-wafer-fab steps are: back-side (or bottom-surface) preparation, final wafer test, separation of the wafer into chips, die bonding of the chips to the package, lead bonding between chip pads and package leads; package closure, and final testing. Consider each of these briefly.

The first step here is to grind off and lap the back side of the wafer. Again, batch processing is used here, so several wafers are mounted with wax onto a heavy metal plate. This is placed in a grinding/lapping machine in which the wafers are made to move in a quasi-random pattern against a bottom plate. The lapping medium is a slurry of water and alumina (Al_2O_3) grit that is available in various sizes. Grit diameter of 0.3 μm is typical of a fine powder. As stated earlier, this operation is used to thin the wafer and to condition the back side for making an ohmic contact. Recall that removal of back-side material also removes doped layers that were created when diffusions were made on the front surface.

Back-side grinding is one of the two "dirty" fabrication steps in processing wafers and really is incompatible with the extreme cleanliness maintained in silicon device processing. The search continues for a better, cleaner method of accomplishing the same results. The second dirty process is dicing.

The last processing step on the wafer as a single unit is the deposition of a gold layer on the backside by vacuum evaporation. This layer serves three functions: it aids in the formation of the back-side ohmic contact[6]; it is the back-side contact; and it aids in the subsequent die bonding of the chips to their mounting cases, or *headers*. The 100% probe of the wafer as described in the last section would be made at this point.

The next step is *dicing*, where the wafer loses its identity as it is cut up, each die becoming a chip. If the first mask was aligned properly with respect to the crystal axes of the silicon substrate, the natural "cleavage" planes of the wafer will lie parallel to the edges of the chips (Section 2.6). Then, if the surface is *scribed*, sawed,[7] or laser-burned along the dice edges, the wafer subsequently may be broken up easily into separate chips.

Headers and mounting/bonding costs often are more expensive than a chip; therefore, defective chips must be culled before mounting takes place. The good dice are stored for future mounting in an appropriate header. Since a given device may be sold in more than one package type, a proper stock balance must be maintained.

---

[6] In die bonding the temperature is raised well above room value and some of the gold diffuses into the silicon, where it greatly reduces the minority carrier lifetime by trapping action. If all minority carriers were removed, a perfect ohmic contact would result.

[7] While the diamond saw method has had widespread use in the semiconductor industry, it ranks with back-side grinding as a "dirty" process. The *kerf* (material that is removed by the saw) and the cooling water form a slurry. Many companies now favor the laser scribe because it provides a very clean and dry operation.

Final assembly is highly labor intensive and affects sale price considerably. As a result, finished chips often are shipped to *offshore* (overseas) assembly plants, where labor costs are lower.

Every package configuration has a tab, plate, or other metal surface to which the die is affixed by a process known as *die bonding*. Where plastic cases are to be used, the tab is part of a metal *lead frame* as shown in Figure 1–14. The process usually involves heating, so it is important to ensure that bonding temperatures remain low enough that (1) junction depths in the device remain unchanged (i.e., no further diffusion of the selectively doped regions takes place), and (2) no structures or materials already existing in or on the chip begin to melt. The Al/Au eutectic temperature is 370°C, which is lower than the Al/Si value of 577°C (Section 1.3).

Die bonding consists of heating the header tab to slightly above 363°C and applying the chip, bottom-side down, on the tab with a slight scrubbing action to insure good contact. Melting occurs at the Au/Si interface; then the temperature is lowered with chip and tab held in close contact. When freezeout occurs, die bonding is completed. Air should not reach the joint

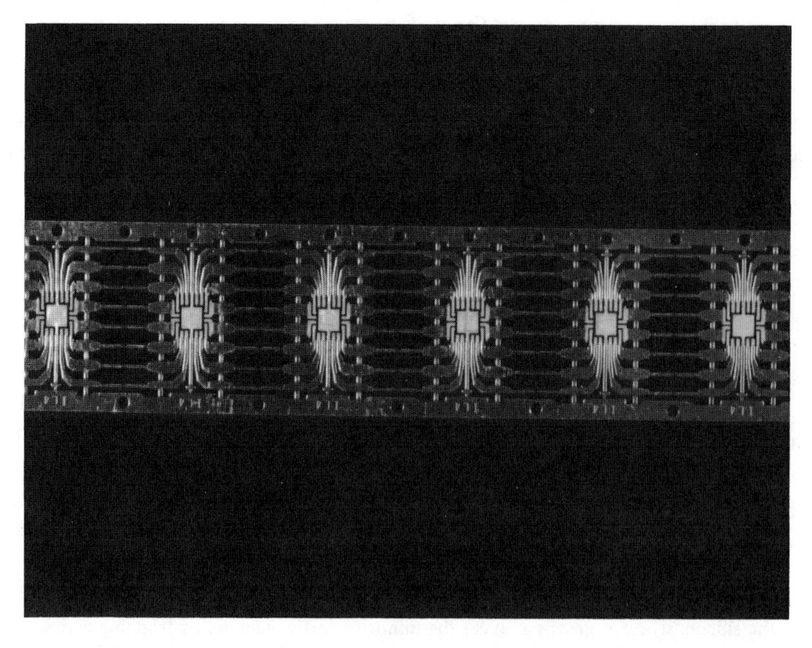

Figure 1–14. A strip of lead frames. The tab on which the die is mounted is at a frame center. Note how the lead size and spacing change: small near the tab, where lead bonding joins frame lead and die contact; large at the periphery, to interface with a socket. The connections between the output leads of the frames are severed after the unit is sealed in plastic.

during the heat/cool cycle; therefore, the bonding operation is often carried out in an inert ambient of nitrogen or in a reducing atmosphere of hydrogen. Other methods of die bonding also are used, the simplest being the use of a conducting epoxy to fasten chip to tab.

The procedure described in the foregoing paragraph relies on the tab gold plating to aid in the bonding action. If the plating is too thin (remember, gold is very expensive) one often uses a very small piece of Au/Si eutectic ($\sim 6\%$ by weight of Si) as a source of additional gold (Chapter 4). The small amount of Si, which gives the eutectic composition, insures the low melting point.

Once the chip is bonded to the header tab, interconnections must be made between the chip contact pads and header leads. This is the region of interfacing between the microworld of the chip, where interconnect widths may be measured in microns (= micrometers = μm), and the macroworld of the header where lead widths average around 15 mils. This represents a change in scale of around 300 : 1! As we shall discuss shortly, some of this difference is made up by the bonding pads on the chip. The actual interfacing between chip and header is provided by gold or aluminum wire about 0.8 to 3 mils in diameter, although with power devices larger diameters, or even two wires in parallel, may be used. These interfacing leads are *welded* to the chip bonding pads and to the header leads in the process known as *lead bonding*.

The welding process is carried out in a *lead bonder*. The header is held in a movable chuck so that the chip bonding pads or header leads may be positioned under the welding head that holds the bonding wire. Pad and bonding wire are brought into close contact. Heat for the weld is furnished in one of two ways.

In thermocompression (TC) bonding, heat is supplied to the header through the electrically heated chuck and head. The metals involved, such as Au/Au or Au/Al, can be welded at 280 to 300°C, temperatures that are well below the Al/Si eutectic temperature. A nitrogen ambient is desirable. The gold bonding wire is held under sufficient pressure between the chip pad and the head for plastic flow to take place, and the weld is formed. The chuck then is moved until the head, still connected to the bonding wire, is aligned with the proper header lead. Pressure is applied again, to form the second weld, and the bond wire is cut free from the head.

The second form of lead bonding is much the same in physical setup. The heat is supplied mechanically, however, in the form of vibrations at ultrasonic frequencies. The head in this case is the vibrating element and the heat, as well as a mechanical cleaning action, results when the bonding wire is scrubbed against the chip bonding pad or header lead. In ultrasonic (US) bonding, either aluminum or gold may be used for the bonding wire.

In the early days of the semiconductor industry, lead bonding was performed manually—in that motion of the header chuck, visual check of the alignment, and application of pressure were all controlled by an operator. Now, automation has taken over.

It is interesting, though distressing, that the minimum size of the chip bonding pads is set by the mechanical requirements of bonding, whether of the TC or US type, rather than by the electronic design or processing of the chip itself. The minimum dimension required to bond the wire to the chip is limited to roughly 3 mils (about the diameter of a human hair) so we expect rectangular bonding pads on ICs to be no smaller than 3 mils × 3 mils. Rectangular pads require no more useful chip space than do circular ones of the same diameter, and rectangles are easier to lay out in mask preparation than circles.

As pointed out earlier, typical IC interconnects tend to have widths in the micron range. With dimensions in the mil range, bonding pads are the most visible features on an IC chip when viewed under low magnification; a lot of real estate on a chip is devoted to the needs of micro-to-macro-world interfacing.

The final assembly step involves closing and sealing the body of the package. The actual processes that are required depend upon the type of package being used. The hermetically sealed types require placement of a lid or cap, which then is welded or soldered to the main body under vacuum. For the more common rectangular plastic types, the metal lead frame is inserted into a hot-molding press and plastic is injected. This is then pressed to shape and cured to form the body around the chip. Packages of this general shape with up to 50 external leads have been used.

Finally, the packaged devices are tested. If some test bad at this point, and inevitably some will, it is depressing and expensive. Notice that batch processing has been used through the 100% wafer probe. Once dicing takes place, however, each chip must be handled individually as far as die- and lead-bonding are concerned, so that cost per unit goes up. Failure of a unit in final test means discarding a unit whose cost has peaked. This is particularly critical when the package cost may exceed the cost of the chip inside the package.

1.9 Summary

In this chapter we briefly have considered basic semiconductor devices and the principal fabrication processes for making them. Some mention was made of the sequence in which the processes are used. As a review, Table 1–1 lists the basic processes that are available to the engineer concerned with wafer fabrication.

Not all the steps required to make a complete diode, transistor, or IC have been covered, but all the necessary information has been given. There really are two types of problems here. (1) Given the masks, what sequence of which processes are required? (2) Given the device and the process sequence, what masks are required? These are addressed in the problems for this chapter.

Table 1–1. The Process Engineer's Arsenal

1. Thermal Oxidation, $Si\downarrow + O_2\uparrow \rightarrow SiO_2\downarrow$, $T \geq 900°C$
2. Chemical Vapor Deposition (CVD)
   a. $SiH_4\uparrow + O_2\uparrow \rightarrow SiO_2\downarrow + 2\ H_2\uparrow$, $T \approx 400–500°C$
   b. $3\ SiH_4\uparrow + 4\ NH_3\uparrow \rightarrow Si_3N_4\downarrow + 12\ H_2\uparrow$
   c. $SiH_4\uparrow \rightarrow Si\downarrow + 2\ H_2\uparrow$
3. Mask Making
   a. Artwork
   b. Photoreduction
   c. Step and Repeat
   d. Photomasks
4. Photolithography
   a. Photoresist
   b. Selective Etching
      1) Liquid
      2) Plasma
   c. Oxide Masking
   d. Mask Alignment
5. Selective Doping
   a. Diffusion, $T \geq 900°C$
   b. Ion Implantation, Room T
6. Metallization

Problems

1–1 Commonly used units in the semiconductor industry are a mix from English, CGS, and MKS systems. You must be able to convert among the several units. See Appendix G.

(a) Fill in the blanks for these conversions:

$1\ \mu m =$ ____________ cm = ____________ in. = ____________ Å

1 mil = ____________ μm, $1\ \mu in. =$ ____________ μm,

1 Å = ____________ $\mu in.$

Note: 1 mil = 10^{-3} in.

(b) $1.5 \times 10^{-15}\ cm^2/s =$ ____________ $\mu m^2/min$

$10\ mil^2 =$ ____________ μm^2

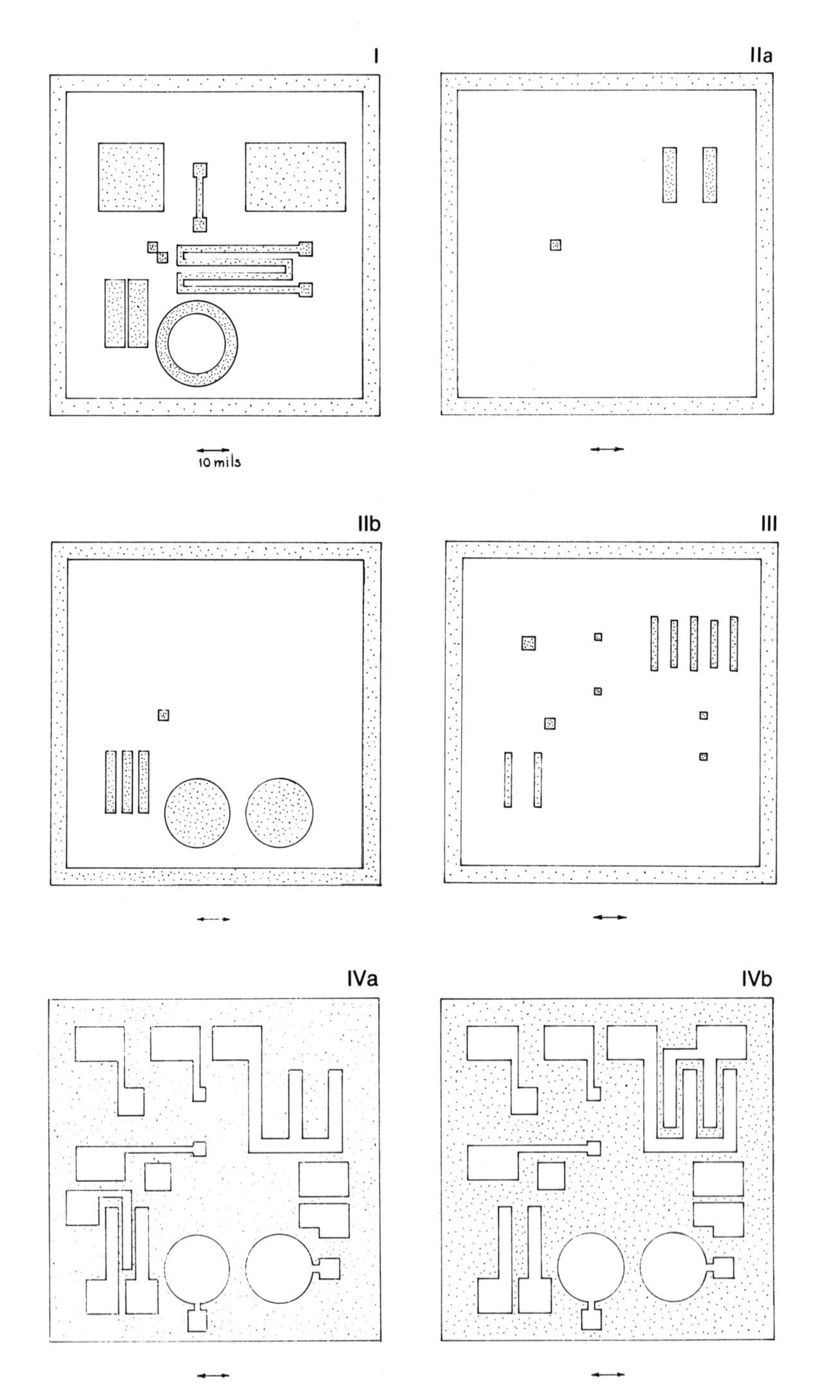

Figure 1–P-1. Masks for Problem 1–2.

1–2 Refer to Figure 1–P-1. These masks can be used to fabricate several devices. Assume an oxidized n-type water. Negative PR is used throughout. Masks I, IIa, and IIb are prediffusion masks for etching windows in the oxide. Mask III is the preohmic mask for etching windows in the oxide prior to metallization. Masks IVa and IVb are used to etch away excess metal. Diffusions are followed by deglaze and reoxidation.

(a) Using the mask sequence I, IIa, III, and IVa, sketch cross sections of each device and identify them.
(b) Repeat for the sequence I, IIb, III, and IVb.

1–3 Sketch the mask set for all the devices in Figure 1–3. Assume a rectangular top view for all of them.
1–4 Sketch the top view of two p-type resistors in a single n-epi tub. Explain how the two are electrically isolated from each other. Are certain d-c bias voltages *essential* to isolation here? Explain.
1–5 Sketch external biasing connections for the devices in Figure 1–3 so that between-tub isolation is guaranteed, irrespective of other applied d-c voltages. Explain.

References

[1] Several reviews of the overall procedure are available. One with several photographs: "Wafer Processing," *Circuits Manuf.*, **20**, (7), 21–34, July 1980.
[2] B. G. Streetman, *Solid State Electronic Devices*, 2nd ed. Englewood Cliffs: Prentice-Hall, 1980, pp. 187–190.

Chapter 2
Wafers

A polished, monocrystalline silicon wafer is the starting material for solid state devices produced by silicon planar processing. In this chapter we consider four main topics concerning these wafers: (1) Why are they monocrystalline? (2) What are their crystalline properties? (3) How is their electrical resistivity related to doping? (4) How are wafers prepared from the raw source material of sand? In the next section we consider the crystal structure of silicon, the properties of its principal planes, and the reasons why certain of the planes are chosen for wafer faces.

2.1 Crystallographic Considerations

If silicon is melted in a suitable container and then cooled until it freezes out into a solid again, the new solid will be composed of a number of small crystals joined together in a single mass. All of these have the same lattice structure, but are randomly oriented relative to each other. This material is called *polycrystalline silicon* or simply *polySi*, and for the most part is not satisfactory for making semiconductor devices. Solar cells may be an exception, but most devices require single-crystal material. Consider the reasons for this.

2.2 Mono- versus Polycrystals

The key to success in semiconductor fabrication is reproducibility: wafer after wafer, device after device should be the same. This requires uniform starting material with known and constant properties. Monocrystalline silicon (*monoSi*) has a uniformity not afforded by polySi. The reasons for this may be seen with help from Figure 2–1, where the sketches diagram wafers with monocrystal and polycrystal structures. The single crystal has long-range order in all directions throughout its entire bulk. This translates into pre-

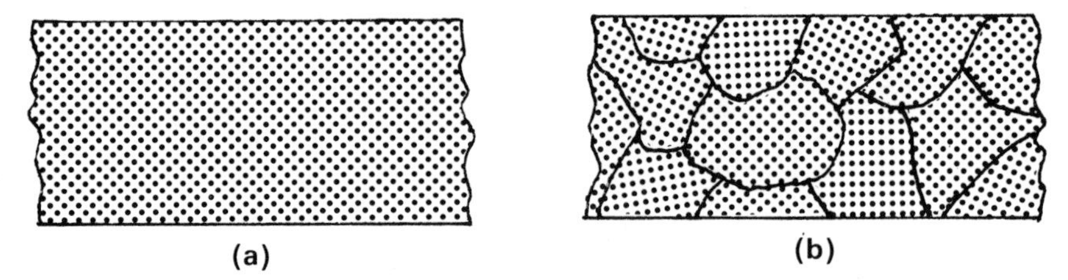

Figure 2–1. Crystal types. (a) Single crystal. (b) Polycrystal.

dictable values of bandgap E_g, carrier mobility μ, bulk resistivity ρ, and other properties that are of concern in solid state devices.

The polycrystalline sample in Figure 2–1(b), however, has only short-range order—since it is broken up into many small single crystals or *grains* that are separated by *grain boundaries.* This type of structure results when silicon is allowed to freeze out from a melt or condense out from a gas phase in a random fashion. (Condensation from the gas phase to produce polySi is discussed in Chapter 9.) In either process the crystals may begin to form on a number of different sites, but their crystallographic orientations will not necessarily be the same. The crystals continue to grow, with each small crystal maintaining its original orientation. Eventually, adjacent grains grow together and touch, forming the grain boundaries. These are discontinuities in the crystal lattice structure, as shown in Figure 2–1(b), and affect the electrical properties of the silicon. For example, values of resistivity ρ may not match the values given by monocrystal samples.

Grain boundaries also change the predicted values of dopant penetration into the wafer during diffusion, because diffusion in each small crystal is slower than the nondiffusion motion of the dopant along a grain boundary. As a result, the incoming dopant penetrates deeper into the wafer along the grain boundaries, causing *spikes* in the otherwise flat junction that lies parallel to the surface.

We shall see in Section 2.8 that the key to single-crystal growth is to provide only one point from which the crystal freezes out from the melt. This point is established by a *seed crystal* of single-crystal silicon, whose orientation sets the orientation of the entire solid crystal being formed.

2.3 Structure, Planes, and Directions

Silicon crystallizes with covalent bonding in the cubic, diamond lattice illustrated in Figure 2–2 [1]. There are three principal crystal planes or directions of interest in silicon technology. These planes exhibit different properties and are defined in terms of the *unit cell,* which in this case is a cube of side dimension *a.* The first plane of interest is defined by any side of the unit cell

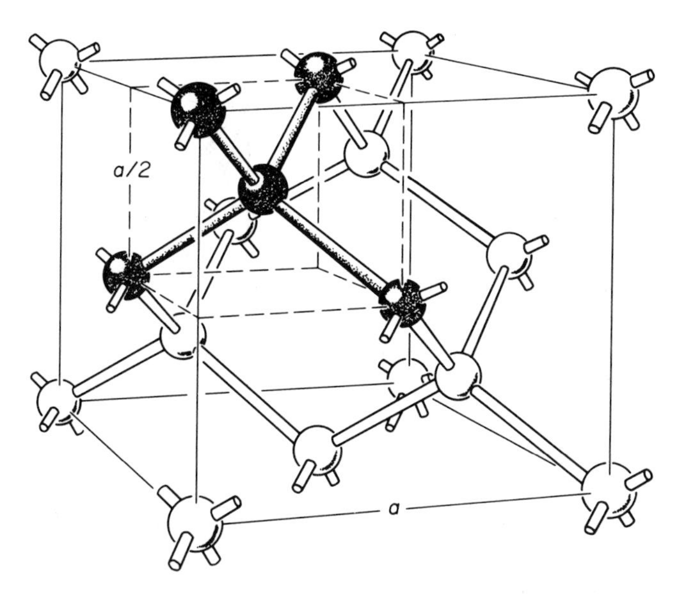

Figure 2–2. The diamond lattice. (From ELECTRONS AND HOLES IN SEMICONDUCTORS by William Shockley © 1950, renewed © 1977 by Wadsworth Publishing Company, Inc. Reprinted by permission of the publisher.)

which, if viewed as an isolated cube, contains five atoms.[1] The plane is shown shaded in Figure 2–3(a) and the normal to the plane is defined as its *direction.* This direction is designated [100] in the figure and the plane as (100). The three ordered digits "1, 0, 0" are called *Miller indices* and are used to identify the planes and directions in cubic crystal structures. They are derived on the following basis. The first number is the reciprocal, normalized with respect to a, of the intersection of the shaded plane and the x-axis. The intersection is a; dividing by a we get $a/a = 1$, and the reciprocal is 1, the first Miller index.

The second number or Miller index is the reciprocal of the y-value, normalized with respect to a, where the plane intersects the y-axis. The intersection is at ∞, since the plane is parallel to the y-axis, so the second Miller index is the reciprocal of ∞/a or 0. The third index is handled in like manner, except that the intersection of the plane with the z-axis is involved. Since the plane is parallel to the z-axis, the third index is also 0. The three indices are placed

[1] Actually in the solid structure each unit cell is surrounded on all sides, the top, and the bottom by other similar cells. As a result, each corner atom is shared by eight cells and each atom in the center of the cube face is shared by two cells. Thus the number of atoms "belonging" to the face of a single cell is: $(1/8)(4) + (1/2)(1) = 1$ atom.

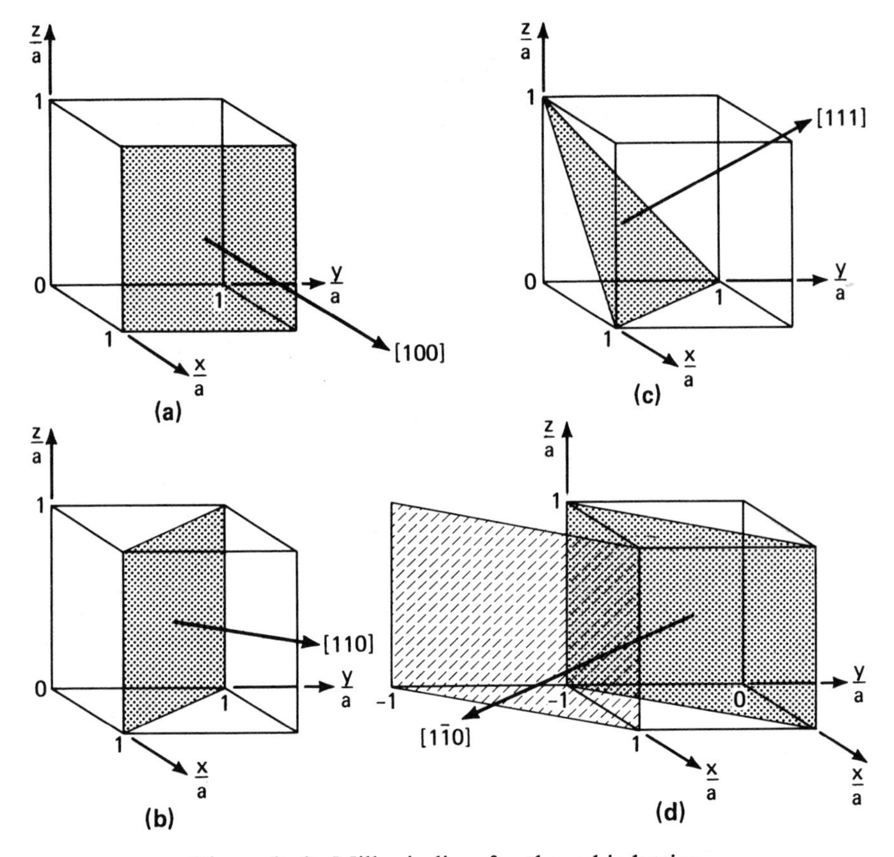

Figure 2–3. Miller indices for the cubic lattice.

in parentheses in the (x, y, z) sequence as (100) to designate the plane. (Note the mnemonic: **p**lane and **p**arentheses go together.) Since the choice of the origin is arbitrary in a periodic structure, all the surfaces of the cube and all planes parallel to them have identical properties. Braces around the indices thus, $\{100\}$, indicate the *family* of these equivalent (100) planes.

The direction normal to the plane [Figure 2–3(a)] is indicated by the same Miller indices, but they are enclosed in square brackets, viz, [100]. The family of equivalent directions is indicated by $\langle 100 \rangle$. Notice that if the direction is considered as a vector, the Miller indices are its three coordinates, normalized with respect to a, in the sequence (x, y, z). Some students find it easier to consider the direction first and then the plane to which it is normal.

The second plane of principal interest is the (110) plane, which lies at 45° relative to the (100) plane of Figure 2–3(a). The (110) plane and its direction [110] are shown in Figure 2–3(b). Let us check the Miller indices: The plane intersects both the x- and y-axes at a. The corresponding normalized values are $a/a = 1$, as are the reciprocals. The plane is parallel to the z-direction and so intersects the z-axis at ∞, so this normalized reciprocal is a/∞ or 0. Thus, the plane is (110) and its direction [110].

The shaded plane of Figure 2–3(d) shows a variation of this example. The situation is the same as at (b), except that the origin, whose location is arbitrary, has been shifted a units in the positive y-direction. As a result, the shaded plane now intersects the y-axis at $-a$. The normalized intersection value thus is $-a/a = -1$. Such a point (with a negative value along one of the axes) is indicated by placing a bar over the corresponding Miller index thus, $\bar{1}$. The bar over the y index replaces the negative sign. The Miller designation for the shaded plane in Figure 2–3(d) is, therefore, $(1\bar{1}0)$, and for the direction, $[1\bar{1}0]$.

For both of the last two examples, (110) and $(1\bar{1}0)$, there are 8 atoms in those planes in the unit cube. Note that this number is not calculated for the *shared* basis, and remember that the unit cell diagonal is longer than the side.

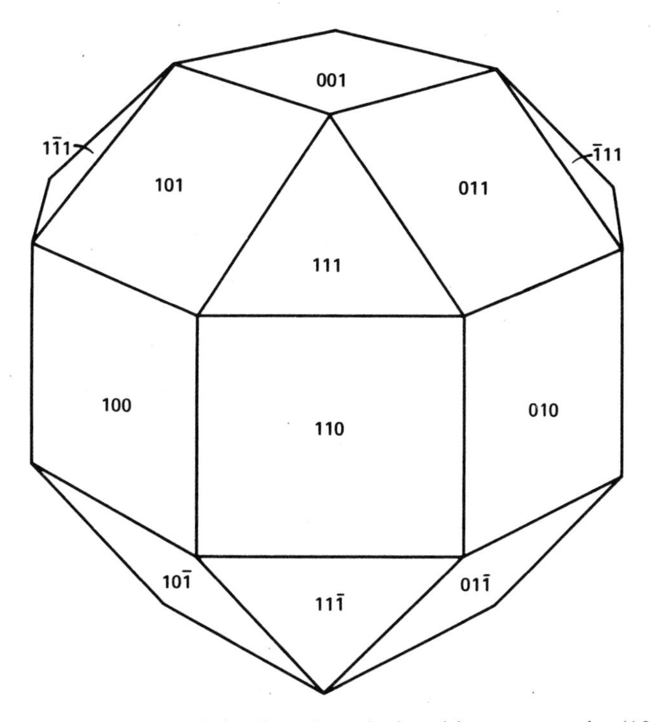

Figure 2–4. One means of viewing the relationships among the (100), (110), and (111) planes in a cubic lattice. Townley [2]. (Courtesy of *Solid State Technology*, PennWell Publishing Company, Copyright 1973.)

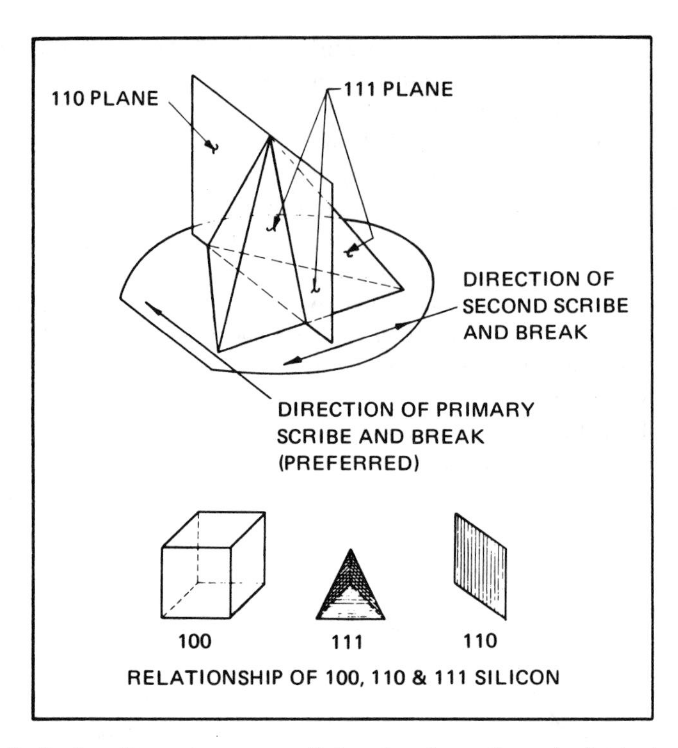

Figure 2–5. An alternate means of showing how the principal planes of a cubic lattice are related. Wosnock [3]. (Courtesy of *Circuits Manufacturing*, © 1973, Miller Freeman Publications.)

The third principal plane of interest is (111). Its corresponding direction [111] is normal to that plane. Working backwards, we can see that the plane will intersect all three axes at normalized values of 1. Both the plane and direction are shown in Figure 2–3(c). The (111) plane in the unit cube contains 9 atoms, although they really lie in 2 levels that are slightly displaced from each other by a small fraction of a, the cube side length. (Here again, the number 9 neglects sharing with adjacent cells.) Remember that the plane considered here is triangular rather than rectangular, as were those in the two previous cases. A three-dimensional model of the diamond lattice greatly clarifies these relationships.

It is important to get a good physical picture of how the three principal planes are inclined with respect to each other. One way of visualizing the relationships is shown in Figure 2–4 [2], which also shows how the various members of a plane family—e.g., (100), (001), and (010)—are related as variations of the {100} family. Another way is shown in Figure 2–5 [3].

2.4 Principal Plane Properties

The three principal planes have different structural and electrical properties and, depending upon how the wafer is oriented relative to the silicon lattice, can affect the devices to be made. The properties are summarized in Table 2–1, yet a fuller understanding of how these differences arise and their effects on processing and device behavior is desirable. Many of the properties are based on geometrical relationships of the Si atoms in the crystal, and a three-dimensional model of the diamond lattice, Figure 2–2, is invaluable for seeing them. The following three sections will clarify these points on a plane-by-plane basis for the interested student.

2.4.1 {100} Planes

There are five clearly defined {100} atomic layers in the unit cell. Adjacent layers are separated by $a/4$, where a is the *lattice constant* or length of the unit cell bounded by {100} faces. These layers are shown in Figure 2–6, where the dashed lines define the unit cell. Figures 2–6(a) to (d) show successively lower planes, spaced at $a/4$ intervals. The bottom plane, which is located a below the top, is identical to the top plane—shown at (a) in the figure and labeled 0 for reference. The circles represent Si atoms, and are shaded for identification in three of the layers. The small arrows pointing toward or away from an atom symbolize bonds to that atom; there always are four of these, a characteristic of the *covalent* bond. In these sketches the arrowheads point from an atom in one layer to an atom in the next layer $a/4$ lower.

Figure 2–6(e) shows the projection of the atoms onto the top surface. Observe that the hatching for the several planes is maintained for easy identifi-

Table 2–1. Comparison of the Principal Planes

Property	{100}	{110}	{111}
Orthogonal planes	yes	yes	no
Plane shape	square	rectangle	hexagon
Plane area of unit cell	a^2	$\sqrt{2}\,a^2$	$\sqrt{3}\,a^2/2$
Atoms in unit cell area	2	4	4
Bonds in unit cell area	2	4	4 (γ) 3 (s)
Atoms/cm^2	6.78 E14	9.59 E14	7.83 E14
Bonds/cm^2	6.78 E14	9.59 E14	7.83 E14 (γ) 2.35 E15 (s)
Spacing between planes, Å	1.36	1.92	2.35 (γ) 0.784 (s)
Channel cross section	square	hex	triangular

Note: $a = 5.43$ Å for silicon.

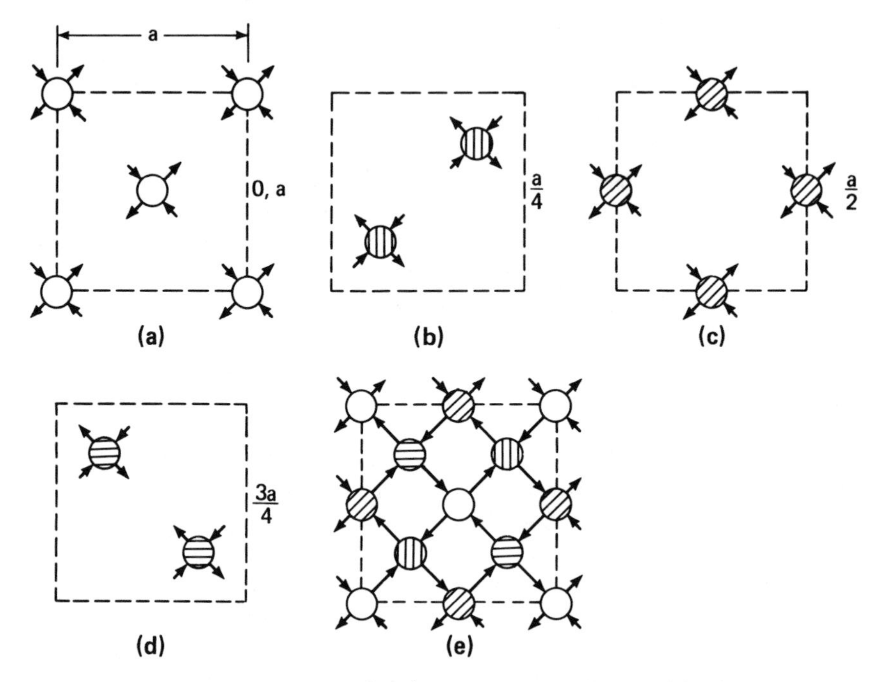

Figure 2–6. Several {100} planes in the diamond lattice.

cation. An arrow touching only one atom in the unit cell indicates a bond between that atom and another lying outside the unit cell.

Note these features of the {100} family:

1. All {100} planes are either parallel or normal.
2. Adjacent {100} planes are spaced by $s_{100} = a/4$. This value may be checked from Figure 2–6(e). Since $a = 5.43$ Å for silicon, we have

$$s_{100} = \frac{a}{4} = \frac{5.43}{4} = 1.36 \text{ Å} \tag{2.4-1}$$

3. The planes normal to those of Figure 2–6(e) and parallel to the diagonals of the unit cell are {110} planes. Reading along a corner-to-corner diagonal, of length $d = \sqrt{2}\,a$, in Figure 2–6(e), we see that the spacing between adjacent {110} planes is

$$s_{110} = \frac{\sqrt{2}\,a}{4} = \frac{\sqrt{2}}{4}(5.43) = 1.92 \text{ Å} \tag{2.4-2}$$

4. Corner atoms in the plane of Figure 2–6(a) are shared among four unit cells *in that plane*, while the center atom is not shared. Thus the number of atoms in the {100} face of the unit cell is (1/4)4 + 1 = 2. This checks with Figures 2–6(b) and (d). Then η_{100}, the number of atoms per cm^2 of a {100} plane, is

$$\eta_{100} = \frac{2}{a^2} = \frac{2}{(5.43 \times 10^{-8})^2} = 6.78 \times 10^{14}\ \text{cm}^{-2} \qquad (2.4\text{-}3)$$

   This is important in choosing wafer orientation for a particular device type (Section 2.5).
5. Atoms projected on a {100} plane as in Figure 2–6(e) show a network of square *channels*, a significant point in Chapter 8, where ion implantation *chanelling* is considered.
6. Adjacent {100} planes in Figure 2–6(e) are joined by four bonds in the unit cell. This and the values of s_{100} and s_{110} are important when cleaving silicon or scribing wafers for subsequent dicing. These are subjects in Section 2.6.

2.4.2 {110} Planes

The dashed rectangle in Figure 2–7 delineates the (110) plane of the unit cell. The figure itself illustrates the projection of the atoms onto a (110) plane located at an arbitrary zero reference level in the lattice. The unshaded atoms lie in the reference level and in the two levels lying at distances $\sqrt{2}\,a/2$, and $\sqrt{2}\,a$ below the reference level. The shaded atoms are in planes located $\sqrt{2}\,a/4$, and $3\sqrt{2}\,a/4$ below the reference level.

The fourth bond to each atom is hidden behind the bond that is vertical in the figure. The arrowheads of these bonds are in opposite directions: one

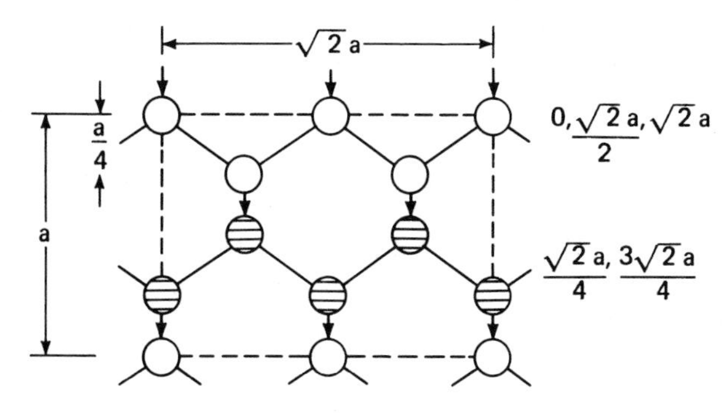

Figure 2–7. The (110) plane of the diamond lattice.

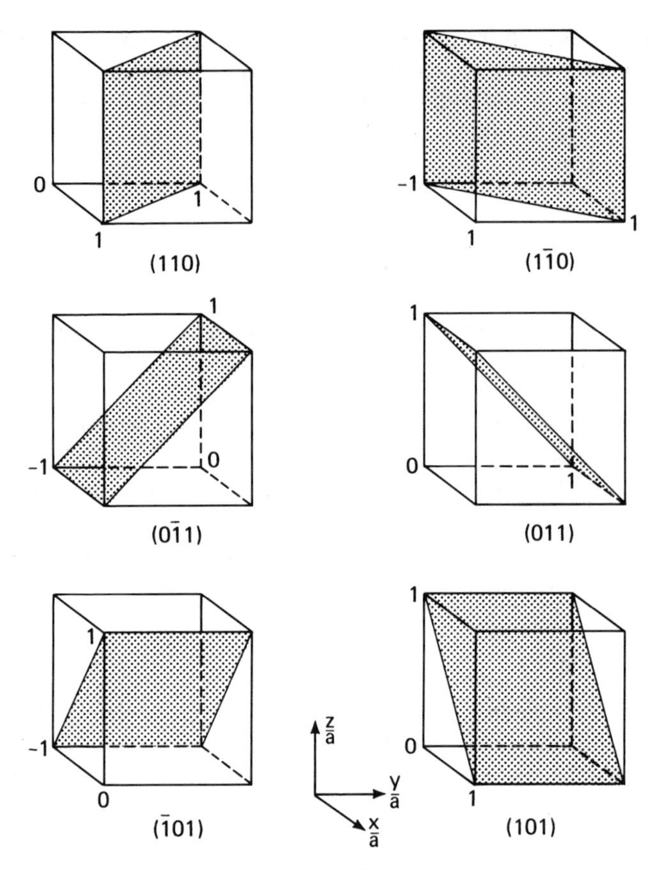

Figure 2–8. Orthogonal pairs of the $\{110\}$ planes of the diamond lattice.

points toward, the other away from the atom. As before, arrowheads on bonds between two atoms point from an upper to a lower plane. Bonds without arrowheads connect two atoms in the same plane. We note the following points:

1. The three pairs of orthogonal $\{110\}$ families in Figure 2–8 are: (110) and (1$\bar{1}$0); (101) and ($\bar{1}$01); and (011) and (0$\bar{1}$1). Each of the planes has a parallel family. Every $\{100\}$ face of the unit cell is cut by an orthogonal pair of $\{110\}$ planes. This is important in scribing wafers (Section 2.6).
2. From Eq. (2.4-2), the spacing between adjacent $\{110\}$ planes is $s_{110} =$ 1.92 Å.
3. The unshaded atoms of Figure 2–7 lie in one plane. Each of the 4 corner atoms is shared among 4 unit cells in that plane; each of the 2 atoms located midway on the sides is shared between 2 unit cells. There are 2

unshaded atoms in the cell, which are not shared in that plane. So, the number of atoms in the $\{110\}$ face is $(1/4)4 + (1/2)2 + 2 = 4$, and η_{110}, the number of atoms/cm^2 of a $\{110\}$ plane, is

$$\eta_{110} = \frac{4}{\sqrt{2}\,a^2} = \frac{4}{\sqrt{2}(5.43 \times 10^{-8})^2} = 9.59 \times 10^{14}\ \text{cm}^{-2} \quad (2.4\text{-}4)$$

4. The $\{110\}$ face in Figure 2–7 shows a network of hexagonal channels, but these are not equilateral. Four sides are of "actual" bond length $\gamma = \sqrt{3}\,a/4 = 2.35$ Å; the other two have length $a/4 = 1.36$ Å.
5. Reading from Figure 2–7, we note that adjacent $\{110\}$ planes are joined by four bonds per unit cell.

2.4.3 $\{111\}$ Planes

The $\{111\}$ plane family is more complicated and difficult to visualize, due partly to the lack of orthogonal sets (see Figures 2–4 and 2–5), and partly to the nonuniform spacing between adjacent parallel planes.

Some basic units of the $\{111\}$ set will be determined. From Figure 2–3(c), the basic cell plane is an equilateral triangle whose side is the diagonal of the unit cube face; hence, the triangle side has length $\sqrt{2}\,a$.

Another basic unit is the "actual" bond length γ, which is the diagonal of the cube whose side is $(a/4)$ (Figure 2–2); hence,

$$\gamma = \text{“actual” bond length} = \sqrt{3}\,a/4 = 2.35\ \text{Å} \quad (2.4\text{-}5)$$

Figure 2–9 presents sketches of two adjacent $\{111\}$ planes, bounded by the triangle mentioned above. The figures in the left-hand column, labeled (a) and (c), show views normal to $\{111\}$ planes; the right-hand figures, labeled (b) and (d), show corresponding side views to illustrate the bonding patterns between two $\{111\}$ planes.

Figure 2–9(a) shows the atoms in a reference level as shaded, and the dashed line delineates the equilateral triangle of the basic cell plane. The convention for the bond arrowheads is the same as before: they point to a lower level. At (b) the side view is shown for the front row of atoms (1, 2, 3) in (a) and the unshaded center atom (6) of the next lower plane [Figure 2–9(c)]. The two dashed-line atoms are projections of the atoms (4, 5) lying halfway up the triangle sides at (a).

Figure 2–9(c) shows the unit triangle of the next lower $\{111\}$ level and (d) shows the atom (6) in cross section, plus the bonding to the layers above and below. This view shows an interesting and important point: adjacent $\{111\}$ planes are not spaced uniformly. Rather, the spacing is $s, \gamma, s, \ldots,$ throughout six successive planes that lie within d, the length of the unit cell's principal diagonal, which is

$$d = \sqrt{3}\,a = 9.41\ \text{Å} \tag{2.4-6}$$

There are three short spacings s_{111} and three long spacings γ along the principal diagonal of the unit cell; hence,

$$d = 3(s_{111} + \gamma)$$

or

$$s_{111} = \frac{d}{3} - \gamma = \frac{\sqrt{3}\,a}{3} - \frac{\sqrt{3}\,a}{4} = \frac{a}{\sqrt{3}\,4} = 0.784\ \text{Å} \tag{2.4-7}$$

Recall from Eq. (2.4-5) that $\gamma = 2.35$ Å.

The sketches of Figure 2–9 show atoms lying in the {111} unit cell *triangle*, but a clearer concept of the {111} structure is shown in Figure 2–10. First, we see a larger, hexagonal projection of the unit cell onto a {111} plane, rather

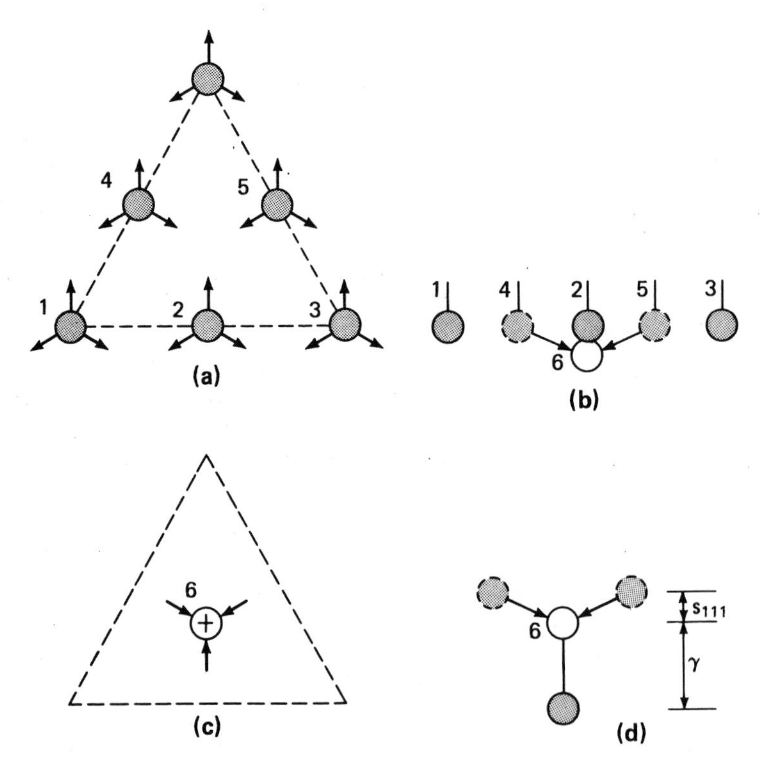

Figure 2–9. Two {111} planes of the diamond lattice. Atoms lying in the unit triangle are shown.

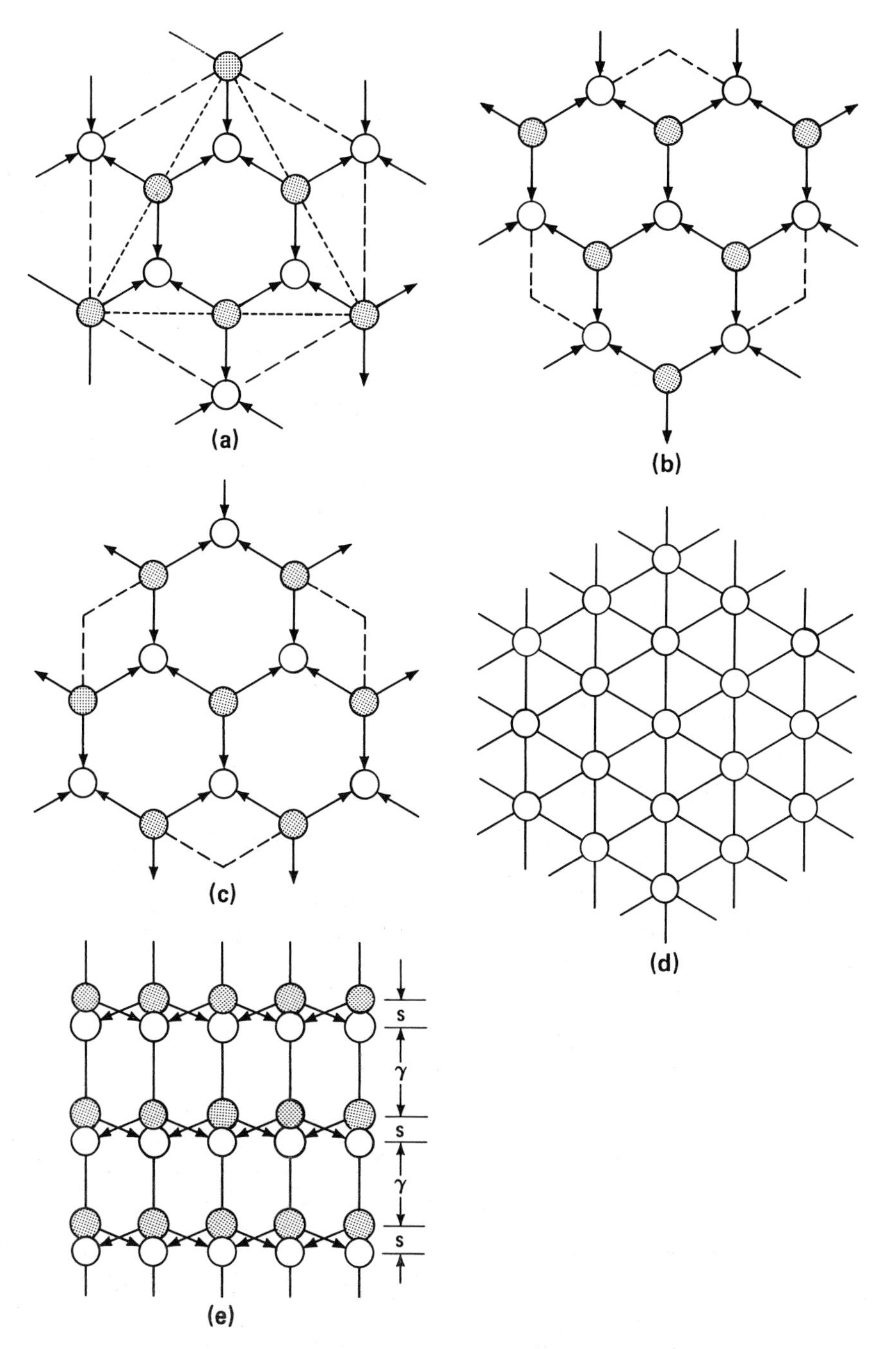

Figure 2–10. Some {111} planes of the diamond lattice. The hexagonal projection of the unit cell is used.

than the triangle of Figure 2–9. The larger area shows more atoms and their bonds, and reveals the $\{111\}$ structure more clearly. Second, since s_{111} is smaller than γ by a factor of one-third, it is convenient to think of two adjacent planes that are separated by s_{111} as a single *super layer* or *super plane* in our sketches. We shall, however, maintain the planes' separate identities by using the shaded/unshaded convention of Figure 2–9: shaded atoms are in the upper layer of the super plane and unshaded atoms are in the lower. Figure 2–10(a) shows the relationship between the hexagonal and triangular areas, as well as the atoms and bonds in a reference $\{111\}$ super layer.

Figures 2–10(b) and (c) show the next two successive super planes. Remember that the unshaded atoms in each of these actually lie s_{111} below the shaded ones and that adjacent super planes are separated by γ. The γ-length bonds joining atoms of adjacent super planes are normal to the planes of the sketches—i.e., normal to the $\{111\}$ planes. The next super plane is not shown because it is identical to the one in (a).

Consider some features that are illustrated by the drawings. In moving down through the structure from (a) to (b) to (c), unshaded atoms are replaced by shaded—i.e., the γ-length bonds that are normal to the $\{111\}$ planes are directed downward from unshaded to shaded atoms. Note the shift of the atom positions from one super plane to another. An hexagonal pattern is clear in Figures 2–10(a) to (c). The projection of these figures onto a single $\{111\}$ plane as in Figure 2–10(d) shows, however, that the entire structure has equilateral triangular channels when viewed along any $\langle 111 \rangle$ direction. This fact is of importance in ion implantation, as discussed in Chapter 8.

Figure 2–10(e) shows the vertical structure—i.e., normal to the $\{111\}$ planes. The patterns of (a) to (c) are projected onto the plane of the three shaded atoms along the base of the triangle in (a). Note the alternate inter-plane spacings of s_{111} and γ as predicted earlier; also note that channels in the planes normal to the $\{111\}$ planes are elongated hexagons separated by two closely spaced planes. The sketches show that the atomic patterns normal to, and in, the $\{111\}$ planes are different; therefore, there are no orthogonal $\{111\}$ planes.

We summarize the observations about the $\{111\}$ family:

1. There are no orthogonal sets of $\{111\}$ planes.
2. There are two spacings between adjacent planes: s_{111} and γ projected onto the normal to the (111) plane. These values alternate, as may be seen in Figure 2–10(e).
3. In Figure 2–10(a), an atom at an apex of the equilateral hexagon is shared among three hexagons in the plane; there are 3 such shaded apex atoms in the upper layer of the super plane. Also, there are 3 shaded atoms lying wholly within the hexagon; therefore, the number of atoms in the shaded layer bounded by the hexagon is $(1/3)3 + 3 = 4$. This is true for all the $\{111\}$ levels. The hexagon area is

$$A_{hex} = \sqrt{3}\, a^2 \tag{2.4-8}$$

Four atoms lie in this area, so η_{111}, the number of atoms per cm^2 of a $\{111\}$ plane is

$$\eta_{111} = \frac{4}{\sqrt{3}\, a^2} = \frac{4}{\sqrt{3}(5.43 \times 10^{-8})^2} \tag{2.4-9}$$
$$= 7.83 \times 10^{14}\ cm^{-2}$$

4. There is one bond per atom of length γ in the hexagon; thus, the number of bonds per cm^2 also is

$$B_{111} = 7.83 \times 10^{14}\ cm^{-2} \tag{2.4-10}$$

5. Within the hexagon, each of the two planes, one shaded and one unshaded, that comprise a single super plane contains 4 atoms. Each of these atoms has 3 bonds in the short-bond direction. Therefore, the number of bonds per cm^2 between two closely spaced adjacent layers is

$$B_{s111} = 12/A_{hex} = 2.35 \times 10^{15}\ cm^{-2} \tag{2.4-11}$$

6. Figure 2–10(d) shows that the channels in the $\langle 111 \rangle$ directions have equilateral triangular cross sections.

The principal numerical results of the foregoing sections are summarized in Table 2–1 (page 45).

2.5 Crystal Structure Implications

The crystallographic orientation of the wafers from which devices are made can affect: (1) the growth properties of the source crystals from which the wafers are cut, (2) the growth rate of SiO_2 on the wafers, (3) the growth properties of epitaxial silicon layers that are deposited on the wafer during device fabrication, (4) the electrical performance of those devices, and (5) wafer dicing—where the several finished electronic parts are separated into individual chips. Several of these effects are related to B, the number of available bonds per square centimeter.

Consider the effect of B when silicon is oxidized. A number of the *dangling bonds* at the surface—i.e., those that have no silicon atoms above the surface to join—are available to link up with oxygen atoms to form SiO_2. From Table 2–1 we see that B is larger for $\{111\}$ planes than for the $\{100\}$s; therefore, other things being equal, the rate of oxide growth during thermal oxidation will be larger for $\{111\}$ than for $\{100\}$ planes.

Table 2–1 shows that η, the atomic density in a plane, is greatest for the {111} super planes. Thus, during crystal growth we would expect growth to take place at the highest rate along the {111} planes. In the epitaxial process, growth starts at random nucleation sites and, theoretically at least, spreads laterally along the {111} plane. Then another layer starts by nucleation, and so on. As a practical matter the growth does not proceed in such an orderly, layer-by-layer manner.

It has been found that if the substrate layer on which the epi is grown has faces inclined 3 to 4° off the [111] axis toward the nearest [110] direction, the corners of the {111} planes tend to become nucleation sites and the growth process is speeded up. It is common practice, therefore, to use wafers that are really about 4° off from the actual {111} planes, if subsequent epi growth is required in device processing as with bipolar integrated circuits.

Note that 4°-off-{111} crystal boules may be grown just as well as actual {111} boules simply by adjusting the orientation of the seed crystal used in pulling the boule from the silicon melt (Section 2.10). These off-axis wafers also may be obtained, however, by sawing them from an actual {111} boule at the 4° angle.

For the reason mentioned above relative to η and the fast growth of {111} planes, single-crystal growth from the melt will be greatest if the crystal is grown in or near to the [111] direction. This direction of crystal growth also tends to minimize a defect in crystal structure known as *twinning*, whereby part of a growing crystal is related to an earlier part by a change in symmetry [4]. For these reasons, crystal growers prefer to grow {111} crystal boules over other orientations.

On the other hand, certain requirements for MOS devices favor the {100} orientation. Recall that in the oxidation process some of the dangling bonds at the wafer surface link up to oxygen atoms to form SiO_2. The remaining surface bonds accumulate charge and so contribute to the *surface density of states*, N_{SS}, which affects surface potential. This N_{SS} has an important effect on the threshold voltage of MOS devices, so it should be as small as possible and of controlled value. Table 2–1 shows that of the three principal planes, the {100} family has the smallest number of atoms per cm^2, η, and the smallest number of available bonds per cm^2, B. This has special significance in MOS devices. With B smallest, N_{SS} will be smallest also; therefore, {100} wafers are invariably used for MOS technology, thereby creating a large demand for wafers of {100} orientation as well as for {111}s, which are preferable for bipolar ICs.

The properties of {110} planes lie between those of the other two, but {110} wafers are not used commonly in the silicon industry. The {100}s and {111}s are the workhorses; silicon boules are almost invariably grown with either [100] or [111] (or 4°-off-[111]) axes, so that after sawing the resulting wafers will have the desired orientations.

2.6 Scribing Considerations

After electronic parts are fabricated in a silicon wafer, a *dicing* process is required to break up the wafer into individual chips, each of which contains one of the fabricated parts. To define the chip boundaries, the surface of the wafer first is scored with a diamond-tipped scribe, a laser beam, or a diamond-coated saw. The saw does not cut completely through the wafer but cuts a kerf extending only partway through. After scoring, the wafer is broken apart. If the scored lines have been laid out properly relative to the silicon lattice, the breaks will follow them. This requires that the first mask be aligned properly relative to the lattice.

Silicon exhibits *cleavage*, a property that is peculiar to certain single crystal materials. Cleavage is observed when a crystal breaks apart with relative ease along certain specific crystallographic planes. In covalent crystals such as silicon, the cleavage planes are those with comparatively weak bonding, so for successful dicing the score lines should run parallel to cleavage planes.

Table 2–1 shows that the longest bond lies between adjacent $\{111\}$ super planes and is γ long. Experimentally, the longest bonds give the weakest bonding, so the $\{111\}$ planes are optimum cleavage planes for silicon. Additional factors must be considered for wafers, however. For example, silicon chips are invariably rectangular; hence, orthogonal cleavage planes are preferred for scribing. Consider what problems arise in this regard for $\{100\}$ and $\{111\}$ wafers.

Figure 2–6(e) shows the bonding pattern of the (100) plane—i.e., the face of a (100) wafer. Even though $\{111\}$ planes are easiest to cleave, they do not occur in orthogonal sets; hence, they are not helpful for dicing rectangular chips. Consider two alternatives for $\{100\}$ wafers: we could scribe parallel to the dashed lines, which are $\{100\}$ planes, or at 45° relative to those dashed lines, along $\{110\}$ planes. Consideration of the relative bond lengths in either the figure or Table 2–1 shows that the $\{110\}$ planes are separated by longer bonds and so cleave better. Therefore, when (100) wafers are used, processing steps should be arranged so that scribing will be along $\{110\}$ planes as in Figure 2–5. Section 2.12.3 shows how a wafer *flat* is used to mark the appropriate (110) plane so it may be identified.

Wafers with the $\{111\}$ orientation present a more complicated problem. Consider Figure 2–11(a), which shows the face view of a (111) wafer. There are no orthogonal planes in the $\{111\}$ family; therefore, we cannot use $\{111\}$ planes to scribe for rectangular chips. A satisfactory compromise is to use the $\{110\}$ planes for scribing again. The planes of choice are those parallel to an altitude of the dashed line triangle and their orthogonals, which are parallel to the corresponding base of the triangle. Both Figures 2–5 and 2–11(a) show the relationships among the planes, but the latter relates to a wafer better.

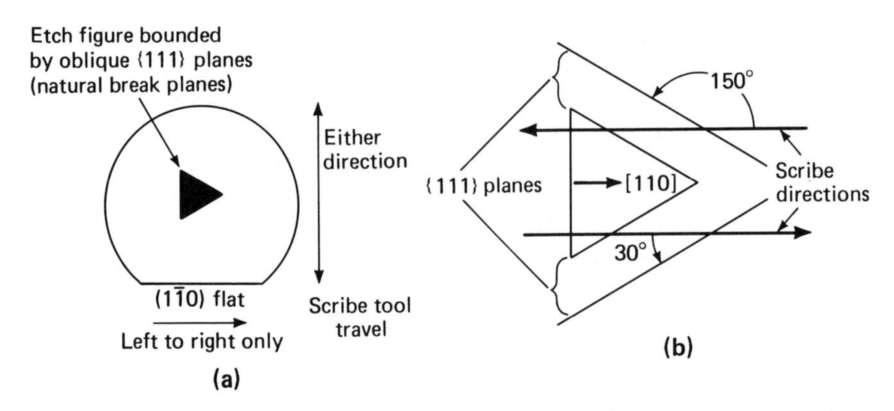

Figure 2–11. Optimum scribing directions for $\{111\}$ silicon wafers. (a) Relationship to the $\{111\}$ triangle. Townley [2]. (Courtesy of *Solid State Technology*, PennWell Publishing Company, Copyright 1973.) (b) Showing the different angles between scribe lines and the $\{111\}$ planes for scribes in opposite directions.

Consider the first scribe, the one parallel to the triangle altitude. Shall the scribe direction be from the base to the apex of the triangle, or from apex to base? Inspection of Figure 2–11(b) shows a difference in the angle between the scribe direction and the $\{111\}$ triangle side on the *right* of the scribe direction. For the base-to-apex scribe this angle is 30° whereas it is 150° for the opposite direction. It has been observed in practice that for both scribe directions the scored line is fringed with tiny microcracks that tend to follow the $\{111\}$ planes. These cracks are smaller for the base-to-apex scribe; this is the preferred scribe direction, as it gives better results when the wafer is broken into chips.

No triangle is visible on a (111) wafer surface to act as a scribing guide. One may be revealed, however, by placing a drop of Sirtl etch on the surface and observing the triangular etch pit shown in Figure 2–11(a) [5].[2] It would not be practicable to use this technique on every wafer, however, especially since the etch pit would waste wafer area that should be used to make parts. The practical alternative is to locate the proper [110] direction by nondestructive x-ray measurements on a silicon boule (Section 2.10) and to grind a flat along the length of the boule parallel to this direction. Subsequently, when the boule is sawed into wafers, each wafer will have an identifying flat as depicted in Figure 2–11(a).

[2] Sirtl etch acts *anisotropically* in silicon in that it attacks different crystallographic planes at different rates and reveals the $\{111\}$ triangular structure clearly. A typical formulation is: by volume, 2 parts of HF and 1 part of 33% (by weight) aqueous solution of CrO_3 (chromic acid) [5]. The etch is named after E. Sirtl.

Townley argues that the wafer flat should be along a $(1\bar{1}0)$ plane so that if the flat is placed at the 6 o'clock position, scribing from left to right will be correct no matter which side of the wafer is scribed [2]. This is in contrast to the situation where the correct scribe direction, left to right or right to left, becomes wafer-side dependent if the flat is a (110) plane.

The second scribe [Figure 2–11(a)], normal to the first, is uncomplicated by comparison; it is along a $\{110\}$ plane that is parallel to an oblique $\{111\}$ cleavage plane (Figure 2–4). The second scribe therefore may be made in either direction.

2.7 Doping Considerations

The electrical parameters of semiconductors can be modified greatly by adding very small amounts of certain dopant atoms. For example, the electrical resistivity ρ of pure or *intrinsic* silicon is 2.5×10^5 Ω-cm, but if only one of every one million silicon atoms is replaced by one atom of arsenic ρ drops to 0.2 Ω-cm! Also the addition of dopants causes the intrinsic semiconductor to become either p- or n-type depending upon the dopant used. It is this property (that added dopants change its electrical parameters) that makes a semiconductor useful for electronic devices.

The unshaded region of Figure 2–12 shows the part of the periodic table that is of interest in silicon technology. In the manufacture of wafers, doping occurs when controlled amounts of the chosen dopant are added to the molten silicon from which the single crystal solid, called a *boule* or *ingot*, is formed. These commonly used dopants have the property of being *substitutional* (i.e., as the solid forms from the molten state, the dopant atoms can substitute for

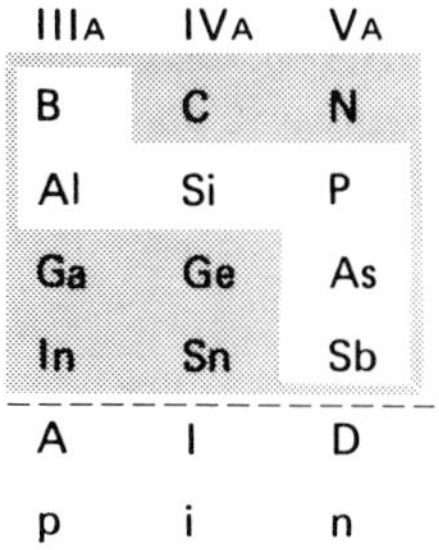

III_A	IV_A	V_A
B	C	N
Al	Si	P
Ga	Ge	As
In	Sn	Sb
A	I	D
p	i	n

Figure 2–12. The portion of the periodic table of concern in semiconductor doping. The unshaded portion shows the elements used in silicon technology. The two lowest lines are mnemonics: "AID" and "pin." The letters A and p indicate that the column III_A elements are Acceptors and make silicon p-type. Letters I and i indicate intrinsic material. Letters D and n remind that column V_A elements are Donors and result in n-type silicon.

and replace silicon atoms in the solid lattice). Dopants from Column IIIA are *acceptors* and cause the silicon to become p-type, with holes being the majority carriers. On the other hand, dopants from Column VA are *donors* and make the semiconductor n-type, with electrons as the majority carriers. In general, the bulk resistivity drops as more dopant is added, up to the limit of solid solubility.

2.7.1 Doping Concentration

The quantity of dopant in silicon can be expressed as a fraction—e.g., one part per million (ppm), or 1 part in 10^6 can be written $1/10^6 = 10^{-6}$. More commonly, the quantity of dopant is expressed as the *concentration* or *number density*, N (often with identifying subscript), in units of cm^{-3}. In semiconductor parlance *density* is often used for concentration even though this term implies different units. In the industry, *doping concentration*, *doping density*, and even *doping level* are used synonymously. The units cm^{-3} are used universally in the semiconductor industry, rather than m^{-3} as might be expected for MKS units.

The doping concentration of a silicon *wafer* often is called the *background doping level*, symbolized by N_b. In some cases, where the type of dopant is known, an alternate notation may be used: N_d if the dopant is a donor, or N_a when the doping is by an acceptor. Two good reference numbers that serve as yardsticks when dealing with doping in silicon are (1) the number of atoms per cubic centimeter of solid intrinsic silicon, namely $N_{\text{Si}} = 5 \times 10^{22}\ \text{cm}^{-3}$, and (2) the *intrinsic carrier concentration* in silicon at room temperature, $n_i = 1.5 \times 10^{10}\ \text{cm}^{-3}$. Typical *background* doping concentrations in silicon wafers run from 10^{14} to $10^{17}\ \text{cm}^{-3}$.[3] This means that the intentional doping level in silicon wafers ranges from one part in 10^8 to one part in 10^5. These numbers illustrate why the purity of silicon must be so high before dopants are added intentionally: the electrical properties must be controlled by the dopants and not by spurious impurities.

2.7.2 Carrier Densities

At or above room temperature, virtually all common donors (or acceptors) in silicon are *ionized* so that the *majority carrier density* N_M is the same as the dopant density N—i.e.,

$$N_M = N \tag{2.7-1}$$

[3] Note that these are *wafer* doping levels. In subsequent doping by ion implantation or diffusion, doping levels may run as high as $10^{21}\ \text{cm}^{-3}$, depending upon the particular doping species and the requirements of the device or part being fabricated.

Now, let N_m be the minority carrier concentration.[4] At thermal equilibrium the actual and intrinsic carrier concentrations are related by [6]

$$N_M N_m = n_i^2 \tag{2.7-2}$$

At or above room temperature, then,

$$N_m = n_i^2/N_M \approx (2.25 \times 10^{20})/N$$

The lowest practical dopant level in typical wafers and devices is $N \approx 10^{14}$ cm^{-3}. Thus, in silicon the typical minority carrier concentration will be $N_m \leq 2.25 \times 10^6$ cm^{-3}, while the majority concentration will be $N_M \geq 10^{14}$ cm^{-3}. For typical silicon device numbers, the minority carrier density can be no more than $(2.25 \times 10^{-8})N_M$—in silicon minority carriers are truly a minority.

2.7.3 Resistivity and Doping

From fundamental considerations, the resistivity ρ of a uniform semiconductor sample is given by [6] or [7]

$$\rho = \frac{1}{\sigma} = \frac{1}{q(\mu_M N_M + \mu_m N_m)} \tag{2.7-3}$$

where σ = electrical conductivity, q = electronic charge magnitude, μ = carrier mobility, and N = carrier concentration.

For typical silicon doping densities $N_m \ll N$. Furthermore, μ_M and μ_m are of the same order of magnitude at any given doping level in silicon; therefore, the second term in the denominator of Eq. (2.7-3) is negligible relative to the first, and we have, to an excellent approximation,

$$\rho = \frac{1}{\sigma} \approx \frac{1}{q\mu_M N_M} = \frac{1}{q\mu_M N} \tag{2.7-4}$$

the last step being true by virtue of Eq. (2.7-1). Some typical values are given in Table 2–2, where the N values are the usual minimum and maximum values used for Si wafers.

The majority carrier mobility μ_M is dependent on N, so Eq. (2.7-4) does not plot linearly on log–log coordinates as might be expected. Empirical curves

[4] New notation is being introduced here. N is the number density or concentration in number of atoms per cm^3. The subscript M indicates **M**ajority carrier, while m indicates **m**inority carrier. This convention avoids the need for knowing whether holes or electrons are the majority carriers.

Table 2–2. Min and Max Values of Si Wafer Doping and Resistivity

N, cm^{-3}	ρ, Ω-cm	
	p-type	n-type
10^{14}	130	50
10^{17}	0.25	0.09

Note: The difference between ρ values for p- and n-types is due to the difference between hole and electron mobilities.

of ρ versus N are given in Appendix F.2.1, the Irvin resistivity curves; corrections for high phosphorus doping levels are plotted in Figure 7–20.[5]

Difficulties may arise in using the ρ versus N curves for *counterdoped* samples—i.e., samples having at least two dopants of the opposite types. A typical example is the base of a discrete, diffused, n–p–n transistor that has donors from the original n-type collector region, plus acceptors that form the base. At any distance x measured from the surface, the effective or net concentration will be[6]

$$N = |N_d| - |N_a| \tag{2.7-5}$$

but the carrier mobilities, which depend upon scattering effects due to *all* the dopants present, will have the form

$$\mu_M = f[|N_d| + |N_a|] \tag{2.7-6}$$

where the function f is unknown.

Note that in computing resistivity the *majority* carrier mobilities must be used. Even if N_d and N_a are constant and of nearly the same value, the curves of Appendix F.2.1 are in severe error; in fact, no general ρ versus N curves are available for such cases. Where N_d and N_a are not constant, other methods are required; these are discussed in Chapter 3.

The emitter region of a diffused transistor presents an even more complicated situation. It is counterdoped twice, so three dopants are present (although emitter and collector dopants may be the same). Typical emitters may be

[5] The data of these curves often are needed for computer calculations. To make this feasible, Irvin breaks the curves into a number of piecewise, power-law segments of the form $\sigma = BN^{\alpha}$ and gives values of B and α for ranges of N. See Section 6.5.2.

[6] Ionized donors have positive charge and ionized acceptors negative charge. In this book we use the convention that if N is signed, N_d is positive and N_a is negative—i.e., they take the sign of the ionized form of the corresponding charge carrier.

doped in excess of 10^{19} cm^{-3}, however, and for such high levels the carrier mobilities become essentially constant and independent of N, and are known as the *degenerate values*; these may be used in Eq. (2.7-4). Typical values of the *majority* carrier mobilities in this range are [8][7]

$$\mu_n \approx 40 \text{ cm}^2/\text{Vs}, \qquad \mu_p \approx 35 \text{ cm}^2/\text{Vs}$$

The presence of impurities such as oxygen also can make the values read from the curves of Appendix F.2.1 incorrect.

2.7.4 Solid Solubility Limit

Equation (2.7-4) implies that ρ can be made vanishingly small if the dopant concentration is raised sufficiently. There is a limit, however, to the amount of any dopant that may be added substitutionally to silicon, called the *limit of solid solubility* and symbolized in this book by N_{SL} with units of cm^{-3}. The value of this limiting concentration depends upon the dopant used and the temperature at which the dopant was introduced into the silicon. *Maximum* possible values of N_{SL} for P and B are 1.5×10^{21} and 6.0×10^{20}, respectively. The concept of the solid solubility limit is discussed in Chapter 4. Curves of N_{SL} versus T, the Trumbore curves, are given in Figure 4–9 [10].

The remainder of this chapter discusses the manufacture of silicon wafers. In the next chapter, we shall consider some electrical tests that are made on wafers for measuring the value of ρ. This value, along with the curves of Appendix F.2.1, allows calculation of the doping density N.

2.8 Polysilicon Preparation

The source material for commercial silicon is impure silicon dioxide (SiO_2) in the form of sand, a very abundant material indeed. Sand is chemically reduced in an electrical furnace to metallurgical grade silicon with approximately 2% impurities. Treatment with hydrochloric acid (HCl) follows, to give one of the five compounds: silane (SiH_4), chlorosilane (SiH_3Cl), dichlorosilane (SiH_2Cl_2), trichlorosilane ($SiHCl_3$), or silicon tetrachloride ($SiCl_4$). Under standard conditions of pressure and temperature the first three are gases. Both $SiHCl_3$ and $SiCl_4$ are liquids, but both may be converted to the gas phase by raising their temperatures.

Further impurities may be removed by chemical means and distillation to a practical limit of approximately one part in 10^5—or as it is often called 5-9 *s*

---

[7] These values are extrapolated from Sze [8, p. 40]. It is not at all uncommon, however, to find different values given for these degenerate mobilities. For example, Glaser and Subak-Sharpe [9, p. 218] cite 75 cm^2/Vs for μ_n. It is difficult to find "correct" parameter values for semiconductor computations. Degenerate mobility values can be dopant-dependent.

(or 99.999%) pure. The compound then is reduced to solid polycrystalline silicon. In some older facilities $SiCl_4$ was the compound of choice for conversion to Si, but trichlorosilane ($SiHCl_3$) is preferred now.

The actual reduction of the silicon compound is carried out in a very high temperature enclosure in the presence of hydrogen, a strong reducing agent. One or more tungsten filaments (or even better, polysilicon rods) are supported in the reducing chamber or *decomposer*. Hydrogen and, say, $SiHCl_3$ are introduced into the chamber while the rods are heated white-hot by passing large electrical currents through them. The highly energy-intensive reduction reaction is

$$SiHCl_3\uparrow + H_2\uparrow \xrightarrow{\text{heat}} Si\downarrow + 3\ HCl\uparrow \tag{2.8-1}$$

with the reduced silicon forming on the rods. There is no single seed crystal on which the reduced silicon can freeze out; hence poly- rather than monosilicon forms on the rods. While the rods are heated white-hot, their temperature must remain below the melting point of silicon—1410°C.

Typical impurity levels at this point should run no more than 1 part in 10^9 or 10^{10}. If further purification is needed, zone refining may be used (Section 4.13). The polySi is the raw material that will be regrown in doped monocrystalline form and eventually will be finished into polished wafers.

2.9 Single-Crystal Growth

Monocrystal silicon for wafers is converted from polySi by proper application of either the Czochralski (CZ) or float zone (FZ) methods of crystal growth. These methods will be considered next [11].

2.10 Czochralski Method

The skeletal setup for growing a single crystal by the Czochralski method is diagrammed in Figure 2–13. Polysilicon is melted in a crucible by resistance or induction heating. Recall that silicon melts at 1410°C, so the crucible material certainly must be able to withstand that temperature. Controlled amounts of dopants (Figure 2–12) are added to the melt.

A *seed* crystal of monocrystalline silicon oriented in the desired direction—e.g., [100] or [111]—is affixed to the lower end of the puller rod. The puller rod and crucible rotate in opposite directions to provide thorough stirring, and the rod is lowered until the seed touches the melt. The rod and seed are raised and the melt surface is pulled up slightly by surface tension at the seed. Cooling occurs at the melt–seed interface, and the melt freezes out following the crystal orientation of the seed. As the rod moves upward, the freezing at the liquid–solid interface continues and a single crystal *boule* or *ingot* is

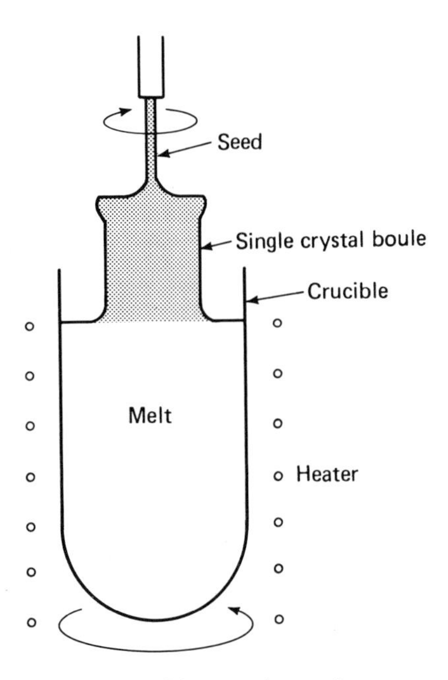

Figure 2–13. Skeletal Czochralski crystal growing apparatus. Note that the melt is held in a crucible.

formed. The dopant is incorporated into the silicon as it freezes out. The boule diameter is determined by the spin and pull rates and by the temperature; the overall size is limited by the *charge* or original amount of silicon in the melt. This is not a problem, however, in some of the modern crystal pullers, which have provisions for replenishing the melt from an auxiliary reservoir as the crystal is being grown [12].

2.10.1 Oxygen and Carbon Contamination

Defects can occur during crystal growth. Defects are any irregularities in the crystal structure, such as twinning (Section 2.5), or, more often, *point defects* (irregularities that occur at isolated points). Atoms located interstitially (between lattice sites in the crystal) or vacancies are common examples. The interstitial atoms may be of the desired dopants or of other unwanted impurities.

Oxygen atoms can locate interstitially during crystal growth, with typical concentrations in the order of 10^{17} to 10^{18} cm^{-3}, and can affect doping. The source of oxygen is usually the ambient air or the crucible that holds the melt. Incorporation of oxygen from the air may be reduced by evacuating the chamber surrounding the Czochralski apparatus and backfilling it with an inert gas such as argon. Further reduction results from using a crucible made of *fused quartz*, a very pure grade of silica (SiO_2) having a very high softening point—something over 1500°C.[8] This is higher than the melting point of silicon.

Where I^2R heating of the crucible is used, the heating elements are graphite, a form of carbon (C). Even in the inert chamber atmosphere there are a few parts per million of oxygen, which can react with the carbon to form CO or CO_2. These gases are soluble in the melt, so carbon also may incorporate into the single crystal as it freezes out. Carbon has a maximum limit of solid solubility in silicon of 5×10^{17} cm^{-3}; since carbon and silicon have similar crystal structures, the carbon goes into the silicon lattice substitutionally and is bonded covalently. This can affect doping by making lattice sites unavailable to dopant atoms. The presence of these impurities may be detected by infrared absorption measurements; oxygen exhibits an absorption band at a wavelength of 9 μm, carbon at 16 μm [13].

2.10.2 Annealing

Oxygen dissolved in silicon acts as a scattering defect for holes and electrons and so affects carrier mobility. It also can act as a donor. Both of these effects make incorrect the calculation of dopant concentration from measured values of bulk resistivity and use of the Irvin curves in Appendix F.2.1. They also may affect junction breakdown voltage and other device parameters [14].

The effects of oxygen may be reduced by heat-treating or *annealing* the boule after crystal growth is completed. The anneal is carried out in a furnace separate from the Czochralski puller, usually working in the 800 to 990°C temperature range. Another treatment method is considered in the next section.

2.10.3 Oxygen Gettering

In the past few years, annealing techniques have been developed for treating already finished silicon wafers so that a *denuded zone* is developed near the surface. This essentially oxygen-free zone is about 15 μm deep and Si devices, which usually lie within 10 μm of the surface, would lie within it.

The wafers are treated with a special thermal anneal in a nitrogen ambient. During the heating process the oxygen atoms are drawn away from the surface into the wafer interior (they move from hotter to cooler regions), where they tend to lock in and cluster on other defect sites—the *gettering process*. Carbon in the silicon may act as nucleation sites for clustering [15].

---

[8] Glasses, including SiO_2, are noncrystalline and so do not exhibit clearly defined melting points. Rather, the phase change from solid to liquid covers a considerable temperature range. Softening points are temperatures at which noticeable sagging occurs (Figure 5–3).

2.10.4 Summary

Over the years, considerable improvement has been made in growing CZ single-crystal boules of large size, uniformity, and perfection. By the mid-1980s CZ boules could be pulled with diameters up to 160 mm (≈ 6 in) with polySi charges up to about 100 kg [16].[9] This means that boules with diameters of, say, 100 mm could be grown to lengths up to about 140 cm. Bulk resistivity, ρ, could be obtained in the range 0.01 to 50 Ω-cm.

Doping is not perfectly uniform during crystal growth and, in fact, N_b may vary both axially and radially in the boule. The first variation depends on the ingot length and dopant type. Herring cites a variation from 7 to 9 Ω-cm over 20.6 cm of length for antimony [17]. The variation is smaller for the other common silicon dopants, boron having the smallest value. Typical radial variations in boules 100 mm in diameter are about 12 to 15% for phosphorus as the dopant, and about 5% for boron. Means for checking such radial variation in finished wafers are considered in Chapter 3.

2.11 Float Zone Method

The second method of growing single crystals is the float zone (FZ) system, which is shown in schematic form in Figure 2–14. A rod of polySi is clamped at both ends, the bottom end being in contact with a single-crystal seed of the desired orientation. There is no crucible for the rod to contact as there was in the CZ apparatus.

Heat is furnished by induction from a short coil supplied with radio-frequency (r-f) energy. The r-f voltage induced in the rod causes large currents to flow, with resulting I^2R heating within the rod. Sufficient power is supplied to melt a short section of the rod within the coil. The melted region is held in place between the two solid regions by surface tension and by the *levitation effect*, which comes from the interaction of the coil field and the currents flowing in the melted region. The coil starts at the bottom, or seed end, of the rod, melting that portion of the rod in contact with the seed crystal. The coil then moves upward, as does the melted region, leaving behind a solid, single crystal that freezes out, layer by layer, while propagating the structure and orientation of the seed.

There is no crucible oxygen contamination because there is no crucible. Also, induction heating eliminates the need for graphite rods for I^2R heating. As a result, the oxygen and carbon content in FZ crystals is much lower than in those grown by the Czochralski method. In current practice, however, the CZ method can provide larger boules of the single-crystal material.

---

[9] Fiegl gives an excellent summary of current technology—including means of replenishing the melt as the crystal is pulled and the use of magnetic fields to reduce swirling in the melt as the crucible and pull rod rotate. These techniques can make boule doping more uniform [16].

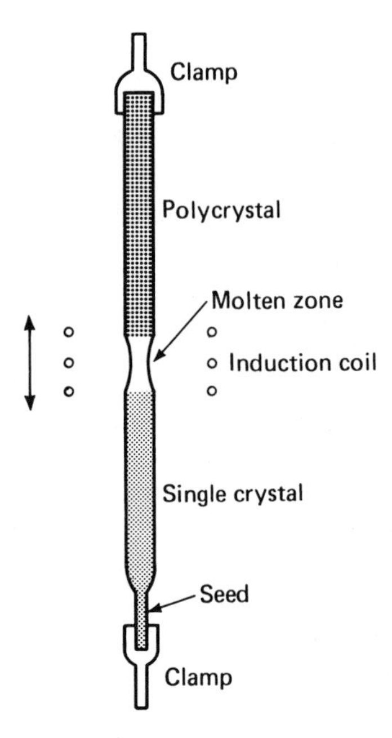

Figure 2–14. Skeletal Float Zone crystal growing apparatus. Note that the silicon does not contact a crucible.

The float zone apparatus closely resembles the equipment used in zone refining (Section 4.13). In fact, if the starting polySi rod were uniformly doped with an acceptor or donor, the conversion of the structure from poly- to monosilicon would be accompanied by a redistribution of the dopant, the concentration of which would no longer be uniform over the rod length. To overcome this problem in FZ crystal pulling, *zone leveling*, whereby the dopant for the entire rod is placed at the bottom end of the initial poly rod, is used. Then, as the induction coil is moved, the dopant is swept upward along with the molten region and so is redistributed throughout the entire rod length. With proper operating conditions, very uniform doping may be obtained with a single pass of the coil [18].

By the early 1980s single crystals up to about 110 mm in diameter could be accommodated by FZ apparatus with a charge of approximately 65 kg [16]. Bulk resistivity, ρ, could be obtained in the range from 10 to 200 Ω-cm [19].

2.12 Wafer Preparation

This section gives an overview of the steps that convert the single-crystal CZ or FZ ingot into finished wafers. More steps are necessary than might be imagined.

2.12.1 Wafer Dimensions

The concept of batch processing teaches that the unit cost of the parts on a wafer decreases as the number of parts per wafer increases. The marked increase in IC size over the years creates a need for ever larger wafer diameters. Wafers with diameters of 6 in. are quite common, particularly on new production lines. The increase in wafer diameter has been accompanied by greater wafer thickness, made necessary by automatic wafer handling on the production line.

In the earlier days of wafer processing, production-line personnel used tweezers to handle wafers. It was found, however, that tweezers produced surface damage and caused contamination by carrying over liquids from one processing position to the next. Traces of liquid left on the wafer surface would dry and cause stains. To minimize these problems and to increase throughput, modern fabrication lines use automatic (machine) handling rather than manual (tweezer) handling. Automatic handling has increased tolerance requirements on wafer diameter: a wafer 4 in. (101.6 mm) in diameter can cause a machine designed to handle 100-mm wafers to malfunction.

Wafer thickness is a function of wafer diameter in that larger wafers tend to break more easily in automatic handling and so are usually thicker. The need for mechanical strength increased markedly with the introduction of diameters of 3 and 4 in., since automatic handling was introduced at the same time.

Larger wafer thickness has raised problems of warpage in high-temperature operations. If large, relatively thick wafers are moved too rapidly out of a high-temperature furnace, they tend to warp: their mass is too large to equilibrate quickly and evenly. This problem and its solutions are discussed in Section 4.11.2.

2.12.2 Ingot Preparation

Precise wafer size is important for automatic wafer handling. While modern computer-controlled CZ and FZ crystal growers give good control of boule size, they cannot maintain constant diameter over the entire boule length. Therefore, boules are grown slightly oversize and then are ground to a cylindrical form with the correct diameter. Grinding may be followed by an etching step to remove grinding damage on the cylinder surface. Remember that the cylinder surface will become the wafer edges.

A quality control check of resistivity then is made on both ends of the cylinder to see how uniform the doping is. This also provides resistivity limit values that will help in sorting the finished wafers by ρ values. Sometimes, boule ends will be cut off and returned for regrowing to insure that ρ in the remaining piece is within specified limits.

2.12.3 Orientation Identification

Subsequent scribe-and-break requires that a [110] direction in the cylinder be identified. To accomplish this, x-ray diffraction methods are used [20]. The (110) plane is marked permanently on the cylindrical boule by grinding a flat surface or *flat* along its length.

In some cases, particularly for diameters of 100 mm or larger, another, shorter, *secondary flat* may be ground; its position relative to the (110) or *primary* flat then indicates whether the wafer is p- or n-type and of (111) or (100) orientation. Details for this marking system are given in Section 2.12.7.

2.12.4 Wafer Sawing

At this stage, the ingot loses its identity, since it is sawed into wafers. Two types of saws are used for this: diamond blade and wire. The diamond saw has a blade impregnated with diamond dust that serves as the cutting medium. In the 1960s, a circular blade with its periphery as the cutting edge was used. A blade of this type must be reasonably thick to provide adequate mechanical support for the cutting edge. As a result the *kerf* (the channel made by the saw) will be even wider, so considerable single-crystal silicon is lost in the sawing process.

To reduce this loss, inside-cutting-edge saws are now used. The blade is annular and the central hole surrounds the cylindrical boule; cutting is done with diamond dust on the inside edge of the annulus. The outer edge now may be quite thick and serve as a support member, so the cutting edge can be thinner than in the older type. A smaller kerf (about 10 mils) results, but even so, the loss is significant; if wafers 20 mils thick are sawed with a kerf of 10 mils, one-third of the single crystal, prepared with a high expenditure of energy, is lost. With both types of diamond saws, water cooling is required.

A second type of saw uses an endless loop of tungsten wire running through a pasty slurry of silicon carbide (SiC) suspended in a liquid vehicle, such as water and glycerin. Since the wire is under tension, it can be quite thin. More than one wafer at a time may be sawed by having the wire make several turns around the supporting mandrels and boule.

Typical finished wafer thickness runs about 10 to 12 mils for 2-in. diameters and around 20 to 22 mils for the 100-mm size. To allow for loss during finishing operations, wafers are sawed about 2 to 5 mils thicker than their final thickness.

Usually, wafers are sawed normal to the axis of the silicon boule, so a (111) wafer results from sawing a (111) cylinder, and a (100) wafer from a (100) boule. This points out the need for proper seed orientation when the single crystal is grown. Typical sawing methods can hold the wafer surface parallel to the desired crystal plane to within $\pm 1^\circ$.

Off-normal sawing also may be used. If the wafer is to be used for bipolar ICs having buried and epi layers, the cut will be made at $3 \pm 0.5^\circ$ off the normal (Section 2.5). Also, if a cylinder with its axis along a [100] direction is sliced at 45° off axis, (111) wafers will result—but note that they will be elliptical rather than circular. This shape can make several processing steps (e.g., photoresist spin on) far from optimum. It does make larger wafers, though, which are useful for experimental purposes.

2.12.5 Backside Treatment

Sawing the boule to produce wafers results in crystal damage to both faces. This damage may be removed with an etch having hydrofluoric acid (HF) and hydrochloric acid (HCl) as the principal ingredients. The HCl is a strong oxidizing agent that converts Si to SiO_2, which in turn dissolves in HF. The speed of reaction may be moderated by adding small amounts of acetic acid (CH_3COOH).

A proper etch usually will provide adequate treatment for the backside or bottom surface of the wafer, but recall from Chapter 1 that backside grinding will be performed *after* device fabrication. Proper etching procedures also will round the edges of the wafer, a good condition for automatic handling because it minimizes chipping and cracking.

By this stage in production, the wafers should be checked to see if they meet specifications for mechanical properties such as *bow*, or departure from flatness, and parallelism of the two surfaces.

2.12.6 Topside Polishing

Recall that devices are confined generally to a layer within 10 μm of the upper wafer surface. It is clear, then, that damage should not extend this far into the wafer, so smoothness of the upper surface is of great importance. At this point, surface damage due to sawing has been removed by etching, but now the wafer topside must be polished to a mirror finish.

This is accomplished by *slurry* polishing, which combines chemical and mechanical means. The wafers are waxed, topside up, to a metal plate or *platen*. Another plate, covered with a polishing pad, is rotated against the wafers. The pad is wet with a slurry, consisting typically of extremely fine SiO_2 particles in an alkali solution with a pH in the 10 to 12 range [17]. Proprietary polishes such as Syton® (Monsanto Company) also are used. As much as 1.5 to 2.0 mils of wafer may be removed during the finishing operation. Wafers may be released from the platen by placing it on dry ice. The platen shrinks and the wafers pop off. This simple and ingenious method avoids the mechanical damage that might result if the wafers were pried off.

The wafers then are cleaned and rinsed. If the wafer finish is proper, the rinsing water will run off completely—no water will remain on the surface, which is said to be *hydrophobic*. The surface also will appear haze-free, and mirrorlike.

Finished wafers may be handled with vacuum pencils. The lift is provided by Bernoulli's principle without the pencil parts touching the wafer surface. This avoids tweezer contact damage before the wafers are loaded into automated carriers. Wafers at this stage usually are stored for shipment or additional processing.

2.12.7 Wafer Marking

When 2-in. wafers were in common use, little information was coded on the wafer; the flat gave the (110) plane location. With the advent of sizes of 3 in. and larger, the trend has been to provide more information by adding another flat. Standards for coding this information have been set up by the Semiconductor Equipment and Materials Institute (SEMI). Their system of marking through the use of primary and secondary flats is shown in Figure 2–15 [17].

Other systems for wafer marking, which use a laser beam as a marking tool, provide still more information about the wafer. A sharply focused laser beam is deflected by two galvanometer-driven mirrors at right angles. A visible mark is burned in where the beam hits the wafer surface. Driving the mirror galvanometers from a computer causes the beam to write alphanumeric characters on a narrow strip near the wafer edge, where devices usually are not made. Typical information given includes growth process (CZ or FZ), dopant species, bulk resistivity, and manufacturer. Details of one such coding system, which has been proposed by SEMI for 125-mm wafers, are shown in Figure 2–16 [21].

2.12.8 Epi and Oxidation

Vendors will furnish wafers with epi layers already grown to the users' specifications. Wafers for bipolar ICs require special considerations, however, because the locations of buried layer (BL) regions vary from one circuit design to another. If the BLs are to be produced by diffusion, this diffusion must be performed before epi deposition, and epi then grown *in house* by the IC fabricator. In contrast, BLs may be implanted through an epi layer (Section 8.14.5), so wafers having epi in place may be used.

With or without epi, finished wafers usually are thermally oxidized before storage and subsequent shipment to the user. Photolithographic processing of wafers requires an initial oxidation anyway, and the oxide will protect the upper surface finish, so the wafer manufacturer provides the oxidation step. Oxide thicknesses can be tailored to users' specifications, and are typically in the 1000 to 1200 Å range.

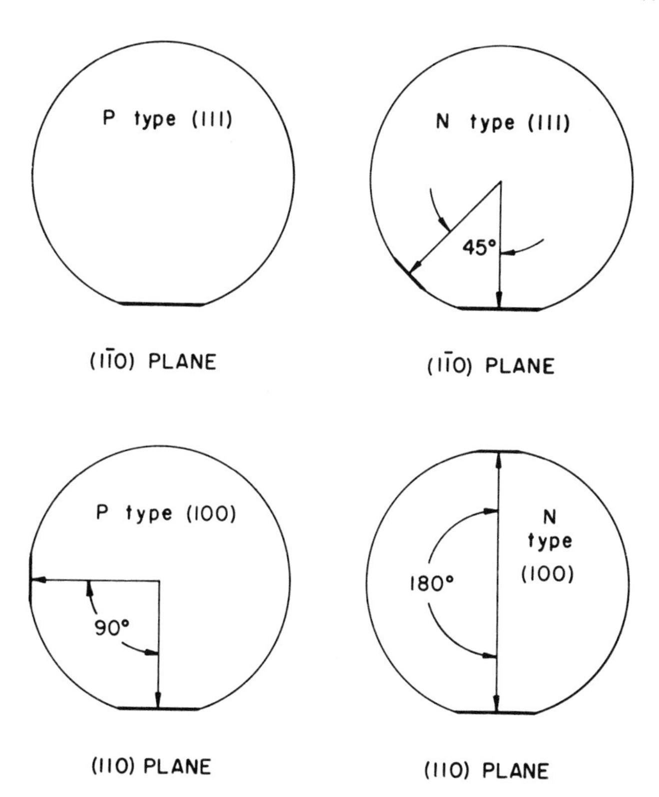

Figure 2–15. The two-flat system of marking silicon wafers with 3-in. or larger diameters. (a) p-type (111). (b) n-type (111). (c) p-type (100). (d) n-type (100). Herring [17]. (Courtesy of *Solid State Technology*, PennWell Publishing Company, Copyright 1976.)

2.12.9 Wafer Recycling

Wafer salvage and recycling may not seem to be germane to the discussion of wafer manufacture, but they are related to wafer preparation and are a concern in the semiconductor industry. Wafers with a large number of defective parts are not uncommon.

Usually, not much is said about manufacturing defects, but the fact remains that wafer-level fabrication (i.e., the fabrication of devices or ICs on wafers before dicing) is not a perfect, trouble-free procedure. This is not surprising, considering the large number of steps, their complexity, and the very small dimensions that are involved. The rejection rate of wafers will depend to a large extent on the complexity and size of the device or IC being processed (often called the *part*) and on how long the production line for that part has been in operation. A figure of 20 to 25% is often cited as an average reject rate at the wafer level for a part that has been manufactured for a long time. Typical

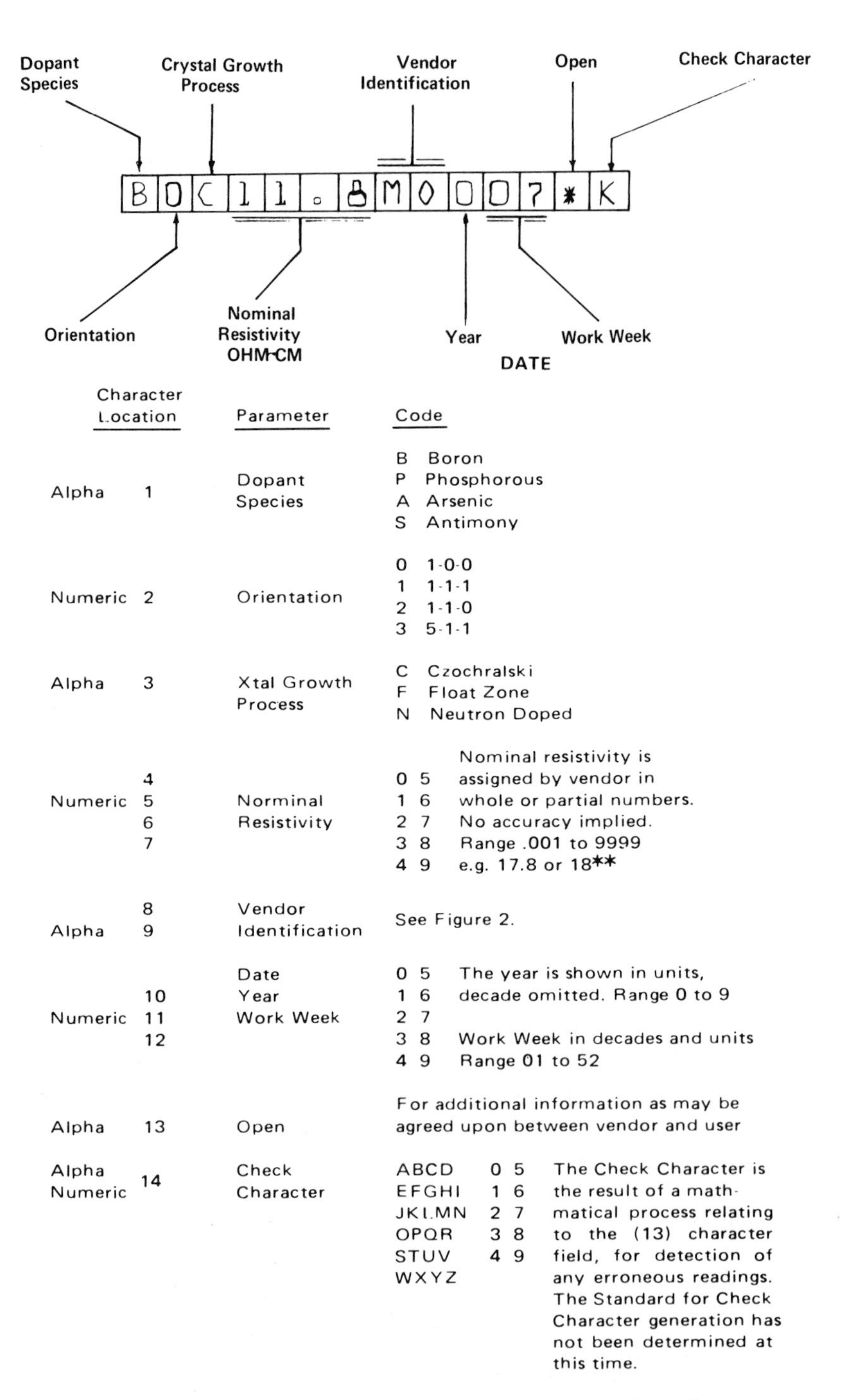

Character Location		Parameter	Code
Alpha	1	Dopant Species	B Boron P Phosphorous A Arsenic S Antimony
Numeric	2	Orientation	0 1-0-0 1 1-1-1 2 1-1-0 3 5-1-1
Alpha	3	Xtal Growth Process	C Czochralski F Float Zone N Neutron Doped
Numeric	4 5 6 7	Norminal Resistivity	0 5 1 6 2 7 3 8 4 9 Nominal resistivity is assigned by vendor in whole or partial numbers. No accuracy implied. Range .001 to 9999 e.g. 17.8 or 18**
Alpha	8 9	Vendor Identification	See Figure 2.
Numeric	10 11 12	Date Year Work Week	0 5 1 6 2 7 3 8 4 9 The year is shown in units, decade omitted. Range 0 to 9 Work Week in decades and units Range 01 to 52
Alpha	13	Open	For additional information as may be agreed upon between vendor and user
Alpha Numeric	14	Check Character	ABCD 0 5 EFGHI 1 6 JKLMN 2 7 OPQR 3 8 STUV 4 9 WXYZ The Check Character is the result of a mathmatical process relating to the (13) character field, for detection of any erroneous readings. The Standard for Check Character generation has not been determined at this time.

Notes: In the absence of information at any of the assigned locations, an asterisk (*) must be used at each such location.
**As in the case of the other character locations, an asterisk (*) must be used in the absence of information.

causes of rejection are photoresist problems, mask misalignment, problems in depositing insulating and metallic layers, wafer cleaning problems, and wafer breakage.

Since wafer rejects exist, what can be done with them? In the United States we tend to be profligate, we tend to simply throw things away. Since gold is very expensive, however, any gold on a rejected wafer will be reclaimed before the basic wafer is cast aside. Another procedure is to strip the defective wafer so the silicon can be cleaned, melted, and regrown into single crystal form. The reprocessing of silicon this way is very energy-intensive, however, because of the monocrystal growing process.

Wafer recycling is another alternative. After all, under the doped regions, some 10 μm below the top surface, lies single-crystal silicon. Thus, if the wafers are not warped, cracked, broken, or otherwise severely physically damaged, and/or if they have not been subjected to a deep gold diffusion extending through the entire wafer, it should be possible to remove the unwanted regions, refinish the top surface, and have a new, but slightly thinner, wafer [22]—an idea that has been used in practice.

The layers above the surface can be removed by grinding or by successive chemical etches. Grinding is bad because it results in damage to the crystal that will have to be removed by subsequent polishing. This entails more loss of wafer thickness, so the multiple-etch method is preferred. Since selective etches are required, it is a big help to know what layers are present and in what order. The HCl/HF combination described earlier can be used to remove the doped silicon region near the surface by etching. This etch also will remove the defect cluster layer lying beneath the denuded zone (Section 2.10.3).

The final steps after layer removal consist of polishing the topside, epi growth (if required), and oxidation. Proponents of wafer recycling claim that the reprocessed wafers are as good as, if not better than, the first-run wafers, although they are thinner by some 10%. The claim for better electrical characteristics argues that all of the high-temperature steps on the original wafer and the subsequent anneals give a new, denuded layer under the new polished surface.

Problems

2–1 The indices (2, 1, 4) represent a plane in a cubic crystal. What are the indices of a second plane, normal to the first, lying in the x, z plane, and with the same origin? Show your work with sketches.

◀ Figure 2–16. Coding system for laser-marking silicon wafers. Duncan [21]. (Courtesy of *Solid State Technology*, PennWell Publishing Company, Copyright 1980.)

2–2 What are the Miller indices of the hatched plane in Figure 2–3(d)?

2–3 Calculate the channel cross-sectional areas in $Å^2$ for the three principal planes in silicon. Show your work.

2–4 (a) Convert $N_d = 3 \times 10^{18}$ cm^{-3} to m^{-3}. Why does the semiconductor industry use the former rather than the MKS version?

(b) How many ppm of As in Si will dope to the maximum limit of solid solubility?

2–5 (a) What is ρ of a Si wafer doped at 10^{17} cm^{-3} with As? If P were used in place of As, what would be the effect on ρ? Explain.

(b) A p-type wafer of 0.01 Ω-cm is desired. What concentration of B is required?

(c) What is μ_M for the boron in part (b)?

2–6 A p-type boule of single-crystal Si measures $\rho = 0.75$ and 1.0 at the two ends. Calculate the percent changes in ρ and N_a. Are they the same? Explain.

2–7 Translate this code marked on the edge of a wafer:

A 1 F 0.02 MO 7 10 * *

2–8 In preparing a wafer for dicing, why isn't the diamond saw allowed to cut completely through the wafer?

References

[1] W. Shockley, *Electrons and Holes in Semiconductors.* Princeton: Van Nostrand, 1950.

[2] D. O. Townley, "Optimum Crystallographic Orientation for Silicon Device Fabrication," *Solid State Technol.*, **16**, (1), 43–47, Jan. 1973.

[3] K. Wosnock, "How to Select the Right Scribing-Tool for Your Application," *Circuits Manuf.*, **15**, (1), 75–80, Jan. 1973.

[4] E. A. Wood, *Crystals and Light.* Princeton: Van Nostrand, 1964.

[5] B. M. Berry, "EPITAXY." In *Fundamentals of Silicon Integrated Device Technology, Vol. 1*, ed. R. M. Burger and R. P. Donovan, p. 435. Englewood Cliffs: Prentice-Hall, 1967.

[6] B. G. Streetman, *Solid State Electronic Devices, 2nd Ed.* Englewood Cliffs: Prentice-Hall, 1980.

[7] J. C. Irvin, "Resistivity of Bulk Silicon and of Diffused Layers in Silicon," *Bell Syst. Tech. J.*, **41**, (2), 387–420, March 1962. Figure 1. This paper has been reprinted in *Micro- and Thin-Film Electronics Readings*, ed. S. N. Levine, pp. 213–236. New York: Holt, Rinehart and Winston, 1964.

[8] S. M. Sze, *Physics of Semiconductor Devices.* New York: Wiley, 1969.

[9] A. B. Glaser and G. E. Subak-Sharpe, *Integrated Circuit Engineering.* Reading: Addison-Wesley, 1979.

[10] F. A. Trumbore, "Solid Solubilities of Impurity Elements in Germanium and Silicon," *Bell Syst. Tech. J.*, **39**, (1), 205–233, Jan. 1960. This paper has been reprinted in ed. S. N. Levine *op. cit.* [7].

[11] J. Lenzing, "Survey of Semiconductor Crystal-Growing Processes and Equipment," *Solid State Technol.*, **19**, (2), 34–39, 43, Feb. 1975.

[12] P. Burggraaf, "Si Crystal Growth Trends," *Semicond. Int.*, **7**, (10), 54–59, Oct. 1984.

[13] P. Stallhofer and D. Huber, "Oxygen and Carbon Measurements on Silicon Slices," *Solid State Technol.*, **26**, (8), 233–237, Aug. 1983.
[14] H. M. Liaw, "Oxygen and Carbon in Silicon," *Semicond. Int.*, **2**, (8), 71–82, Oct. 1979.
[15] R. A. Craven and H. W. Korb, "Internal Gettering in Silicon," *Solid State Technol.*, **24**, (7), 55–61, July 1981.
[16] G. Fiegl, "Recent Advances and Future Directions in CZ-Silicon Crystal Growth Technology," *Solid State Technol.*, **26**, (8), 121–131, Aug. 1983.
[17] R. B. Herring, "Silicon Wafer Technology—State of the Art 1976," *Solid State Technol.*, **19**, (6), 37–42, June 1975.
[18] S. K. Ghandhi, *The Theory and Practice of Microelectronics.* New York: Wiley, 1968. Chapter 2.
[19] J. H. Matlock, "Crystal Growing Spotlight," *Semicond. Int.*, **2**, (8), 33–44, Oct. 1979.
[20] B. D. Cullity, *Elements of X-Ray Diffraction.* Reading: Addison-Wesley, 1968.
[21] G. Duncan, "Proposed Alphanumeric Marking Standards for 125 mm Silicon Wafers," *Solid State Technol.*, **23**, (7), 54–55, July 1980.
[22] J. E. Lawrence, "The Alchemist's Dream: Wafer Recycling," *Circuits Manuf.*, **23**, (5), 25–30, May 1983.

Chapter 3
Wafer Measurements

Several measurements on wafers are made routinely. One group concerns the determination of wafer type—i.e., p- or n-type—background doping level N_b, resistivity ρ, and sheet resistance R_S, and is made on incoming wafers. A second group measures junction depth, doping profiles within the silicon, layer profiles on the top surface, and film layer thickness. The latter group is made during fabrication.

3.1 Type Determination

Two basic phenomena are available to determine if a silicon wafer is p- or n-type: (1) diffusion of charge carriers under a temperature gradient to form a voltage, and (2) the Hall effect. The former is more common for production line work and uses a *hot point probe.*

3.1.1 Hot Point Probe (HPP)

The basic hot point probe (HPP) comprises two test probes, C and H, connected by a microammeter. The cool probe C is at room temperature, and H, the hot probe, is heated to a higher temperature electrically. The circuit operates roughly in this fashion. When the probes are in contact with the sample, charge carriers tend to diffuse through the sample from H toward C, causing a current to flow. The net current will depend primarily on the majority carriers, electrons for n-type and holes for p-type samples. Since these carriers are of opposite sign, the direction of net current flow will differ for the two types. If H is positive wrt C, the sample is n-type, otherwise it is p-type. Thus, the type of the sample can be determined by the direction of deflection of a zero-center current meter. More elegant explanations of the operation are available [1]. Notice that the HPP test is nondestructive. In practice, only one wafer in an incoming batch is checked with the HPP to verify the information coded on the wafers by either the two-flat or laser writing system (Section 2.12.7).

Hot point probes have been incorporated into rather sophisticated equipment, such as spreading resistance probes.

3.1.2 Hall Effect

The second type-determination method is based on the Hall effect: when charge carriers drift in a direction perpendicular to an externally applied magnetic field, they tend to deflect perpendicularly to both the magnetic field and their original drift direction.

The basic experimental setup is shown in Figure 3–1. The sample *edge* is contacted at four points—a, b, c, and d. It is convenient if these are spaced roughly at 90° intervals. The sample is placed in a z-directed magnetic field of magnetic flux density $\mathscr{B}_z$. A current from an external source flows from contact c to contact d in the *x*-direction. The voltage V_{ab} is measured with a high-impedance, d-c electronic voltmeter to avoid drawing current.

It can be shown from theoretical considerations that V_{ab} will be positive if the sample is p-type, and negative if the sample is n-type, so the semiconductor type is determined simply by noting the polarity of V_{ab}. If many wafers are to be sorted by type, a zero-center meter is of great help. The voltage *magnitude* and the sample shape are not required for *type* determination. A wafer with flats works quite well.

The Hall effect method raises some problems in making contacts to the wafer edges and requires a magnetic field, so the HPP method is preferred for production line use.

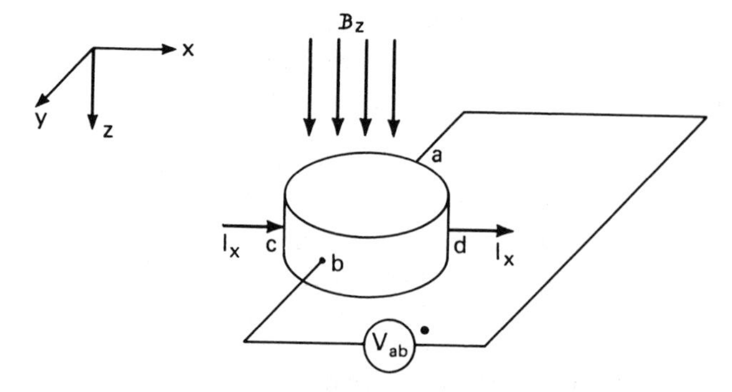

Figure 3–1. The Hall effect used for determining wafer type. If V_{ab} is positive, the sample is p-type.

3.2 Resistivity and Doping Measurements

Wafer doping concentration N_b is determined indirectly from a measured value of bulk resistivity ρ given by Eq. (2.7-3). This value depends on the majority and minority carrier concentrations and hence on the *ionized* dopant concentrations. Then we assume that $N_M \gg N_m$, a typical condition in Si, and make the measurement at or above room temperature where most *dopant* atoms are ionized. Thus, $N_M \approx N_b$, so Eq. (2.7-4) is valid, namely $\rho = 1/(q\mu_M N_b)$. Then N_b is read from the appropriate Irvin curve in Appendix F.2.1.

Since optical injection can make $N_M > N_b$, semiconductor resistivity measurements should be made with the sample in darkness. For example, if the sample is illuminated with light of wavelength $\lambda \leq hc/E_g$, where h = Planck's constant, c = velocity of light, and E_g = sample band gap, additional covalent bonds in the semiconductor will be broken, so both N_M and N_m will increase. For silicon $\lambda_g \approx 10{,}000$ Å, so visible light will cause optical injection. The enhanced values of the two carrier concentrations will invalidate the assumption that $N_M \approx N$, the doping concentration, and Irvin's ρ versus N curves no longer may be used. The need for keeping the sample dark applies to *all* forms of semiconductor resistivity measurement.

As a general rule, resistivity measurements are based on a V/I ratio (often called the *spreading resistance* R_{Sp}), where V and I are measured values of voltage and current. The ratio then is converted to bulk resistivity ρ (or sheet resistance R_S—Section 3.8) by correcting for the geometry of the sample itself and of the probing contacts used for measuring V and I. On semiconductor production lines it is desirable to use a measurement system that requires no correction factor, particularly where a whole wafer is being measured.

We consider two types of resistivity measurements: (1) van der Pauw, and (2) four point probe.

3.3 Van der Pauw Measurement

Van der Pauw has developed a means of measuring resistivity that is applicable to entire wafers [2]. Through the use of field theory he has shown that ρ of a uniformly thick sample of any arbitrary *shape*, can be determined from five quantities: two voltages, two currents, and the sample thickness t.

The basic setup for making the four electrical measurements is shown in Figure 3–2. Four contacts are made to the *edge* of the sample at points a, b, c, and d. The points may be located arbitrarily along the edge but are labeled in sequence for convenience. The first measurement configuration is shown in (a). An external current source drives the known current I_{ab} from a to b through the sample. Voltage V_{cd} is measured, and the equivalent resistance

$$R_1 = R_{ab,cd} = \frac{V_{cd}}{I_{ab}} \tag{3.3-1}$$

is calculated.

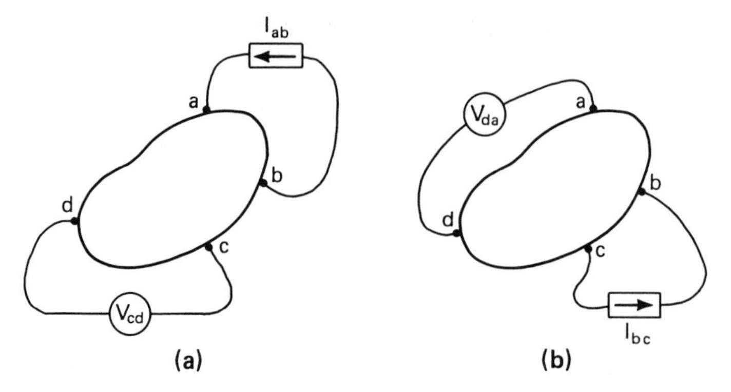

Figure 3–2. Basic connections for the van der Pauw measurement. The sample has uniform thickness t.

Then the current source and voltmeter are indexed, usually by a four-pole, two-position switch, to the second position as shown in Figure 3–2(b). The values of I_{bc} and V_{da} are measured, and the effective resistance

$$R_2 = R_{bc,da} = \frac{V_{da}}{I_{bc}} \tag{3.3-2}$$

is calculated. Van der Pauw has shown that the bulk resistivity of the sample then is given by

$$\rho = \frac{\pi}{\ln 2} t \left(\frac{R_1 + R_2}{2} \right) f\left(\frac{R_1}{R_2} \right) \tag{3.3-3}$$

The factor $\pi/\ln 2$ (≈ 4.53) will be encountered again in Section 3.5.3. The function $f(R_1/R_2)$ is plotted in Appendix F.3.1 and is a correction factor.[1] Note that the minimum value of the abscissa scale for the curve is unity; therefore, if the resistance ratio is less than one, its reciprocal must be used to determine f.

If the sample is circular and of uniform ρ, and if adjacent contacts are spaced 90° apart, $R_1 = R_2 = R$, $f = 1$, and

$$\rho = \frac{\pi}{\ln 2} tR \tag{3.3-4}$$

Since most semiconductor wafers are nearly circular (remember the *flats*), the last equation often can be used for a good approximation.

---

[1] The f curve was evaluated incorrectly in van der Pauw's original paper. The author is indebted to John S. Moore, then of the Electrical Engineering Department, University of Illinois, for the corrected version shown in the appendix.

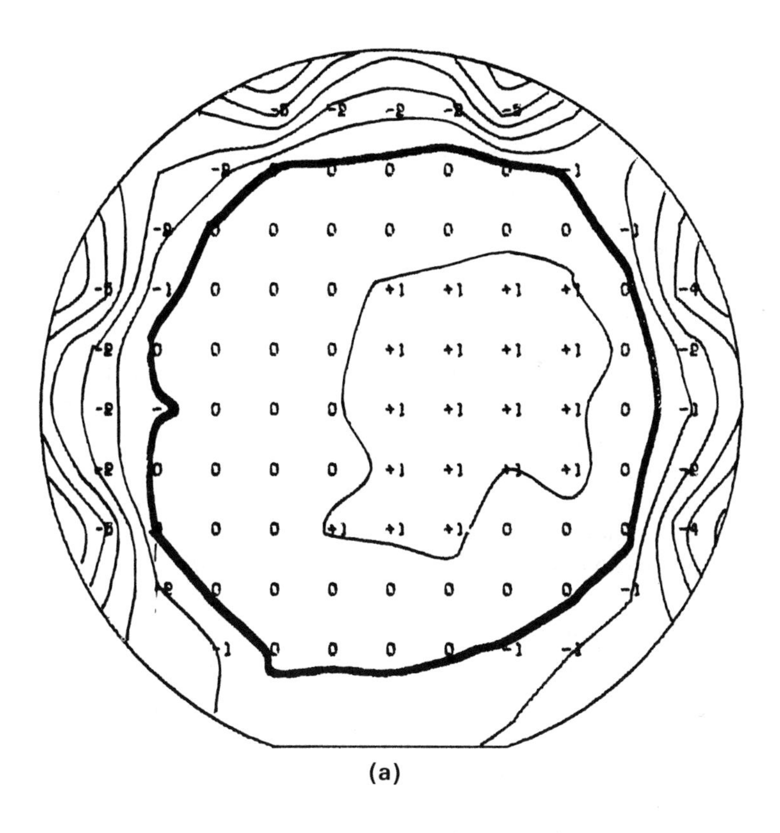

(a)

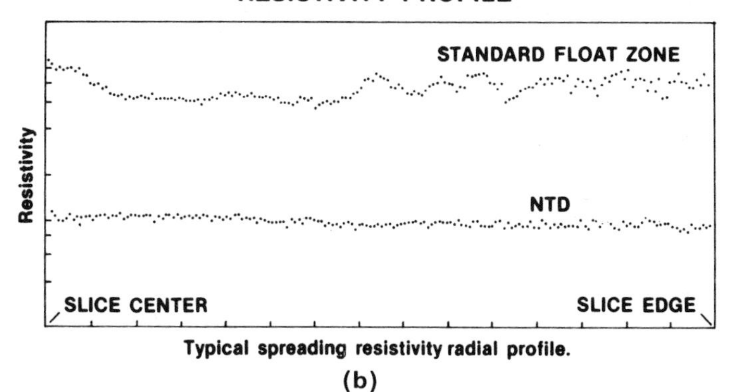

(b)

Figure 3–3. Plots showing variation of ρ. (a) Contours of constant ρ on a CZ wafer. Burggraaf [3]. (b) Variation of ρ with radial distance from the center of an FZ wafer. The dopant is phosphorus. The results labeled NTD are for a wafer doped by neutron transmutation doping. Matlock [33]. (Courtesy of *Semiconductor International Magazine.*)

The need for contacting the wafer *edge* at the four contact points can raise some practical problems. For standard-size wafers, it is feasible to use a jig having four spring-loaded, knife-edge contacts that are forced against the wafer edge.

The van der Pauw method assumes a uniform value of ρ over the entire wafer. If ρ does vary with location, the measured value is some sort of an average; the method cannot give point-by-point values. Actual wafers can show variations in ρ, particularly as a radial change, the inevitable result of different doping near the edges of Czochralski ingots. Figure 3–3(a) shows a typical map of ρ variation on a CZ wafer [3]. The variation of ρ along a radial line for an FZ wafer is shown in (b) [33].[2] Because of these variations it is desirable to have still another method of measuring resistivity that would allow ρ checks, if not on a point-by-point basis, at least over small regions of the wafer. This is particularly desirable for wafers of 3 in. or more in diameter. The four and two point probes of Sections 3.4 and 3.12.1 provide this capability.

3.4 Four Point Probe (FPP)

The four point probe (FPP) is commonly used in the semiconductor industry for measuring bulk resistivity and sheet resistance for good reasons: it does not require a special sample shape; the probes may contact the top surface, rather than the edges, of the sample; and it may be moved about on the sample surface to give readings on a small-area-by-small-area basis. This last property allows mapping of the sample resistivity, as shown in Figure 3–3(a).

We shall consider here the theory of the measurements in order to understand the bounds placed on the several resistivity equations that relate measurements to sample geometry. Mathematical details are in Appendix A.

3.4.1 Collinear FPP Configuration

The basic configuration of the probe proper and its associated measuring circuit are shown in Figure 3–4(a); a cutaway view of a typical commercial probe is shown in (b) [3].

The four probe tips are collinear, equispaced, and spring loaded. The tips typically are made of osmium, tungsten, or tungsten carbide. Probes with

---

[2] The lower curve in Figure 3–3(b) is labeled NTD, for *neutron transmutation doping* [15]. Normal silicon has an atomic structure consisting of a nucleus of 14 neutrons and 14 protons (mass number = 28), surrounded by 14 orbital electrons (atomic number = 14)—i.e., $_{14}Si^{28}$ in standard notation. In this unusual doping method a beam of neutrons bombards the silicon and some atoms are converted to the $_{14}Si^{31}$ isotope by incorporating neutrons. Incorporation of an electron results in transmutation to a normal phosphorus atom $_{15}P^{31}$. These transmuted phosphorus atoms act as donors in the usual manner. Neutron bombardment also causes radiation damage in the silicon (i.e., it upsets the lattice structure), but this may be annealed out thermally (Sections 8.12 and 8.13).

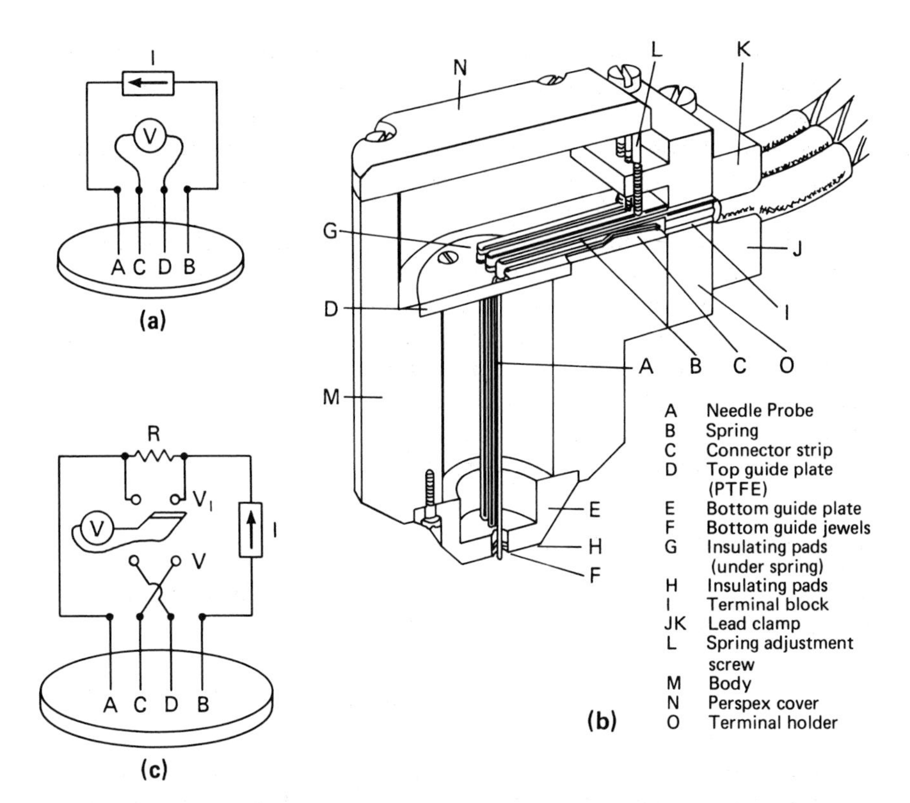

Figure 3–4. Four point probes (FPP). (a) Basic configuration. (b) Cross section of a commercial FPP head. Burggraaf [3]. (Courtesy of *Semiconductor International Magazine*.) (c) Modification of the basic FPP connections so a single voltmeter is used to measure both V and $V_I = IR$.

intertip spacings of 10 to 62.5 mils $\pm$ 1% are available commercially, and springs provide loading in the order of several grams per tip. In use, the four probe tips are made to contact the sample surface, and a constant-current source drives a known current I through the sample between the two outer tips, A and B [Figure 3–4(a)]. The resulting voltage drop between the two inner contacts, C and D, is measured with an high-impedance electronic voltmeter. Two variations of the basic FPP and circuit are considered in Sections 3.10.1 and 3.10.2.

Usually, to minimize meter errors, a single meter is used to read both V and I. The modified circuit of Figure 3–4(c) facilitates this. A current sampling resistor R has been added in the current supply loop so $V_I = IR$, and a dpdt switch allows both voltages to be measured. The leads between C and D and the switch are crossed so that V and V_I have the same polarity on the voltmeter.

Polarity-reversing switches for V and I often are provided so that two sets of readings, of opposite polarity, can be made. This tends to cancel out any voltage offsets that arise due to poor contacts between the probe tips and the sample. Ideally, the two readings should be identical, although in practice small differences may be tolerated if the average value of the two voltmeter readings is used. If the discrepancy is large, the probe location on the wafer should be changed slightly. As a general rule, it is easier to get good contacts on the rougher, back side of the wafer than on the mirror-finish front side, but this is not satisfactory if doping is different on front and back.

A given value of the V/I ratio (or *spreading resistance*) is not uniquely related to ρ or R_S; rather, the relation depends on the probe and sample geometries. Consequently, there are several equations relating the quantities. Even though the derivations are relegated to Appendix A, they are important because they show the limitations on the several equations.

3.4.2 Semi-Infinite (S-I) Sample

The first result is for a semi-infinite (S-I) sample—one of infinite extent in all directions *on* and *below* the surface of the sample.

$$\rho_\infty = 2\pi s \frac{V}{I} = 2\pi s R \frac{V}{V_I} \qquad \text{(semi-infinite sample)} \qquad (3.4\text{-}1)$$

where s is the intertip spacing, R is the sampling resistance, and the subscript ∞ on ρ serves as a reminder that the equation is valid only for a sample that is S-I, or at least "large" relative to the probe tip spacing. We shall define "large" more precisely later, and shall consider correction factors for finite samples of different shapes. Equation (3.4-1) is derived in some detail in Appendix A.1.

A commonly used intertip spacing is $s = 62.5$ mils $= (62.5 \times 10^{-3})(2.54)$ cm. The conversion from mils to cm is used so that ρ_∞ will have the common units of Ω-cm, rather than Ω-mil. Say the sampling resistor has a typical value of R = 100 Ω. Then

$$\rho_\infty = 2\pi(62.5 \times 10^{-3})(2.54)(100)\frac{V}{V_I}$$

$$= 99.7\frac{V}{V_I} \qquad \text{(S-I sample, constant } s = 62.5 \text{ mils, } R = 100\ \Omega)$$

The units of ρ_∞ are Ω-cm if the two voltages have the same units (e.g., mV). The numerical forefactor may be taken as 100.

Many variations of the basic circuit are available [3], with preset current values and digital readout of V or ρ_∞. We next consider some factors that cause the sample ρ to differ from ρ_∞.

3.5 Thickness Correction Factors

Production line wafers are neither infinitely thick nor infinitely large horizontally; therefore correction factors are needed to convert ρ_∞ of Eq. (3.4-1) to actual ρ of the wafer. These factors may be derived by using the potential analog of Appendix A.1 with appropriate boundary conditions as in Appendix A.2. One important aspect of the results is that the terms "thick" or "large" wafer are clarified. We first consider the effect of finite sample thickness.

It is shown in Appendix A.2 that the FPP solution for a sample of uniform thickness t and extending to infinity in all directions in the horizontal plane (I-t sample) is

$$\rho = a\left(2\pi s\frac{V}{I}\right) = a\rho_\infty \qquad \text{(I-t sample)} \tag{3.5-1}$$

where a is the thickness correction factor. It is defined in Eq. (A.2-3) and is plotted in Appendix F.3.2. Use of the correction factor is illustrated in the following example.

Measurements are made on a silicon sample 15 mils thick and of horizontal size very large wrt the probe intertip spacing. (We shall assume that the sample is I-t.) The following numbers are determined: $s = 40$ mils, $V = 5.81$ mV, and $V_I = 100$ mV measured across a sampling resistance $R = 100\ \Omega$.

Determine the bulk resistivity:

$$\frac{t}{s} = \frac{15}{40} = 0.375$$

From Appendix F.3.2, $a = 0.27$, so

$$\begin{aligned}\rho = a\rho_\infty &= a2\pi sR\frac{V}{V_I}\\ &= (0.27)(2\pi)(40 \times 10^{-3})(2.54)(100)(5.81/100)\\ &= 1\ \Omega\text{-cm}\end{aligned}$$

Given the curve of a we now can define "thick" and "thin" in the context of the FPP.

3.5.1 Thick Samples

The curve of thickness correction factor a in Appendix F.3.2 shows that for all practical purposes if $t/s \geq 5$, then $a = 1$ or $\rho = \rho_\infty$. Thus *if the sample thickness is at least 5 times the probe intertip spacing, the FPP views the sample as being infinitely thick.* This, then, is the definition of a thick sample for the FPP.

Wafers seldom are more than 30 mils thick, and the smallest intertip spacing on commercial FPPs is 10 mils; hence the $t/s \geq 5$ criterion usually is not met and thickness correction is needed. Note, however, that practically, "infinite thickness" for the FPP, is defined by $t/s \geq 5$. We shall find later that a similar criterion is available for the "infinite horizontal plane."

3.5.2 Moderately Thick Samples

The curve of Appendix F.3.2 exhibits no special simplifying features in the range

$$0.5 \leq \frac{t}{s} \leq 5 \qquad \text{("moderately thick," I-t)}$$

so ρ must be evaluated using a as read from the curve. We define a sample in this range as being moderately thick.

3.5.3 Thin Samples

The curve of Appendix F.3.2 is plotted on log–log coordinates and is linear in the range $t/s \leq 0.5$. This implies that the data obey a power law of the form

$$a = K\left(\frac{t}{s}\right)^m \quad \text{for } \frac{t}{s} \leq 0.5 \qquad \text{("thin," I-t)}$$

where m is the slope of the linear portion of the curve. In this range the slope is unity, so $m = 1$. To find K we extrapolate the linear region to its intersection with $(t/s) = 1$, and read $K = 0.72$. Hence, for infinite-plane samples (I-t) whose thickness is one half or less of the probe intertip spacing,

$$a = 0.72\left(\frac{t}{s}\right) \quad \text{for } \frac{t}{s} \leq 0.5 \qquad \text{("thin," I-t)}$$

When substituted into Eq. (3.5-1) this yields

$$\rho = a(2\pi s)\frac{V}{I} = 0.72\left(\frac{t}{s}\right)(2\pi s)\frac{V}{I}$$

or

$$\rho = 4.53t\frac{V}{I} \quad \text{for } (t/s) \leq 0.5 \qquad \text{("thin," I-t)} \qquad (3.5\text{-}2)$$

Notice that in this "thin" range the probe spacing s cancels out of the basic equation for ρ.

The numerical forefactor of 4.53 in the last equation is the approximate numerical value for $\pi/(\ln 2)$ that appeared in the van der Pauw formula, Eq. (3.3-3). The value 0.72 is the approximate numerical value for $1/(2\ln 2)$. The 4.53 constant may be clarified if Eq. (3.5-4) is derived by use of the logarithmic potential, as in Appendix A.3.

3.6 Diffused or Implanted Layers

Generally wafers will be no thinner than 10 mils (254 μm) to avoid mechanical breakage problems. As a practical matter, however, the major portion of FPP measurements are made on samples that are very much thinner—on diffused or implanted layers at one surface of a much thicker substrate. These layers typically are 10 μm or less in thickness. We now consider measuring the resistivity of such very thin layers.

3.6.1 Type n–p or p–n Junctions

Generally, FPP measurements on diffused or implanted layers can be made conveniently only if the layer is of opposite type from the substrate, either p on n, or n on p. This implies that a junction is present that essentially prevents probe current flow between the two regions.

If current is to traverse the lower layer, it must cross the junction twice—once in the reverse direction—so it is limited to a small leakage value, essentially negligible wrt the main flow in the upper layer. Therefore, with the p–n junction present, probe readings made on the upper surface are unaffected by the substrate, which effectively looks like an insulator. (Isolation between the layer and the opposite type of substrate can be increased by applying a reverse bias across them, independent of the FPP circuitry.) This condition was assumed in deriving the basic FPP thin sample equation; hence Eq. (3.5-2) is valid.

3.6.2 Type n–n^+ or p–p^+ Junctions

Particularly in the manufacture of certain discrete semiconductor devices there are cases where the layer to be tested is diffused or implanted in a highly doped wafer of the *same* type. This results in an n on n^+ or p on p^+ junction (the superscript "+" indicates high doping concentration) that does not have the rectifying and isolating properties of its p–n counterpart. The previously derived equations are not valid for these cases and other ways of handling resistivity measurement are necessary.

One solution to the no-isolation problem is to include a special test wafer of opposite type along with a group of wafers that are to be batch-diffused, or implanted. Then, when the process results in layer formation, the test wafer will have a p–n junction and FPP measurements can be made on it. The corresponding doping level is then calculated for the test wafer and assumed to be the same for all the other wafers of the batch. This often is referred to as the *control-wafer method*.

Alternatively, equations can be derived for probe measurements made on a sample resting on a conducting rather than an insulating surface [4], although this approach is not used commonly in the semiconductor industry. Another method is discussed in Section 9.12.

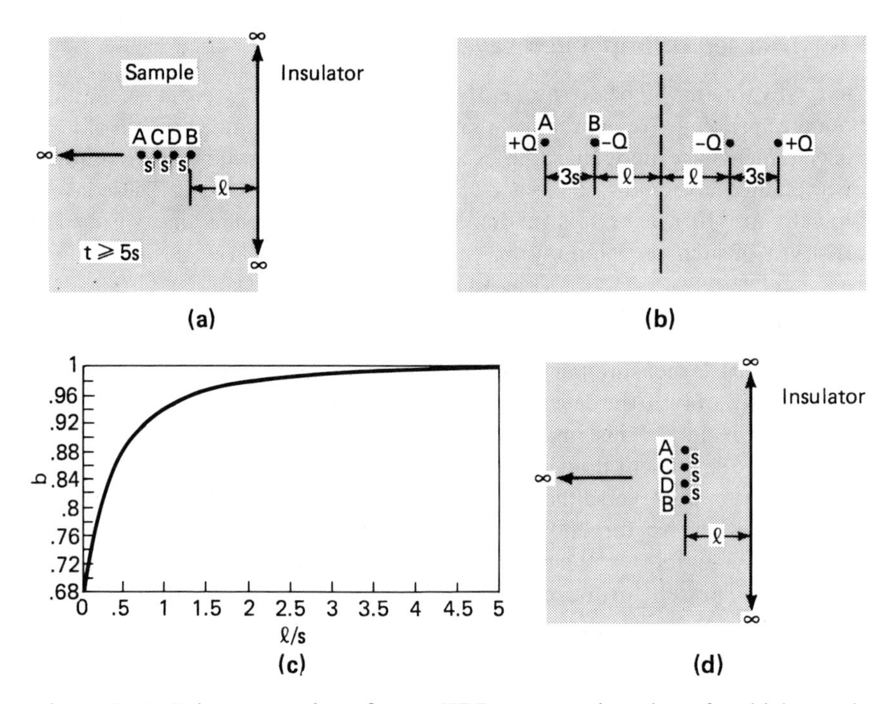

Figure 3–5. Edge corrections for an FPP near to the edge of a thick, semi-infinite sample. (a) The probe axis is normal to the sample edge. (b) The charges required to satisfy boundary conditions in the electrostatic analog of (a). (c) The correction factor *b*. (d) The probe axis is parallel to the sample edge.

3.7 Size Correction Factors

We have seen that the FPP views an I-t sample as being infinitely thick if $(t/s) \geq 5$. Can a similar rule be found for samples of *finite* horizontal size, or failing this, can we determine size and shape correction factors for samples of different sizes and shapes? We first consider infinitely thick samples with the probe tips near a sample edge.

Consider an infinitely thick sample (or at least with $t \geq 5s$) but semi-infinite in the horizontal plane. Say the FPP is placed so that its line of tips is normal to the boundary of the semi-infinite plane, with tip B a distance ℓ from that boundary, as shown in Figure 3–5(a). We seek a correction factor b such that the resistivity of the sample is given by

$$\rho = b(2\pi s)\frac{V}{I} = b\rho_{\infty} \tag{3.7-1}$$

where V and I are the FPP voltage and current as before. We must evaluate b as a function of ℓ/s. Notice in the figure that the medium to the right of the semi-infinite plane boundary is an insulator, so no current can flow across the boundary. This implies that the *normal* component of the $\mathscr{E}$ field is identically zero along that boundary, a condition that must be satisfied in our solution.

We use the electrostatic analogy again. Charges $+Q$ at A and $-Q$ at B (which are necessary for modeling the FPP) in Figure 3–5(b) cannot satisfy the boundary condition, so we shall add image charges. It is easy to argue from physical considerations that only two such charges are required and must be placed as shown in the figure—both to the right of the boundary and in line with the probe tips. We need $-Q$ at ℓ and $+Q$ at $(\ell + 3s)$, both measured from the boundary. The voltage between C and D, corresponding to the locations of the inner tips of the actual probe, is then calculated. Details of the calculation are left as an exercise, see Problem 3–5. A rough plot of b versus ℓ/s is shown in Figure 3–5(c). This indicates that the probe reads high—i.e., $\rho_\infty = 2\pi sV/I > \rho$, the correct value—as the probe approaches the sample edge (ℓ/s decreases). For example for $\ell = s$, ρ_∞ reads roughly 5% high, even though the actual resistivity is independent of the probe location.

Evaluation of b shows that for $\ell/s \geq 5$, $b \geq 0.997 \approx 1$, so for practical purposes, if the probe is normal to the boundary and away from it by at least $5s$, the FPP sees no boundary at all. This defines a large sample, at least for this probe orientation.

Another configuration is shown in Figure 3–5(d), where the line of the probe tips is parallel to the boundary. This situation is studied in Problem 3–6, where it may be shown that the solution is of the form

$$\rho = c(2\pi s)\frac{V}{I} = c\rho_\infty \tag{3.7-2}$$

Again, evaluation of c will show that the sample looks infinitely large if $\ell/s \geq 5$. We must be careful, however, not to infer that this *5 times s* criterion for *infiniteness* applies to all cases.

It is tempting to extrapolate these single-edge results. If the probe is centered on a finite rectangle as in Appendix F.3.3, is it valid to write

$$\rho_{\text{rect}} \stackrel{?}{=} (2b)(2c)(2\pi s)\frac{V}{I}$$

The answer is no! Consider the following partial argument. The right-hand edge of the rectangle was handled as a *single, isolated* boundary in Figure 3–5(b). Only two image charges were required to make the normal component of $\mathscr{E}$ identically zero on that boundary and the corresponding correction factor was designated c.

Now we add the left-hand edge of the rectangle. As an *isolated* boundary it would require two image charges properly placed to the left of the rectangle. But these two new images would upset the $\mathscr{E}_{\text{normal}} = 0$ condition on the right-hand edge and we are in trouble right away. It turns out that the correct solution for such a finite, closed boundary problem requires a two-dimensional, infinite array of image charges. Some solutions for this class of problem are given in later sections.

3.8 Sheet Resistance (R_S)

Most samples, particularly diffused and implanted layers, satisfy the thin criterion whose solution is given by Eq. (3.5-2). It is convenient to define the *sheet resistance* R_S for such layers.[3] This is obtained by dividing both sides of Eq. (3.5-2) by t, viz

$$R_S = \frac{\rho}{t} = 4.53\frac{V}{I} \tag{3.8-1}$$

Even though R_S is defined in the preceding equation in terms of an FPP reading, sheet resistance has significance in its own right. First, observe that the RHS of the equation is independent of any sample geometry and dimension, except that infinite extent in the horizontal plane is assumed; hence we might expect that R_S is a property of the sample material alone. Its significance may be clarified by noting that the end-to-end resistance of a rectangular parallelepiped is

$$R = \rho\frac{\ell}{wt} \tag{3.8-2}$$

where ℓ = sample length, w = sample width, and t = sample thickness.

Now, if the sample is made *square* (normal to the direction of current flow), by setting $w = \ell$, then the sample resistance reduces to ρ/t, which we already have defined to be the sheet resistance R_S. Therefore, R_S may be interpreted as the end-to-end resistance of a square sample. For this reason the units of R_S often are given as *ohms per square* or simply Ω/sq. Dimensionally this is the same as ohms, but it is convenient to use it as a reminder of the geometrical significance of sheet resistance.

It is apparent from the equation that R_S (being independent of the side length of the square) is a property of the material and its thickness. We can, however, reinforce this idea by a simple exercise. A square sample of length and width w has an end-to-end resistance R_S. If two such samples are joined

[3] Some authors use ρ_S, which has implied dimensions Ω-cm. We use R_S with dimension Ω.

in series, they give an end-to-end resistance $2R_S$. Then, if two of the two-square units are placed in parallel, the end-to-end resistance becomes $\frac{1}{2}(2R_S) = R_S$. Thus we have quadrupled the area of the square sample but the sheet resistance has remained unchanged; R_S is independent of the length of the square's side.

A useful property of sheet resistance is that it may be defined for, and measured on, samples that are of either uniform or varying resistivity in the t-direction—i.e., normal to the surface. For this reason it is applicable, say, to diffused layers as well as to uniformly doped substrates. Calculation of R_S for diffused layers is considered in Section 6.5.

3.9 R_S Shape Correction Factors

In Section 3.7 we saw that a two-dimensional infinity of image charges is necessary to meet the boundary conditions at the edges of a sample that is finite in the horizontal plane. We shall omit the details of solving such problems to get shape correction factors, since a number of references are available that explain the details [5–9]. Rather, we shall present results for a few of the more commonly encountered shapes.

The large majority of FPP measurements are made on very thin diffused or implanted layers so that indeed $t/s \leq 0.5$, and Eq. (3.8-1) is valid if the sample is infinite in the horizontal plane. Then a factor, say k, can correct the basic R_S equation for the particular sample shape and size in the horizontal plane. Thus,

$$R_S = 4.53 \frac{V}{I} k \tag{3.9-1}$$

Many authors prefer to define a single factor, $C = 4.53k$, such that

$$R_S = C \frac{V}{I} \tag{3.9-2}$$

Thus, if $C = 4.53$, $k = 1$, which implies that the FPP sees the sample as being of infinite extent in the horizontal plane and no correction to the basic equation is required.

Correction curves of C for an FPP centered on rectangular and circular samples are shown in Appendix F.3.3. These curves are adapted from Smits [6]. To illustrate their use consider the following example.

A rectangular p-type region, 250 × 125 mils in size, is diffused into an n-type wafer. An FPP with $s = 62.5$ mils is centered on the p-region. The spreading resistance (V/I) is measured to be 21 Ω. Calculate R_S of the p-layer.

Using the notation of Appendix F.3.3 (a is not the thickness correction factor!) we see that

$$a = 250 \text{ mils}, \qquad d = 125 \text{ mils}, \qquad \frac{a}{d} = \frac{250}{125} = 2$$

Therefore the curve is valid for this case. Also, $d/s = 2$. Then, reading from the rectangle curve, we get $C = 1.95$. Hence

$$R_S = CR_{Sp} = 1.95(21) = 41 \ \Omega/\text{sq}$$

If the same p-layer were diffused into the entire upper surface of a circular n-type wafer 3 in. in diameter, what R_{Sp} would be read by the same FPP centered on the wafer?

The wafer diameter is 3 in. = 3000 mils, so $d/s = 3000/62.5 = 48$. From the circle curve of Appendix F.3.3 we read that C has the asymptotic value of 4.53. Thus $R_S = 4.53 R_{Sp}$. Since R_S is still the same for the p-layer

$$R_{Sp} = \frac{41}{4.53} = 9.1 \ \Omega$$

These results show that for the same p-layer sheet resistance we read two different V/I ratios, or spreading resistance values, simply because the probe was on differently shaped samples of different sizes.

Note that in the second case the wafer is infinitely large as far as the probe is concerned, and that the criterion for infinite size in this case is $d/s \geq 20$, rather than 5, the value we have encountered in all the earlier cases.

Inspection of the curve for rectangular samples in Appendix F.3.3 shows that the log–log plot approaches linearity with slope m approaching unity and $K = 1$ for $d/s < 1.5$. In this narrow sample range, then, we may write

$$C = K\left(\frac{d}{s}\right)^m = 1\left(\frac{d}{s}\right)^1 = \frac{d}{s} \quad \text{for } \frac{d}{s} < 1.5$$

and by Eq. (3.9-2), R_S, which is ρt, becomes

$$R_S = C\frac{V}{I} = \frac{d}{s}\frac{V}{I} = \rho t$$

The ratio V/I is just the total resistance, say R, between the voltage sampling points that are separated by distance s. Furthermore, d is the conventional sample width w. Thus

$$R = \rho \frac{s}{wt}$$

This means that in the asymptotic range for small d, the probe views the narrow sample as a rectangular parallelepiped from geometrical considerations, an unsurprising result.

The curves of Appendix F.3.3 assume the probe to be centered on the sample. In contrast, Appendix F.3.4 shows C curves for an FPP aligned radially on a circular sample. The Greek δ is the distance between the sample and probe centers. The curves are calculated from the following equation, cited by Swartzendruber for this configuration [10]:

$$\eta = \left(\frac{1}{2\ln 2}\right)\ln\frac{[1-(x+0.5S)(x-1.5S)][1-(x-0.5S)(x+1.5S)]}{[1-(x-0.5S)(x-1.5S)][1-(x+0.5S)(x+1.5S)]} \tag{3.9-3}$$

where $x = \delta/r, S = s/r, r =$ radius of the circular sample, and $C = 4.53/(1+\eta)$.

The curves are of great significance when a probe is moved about the surface of a wafer to check for uniformity of R_S. Consider the following illustration of this point.

Assume a *uniform* n-layer implanted on the upper surface of a circular p-type wafer of 5 in. diameter. The FPP intertip spacing is 62.5 mils. Then we have

$$S = \frac{s}{r} = 62.5/2500 = 0.025$$

With the probe centered on the wafer, $\delta = 0$, and we calculate from Eq. (3.9-3) that $C = 4.524$, so $R_S = 4.524 R_{Sp}$.

Now say that the probe is moved until $\delta = 0.9r$, so that $x = \delta/r = 0.9$. Then we calculate that $C = 4.361$ so $R_S = 4.361 R_{Sp}$. Thus, even though R_S is perfectly constant over the entire wafer surface, the probe's reading will change about 3.6% just because of its change in location. The change is 10% for the same s but $D = 3$ in. Note that when $x = 0.9$ and $D = 3$ in., the outer tip of the probe is just under one s from the wafer edge.

The change in probe reading illustrated by this example can be misleading, especially when the FPP is moved around on the sample surface to map the variation in R_S. If nonuniformities appear in R_{Sp}, the experimenter must know how much of the change is due to the variation in doping, and how much to the edge effect of the FPP.

From earlier equations we have seen that

$$R_{Sp} = \frac{V}{I}, \qquad R_S = C R_{Sp},$$

$$\rho_\infty = 2\pi s R_{Sp}, \qquad \rho = R_S t = C t R_{Sp}$$

Thus, for all cases of calculating either ρ or R_S, the answer is obtained by multiplying the spreading resistance (R_{Sp} = measured V/I ratio) by one correction factor C, or at most by two—C and the thickness t. A number of commercial FPP measuring sets are available that allow the entering of appropriate multipliers, perform the multiplications internally, and give digital readout [3].

3.10 Modified FPP Configurations

The connections for the collinear four point probe (FPP) shown in Figure 3–4(a) are the most common encountered in practice. Two other useful configurations are shown in Figure 3–6. We shall consider these briefly, concentrating on the advantages or particular uses of each. Mathematical details are covered in Problems 3–7 and 3–8.

3.10.1 Modified Collinear FPP

Consider the modified collinear FPP configuration shown in Figure 3–6(a). The probe head is the same as before, but the current source and voltmeter connections are changed from what they were in the conventional form. This configuration also may be used for ρ and R_S measurements. It finds particular application in conjunction with the usual connection, as considered next.

No matter how carefully a probe is built, there is always some chance that one or more of the tips will be bent out of collinearity when they are pressed into contact with the sample, which causes errors in the probe readings. Brennan and Dickey have discussed a method of compensating for such probe geometry errors that involves readings taken with both the conventional and modified collinear connections [11].

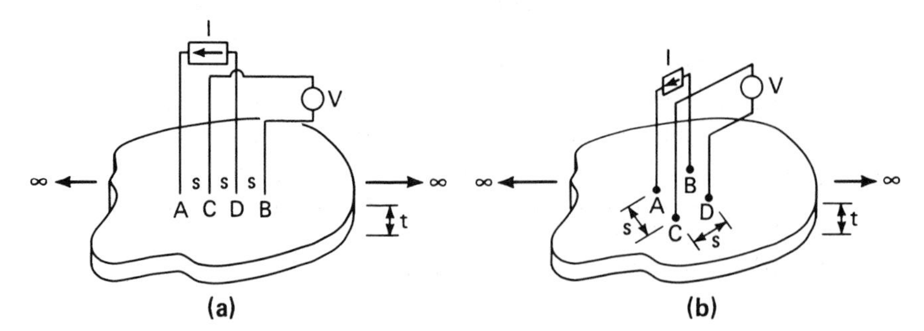

Figure 3–6. Other FPP configurations. (a) Modified collinear probe. (b) Square tip probe.

The probe is placed so that its tips make contact with the sample surface. Then, with no intervening tip movement, two readings of spreading resistance are made: $R_{S_{pc}}$ with the current source and voltmeter in the conventional circuit, and $R_{S_{pm}}$ with the source and voltmeter in the modified positions. The sheet resistance then may be written as

$$R_S = C_r R_{S_{pc}}$$

where C_r is a function of the spreading resistance ratio $R_{S_{pc}}/R_{S_{pm}}$. It is claimed that this method can give accuracies of about 0.5%.

3.10.2 Square Tip Probe

Another fairly common form of FPP head, shown in Figure 3–6(b), has the tips located at the corners of a square of side s. The current source is connected, not to diagonally opposite tips, but to tips on one side of the square, while voltage is measured between the tips on the opposite side of the square. It is easy to show (Problem 3–8) that ρ_∞, the resistivity of a semi-infinite sample, is given by

$$(\rho_\infty)_\square = \frac{2\pi s}{(2 - \sqrt{2})} \frac{V}{I} = \frac{2\pi s}{0.586} \frac{V}{I} \tag{3.10-1}$$

and that for the I-t sample

$$(R_S)_\square = \frac{2\pi}{\ln 2} \frac{V}{I} = 2(4.53)\frac{V}{I} \tag{3.10-2}$$

or twice the value for the conventional collinear probe head [see Eqs. (3.4-1) and (3.5-2)]. Correction factors $C_\square$ for the equation

$$R_S = C_\square \frac{V}{I} \tag{3.10-3}$$

applicable to the square probe have been calculated by several authors [5, 7, 10, 12]. Curves of normalized correction factors for the conventional collinear probe and square probe centered on a square of side d and centered on a circle of diameter d are shown in Appendix F.3.5 [12].

Advantages of the square FPP relative to the conventional collinear form are: (1) it is more compact for a given tip spacing; (2) for a given driving current I, it gives higher voltage, and hence spreading resistance readings; and (3) it can be moved closer to the sample edges without the need for correction factors. The last two points may be verified from Appendix F.3.5.

3.11 Temperature Correction

The foregoing results have assumed that at room temperature essentially all the dopants in a Si sample are ionized, and that $N_M \gg N_m$; therefore $N_M = N_b$. Nevertheless, resistivity can be temperature dependent in the room-temperature range, especially in high-ρ samples. Consider the reason for this.

Equation (2.7-3), namely $\rho = 1/q(\mu_M N_M + \mu_m N_m)$, shows that ρ depends upon the concentrations and mobilities of both majority (M) and minority (m) carriers. At low resistivities N_M completely dominates N_m in the room-temperature range. For higher resistivity silicon, say at 10 Ω-cm or more, N_M becomes lower ($\leq 5 \times 10^{14}$ cm^{-3} for n-type) and N_m rises, because $N_m = n_i^2/N_M$.

Furthermore, n_i^2 increases exponentially with absolute temperature in Kelvins; therefore, in silicon ρ can change noticeably as temperature rises even though the doping $N \approx N_M$ holds constant. Moreover, due to increased scattering at higher temperature, both majority and minority carrier mobilities decrease.

These temperature effects can cause difficulties in monitoring fabrication processes. For example, if wafers into which a layer has been diffused are subject to FPP sheet resistance measurements while located very close to the diffusion furnace, where the temperature might be rather high, say around 30°C ($\approx$86°F), the measured value would read high compared to the value at the ASTM standard temperature of 23°C ($\approx$73.4°F). From a practical viewpoint, if the measured value of ρ were used to monitor the diffusion process, the results would be incorrect. Compensation may be obtained by the ASTM method outlined by Germano [13][4]:

Let

$$F_T = 1 - C_T(T_C - 23) \tag{3.11-1}$$

where T_C is the measurement temperature in °C, and C_T is the temperature coefficient determined from Table 3–1 (whose values were obtained by curve-fitting data). Then the resistivity at the standard temperature of 23°C is given by

$$\rho_{23} = \rho_T F_T \tag{3.11-2}$$

where ρ_T is the measured resistivity at temperature T_C.

By way of illustrating the magnitude of this temperature effect, say ρ_T is measured as 10 Ω-cm at 30°C. Using data from the table for 10 Ω-cm resistivity

[4] The American Society for Testing Materials (ASTM) prepares testing procedures for many segments of American industry. For example, a principal publication for the semiconductor industry is Part 43 of the Annual Book of ASTM Standards, 1975, "Measuring Resistivity of Silicon Slices with a Collinear Four-Probe Array."

Table 3–1. Temperature Compensation Coefficient for Silicon

Resistivity Range	C_T
<0.0008	0.00192
0.0008 to 0.249	$(1.94\ \text{E}-3) - (2.54244\ \text{E}-1)\ \rho + (1.173\ \text{E}1)\ \rho^2 - (1.83887\ \text{E}2)\ \rho^3 + (1.35035)\ \rho^4 - (4.69242\ \text{E}3)\ \rho^5 + (6.20569\ \text{E}3)\ \rho^6$
0.25 to 7.98	$(5.54\ \text{E}-3) + (2.326\ \text{E}-3)\ \rho - (8.535\ \text{E}-4)\ \rho^2 + (1.32\ \text{E}-4)\ \rho^3 - (7.1\ \text{E}-6)\ \rho^4$
7.99 to 500	$(8.138\ \text{E}-3) + (1.037\ \text{E}-5)\rho - (9.778\ \text{E}-8)\rho^2 + (4.227\ \text{E}-10)\rho^3 - (8.33\ \text{E}-13)\ \rho^4 + (6.075\ \text{E}-16)\ \rho^5$
>500	0.00864

Note: $\rho = \rho_T$.
Germano [13]. Courtesy of *Solid State Technology*, PennWell Publishing Company, Copyright 1976.

and substituting, C_T is calculated to be $8.23 \times 10^{-3}\ (°\text{C})^{-1}$ to three significant figures. Then F_T is 0.942, and the resistivity at standard temperature is $\rho_{23} = 9.42\ \Omega$-cm. The initial measured value, then, was about 6% high. The error increases as ρ_T gets larger, as may be seen from the table. On the other hand, if ρ_T measured out as 1 Ω-cm at 30°C, the error would be less than 0.01% relative to the 23°C value.

3.12 Other Resistivity Measuring Systems

A number of other resistivity measuring systems are available [3, 14], but they are used primarily for diagnostic reasons; they cannot compete with the FPP for normal production line use. We shall consider two different probe methods.

3.12.1 Two Point Probe (2PP)

As its name implies, the 2PP has only two tips, which serve simultaneously for current injection and voltage measurement. Again, the V/I ratio is the spreading resistance R_{Sp}. Calibration curves used to relate sample ρ to R_{Sp} are obtained by measuring R_{Sp} on standard silicon samples of known resistivity. Such samples are available in a range covering five decades of resistivity beginning at 0.001 Ω-cm at the low end, for ⟨100⟩ and ⟨111⟩ orientations in both p- and n-type silicon.

There is no advantage of the 2PP over the FPP for usual ρ and R_S measurements. It is used primarily for ρ-contour measurements [Figure 3–3(a)] and in *profiling*, where the variation of resistivity with distance into the wafer,

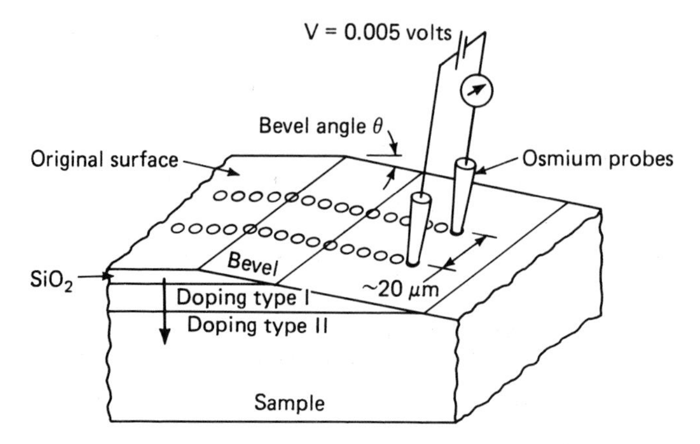

Figure 3–7. Profiling with a 2PP on the beveled edge of a sample. Brennan and Dickey [11]. (Courtesy of *Solid State Technology*, PennWell Publishing Company, Copyright 1984.)

x, is measured. The small size of the 2PP is of advantage in both applications. Profiling requires beveling the wafer so a series of measurements into the wafer can be made as in Figure 3–7 [11]. Beveling is discussed in Section 3.14.1.

A typical bevel angle of 12° relative to the top surface will spread out a 10-μm layer to 48.1 μm (1.89 mils). It is apparent that a very precise positioning mechanism is needed if the probe is to be moved to several measuring points along a path only a few micrometers long, with the same force being applied for each measurement. The probe itself is small—2 μm being a typical intertip spacing, s. Typical tip loading is from 2.5 to 10 g per tip, depending upon the thickness of the layer being probed. The construction of such a probe is not trivial [16].

Considerable calculation is needed to convert the R_{Sp} readings as a function of probe position to a plot of $N(x)$. A principal disadvantage of 2PP profiling is that beveling of the sample is destructive. An alternative, but still destructive method, is to successively etch away thin layers of the wafer, with a resistivity measurement being made on each exposed surface.

3.12.2 Three Point Probe (3PP)

The three point probe (3PP), shown in Figure 3–8, is used for determining the reverse breakdown voltage, BV, between a metal point and a lightly doped, ion implanted or epi layer on a highly doped substrate of the same type—i.e., n on n^+ or p on p^+. The breakdown voltage of a reverse biased point contact depends on the doping (and so on ρ) of the semiconductor; the lighter the doping, the higher is BV. The layer resistivity may be determined from the BV value.

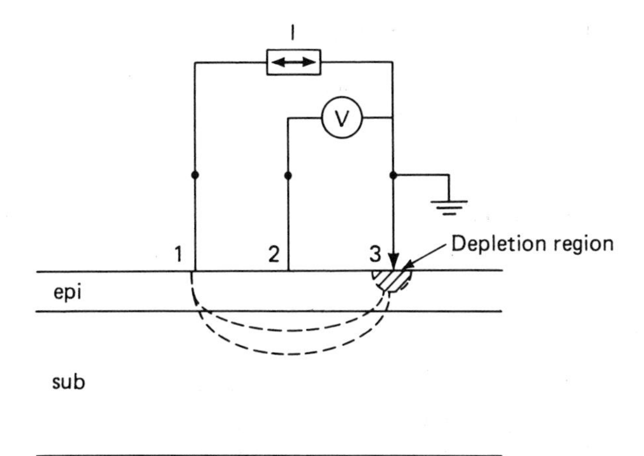

Figure 3–8. The 3PP on a sample. A rectifying contact is formed by cone-pointed tip 3 contacting the lightly doped epi layer. Current flows primarily in the more heavily doped substrate.

The breakdown characteristic of a point contact under reverse bias is very soft as compared to the sharp breakdown typical of a reverse biased silicon p–n diode, so the exact value of BV for the point contact is hard to define and to measure. The measurement is made more precise by defining BV, for a given point size, as the magnitude of the reverse bias voltage at a specified value of reverse current. Empirically determined calibration curves are required to convert the BV readings to doping concentration N or to resistivity ρ of the semiconductor layer [17, 18].

Tip 3 of Figure 3–8 is a truncated cone that terminates in a flat tip of, say, 2.5 mils in diameter. It forms the point contact diode with the silicon layer and also is the common grounded lead. Tips 1 and 2 are blunt and typically will be 10 mils in diameter in order to serve as ohmic contacts on the epi layer.

The current source is polarized to reverse bias the point contact diode: tip 3 is positive for a p-type epi layer, and negative for an n-layer. The epi layer has higher resistivity and is thinner than the substrate; thus, current flows almost entirely in the substrate (dashed lines in the figure) so IR drops in the epi layer are negligible. A high-impedance voltmeter reads the voltage across the point contact diode. The current source prevents large currents that might damage tip 3.

Many variations of the basic d-c circuit have been designed for use with the 3PP. A 60-Hz system that uses an oscilloscope for readout has been described [17]. Schumann et al. discuss several systems and show an electronic measuring system that uses a ramp input signal [18].

3.13 Depletion Layer Capacitance Profiling

The measurement of junction capacitance C as a function of reverse bias voltage magnitude V is the basis of a powerful method for determining semiconductor doping profiles nondestructively. The $C-V$ measurements are simple to make and relatively easy to convert to $N(x)$ versus x [19, 20]. Other methods are based on harmonic generation in the nonlinear junction capacitance. Both are known as *profiling*. Two variations are considered.

3.13.1 *C–V* Method

Consider a one-sided junction—that is, one in which the depletion or transition region, of width w, is confined to the semiconductor on one side only of the junction. This condition may be realized by three structures: (1) a p–n junction with one side very highly doped relative to the other and the depletion region primarily on the lightly doped side; (2) a metal/semiconductor (MS) or Schottky junction; or (3) a metal/insulator/semiconductor (MIS) junction. In the two latter types the depletion region is confined to the semiconductor.

With the junction reverse biased, mobile carriers are drawn away from the transition region, depleting it of mobile carriers, and *uncovering* ionized donors or acceptors that are locked in the lattice. As the reverse bias is changed, so are the depletion region width w and the total uncovered bound charge Q in the depletion region. The latter has the magnitude

$$Q = qA \int_0^w N(x)\,dx \qquad (3.13\text{-}1)$$

where

q = electron charge magnitude

A = cross section area of the junction

We seek a $w-V$ relationship. Differentiating wrt w,

$$\frac{dQ}{dw} = qAN(w) \qquad (3.13\text{-}2)$$

But by definition the voltage dependent depletion layer capacitance is

$$C = \frac{dQ}{dV} = \frac{dQ}{dw}\cdot\frac{dw}{dV} \qquad (3.13\text{-}3)$$

Substituting from Eq. (3.13-2)

$$C = qAN(w)\frac{dw}{dV} \tag{3.13-4}$$

Even though the bound charge is *distributed* throughout the depletion layer, it may be shown that the related capacitance is the same as for a parallel plate capacitor of the same area and of plate separation w, so

$$C = \frac{\varepsilon A}{w} \quad \text{or} \quad w = \frac{\varepsilon A}{C} \tag{3.13-5}$$

where ε is the permittivity of the semiconductor. Then

$$\frac{dw}{dV} = -\frac{\varepsilon A}{C^2}\frac{dC}{dV} \tag{3.13-6}$$

so

$$N(w) = \frac{C^3}{q\varepsilon A^2(dC/dV)} \tag{3.13-7}$$

The negative sign has been dropped because both voltage and concentration have been defined as magnitudes and hence are positive. Data for this equation are obtained by measuring C (with a small a-c signal superimposed on the d-c reverse bias) at several bias values.

The principal problem in using the equation is determining the derivative dC/dV from C–V data. One method is to take two data pairs (C_2, V_2) and (C_1, V_1) close together and approximate the slope by

$$\frac{dC}{dV} \approx \frac{C_2 - C_1}{V_2 - V_1} \tag{3.13-8}$$

Slide rules are available for evaluating Eq. (3.13-7) using this approximation [21], and others include temperature correction as well [22].

An alternative method more suitable to computer solution is based on the fact that the C–V relationship can be expressed as a power law, viz

$$C = C_o V^m, \qquad -1/2 \leq m \leq -1/3 \tag{3.13-9}$$

and Eq. (3.13-7) becomes

$$N(w) = \frac{C_o^2 V^{(2m+1)}}{q\varepsilon A^2 m} \tag{3.13-10}$$

and from Eqs. (3.13-5) and (3.13-9)

$$w = \frac{\varepsilon A}{C_o V^m} \tag{3.13-11}$$

Since w in the last two equations is measured from the junction at $x = 0$ it may be replaced by x, giving two parametric equations for obtaining $N(x)$ versus x that may be handled easily with a simple computer program. A least squares fit of the $C-V$ data allows calculation of C_o and m, and also has the advantage of smoothing data. Given these, the plot may be made readily. Alternate formulation of the equations is possible, too.

Several commercial profilers that perform all the functions automatically and deliver a profile plot are available. To speed up work, the stepwise varied d-c reverse bias is replaced by a d-c ramp and the a-c capacitance measurement is made "on the fly" [20].

3.13.2 Harmonic Method

Another related method, capacitance inverse profiling (CIP), essentially yields a plot of $1/N(x)$ versus x—i.e., an *inverse* profile [23, 24]. Again, the depletion region width w is changed by applying a d-c ramp signal, and a sinusoidal r-f *current* signal of constant amplitude I is applied to the junction as well.

The diode capacitance is nonlinear in voltage, so a significant second harmonic, in addition to a fundamental, voltage component appears across the junction. It may be shown that the r-f output voltage is [23]

$$V = \frac{I\cos(\omega t)}{\varepsilon A\omega} w + \frac{I^2[\cos(2\omega t) + 1]}{4q\varepsilon A^2\omega^2} \frac{1}{N(w)} \tag{3.13-12}$$

so the fundamental component is directly proportional to w, and the second harmonic is *inversely* proportional to $N(w)$. These components may be separated by filters and processed to give the desired plot. The use of a logarithmic amplifier allows a noninverted plot of $N(w)$ versus w over a two-decade range of doping concentration. Details of the measuring equipment are available [24–26].

3.13.3 Mercury Probe

If a bare or oxidized silicon wafer is to be profiled, a temporary metal contact of known area must be provided (temporary so that the wafer is left un-

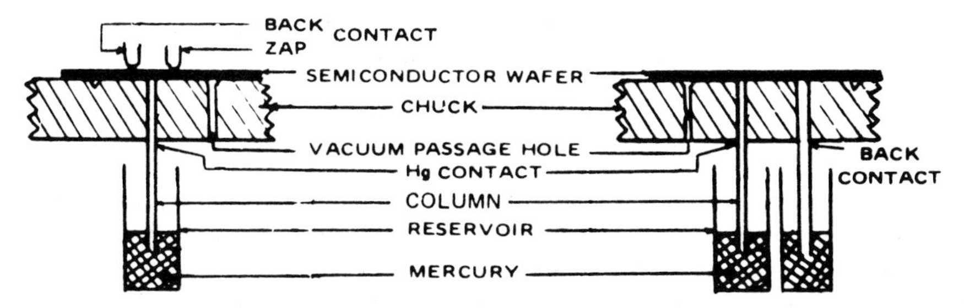

Figure 3–9. Forms of Hg probes. (a) Single Hg column. (b) Double Hg column [27]. (Courtesy of MSI Electronics Inc.)

harmed). Mercury (Hg) is well suited to this purpose because of its nonwetting properties and its noninteraction with Si and the common compound semiconductors. Two forms of commercial Hg probes for making such contacts are shown schematically in Figure 3–9 [27–29]. The wafer is placed measure-side down on the chuck, and Hg is pulled up from a reservoir by vacuum. The contact area (A in the foregoing equations) is determined by a standard orifice in a 10-mil-thick Mylar sheet located over the Hg siphon.

The unit in Figure 3–9(a) uses a metal back contact on the rear (upper) side of the wafer. A thin native oxide layer there hinders good electrical contact, so it is "zapped" away by application of a current pulse between the two top-side electrodes. This breaks down the oxide before measurements are made.

In some layered samples (e.g., n–n^+), both contacts are made on the same surface, Figure 3–9(b). Orifices of different sizes are used, with the larger defining the "back" ohmic contact.

If the silicon surface concentration exceeds 10^{17} cm^{-3}, the Hg no longer can make a rectifying junction, and depletion-layer-based measurements cease to be possible. This may be handled by using an oxide layer between the Hg and Si, giving an MIS contact. Then C of the equations is the oxide capacitance in series with the depletion-layer capacitance. Compensation must be made for this in the calculations. Several references are available on this and MIS contact measurements in general [30, 31].

3.14 Junction Depth Determination

The need frequently arises for measuring the depth of a junction x_j within a wafer. Multiple junctions also may be involved—e.g., the emitter and base junctions of a bipolar transistor, or the source and drain junctions in a MOST. These must be checked during production to provide process control. This section considers these measurements. Three separate steps are involved:

junction exposure, junction decoration (or staining) to reveal the p- and n-type layers separately, and actual measurement of x_j.

3.14.1 Junction Exposure

Typically, a junction lies a few microns (μm) below the wafer surface and hence is not accessible from the surface. A portion of the upper layer is removed so the junction is exposed. Because x_j is so small, we try to expose the junction in such a way as to spread out the region from the surface to x_j over a longer path. This concept was used in two-point-probe profiling. We consider two different methods of junction exposure to meet this requirement. One method sections the wafer on a plane inclined at a small angle to the surface. The second sections the wafer in the arc of a circle.

The first method uses a special grinding jig to hold the sample as a bevel is ground at some region near the wafer's edge. The jig is comprised of a cylinder, beveled on one end at 10 to 12°, mounted within an outer sleeve. The cylinder is free to move vertically but cannot rotate within the sleeve. A typical design for hand operation is given by Bond and Smits [32].

The sample is mounted on the central cylinder's face with wax, such as Apiezon "black wax," so that a small region of its surface projects above the top of the jig. The jig and sample are placed upside down on a flat glass working plate, and a slurry of fine abrasive particles, usually of alumina (Al_2O_3), in distilled or deionized water (DI) is applied to the plate. The jig is moved around in a figure-eight pattern by hand to lap off the desired bevel at the region of interest. The bevel angle will be determined by the jig, and the weight of the jig cylinder applies force for the grinding action. Sometimes a drop of detergent is added to the slurry to prevent the sample from "hanging up" on the glass. Frequent inspection of the sample is necessary to determine when the beveling action can be stopped.

A succession of slurries is used, with particles decreasing from roughly 0.5 to 0.1 μm in size. If a high polish is desired, final polishing is done on a beeswax-coated plate. Careful cleaning is required when slurries are changed, so larger particles will not carry over and scratch the surface.

With beveling finished, the sample is removed by melting the wax. Any excess can be dissolved in suitable organic solvents such as trichloroethylene (TCE) or trichloroethane (TCA). The latter is less noxious, but both of these chlorinated solvents must be used properly—in a vented fume hood to remove vapors, and with hands protected with good plastic or rubber gloves. As always in any laboratory work, proper eye protection should be worn.

After the wafer has been cleaned with TCE or TCA, it should be rinsed in acetone, in either ethyl or methyl alcohol, and finally in DI—in that sequence. The two chlorinated solvents do not mix with water but do with acetone. Acetone mixes with alcohol but not with water. Alcohol mixes with water.

The final DI rinse leaves the sample free of all organics and particulate matter.

The semiconductor industry has been eliminating, or at least reducing, the use of chlorinated solvents because of their hazardous nature. To affix the sample to the jig cylinder for example, black wax may be replaced by glycol phthalate. The latter's melting point is lower, but more important, it is soluble in acetone, thus eliminating the need for TCE or TCA.

The linear bevel method has some shortcomings. Practically, its use is confined to small samples—otherwise, a very large jig is required. Often a small piece of a broken wafer that was batch-processed with a run of wafers is used for this purpose. Second, the bevel must be at the edge of the sample. Third, the process is slow.

A second method of junction exposure removes material on a circular part-cylinder to expose the junction. The grinding is done by a motor-driven, circular metal wheel, typically about 0.75 in. in diameter and up to 0.125 in. thick. The sample is held against the rotating wheel by a spring-loaded holder in the region where x_j is to be exposed, and a slurry is applied to grind a groove in the sample. The sample then is removed and cleaned in the manner already described. At this stage it is not possible to visually distinguish the p- from the n-type material at the junction, so *decoration* or selective staining is necessary. The decoration method is described in Section 3.14.2, and colors the two regions differently.

The top view of the groove (after decoration) is illustrated in Figure 3–10(a). In general, the groove edges will not be perfectly straight, so some averaging by eye is necessary when the groove dimensions are subsequently measured. Part (b) of the figure shows a cross section of the groove, which ideally is a sector of a circle.

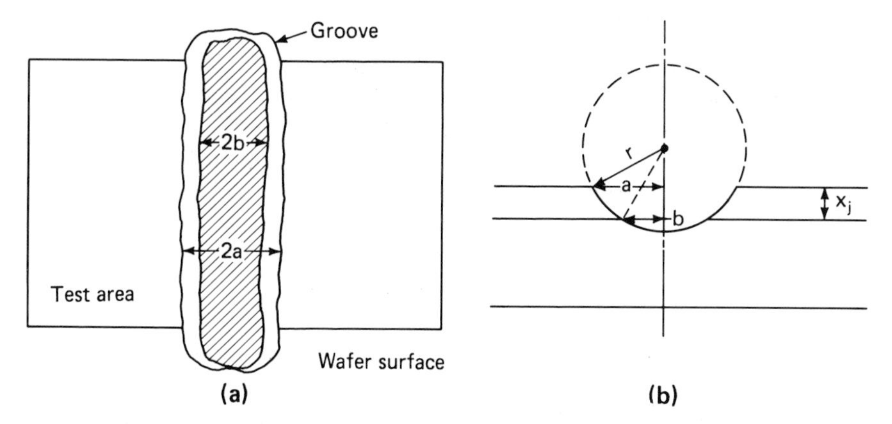

Figure 3–10. Junction measurement from a decorated, cylindrical groove in a sample. (a) Top view. (b) The geometry for calculating x_j.

3.14.2 Junction Decoration

The exposed junction must be *decorated* (i.e., stained selectively), so that the two types of silicon on either side of the junction appear different. Then x_j will lie where the color changes. Excellent results are provided by proprietary solutions.[5] Instructions furnished with these solutions should be followed strictly.

Lacking a proprietary stain, one can make do with the home-grown variety. The simplest stain is concentrated hydrofluoric acid (HF). A drop of this is placed on the exposed junction region and the p-region darkens. Extreme care must be used with HF since it attacks human skin slowly, but devastatingly, without causing any sensation of pain; proper acid-proof gloves in good condition must be used and proper eye protection must be worn. Remember that HF fumes etch glass! Since the reaction is viewed through a microscope, the objective lens should be protected by a piece of optically flat glass. When the staining is adequate, the action is stopped by adding DI to the drop of acid. The whole sample then should be flushed thoroughly in DI to insure that no HF is left. The sample may be blow-dried gently with dry nitrogen.

Sometimes the HF stain fails to work, in which event these tricks are available. (1) Shine bright white light on the junction region during staining. (2) Touch the tip of a stainless steel tweezer to the drop of HF. Be sure to rinse off the tweezer properly after this operation. (3) Modify the stain by adding a small amount of hydrochloric acid (HCl), say 1 drop of HCl in 10 cm^3 of HF.

A variation of the procedure places the junction under reverse bias during staining. This makes the n- rather than the p-region stain dark.

3.14.3 Measurement of x_j

Once the sample junction has been exposed and decorated by selective staining, the final step is to measure the junction depth. First, consider how this is done for the exposure method of Figure 3–10.

The geometry of the figure allows calculation of the upper-layer thickness, or junction depth x_j, from measured values of a and b and the known wheel radius r.

$$x_j = \sqrt{r^2 - b^2} - \sqrt{r^2 - a^2} \qquad (3.14\text{-}1)$$

The quantities $2a$ and $2b$ are viewed through a microscope and measured with

[5] Safe-t-Stain, available from Philtec Instrument Co., Philadelphia, PA, comes in five different formulations—each of which stains a different color. The stains also will work for n–n^+ and p–p^+ junctions, with the higher doped side staining darker.

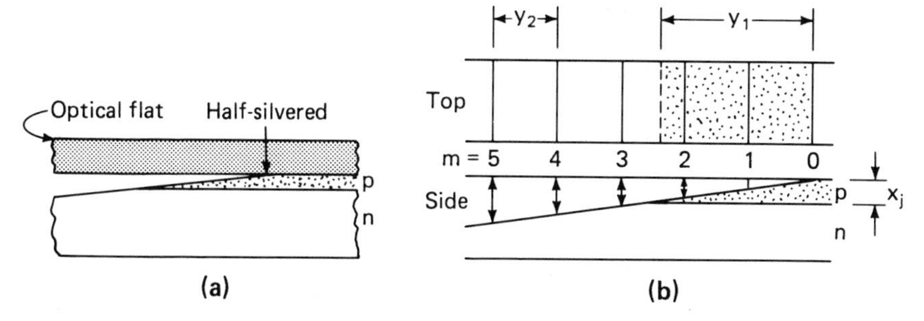

Figure 3–11. Measurement of junction depth by optical destructive interference. (a) Cross section of the physical setup. (b) Locations of the dark lines.

a filar eyepiece (one that has a movable, calibrated cross hair to measure distances accurately).

A different, interferometric method is used with the beveled, exposed junction. These measurements are made with monochromatic light and are well adapted to the small x_j values of a few micrometers. A brief review of optical interference is given in Appendix A.4 as background material.

Consider that a half-silvered optical flat (actually, a half-silvered microscope slide will serve nicely) is placed on the top surface of a linearly beveled sample. As can be seen in Figure 3–11(a), the flat and the bevel enclose a thin wedge of air. If this arrangement is illuminated with monochromatic light, we might expect that certain positions along the air wedge will satisfy Eq. (A.4-2) for destructive interference, and dark lines would show up at these positions. Consider how these dark lines may be used to measure x_j.

The setup is viewed with a metallurgical microscope, as shown in Figure 3–12. The monochromatic source is a sodium vapor lamp that has a wavelength in air of $\lambda_a = 5896$ Å. Notice that the lighting arrangement makes the incident light arrive normal to the surface of the flat, so $\cos \phi = 1$ in Eq. (A.4-2), and destructive interference will occur at those places at which the air wedge thickness is

$$t = \frac{m\lambda_a}{2}, \qquad m = 1, 2, 3, \ldots \tag{3.14-2}$$

As viewed through the microscope eyepiece, these positions show up as dark lines that provide a horizontal distance scale for determining the wedge thickness at any location.

This idea is illustrated in Figure 3–11(b), so we can write the proportion

$$\frac{x_j}{\lambda_a/2} = \frac{y_1}{y_2} \quad \text{or} \quad x_j = \frac{y_1}{y_2}\frac{\lambda_a}{2} \tag{3.14-3}$$

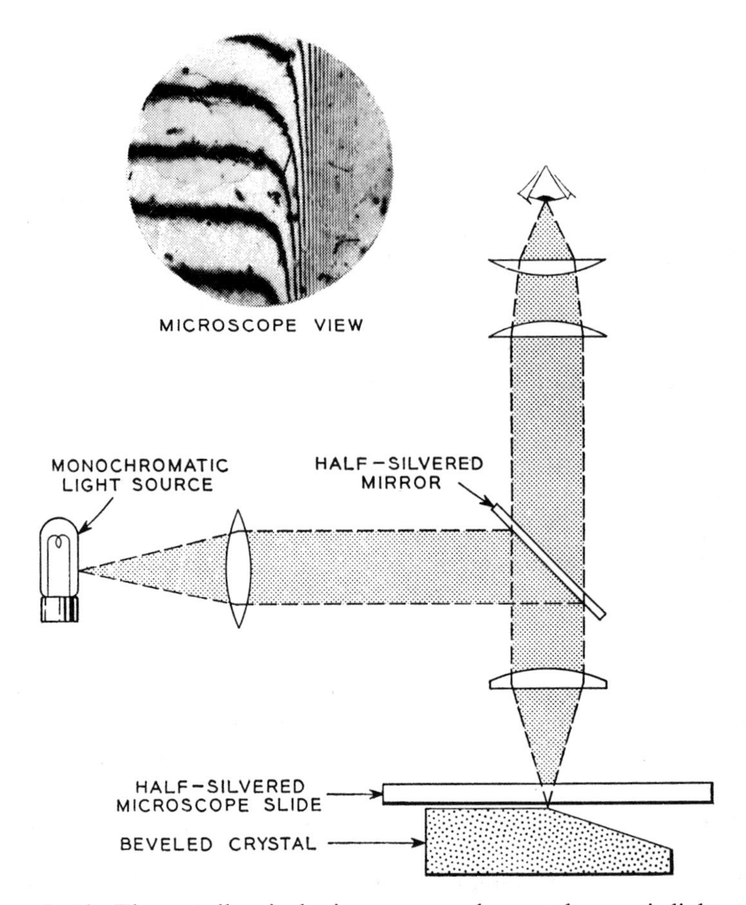

Figure 3–12. The metallurgical microscope and monochromatic light source are used to observe the locations of the dark bands. Bond and Smits [32]. (Courtesy of AT&T Bell Laboratories.)

A filar micrometer eyepiece may be used to measure y_1 and y_2, or a photograph may be made and the two distances measured with a scale. Since a ratio is used rather than an absolute length, the angle of the bevel is not involved in the calculation. However, a small bevel angle does spread the interference bars farther apart, making measurement easier.

It often happens in practice that the interference lines are not parallel to the junction edge. This means that the optical flat is not level on the upper surface of the sample, but it may be moved around slightly until the condition improves. Commercial equipment that simplifies the flat adjustment by means of thumbscrews is available.

Commercial equipment of recent design shines an infrared beam, to which silicon is transparent, on the top surface. The reflected IR components from the wafer surface and the junction combine to give an interference pattern that allows determination of x_j. This method is used routinely for checking epi layer thickness; it requires a significant difference in epi and substrate doping, but it is nondestructive.

3.15 Film Thickness Measurement

Insulating and/or conducting films grown or deposited on the upper surface of a wafer have typical thicknesses in the range from 0.1 μm (1000 Å) to a few micrometers. Yet gate oxides for modern MOS devices may be in the 200 to 400 Å range, while tunnel oxides for EEPROMS (electrically erasable programmable read only memories) may be below 100 Å! The interferometer method may be adapted to such films. An edge of the film must be accessible; if not, one can be etched for test purposes.

To insure adequate reflection, particularly with transparent films, a thin, uniform coating of silver or aluminum about 1000 Å thick is evaporated over the sample in the region of the film edge. A half-silvered optical flat is laid over the step and illuminated with monochromatic light. In Figure 3–13(a) you are looking head-on at the step caused by the film edge, so the circled dimensions are measured in *front* of the step. Two air wedges are formed, separated by the step: one in front between the flat and the substrate, another in back between the flat and the top of the film. Two sets of interference lines are formed, the lines of one set being displaced from the lines of the other by a distance proportional to the film thickness τ.

The flat is adjusted so that it lies at a slight angle to the film surface *in the direction parallel to the step edge*, and the dark interference lines are perpendicular to the step as shown in Figure 3–13(b). Then, by analogy to our earlier work,

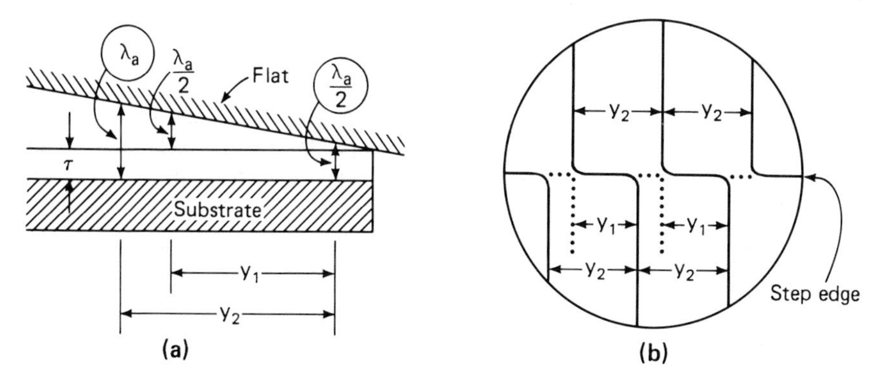

Figure 3–13. Film thickness measurement. (a) Head on view of the step in the film being measured. (b) Appearance of the dark lines. Note the displacement at the film step.

$$\tau = \frac{\lambda_a}{2} \cdot \frac{y_1}{y_2}, \qquad (\lambda_a = 5896 \text{ Å for sodium}) \qquad (3.15\text{-}1)$$

If the dark lines are not normal, or nearly so, to the step, the optical flat is not slanted properly and should be reoriented.

In some commercial equipment the half-silvered optical flat is jig-mounted, so its slant may be adjusted easily with thumbscrews. In others the optical flat is replaced by a Fizeau plate that also is screw-adjusted. These refinements contribute to the ease of adjustment and use.

Another optical system for measuring the thickness of transparent films, such as silicon dioxide or silicon nitride, utilizes the *ellipsometer*, whose operation is based on the reflection of *polarized* light from the film, rather than on the interference of light. It is discussed in Section 5.10.2.

3.16 Surface Profilers

Incredible as it may seem, electromechanical devices are available commercially that can scan across a wafer surface and read out vertical variations with a resolution of better than 5 Å! Two such instruments are the DEKTAK[6] and the Alpha-Step.[7]

In principle, these instruments use a linear variable differential transformer (LVDT) to convert vertical mechanical motion of the core to electrical output. The transformer core is connected to a diamond stylus that contacts the sample surface. The stylus is moved in a line across the sample surface, following the contour, and the LVDT produces a corresponding electrical signal. Readout options of height against scan position are available—e.g., an oscilloscope or x–y recorder. Thus, step heights of surface layers can be determined quickly and accurately. Step measurement is independent of the optical properties of the film and substrate; hence, the profiler will measure both dielectric and metal films.

Problems

3–1 A phosphorus-doped Si wafer is 4 in. in diameter (consider it to be an infinite plane) and 35 mils thick. There is a boron doped layer 1 μm thick at the upper surface. Measurements are made at the center of both surfaces with an $s = 25$ mil FPP.

(a) The sheet resistance on the B side is 20 Ω/sq. What V is expected if $I = 10$ mA?

(b) On the P-doped side, $I = 1$ mA, and $V = 0.627$ mV. What is the P doping concentration?

---

[6] Sloan Technology Corp., Santa Barbara, CA.

[7] Tencor Instruments, Mountain View, CA.

(c) The same probe is used on a large P-doped wafer 10 mils thick. Again, I and V are the same as in part (b). Calculate the P concentration.
(d) Calculate R_S for parts (b) and (c). Comment on your results.

3–2 (a) Using appendix data determine if, in Problem 3–1, it is valid to consider a 4 in. in diameter wafer as an infinite plane when a 25-mil FPP is used.
(b) If the wafer were 2 in. in diameter would the same assumption be valid? Explain.
(c) If the probe were moved off center, toward the wafer edge, would the V reading go up or down? The wafer is uniformly doped radially. Explain from physical considerations.

3–3 An outer point of a four point probe assembly is bent from its normal position, but all four points remain collinear. The top view of the points is shown in Figure 3–P-1. Measurements are made on a semi-infinite sample.

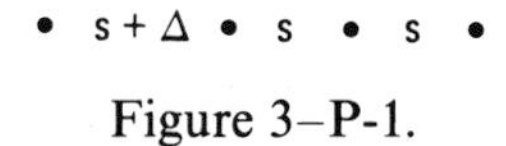

Figure 3–P-1.

(a) Derive an expression for V^*, the voltage between the two inner points of the bent assembly. Use $\delta = \Delta/s$.
(b) Normalize your expression for V^*/V, where V is the value for $\delta = 0$ (i.e., for the undamaged probe).
(c) What is the percent error in measured ρ_∞ if $\delta = 0.05$?

3–4 An inner point of an FPP is bent from its normal position as shown in Figure 3–P-2, but remains collinear with the other three. Repeat Problem 3–3 for this situation.

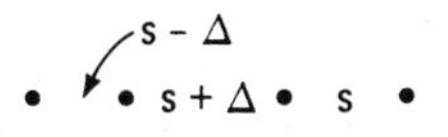

Figure 3–P-2.

3–5 (a) Derive the correction factor b for Figure 3–5(a).
(b) Check your results with Figure 3–5(c).

3–6 (a) Calculate the correction factor c for the situation of Figure 3–5(d).
(b) Calculate and plot c as a function of ℓ/s.
(c) What is the effective definition of "infinity" for this configuration?

3–7 Derive the equation for ρ_∞ for the FPP of Figure 3–6(a) and compare your results with those of the conventional FPP.

3–8 Verify Eq. (3.10-2) for the square FPP of Figure 3–6(b).

3–9 The connections to a square FPP are changed so that I is connected between A and D, and V between B and C. Analyze the situation.

3–10 A Schottky diode yields these data:

C, pF	V, V	C, pF	V, V	C, pF	V, V
8.06	5.00	6.19	8.50	5.44	11.0
7.36	6.00	6.01	9.00	5.32	11.5
6.82	7.00	5.85	9.50	5.21	12.0
6.59	7.50	5.71	10.00	5.10	12.5
6.38	8.00	5.57	10.50	5.01	13.0

The diode has a junction area $A = 9 \times 10^{-5}$ cm^2. Calculate $N(x)$ for the semiconductor side of the diode. (Note: Round off intermediate and final results to three significant figures.)

3–11 A varactor diode is to have a C–V relationship given by $C = C_o V^{-2}$ in order to be used for "tuning" a radio receiver.

(a) Derive the expression for $N(x)$ versus x for the lightly doped side of the diode.
(b) Sketch $N(x)$.
(c) If the varactor is used as the variable capacitor, shunting an inductor, in a tuned circuit, how will the resonant frequency f_o vary with V?

3–12 (a) Verify Eq. (3.14-1).
(b) If $a, b \ll r$, derive a simplified equation. Use the binomial expansion to obtain your answer.
(c) Calculate x_j for these numbers: $r = 250$, $a = 63.0$, and $b = 62.1$ mils. What is a possible difficulty in measuring very shallow junction depths? Should r be large or small in this regard? Explain.

References

[1] R. B. Adler, A. C. Smith, and R. L. Longiní, *Introduction to Semiconductor Physics, SEEC Vol. 1*. New York: Wiley, 1982. p. 197.
[2] L. J. van der Pauw, "A Method of Measuring Specific Resistivity and Hall Effect of Discs of Arbitrary Shape," *Philips Research Reports*, **13**, (1), 1–9, Feb. 1958.
[3] P. S. Burggraaf, "Resistivity Measurement Systems," *Semicond. Int.*, **3**, (6), 37–52, June 1980.
[4] L. B. Valdes, "Resistivity Measurement on Germanium for Transistors," *Proc. IRE*, **42**, (2), 420–427, Feb. 1954.
[5] A. Uhlir, Jr., "The Potentials of Infinite Systems of Sources and Numerical Solutions of Problems in Semiconductor Engineering," *Bell Syst. Tech. J.*, **34**, (1), 105–128, Jan. 1955.
[6] F. M. Smits, "Measurement of Sheet Resistivities with the Four-Point Probe," *Bell Syst. Tech. J.*, **37**, (3), 711–718, May 1958.
[7] M. A. Green and M. W. Gunn, "The Evaluation of Geometrical Effects In Four Point Probe Measurements," *Solid State Electron.*, **14**, (11), 1167–1177, Nov. 1971.
[8] R. Hall, "Minimizing errors of four point probe measurements on circular wafers," *J. Sci. Instrum.*, **44**, (1), 53–54, Jan. 1967.

[9] M. A. Logan, "Sheet Resistivity Measurements on Rectangular Surfaces—General Solution for Four Point Probe Conversion Factors," *Bell Syst. Tech. J.*, **46**, (10), 2277–2322, Dec. 1967.

[10] L. J. Swartzendruber, "Four-Point Probe Measurement of Non-Uniformities in Semiconductor Sheet Resistivity," *Solid State Electron.*, **7**, (6), 413–422, June 1964.

[11] R. Brennan and D. Dickey, "Determination of Diffusion Characteristics Using Two- and Four-Point Probe Measurements," *Solid State Technol.*, **27**, (12), 125–132, Dec. 1984.

[12] J. S. Glick, M.S. Thesis, Electrical Engineering Department, University of Illinois, 1974.

[13] C. Germano, "Simplified Temperature Corrections for Silicon Resistivity Measurement," *Solid State Technol.*, **19**, (12), 53, Dec. 1976.

[14] G. L. Allerton and J. R. Seifert, "Resistivity Measurements for Semiconductors," *Western Electric Eng.*, **V**, (3), 43–49, July 1961.

[15] J. M. Meese, D. L. Cowan, and M. Chandrasekhar, "A Review of Transmutation Doping In Silicon," *IEEE Trans. Nucl. Sci.*, **NS-26**, (6), 4858–4867, Dec. 1979.

[16] P. A. Schumann, Jr., E. F. Gorey, and C. P. Schneider, "Small Spaced Spreading Resistance Probe," *Solid State Technol.*, **15**, (3), 50–52, March 1972.

[17] E. E. Gardner, J. F. Hallenback, and P. A. Schumann, Jr., "Comparison of resistivity measurement techniques on epitaxial silicon," *Solid State Electron.*, **6**, (3), 311–312, May–June 1963.

[18] P. A. Schumann, Jr., M. R. Poponiak, J. F. Hallenback, Jr., and C. P. Schneider, "Measurement Electronics for the Three-Point Probe," *Solid State Technol.*, **11**, (11), 32–36, Nov. 1968.

[19] D. C. Gupta, "Diode Voltage-Capacitance Method of Measuring Resistivity and Impurity Profile in a Silicon Epitaxial Layer," *Solid State Technol.*, **11**, (2), 31–34, Feb. 1968.

[20] P. S. Burggraaf, "C–V Plotting, C–T Measuring and Dopant Profiling: Applications and Equipment," *Semicond. Int.*, **3**, (9), 29–42, Oct. 1980.

[21] I. Amron, "A Slide Rule for Computing Dopant Profiles in Epitaxial Semiconductor Films," *Electrochem. Technol.*, **2**, (11–12), 1–2, Nov.–Dec. 1964.

[22] R. F. Pierret, T. L. Chiu, and C. T. Sah, "A Slide Rule for Computing U_F and the Bulk Doping Density from MIS-Capacitor High-Frequency C–V Curves," *IEEE Trans. Electron Devices*, **ED-16**, (1), 140–147, Jan. 1964.

[23] J. A. Copeland, "A Technique for Directly Plotting the Inverse Doping Profile of Semiconductor Wafers," *IEEE Trans. Electron Devices*, **ED-16**, (5), 445–449, May 1969.

[24] N. I. Meyer and T. Guldbrandsen, "Method for Measuring Impurity Distributions in Semiconductor Crystals," *Proc. IEEE*, **51**, (11), 1631–1637, Nov. 1963.

[25] D. Leenov and R. G. Stewart, "A Proposed Method for Rapid Determination of Doping Profiles in Semiconductor Layers," *Proc. IEEE, Letters*, **56**, (11), 2095–2096, Nov. 1968.

[26] R. R. Spiwak, "Design and Construction of a Direct-Plotting Capacitance Inverse-Doping Profiler for Semiconductor Evaluation," *IEEE Trans. Instrum. Meas.*, **IM-18**, (3), 197–202, Sept. 1969.

[27] "Mercury Probes to Prepare Non-Destructive Semiconductor Contacts," *Bulletin MP 4/80*, MSI electronics inc., 1980.
[28] A. Lederman, "Vacuum Operated Hg Probe for *CV* Plotting and Profiling," *Solid State Technol.*, **24**, (8), 123–126, Aug. 1981.
[29] D. L. Rehrig and C. W. Pearce, "Production Mercury Probe Capacitance-Voltage Testing," *Semicond. Int.*, **3**, (5), 151–162, May 1980.
[30] K. H. Zaininger and F. P. Heiman, "The *C–V* Technique as an Analytical Tool, Part 1," *Solid State Technol.*, **13**, (5), 49–58, May 1970. "Part 2," **13**, (6), 46–55, June 1970.
[31] A. Goetzberger, "Ideal MOS Curves for Silicon," *Bell Syst. Tech. J.*, **45**, (7), 1097–1122, Sept. 1966.
[32] W. L. Bond and F. M. Smits, "Interference Microscope for Measurement of Extremely Thin Surface Layers," *Bell Syst. Tech. J.*, **35**, (5), 1209–1222, Sept. 1956. This paper is reprinted in *Bell System Monograph 2682*.
[33] J. H. Matlock, "Crystal Growing Spotlight," *Semicond. Int.*, **2**, (8), 33–44, Oct. 1979.

Chapter 4
Equilibrium Concepts

In later chapters, we shall need certain concepts that occur when systems of different materials interact with each other under different conditions of temperature, pressure, and composition. Fabrication processes where this occurs are oxidation, diffusion, ion implantation, and contact formation. Some of these concepts are the limit of solid solubility, the segregation coefficient, eutectic temperature, and eutectic composition. Such systems are easiest to consider when they are at thermal equilibrium—when all parts of the system are at the same temperature and undergo no changes. Our study will be brief, providing only a working background for the concepts that are used later.

4.1 Definitions

Consider a closed system at thermal equilibrium. This system will be composed of one or more chemical species in their various states of aggregation.[1] A *component* in the system is any one of the pure chemical substances, either element or compound, that make up the system. Let

$$C = \text{number of components in the system}$$

A *phase* in the system is any homogeneous, physically distinct, and *mechanically separable* portion of the system. Phases may be liquid, gas (vapor), or solid. Note that a single solid may be comprised of two or more solid phases that are distinguishable by chemical composition, crystal structure, and even appearance. For example, granite appears as a solid matrix in which clearly

[1] Contrary to popular usage, the word "species" is both singular and plural in this sense. "Specie" is a coin, which obviously is not part of a system under consideration here.

visible crystals of different composition and appearance are embedded. Let

$$P = \text{number of phases in the system}$$

If the system has more than one component, we must know the *composition* which tells us the relative portions of the components. We shall designate weight composition by x, with suitable subscripts where necessary. Our study will be confined largely to *binary systems*—i.e., to those systems that have only two components. If we designate the two components as A and B, then the composition by weight (or weight fraction) will be defined by

$$x = \frac{\text{weight of B}}{\text{weight of A} + \text{weight of B}} = \frac{W_B}{W_A + W_B} \tag{4.1-1}$$

where W = weight.

In certain cases composition may be expressed in terms of *atomic* or molecular fraction rather than weight fraction, in which event we define

$$x' = \frac{\text{no. of atoms of B}}{\text{no. of atoms of A} + \text{no. of atoms of B}}$$

$$= \frac{n_B}{n_A + n_B} \tag{4.1-2}$$

where n = total number of atoms in the system.

The two values of composition are related in this manner. Let

$$M_j = \text{molecular weight of species j}$$

$$= \text{weight per mole of species j}$$

$$n_a = \text{Avogadro's number} = \text{no. of atoms per mole}$$

$$= 6.02 \times 10^{23} \text{ molecules/mole}$$

$$n_j = \text{number of atoms of species j}$$

$$W_j = \text{weight of } n_j \text{ atoms of species j}$$

Then

$$W_j = \frac{n_j M_j}{n_a}$$

Whence

$$\frac{1}{x} = 1 + \frac{M_A}{M_B}\left(\frac{1}{x'} - 1\right) \tag{4.1-3}$$

or

$$\frac{1}{x'} = 1 + \frac{M_B}{M_A}\left(\frac{1}{x} - 1\right) \qquad (4.1\text{-}4)$$

The *degrees of freedom* of the system are the number of external variables that may be controlled independently by the observer—e.g., temperature, pressure, and composition. We will let F = number of degrees of freedom.

Another term we encounter later is *constituent*, an association of two or more phases in a distinctly recognizable form.

4.2 Phase Rule

The phase rule of Willard Gibbs gives the relationship among the variables of a system at thermal equilibrium. This rule states that

$$P + F = C + 2 \qquad (4.2\text{-}1)$$

and is of great help in interpreting *phase diagrams* or maps of systems in equilibrium. The rule tells how many external variables may be controlled independently in a system. For example, we calculate that in a two-component, three-phase system there is only one degree of freedom—only one of the quantities, temperature, pressure, or composition, can be varied independently.

We shall consider a few different systems to see how the phase rule and phase diagram are applied and what type of information may be obtained from them.

4.3 Unary Systems

The simplest type of system is the one-component or *unary* system. In this case $C = 1$ and the phase rule reduces to

$$P + F = 3 \quad \text{for } C = 1 \qquad (4.3\text{-}1)$$

Typical phase diagrams for two types of unary systems are shown in Figure 4–1, the one at (a) being typical of water, while (b) shows a more general type. Note that the two diagrams differ primarily in the sign of the slope of the boundary line, separating the L and S regions, to the right of the point a. There are three regions in both diagrams, each of which corresponds to a single phase and is labeled: S for solid, L for liquid, and V for vapor (or gas). It must be emphasized that phase diagrams are drawn from experimental data; they are not postulated by a latter-day Aristotle.

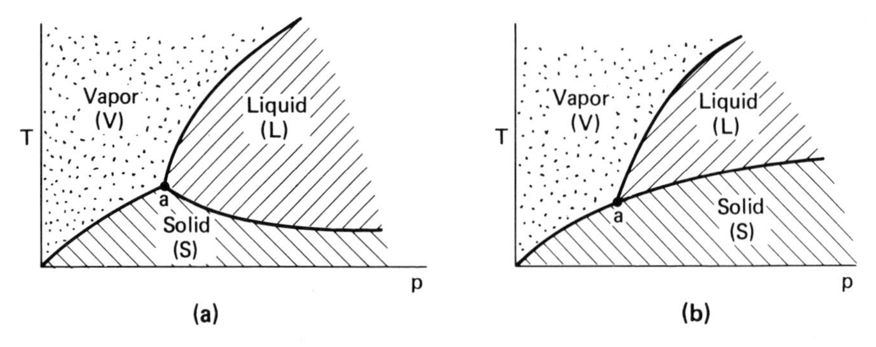

Figure 4–1. Unary or one-component phase diagrams. (a) Typical of water. (b) General type where solidus slope remains positive.

Let us apply the phase rule to various parts of a unary diagram to see how it may be used. In any one of the three single-phase regions $P = 1$; hence from Eq. (4.3-1) we have that $F = 2$, so there are two degrees of freedom. This statement checks with the diagram, for in any single-phase region we can independently vary both p and T.

Two phases coexist along any one of the three lines that separate adjacent single-phase regions. This means that on these lines $P = 2$ and $F = 1$, by Eq. (4.3-1). This is easy to verify on the diagram: if one variable, either p or T, is chosen arbitrarily, the other is fixed if the system is to remain on the line.

Lastly, consider the point marked a in Figure 4–1, the *triple point* at which all three phases coexist. At the point, $P = 3$ so that $F = 0$ and there is no degree of freedom. The triple point is invariant for a given system and is uniquely determined by nature. If we change either p or T or both, at least one of the phases must disappear.

These examples show the relationship between the phase rule and the experimentally derived phase diagram for a unary system. Water is such a one-component system—with ice, liquid water, and water vapor as the three phases. The triple point for water lies at a pressure of 4.6 torr (= 4.6 mm Hg = 0.006 at.) and a temperature of 0.0072°C.

4.4 Binary Systems

The unary system is of interest in our work only as a vehicle for introducing concepts. In semiconductor work *binary* or two-component systems such as B/Si, P/Si, Au/Si, or Au/Ge are common. We shall use phase diagrams to determine what happens as two such components interact with each other under different conditions. The phase rule will guide us in interpreting some of the results.

4.4.1 The Reduced Phase Rule

In constructing the phase diagram of any binary or two-component system, *three* variables must be specified: in addition to p and T the composition x is needed. It appears, then, that three-dimensional phase diagrams are needed for binary systems. Simplification is possible, however. Most device processing is done at constant pressure (usually at atmospheric value), so pressure can be eliminated as a degree of freedom. This reduces the diagram to a two-dimensional plot again, with T and x as the axes. To reflect this change in the phase rule, let

$$F' = \text{reduced number of degrees of freedom}$$
$$= F - 1, \qquad p \text{ constant} \tag{4.4-1}$$

and define a reduced form of the phase rule, namely

$$P + F' = C + 1, \qquad p \text{ constant} \tag{4.4-2}$$

Furthermore, in binary systems $C = 2$; hence we have[2]

$$P + F' = 3, \qquad \text{binary system, } p \text{ constant} \tag{4.4-3}$$

This reduced form and the two-axis phase diagram will be used for the remaining binary systems to be discussed.

4.4.2 Solid Solubility

Binary systems fall into several different categories depending upon the particular components that are involved. For example, the B/Si system behaves quite differently from that of Cd/Bi. These systems differ fundamentally in their *degree of solid solubility*. This concept is best approached by analogy to liquid solubility, which is quite familiar.

It is a matter of common knowledge that alcohol and water are completely soluble in all compositions: they have *complete* liquid *solubility* in each other. Oil and water, on the other hand, are completely insoluble in any portions or composition and remain separate; they exhibit complete liquid *insolubility*. As a third example, consider a sugar/water solution. At a given temperature a liquid-only single phase is formed as sugar is added to the water. Finally, at some composition the water can hold no more sugar, and any additional sugar remains in solid phase at the bottom of the liquid. This is an example of partial solubility.

---

[2] There is an apparent exception if $x = 0$ or 1. This is considered later.

Analogous behavior can be observed in some solids. For example if fixed weights of germanium and silicon are melted together and then allowed to freeze out, only one solid phase results; it is a single homogeneous solid from which we cannot separate out the Ge and Si. This is true no matter what proportions of Si and Ge were originally melted together. Germanium and silicon form an *isomorphous* system (one that exhibits complete solid solubility).

Bismuth and cadmium, on the other hand, behave in an entirely different manner. No matter what proportions of the original two materials are melted together, Bi freezes out alone and Cd freezes out alone as the temperature of the melt is lowered. These two materials form a system with complete solid insolubility.

A word of caution in interpreting the last paragraph: the two components will not freeze out from a melt into two nicely separated piles, Bi on the left and Cd on the right. Rather, a general matrix of the more abundant component is present with small crystals of the other component within that matrix. If a sample of the solid is polished and etched properly, the regions of the two components can be seen quite clearly when viewed under a microscope.

Between the two extremes of complete solid solubility and insolubility is a wide range of systems that show *partial* or *limited solid solubility*. Most of the systems of concern to us here fall into this latter category. We shall devote some time to the other two types, however, for they have simpler phase diagrams and can be used to illustrate some properties with less clutter.

4.5 Isomorphous Binary Systems

Typical binary systems that fall into the isomorphous category because they they exhibit complete solid solubility are Ge/Si, Au/Pt, and Cu/Ni. A typical corresponding reduced-phase diagram, for assumed constant pressure, is shown in Figure 4–2(a). Recall that these diagrams are based on experimental data, as are the identifications of the phases in the three regions.

We use A and B to designate the two components unless a particular system is being considered. Component A has the lower of the two melting points, T_A and T_B, respectively, and is at the left-hand side of the phase diagram, where $x = 0$. The composition x, then, gives the weight fraction (or percentage) of B as defined in Eq. (4.1-1). If $x = 0$, the system is all A with no B present; if $x = 1$ (or 100%) the system is all B.

The upper of the two lines joining T_A and T_B serves as the lower boundary of the liquid-phase region, and is called the *liquidus*. (Every priesthood must have its own mumbo jumbo, Latin preferred.) The lower line joining the two melting points serves as the upper boundary of the solid-phase region, and is called the *solidus*. In both single-phase regions, which are marked L and S, $P = 1$. Then by Eq. (4.4-3), the reduced-phase rule for a binary system, we

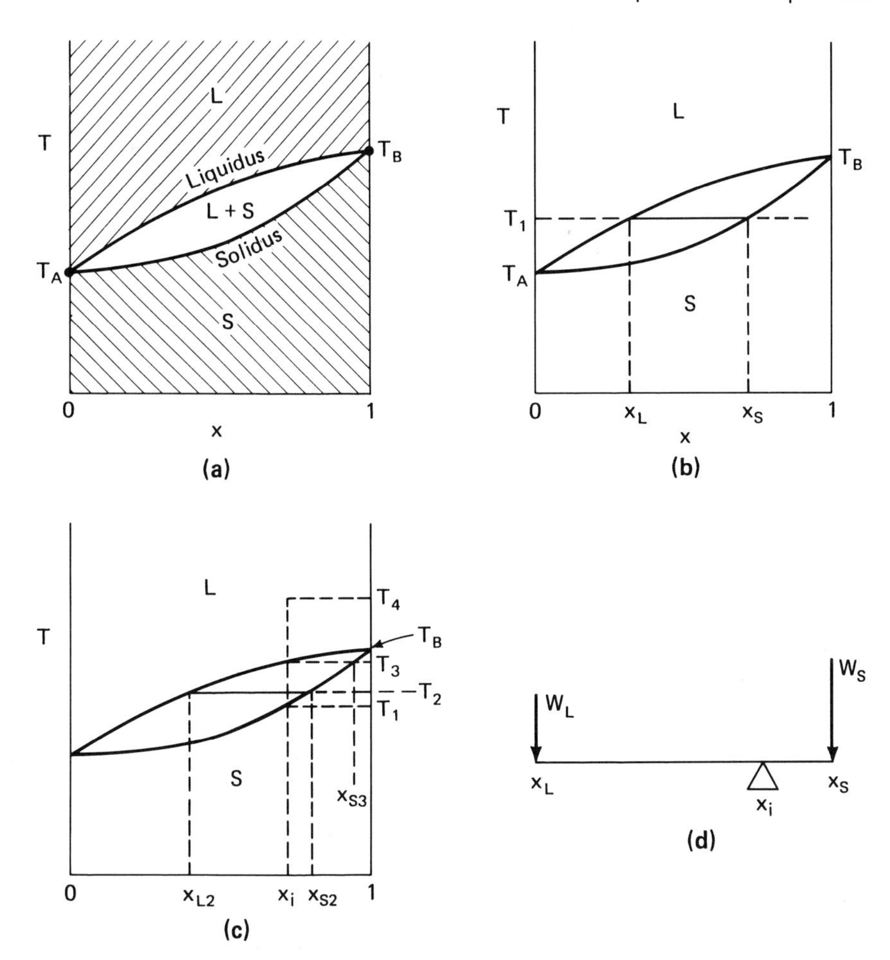

Figure 4–2. Isomorphous binary system phase diagrams. The two components form a single solid solution in any proportions. (a) Basic diagram. (b) A tie line between the liquidus and solidus at temperature T_1. (c) Illustrating freezeout. (d) Illustrating the lever rule.

calculate that $F' = 2$. This checks with the diagram because in such a region we may vary T and x independently.

The unshaded region (marked L + S) presents some logical problems and must be considered separately. Remember that experiment says two phases, liquid and solid, are present, so $P = 2$. With this value Eq. (4.4-3) gives $F' = 1$. Our previous experience, however, has indicated that $F' = 1$ implies location on a line, yet the phase diagram seems to indicate a region which has $F' = 2$. How do we reconcile this apparent discrepancy?

Choose a temperature T_1 lying between T_A and T_B (there goes the one degree of freedom). Now draw a horizontal *tie line* at T_1 between the liquidus and solidus [Figure 4–2(b)]. The points of intersection correspond to the compositions labeled x_L and x_S. Gibbs says that at T_1 all points on the tie line between x_L and x_S are forbidden, because the one allowed degree of freedom was used up in choosing the temperature; therefore composition(s) must be fixed. There *are* two phases, a liquid and a solid, whose compositions must be read as x_L and x_S at the intersections of the tie line with the liquidus and solidus, respectively. This tie line concept must be used whenever an apparent two-phase region is encountered on a phase diagram.

All we have left to consider now are the two melting points T_A and T_B. There is a subtle point involved here: even though we are dealing with a binary system, $C = 1$ at these two points where x is either 0 (all A) or 1 (all B). At these two special points the system is really unary, either only A or only B. With this point clarified, we can use the reduced phase rule, Eq. (4.4-2), to find that $F' = 0$; there are zero degrees of freedom, so the melting points of A and B (for the constant pressure chosen) are invariant and are unique properties of the materials themselves.

4.5.1 Freezeout Compositions

As a final exercise to gain familiarity with the isomorphous type of diagram consider Figure 4–2(c). Portions of solid A (S_A) and solid B (S_B) to form an initial composition x_i are placed in a crucible and melted by raising the temperature to some T_4 greater than T_B. We then begin a very slow cooling process (very slow to allow the system to equilibrate at any temperature in question). The entire mass remains liquid until T_3 (where the x_i line intersects the liquidus) is reached, when initial freezeout into solid phase begins. The composition of this initial solid will be x_{S3}, that of the liquid will be x_i. Note that both values are read at the ends of the tie line: x_i at the liquidus end, and x_{S3} at the solidus end.

On further cooling to, and equilibration at, T_2, there are still two phases present: liquid with composition x_{L2} and solid with composition x_{S2}. This procedure of reading the two compositions is continued as T is lowered until it is just below T_1 when the entire mass will be solid again at composition x_i. Notice that the initial and final compositions of the solid are at the same composition x_i (this must be so since neither A nor B has been subtracted or added during the melt and freezeout cycles), but they do differ in physical form. Before melting, there were two separate lumps of A and B, whereas finally there is only a single phase, a solid *solution* of A and B with overall composition x_i. If the freezeout is too rapid for equilibration, the composition may vary throughout the solid.

4.5.2 Lever Rule

Tie lines also may be used to calculate the weight of each phase. Consider the situation shown in Figure 4–2(c). Let W_A = weight of component A, and W_B = weight of component B in the system. Say that the system has reached equilibrium at some temperature, say T. Let W_S = weight of the solid phase, and W_L = weight of the liquid phase. As we raise or lower temperature, the total weight in the system remains unchanged since no material is added or subtracted; therefore, it must be true that at all temperatures

$$W_A + W_B = W_L + W_S \tag{4.5-1}$$

Then by the definition of composition we can write

$$\begin{aligned} x_i(W_A + W_B) &= x_i(W_L + W_S) \\ &= \text{weight of B in original melt} = W_B \end{aligned}$$

Also,

$$x_L W_L = \text{weight of B in the liquid phase at } T \tag{4.5-2}$$

$$x_S W_S = \text{weight of B in the solid phase at } T \tag{4.5-3}$$

Since the total weight of B remains unchanged and can only distribute itself between the liquid and solid phases, we can write

$$x_L W_L + x_S W_S = x_i(W_L + W_S)$$

whence

$$W_L(x_i - x_L) = W_S(x_S - x_i) \tag{4.5-4}$$

Figure 4–2(d) shows the relationship expressed by this equation, and it is apparent why this equation is called the *lever rule*. Note that once W_L and W_S have been determined, the weight of B in each of the two phases may be calculated from Eqs. (4.5-2) and (4.5-3). The corresponding weights of A also can be calculated quite easily.

The lever rule and other concepts that have been discussed relative to the isomorphous system phase diagram may be transferred over to the more complicated binary systems that are considered next.

4.6 Eutectic Binary Systems

There is a large class of binary systems that, on freezing out, exhibit a *constituent* called the *eutectic*. This is an association of phases in a recognizable form and having a unique melting point that is less than the melting points of either component. We shall consider two types of eutectic binary systems. One has complete solid insolubility and the other limited or partial solid solubility.

4.7 Binary Systems; Complete Solid Insolubility

Figure 4–3(a) shows a phase diagram typical of binary systems having complete solid insolubility. As stated earlier, bismuth/cadmium forms such a system. Certain features that distinguish this diagram from that of the isomorphous systems should be noted. There are two two-phase regions, labeled $L + S_A$ and $L + S_B$, where the tie line concept and lever rule must be used.

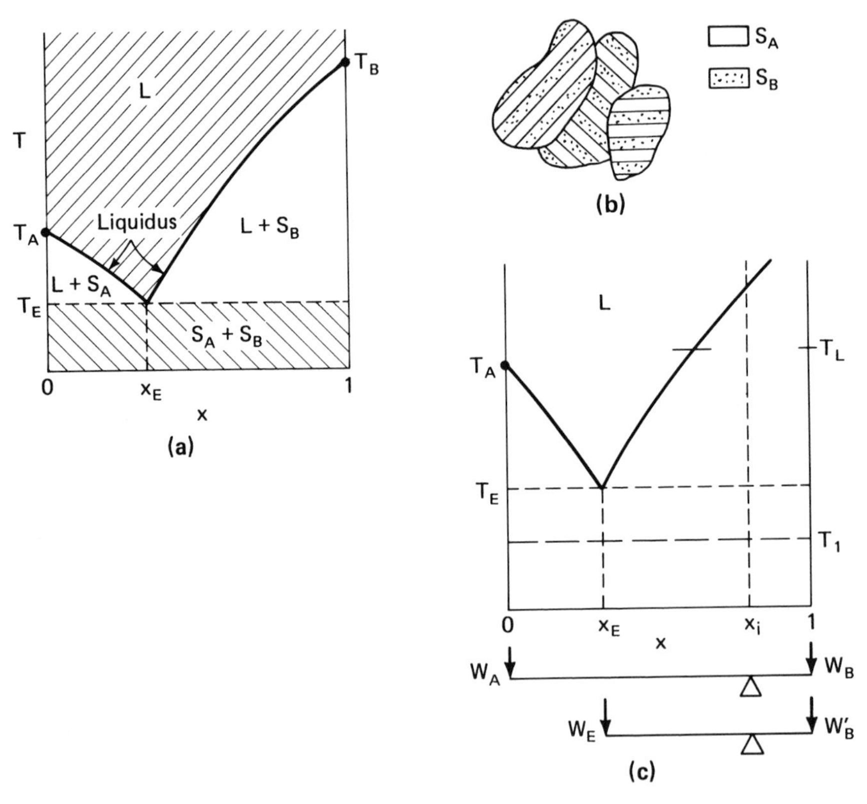

Figure 4–3. Eutectic binary system phase diagrams where the solid components are completely insoluble. (a) Basic phase diagram. Pure A and B are the only admissible solids. (b) Illustrating the eutectic. Each single crystal comprises a laminar association of A and B. (c) Alternate lever rule forms for temperatures below T_E.

The solid region here is marked $S_A + S_B$ because experiment shows that the frozen solid indeed has two phases, pure A and pure B, separate and not in solid solution; the property of complete solid insolubility forbids the formation of a single-phase solid solution. Since this is a two-phase region, the tie line and lever rule must be used here also, not to determine compositions, which are fixed at 0 and 1, but to get weight data. The only single-phase region on the diagram lies above the liquidus, which has two branches, and is labeled L (as might be expected since the entire mass is liquid).

4.8 The Eutectic

Another distinguishing feature here as compared to the isomorphous diagram is the existence of the *eutectic point* whose coordinates are the eutectic temperature T_E and the eutectic composition, x_E. The eutectic is not another solid phase, but rather a constituent of the system—i.e., an *association of phases* in a distinctly recognizable fashion, and having a distinct melting point. Notice that the eutectic melts at T_E, the eutectic temperature, which is lower than both T_A and T_B.

The eutectic is an association of S_A and S_B in the unique composition x_E. The two eutectic solid phases often are associated in laminar form as indicated by the sketch of Figure 4–3(b). When a melt is frozen out, S_A and S_B tend to associate in this special eutectic form. To further illustrate this idea, consider the situation shown in Figure 4–3(c). A melt of initial composition x_i has been cooled slowly and allowed to equilibrate at a temperature T_1 that is lower than T_E. All of the material will be either S_A or S_B and by using the lever rule illustrated in the upper lever diagram we can calculate W_A and W_B. All of A is associated with the eutectic but B can be identified in two guises: one as solid B associated with solid A in the eutectic and having weight ($W_B - W_B'$), and the other as solid B *not* in the eutectic and having weight W_B'. This latter weight, W_B', of the noneutectic B can be found by applying the lever rule to the lower lever diagram in Figure 4–3(c). The interpretation associated with this lower diagram is quite realistic. If a sample of the frozen composition x_i were properly prepared by polishing and staining and then observed under a microscope, it would appear as regions of the eutectic (of total weight $W_A + W_B - W_B'$) in a matrix of solid B only, having a total weight of W_B'. If the initial composition lay to the left of x_E in Figure 4–3(c), more A would be present and we would expect the matrix to be eutectic, with regions of pure A scattered about in it.

To consider another feature of the eutectic, say that in Figure 4–3(c) heat is removed from a melt of composition x_i until the temperature decreases below T_L. We know from the diagram that two phases are present: S_B and a liquid of composition determined by the tie line intersection with the liquidus.

On further removal of heat, the temperature continues to drop until T_E is reached. Now observe that at the eutectic point there are three phases coexisting: L of composition x_E, and solids S_A and S_B. Then by the reduced form of the phase rule for a binary system, Eq. (4.4-3)

$$F' = 3 - P = 3 - 3 = 0$$

that is, the eutectic is an invariant point and the transformation from liquid to the two solid phases is an invariant transformation and takes place isothermally; all of the liquid of composition x_E freezes out completely to S_A and S_B at constant temperature T_E. In fact, an invariant phase transformation of the general form

$$L \xrightarrow{\text{cool}} S_1 + S_2 \tag{4.8-1}$$

is designated as a *eutectic transformation.* Here S_1 and S_2 are any solid phases—which may be pure component, solid solution, or solid compound. Once the transformation from liquid to solid is completed, the temperature continues to lower as more heat is removed from the system.

From the diagram we see that a eutectic point of transformation occurs at a downward-pointed cusp in the liquidus which lies at the top of a two-phase solid region. Further examples of this will appear in later sections. We shall also see another type of invariant transformation—the peritectic.

4.9 Binary Systems; Partial Solid Solubility

The second type of eutectic binary system to be considered is characterized by partial or limited solid solubility. A typical form of the corresponding phase diagram is shown in Figure 4–4(a) and applies to several systems comprised of silicon and any one of several of the common dopants.

This diagram is different in several respects from ones examined earlier. First, we note a new double-branched boundary line, the *solvus.* These branches, one on the left and one on the right of the diagram, join up with the lower ends of the solidus branches at T_E. Second, we see no S_A or S_B designations; there are no pure-A and pure-B regions. Third, there are two new single-phase regions labeled S_α and S_β. These regions represent *single-phase solid solutions* of temperature dependent composition. The limiting values of x for the α and β solutions are delineated by the solidus above T_E and by the solvus below T_E. The solid phase α, having maximum x-values near zero, comprises a solid of a little B in mostly A. The solution β, on the other hand, is characterized by compositions close to unity and has a small fraction of A in mostly B.[3]

[3] In some systems the S_α and/or S_β regions are so small as to be invisible on the phase diagram; this is true of binary systems—for example, of Si with Ga, In, or Sb. To handle this problem an alternate form of plotting and a different scale must be used, as in the Trumbore curves, which are discussed in Section 4.11.1.

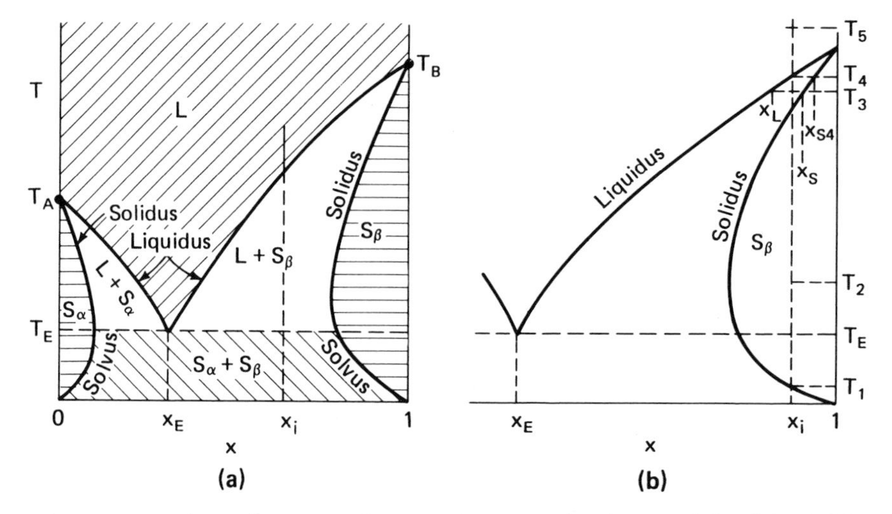

Figure 4–4. Phase diagrams of a eutectic system having partial solid solubility. (a) Complete phase diagram. (b) Enlarged view for x_i near unity to illustrate the limit of solid solubility.

As in the previous examples, the single-phase liquid region L lies above the liquidus. There are now three two-phase regions: ($L + S_\alpha$), ($L + S_\beta$), and ($S_\alpha + S_\beta$). The tie line and lever rule must be used in all of them. Take note that S_α (alone) and S_β (alone) correspond to single-phase regions where the tie line and lever rule must *not* be applied. Note further that S_α and S_β are not of fixed composition; their compositions are temperature dependent.

It must be emphasized that the limiting composition for each solid solution phase is given by the solidus and solvus. To illustrate, consider Figure 4–4(b), which shows the β side of the diagram. At T_3 an S_β solid of any composition lying between x_S and 1 is admissible. A β-phase of any composition lying between x_L and x_S cannot exist at T_3. At that temperature the value of x_S represents the *limit of solid solubility* for solid solution β; if any more A were present in β (i.e., if $x < x_S$) the liquid phase of composition x_L would start to form. We shall have more to say about this limit of solid solubility shortly.

Perhaps the situation may be made more clear if we consider another example relative to Figure 4–4(b). Say we start with solid A and solid B of weights corresponding to a composition x_i as shown. We raise the temperature to some T_5 above the intersection of the liquidus and the tie line at T_4. This temperature increase is represented on the phase diagram by moving upward along the vertical dotted line at x_i until T_5 is reached.

We now start the slow cool-and-equilibrate process. Freezout begins at T_4, with the solid phase comprising β at its T_4 limit of solid solubility, which is marked x_{S4}. With the temperature lowered further to T_3 the liquid is less rich in B (x_L has decreased) and the solid phase is also less rich in B, since its composition has dropped from x_{S4} at T_4 to the lower value x_S at T_3.[4]

At the still lower temperature T_2 the entire weight of the system is frozen out into the β phase with the initial composition x_i. (Remember, do *not* use a tie line in the single-phase S_β region.) This is possible, with no liquid present, because the system has a composition that is greater than the limit of solid solubility at T_2 and hence is admissible as β phase alone.

With further cooling the composition remains constant at x_i and in the β phase until T_1 is reached. T_1 and x_i are the coordinates for a point on the solvus, so just below this temperature the β phase starts to reject B and an α phase starts to form. For such a condition, one that lies below the solvus at initial composition x_i, the system is in the two-phase region with both S_α and S_β present. Compositions and weights are therefore determined by the tie line and lever rule.

On the other hand if, as in Figure 4–4(a), x_i lay between x_E and the solvus intersection with T_E, a different set of phase transformations would result on cooling. Cooling the liquid phase below the liquidus would give L and S_β as before. With further heat removal, however, the temperature eventually would drop to T_E, the eutectic value. Then an isothermal transformation would convert liquid of composition x_E to two solid phases, α and β, in compositions and weights determined by the tie line and lever rule.

Notice that the solid phases α and β are solid *solutions* because the system has limited solid solubility. Contrast this to the situation in Figure 4–3(a) where A and B are the solid phases; solid solutions could not form because the system exhibited complete solid insolubility.

Phase diagrams of three binary systems that are common in silicon technology are shown in Figures 4–5 to 4–7 [1]. In all of these diagrams the silicon host has the higher melting point, and so is on the RHS of the diagram, in accord with our convention. It follows that we were quite realistic in concentrating attention on the right-hand or β side of the diagram; we usually are interested in a large amount of silicon with a very small amount of dopant, such that the composition is close to unity.

Figures 4–5 and 4–6 have some special features of interest. First, they have scales of both x (weight percent, at the top), and x' (atomic percent, at the bottom). Second, the correspondence between x and x' is not the same for the two diagrams. For example, $x = 0.5$ (by weight) corresponds to $x' \approx 0.88$ in the Au/Si system, but to $x' \approx 0.81$ in the Sb/Si case. The reason for this may be seen from Eq. (4.1-4); the conversion from x to x' depends on the molecular weight ratio of the system components, and Au and Sb have different molecular weights.

[4] How can the percent of B decrease in *both* phases? Remember that weights change, too, and in such a way as to take up the slack.

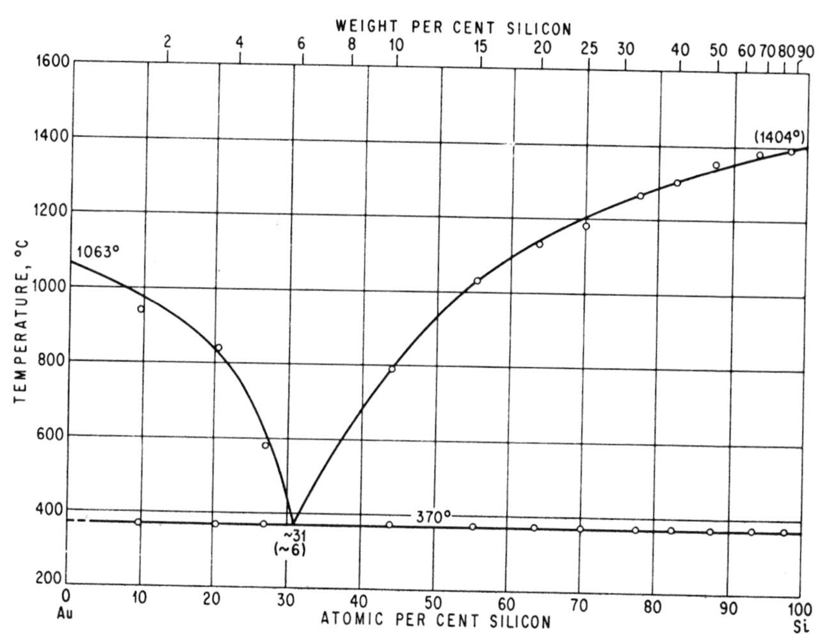

Figure 4–5. Phase diagram of the Au/Si system. Hansen and Anderko [1]. (Courtesy of McGraw-Hill Book Company.)

Third, the eutectic composition for Sb/Si is so small (almost all Sb) that it cannot be read from the phase diagram. The fact that the value is so small is fortunate when Sb is diffused into Si as a dopant. We shall see later that at diffusion temperatures, which typically are well above the 630°C eutectic value, the limit of solid solubility of Sb in Si does not exceed about one part in 500—i.e., the system prevents the formation of eutectic; we need not worry about eutectic melting during the higher temperature diffusion cycle. Stated differently, during diffusion the system confines itself to $x \geq 500/501 \approx 0.998$, which is nowhere near small enough for eutectic to form.

The Al-Si phase diagram is shown in Figure 4–7. Notice that the α and β solid phases show up clearly, indicating that this system has partial solid solubility. It is difficult, however, to read values of x along the solidus and solvus branches, since they lie so close to the vertical axes, so insets are provided. The solid phase regions of α and β cannot be seen at all in the two preceding figures! We shall see later how the Trumbore curves solve this problem. The diagram shows no evidence of chemical reaction between the components, but recent work clearly indicates the formation of aluminum silicides.

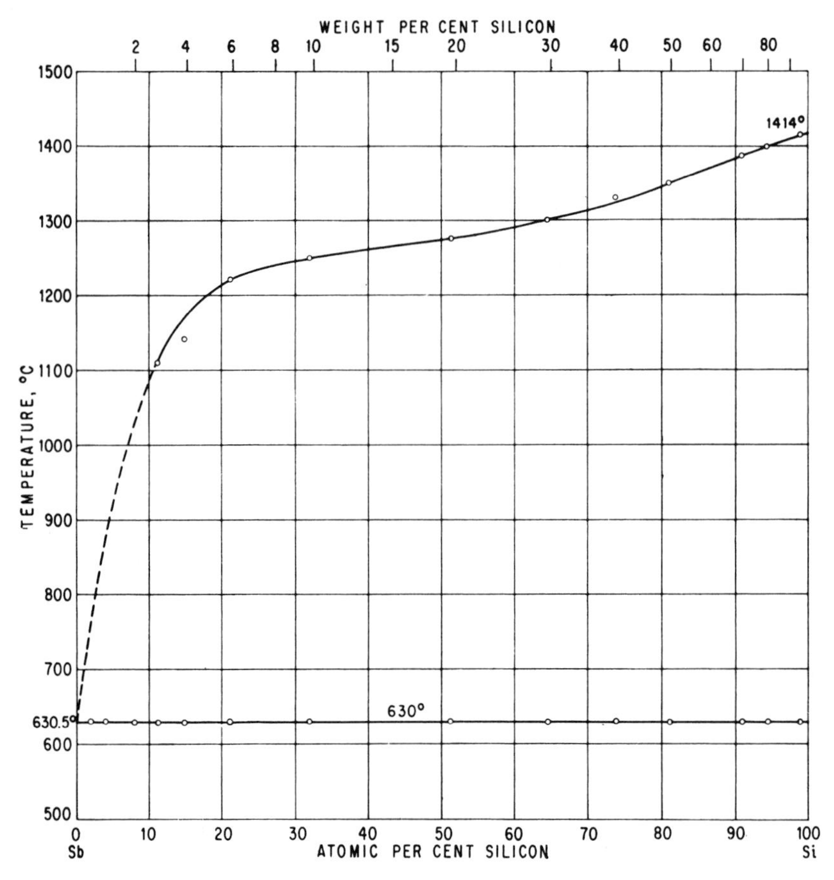

Figure 4–6. Phase diagram of the Sb/Si system. The eutectic composition is so small, it cannot be read from the diagram. Hansen and Anderko [1]. (Courtesy of McGraw-Hill Book Company.)

4.10 Contact Formation Temperatures

The formation of eutectics has important implications in processing silicon devices, especially when metal-to-silicon contacts are involved. Consider some of the temperatures of concern. If temperatures remain below roughly 800°C, the common dopants do not diffuse in silicon significantly. Of the commonly encountered pure materials Si, SiO_2, Al, and Au, aluminum has the lowest melting point, 660°C. Melting points may be lowered when two materials are in contact and have formed an alloy (a substance having two or more components that are soluble in each other in the solid phase). For example, Figure 4–7 shows that Al/Si contacts can melt at less than 660°C. Thus post-contact-

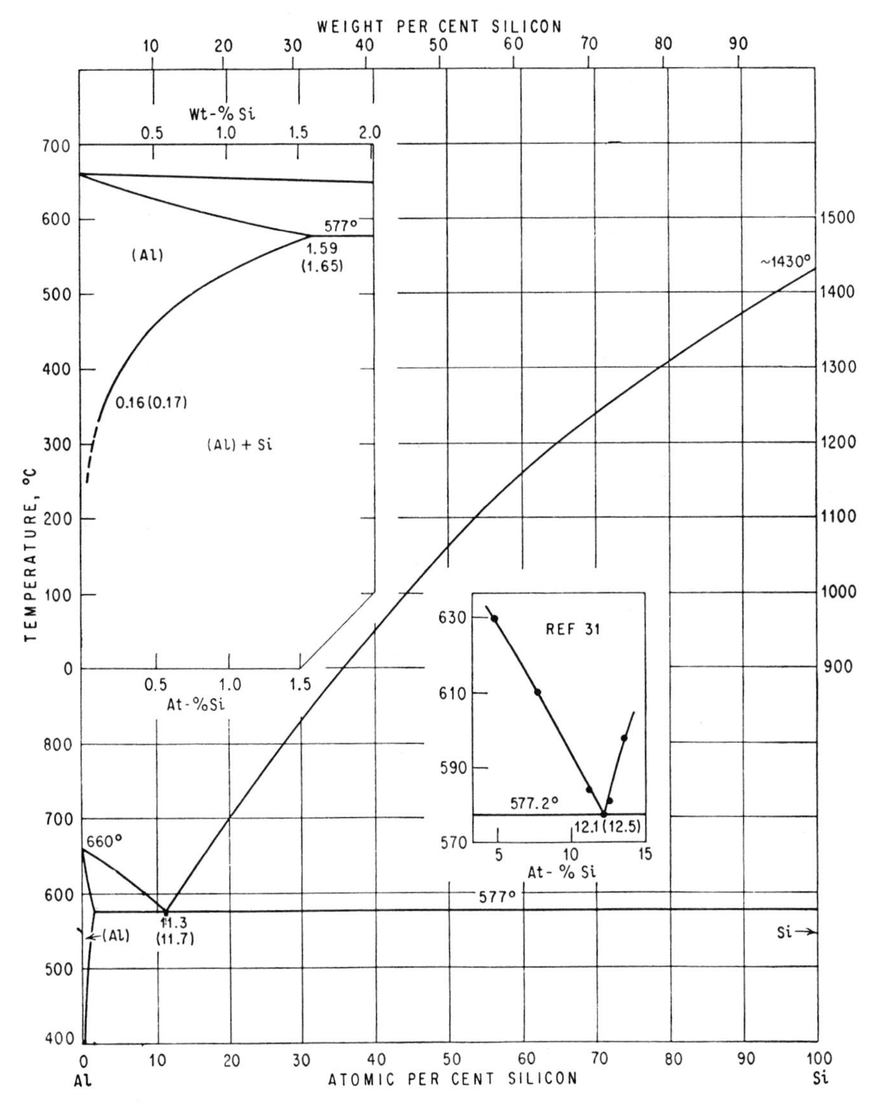

Figure 4–7. The Al/Si phase diagram. The insets are needed to read values of x near 0 and the eutectic point. Hansen and Anderko [1]. (Courtesy of McGraw-Hill Book Company.)

forming processes must be carried out below the 577°C eutectic value to avoid melting.

In die bonding the bottom side of a chip to a header, silicon and gold are in contact. This binary alloy has a T_E value of 370°C, so die bonding may take place after Al/Si contacts are made.

The eutectic of the Au/Ge system (12% Ge by weight) has a melting point of 356°C. This material may be used in place of the Au/Si eutectic for die bonding, if the lower temperature is desirable. Since Si and Ge are completely soluble in each other, the presence of a small amount of germanium in the bond causes no difficulties.

4.11 Limit of Solid Solubility

Consider in more detail a subject that has been mentioned earlier: the *limit of solid solubility*. On the S_β side of the phase diagram this represents the maximum amount of dopant (component A) that can go into solid solution with the host material (component B) without a liquid phase forming. Perhaps we can make this concept more clear by using the equilibrium phase diagram as a guide to what happens in a nonequilibrium situation. Say that phosphorus is being diffused into silicon at some temperature near 1000°C.[5] We can assume an infinite dopant source external to the host wafer in Figure 4–8, infinite in the sense that it does not deplete during the diffusion process, and that eventually it is large enough to produce a doping concentration at the

[5] In previous work we tacitly assumed that W_A and W_B were constant, so the overall system weight did not change. Here, where we consider indiffusing phosphorus (component A) into silicon (component B), W_B is constant, but W_A increases as phosphorus is added from an external source. The change in total weight is extremely small, however.

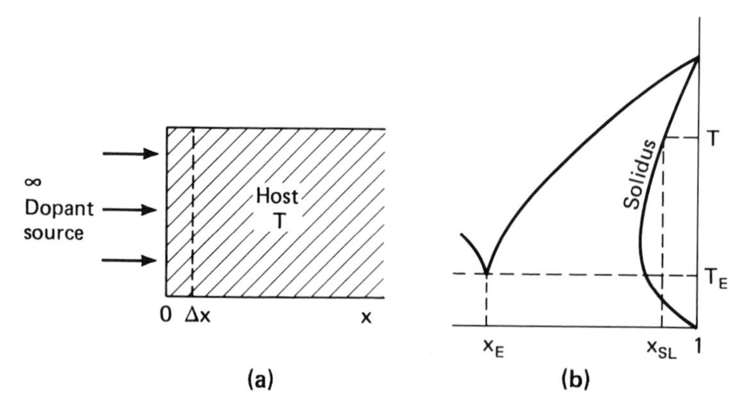

Figure 4–8. Indiffusion from an infinite source. (a) The indiffusant enters the host through a very thin layer of thickness Δx. (b) The corresponding phase diagram.

surface of the host of N_{SL} (the solid solubility limit in atoms/cm^3) at the diffusion temperature. The actual mechanisms involved in this process will be discussed in Chapter 6, but for the moment let us examine what happens qualitatively in terms of Figure 4–8.

Consider a thin layer of thickness Δx just within the surface of the host wafer. At $t = 0$ this layer is undoped; only the host is present, so $x = 1$. As the diffusion process begins, dopant starts moving into the Δx layer so that a β solid phase starts to form as the concentration of dopant (component A) builds up. On the phase diagram this roughly corresponds to motion along the constant-temperature line at T, with x decreasing from 1 toward x_{SL}. Finally, at some time the composition of β phase in the Δx layer becomes just slightly greater than x_{SL}, the solid solubility limit.

Now, if the very thin Δx layer were to accept any more dopant, x would decrease further, and moving along the constant-T line on the phase diagram, would enter the two-phase region of (L + S_β) and liquid would start to form (i.e., melting would begin). As a matter of fact, this does not happen during diffusion into silicon. Any indiffusing dopant, beyond that required to reduce the composition to x_{SL}, continues on through the Δx layer and builds up in silicon layers farther in from the surface. Thus, there is a limiting mechanism that does not allow the doping density to drop the composition below the solid solubility limit, and the entire system remains solid. The composition x_{SL}, then, sets the maximum amount of dopant that can be introduced into any region of the host at a particular temperature by diffusion or any thermally controlled or thermodynamic process. Ion implantation, however, is not such a process and can produce doping levels in excess of N_{SL}, which is the doping *concentration* at the limit of solid solubility.

It follows that x_{SL} for any dopant-host combination sets a maximum doping limit for the diffusion process and hence is an important parameter for both the device designer and the process engineer. In principle, x_{SL} can be read off directly from the appropriate phase diagram, but as we have seen this may not be easy to do. First, x_{SL} is temperature dependent; second, it often has values so close to unity that it is extremely difficult to read. (Recall that on some diagrams of limited solid solubility, the solidus and solvus did not even show up.) It would be useful to have the solid solubility limit data presented in a more readable form, the Trumbore curves.

4.11.1 Trumbore Curves

Trumbore has provided plots more suitable for reading solid solubility limits. He retains the temperature ordinate, but replaces x (or x') of the abscissa by N_{SL}, the dopant concentration in atoms/cm^3, at the solid solubility limit [2]. The so-called Trumbore curves for silicon as the host material are shown in Figure 4–9. Notice that the N_{SL} values are expanded by use of a logarithmic scale. These curves are used widely in the semiconductor industry.

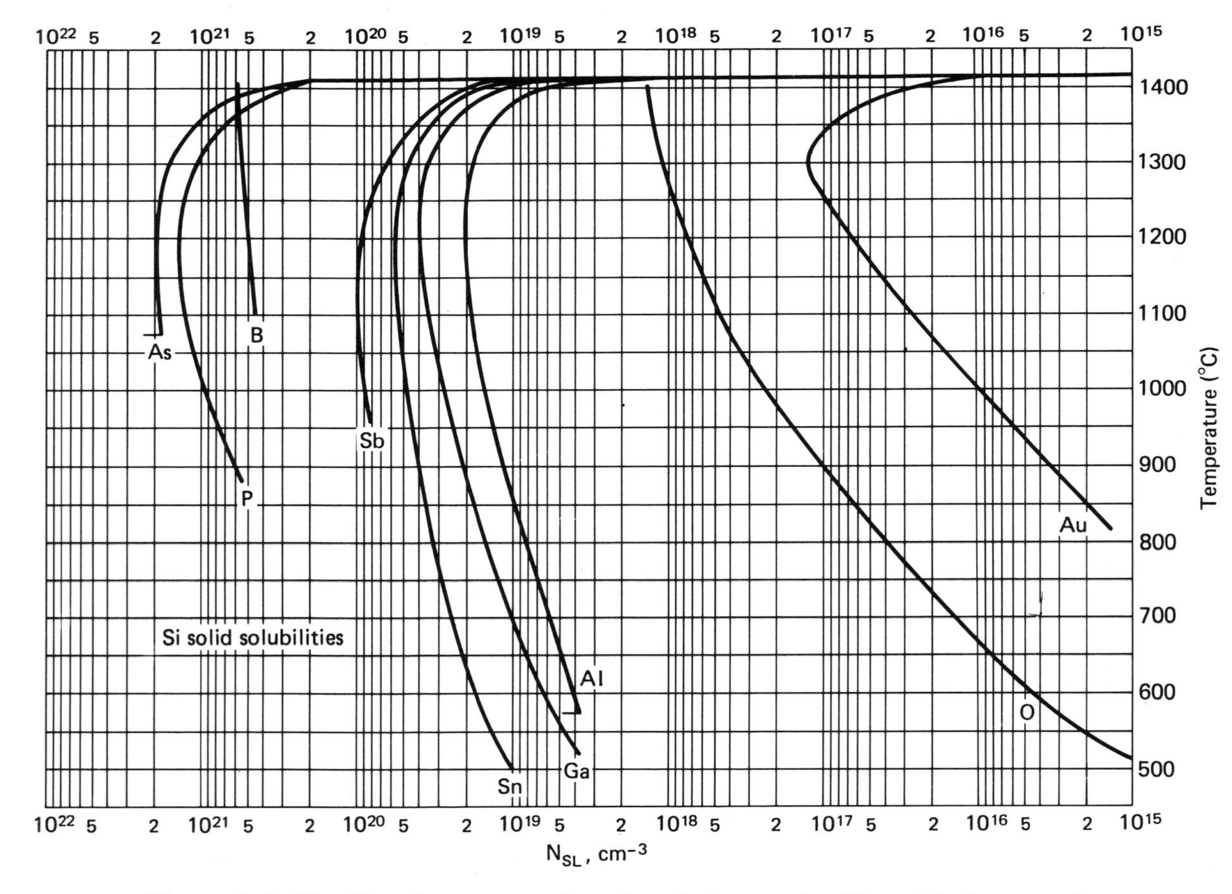

Figure 4–9. The Trumbore curves for selected dopants in silicon. Updates are included. After Trumbore [2]. (Courtesy of AT&T Bell Laboratories.)

4.11.2 Quenching

Semiconductor device fabrication often involves nonequilibrium processes so that phase diagrams and the Trumbore curves sometimes must be interpreted rather freely. For example, after a diffusion has been performed on a silicon wafer at, say, 1200°C, the wafer temperature often is not lowered to room value slowly enough for all the changes indicated by the phase diagram to take place. We might think of this as a *quenching* process whereby the dopant is frozen into the silicon lattice relatively quickly, without having a chance to outdiffuse during the short cooling cycle.

Say P is diffused into Si at 1200°C until the *surface* concentration of the dopant $N_{x=0} \equiv N_0$ is at N_{SL}, the solid solubility limit value. Reading from the appropriate Trumbore curve we note that $N_{SL} = 1.5 \times 10^{21}$ cm^{-3}. Then the wafer is cooled rather quickly. If the cooling rate were slow enough, Trumbore says that at 900°C N_0 would drop to 6×10^{20} cm^{-3}. As a practical matter, quenching allows the wafer to be brought back to room temperature with N_0 remaining near the 1200°C values of 1.5×10^{21} cm^{-3}. Since negligible diffusion takes place in silicon below about 800°C, any dopant adjustment away from the surface cannot take place below that value.

There are some practical limitations on how fast quenching may take place. First, at the typically high diffusion temperatures, many vacancies (which are necessary for substitutional dopant diffusion to take place) form in the host crystal. If these are frozen in by too-rapid cooling, the behavior of the finished devices will be affected adversely. Second, the widespread use of 4 in. and even larger wafers has led to problems of wafer *warping* if the temperature is decreased too rapidly.[6]

Two general methods of controlling the cooling rate and schedule are used to prevent warping. First, a *slow pull* may be used to move wafer-loaded boats from the furnace hot-zone gradually, until the boat is outside the furnace. The slow pull may be done by hand, but commercially, where boats are large and heavy, uniform-pull-rate motors are used. By this means the wafer temperature is gradually lowered from the high diffusion value, through the cooler furnace end-zone temperature, to room value. Typical silicon dopants practically cease diffusing once T drops below roughly 800°C, so the final cool outside the furnace has little effect on the dopant distribution. The overall temperature cycle can be made to eliminate warpage completely.

Wafers enter and leave the furnace from the same end. Hence, the first wafers into the furnace hot-zone are the last to leave the hot-zone (the FILO problem) and are exposed to the high diffusion temperature for the longest

---

[6] The effect here is indirect. As wafer diameter has increased, wafer thickness also has had to be increased to reduce chances of breakage during handling. Wafers with thicknesses of twenty or more mils are very subject to warpage on fast cooling. In the days of thin wafers with diameters of 2 in. or less this was not a problem.

time. All wafers in the batch do not have the same time/temperature product, which makes for nonuniformity.

A second and widely used means of lowering wafer temperature at a controlled rate is furnace *ramping*: the loaded boat is left in the furnace and the furnace temperature is reduced, usually under microprocessor control. The boat and wafers are removed when the furnace temperature is well below diffusion values. Final cooling takes place at room temperature. The FILO (first in, last out) problem is eliminated.

4.12 Distribution Coefficient

The *distribution coefficient* is another important quantity that can be read from the phase diagrams. Unfortunately it is defined in different ways in the literature, but they all give about the same numerical value. First, consider the definition used by Trumbore that we designate by the symbol k_d, rather than by the more commonly used k, which can be confused with Boltzmann's constant [2]. Where there is no chance for confusion, we may drop the "d" subscript. Following Trumbore we write

$$k_d = \text{distribution coefficient}$$

$$= \frac{\text{atomic fraction of A in the solidus alloy}}{\text{atomic fraction of A in the liquidus alloy}} \quad \text{at } T_B$$

where A is the impurity or dopant.

This may be written in more compact form as

$$k_d = \left. \frac{(\text{atomic fraction of A})_{\text{solid}}}{(\text{atomic fraction of A})_{\text{liquid}}} \right|_{T_B} \tag{4.12-1}$$

where T_B is the melting point of the host component B.

We convert this definition into notation that has been used earlier. Since x' = atomic fraction of the *host*, B, then $(1 - x')$ = atomic fraction of the *impurity*, A; hence

$$k_d = \left. \frac{(1 - x_S')}{(1 - x_L')} \right|_{T_B} \tag{4.12-2}$$

Since k_d is defined at the host melting point T_B, it is temperature independent. Some typical silicon technology values are shown in Table 4–1.

The coefficient may be defined in another but equivalent form by reference to Figure 4–4(b). If $T_3 \approx T_B$, so the liquidus and solidus lines may be considered to be straight near T_B, simple trigonometry shows that

$$k_d = \left. \frac{\text{slope of liquidus}}{\text{slope of solidus}} \right|_{\text{near } T_B} \tag{4.12-3}$$

Table 4–1. Distribution Coefficient for Certain Impurities in Silicon at the Melting Point of Silicon

Impurity	Al	As	B	P	Sb	Au
k_d	0.002	0.3	0.8	0.35	0.023	2.5×10^{-5}

After Trumbore [2]. Courtesy of AT&T Bell Laboratories.

Remember that the slopes must be evaluated in terms of x', the *atomic* fraction, rather than in terms of x, the composition by weight.

Still another definition of the distribution coefficient, call it K_d, is

$$K_d = \left.\frac{N_S}{N_L}\right|_{\text{A, near } T_B} \tag{4.12-4}$$

where N_S and N_L are respectively the *concentrations* of the *impurity* (A) in the solid and liquid phases, near the melting point of B. This is the form most often found. The relation between the two definitions may be calculated.

Proof of Eq. (4.12-5) is left to the student. Here are a few hints. A volume V of the alloy contains n_A (not n_a, Avogardo's number) atoms of the impurity component A, and n_B atoms of the host component B. By the definition of concentration, $N_A = n_A/V$ and $N_B = n_B/V$, whence

$$\frac{n_A}{n_B} = \frac{N_A}{N_B} \quad \text{and} \quad x' = \frac{N_B}{N_A + N_B}$$

If $N_A \ll N_B$, $1 - x' \approx N_A/N_B$. Then

$$k_d = \frac{1 - x'_S}{1 - x'_L} = \left(\frac{N_S}{N_L}\right)_A \left(\frac{N_L}{N_S}\right)_B$$

But $N = \rho\, n_a/M$, where ρ = density of the phase, n_a = Avogadro's number, and M = molecular weight. Since for a given material M is independent of the phase, and Avogadro's number is constant, N is proportional to ρ. Finally

$$K_d = (\rho_S/\rho_L)_B k_d \tag{4.12-5}$$

Since the solid-to-liquid density ratio of silicon near its melting point is nearly one, the two definitions, k_d and K_d are taken to be equivalent.

It is possible to set up an equation for k_d in terms of the weight compositions x_S and x_L. This is considered in one of the problems.

4.13 Decant Refining

Zone refining is the process that allowed silicon to be purified enough for the manufacture of semiconductor devices. This process was developed by Pfann of Bell Telephone Laboratories, and utilizes the fact that common impurities in silicon have k_d values less than unity [3, 4]. We shall not consider the zone refining process per se, but rather a related process, *refining by decanting*, in order to illustrate how the value of the distribution coefficient is related to the refining process. Consider the situation shown in Figure 4–10.

The initial composition of the material is x_i. This material has been chemically purified as much as is economically feasible, so x_i is very close to 1. The mass is melted by raising the temperature to T_3 and then cooled to T_1. A certain portion of the total weight (you know how to calculate it by using the lever rule) freezes out with composition x_{S1}, whose value is closer to 1 than was x_i. The solid now is purer B, a result of $k_d < 1$. We decant, or pour off, the liquid, leaving behind a smaller weight of purer solid. The temperature is raised again to T_3, and cooled to $T_2 > T_1$. Solid freezes out at a still higher composition, x_{S2}, but of even smaller weight. This process can be continued until a very small amount of very pure β-phase solid is left. Of course the decanted liquid that is poured off after each partial cooling step is not thrown away but saved for another pass through the whole process.

Zone refining provides a very ingenious alternative to the decanting process: a portion of the system is melted, and the melted region is made to pass through the entire sample, usually in the form of a long rod (Figure 2–14). Several passes of the molten region through the rod effectively sweep impurities from the starting end toward the finishing end of the passes, all without the need of pouring off the liquid. The overall effect is the same as in decanting.

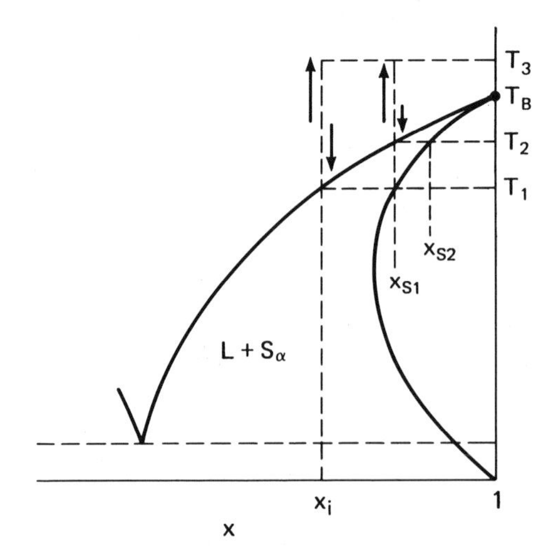

Figure 4–10. Refining by decanting.

The repeated melting and freezing of any given section of the rod results in purification. Since $k_d < 1$, each melt/freeze cycle favors impurities leaving the solid for the liquid, and purer solid is left behind.

4.14 Congruent Transformations

Another invariant transformation that may be encountered in typical doped-semiconductor systems is the *congruent* type, which yields a solid *compound* on freezeout from a melt. The congruent transformation may be symbolized by

$$L(A + B) \xrightarrow{\text{cool}} (A_n B_m)_S \tag{4.14-1}$$

If the compound forms stoichiometrically (n and m fixed integers), it may be represented by a vertical line on the phase diagram corresponding to a fixed value of x' (or x, but x' is easier to calculate). For example, in Figure 4–11 we see two congruent transformations producing the compounds $SiAs_2$ at $x' = 66.7\%$ and SiAs at $x' = 50\%$.[7]

When a compound is formed in a binary system it effectively divides the phase diagram into two separate regions. By way of example, in Figure 4–11 the half-diagram to the left of $x' = 50\%$, where SiAs is formed, can be considered as a separate phase diagram with As and SiAs as the components. The new x' scale would start at zero, corresponding to Si, and continue up to 0.5 at SiAs. In the same way, the right-hand side of the diagram can be considered also as representing a system with As and $SiAs_2$ as the components. Still a third system could be considered with SiAs and $SiAs_2$ as the components. This technique of using compounds formed by congruent transformations to break down phase diagrams into simpler ones often is used to simplify complicated diagrams. It also may be applied to eutectics.

This concept is used extensively in the study of ternary (three-component) compound semiconductors. For example, a system of wide band gap, and therefore of interest for optoelectronic applications, is made up of the three components Al, As, and P. Two stoichiometric compounds are formed: AlAs (aluminum arsenide) and AlP (aluminum phosphide). (In the periodic table Al lies in column III_A while As and P are in column V_A, so both compounds are of the III-V type.) Thus, a "pseudo-binary" phase diagram may be formed, with AlAs and AlP as the A and B components of our earlier diagrams. Because of the complete solid solubility of the components, the diagram will have the general shape of Figure 4–2(a). This diagram is simpler to use and focuses attention on the range of nonstoichiometric, ternary alloys that may exist because the components have complete solid solubility. These solid solutions may be designated $AlAs_{(1-x')}P_{x'}$ where x' is the atomic fraction in

---

[7] Hansen and Anderko [1] draw the As/Si phase diagram with Si as the A component on the left side of the diagram. This is the reverse of the convention used in earlier diagrams here and means that the higher melting point is on the left also.

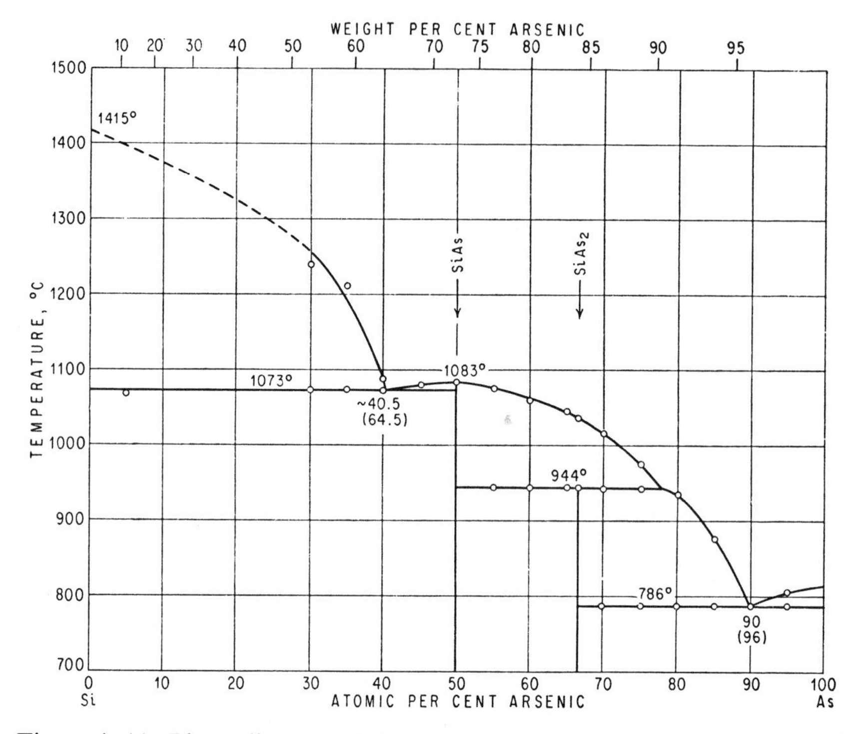

Figure 4–11. Phase diagram of the As/Si system. Hansen and Anderko [1]. (Courtesy of McGraw-Hill Book Company.)

accordance with the notation that we have been using. (In compound semiconductor work the atomic fraction typically is designated x.)

Some other binary components that fall into this category are AlSb/GaSb, InAs/GaAs, AlSb/InSb, InSb/GaSb, and InP/GaP. In these examples the column V_A element is common to the two binary components. The prior example, AlAs/AlP, and InAs/InP share a common column III_A element.

4.15 Peritectic Transformations

Still another invariant transformation encountered in some doped-semiconductor systems is the *peritectic* type. This transformation may be symbolized by

$$L + S_1 \xrightarrow{\text{cool}} S_2 \qquad (4.15\text{-}1)$$

In this situation only a portion of S_1 may combine with the liquid-phase L to form S_2 with some remaining as S_1. The solid phases S_1 and S_2 may be elements, compounds, or solid solutions.

An example of the peritectic transformation may be seen in Figure 4–11. Consider a melt of atomic composition $x_i' = 0.6$ (or 60%) As. As the temperature is lowered below the liquidus line, freezeout occurs with the solid composition fixed at $x' = 0.5$ (SiAs) and the liquid composition determined by the tie line intersection with the liquidus. As T decreases even more, x' increases along the liquidus which indicates that the melt is becoming more rich in As (or less rich in Si). Just above 944°C, the peritectic temperature, the liquid composition is $x_L' \approx 0.78$. At 944°C and below, the tie line terminates at the vertical lines corresponding to SiAs and $SiAs_2$, so the peritectic reaction converts the combination of liquid at $x_L' \approx 0.78$ and some of the solid at $x_S' = 0.5$ to the new solid phase at $x_L' = 0.667$, corresponding to $SiAs_2$.

The Si/As diagram of Figure 4–11 provides a good summary of our work since it shows two congruent transformations resulting in compounds, two eutectic transformations (at temperatures of 1073°C and 786°C), and a peritectic transformation at 944°C. All five of these are invariant transformations.

Reference to the Trumbore curves, Figure 4–9, shows that the maximum limit of solid solubility for As in Si is $(N_{SL})_{max} \approx 2 \times 10^{21}$ cm^{-3}. This corresponds to an x' value of

$$x' = \frac{(N_{SL})_{max}}{N_{Si}} = \frac{2 \times 10^{21}}{5 \times 10^{22}} = 0.04$$

Note that N_{Si} is the concentration of Si. This means that $SiAs_2$ does not form when As is indiffused into Si. After the diffusion cycle ends and the wafer cools down to room temperature a very minute fraction of the solid will be SiAs. The solid β phase does not show up on the phase diagram.

4.16 Contact Systems

The formation of ohmic contacts on semiconductor devices represents one of the more arcane aspects of semiconductor processing. Generally, three types of ohmic contacts are involved: (1) the aluminum-to-silicon contact on the chip proper, (2) the silicon-to-header contact, the header being metal, and (3) the gold-to-aluminum interface between the gold bonding wire and the aluminum contact pads on the chip. We shall consider these in order.

As to the joining of aluminum contact pads to silicon, we know from Chapter 1 that aluminum is deposited on the entire upper surface of the wafer and makes contact to the silicon where oxide windows are provided. If the silicon there is very highly doped, a proper ohmic contact is formed by subjecting the wafer to a *sintering* or *annealing* step, in which the wafer temperature is raised to just *under* the eutectic value of 577°C for the Al/Si system. This prevents melting at the contact.

The Al/Si system is well behaved and compounds do appear. As stated earlier these typically will be silicides, and will be of the β phase, of composition so near to unity that they do not show on the basic phase diagram, Figure 4–7, since the x scale is not expanded enough. It is now believed that the presence of the silicides plays an important role in the formation of the ohmic contacts.

Next, the silicon-to-header contact is formed when the chip or die is *die bonded* to the header. The latter usually is made of gold-plated Kovar.[8] The surface between the die and header after bonding is rather large so that if a junction is formed it will have a large leakage current and tend to behave like an ohmic contact. The die-bonding temperature certainly must be lower than 577°C so that eutectic melting will not take place at the already-formed Al/Si contacts (Section 4.10).

With both gold and silicon present, *eutectic bonding* between chip and header may be used. This takes advantage of the formation of the Au/Si eutectic of composition $x \approx 6\%$ Si by weight at $T_E = 370$°C. Thus, if the header gold plating is thick enough to supply an adequate amount of gold, and if the chip is scrubbed slightly on the header with temperature just above T_E, some melting occurs. On cooling, the melt solidifies to complete the bond.

Since the cost of gold is very high, plating tends to be thin, so the practice usually is to furnish some Au/Si eutectic in the form of a thin, small tab or *preform* that is placed between the chip and header. This insures that some eutectic will be present, which will melt at 370°C when the temperature is raised. Cooling again results in the *eutectic bond*. Eutectic bonding at a slightly lower temperature may be obtained by using a preform of Au/Ge, rather than Au/Si, and a eutectic of composition $x \approx 12\%$ Ge by weight. In this case $T_E =$ 356°C.

When die bonding is completed the gold/aluminum system comes into play when a gold wire, typically 0.8 to 1 mil in diameter, is bonded to the chip aluminum contact pads by thermocompression (TC). These thin gold wires effectively serve as the links between the microworld of the chip and the macroworld where the header-mounted chip is used. Any elevated temperature used during lead bonding must be lower than the die-bonding temperature so that the die bond maintains its integrity.

During TC lead bonding the chip temperature is raised to about 300°C and the Au wire is pressed in tight contact to the aluminum pad, thereby forming a weld between the wire and the aluminum. Usually under modern controlled conditions this system is well behaved. The Al/Au phase diagram does show, however, that a number of intermetallic compounds may form

[8] Kovar is the trade name of an alloy made by the Westinghouse Electric Corporation. It has a thermal expansion coefficient that can be matched by certain glasses (used to provide seals for external leads passing through the header) and is composed of 17% Co, 29% Ni and the remainder mostly Fe. A more precisely specified formulation is the ASTM 15-68 alloy which has the composition: 17% Co, 29% Ni, 0.02% C, 0.30% Mn, 0.20% Si, and 53.48% Fe. In addition to matching glass thermal expansion well, these alloys plate well with gold, are good electrical conductors, and bond well.

under certain conditions. In particular, congruent transformations occur at 1060°C, where $AuAl_3$ is formed, and at 624°C, where a transformation yields Au_2Al. The latter is tan-colored and a poor conductor that can cause contact failure, while $AuAl_2$ is a purple-colored good conductor [5]. For many years the purple $AuAl_2$, which obscured the presence of the trouble-causing Au_2Al, was blamed for contact failure, which was termed the *purple plague*. Even now, when finished devices are stored at temperatures that are too high this failure mechanism may show up. There are also other plagues that can cause contact failure on semiconductor chips [6].

Our discussion has been limited to binary or two-component systems, yet it is clear that if we have a gold lead wire welded to an aluminum bonding pad that makes contact to a region of boron doped silicon, we are dealing with a many-component system. It is fortunate that most of these can be considered on a two-at-a-time basis so the methods that have been discussed in this chapter are usually adequate.

Problems

4–1 A 100-gram sample of 60%Pb/40%Sn (by weight) alloy is melted and held at 500°F. The sample temperature is lowered slowly to 400°F and held there. The phase diagram is shown in Figure 4–P-1.

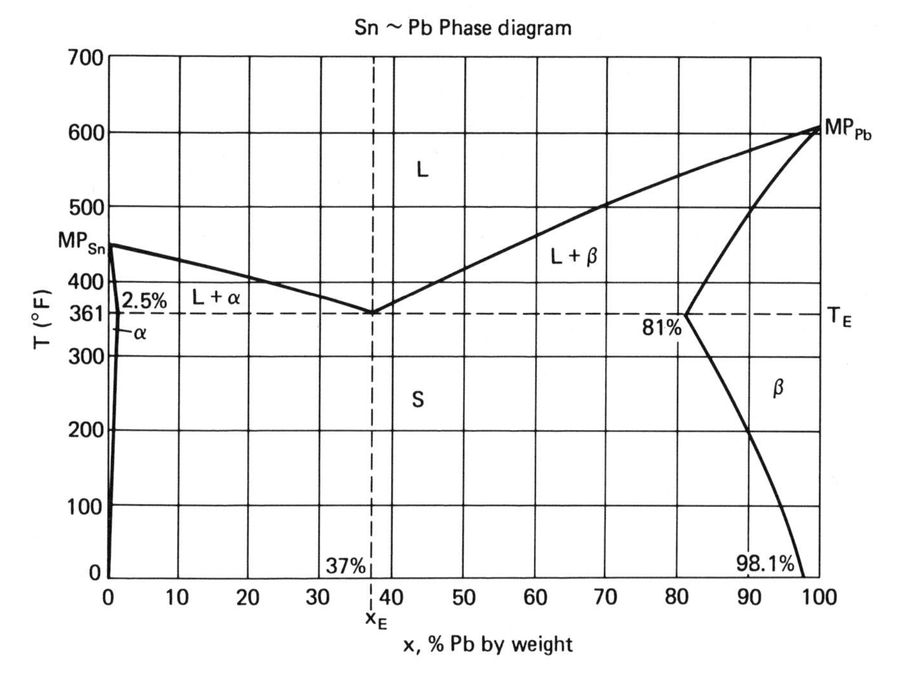

Figure 4–P-1. The Sn/Pb phase diagram.

(a) Describe the phases present, giving the composition and weight of each phase.
(b) What is the "pasty" temperature range of this alloy—i.e., what is the range of temperature where both liquid and solid phases are present?
(c) What is the pasty range if the weight percentages of Pb and Sn are interchanged?
(d) Is there any other Pb/Sn alloy that has the same melting point as that in part a? Explain.

4–2 Repeat Problem 4–1 but for an initial composition of 0.5.
4–3 Verify Eq. (4.12-5). Show all your steps.
4–4 Describe what would happen in decant refining if $k_d > 1$.
4–5 Consider a eutectic system as in Figure 4–P-1. The initial composition $x_i > x_E$. Consider the eutectic/B system as pseudo-binary.

(a) Show that the lever rule for the pseudo-binary system [Figure 4–3(c)] may be written as

$$W_E(x_i - x_E) = W_B''(1 - x_i)$$

where W_E = eutectic weight and W_B'' = weight of B *not* in the eutectic.
(b) Show that this is equivalent to the A/B lever rule

$$W_A(x_i - 0) = W_B(1 - x_i)$$

References

[1] M. Hansen and K. Anderko, *Constitution of Binary Alloys*. New York: McGraw-Hill, 1958.
[2] F. A. Trumbore, "Solid Solubilities of Impurity Elements in Germanium and Silicon," *Bell Syst. Tech. J.*, **39**, (1), 205–233, Jan. 1960.
[3] W. G. Pfann, "Principles of Zone-Melting," *J. Metals*, **4**, (7), 747–752, July 1952.
[4] W. G. Pfann, "Zone Refining," *Sci. Am.*, **217**, (6), 66–72, Dec. 1967. This is an excellent review article.
[5] S. K. Ghandhi, *The Theory and Practice of Microelectronics*. New York: Wiley, 1968.
[6] M. Fogiel, *Modern Microelectronics*. New York: Research and Education Association, 1972. p. 816.

Chapter 5
Oxidation

Silicon dioxide plays essential roles in silicon planar processing. It serves as the most common insulator and dielectric in the technology; it serves to mask silicon against the indiffusion of dopants; and it provides two types of passivation, first as a tough glassy covering that affords mechanical protection to completed dice, and second, as a means of saturating the dangling bonds at the surface of the wafer. By this process the wafer surface becomes well behaved, making for better devices.

In this chapter we consider the nature of SiO_2, what forms its structure may take, and how it may be *grown* by combining externally supplied oxygen in some form to react with the substrate silicon. We shall concentrate on the high-temperature process of thermal oxidation. Oxide also may be *deposited* from an external source (Chapter 9).

5.1 Structure

Even pure silica or silicon dioxide (SiO_2) can appear in many forms. Some 22 phases of silica have been identified, most of which are crystalline [1]. Silica thermally grown on silicon is *amorphous*, however. These forms often are called *vitreous*, which means *glasslike* or noncrystalline, but amorphous does not imply an atomic arrangement with no order. Rather, short-range order is present in the form of a tetrahedral atomic arrangement, but long-range order is absent because the tetrahedra are not arranged in a regular, three-dimensional array as in a crystal.

5.1.1 Basic Tetrahedron

The basic subunit of silica structure, whether amorphous or crystalline, is made up of a silicon atom surrounded by four oxygen atoms, as illustrated in

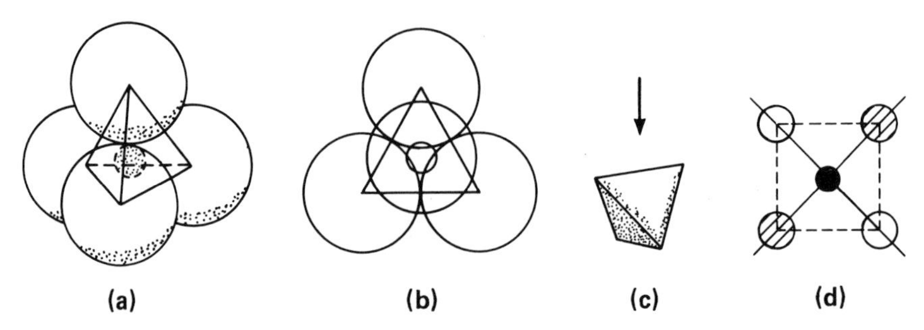

Figure 5–1. Silicon dioxide tetrahedron. (a) Basic tetrahedron. (b) Top view of two-dimensional projection. (c) Another view of the coordination tetrahedron. (d) Projection of (c) onto a plane from the arrow direction indicated.

Figure 5–1(a). The relative radii of silicon and oxygen are 0.42 and 1.32 Å, respectively, so that the silicon atoms fit nicely among the four oxygen atoms, which touch each other [2]. As illustrated in the figure, if straight lines are drawn between centers of touching oxygen atoms, a regular tetrahedron is formed with the silicon atom located at its center. Notice that this structure, called a *coordination tetrahedron*, does not exist but is a useful construct for illustrating structure without showing individual atoms. The similarity to the basic tetrahedron of silicon is apparent, except that two elements are involved rather than only one, and the sizes are different. The side lengths of the SiO_2 and Si tetrahedra are 2.65 Å and 7.68 Å, respectively.

If the tetrahedron of Figure 5–1(a) is viewed from above, its two-dimensional projection is an equilateral triangle, as shown in (b). Even though this projected view is two-dimensional, remember that the actual atomic array is three-dimensional as described above. Both the atoms and the triangular projection of the coordination tetrahedron are shown in Figure 5–1(b).

A different view of the coordination tetrahedron is shown, without the atoms, in Figure 5–1(c). When viewed from above in the direction of the arrow the square two-dimensional projection appears as shown in Figure 5–1(d), where the atoms are not drawn to scale. The two diagonally opposite shaded oxygen atoms are in one plane, while the two unshaded ones will be in a plane one lattice constant below (or above). The four short lines projecting outward from the corners indicate covalent bonds from the oxygen atoms that can link to other coordination tetrahedra or to subunits of other kinds to be discussed shortly.

Consider that the square projection of Figure 5–1(d) is bonded to four other tetrahedra, as in typical crystalline forms. Then each corner oxygen is shared covalently with one other tetrahedron: a given tetrahedron "owns" only one half of four oxygen atoms for a total of two oxygen atoms. Silicon atoms are not shared by the polyhedra: thus, each tetrahedron contains or owns one silicon atom. The total per tetrahedron is one silicon and two oxygen atoms, which reconciles the SiO_2 formula.

5.1.2 Crystalline Silica

One possible crystalline phase of silica has the form represented in Figure 5–2(a). Remember that the atoms (not shown) do not all lie in the same plane. The small arrowheads and tails at the centers of the triangles represent the direction in which the fourth oxygen atom lies relative to the plane of the other three oxygens. Thus in tetrahedra 1, 3, and 5 the extra oxygen is above the plane, while in 2, 4, and 6 it lies below. These fourth oxygen atoms provide covalent bonding to tetrahedra in other similar atomic planes, leading to a repetitive three-dimensional structure. The tetrahedra provide local or short-range order, and their regular, repeated organization in three dimensions gives the long-range order associated with the crystalline state. Other bonding arrays are possible that give other crystalline phases or *allotropic* forms.[1] The manner in which the tetrahedra bond to each other determines the structure and properties of the particular phase.

---

[1] Three principal forms of crystalline silica and their lattice forms are quartz, hexagonal; tridymite, triclinic; and cristobalite, tetragonal. It is unfortunate that many forms of vitreous silica are incorrectly referred to as quartz.

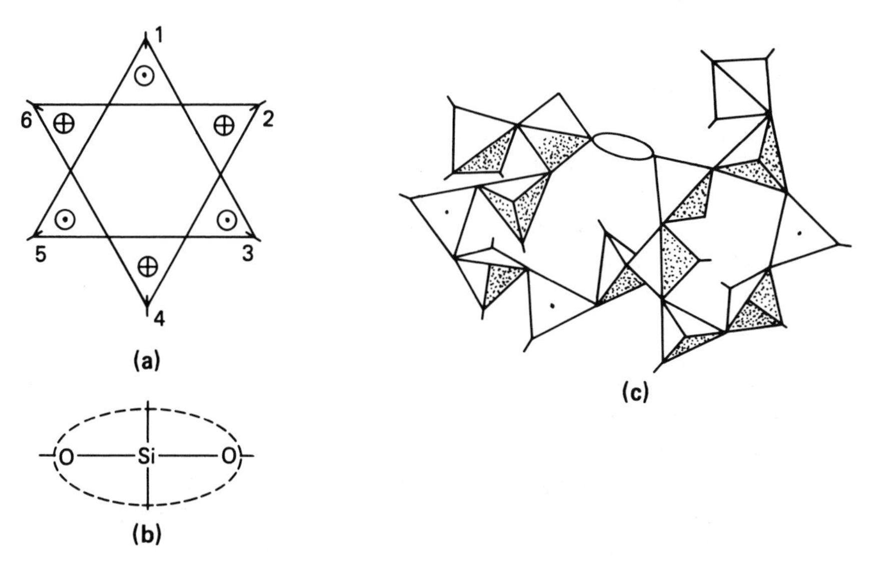

Figure 5–2. Silica structure. (a) The hexagonal structure of quartz, a crystalline form. (b) The SiO_2 radical. (c) A typical array of subunits in amorphous silica. The actual network is three dimensional.

5.1.3 Amorphous Silica

In amorphous or vitreous silica, $a\mathrm{SiO}_2$, the phase of principal interest here, the individual tetrahedra do not link up in a regular three-dimensional pattern. Also, the oxygens may not all be shared by two tetrahedra; some may link up to other subunits such as the one shown in Figure 5–2(b), whose bonds allow it to replace a tetrahedron structurally. Some idea of this may be obtained from the sketch in Figure 5–2(c), which attempts to show possible structures of $a\mathrm{SiO}_2$. No extended regular structure can be seen because there is none. In studying this figure, remember that you are looking at a representation of a three-dimensional structure; hence not all of the bonding can be shown. The small lines appearing at the apexes in the diagram indicate where the structure might continue to tetrahedra or other subunits that are not visible.

As compared to the crystalline forms, $a\mathrm{SiO}_2$ has a relatively open structure—comparative relative densities are 2.2 g/cm^3 for the glass versus 2.65 g/cm^3 for quartz.

Another method of restoring charge balance and bonding when a tetrahedron is incompletely bonded to adjacent tetrahedra can occur if positively charged ions of another element are present. A frequently occurring cation of this type in glasses is Na^+. Adding a third element to Si and O results in impure or *extrinsic* silica. These cations are called *network modifiers*—in contrast to the designation *network former* given to Si or other atoms that, with oxygen, form the basis of the particular glass structure.

5.2 Phase Changes

The structural difference in crystals and vitreous materials shows up in the phase change between liquid and solid as temperature is changed. This is illustrated in the typical cooling (or heating) curves of Figure 5–3. With crystals an increase in T raises the total heat content; and melting takes place isothermally at a distinct melting point (MP). This seems reasonable, because every atom in crystals is in the same structural environment (due to the regularity of the structure). Melting requires breaking bonds for all the atoms, so they all break at the same temperature. As the heat content of the system is reduced and temperature drops, a similar isothermal transition from liquid to solid (freezing) occurs (and at the same MP temperature).

On the other hand, with amorphous glasses, the atoms are not all in the same environment (due to the random structure), so on heating, the bonds do not all break at the same temperature. Hence, there is no definite melting point and the transition from solid to liquid takes place gradually over a considerable temperature range. In this situation, one speaks of a glass *transition temperature*, T_g, or *softening point*, at which—on cooling, for example—the material first appears to be solid [3].

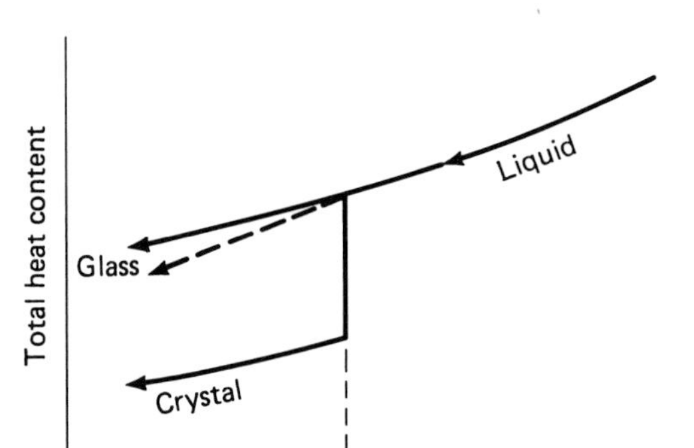

Figure 5–3. Illustrating the difference in cooling/heating curves between amorphous (vitreous) and crystalline materials. Two curves are shown for glass to illustrate how its curve depends on the cooling rate.

When a glass is cooled, the heat-content–T curve depends upon the cooling rate (as illustrated in Figure 5–3 by the two curves for the same amorphous material). The lower, dashed curve represents a slower cooling rate, which tends to shift the solid amorphous structure more closely toward a crystalline form [4].

Since the glass structure can be modified by impurities we may expect that the type and quantity of the impurity can affect the softening point. For example, "Vycor" (Corning No. 7900, with 96% silica) has a softening point of 1500°C, while "fused silica" (Corning No. 7940, with 99.8% silica) has a softening point of 1580°C [5]. In contrast, crystalline silica melts at 1715°C in its cubic form of cristobalite, and at 1610°C as hexagonal quartz [6]. "Pyrex" (Corning No. 7740, comprised of 80% SiO_2 and 14% B_2O_3, with traces of Al_2O_3, Na_2O, and K_2O) is an extrinsic glass and softens at 820°C.

5.3 Other Processing Glasses

A few other glasses are encountered in silicon processing, the principal ones being boron trioxide, B_2O_3, and phosphorus pentoxide, P_2O_5. Also, B and P can incorporate into SiO_2 on a silicon wafer when the latter is exposed to them in the vapor phase, say, during predep diffusion. The result will be borosilicate or phosphosilicate glass, respectively, forming a thin glazed skin on the SiO_2 layer.

5.3.1 Boron Trioxide

Boron trioxide can form a properly coordinated polyhedron that is able to produce a three-dimensional glass structure. In contrast to silica, which has

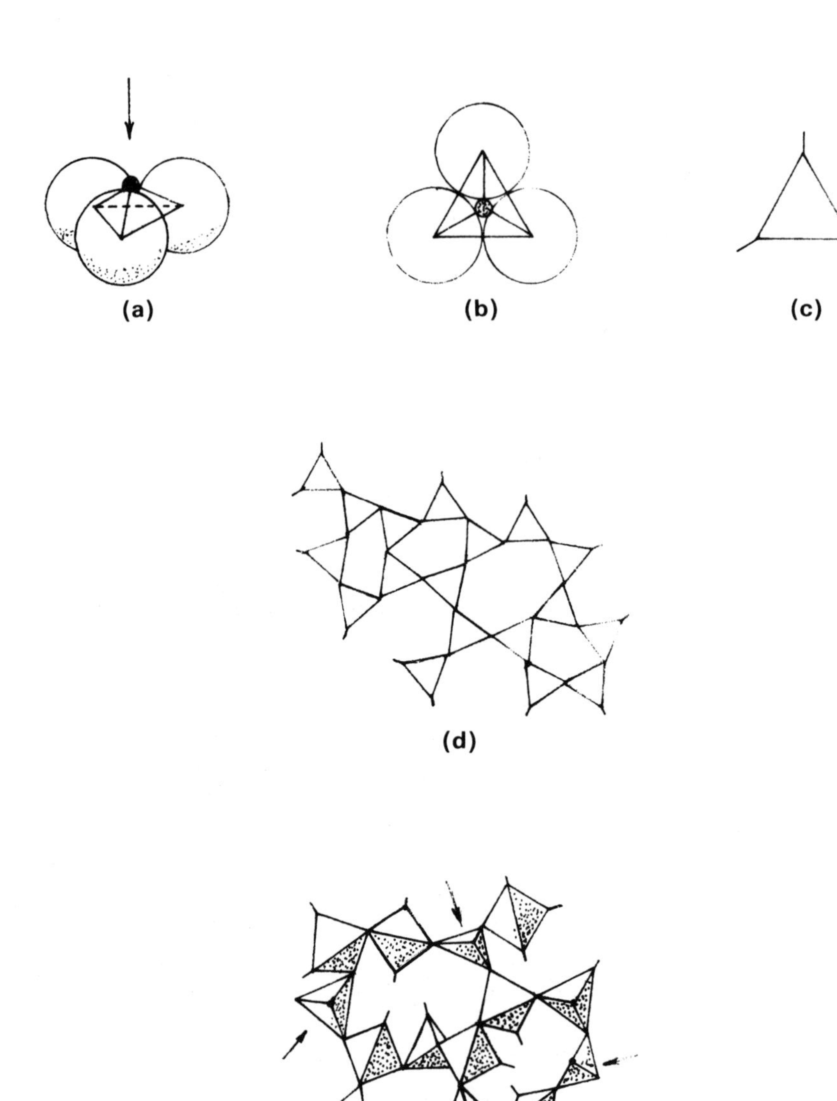

Figure 5–4. Structure of boron trioxide, B_2O_3. (a) Basic shallow pyramidal unit. (b) Top view of (a) projected onto a plane. (c) The coordination pyramid projected onto a plane. (d) A two-dimensional representation of a typical B_2O_3 network using the coordination pyramid. (e) Possible network of borosilicate glass. Arrows point to the B_2O_3 pyramids. Note the absence of long-range order in (d) and (e).

a coordination tetrahedron, B_2O_3 coordinates as a shallow pyramid with the boron atom at one apex and oxygen atoms at the other three, as shown in Figure 5–4(a) [3]. The arrow indicates the sighting direction for the top view shown in (b). The coordination pyramid projects onto the plane of the oxygens as the equiliateral triangle shown in Figure 5–4(c). The small lines extending from the triangle apexes represent covalent bonds to other coordination triangles with which they are shared. The B atom bonds to no other polyhedra since its valence is three. It is easy to reconcile the pyramid with the B_2O_3 formula (Problem 5–1).

The covalent oxygen bonding between pyramids allows the formation of an amorphous, three-dimensional array of the shallow pyramids. Figure 5–4(d) shows the projection of such a glass structure onto a plane.[2]

5.3.2 Borosilicate Glass

During boron predep diffusions, an oxidized silicon wafer is exposed to B_2O_3 vapor at temperature generally greater than or equal to 900°C. Gas-phase B_2O_3 can diffuse into the existing SiO_2 masking layer and convert some of it into a *borosilicate* glass that involves Si, B, and O_2.

In the unmasked regions the silicon also is exposed to B_2O_3 vapor. As the boron diffuses into the silicon, a thin layer of SiO_2 doped with the B_2O_3 simultaneously will be growing on that silicon. As a result, the entire wafer surface will be covered by borosilicate glass. A possible bonding pattern in the amorphous structure is shown in Figure 5–4(e). The small arrows point to the B_2O_3 coordination pyramids, which have only three apexes that bond.

Borosilicate glass, formed during boron predep, etches much more slowly in conventional HF-based oxide etches than does pure silica. Fortunately, its thickness on silicon is much less than that of the masking oxide surrounding the window. For this reason, after predep the mask window regions may be cleared of borosilicate glass by a short dip into an etch without completely removing the SiO_2 mask regions.

In the etching process some elemental boron will be left on the exposed silicon in the mask windows. This boron may be removed by first oxidizing it for 10 sec in a solution such as 1 H_2SO_4 : 1 HNO_3, followed by a 5-sec dip in the oxide etch solution. The etching of SiO_2 and other glasses is covered in Chapter 10.

---

[2] This is the simplified form of the diagram generally used to illustrate the amorphous phase of silica, SiO_2. Actually the form shown in Figure 5–2(c) is better, since it shows the *four* oxygens at the apexes of a tetrahedron and not in the same plane. For B_2O_3, the *three* oxygens are in a single plane and are located at apexes of an equilateral triangle, as in Figure 5–4(d).

The conventional silica diagram usually shows additional cations such as Na^+ and Ca^{2+} in the interstices of the structure. This conventional diagram is based on the work of Zachariasen which is summarized concisely by Van Vlack [7]. We consider it later.

5.3.3 Phosphorus Pentoxide

Phosphorus pentoxide (P_2O_5) also can form an amorphous, three-dimensional glass. The coordination polyhedron is tetrahedral, with an oxygen at each of the four apexes and a phosphorous atom at the center. This superficially resembles SiO_2, but the valence of 5+ for phosphorus, as against 4+ for silicon, requires that one of the four oxygens be double-bonded to the phosphorous, leaving only three oxygens available for bonding to other tetrahedra. The general configuration is illustrated in Figure 5–5 [8].

The basic tetrahedron is at (a); nothing shows the presence of the double bond to the *nonbridging* oxygen, which is unable to link or bridge to another tetrahedron. Figure 5–5(b) shows the square projection of the tetrahedron (see Figure 5–1 for comparison). Note the convention of a double line from the phosphorus to one oxygen, the nonbridging one. Single lines join the remaining oxygens, which are free to bridge to adjacent tetrahedra. Remember that in P_2O_5 only *three* of the four oxygens are available for bridging. Reconciling the tetrahedral model with the chemical formula is easy (Problem 5–1).

A typical three-dimensional array of amorphous P_2O_5 is shown in Figure 5–5(c). The small crosses indicate the nonbridging apexes; only three apexes in each tetrahedron are used for bonding.

5.3.4 Phosphosilicate Glass

When P dopes SiO_2 during the phosphorus predep diffusion, the processes involved are similar to those of Section 5.3.2, but with phosphorus and P_2O_5 replacing boron and B_2O_3. A sketch illustrating a possible form of the phosphosilicate glass is shown in Figure 5–5(d). Again, the small crosses indicate apexes with a nonbridging oxygen.

In contrast to borosilicate glass, phosphosilicate glass etches much more rapidly than intrinsic silica, a fact that raises problems in clearing mask windows after a phosphorus diffusion. Particularly in n–p–n transistors with P-doped emitters, care must be used not to etch away the entire oxide mask in this window-clearing step. Usually the problem is solved by using a more dilute etch, or increasing the buffering (Section 10.3). Special etches also are available (e.g., Pliskin etch).

Contrary to the case with boron trioxide, we sometimes are interested in P-doped glass (PSG) per se because it is used to cover finished IC chips as a passivating and "scratch-protecting" agent. Some years ago it was used alone for this purpose. Since H_2O vapor readily diffuses through the PSG, however, it can generate H_3PO_4. This leads to corrosion of the Al-based metallization, and so can adversely affect product reliability.

Current practice replaces PSG with silicon oxynitride, which is a better diffusion barrier against H_2O. The PSG still is used as a stress-relief layer

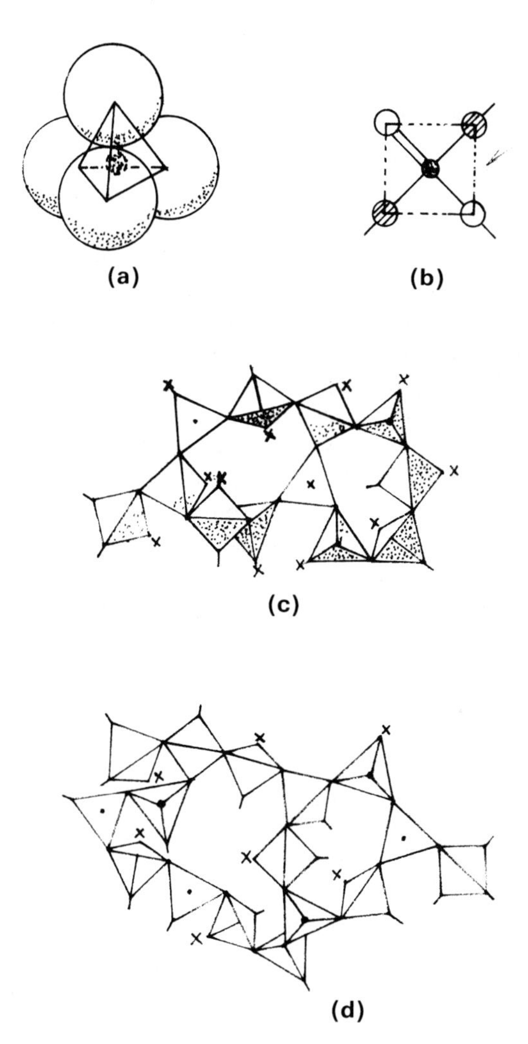

Figure 5–5. Structure of phosphorus pentoxide, P_2O_5. (a) Basic tetrahedron. (b) Square projection of (a) showing how the phosphorus atom is double-bonded to one of the oxygen atoms. The three other oxygens can link up to other units. (c) Possible three-dimensional network of amorphous P_2O_5. The small crosses indicate nonbridging oxygens. (d) A possible network of phosphosilicate glass.

between the metallization and a multilayer passivation of silicon nitride and silicon oxynitride. These materials are deposited by chemical vapor deposition (Sections 9.13.2 and 9.13.3).

5.3.5 Silicon Monoxide

Silicon monoxide (SiO) is another oxide of silicon, used rarely in silicon processing but quite often in thin film, hybrid IC manufacture. It is useful there because no silicon is available for *in situ* growth of the dioxide, and it is deposited easily by physical vapor deposition (Section 12.2.4), a process quite similar to the vacuum deposition of aluminum described in Section 1.8.6. Usually the condensed oxide is nonstoichiometric, being a mixture of SiO, SiO_2, and Si; often, it is designated SiO_x. The value of x, which ranges between 1 and 2, is influenced by the deposition conditions.

The SiO source material is prepared by heating SiO_2 with Si, carbon, and hydrogen, or a hydrocarbon, to reduce the dioxide to the monoxide. Reduction takes place at roughly 1000°C, and SiO is formed in the vapor phase. This temperature is much too low to produce a significant vapor pressure of SiO_2, so the two oxides may be separated. The monoxide vapor is condensed out and subsequently powdered [1]. This material, which is suitable for evaporation, is available commercially and need not be prepared by the IC manufacturer.

5.3.6 Water and Silica

Silicon wafers frequently are oxidized in the presence of water vapor or steam rather than in dry oxygen, because the water vapor gives a higher growth rate. The phase diagram for the H_2O–SiO_2 system is very complex and exhibits at least 15 triple points. In silicon processing, if H_2O, in either liquid or vapor phase, is in contact with silica (particularly at high thermal oxidation temperatures) a reaction takes place within the silica and a very thin H_2O film may form on its surface [1].

Actually the hydroxyl radical OH^-, rather than H_2O itself, reacts with the silicon. The net effect is illustrated in Figure 5–6, where the square projection of an SiO_2 tetrahedron is shown in part (a). A possible variation is shown in part (b), where each of two bridging oxygens has been replaced by a hydroxyl radical. Each such radical has a valence of 1^-, rather than the 2^- of the replaced oxygens, and is bound to the Si alone. This means that the OH^- radicals reduce the amount of bridging in the original silica structure, which in turn leads to a less-dense material. This is a penalty paid for the faster growth rate of wet oxidation.

Improvement of the silica structure and increase in its density can be brought about after oxidation by heat treating or *annealing* the oxidized

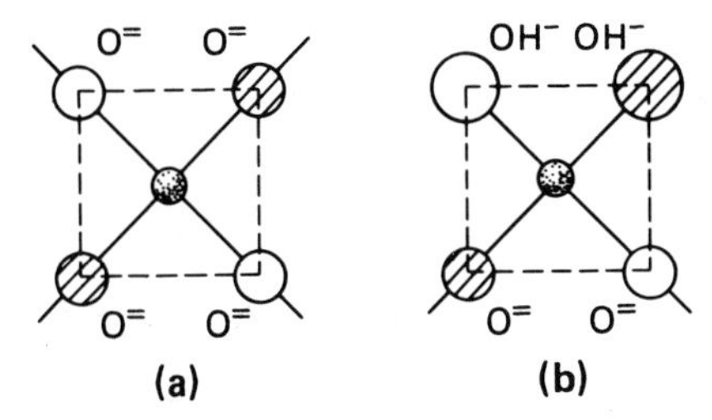

Figure 5–6. Modification of a silica tetrahedron by hydroxyl radicals. (a) Square projection of a silica tetrahedron. (b) Two of the bridging oxygens are replaced by hydroxyl radicals, OH^-.

sample. This is referred to as a *densification anneal*, and is discussed further in Section 5.8.2. The anneal has a negligible effect on the overall oxide thickness but does markedly improve the oxide structure by eliminating the OH^- radicals.

5.3.7 Sodium in Silica

Sodium is one of the most common components in commerical glasses; it usually is added in the form of sodium monoxide, Na_2O. In contrast to B_2O_3 and P_2O_5, sodium cannot form a glass by itself; it is a network *modifier* rather than a network *former*. It is easy to see the reason for this. Sodium is monovalent, so the basic structural unit of the oxide has the form shown in Figure 5–7(a), where a single oxygen is bonded to two sodiums. All the bonds are saturated and none is available to bridge to other similar units (a requirement for forming a glass network). The same is true of sodium peroxide, Na_2O_2, whose bonding unit is shown in Figure 5–7(b). Again, all bonds are saturated, so no bridging oxygens are available for network formation.

The sodium/oxygen bond is a weak ionic bond that can break at lower temperatures than the covalent bond, say between silicon and oxygen. Thus, adding sodium oxide to silica in commercial glass manufacture significantly lowers the softening point [4].

Structurally, the effect of adding sodium oxide to silica is quite different from that of B_2O_3 or P_2O_5. Since sodium is not a network former, it does not link into the basic silica network. Rather, it enters the network interstitially and is loosely bonded ionically to an unsaturated, bridging oxygen in a nearby SiO_2 tetrahedron. A typical structure is shown in Figure 5–7(c) where for simplicity the SiO_2 tetrahedra are all shown in their square projections. The large, shaded circles represent the sodium ions. Two of the silica tetrahedra, which are indicated by the small arrows, have only three oxygens bridging to other tetrahedra and one oxygen to complete the weak ionic bond to a sodium

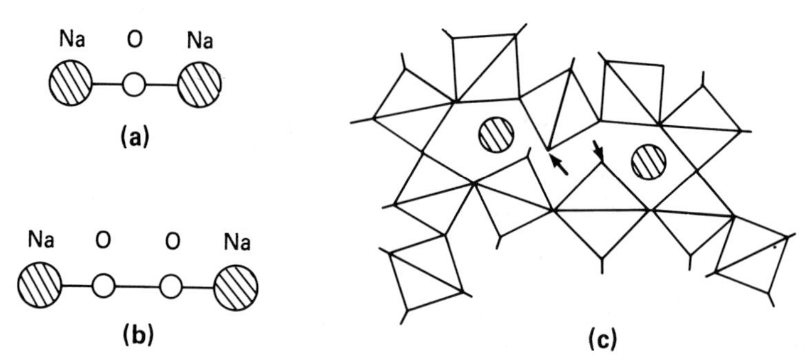

Figure 5–7. Sodium in glass. Neither (a) Na_2O nor (b) Na_2O_2 has bonds available for forming a glass network. (c) Sodium atoms (shaded) locate interstitially and have weak ionic bonds to silica tetrahedra which have a free bond (arrows).

cation. The bond is weak enough that the cations may drift if an external electric field is applied. Their mobility is small, however, compared to that of electrons and holes in silicon.

The cations locate themselves subject to these two principles: first, they cannot be too close together since they repel each other by virtue of their positive charges, and second, they must remain in interstitial sites [7].

Sodium is not introduced into the oxide on silicon wafers intentionally. On the contrary, every reasonable effort is made to prevent its presence, particularly in the oxides of MOS devices, where mobile sodium cations can cause threshold voltage shifts in response to changes in gate-to-source voltage. In fact, a check of MOS threshold drift is an important means for monitoring the presence of Na contamination on a production line.

Sodium can be introduced inadvertently into semiconductor oxides during processing from at least the following sources.

1. Water used for rinsing, and for growing wet or steam oxides thermally. This dictates that all water used should be clean and thoroughly deionized. A general criterion for processing water is that its resistivity be at least 14, or preferably 18 MΩ-cm. (In the trade this is called *18 Meg water.*)

Sodium-free water for oxidizing silicon can be produced by burning hydrogen and oxygen together (very carefully). Very pure, sodium-free gases are available commercially. Special sodium-free containers are required for this water. (See item 3 below.)

2. Liquid chemicals, particularly acids and bases that are used for wet-processing silicon, are possible sources of small amounts of sodium. For example the better-than-common reagent grade of glacial acetic acid (such as the "instrument analyzed reagent grade") assays sodium at 0.5 ppm (parts per

million) and the "electronic grade" at 0.02 ppm of sodium [9]. The latter is used for MOS processing. These low-sodium grades come at premium prices, as might be expected.

3. Glassware. Pyrex glassware (Corning No. 7740)—used in typical chemistry beakers, graduates, and flasks—has a sodium content of 4% by weight. If this or other sodium-bearing glass is used for holding silicon wafers or chemicals for processing them, sodium may be transferred to the wafer oxides. For this reason, sodium-free "fused silica" (Corning No. 7940) or "Vycor" (Corning No. 7900) are used for vessels in silicon processing. They also have higher softening points, as we have seen earlier. In recent years, the use of vessels made of fluorocarbons such as Du Pont's Teflon has become common.

4. Humans. Skin oils are sodium-rich. To avoid contaminating device materials, operators should never touch wafers with their bare hands, but should wear special gloves and use tweezers or vacuum pencils for moving wafers. Needless to say, proper care and cleaning of such wafer-handling devices is essential.

5. Swabs and cleanup devices. In early processing swabs were used to lightly scrub wafer surfaces. Common cotton-tip swabs, particularly those that have been dipped in a binder to make the swab heads hold shape when wet, should not be used in silicon processing, and are not in current commercial practice. Mechanical swab cleaning has been replaced by other scrubbing methods (Section 10.10.1).

In summary, avoid sodium contamination, it causes trouble. Small amounts of Na in MOS oxides may be stabilized by *gettering* with chlorine, however (Section 5.8.1). This technique was the key to commercial MOS devices. But play safe; think sodium-free operation.

5.4 Thermal Oxidation; Diffusion

The physical setup for oxidizing silicon has been described in Chapter 1; we now consider the oxidation mechanism.

The *growth* of silicon dioxide on silicon is the direct result of oxygen and/or water vapor chemically oxidizing the silicon. The basic oxidation equations are

$$\mathrm{Si}\downarrow + \mathrm{O_2}\uparrow \rightarrow \mathrm{SiO_2}\downarrow \tag{5.4-1}$$

and

$$\mathrm{Si}\downarrow + 2\,\mathrm{H_2O}\uparrow \rightarrow \mathrm{SiO_2}\downarrow + 2\,\mathrm{H_2}\uparrow \tag{5.4-2}$$

The rate of oxidation is temperature dependent, with the rate increasing rapidly as temperature is raised.

When a silicon wafer is exposed to an oxidant the two can form the desired oxide where they are in direct contact. In short order, however, oxidant and silicon will be separated by a layer of SiO_2 and the oxidation will stop unless the two can continue to come in contact, either by one or both moving through the already-grown oxide layer.

Studies with radioactive tracers have shown that in thermal oxidation it is the oxidant that moves through the SiO_2 layer, so oxidation continues at the SiO_2/Si interface [10]. This idea is basic to developing the mathematical model for the oxidation process and leads us to expect that the growth rate of the oxide will decrease as the oxide thickness increases, because the oxidant has farther to travel to reach the silicon. The oxidant moves through the oxide by the process of diffusion that is introduced here; it is covered in more detail in Chapter 6.

Diffusion is a thermally dependent particle transport mechanism that does not depend on particle charge and an electric field. It places two conditions on the diffusing species: (1) the species must have a nonuniform spatial distribution, and (2) it must have random thermal motion. When these requirements are met, thermal motion drives the species particles from regions of high to those of low concentration.

Even though diffusion is a three-dimensional process, in semiconductor work we usually confine our attention to the x-directed component (i.e., the one normal to the wafer surface), because that is the direction in which principal modifications to the wafer are made. For example, the t_a curve in Figure 7–2 shows an initial x-directed concentration distribution of, say, the α species. Because of thermal motion the particles tend to move away from the region of peak concentration as time goes on. If the process is allowed to continue long enough, the profile tends to become uniform as indicated by the curves for $t > t_a$.

Diffusion is described mathematically by Fick's First Law (FFL) which in one dimension is

$$\mathscr{F}_\alpha(x) = -D_\alpha[T, N_\alpha(x)]\frac{\partial N_\alpha(x)}{\partial x} \tag{5.4-3}$$

where

$$\mathscr{F} = \text{particle flux density, cm}^{-2}\text{s}^{-1}$$

$$D = \text{particle diffusivity, cm}^2\text{s}^{-1}$$

$$T = \text{temperature, K}$$

$$N = \text{particle concentration, cm}^{-3}$$

$$\frac{\partial N(x)}{\partial x} = \text{particle profile gradient, cm}^{-4}$$

The flux density is the number of particles that in unit time cross a unit area, which is normal to the diffusion direction x. The diffusivity depends on T and the concentration $N_\alpha(x)$ as well as on the diffusing species and the material through which it diffuses, the *host*. For example, the diffusivities of boron and phosphorus in silicon dioxide will differ, but are the same in silicon. Also, each of them has diffusivities of different value in Si and SiO_2.

The diffusivities of the gases H_2, O_2, and H_2O in SiO_2 are important in oxidation and have the values 4×10^{-6}, 4×10^{-9}, and 5×10^{-10} cm^2/s, respectively, at 1200°C. Older references show D greater for H_2O than for O_2, but the above values are widely accepted now. Values for other temperatures may be calculated from

$$D = 10^{-(d_1 + d_2/T)} \qquad (T \text{ in Kelvins}) \tag{5.4-4}$$

given values for the parameters d_1 and d_2. See, for example, Donovan [2]. The diffusivities of B and P in Si are orders of magnitude smaller than those cited for the gases.

In semiconductor work $N_\alpha(x)$ often is much less than the limit of solid solubility and we can assume that $D_\alpha[T, N_\alpha(x)]$ is independent of $N_\alpha(x)$. At a given temperature, the diffusivity then becomes constant, D_α, and Fick's First Law becomes

$$\mathscr{F}_\alpha(x) = -D_\alpha \frac{\partial N_\alpha(x)}{\partial x} \quad \text{for constant } T \text{ and } N_\alpha(x) \ll N_{SL} \tag{5.4-5}$$

This is a linear differential equation with constant coefficients, so superposition applies. Thus two different species, say α and β, may diffuse simultaneously through the medium, independently of each other, and the diffusion equation may be solved for one species without considering the presence of the other. Equation (5.4-5) is the cornerstone for developing the oxide-growth-rate equation.

Note the significance of the negative sign in Eq. (5.4-5). If, for example, $[\partial N_\alpha(x)/\partial x]$ is negative, the flux density and direction of particle flow will be positive (i.e., FFL states that the motion due to diffusion is downhill in concentration).

5.5 Oxidation Model

The model here for thermal oxidation of silicon follows the work of Deal and Grove [11]. We assume that some initial oxide of thickness x_i is present at the beginning of the particular oxidation cycle under investigation. The model has three regions, shown in Figure 5–8(a). The interface between the gas-phase oxidant and the oxide lies at $x = 0$ by definition. The oxide thickness is $x_{ox}(t)$,

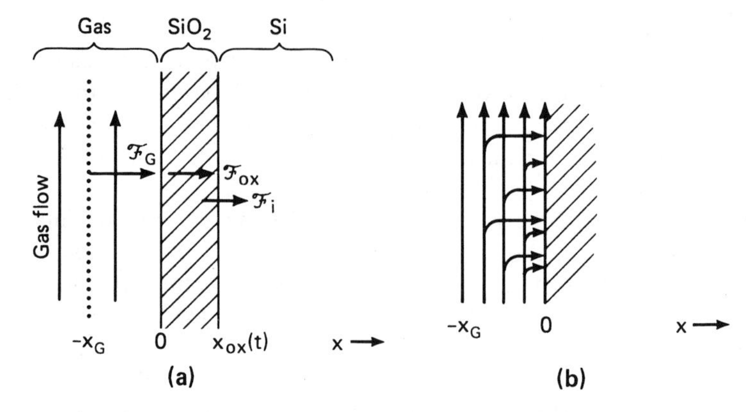

Figure 5–8. The Deal and Grove model for thermal oxidation. (a) Three regions, gas, oxide, and silicon, are identified. (b) Defining $-x_G$ where the oxidant gas is in steady state and does not supply the oxidation process. After Deal and Grove [11]. (Courtesy of the *Journal of Applied Physics*.)

which changes with time as oxide growth progresses. This places the oxide/silicon interface at $x = 0 + x_{ox}(t) = x_{ox}(t)$. The region in the diagram lying to the left of $x = 0$ is the gaseous oxidant source region. It is assumed that the oxidant flows from the bottom to the top of the diagram, parallel to the silicon surface. The gas region lying between $x = 0$ and $-x_G$ is furnishing the oxidant that diffuses through the oxide to support the chemical reaction at $x_{ox}(t)$. Figure 5–8(b) illustrates schematically how the gas near the solid interface is borrowed more heavily than gas flowing nearer to $-x_G$. The value of x_G is defined so that the gas region to its left is far enough away that it does not supply oxidant to the solid portion of the system.

Figure 5–8(a) also defines three *oxidant* flux densities: $\mathscr{F}_G$ in the gas phase, $\mathscr{F}_{ox}$ in the oxide, and $\mathscr{F}_i$ across the oxide/silicon interface. All three are functions of position and time. The initial step in developing the mathematical model of oxide growth considers the steady state, in which the three flux densities must be equal

$$\mathscr{F}_G = \mathscr{F}_{ox} = \mathscr{F}_i \qquad \text{(in steady state)} \qquad (5.5\text{-}1)$$

since there can be no accumulation of oxidant at any point. Consider the flux densities for each of the three regions.

5.5.1 Gas Region

Details of the gas region that supplies oxidant to the system are shown in Figure 5–9(a). The concentrations N_G and N_S are, respectively, the gas-phase

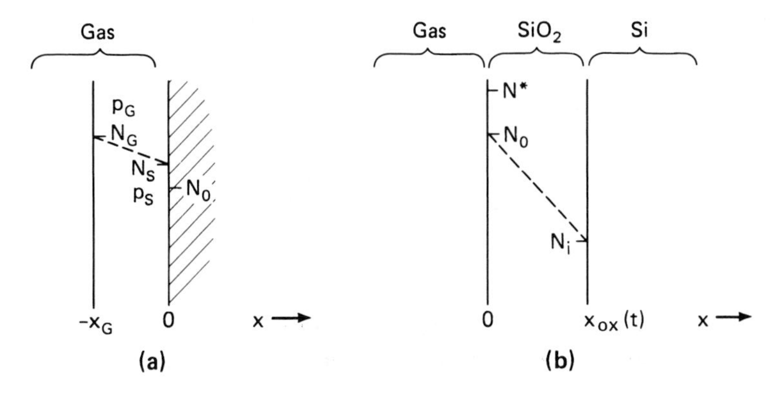

Figure 5–9. Notation for the model. (a) The gas region. (b) The oxide region. A linear oxidant gradient is assumed between N_0 and N_i, both of which change as the oxidation progresses.

oxidant concentrations at $x = -x_G$ and $x = 0^-$. The corresponding oxidant partial pressures are p_G and p_S. Note that N_S and p_S are the values at the *gas* side of the interface—i.e., at $x = 0^-$ or just outside the silicon surface. If x_G is small, as assumed by Deal and Grove, the oxidant profile may be assumed to be linear between $-x_G$ and $x = 0$. (This assumption is verified subsequently by how well the model checks the real world.)

Applying FFL we write

$$\begin{aligned}\mathscr{F}_G &= \text{gas-phase oxidant flux density}\\ &= -D_G \frac{\partial N_g(x)}{\partial x} \approx -D_G \frac{N_G - N_S}{-x_G - 0} \qquad (5.5\text{-}2)\\ &= \frac{D_G}{x_G}(N_G - N_S)\end{aligned}$$

Note that the g or G subscripts refer to the oxidant in the *gas* phase, so D_G is the diffusivity of the oxidant in that region.

Equation (5.5-2) is a perfectly good statement of FFL for this particular region, but in dealing with the gas phase it is more natural to use gas pressure than concentration. Therefore the equation is recast in terms of pressure by the general gas law

$$p = NkT \qquad (5.5\text{-}3)$$

where k = Boltzmann's constant and N is concentration. Thus we have

$$N_G = \frac{p_G}{kT} \quad \text{and} \quad N_S = \frac{p_S}{kT} \qquad (5.5\text{-}4)$$

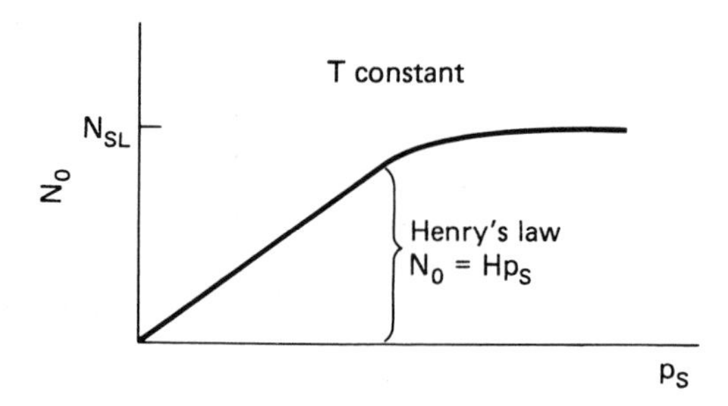

Figure 5–10. Illustrating the relationship between N_0 just inside the host surface and p_S just outside. Henry's law applies in the linear portion.

so Eq. (5.5-2) may be written as

$$\mathscr{F}_G = \frac{D_G}{x_G kT}(p_G - p_S) \tag{5.5-5}$$

and FFL is in terms of pressures. Still another transformation simplifies later work.

Consider a system in which a gas interfaces with a solid. For such a system there is an empirical relationship between the concentration N_0 of a gas species just *within* the solid surface ($x = 0^+$) and the partial pressure p_S of that species in the gas phase just *outside* the solid surface ($x = 0^-$). The relationship for constant T has the form shown in Figure 5–10. Other temperature values will yield curves of the same shape but displaced from the one shown. Observe that the curve saturates above some value of p_S (corresponding to the limit of solid solubility discussed in the last chapter). A linear region below the knee of the curve is described by Henry's law, namely

$$p_S = \frac{N_0}{H} \tag{5.5-6}$$

where H is Henry's constant, a function of the species, the solid material or host, and the temperature. Deal and Grove assume that during oxidation the oxidant partial pressure will be such that Henry's law will apply. Utilizing this idea we may replace p_S by N_0/H.[3]

There is a slight problem in making a similar substitution for p_G, since p_G is not a partial pressure *just outside* the solid surface. Deal and Grove surmount this problem by defining a *virtual* concentration of the oxidant *just within* the solid surface, N^*, such that N_0 would become N^* if p_S were to

[3] Even if this is not so, we can assume that p_S is N_0/η where η is some proportionality constant (or "answer coefficient") that may be determined experimentally. Practically, its separate evaluation is unnecessary. See after Eq. (5.5-8).

become equal to p_G.[4] Thus, we can write

$$p_G = \frac{N^*}{H} \tag{5.5-7}$$

Using these last two relationships we finally can write for FFL in the gaseous oxidant feed region

$$\mathscr{F}_G = h(N^* - N_0) \tag{5.5-8}$$

where

$$h = \frac{D_G}{x_G kTH}$$

This last quantity often is called the *solid-phase, mass transfer constant.* Note that h absorbs four parameters that consequently do not need individual evaluation; one determination of h suffices.

5.5.2 Oxide Region

Figure 5–9(b) shows the conditions assumed in the oxide. The oxide concentrations N_0 and N_i are defined at the *oxide* sides of the two interfaces that lie at $x = 0$ and $x_{ox}(t)$. A linear oxidant profile is assumed between those two values, but they both change with time as the oxide grows thicker.

Applying FFL to the *oxidant* flux density in the oxide yields

$$\begin{aligned} \mathscr{F}_{ox} &= -D_{ox}\frac{\partial N(x)}{\partial x} = -D_{ox}\frac{N_0 - N_i}{0 - x_{ox}(t)} \\ &= +\frac{D_{ox}}{x_{ox}(t)}(N_0 - N_i) \end{aligned} \tag{5.5-9}$$

where D_{ox} is the diffusivity of the *oxidant* in the oxide. The equation is cast in terms of concentrations within the oxide and thus is compatible with Eq. (5.5-8).

---

[4] As a practical matter, in thermal oxidation N^* will be near the solubility limit of the oxidant in SiO_2 at the diffusion temperature. Deal and Grove cite these solubilities in SiO_2 at 1000°C:

$$\text{for } O_2 \quad 5.2 \times 10^{16}, \quad \text{for } H_2O \quad 3 \times 10^{19}\ \text{cm}^{-3}$$

These values will be very important in a later discussion.

5.5.3 Oxide/Silicon Interface

The equation for the flux density of the oxidant across the oxide/silicon interface at $x = x_{ox}(t)$ is a simple statement that its flow across the boundary is limited by the supply source of concentration N_i. Thus

$$\mathscr{F}_i = k_S N_i \tag{5.5-10}$$

where k_S is a chemical reaction rate constant.

5.5.4 Steady State Equations

Equations for the three flux densities are in hand, and in the steady state they are all equal. Considerable algebraic manipulation yields

$$N_i = \frac{N^*}{\left(1 + \frac{k_S}{h}\right) + \frac{k_S}{D_{ox}} x_{ox}(t)} \tag{5.5-11}$$

and

$$N_0 = \left[1 + \left(\frac{k_S}{D_{ox}}\right) x_{ox}(t)\right] N_i \tag{5.5-12}$$

Some insight into the oxidation process may be had from these equations. Say that time increases, so then $x_{ox}(t)$ must increase. Equation (5.5-11) shows that N_i will then decrease, and from Eq. (5.5-10), $\mathscr{F}_i$ also will decrease, as will the two other flux densities (they are all identical in the steady state), so the growth process is self-limiting—the oxide growth rate slows down as the oxide grows.

By differentiating Eq. (5.5-12) it is easy to show that $\partial N_0/\partial x$ is positive; therefore, with increasing time N_0 increases toward the limiting value of N^*. Taken at face value, this statement apparently leads to a paradox: if as time goes on N_0 increases and N_i decreases, the oxidant gradient in the oxide would seem to increase, causing flux density to increase. This contradicts earlier results, but remember that the *gradient* varies inversely with $x_{ox}(t)$, which also is increasing; thus, the flux density will indeed decrease as predicted earlier.

The mathematical model thus far does not include time explicitly, so it does not yield the oxide growth rate. That quantity is derived next.

5.5.5 Growth Rate

Derivation of the oxide growth rate equation requires that time be introduced explicitly into the steady state solution. This may be done in the following manner.

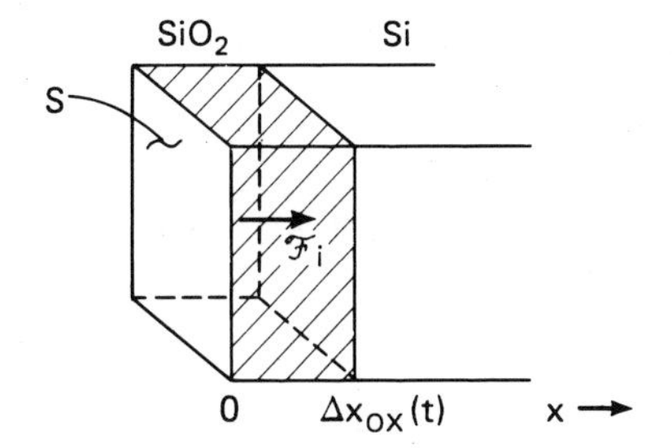

Figure 5–11. An incremental volume of oxide grows in time Δt.

Consider the detail shown in Figure 5–11, where the shaded region is an incremental volume of oxide $S\Delta x_{ox}(t)$ that has growth in time interval Δt. Let S represent the cross-sectional area, and

$$\begin{aligned} N_{ox} &= \text{number of molecules SiO}_2 \text{ per cm}^3 \\ &= 2.3 \times 10^{22}\ \text{cm}^{-3} \end{aligned} \qquad (5.5\text{-}13)$$

Then we may write

$$\mathscr{F}_i S\Delta t = \text{number of oxidant molecules reaching area } S \text{ of the interface in time interval } \Delta t \qquad (5.5\text{-}14)$$

and

$$N_{ox} S\Delta x_{ox}(t) = \text{number of oxidant molecules incorporated into the incremental volume grown in time } \Delta t \qquad (5.5\text{-}15)$$

Assume that all of the oxidant arriving at the SiO_2/Si interface is consumed in growing the oxide, and equate these two quantities. The elemental volume cross-sectional area S cancels out, and we may solve for $\Delta x_{ox}(t)/\Delta t$. In taking the limit $\Delta t \to 0$ we get

$$\frac{dx_{ox}(t)}{dt} = \frac{\mathscr{F}_i}{N_{ox}} = k_S \frac{N_i}{N_{ox}} \qquad (5.5\text{-}16)$$

Equation (5.5-11) allows the elimination of N_i, yielding

$$\frac{dx_{ox}(t)}{dt} = \frac{k_S N^*/N_{ox}}{\left(1 + \dfrac{k_S}{h}\right) + \dfrac{k_S}{D_{ox}} x_{ox}(t)} \qquad (5.5\text{-}17)$$

If numerator and denominator are multiplied by $2D_{ox}/k_S$, the equation may be reduced to

$$\frac{dx_{ox}(t)}{dt} = \frac{B}{A + 2x_{ox}(t)} \tag{5.5-18}$$

where

$$B = 2D_{ox}\frac{N^*}{N_{ox}} \tag{5.5-19}$$

and

$$A = \frac{2D_{ox}}{k_S}\left(1 + \frac{k_S}{h}\right) \tag{5.5-20}$$

This equation contains the required growth rate information in derivative form. To find x_{ox} directly, we separate the variables and integrate.[5] Recall that an initial oxide thickness x_i at $t = 0$ has been assumed. Thus

$$\int_{x_i}^{x_{ox}} [A + 2x_{ox}(t)]\, dx_{ox} = \int_0^t B\, dt \tag{5.5-21}$$

or

$$(Ax_{ox} + x_{ox}^2) - (Ax_i + x_i^2) = Bt \tag{5.5-22}$$

It is convenient to eliminate x_i in terms of an equivalent time τ such that if the oxide were grown under the present conditions starting from zero thickness, the oxide would be x_i thick at time τ. We can do this by noting that dimensional homogeneity must prevail in Eq. (5.5-22); hence, we define τ such that

$$B\tau = (Ax_i + x_i^2) \tag{5.5-23}$$

and rewrite Eq. (5.5-22) as

$$x_{ox}^2 + Ax_{ox} - B(t + \tau) = 0 \tag{5.5-24}$$

Then, solving by the quadratic formula and taking only the admissible root, we finally have for the growth rate equation

$$x_{ox} = \frac{A}{2}\left[-1 + \sqrt{1 + \frac{4B(t + \tau)}{A^2}}\right] \tag{5.5-25}$$

[5] Having established that $x_{ox}(t)$ is indeed time dependent, we use x_{ox} to indicate the final oxide thickness at the end of the oxidation cycle.

Note that the original model invoked a number of parameters—viz, D_{ox}, N^*, k_S, D_G, x_G, and T—and that all these have been reduced to only two parameters, A and B (plus T), that need be evaluated from experimental data for the particular oxidant, temperature, and host being used.

The general form of this equation is shown in Figure 5–12. Using the dimensionless, normalized variables $x_{ox}(t)/(A/2)$ and $4B(t + \tau)/A^2$ and logarithmic scales allows a single plot suitable for covering more than one oxidant and a wide range of temperatures.

Note that although logarithmic scales are used on both axes in Figure 5–12, the scale calibrations are *not* equal; one decade on the horizontal axis has the same length as a half-decade on the vertical axis. This means that a line of mathematical slope 1/2 plots at 45° on the coordinates, and is an artifice to allow presentation of data over wider ranges.

Deal and Grove have presented experimental data for checking their model over a temperature range from 700 to 1300°C and for both oxygen and water vapor as oxidants. They have plotted the data on curves like those of Figure 5–12 (except they use a still different scale ratio) and the agreement of experiment and theory is excellent [11, 12].

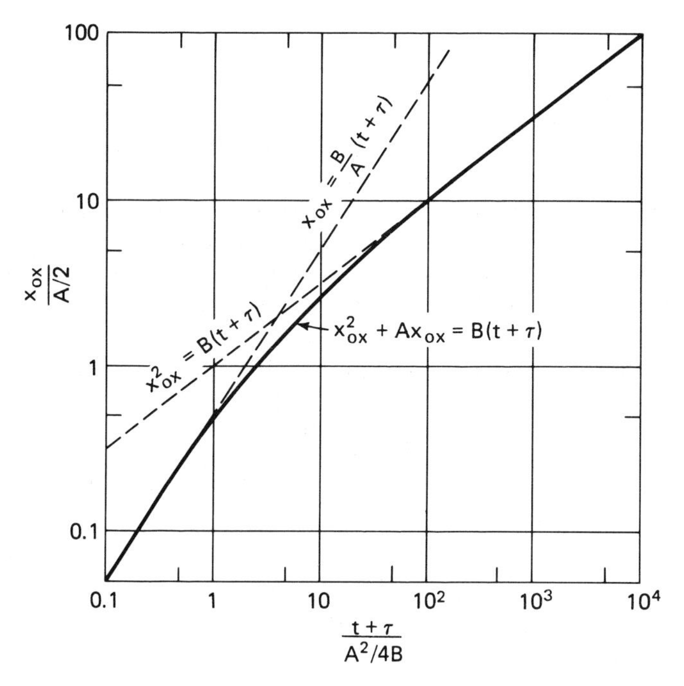

Figure 5–12. Plot of Eq. (5.5-25) and its asymptotes. Note the unequal scales. After Deal and Grove [11]. (Courtesy of the *Journal of Applied Physics.*)

5.5.6 Asymptotic Solutions

Figure 5–12 also shows the short-time and long-time asymptotes of Eq. (5.5-25). These warrant further study. (Remember that they appear not to have slopes of 1 and 1/2 on the unequal coordinate scales of the figure.)

1. Short-Time or Linear Asymptote: This condition is defined by

$$\frac{4B(t+\tau)}{A^2} \ll 1$$

Recall the binomial expansion

$$(1+u)^{1/2} = 1 + \frac{u}{2} - \frac{u^2}{8} + \cdots$$

Then if $u \ll 1$, $(1+u)^{1/2} \approx 1 + u/2$. Applying this approximation to Eq. (5.5-25) we get the short-time asymptote

$$x_{ox}(t) = \frac{A}{2}\left[-1 + 1 + \frac{2B(t+\tau)}{A^2}\right]$$

$$\approx \frac{B}{A}(t+\tau) \qquad \text{(short time)} \qquad (5.5\text{-}26)$$

Since this defines a linear oxide buildup range, $B/A = D$ may be defined as the *linear growth rate constant*. Practically, linear growth can be observed only when an unoxidized silicon sample is undergoing initial oxidation in, say, air at room temperature. This can be rationalized as follows: From Eqs. (5.5-19) and (5.5-20) we observe that the forefactor B/A is independent of D_{ox}. This implies that the oxide is so thin that it does not impede the flow of oxidant to silicon.

2. Long-Time or Parabolic Asymptote: This condition is defined by

$$\sqrt{\frac{4B(t+\tau)}{A^2}} \gg 1$$

for which Eq. (5.5-25) becomes

$$x_{ox}(t) \approx \frac{A}{2}\sqrt{\frac{4B(t+\tau)}{A^2}}$$

$$\approx \sqrt{B(t+\tau)} \qquad \text{(long time)} \qquad (5.5\text{-}27)$$

and the parameter B is defined as the *parabolic growth rate constant*.

Many silicon oxidation cycles fall near or within this parabolic range. The factor B is proportional to D_{ox}, so the oxide is present in sufficient thickness to affect the flow of oxidant to the silicon.

5.5.7 Silicon Consumption

The basic oxidation reactions, Eqs. (5.4-1) and (5.4-2), show that silicon is consumed in the oxidation process of forming SiO_2. This silicon is furnished by the substrate or any other silicon-bearing material [for example, silicon nitride (Si_3N_4)] upon which the oxide grows. The ratio of oxide volume grown to the volume of solid material, in this case silicon, consumed in the growth is known as the Pilling-Bedworth ratio (PBR). Since cross-sectional area is the same for both materials, we can write

$$\text{PBR} = \frac{\text{thickness of SiO}_2\text{ grown}}{\text{thickness of Si consumed}} \tag{5.5-28}$$

The numerical value for SiO_2/Si is 2.2; if 2000 Å of oxide are grown, the silicon under the oxide is thinned by 909 Å. Usually, this substrate thinning is not shown on device cross section sketches as in Figure 1–1, because, in part, of relative magnitudes. A typical bipolar emitter junction might lie 2 μm below the substrate surface. If 2000 Å of oxide are grown, the 909 Å thinning is difficult to show because of its small size. For MOS and LSI (large scale integration) devices that have shallower junctions, however, the silicon loss can become significant.

A second reason for not showing the thinning is to simplify already complex diagrams. The oxidation of Si_3N_4 is discussed later; it, too, is consumed in being converted to SiO_2.

5.6 Oxidation Plots

A number of plots of experimental data that relate x_{ox} to t are available in the literature [2, 12]. Some examples are shown in Appendices F.5.1 through F.5.6 where the units are:

$$x_{ox}\,(\mu\text{m}), \qquad t\,(\text{min}), \qquad T = T_c + 273\,(\text{K}), \qquad T_c\,(^\circ\text{C})$$

Other units may be encountered on occasion, however. These curves were calculated from parameters listed in Appendix F.5.7 and are traceable to measured data.

Published oxidation curves show considerable differences, even though they all are based on experimental data. One reason is that the accuracy of measuring oxide thickness has improved. There is a lesson to be learned here:

do not expect to read three-significant-figure data from the curves ("micrometer eyeballs" not withstanding), it is not justified. Consider a few features of these curves.

5.6.1 Oxidants

Published oxidation curves usually are drawn for three principal oxidants as shown in Appendices F.5.1 to F.5.3: dry oxygen, wet oxygen, and steam. It is important to distinguish among them. Consider the schematic oxidation system of Figure 5–13. The bubbler water must be clean (Section 5.3.7).

If the petcocks P_L and P_R are in position 1, dry oxygen is fed to the furnace, giving *dry oxidation*. With the petcocks in position 2, supply oxygen passes through the *bubbler* and picks up water vapor that is carried to the furnace. If T_L is 95 to 97°C, *wet oxidation* results; if the bubbler water boils, *steam oxidation* takes place. The U-tube and petcock in the water vapor line in Figure 5–13 serve as a trap and drain for condensed water vapor.

Water vapor pressure increases as T_L is raised, so we expect steam to have higher values of p_G and so of N^* than wet oxygen. The resulting oxide

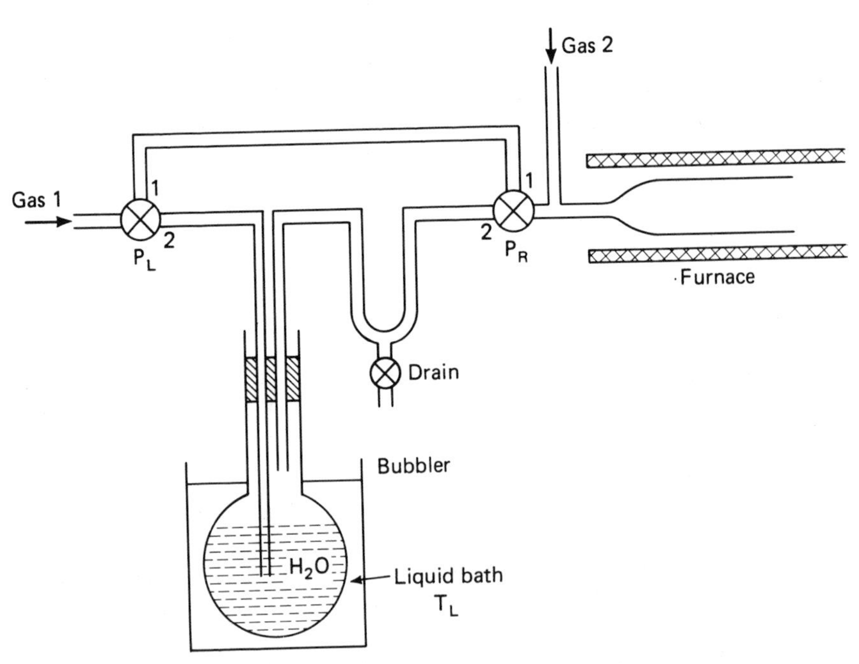

Figure 5–13. A generalized oxidant supply system for furnishing dry oxygen, wet oxygen, or steam.

thicknesses for a given time and temperature are so nearly equal for steam and wet oxidation, however, that the former is seldom used now (Appendices F.5.2 and F.5.3).

Comparison of Appendices F.5.1 and F.5.2 shows that water vapor provides oxide growth rates spectacularly higher than those of dry oxygen. This requires some explanation because the data cited before Eq. (5.4-4) and in Appendix F.6.1 show that $D_{O_2} > D_{H_2O}$ in SiO_2, which appears contradictory.

For typical oxidation times, where the growth is parabolic, x_{OX} is proportional to $B^{1/2}$. From Eq. (5.5-19) we see that $B \propto D_{ox}N^*$. Thus the $D_{ox}N^*$ *product*, rather than just D_{ox} alone, must be considered. Typically, N^* approaches the solubility limit of the particular oxidant in SiO_2. Deal and Grove data predict that $N^*_{H_2O} \approx 3000\ N^*_{O_2}$ at 1000°C. Thus, the smaller value of D_{H_2O} is completely overridden by the orders-of-magnitude-larger value of $N^*_{H_2O}$, so the buildup of oxide in wet oxidation can be faster than with dry oxidation.

The numbers given before Eq. (5.4-4) and in Appendix F.6.1 show that hydrogen, which is the by-product when H_2O oxidizes silicon [Eq. (5.4-2)], does not pile up at the oxide/silicon interface since it outdiffuses back toward the gas region faster than the water vapor diffuses in.

5.6.2 Silicon Orientation

In Chapter 2 we saw that a clean (111) surface has more dangling bonds than a clean (100) surface of silicon; therefore, the (111) face should oxidize faster than the (100), other things being equal. Appendix F.5.1 shows this to be the case. The effect shows up more for thin oxides. The effect decreases as x_{ox} increases because the difference in the number of dangling bonds, (111) versus (100), at the silicon surface becomes negligible, as the bonds have had ample opportunity to bond with oxygens during the oxidation process. Most of the older published oxidation curves neglect this effect and presumably are plotted for (111) wafers.

5.6.3 Oxidant Gas Pressure

Comparison of Appendices F.5.1 and F.5.4 shows that the oxidant bulk gas pressure p_G has a marked influence on the oxide growth rate, at least for the dry oxidation case. This is predictable from the oxidation model and is consistent with results in the following sections. We shall see that both rate constants D and B are directly proportional to pressure; hence, we expect that an increase in p_G will give an increase in x_{ox} for a given temperature, time, and silicon orientation. This effect is exploited in HiPOx (see Section 5.8.3).

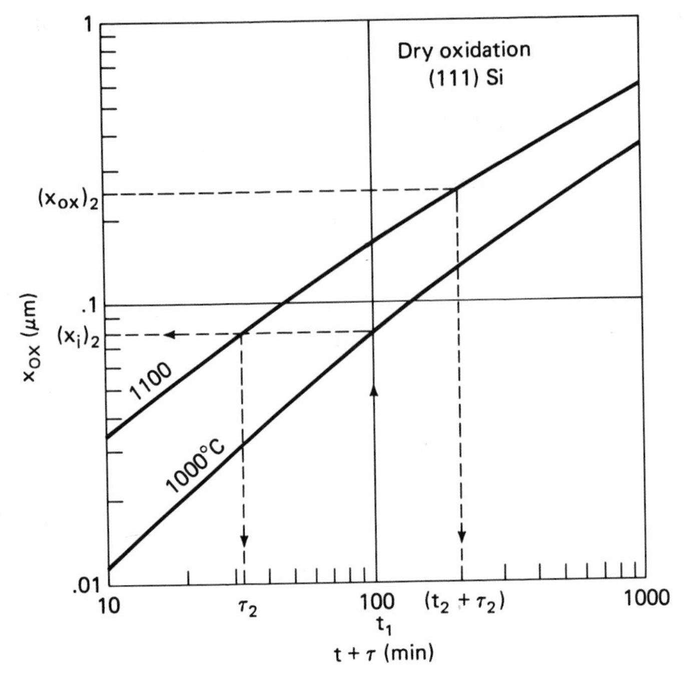

Figure 5–14. Dry oxidation curves for the illustrative example in the text.

5.6.4 Graphical Calculations

The use of oxidation curves for typical oxidation-time calculations is illustrated in the following example.

An unoxidized, (111) silicon wafer is subjected to a double dry-oxidation cycle. The first is for 100 minutes at 1000°C. The second is at 1100°C for a time t_2 to give a final total oxide thickness of 2500 Å. Find t_2. Follow the computations on Figure 5–14.

Consider the first cycle with a duration of $t_1 = 100$ min at 1000°C. Since the sample is unoxidized, $(x_i)_1$ and also τ_1 are identically zero. Enter the curves at 100 min and find the vertical intersection with the 1000°C curve. The point has an ordinate of $(x_{ox})_1 = 0.08\ \mu m = 800$ Å. This becomes the initial thickness $(x_i)_2$ for the second cycle.

For the second cycle, enter the chart at $(x_i)_2$ and find the horizontal intersection with the 1100°C curve. This corresponds to an abscissa value of 34 min and is τ_2, the time that would have been required at 1100°C to grow $(x_i)_2$. (Note the effect of T on t_1 and τ_2.)

Enter the chart at 2500 Å $= 0.25\ \mu m = (x_{ox})_2$, the final total thickness. Moving horizontally to the intersection with the 1100°C curve, read the corresponding abscissa value of $(t_2 + \tau_2) = 208$ min. Thus

$$t_2 = (t_2 + \tau_2) - \tau_2 = 174 \text{ min} = 2 \text{ h } 54 \text{ min}$$

Note, in passing, that dry oxidation is a rather slow process. Also, we shall see in Section 5.7 that problems of this sort may be solved by equations, if computers or proper pocket calculators are available.

Invariably, oxidation curves in the literature label the abscissa as "time" or simply (t). Actually, as we have seen in the foregoing example, the time axis really is ($t + \tau$). You will avoid confusion if you remember this distinction.

5.7 Computer Calculations

It is much easier to perform oxidation calculations by solving Eq. (5.5-25) with a computer or calculator. These computations require the rate constants B and $D = B/A$, which must be extracted from experimental plots. Given these values, the curves serve little purpose. Next, consider how the B and D values may be determined from experimental data. We assume that data are measured at the nominal bulk oxidant pressure of 1 atmosphere; this restriction will be removed later. Data should be available for both (111) and (100) silicon faces, for different oxidants, and for a number of different temperatures.

5.7.1 Determination of B

The long-time asymptote is given by Eq. (5.5-27). Since this equation contains B alone as a parameter, it may be used to evaluate B from data. Taking logs yields

$$\log x_{ox} = \tfrac{1}{2}\log B + \tfrac{1}{2}\log(t + \tau) \tag{5.7-1}$$

This equation is valid on the linear, long-time asymptote. Then B may be evaluated by linear regression or by slope/intercept methods. Data of B versus T may be accumulated by applying either procedure to several T curves.

Next, B must be separated into a constant and a temperature dependent factor. From Equations (5.5-19) and (5.5-7)

$$B = \frac{2D_{ox}Hp_G}{N_{ox}} \tag{5.7-2}$$

where H is Henry's constant, which is condition dependent, and p_G is the partial pressure of the gas-phase bulk oxidant source, which is assumed for the moment to be 1 atm.

In Chapter 6 we shall learn that the temperature dependence of diffusivity is exponential and may be expressed in the form

$$D_{ox} = D_{\infty}10^{-e/kT}$$

where e is some activation energy, and k is Boltzmann's constant. Henry's constant also has a temperature dependence, so we group the factors and finally write B as

$$B = C_1 10^{-E_1/kT} \tag{5.7-3}$$

that is, the temperature dependence of B may be handled by the two parameters C_1 and E_1, the latter being an equivalent activation energy. [Note that other forms may be used; see, for example, Eq. (5.4-4).]

Taking logs again, we get

$$\log B = \log C_1 - \frac{E_1}{kT} \tag{5.7-4}$$

Then linear regression applied to the B versus T data allows extraction of C_1 and E_1 for a given set of oxidation conditions such as p_G, silicon orientation, and type of oxidant. Remember that p_G is assumed to be 1 atm.

5.7.2 Determination of *A*

The parameter A may be determined in a similar manner by using the short-time asymptotic solution, Eq. (5.5-26). Define $D = B/A$ for convenience. The reduction of D into two factors closely follows the procedure we used for B. It may be shown that

$$D = \frac{B}{A} = \frac{Hp_G}{N_{ox}} \frac{k_S}{(1 + k_S/h)} = C_2 10^{-E_2/kT} \tag{5.7-5}$$

Notice that an exponential dependence on temperature is present, but E_2 will be greater than E_1 since D_{ox} is absent. The parameters C_2 and E_2 may be extracted by linear regression, D may be evaluated as before, and A may be determined from $A = B/D$.

Data are needed for the same conditions as before: dry oxygen, wet oxygen, and steam on (111) and (100) silicon faces. Thus far, all data were assumed to be taken at atmospheric pressure. This is easy to extend. Foregoing equations show that both D and B are directly proportional to p_G; therefore, if the original data were measured at 1 atm. as specified, the original D and B values need only be multiplied by the new p_G value in *atmospheres*. This theoretical means of obtaining rate constants for high p_G may give slightly different values from measurements. Other methods are considered in Section 5.8.3.

5.7.3 Formulary

The appropriate equations and parameters for oxide calculations are as follows:

$$x_{ox} = \frac{A}{2}\left(-1 + \sqrt{1 + \frac{4B(t + \tau)}{A^2}}\right) \tag{5.5-25}$$

$$t = \frac{x_{ox}}{B}(x_{ox} + A) - \tau \tag{5.5-24}$$

$$\tau = \frac{x_i}{B}(x_i + A) \tag{5.5-23}$$

$$B = C_1 p 10^{-E_1/kT} \tag{5.7-3}$$

$$D = \frac{B}{A} = C_2 p 10^{-E_2/kT} \tag{5.7-5}$$

Notice that the oxygen pressure p *in atmospheres* is included in the definitions of the A and B parameters.

Several programs have been devised for oxidation calculations on a computer or a programmable hand-held calculator with considerable memory. Two examples are SUPREM II and PRIDE [13, 14]. Both of these programs handle much more than just oxidation calculations, however. A simple program of this type in MBASIC is listed in Appendix F.5.7 for a pressure of 1 atm. The parameters for calculating B and D of the formulary are from these two programs and are listed on lines 110 to 150, and 190. Appendix F.5.7 allows calculation either of times to get a given oxide thickness, or thickness for a given time and other operating conditions. It is interesting to verify the values given in earlier graphical examples by means of such a program.

5.8 Special Oxidations

Sometimes the oxidation furnace tube is fed gases other than those already discussed, or at pressures much higher than 1 atm. We shall consider three such situations: (a) chlorine-bearing gas species, (b) nitrogen ambient, and (c) dry or wet oxidation at elevated pressures (HiPOx).

5.8.1 Chlorine-Doped Oxides

The incorporation of chlorine into silicon dioxide during its growth gives several advantages over standard oxides [15, 16]. Adding chlorine

1. reduces the incorporation of cations from the furnace tube into the growing oxide. Probably Cl^- combines with the cations to form volatile compounds that leave the tube as effluent.
2. stabilizes MOS threshold voltages (Section 5.3.7). The Cl^- chlorine ion can *getter* the cations by making them immobile in the oxide.
3. stabilizes dielectric breakdown voltage.
4. reduces the density of surface states on the oxide side of the SiO_2/Si interface, which are due in part to the excess Si^{2+} in the nonstoichiometric SiO_x in that region. Apparently Cl^- neutralizes some of this positive charge.
5. cleans the furnace tube.
6. greatly enhances the oxide growth rate.

The presence of chlorine has negligible effect on the density, permittivity, and index of refraction of the oxide.

Early work on incorporating chlorine into oxides was done by feeding dry oxygen and either chlorine or vapor-phase HCl simultaneously into the oxidation furnace tube. It was determined that Cl *atoms* were the active species in bringing about favorable results. Chlorine is extremely toxic and HCl vapor in the presence of water vapor is highly corrosive, attacking even stainless steel gas supply lines and exhaust stacks of the oxidation system. It does enhance the oxide growth rate, however; curves showing typical effects on oxidation are given in Appendix F.5.5.

A less corrosive chlorine source is trichloroethylene (TCE). This chemical is a liquid at room temperature and so is easier to handle, even though its vapor is very hazardous to humans. It is available commercially in small sealed containers in hundreds-of-cm^3 sizes and in "4-9s-5" (i.e., 99.995%) purity [17].

The plumbing for the TCE System is shown in Figure 5–15. The bubbler flask, which holds the TCE, is immersed in a water bath held at a constant temperature T_L, usually around room value to minimize condensation of TCE vapor in the furnace feed line. At a typical T_L of 26°C (78.8°F), the TCE vapor pressure is roughly 90 torr, so the volume of TCE vapor entering the furnace tube and hence the oxidation rate are determined by the nitrogen flow rate (Appendix D.3.5).

Aside from ease of handling, TCE also produces less by-products than both Cl_2 and HCl, yet its effects on the oxide are much the same except for a somewhat smaller growth rate. Typically, at 1100°C with flow rates of 50 cm^3/min of N_2 and 500 cm^3/min for O_2 the TCE gave an oxide roughly 45% thicker than dry O_2 alone [18].

A by-product of TCE oxidation is CO_2. If insufficient oxygen is present, a black carbon deposit can form at the effluent end of the furnace tube. This

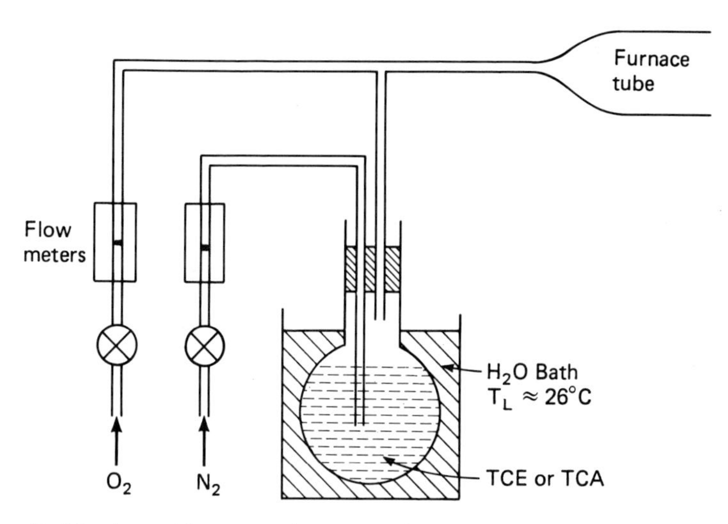

Figure 5–15. An oxidant supply system for growing chlorine-doped oxides using either TCE or TCA.

may be eliminated by substituting 1,1,1–trichloroethane (TCA) for TCE as the chlorine source. It is believed that the single bonds in CH_3CCl_3 (TCA) rather than the double bond in $ClCH{:}CCl_2$ is the reason for this.

Of the three chlorine sources considered for oxide doping, TCA is the one of choice for safety reasons. Chlorine gas and TCE are more hazardous and TCE also is carcenogenic.

5.8.2 Oxide Anneal

The glass $a\mathrm{SiO}_2$ has a number of properties that affect semiconductor devices. Principal ones are the specific surface charge (Q_{SS}, $\mathrm{C/cm^2}$) at the SiO_2/Si interface, dielectric breakdown voltage (BV, V/μm), density (ρ, $\mathrm{g/cm^3}$), permittivity (ε, F/cm)—and in some cases, index of refraction (n, numeric).

The nominal density value for silica glass is 2.2 $\mathrm{g/cm^3}$, but Deal cites variations from 2.05 for steam oxidation at 1200°C to 2.27 for dry oxidation at 1000°C [19]. He also shows that dielectric breakdown increases linearly with increasing density. Apparently, fast growth (wet versus dry, high versus low temperature) usually makes for a less dense silica network due to oxygen deficiencies, OH^- radicals, and the like. As a general rule, if oxidation is followed by an annealing step where the silicon wafers remain in the oxidation furnace, but with N_2 or Ar replacing the usual oxidant gas or gases, the density increases. This is the *densification* process referred to earlier. Such treatment tends to produce densities at the higher end of the range near 2.2 $\mathrm{g/cm^3}$, and

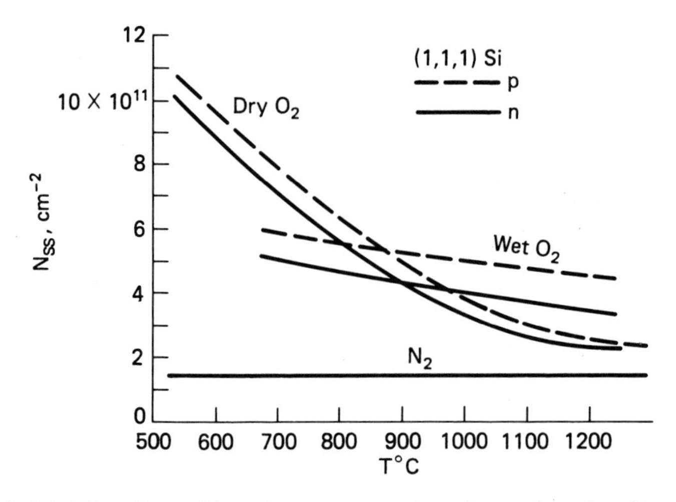

Figure 5–16. The effect of dry nitrogen anneal on the surface density of states. After Deal et al. [20]. (Reprinted by permission of the publisher, The Electrochemical Society, Inc.)

a dielectric breakdown voltage of 600 V/μm is readily obtained in commercial practice [19].[6]

The post-oxidation anneal process in dry N_2 or Ar also has a marked effect on reducing the surface density of states N_{SS} ($= Q_{SS}/q$, cm^{-2}, where q = electronic charge magnitude) which is of particular importance in MOS devices. Figure 5–16 shows the effect of nitrogen in this regard. Raising the oxidation temperature in both wet and dry oxidations lowers N_{SS} [20]. A nitrogen anneal after oxidation further reduces N_{SS} greatly and the effect is independent of the anneal temperature, as long as that remains in the oxidation range. The anneal is performed in the oxidation furnace but with the gases changed, and can be quite short, say 10 to 15 min.

In current MOS fabrication, where minimum N_{SS} is very important, Ar has replaced N_2 as annealing ambient. (1) It acts more efficiently in the annealing process. (2) It gives significantly lower background contamination when furnished from a cryogenic (i.e., low-temperature) source, the common practice. (3) Argon also eliminates a significant problem of N_2 anneal—namely, that N_2 diffuses through the oxide layer during anneal and forms a nitride layer at the oxide/silicon interface, causing N_{SS} to increases with further annealing. In processing bipolar devices, where Q_{SS} is not of such great importance, wet

[6] The breakdown voltage must be measured with the metal electrode, which is in contact with the oxide under test, biased so that the semiconductor surface is in accumulation. This means, for example, that the *metal electrode* is negative for dielectric breakdown testing of oxides grown on p-type semiconductors.

or steam oxidations often are followed by a dry O_2, rather than a dry N_2, anneal. Even if the short anneal is carried out in dry oxygen at the oxidation temperature, there will be insignificant additional oxide growth because of the large difference in wet and dry rates.

5.8.3 HiPOx

We have seen that oxide growth rate increases with pressure of the oxidant. We now consider two practical applications of the Hi Pressure Oxidation, or HiPOx, technique [21, 22].

Raising pressure during oxidation allows use of a lower temperature for a given x_{ox} and $(t + \tau)$. A convenient rule of thumb is that an *increase* of 1 atm. (=1.013 Pa = 760 torr) allows a *decrease* of about 30°C in temperature [21]. This is verified by the following example.

From Appendix F.5.4 we read these data:

x_{ox}, μm	$(t + \tau)$, min	p, atm.	T, °C
0.2	40	10	1000
0.2	40	1	1260

The pressure *change* is $\Delta p = -9$ atm., and the temperature change is $\Delta T =$ 260°C.

$$\frac{\Delta T}{\Delta p} = -\frac{260}{9} = -28.9\text{°C/atm.}$$

This checks well with the value of −30°C given by the rule.

This effect has important applications in fabrication when a thermal reoxidation step is used after dopants already have been introduced into the silicon substrate.

Any post-introduction, high-temperature operation such as reoxidation will cause further diffusion of the dopants. Diffusivity increases exponentially with temperature; hence, lowering the reoxidation temperature reduces the post-introduction diffusion effect. This gives much better control of junction depths, and, in fact, is essential in modern VLSI (very large scale integration) processing, where the very small devices have correspondingly shallow junction depths of less than one micron.

Another commercially useful effect of HiPOx is the large decrease in oxidation time at higher pressure for a given x_{ox} and temperature. For example, we read from Appendix F.5.4 that for $x_{ox} = 0.2$ μm, dry oxidation, and T = 1000°C, 390 min (6.5 h) are required at 1 atm., while only 20 min are needed at 20 atm. Even further time reductions can be obtained by using wet or steam oxidation at elevated pressures. The commercial importance here is that the

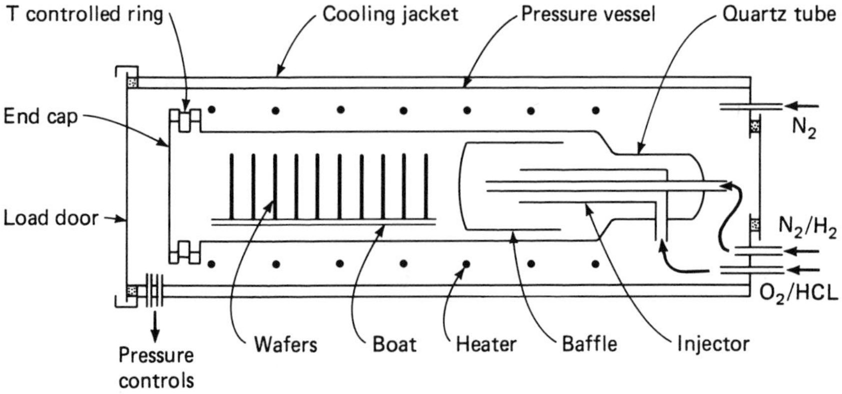

Figure 5–17. A HiPOx oxidation furnace used by GaSonics. After Levinthal [21]. (Courtesy of *Semiconductor International Magazine.*)

throughput (wafers handled per hour) of a single oxidation furnace is increased tremendously; one HiPOx furnace may replace several conventional furnaces.

These advantages are bought with greater cost for a furnace built for high-pressure operation. Commercial units often are designed for safe operation at 10 to 25 atm. There are practical limits to the pressure value. Safety measures aside, a limitation arises because of etching, at least with steam as the oxidant. The oxidant begins to *etch* silicon if temperature is above 1000°C and pressure is above 150 atm. [2]. Fortunately, this is well above the usual HiPOx range.

Figure 5–17 shows the cross section of a typical high-pressure setup, although several variations are in use. The outer pressure vessel must withstand the desired pressure, usually with blowout plugs provided for safety. The pressure differential between inside and outside of the "quartz" tube must be positive to prevent contaminants from reaching the wafers, yet small enough so the tube will not break. Computer control of the gas flow, pressure, and heating elements is standard, as are provisions for both wet and dry oxidation.

5.9 Oxidation of Silicon Nitride

The use of silicon nitride (Si_3N_4) in semiconductor devices and ICs has increased in recent years. Usually deposited by chemical vapor deposition (CVD), the nitride has several uses: e.g., (1) as a dielectric in MOS processing because its relative permittivity ε_r is 7.5 as against 3.9 for SiO_2, and (2) as a mask during thermal oxidation of Si. Regions of silicon covered by a sufficiently thick nitride layer do not oxidize—i.e., they are masked, thereby permitting selective oxidation. We digress briefly on the subject of dielectric isolation to see a use for this application of Si_3N_4.

5.9.1 Dielectric Isolation

Junction isolation (JI), whereby back-to-back p–n junctions provide isolation between adjacent devices on an IC chip, was discussed in Section 1.2. The extra junction to the substrate results in relatively high output and interdevice capacitances. Of the two components of these capacitances, that due to the tub sidewalls is much greater than that on the tub bottom. This comes about because the isolation diffusion, required for JI, causes high doping along the tub sidewalls, whereas the substrate doping under the tub is much smaller. This makes the depletion widths of the sidewalls smaller than of the bottom, causing the sidewall capacitance per unit area to be correspondingly higher.

An alternative is *dielectric isolation* (DI), which nearly eliminates both capacitive components by completely surrounding each tub, bottom and sides, by an SiO_2 layer much thicker than the depletion-layer width in JI. Fabrication of a DI wafer is more expensive because of processing complexity, however, and results in a rather fragile wafer.

The basic steps for producing DI on a wafer are shown in Figure 5–18(a). The starting material is monocrystalline n-type silicon, shown at (1) in the figure. Conventional photolithography produces troughs, t, in the upper surface as in (2), and that surface is oxidized, (3). Then a thick layer of silicon is deposited on the oxide layer by CVD to give the result shown at (4). The deposited layer will be polySi because the underlying $a\text{SiO}_2$ cannot act as a *seed* for monocrystal Si growth. The polySi must be several *mils* thick since it eventually will serve as the "substrate" of the finished water.

The wafer then is *inverted* and the original monocrystalline Si is ground off and polished to the level of the arrows. The final wafer, *inverted again*, is shown at (5). The tubs of original n-type, single-crystal substrate lie between the etched trough regions of (2) and are completely surrounded on sides and bottom by SiO_2, a dielectric (or insulator). This gives rise to the *dielectric isolation* designation. Devices will be fabricated in the tubs, just as they would be for a JI type of wafer. Since they are insulated from the polySi, the latter's type is of no concern.

The DI process provides better isolation than JI, but adds fabrication expense; most of the original wafer, prepared with high energy expenditure, is ground away.

A good compromise between JI and DI results when only the high sidewall capacitance of JI is eliminated by replacing the highly doped, isolation regions of JI by a region of thick SiO_2, called ROx or *recessed oxide*, recessed into the silicon. This reduces the larger capacitance component and costs less to fabricate than conventional DI isolation. The ROx isolation concept is used, for example, in the Fairchild ISOPLANAR and Motorola LOCOS processes [30]. The use of silicon nitride as a mask during the growth of the ROx regions is an essential part of these techniques.

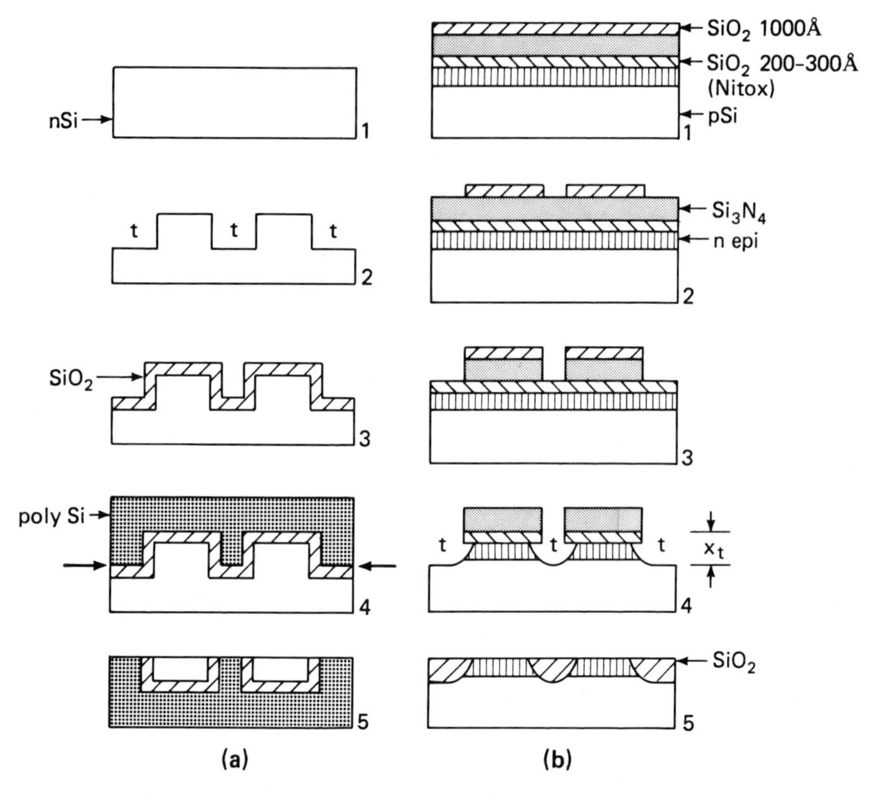

Figure 5–18. Two forms of isolation that use SiO_2 as the isolating material. (a) Dielectric Isolation (DI) where each tub is surrounded by SiO_2 on all sides and the bottom. (b) Recessed Oxide (ROx) Isolation. Each tub is surrounded by SiO_2 on the sides only. JI is used on the bottom. Part (b) after Bassous, Yu, and Maniscalco [23]. (Reprinted by permission of the publisher, The Electrochemical Society, Inc.)

5.9.2 Nitride Masking

The idealized sketches of Figure 5–18(b) show the principal steps in the ROx process [23]. A p-type monosilicon substrate with an n-epi layer is thermally oxidized to a thickness of some 200 to 300 Å. This oxide often is called *Nitox* because it will underlie a layer of silicon nitride that is deposited next by CVD. (The Nitox serves a masking function to be described shortly.) Another layer of SiO_2, about 1000 Å thick, is deposited by CVD on top of the nitride. Both CVD processes are typically run at 800°C. The silicon and its three layers of insulators are illustrated in Figure 5–18(b)(1).

Conventional photolithography and buffered HF are used to etch trenches in the top oxide to define islands as shown at (2) in the figure. Buffered HF does not attack the nitride. The islands are masks for the next step.

Next, the nitride under the troughs is removed by etching in boiling H_3PO_4 (orthophosphoric acid). Since SiO_2 is immune to this acid, the thin Nitox layer protects the underlying Si [Figure 5–18(b)(3)].

A buffered HF etch is used again to remove all the remaining, exposed oxide—top and Nitox layers—but the nitride is unaffected. Then, troughs are etched deeply in the Si with HNO_3/HF to a trench depth x_t of, say, 1.1 μm. (The islands are protected by the nitride.) Note that the etch must go completely through the n-epi layer. The result is shown in Figure 5–18(4), where slight undercutting of the Nitox should be noted. Undercutting can be minimized by using an anisotropic etch and a (100) substrate (Chapter 10).

Next, the wafer is subjected to a long, wet oxidation so that the troughs, t, are oxidized. This will typically require several hours at temperatures in the 1100 to 1150°C range. Values may be determined from the curves of Appendices F.5.2 or F.5.3. Note that the nitride protects the silicon islands during this long oxidation step, and that the thick trough oxide is literally recessed into the silicon and so gives rise to the recessed oxide or ROx notation.

Two features are important here. (1) More silicon will be consumed in the troughs to support the thick oxide growth there. (2) Some of the nitride also will be converted to SiO_2. This second effect is our principal concern here, since it sets a minimum nitride thickness required to protect the silicon islands during the long oxidation.

As a last step the nitride and Nitox layers are etched selectively to leave the wafer with the ideal cross section shown in Figure 5–18(b)(5). If all processing has been done perfectly and if the thicknesses of the several layers have been chosen properly, the top surface of the silicon and oxide will be essentially flat. Notice that none of the silicon wafer was ground away and discarded as in the DI process; the original substrate remains as substrate. The figure shows how the tubs are isolated from each other along the sidewalls by ROx (i.e., by DI), while the tub bottoms are junction isolated between the n-epi tubs and the underlying p-substrate.

An anomalous structural feature, known as the *bird's head and beak*, may arise in practice because of lateral oxidation under the edges of the nitride mask. Consider this. The sketches of Figure 5–18(b) are idealized: they do not show all undercutting resulting from etching operations. A more realistic representation is shown in Figure 5–19 where the cross section in (a) shows the wafer in the region of an island just before the ROx is grown. During this oxidation the growth proceeds vertically, and laterally under the edges of the nitride mask layer; thus, the masking afforded by the nitride is not perfect at its edges. The resulting lateral oxide growth strains the nitride layer, which, if it is thin enough, literally will be bent upward near its edges, as shown in Figure 5–19(b) [23].

Imaginative study of the figure shows that the oxide layer under the edges of the nitride resembles a bird's head and beak. After removal of the nitride

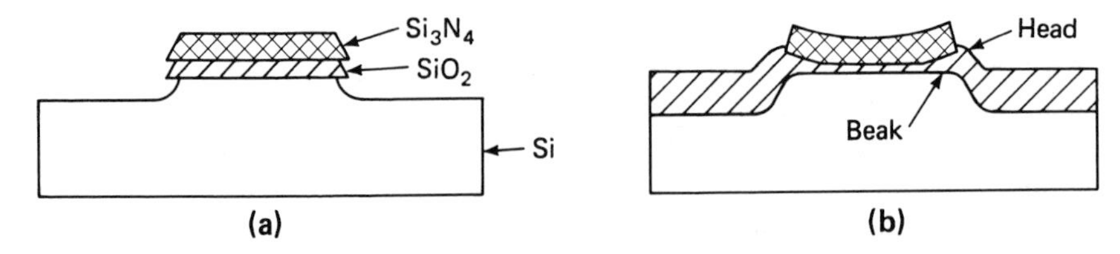

Figure 5–19. Growth of the bird's head and beak during ROx growth. (a) Layers before oxidation showing undercutting from etching. (b) Undercutting of the nitride layer due to lateral oxidation. After Bassous, Yu, and Maniscalco [23]. (Reprinted by permission of the publisher, The Electrochemical Society, Inc.)

layer once its masking function is completed, the head and beak cause irregularities of the oxide upper surface (which ideally should be flat). The bird's beak reduces the effective size of the tub where the device subsequently will be fabricated. Compensation for this loss of area requires a larger initial island size, that finally translates into a lower packing density in the complete integrated circuit.

The size of the bird's head and beak decreases as (1) the nitride thickness increases, (2) the Nitox thickness decreases, and (3) the ROx thickness decreases. The student may argue out the reasons for these trends.

5.9.3 Nitride Oxidation

We have seen that Si_3N_4 may be used to mask Si against thermal oxidation. A critical feature of this procedure is that some of the nitride itself is consumed and converted to SiO_2 during oxidation; we must know to what extent in order to calculate the minimum nitride thickness required to provide adequate masking.

The basic oxidation processes of Si_3N_4 are, for dry oxidation,

$$Si_3N_4\downarrow + 3\,O_2\uparrow \rightarrow 3\,SiO_2\downarrow + 2\,N_2\uparrow \tag{5.9-1}$$

and for wet oxidation, theoretical possibilities

$$Si_3N_4\downarrow + 6\,H_2O\uparrow \rightarrow 3\,SiO_2\downarrow + 6\,H_2\uparrow + 2\,N_2\uparrow \tag{5.9-2}$$

or

$$Si_3N_4\downarrow + 6\,H_2O\uparrow \rightarrow 3\,SiO_2\downarrow + 4\,NH_3\uparrow \tag{5.9-3}$$

The first is preferred on thermodynamic grounds [24].

In all three equations, one molecule of Si_3N_4 is converted into three molecules of SiO_2; hence, the Pilling-Bedworth ratio (PBR) is the same for all three reactions. If we interpret the definition of the PBR from Section 5.5.7 in terms of nitride, rather than silicon, consumed, we have

$$\text{PBR} = \frac{\text{thickness of } SiO_2 \text{ grown}}{\text{thickness of } Si_3N_4 \text{ consumed}} = 2.01$$

This implies that the thickness of consumed nitride is very close to one half of the resulting oxide thickness.

Because of the thick oxide layers required in ROx, wet oxidation is invariably used and the oxidation times will be long, running into several hours, as may be verified by Appendix F.5.6. The oxidation curves for the nitride are plotted in the lower portion of the graph while corresponding oxidation curves for (111) silicon are shown in the upper part for easy comparison.[7] A significant point is that silicon oxidizes at least 15 times faster than silicon nitride.

A dashed line of slope 1/2 is drawn in the lower right-hand corner of Appendix F.5.6. This is the slope anticipated by the Deal and Grove oxidation model for the parabolic domain. Note that the long-time portions of the curves for both media at 1000 and 1100°C conform very well to this model. Above 700 min, however, the 900°C nitride curve shows a definite saturation effect that is not predicted by the model. Consider some typical numbers.

Say 2 μm of recessed oxide are desired. What nitride thickness is required on the islands to provide adequate masking? What silicon trench depth, x_t, is required so that the top surface will be flat?

Read from Appendix F.5.6 that 2 μm of oxide may be grown on (111) Si at 1 atmosphere and 1100°C in 480 min (8 h). Under the same conditions, nitride will convert to 0.11 μm of SiO_2.

Since the PBR for SiO_2/Si_3N_4 is 2.01, this means that $0.11/2.01 = 0.055$ μm of nitride are consumed in the oxidation process. Thus, the initial nitride must be at least this thick to provide adequate masking for the underlying Si island. Usually a 2 : 1 safety factor is allowed, however, so a nitride thickness of 1100 to 1200 Å would be used.

In calculating the silicon trench depth before oxidation we recall that the depth of silicon consumed in oxide growth is $x_{ox}/\text{PBR}_{SiO_2}$. Thus, the required

[7] The nitride curves are based on data from Fränz and Langheinrich [24]. The silicon curves are computed from Appendix F.5.7.

Published curves of x_{ox} versus time for silicon nitride show wide variations. Compare, for example, Glaser and Subak-Sharpe, Figure 6–7 [25]. See also [26]. Differences are due, at least in part, to difficulties in measuring thicknesses of the oxide and the underlying nitride layer, both of which are transparent. Fränz and Langheinrich [24] describe the method used in obtaining their data.

silicon trench depth is

$$x_t = x_{ox}(1 - 1/PBR_{SiO_2})$$

where $PBR_{SiO_2} = 2.2$. Therefore,

$$x_t = 2[1 - (1/2.2)] = 1.09 \ \mu m$$

It should be evident that the long (8-h) recessed oxidation time could be reduced by using a higher temperature or the HiPOx technique.

5.10 Thickness Determination

Oxide thickness must be measured during processing. Some of the methods discussed in Section 3.14 for determining junction depth also can be used to measure oxide (or nitride) film thickness. We now consider two methods that are applicable only to transparent films.

5.10.1 Color Comparison

The simplest method of determining the thickness of transparent films involves observing the apparent film color when it is viewed in white light. Because of destructive interference, one wavelength will not be reflected. When viewed by the eye, the reflected light will be the incident white light minus the nonreflected component, and will have a color roughly corresponding to the *complement* of the rejected wave.[8] Consider the equations involved in this phenomenon.

Destructive optical interference is considered in Appendix A.4 with the defining physical arrangement shown in Figure A–6 for a glass/air interface, but here the oxide replaces the air layer. The condition for destructive interference is that the total phase difference between the two reflected paths, namely $(\psi + \pi)$, must be an *odd* multiple of π radians.

In the current situation the oxide (or nitride) has a *higher* index of refraction than the air, so reflection at the upper surface takes place without a phase shift. The difference in path length is the same as before, and the condition for destructive interference is that ψ alone must equal an odd multiple of π radians. Then, substituting for ψ we have

$$\frac{4tn\cos\phi}{\lambda_a}\pi = (2m + 1)\pi \qquad (5.10\text{-}1)$$

where t is the thickness of the layer in question.

---

[8] Two colors of light that add together to give white light are said to be *complementary* colors—e.g., red and blue-green (cyan), blue and yellow, green and purple, and orange and blue-violet.

If the sample is illuminated and viewed normal to the surface, $\phi = 0$, and the wavelength in air, λ_a, that is eliminated by destructive interference is

$$\lambda_a = \frac{4nt}{2m + 1}, \qquad m = 0, 1, 2, 3, \ldots \tag{5.10-2}$$

where n is the refractive index of the oxide (or nitride). Admissible values of m are those that make λ_a lie in the visible spectrum (i.e., in the range of roughly 4000 to 7000 Å). The colors of the visible spectrum are listed in [6].

Consider an illustrative example. An SiO_2 layer 0.3 μm (=3000 Å) thick is viewed vertically in air under normal incident white light. (Daylight fluorescent lamps usually are used.) What color will it be? The refractive index of SiO_2 is 1.5, correct to two significant figures.

By Eq. (5.10-2) the wavelength rejected by destructive interference is

$$\lambda_a = \frac{4(1.5)(0.3)}{2m + 1} = 0.6\ \mu\text{m} \qquad \text{for } m = 1$$

(Note that $m = 1$ is the only admissible value that makes λ_a lie in the visible range.) A 0.6-μm wavelength corresponds to an orange hue. The complement of orange is blue-violet, so this is the apparent color of the sample. This checks Appendix F.5.8.

By simple extension of the idea illustrated in the example, *color charts*, based on experimental observation of thickness versus color, have been prepared—e.g., Appendix F.5.8 [27].

Color judgement varies considerably among different observers, and as Appendix F.5.8 shows, a given color spans quite a range of wavelengths. Also, the fact that destructive interference may only attenuate, and not completely eliminate, the λ_a component makes color judgement difficult. A chart of standard colors, such as prepared by the National Bureau of Standards, to which the sample may be *matched* simplifies the problem. The best way to eliminate ambiguity, however, is to have actual samples of oxidized silicon wafers where the oxide thicknesses have been determined independently. Their use simplifies accurate color matching.

Appendix F.5.8 shows that certain colors, such as "carnation pink," may correspond to different values of oxide thickness. This indicates the different admissible values of m in Equation (5.10-2). The oxidation history of the sample gives a rough estimate of t, however, that resolves the ambiguity.

A color chart for silicon nitride thickness also is available; however, a simple artifice allows the oxide chart of Appendix F.5.8 to be applied to silicon nitride [28]. Equation (5.10-2) shows that the rejected wavelength is directly proportional to the index of refraction, n, of the transparent layer. Thus we may set up the ratio

$$\frac{(\lambda_a)_{\text{ni}}}{(\lambda_a)_{\text{ox}}} = \frac{n_{\text{ni}} t_{\text{ni}}}{n_{\text{ox}} t_{\text{ox}}}$$

where subscript "ni" refers to the nitride. Then if a nitride and an oxide sample have the same apparent color, presumably they have the same λ_a, whence

$$t_{\text{ni}} = t_{\text{ox}}(n_{\text{ox}}/n_{\text{ni}}) \tag{5.10-3}$$

The indices of refraction, rounded off to two significant figures, are $n_{\text{ox}} \approx 1.5$ and $n_{\text{ni}} \approx 2.0$. Thus for the same color we have $t_{\text{ni}} \approx 0.75 t_{\text{ox}}$, so Appendix F.5.8 may be used for Si_3N_4 if table values of thickness are multiplied by 0.75.

5.10.2 Ellipsometry

Ellipsometry is another method for measuring the thickness of transparent films. The method senses the state of polarization of an elliptically polarized light beam, incident at an angle ϕ, as it is reflected from the film.

A plane electromagnetic (EM) wave has electric and magnetic component waves that are normal to each other and to the direction of propagation. The *direction of polarization* is defined as that of the electric field component. In the general case a light beam comprises two component plane waves, one parallel to the *plane of incidence* and of magnitude and phase M_p and β_p, respectively, and the second normal to the plane of incidence and characterized by M_n and β_n.[9] In the general case of elliptical polarization $M_p \neq M_n$ and $\beta_p \neq \beta_n$.

If such a beam arrives at a transparent film of thickness t and refractive index n situated on an absorbing substrate (e.g., silicon) the reflected wave consists of many components that result from refraction, and multiple reflections at both surfaces of the film being measured. This is illustrated in Figure 5–20. Consequently, there are changes in the magnitudes and phases of the two component plane waves. The total effect may be summarized by two parameters ψ and Δ that are defined in terms of the incident (i) and reflected (r) components [29].

$$\tan\psi = \left(\frac{M_p}{M_n}\right)_r \left(\frac{M_n}{M_p}\right)_i \tag{5.10-4}$$

and

$$\Delta = (\beta_p - \beta_n)_r - (\beta_p - \beta_n)_i \tag{5.10-5}$$

[9] The *plane of incidence* is that plane containing the direction of the incident EM wave and the normal to the reflecting surface. In Figure 5–20 it is the plane in which the angle of incidence, ϕ, is shown.

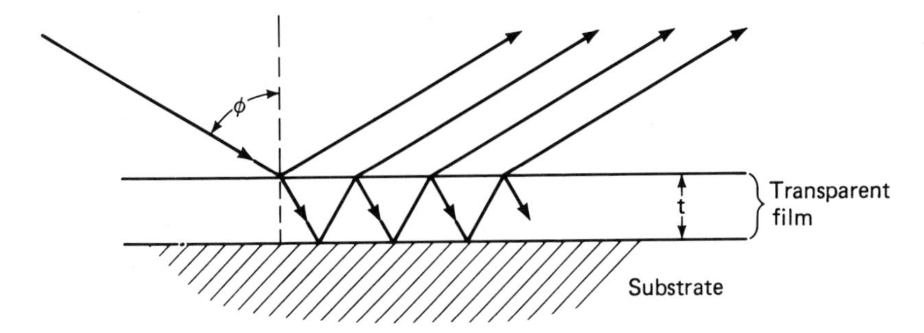

Figure 5–20. Multiple refractions and reflections at a thin film on a substrate. The incident ray arrives at angle ϕ relative to the normal.

Thus ψ and Δ give the effect of reflection on the state of the polarized wave, and are measured with the ellipsometer. Note from Figure 5–20 that optical properties of the *substrate*, as well as those of the film, must be taken into account since the reflected wave depends upon the multiple reflections at the film/substrate interface. In the case of a silicon substrate, which is of primary interest here, the properties are known and may be included in the calculations for the *film* t and n from ψ and Δ.

One basic form of the ellipsometer is shown in Figure 5–21. A collimated monochromatic light source is passed through a *polarizer* and *wave plate*. The polarizer produces an ordinary, linearly polarized wave and the wave plate generates an additional extraordinary polarized component shifted by roughly 90° from the ordinary wave. These two plane wave components produce an elliptically polarized wave that is incident at angle ϕ to the film under test. If the wave plate introduced a shift of exactly 90° it would be a

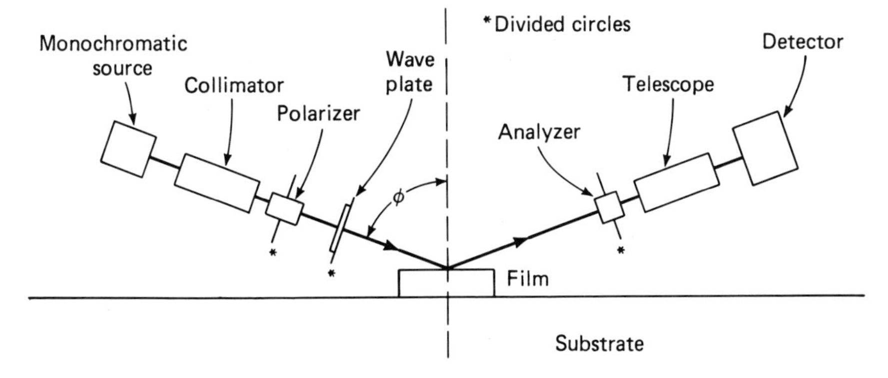

Figure 5–21. Physical layout of an ellipsometer. After Archer [29]. (Courtesy of the *Journal of the Optical Society of America*.)

quarter-wave plate. In practice the shift should be nearly, but not exactly, equal to 90°. Call it · γ.

The reflected wave passes through a polarizing analyzer to a detector (which may be the human eye, but a photodetector is preferable). The wave plate and the two polarizing units are mounted on divided circles so that their angles may be determined to within, say, 0.01°. Angles are measured with respect to a coordinate system in which the z-axis is the direction of the wave, and the z–x plane is the plane of incidence. The graduated circles are normal to the z-direction and the zero angle reference is the plane of incidence.

In use, the wave plate has its axis set at about 45° and the polarizer and analyzer angles, P and A, are adjusted to values (say, P_0 and A_0) that produce extinction of the beam at the detector. Two sets of P_0 and A_0 values may be found, but the following equations assume that P_0 is confined to the first quadrant. Next, ψ and Δ may be determined from P_0 and A_0 by these equations:

$$\tan\Delta = \sin\gamma\tan(90° - 2\mathrm{P}_0) \tag{5.10-6}$$

$$\tan\psi = \cot L\tan(-\mathrm{A}_0) \tag{5.10-7}$$

where

$$\cos 2L = -\cos\gamma\cos 2\mathrm{P}_0 \tag{5.10-8}$$

Next, the film parameters must be determined from ψ and Δ. Archer outlines the derivation and shows typical results for transparent films on silicon [29]. These results are reproduced in Figure 5–22 for a silicon substrate, a 5461-Å source, and an angle of incidence $\phi = 70.00°$. The chart reproduced here really lacks sufficient accuracy for practical use, so computer solutions are used; but it does illustrate salient features of the solution.

First, each solid line is a locus of film thicknesses for a film of specific refractive index n, which is shown as an underlined number. On each such locus the film thickness t is specified in terms of a related angle δ given in degrees where

$$\delta = \frac{360}{\lambda}t\sqrt{n^2 - \sin^2\phi}, \text{ degrees} \tag{5.10-9}$$

For the specified conditions this reduces to

$$t = \frac{15.17\,\delta}{\sqrt{n^2 - 0.8830}}, \text{ Å} \tag{5.10-10}$$

Physically, δ is the phase change (in degrees) of light with wavelength λ passing through a thickness t of the film whose refractive index is n. Notice

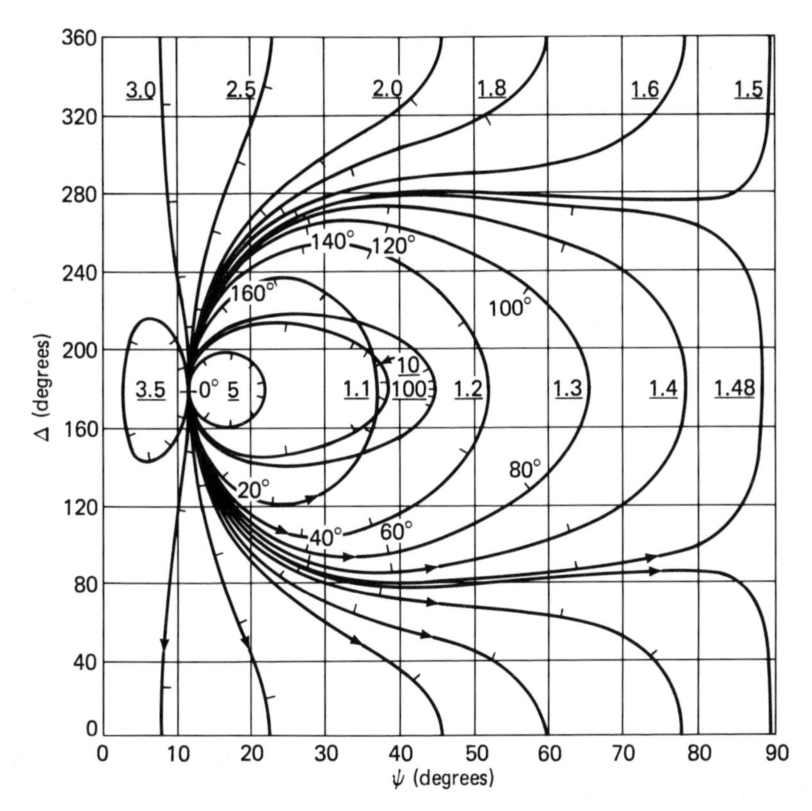

Figure 5–22. Loci of constant refraction index n of the film in the Δ–ψ plane. The film thickness is determined from δ in degrees on the loci. See Eq. (5.10-10). Underlined numbers are values of n. Δ and ψ are determined from ellipsometer readings. Archer [29]. (Courtesy of the *Journal of the Optical Society of America.*)

that increasing film thickness corresponds to increasing δ and hence to counterclockwise motion on a given locus of fixed n.

Second, for practical transparent films and thicknesses, a data pair (ψ, Δ) is uniquely related to a set of film properties (t, n). For required accuracies, computer solutions or tables rather than the curves are used for the transformation.

Use of the method of ellipsometry is illustrated in the following example.

A thin film on a silicon substrate is measured with the ellipsometer shown in Figure 5–21 with $\phi = 70°$ and $\lambda = 5461$ Å. With analyzer and polarizer adjusted for extinction of the detected beam, it is found that $P_0 = 83.0°$ and $A_0 = -62.2°$. $\gamma = 88°$. Find n and t of the film.

We first find ψ and Δ. By Eq. (5.10-8)

$$\cos 2L = -\cos\gamma\cos 2P_0 = -\cos 88\cos 166 = 0.03386$$

or

$$L = 44^\circ$$

By Eq. (5.10-7)

$$\begin{aligned}\psi &= \arctan[\cot L\tan(-A_0)]\\ &= \arctan[\cot 44\tan 62.2] = 63^\circ\end{aligned}$$

By Eq. (5.10-6)

$$\begin{aligned}\Delta &= \arctan[\sin\gamma\tan(90 - 2P_0)]\\ &= \arctan[\sin 88\tan(-76)] = -76 \rightarrow 104^\circ\end{aligned}$$

Reading from Figure 5–22, we have $n = 1.4$ and $\delta = 80^\circ$. By Eq. (5.10-10)

$$t = \frac{15.17\,\delta}{\sqrt{n^2 - 0.883}} = \frac{15.17(80)}{\sqrt{1.4^2 - 0.883}} \approx 1170\ \text{Å}$$

This example shows how tedious it is to convert from a (ψ, Δ) pair to the corresponding (n, δ) pair by using Figure 5–22. Note, also, that accuracy is limited by reading values from the figure so solution by computer is indicated.

Finally, there is the periodicity problem in that δ repeats its cycle around a given n contour of Figure 5–22 in periods of 180°. Thus, for the numbers of the illustrative example, the contour is repeated in thickness multiples of

$$t = \frac{15.17(180)}{\sqrt{1.4^2 - 0.883}} = 2631\ \text{Å}$$

The significance is that the actual film thickness must be known within a multiple of this value. A similar problem was encountered in using the color chart to determine film thickness.

Problems

5–1 (a) By counting atoms "owned" by the pyramid in Figure 5–4(a), verify the formula B_2O_3.

(b) Repeat for P_2O_5 using Figure 5–5(a).

5–2 (a) Derive an equation for the Pilling-Bedworth ratio for the SiO_2/Si system in terms of densities and molecular weights.
(b) Calculate the PBR given these data:

Material	Molecular Weight	Density, g/cm^3
SiO_2	60.08	2.18
Si	28.09	2.33

5–3 (a) Discuss the advantages of the unequal scales in Figure 5–12 for plotting Eq. (5.5-25) over a wide temperature range. Assume that people like square log–log sheets.
(b) If you wished to plot the function $y = ax^2$, where x ranges over many decades, what relative scales would be good for a log–log plot.

5–4 (a) Using the appropriate curves, determine and tabulate the time in minutes required to oxidize bare Si to thicknesses of (1) 1000 Å, and (2) 2000 Å at 1000°C for these oxidants: dry O_2, wet O_2 (95°C H_2O), and steam.
(b) Verify your results using Deal and Grove's equations for wet O_2 and 2000 Å.

5–5 (a) If a Si wafer has an initial oxide thickness of 0.20 μm, how long will it take to grow an additional oxide thickness of 0.10 μm in wet O_2 at 1200°C? A graphical solution is acceptable. Verify by computer calculation.
(b) Deal and Grove's model shows that for long oxidation times $x_{ox} = \sqrt{B(t + \tau)}$, where B is a function of T and the oxidant. One point on the curves for dry O_2 oxidation of Si at 1200°C is: $x_{ox} = 0.30$ μm at time $(t + \tau) = 150$ min. Calculate B for these conditions.
(c) Check your B value using numerical data.

5–6 An SiO_2 layer is 0.25 μm thick.

(a) Calculate λ_a in the visible range.
(b) What is the complementary color?
(c) Check your result against Appendix F.5.8 and explain any discrepancies.
(d) An Si_3N_4 film appears carnation pink when viewed under normal incident white light. What is its thickness? What method did you use?

5–7 Readings are made on a nitride film with an ellipsometer for which $\gamma = 88.0°$ and $\phi = 70.0°$. These values are read: $A_0 = -29.25°$, $P_0 = 75.0°$. Calculate n and t for the film.

5–8 Explain the polarity requirements described in footnote 6, Section 5.8.2.

5–9 Given an oxidized silicon wafer, a capacitance meter, and a mercury probe (Section 3.13.3), explain how to measure the oxide thickness. What d-c voltage polarity should be applied? Why? What assumption must be made about the oxide?

5–10 A 1-μm oxide is to be grown on Si_3N_4 at 1 atm. and 1200°C. What oxidation time is required?

References

[1] R. B. Sosman, "The Phases of Silica." *Am. Ceram. Soc. Bull.*, **43**, (3), 213, March 1964. For greater detail, see R. B. Sosman, *The Phases of Silica*. New Brunswick: Rutgers University, 1965.

[2] R. P. Donovan, "Oxidation." In *Fundamentals of Integrated Device Technology*, ed. R. M. Burger and R. P. Donovan. Englewood Cliffs: Prentice-Hall, 1967.

[3] W. G. Moffatt, G. W. Pearsall, and J. Wulff, *The Structure and Properties of Materials. Vol. 1*. New York: Wiley, 1964. Chap. 5.

[4] R. J. Charles, "The Nature of Glasses." *Sci. Am.*, **217**, (3), 126–136, Sept. 1967.

[5] *Handbook of Tables for Applied Engineering Science, 2nd Ed.*, ed. R. E. Bolz and G. L. Tuve. Cleveland: CRC Press, 1973.

[6] *Handbook of Chemistry and Physics, 51st Ed.*, ed. R. C. Weast. Cleveland: The Chemical Rubber Co., 1966.

[7] L. H. Van Vlack, *Physical Ceramics for Engineers*. Reading: Addison-Wesley, 1964.

[8] V. V. Tarasov, *Physics of Glass* (translation). Jerusalem: Israel Program for Scientific Translation, 1963.

[9] J. T. Baker Chemical Co. catalog data. Also available is an "ultra pure reagent grade" that assays Na at 0.02 ppb (parts per billion or ng/g) in glacial acetic acid!

[10] M. M. Atalla, "Semiconductor Surfaces and Films: The Silicon-Silicon Dioxide System." In *Properties of Elemental and Compound Semiconductors, Vol. 5*, ed. H. Gatos, pp. 163–181. New York: Interscience, 1960.

[11] B. E. Deal and A. S. Grove, "General Relationship for the Thermal Oxidation of Silicon," *J. Appl. Phys.*, **36**, (12), 3770–3778, Dec. 1965.

[12] A. S. Grove, *Physics and Technology of Semiconductor Devices*. New York: Wiley, 1967.

[13] D. T. Antoniadis, S. E. Hansen, and R. W. Dutton, "SUPREM II—A Program for IC Modeling and Simulation," *Technical Report No. 5019-2, Integrated Circuits Laboratory*. Stanford: Stanford University, June 1978. Numerical data here are from the SUPREM manual, and are used with permission.

[14] J. R. Pfiester, "PRIDE—Portable Process and Device Design," Stanford: Integrated Circuits Laboratory, Stanford University, Aug. 1981. PRIDE was developed specifically for the HP-41CV calculator. The entire program, covering many semiconductor calculations, was available in a plug-in ROM module.

[15] R. J. Kriegler, Y. C. Cheng, and D. R. Colton, "The Effect of HCl and Cl_2 on the Thermal Oxidation of Silicon," *J. Electrochem. Soc.*, **119**, (3), 388–392, March 1972.

[16] D. W. Hess and B. E. Deal, "Kinetics of Thermal Oxidation of Silicon in O_2/HCl Mixtures," *J. Electrochem. Soc.*, **124**, (5), 735–739, May 1977. This paper has an interesting discussion regading the effect of silicon orientation on the growth rate constants *B* and *D*.

[17] R. S. Clark, "Thermal Oxidation of Silicon Using Trichloroethylene," *Solid State Technol.*, **21**, (11), 80–82, Nov. 1978.

[18] J. R. Flynn, "Trichloroethylene Oxidation of Silicon," MS Thesis, EE Department, University of Illinois-UC, Dec. 1979.

[19] B. E. Deal, "The Oxidation of Silicon in Dry Oxygen, Wet Oxygen, and Steam," *J. Electrochem. Soc.*, **110**, (6), 527–532, June 1963.
[20] B. E. Deal et al., "Characteristics of Surface State Charge (Q_{SS}) of Thermally Oxidized Silicon," *J. Electrochem. Soc.*, **114**, (3), 266–273, March 1967. The nitrogen anneal data are in part from A. S. Grove [12].
[21] D. J. Levinthal, "Diffusion System Trends," *Semicond. Int.*, **2**, (5), 31–41, June 1979.
[22] W. A. Brown, "High-Pressure Oxidation," *Hewlett-Packard J.*, **33**, (8), 34–36, Aug. 1982. This paper gives considerable data on the *D* and *B* rate constants as a function of pressure.
[23] E. Bassous, H. N. Yu, and V. Maniscalco, "Topology of Silicon Structures with Recessed SiO_2," *J. Electrochem. Soc.*, **123**, (11), 1729–1737, Nov. 1976. The authors also discuss reasons for using the Nitox layer to protect the underlying Si surface.
[24] I. Fränz and W. Langheinrich, "Conversion of Silicon Nitride into Silicon Dioxide Through the Influence of Oxygen," *Solid State Electron.*, **14**, (6), 499–505, June 1971.
[25] A. B. Glaser and G. E. Subak-Sharpe, *Integrated Circuit Engineering*. Reading: Addison-Wesley, 1979.
[26] *Semiconductor Technology Handbook*. Portola Valley: Technology Associates, 1978, Figure OX-9.
[27] W. A. Pliskin and E. E. Conrad, "Nondestructive Determination of Thickness and Refractive Index of Transparent Films," *IBM J. Res. Dev.*, **8**, (1), 43–51, Jan. 1964.
[28] P. E. Gise and R. Blanchard, *Semiconductor and Integrated Circuit Fabrication Techniques*. Reston: Reston, 1979.
[29] R. J. Archer, "Determination of the Properties of Films on Silicon by the Method of Ellipsometry," *J. Opt. Soc. Am.*, **52**, (9), 970–977, Sept. 1961.
[30] J. A. Appels et al., "Local Oxidation of Silicon and Its Application in Semiconductor-Device Technology," *Philips Res. Repts.*, **25**, (2), 118–132, April 1970.

Chapter 6
Diffusion: Predeposition

Diffusion is a process used for forming p–n junctions in semiconductor devices. A particular atomic species, having a nonuniform distribution and undergoing random thermal motion, suffers a directed motion by diffusion from regions of higher to lower concentration of that species. It is a highly temperature dependent process.

Subject to any boundaries and the direction of the concentration gradient, the diffusing species is free to move in any direction. In fabrication, however, interest usually is confined to motion in the x-direction—i.e., through and normal to the wafer surface, so we use one-dimensional equations for first-order models. Lateral diffusion, parallel to the wafer surface, is either neglected, or, where it is important, handled independently (Section 6.6.2).

The student probably is familiar with the diffusion of holes and electrons, since this is basic to semiconductor device behavior. In device fabrication the diffusion of the common elemental *dopants* in silicon—e.g., boron (B), phosphorus (P), and arsenic (As)—is important. While the law of particle flux density is the same for dopants and for holes and electrons, the equations of continuity for the two types of diffusants differ in one respect. Dopants do not *recombine* as do holes and electrons; hence, continuity equations for dopant atoms have no lifetime terms.

In this chapter we look at some of the mechanisms involved in diffusion, and consider the basic diffusion equations, namely Fick's laws, but in one dimension only. Solutions of these equations that are related to the predeposition (or *predep*) diffusion are considered next.[1] Practical aspects of the predep are considered last.

---

[1] Other solutions for diffusion processes used in device fabrication are considered in Chapter 7. These are applicable to special conditions.

6.1 Diffusion in Monocrystals

Consider diffusion in a single-crystal host material. At typical diffusion temperatures of at least 900°C in Si, the diffusant atoms must move through a regular lattice structure whose atoms are vibrating very rapidly about their mean lattice positions. Despite these vibrations, there is space for the diffusing atoms to move between lattice atoms, but there are potential barriers that must be surmounted. Common dopants in silicon diffuse in two principal ways: interstitially and substitutionally.

6.1.1 Interstitial Diffusion

In interstitial diffusion the diffusing atoms move through the lattice by jumping from one *interstice* [the space between normal, adjacent Si lattice sites—shown by open circles in Figure 6–1(a)] to another adjacent one. The diffusant atom (shaded circle) must have at least an activation energy E_a to squeeze through, or, put more elegantly, to *overcome the potential barrier*. Copper (Cu), gold (Au), and silver (Ag) are interstitial diffusers in single-crystal Si and have activation energies of 1.00, 1.12, and 1.60 eV, respectively. The E_a values for all interstitial diffusers in Si are in the 1 to 2 eV range. Effective dopant atomic radii (r_a) in silicon are given by Wolf [1]. It seems that smaller atoms should squeeze between adjacent lattice atoms more easily, and that is true for these three dopants: E_a increases as r_a increases. This trend is not followed by substitutional diffusers, however, so more is involved than just the atom's size.

After diffusion is completed and the wafer has been returned to room temperature, interstitial dopant atoms remain at the interstices rather than on lattice sites, and so do not act as conventional p- or n-type dopants. They occupy *deep levels* in the forbidden band, as shown in the band diagram of

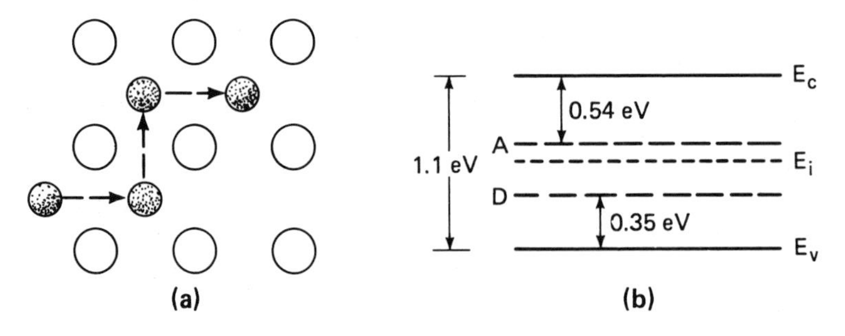

Figure 6–1. Interstitial diffusion. (a) Motion of the dopant atom (shaded) is from interstice to interstice in the silicon (unshaded) lattice. (b) Band diagram for gold in silicon.

Figure 6–1(b), deep level implying that the energy level lies near the *center* of the band gap (E_i) rather than near a band edge, E_c or E_v. Notice that two levels are possible for gold, an A (acceptor) and a D (donor) level.

Gold is the most commonly used interstitial dopant in silicon technology. It is noncorrosive and forms no silicon compounds. Its atoms act as efficient generation/recombination centers, and so can be used to reduce the minority carrier lifetime [2]. In earlier days gold often was diffused into bipolar transistors to reduce switching times, a practice not used today.

6.1.2 Substitutional Diffusion

In substitutional diffusion, a diffusant atom jumps from a lattice site to an adjacent *vacant* lattice site. Thus, two steps are involved for one jump of the atom. (1) A lattice *vacancy* must be created to make room for the dopant atom. Energy is expended in doing this because the atom must break bonds to its neighbors and overcome potential barriers to move to another vacancy or to the wafer surface. (2) The dopant atom also must overcome a potential barrier to move from a lattice site to the vacancy. We expect, then, that the activation energy for substitutional diffusion would be higher than for a typical interstitial event. This is true, as may be seen from Appendix F.6.1, where $E_a > 3$ eV (i.e., much higher than the two-volt maximum for the interstitial types).

Figure 6–2 is a two-dimensional representation of the substitutional diffusion process. In the left of (a) we see a shaded dopant atom (DA) outside the host. Next, a vacancy V_2 is created by a silicon atom moving from the lattice to the surface. Finally the DA moves into that vacancy. Before the DA can move substitutionally in the crystal again, an adjacent vacancy must be created. Say a silicon atom moves into V_1, thereby creating a vacancy V_3. Then, as illustrated by the sketches in Figure 6–2(b), V_4 is created by a silicon atom moving into V_3. Finally, the DA moves into the V_4 location, leaving another vacancy V_5 behind.

In certain situations some of the substitutional diffusant atoms may be lodged temporarily at interstices. If the wafer is *quenched* (i.e., if its temperature is lowered very quickly), these atoms may be frozen into the interstitial locations (Section 4.11.2). In that event, they are not *electronically active* and do not serve as donors or acceptors as the case may be.

Another consequence of thermal quenching is that vacancies can be frozen in the lattice. Subsequently they can act as deep level recombination/generation centers or traps, and affect minority carrier lifetime in a more or less unpredictable manner [2].

The substitutional diffuser atoms listed in Appendix F.6.1 are the usual dopants used in silicon technology. When they are properly located on lattice sites, they contribute shallow levels in the band gap that are located near E_c

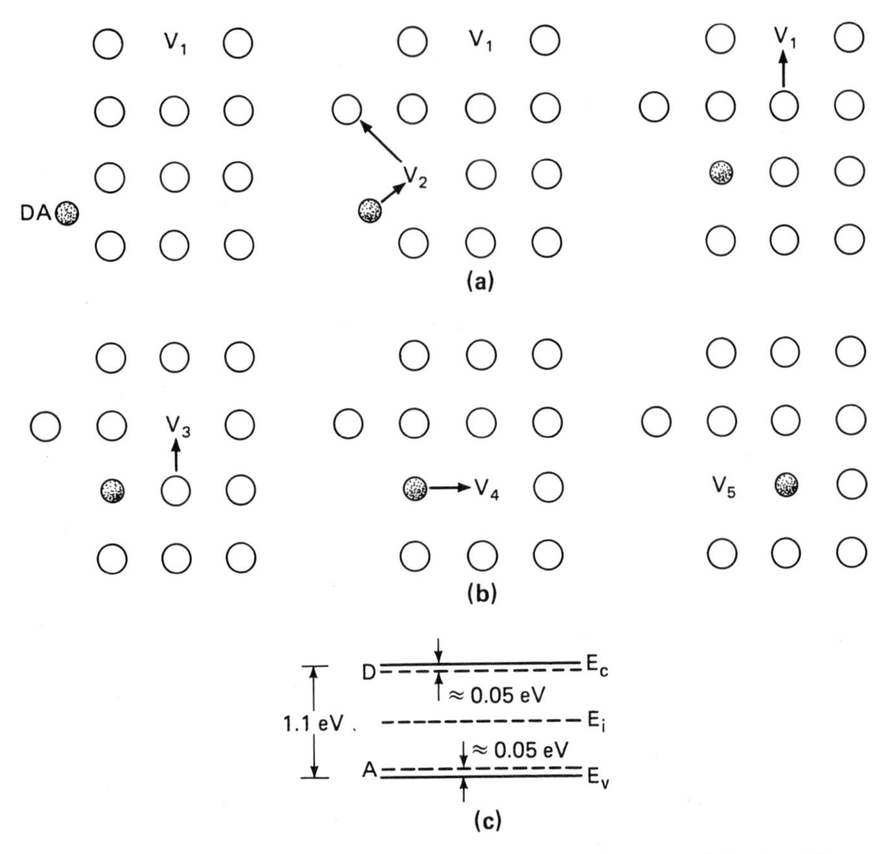

Figure 6–2. Substitutional diffusion. (a) A vacancy is created in the silicon (unshaded) so a dopant (shaded) atom may move into a lattice site, substituting for a silicon atom. (b) Another vacancy is created so the dopant atom moves to another lattice site. (c) Band diagram for substitutional dopants in silicon.

(for donors) or E_v (for acceptors) as shown in Figure 6–2(c). Typical separation from the nearer band edge in a silicon host is roughly 50 meV but differs slightly from dopant to dopant.

6.1.3 Diffusivity and Temperature

The diffusivity D is a key parameter in characterizing the diffusion of a particular atom species in a particular host material. The exponential nature of the diffusivity dependence on absolute temperature is stated in different, but equivalent forms, e.g.,

$$D = 10^{(d_1 - d_2/T)} = D_\infty e^{-b/T} \tag{6.1-1}$$

where $b = E_a/k$, k being Boltzmann's constant $= 8.62 \times 10^{-5}$ eV/K, and temperature T is in Kelvins. Appendix F.6.1 gives D_∞ and b for the common dopants in *monocrystal* Si at *low doping levels.*

The second form is useful in extracting D_∞ and b values from D versus T data, since it serves as the basis of the Arrhenius plot of ln D (or log D) versus $1/T$. Thus, if we take logs in Eq. (6.1-1), we have

$$\log D = \log D_\infty - (b \log e)\frac{1}{T} \tag{6.1-2}$$

Plotted data for this type of exponential dependence may be extrapolated as a straight line on the semilog coordinates to find D_∞; $(b \log e)$ is the slope. Equation (6.1-2) also is the key to extracting E_a and b from data by linear regression. These methods have been discussed in Section 5.7.

It is apparent from Eq. (6.1-1) that

$$10^{d_1} = D_\infty \quad \text{or} \quad d_1 = \log D_\infty \tag{6.1-3}$$

Thus if $D_\infty > 1$, $d_1 > 0$. On the other hand, if $D_\infty < 1$, then $d_1 < 0$. This accounts for the difference in signs in Eqs. (5.4-4) and (6.1-1). For atoms diffusing in *silica* D_∞ is less than one for all the atoms listed in Appendix F.6.1; hence the negative sign has been factored out of the exponent parentheses in the former equation.

Figure 6–3 shows semilog Arrhenius plots of diffusivity versus $1/T$ for the common dopants in silicon. The three interstitial diffusers in the figure show similar slopes, an unsurprising fact since the Arrhenius slope is $-(b \log e)$, and $b = E_a/k$ is nearly the same for these three. Displacement differences among the three curves are due to differences in D_∞ values. The scales in Figure 6–3 were chosen to display many decades of D, but they are difficult to read. Working with reciprocal temperature in Kelvins is also a bother, so a temperature scale in °C is added at the top of the figure, but that scale is nonlinear, making interpolation difficult.

It is easier to read D values in the limited range of diffusion temperatures from a plot of log D versus T°C as in the solid curve of Figure 6–4. Furthermore, many diffusion solutions involve $\sqrt{D}$, so some authors prefer to plot $\log \sqrt{D}$ versus T°C, as in the dashed curve. Note the change in ordinate scales. Practically, it is more satisfactory to calculate diffusivity from D_∞ and b values given in Appendix F.6.1.

Listed diffusivity values for dopants in monocrystal silicon vary considerably in the literature, two principal reasons being that D depends on dopant concentration, and measurement methods. The values given here are

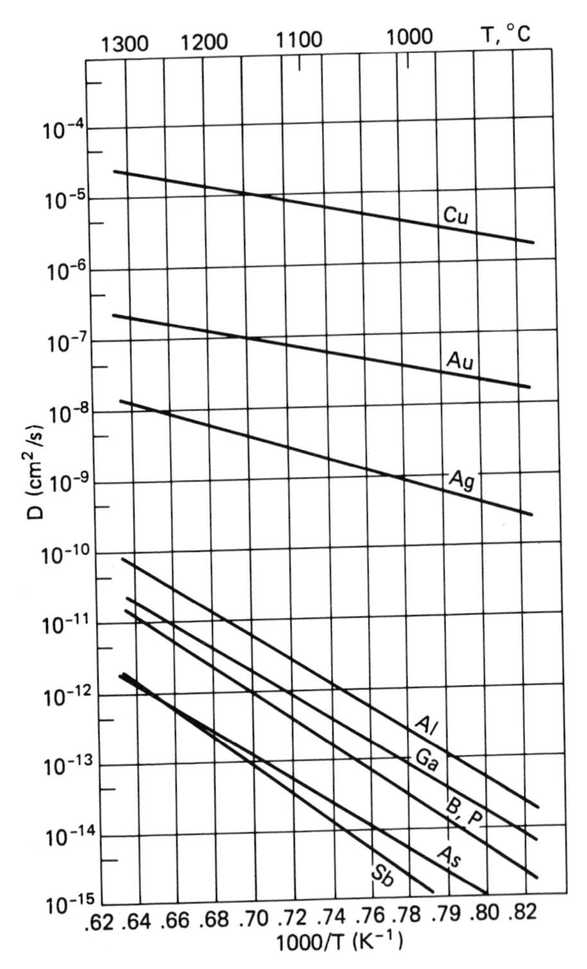

Figure 6–3. Typical Arrhenius plots of diffusivity versus reciprocal absolute temperature for several dopants in silicon. (After *The Theory and Practice of Microelectronics*, S. K. Ghandhi, © 1968 by John Wiley & Sons, Inc. Used with permission.)

called *intrinsic* values; they are valid where doping is low enough that D is concentration independent. At higher levels, where doping levels approach the limit of solid solubility, the effect comes into play. The effect is strong in diffused As and P bipolar emitters. These are studied in Sections 7.10 to 7.12, which show that the doping profile departs from the erfc shape of the first-order theory. Computer programs (such as SUPREM, already cited in Chapter 5) that include many such effects also are available. A fact of importance

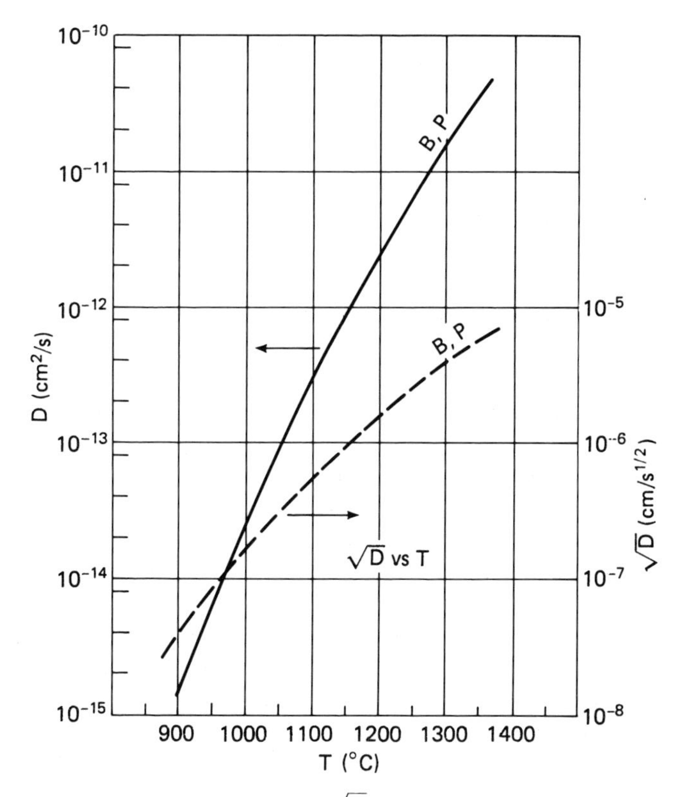

Figure 6–4. Semilog plots of D and $\sqrt{D}$ versus T for boron and phosphorus in silicon.

in annealing after ion implantation (Section 8.13.1) is that dopant diffusivities in poly- or amorphous silicon are higher than in the monocrystal form.

6.2 Fick's Laws of Diffusion

Fick's laws are mathematical statements of the diffusion process. While Fick's first law (FFL) in one dimension was introduced briefly in Section 5.4, we consider it in more detail here. We also consider Fick's second law (FSL), the continuity equation. The solutions of these differential equations, subject to specific initial and boundary conditions that model conditions encountered in practice, permit mathematical handling of diffusion problems in device fabrication.

6.2.1 Fick's First Law (FFL)

Assuming that $N_\alpha(x) \ll N_{SL}$, the solid solubility limit, so D_α is independent of $N_\alpha(x)$, FFL in one dimension is

$$\mathscr{F}_\alpha = -D_\alpha \frac{\partial N_\alpha(x,t)}{\partial x} \tag{6.2-1}$$

where

$\mathscr{F}_\alpha$ = flux density of the α species, $cm^{-2}s^{-1}$

N_α = concentration of the α species, cm^{-3}

t = time, s

x = distance in the direction normal to the wafer surface, cm

D_α = diffusivity of the α species, cm^2/s

Other units may be used (e.g., μm in place of cm, minutes instead of seconds, and so on) if appropriate changes are made in the numerical constants.

The flux density measures how many atoms of the diffusing species cross a unit area in unit time, nominally one square centimeter in one second. The physical significance of the negative sign is that flux density is in the *opposite* direction from the concentration gradient—i.e., the motion is downhill from high to low concentration.

As a general rule, diffusivity D is not a constant, but depends on the diffusing species, the host material, and the temperature. We already have discussed the temperature variation and the concentration dependence of D at very high diffusant concentrations. We neglect this latter effect now, but will consider it in Chapter 7.

6.2.2 Fick's Second Law (FSL)

Fick's second law (FSL) is simply the equation of continuity with no lifetime term, and really is a bookkeeping statement. Members of the diffusing species are neither created nor destroyed, so the rate of change in their number within the small volume of Figure 6–5 must equal the difference between the incoming and outgoing flux densities. If we assume that D_α is independent of $N_\alpha(x, t)$, it is easy to show that in the limit as Δx becomes vanishingly small, this statement reduces to the mathematical form of FSL (Problem 6–1).

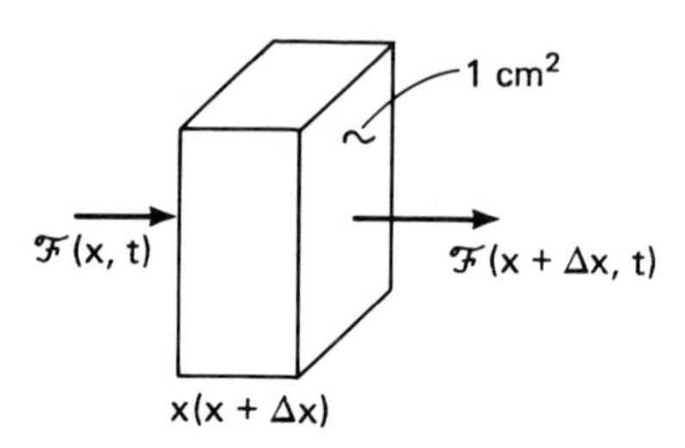

Figure 6–5. The small volume $\Delta x \cdot 1\ cm^3$ used for deriving Fick's second law.

$$\frac{\partial N_\alpha(x,t)}{\partial t} = D_\alpha \frac{\partial^2 N_\alpha(x,t)}{\partial x^2} \tag{6.2-2}$$

If D_α is constant as we have assumed, then both of Fick's laws are *linear* differential equations, and the principle of superposition may be applied to their solutions. This means, for example, that if phosphorus is diffused into a boron-doped wafer, we may solve for the P and B diffusion profiles independently. The algebraic sum of the two is the total or *compensated* profile. The use of superposition, when D_α may be assumed constant, greatly simplifies the calculation of problems when two or more different dopants are diffusing simultaneously in a single host crystal.

The solutions to FSL, with D_α constant, for problems related to the predep are considered in this chapter. We shall not derive the results since these derivations are available in the literature [3, 4].

6.3 Predeposition Diffusion (Case I)

The very nature of the diffusion process is such that any foreign atoms having a nonuniform distribution in the host will diffuse if the temperature is high enough. We can think of this as a *redistribution diffusion*, examples of which are considered in the next chapter.

Diffusion also is used to *introduce* dopant atoms into the host. This *predeposition* or *predep* diffusion, or the now more commonly used ion implantation (Chapter 8), is essential in forming p–n junctions for certain devices, and creates a very thin layer of the dopant just within the surface of the host. If necessary, the atoms may be redistributed or driven into the host by a second or *drive* diffusion. We shall distinguish between the two by using subscripts, 1 for the first or predep and 2 for the second or drive diffusion. In this section we shall consider some of the mathematical aspects related to this predeposition process.

For mathematical purposes the predep diffusion process may be described as "indiffusion from an infinite (constant, or unlimited) source into an undoped host for a limited time." This states the initial and boundary conditions for solving FSL. The term *undoped host* implies only that the host initially contains none of the *indiffusing* dopant species, say α. If there is wafer background doping by another species, say β, it may be ignored under our assumption that superposition is applicable. The basic situation is illustrated in Figure 4–8(a). Let us translate the initial condition (IC) and boundary conditions (BC) into mathematical terms.

The IC, *undoped host*, implies that

$$N_1(x,0) = 0, \qquad x \geq 0 \tag{6.3-1}$$

Notice that the subscript α has been replaced by 1 because it is understood that we are concerned only with the indiffusing species during predep.

The BC, *indiffusion from an infinite souce*, implies that so long as the diffusion continues, the source of dopant never depletes and is able to maintain a constant concentration just *within* the surface of the host. Stated mathematically

$$\begin{aligned} N_1(0,t) &= \text{constant for } t \geq 0 \\ &= N_{01}U(t) \end{aligned} \tag{6.3-2}$$

We are using the symbol N_{01} for the *constant* surface concentration (i.e., at $x = 0^+$). The second subscript 1 refers to the predep or first diffusion of a two-step diffusion process. The unit step function $U(t)$ is necessary since N_{01} is the surface concentration only for $t \geq 0$.

The second boundary condition is a bit more difficult to understand. The statement *for a limited time* implies that the diffusion time or duration is short enough that the indiffusing dopant never diffuses through to the other side of the wafer. It is convenient to handle this mathematically by assuming that the wafer is infinitely thick in the x-direction. Thus, we write for the second BC

$$N_1(\infty, t) = 0 \tag{6.3-3}$$

Solving FSL subject to these three conditions yields Eq. (1.7-1), namely

$$N_1(x, t_1) = N_{01} \operatorname{erfc}\left(\frac{x}{\sqrt{4D_1 t_1}}\right) = N_{01} \operatorname{erfc} u \tag{6.3-4}$$

It is sometimes useful to symbolize the product $D_1 t_1$ by Δ_1.

6.3.1 The Error Function

The function erfc u is known as the *complementary error function* and is related to the *error function* erf u by the identity

$$\left.\begin{aligned} \operatorname{erfc} u &= 1 - \operatorname{erf} u \\ \text{where} \quad & \\ \operatorname{erf} u &= \frac{2}{\sqrt{\pi}} \int_0^u e^{-y^2}\, dy \\ &= \frac{2}{\sqrt{\pi}} \left(u - \frac{u^3}{3 \times 1!} + \frac{u^5}{5 \times 2!} - \cdots\right) \end{aligned}\right\} \tag{6.3-5}$$

Other properties of this function are listed in Appendix F.6.2. Since the error function is definable by a definite integral its value may be tabulated as a function of the upper limit u as in Appendix F.6.3.

Since computer or calculator solutions to problems are desirable, an algorithm or approximation for erf u or erfc u is needed. One commonly encountered approximation is

$$\operatorname{erf} u \approx \sqrt{1 - \exp(-4u^2/\pi)} \qquad \text{for } u \leq 1 \tag{6.3-6}$$

but it is not satisfactory for typical predep diffusions where u lies in the range from 2 to 4. For $u = 2$, the approximation is 34% low, so greater accuracy is required. A better but longer algorithm that requires considerable memory is given in Appendix F.6.4. Still other forms are available in the literature.

The general shape of the erfc u function is shown by the left-hand curve in Figure 6–6(a) with the u abscissa read at the top of the graph. To show how the profile moves in x as the predep time increases, four curves are plotted with t_1 starting at t_a and increasing by factor $\sqrt{2}$. In order to convert from u to x in the figure, $\sqrt{4D_1 t_a}$ was assumed to be 1 μm.

Notice that erfc $(0, t)$, the normalized surface concentration, is independent of t as required by Eq. (6.3-1). The area under any one of the profiles is a measure of the total amount of dopant that has diffused into the host by that

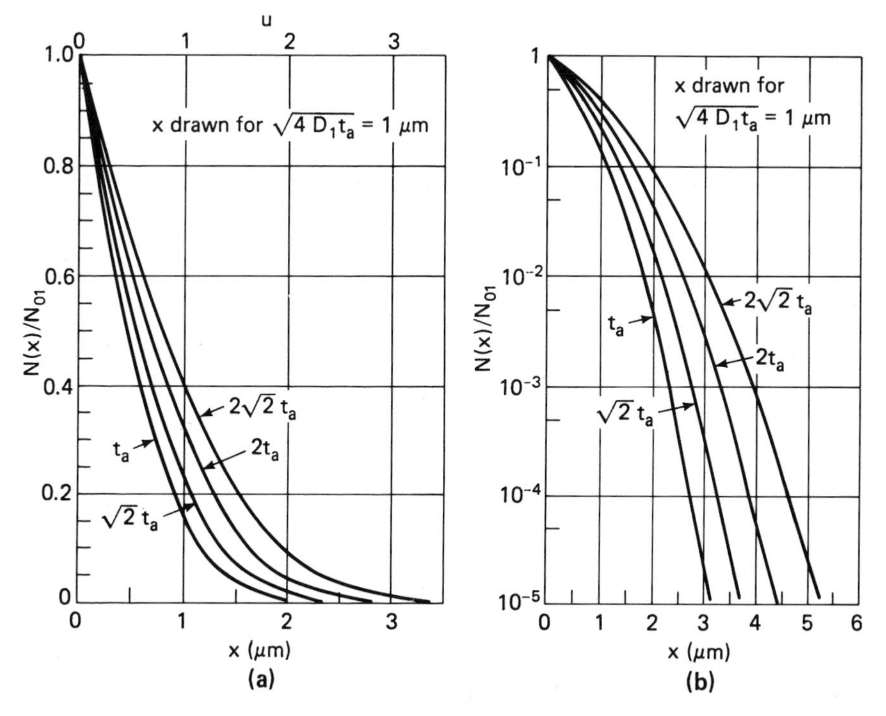

Figure 6–6. Curves of erfc u. (a) Linear scales. The left-hand curve shows erfc u versus u. The x values are based on $\sqrt{4D_1 t_a} = 1$ μm. (b) Semilogarithmic scales.

particular value of time. As might be expected, the areas under successive curves do increase as t_1 gets larger.

For all profiles in Figure 6–6(a) the area under the curve to the right of the erfc $u = 0.01$ point is negligible relative to the area to the left of that point. Furthermore, on the linear scales it is not feasible to read ordinate values less than 0.01, although they may be important in later calculations.

The same data are plotted on semilogarithmic scales in Figure 6–6(b). These curves look very different from those at (a) but ordinate values can be read easily at magnitudes well below 0.01.

The following work uses the *inverse* error function $\text{erf}^{-1} y$ and the *inverse* complementary error function $\text{erfc}^{-1} y$. Consider what these mean and how they are evaluated. By definition

$$\text{If } y = \text{erfc}\, u, \text{ then } u = \text{erfc}^{-1} y \tag{6.3-7}$$

i.e., the -1 exponent does not imply the reciprocal, but rather an *inverse* function just as $\tan^{-1}$ (also symbolized arc tan) is the inverse tangent function. By Eq. (6.3-5)

$$\text{erfc}\, u = 1 - \text{erf}\, u$$

whence

$$u = \text{erfc}^{-1} y = \text{erf}^{-1}(1 - y) \tag{6.3-8}$$

This quantity may be evaluated by reading from right to left in Appendix F.6.3. A numerical example will be considered later.

6.3.2 Predep Properties

The predep diffusion has several properties useful in later studies. First, consider the flux density of the dopant in and into the wafer. This can be obtained directly by substituting Eq. (6.3-4) into FFL

$$\begin{aligned} \mathcal{F}_1(x, t_1) &= -D_1 \frac{\partial N_1(x, t_1)}{\partial x} = -D_1 N_{01} \frac{\partial}{\partial x}\left[1 - \text{erf}\left(\frac{x}{\sqrt{4D_1 t_1}}\right)\right] \\ &= N_{01}\sqrt{\frac{D_1}{\pi t_1}} \exp\left(\frac{-x^2}{4D_1 t_1}\right) \end{aligned} \tag{6.3-9}$$

or we can calculate the flux density at the surface (i.e., the flux density of the atoms entering the surface) by setting $x = 0$, whence

$$\mathcal{F}_1(0, t_1) = N_{01}\sqrt{\frac{D_1}{\pi t_1}} \tag{6.3-10}$$

We next calculate the *dose*, a term of ion implantation notation that we shall use for the predep diffusion also. This will make comparison of the two methods of introducing dopants into a host much easier later in Chapter 8. Let Q_1 = number of dopant atoms that enter the host through one cm^2 of surface in time t_1 = the dose.

$$Q_1 = \int_0^{t_1} \mathscr{F}_1(0, t_1)dt_1 = 2N_{01}\sqrt{\frac{D_1 t_1}{\pi}} \qquad (6.3\text{-}11)$$

Calculations with some typical numbers will show the orders of magnitude involved. We also shall estimate how deep a typical predep diffusion might be, and how many atoms will diffuse into the host during the predep.

Say boron is diffused into silicon from an infinite source for 15 min at 900°C. Conditions are such that the surface concentration is constant at the limit of solid solubility (i.e., $N_{01} = N_{SL}$). Calculate x^*, the value of x at which the concentration is $0.01N_{01}$, and the dose Q_1.

From Appendix F.6.1 for boron at 900°C: $D_1 = 1.5 \times 10^{-15}$ cm^2/s and $N_{SL} = 3.7 \times 10^{20}$ cm^{-3}. Then at x^*

$$N_1(x^*, t_1) = N_{01}\,\text{erfc}\left(\frac{x^*}{\sqrt{4D_1 t_1}}\right) = 0.01N_{01}$$

or

$$x^* = \sqrt{4D_1 t_1}\,\text{erfc}^{-1}(0.01)$$

This can be evaluated from Appendix F.6.3. Thus,

$$\text{erfc}^{-1}(0.01) = \text{erf}^{-1}(0.99) = 1.82$$

or

$$x^* = \sqrt{4(1.5 \times 10^{-15})(15)(60)}\,1.82$$
$$= 4.23 \times 10^{-6}\,\text{cm} = 423\,\text{Å}$$

Notice that the 15-min factor is multiplied by 60 to convert to seconds; this is necessary since D_1 is given in cm^2/s.

By Eq. (6.3-11)

$$Q_1 = 2N_{01}\sqrt{\frac{D_1 t_1}{\pi}} = 2(3.7 \times 10^{20})\sqrt{\frac{(1.5 \times 10^{-15})(15)(60)}{\pi}}$$
$$= 4.85 \times 10^{14}\,\text{atoms/cm}^2$$

Consider the significance of these results. First, most of the indiffused atoms lie within 423 Å of the wafer surface. This represents a very thin layer indeed, in fact much thinner than a wavelength of visible light! We say *most* of the atoms because the profiles in Figure 6–6(a) show that the relative number of atoms lying to the right of x^* is negligible compared to those lying to its left. Secondly, on an atomic scale the value of Q_1, the total number of boron atoms that have diffused through one cm^2 of wafer surface in 15 min, is very small also. Consider the following for clarification.

Single-crystal silicon has 5×10^{22} atoms/cm^3. If we consider a very simple silicon model—that of closely packed spherical atoms crowded into touching layers—we calculate that each layer (with area 1 cm^2) of atom-spheres will contain $(5 \times 10^{22})^{2/3} = 1.36 \times 10^{15}$ atoms. Since this is the number of atoms in a layer one atom thick (i.e., a *monatomic layer* or *monolayer*) and 1 cm to a side, we see that the 15-min boron predep into silicon at 900°C results in a dose of less than one monolayer of atoms! These are quite typical numbers for a boron predep diffusion. Another useful result from this model is that a silicon monolayer has a thickness of roughly $[1/(5 \times 10^{22})^{1/3}]10^8 \approx 2.7$ Å.

6.3.3 Substrate Outdiffusion (Case II)

When a doped substrate is raised to diffusion temperatures, the background dopant may outdiffuse under certain conditions. This effect is related to predep because it may affect the junction depth x_{j1} at the end of the predep process. In this section we consider outdiffusion alone; later we combine its effect with that of an indiffusant. Outdiffusion from the wafer involves another solution of FSL with appropriate initial and boundary conditions. Assume that all background atoms on reaching the surface will leave the host.

This case may be described as "outdiffusion from a uniformly doped host into an infinite sink for a limited time." The general conditions for this case are illustrated in Figure 6–7(a). The initial and boundary conditions may be translated into suitable mathematical terms in this manner: the IC *uniformly doped host* implies that

$$N_o(x, 0) = N_b = \text{constant} \tag{6.3-12}$$

where the subscript o indicates an outdiffusing dopant. The symbol N_b represents the uniform background doping level or concentration in the host.

The *into an infinite sink* BC implies that after $t = 0$, the surface concentration is held at zero:

$$N_o(0, t_o) = 0\, U(t) \tag{6.3-13}$$

The second BC, namely *for a limited time*, implies that the diffusion duration is short enough that the far side of the wafer is not aware of any dopant loss. Again, this may be modeled conveniently by assuming the wafer to be infinitely

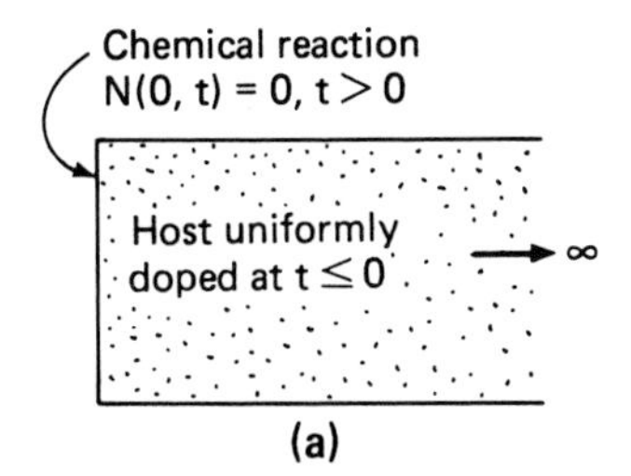

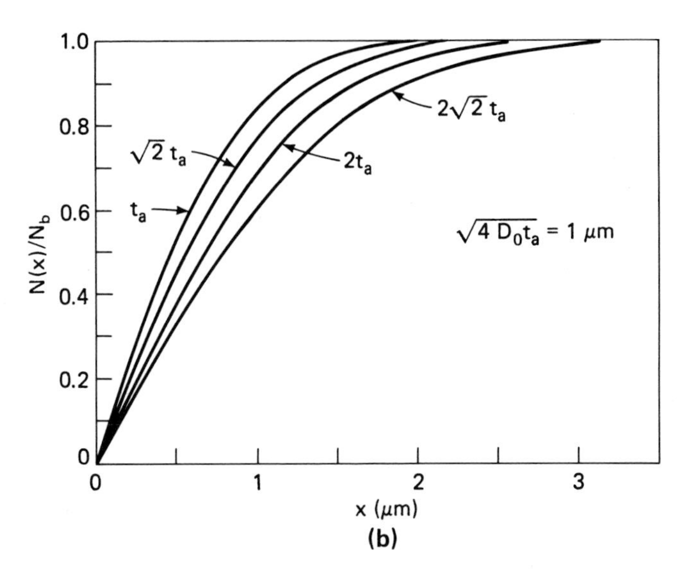

Figure 6–7. Substrate outdiffusion, Case II. (a) Chemical reaction keeps the surface concentration at zero once diffusion starts. The semi-infinite host has uniform doping initially. (b) Outdiffusion profiles for different times. It was assumed that $\sqrt{4D_o t_a} = 1\ \mu\text{m}$.

thick, so we write

$$N_o(\infty, t_o) = N_b \tag{6.3-14}$$

The solution of FSL with these three conditions is

$$N_o(x, t_o) = N_b \operatorname{erf}\left(\frac{x}{\sqrt{4D_o t_o}}\right) \tag{6.3-15}$$

Typical profiles for this function as time varies are shown in Figure 6–7(b).

It is difficult to realize a true infinite sink for the outdiffusant (i.e., one that can hold the surface concentration at zero). A good approximation is afforded, however, if an ambient gas or solid is present at the host's surface that reacts with the outdiffusing species to form a compound outside the silicon. For example, if a boron-doped wafer is heated to diffusion temperatures in a hydrogen ambient, this reaction can take place:

$$2\,B\!\downarrow + 3\,H_2\!\uparrow \rightarrow B_2H_6\!\uparrow$$

Since B_2H_6 is a gas, it removes boron at the surface; the concentration there may drop two or more orders of magnitude below N_b, in which event $N_o(0, t_o)$ is practically negligible, and Eq. (6.3-15) is adequate for profile calculation.

Aluminum-doped silicon also can show the infinite sink effect [5]. If an oxidized wafer of this material is heated to diffusion temperatures, aluminum can outdiffuse into the SiO_2 layer from the silicon and cause the reaction

$$4\,Al\!\downarrow + 3\,SiO_2\!\downarrow \rightarrow 2\,Al_2O_3\!\downarrow + 3\,Si\!\downarrow$$

and the Al concentration in the silicon can drop markedly near the surface. If it does not drop several magnitudes below N_b, Eq. (6.3-15) is a worst-case solution in that it shows the effect of outdiffusion extending the greatest possible distance into the wafer. Other cases of outdiffusion into a solid are discussed in Chapter 7.

6.4 Predep Junction Formation

A p–n junction may be formed by a predep diffusion. For example, if an acceptor is indiffused, a junction may form whether the host donor dopant outdiffuses or not, provided N_{01} of the indiffuser is great enough. No junction will form if both dopants are of the same type, however.

Assume that the conditions under which the processing is carried out can be modeled by the erfc u solution for the indiffuser (Case I) and by Case II for the outdiffuser. We shall use $\pm$ and $\mp$ signs to indicate the opposite types of the two diffusing species.[2] By superposition, the net or compensated profile will be the algebraic sum of the two separate solutions, so

$$N(x, t_1) = \pm N_{01}\,\mathrm{erfc}\left(\frac{x}{\sqrt{4D_1 t_1}}\right) \mp N_b\,\mathrm{erf}\left(\frac{x}{\sqrt{4D_o t_1}}\right) \qquad (6.4\text{-}1)$$

Notice that the subscripts allow for different diffusivity values for the two dopants, yet the time of diffusion for both must be the same because both species diffuse simultaneously, at the same temperature, and in the same wafer.

---

[2] Regarding sign conventions, we use + for the donors since their *ions* have positive charge, and – for acceptors because their *ions* are charged negatively (Section 1.7.3).

Now at x_{j1}, the metallurgical junction after predep, the net doping density must be zero (Section 1.7.3), that is, $N(x_{j1}, t_1) = 0$, so

$$\pm N_{01}\,\mathrm{erfc}\left(\frac{x_{j1}}{\sqrt{4D_1 t_1}}\right) \mp N_b\,\mathrm{erf}\left(\frac{x_{j1}}{\sqrt{4D_o t_1}}\right) = 0 \qquad (6.4\text{-}2)$$

This equation is difficult to solve analytically in the general case. One can calculate and plot each term of the equation as a function of x_{j1}, however, and find the solution at the point of intersection. As an alternative, trial and error can yield a solution rather quickly if the first guess is close to the correct value. Experience helps in making a good first guess, and computers can crunch the numbers rapidly. Some special cases arise in practice for which there are simple closed solutions. Two of these will be considered because they give an understanding of the processes.

6.4.1 Slow Outdiffuser Approximation (SOA)

Consider a special case of junction formation in which $D_o \ll D_1$. Assume for the moment that

$$\mathrm{erf}\left(\frac{x_{j1}}{\sqrt{4D_o t_1}}\right) \geq 0.995 \approx 1 \qquad (6.4\text{-}3)$$

From Appendix F.6.3 we see that this implies that

$$\frac{x_{j1}}{\sqrt{4D_o t_1}} \geq 2 \qquad (6.4\text{-}4)$$

Subject to this assumption, Eq. (6.4-2) can be written and solved for x_{j1}

$$x_{j1} = \sqrt{4D_1 t_1}\,\mathrm{erfc}^{-1}\left(\frac{N_b}{N_{01}}\right) \quad \text{SOA} \qquad (6.4\text{-}5)$$

where SOA stands for "slow outdiffuser approximation."

Define a predep parameter $\mathscr{P}$ as

$$\mathscr{P} = \mathrm{erfc}^{-1}\left(\frac{N_b}{N_{01}}\right) \qquad (6.4\text{-}6)$$

so

$$x_{j1} = \sqrt{4D_1 t_1}\,\mathscr{P} \qquad (6.4\text{-}7)$$

(Equations 6.4-6 and 6.4-7: SOA)

Consider the consequences. Substituting for x_{j1} from Eq. (6.4-7) into Eq. (6.4-4) we have

$$\frac{x_{j1}}{\sqrt{4D_o t_1}} = \sqrt{\frac{4D_1 t_1}{4D_o t_1}}\,\mathscr{P} \geq 2$$

whence we have the criterion for SOA validity:

$$\frac{D_1}{D_o} \geq \left(\frac{2}{\mathscr{P}}\right)^2 \quad \text{SOA criterion} \qquad (6.4\text{-}8)$$

This gives the clue to the designation *slow outdiffuser approximation* (SOA), for it states that if D_o is less than D_1 by factor $(\mathscr{P}/2)^2$, then the approximation is valid.

The physical significance of the slow outdiffuser approximation is shown in Figure 6–8. If the effect of background outdiffusion has not reached x_{j1}, the outdiffusing profile has the original background level N_b *in the vicinity of the junction.*

Consider an illustrative example. Boron, with $N_{01} = 10^{19}$ cm^{-3}, is diffused into an arsenic-doped silicon wafer ($N_b = 10^{18}$ cm^{-3}) for 15 min at 1100°C. (The value of 10^{18} for background doping level is quite high, but is chosen to illustrate a point.) At this temperature: $D_1 = 3 \times 10^{-13}$ cm^2/s and $D_o = 4.4 \times 10^{-14}$ cm^2/s. Using the slow outdiffuser approximation, we have

$$\mathscr{P} = \text{erfc}^{-1}\left(\frac{N_b}{N_{01}}\right) = 1.16$$

Checking the SOA criterion by Eq. (6.4-8) we have

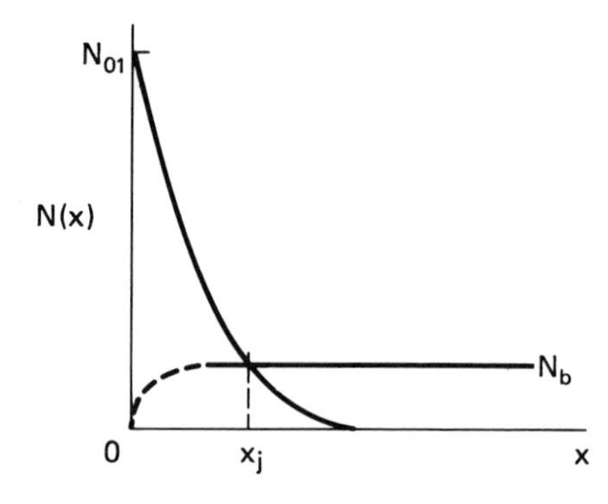

Figure 6–8. The slow outdiffuser approximation. The effect of outdiffusion has not reached x_j, where the background concentration is still at N_b.

$$\frac{D_1}{D_o} \geq \left(\frac{2}{1.16}\right)^2 \approx 3$$

But the actual diffusivity ratio is

$$\frac{D_1}{D_o} = \frac{3 \times 10^{-13}}{4.4 \times 10^{-14}} = 6.8$$

so the criterion is satisfied. Then

$$\begin{aligned} x_{j1} &= \sqrt{4D_1 t_1}\,\mathscr{P} = \sqrt{4(3 \times 10^{-13})(15)(60)}\,(1.16) \\ &= 3.81 \times 10^{-5}\ \text{cm} = 0.38\ \mu\text{m} \end{aligned}$$

The result is rounded off to two significant figures. One cannot justify even three-significant-figure accuracy in this type of work, especially since data are seldom accurate to better than two. The ten-figure capabilities of calculators are tempting, but don't yield to them.

The results of the foregoing example may be generalized to some extent. Typically, silicon predep conditions will be such that $N_b/N_{01} \leq 0.01$ or $\mathscr{P} = \text{erfc}^{-1}(N_b/N_{01}) \geq 1.82$. The SOA criterion then asks that the diffusivity ratio be

$$\frac{D_1}{D_o} \geq \left(\frac{2}{1.82}\right)^2 = 1.2$$

The curves of Figure 6–3 show that the criterion is satisfied easily for most common silicon dopants if an acceptor diffuses into a donor-doped wafer.

Boron and phosphorus are an exception because they have the same intrinsic diffusivity. Since one is p- and the other n-type, they can form a junction if $N_b < N_{01}$. With $D_1 = D_o$ for this unique case, $\mathscr{P}$ would have to be ≥ 2 or $N_b/N_{01} \leq 0.0047$, a condition often met in practice.

Remember that approximations in this section are limited to a predep where the background level is N_b in the vicinity of x_{j1} (i.e., where the SOA is satisfied).

6.4.2 Equal Diffusivities Approximation (EDA)

Boron and phosphorus give a special case of predep junction formation even if the SOA is not satisfied. We consider this for some practice in handling the erf u and erfc u functions and to gain more insight into this unique case. Setting $D_o = D_1$ in Eq. (6.4-2) we get

$$\pm N_{01}\operatorname{erfc}\left(\frac{x_{j1}}{\sqrt{4D_1 t_1}}\right) \mp N_b \operatorname{erf}\left(\frac{x_{j1}}{\sqrt{4D_1 t_1}}\right) = 0 \tag{6.4-9}$$

The two arguments are identical, and erf $u = (1 - \operatorname{erfc} u)$, so

$$x_{j1} = \sqrt{4D_1 t_1}\operatorname{erfc}^{-1}\left(\frac{N_b}{N_b + N_{01}}\right), \qquad \begin{cases} D_o = D_1 \\ \text{EDA} \end{cases} \tag{6.4-10}$$

This equation does not assume the SOA, but only the equal diffusivity approximation (EDA), and so is valid without restriction on the N_b/N_{01} ratio. As might be expected, as the ratio gets smaller, Eq. (6.4-10) approaches the SOA solution. Typically $N_{01}/N_b \geq 10^4$.

6.5 Calculation of Sheet Resistance

The concept and measurement of sheet resistance R_S were considered in Section 3.8. We now relate R_S to the diffused layer near the sample surface. Given values of surface concentration N_0, background concentration N_b, and junction depth x_j, we can *calculate* R_S for certain layer profiles. This allows design of a diffusion schedule to give desired values of R_S and x_j.

The predep diffused profile is modeled by the erfc u function if the diffusivities are concentration independent. In Chapter 7 we shall study a drive diffusion whose profile may be modeled by a gaussian. Both profile shapes will be considered in this section because the calculation *methods* for R_S are the same, even though the curves differ.

6.5.1 Basic Concepts

Consider the highly idealized situation of Figure 6–9(a). A thin, *uniformly* doped layer of concentration N_0 and thickness x_j is just within the surface of a wafer doped with the opposite type of dopant at uniform concentration N_b. Since the two dopants differ in type, x_j is both junction depth and surface-layer thickness. For the uniform singly doped wafer below the surface layer (i.e., for $x > x_j$), we can write the familiar equation for conductivity and its reciprocal, resistivity:

$$\sigma = \frac{1}{\rho} = q\,\mu_M\{N_b\}\,N_b, \qquad x > x_j \tag{6.5-1}$$

where

q = electronic charge magnitude

$\mu_M\{N_b\}$ = majority carrier mobility for the concentration N_b

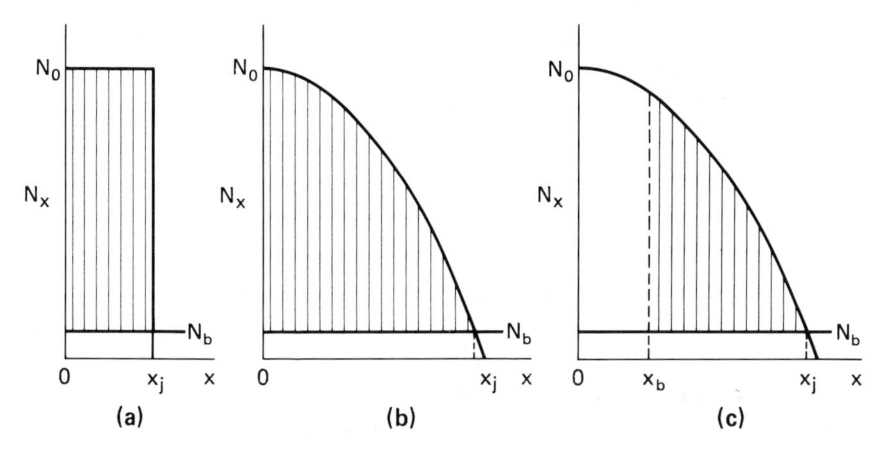

Figure 6–9. Sheet resistance profiles. Sheet resistance depends on the doping in the shaded region. (a) A uniformly doped surface layer. (b) Surface layer with nonuniform doping. (c) A buried layer extending from x_b to x_j.

The surface layer is counterdoped, so conceptually the sheet conductivity of the layer will depend upon the shaded area in the figure—i.e., on the *net* or *compensated* profile $(N_0 - N_b)$ for $0 \leq x \leq x_j$.

Majority carrier mobility, however, is affected by the scattering effects of *both* dopants so it will depend on the *total* or *uncompensated* concentration $(N_0 + N_b)$ and we may symbolize it by $\mu_M\{N_0 + N_b\}$. Then the sheet conductivity and sheet resistivity will become

$$\sigma_S = \frac{1}{\rho_S} = q\,\mu_M\{N_0 + N_b\}\,(N_0 - N_b), \qquad x \leq x_j \tag{6.5-2}$$

It should be clear that if $N_0 \gg N_b$, the curves of Appendix F.2.1 can be used directly to evaluate ρ_S.

Next, consider the situation shown in Figure 6–9(b), where the surface layer has been diffused with a profile $N(x)$, nominally of either the erfc u or gaussian type. To save writing we define

$$N_x = N(x) = \text{value of the diffused profile concentration at distance } x \text{ from the surface}$$

$$= N_0 f(x) \tag{6.5-3}$$

where $f(x)$ is simply the profile normalized wrt the surface concentration N_0.

Since N_x changes between the surface and x_j, both the concentration and the majority carrier mobility become position dependent, as does the sheet conductivity. Therefore, the entire layer for $0 \leq x \leq x_j$ is characterized by an *average* value, indicated by $\langle \sigma_S \rangle$ or $\langle \rho_S \rangle$, which depends upon the shaded region in Figure 6–9(b).[3] By the mathematical definition of the average we have

$$\langle \sigma_S \rangle = \frac{1}{\langle \rho_S \rangle} = \frac{1}{x_j} \int_0^{x_j} q\,\mu_M\{N_x + N_b\}\,(N_x - N_b)\,dx \qquad (6.5\text{-}4)$$

Note that $\{N_x + N_b\}$ is the *argument* of μ_M; it is not a factor.

Direct evaluation of this integral is not feasible. Even though N_x can be replaced by an equation in the nominally assumed cases, a simple analytical curve fit for the mobility is not available.

Irvin has approached this problem by breaking his ρ versus N resistivity curves into a number of piecewise power-law segments of the form [6]

$$\sigma = BN^{\alpha} \quad \text{for } N_L \leq N \leq N_U \qquad (6.5\text{-}5)$$

where N_L and N_U are the lower and upper limits for each segment. Values for B and α are tabulated in the reference. These segments must be forced into Eq. (6.5-4).

6.5.2 Irvin R_S Curves

Assume that all the dopants are ionized, which is a good approximation at room temperature. Multiply the integrand of the last equation by $(N_x + N_b)/(N_x + N_b)$ to get

$$[q\,\mu_M\{N_x + N_b\}\,(N_x + N_b)]\frac{(N_x - N_b)}{(N_x + N_b)}$$

The factor enclosed within brackets may be identified with BN^{α} of Eq. (6.5-5), but with N replaced by $(N_x + N_b)$; hence $\langle \sigma_S \rangle$ becomes

$$\langle \sigma_S \rangle = \frac{1}{\langle \rho_S \rangle} = \frac{1}{x_j} \int_0^{x_j} B(N_x + N_b)^{(\alpha - 1)}(N_x - N_b)\,dx \qquad (6.5\text{-}6)$$

where N_x is $N_0 f(x)$ from Eq. (6.5-3).

As a practical matter, the lower limit of the integral may be changed from zero to any value less than x_j, say x_b (Irvin uses x), to get the average conductivity of the *buried layer* between x_b and x_j. Thus

---

[3] The average value also is indicated by an overbar, viz $\overline{\sigma_S}$ or $\overline{\rho_S}$; either form may be used.

$$\langle \sigma_{Sb} \rangle = \frac{1}{\langle \rho_{Sb} \rangle} = \frac{1}{x_j - x_b} \int_{x_b}^{x_j} B(N_x + N_b)^{(\alpha-1)}(N_x - N_b)\, dx \qquad (6.5\text{-}7)$$

This *buried layer* is indicated in Figure 6–9(c). Layers of this type are of considerable importance in determining the short-circuit, forward current gain β of a BJT. Remember, however, that in this chapter our primary interest is in the predep, where the diffused layer will begin at the surface so that $x_b = 0$ and the profile is of the erfc u shape.

Given Irvin's B and α values, Eq. (6.5-7) may be computer integrated and plotted in families of the Irvin R_S curves. Each curve *family* is plotted on N_0 versus $\langle \sigma_S \rangle$ coordinates and is characterized by fixed values of N_b (Irvin uses N_{BC}), *layer* type (p or n), and profile shape (erfc u for predep or gaussian for drive). Each curve in the family is drawn for a different value of x_b/x_j. A few of the Irvin R_S curves are plotted in Appendix F.6.5, and are used for determining $\langle \sigma_S \rangle$ or $\langle \sigma_{Sb} \rangle$, as the case may be. Great care must be exercised to choose the correct family in using the curves. Given the appropriate conductivity value and x_j, the sheet resistance will be

$$R_S = \frac{1}{\langle \sigma_S \rangle x_j} \quad \text{(surface layer)} \qquad (6.5\text{-}8)$$

or for the buried layer

$$R_{Sb} = \frac{1}{\langle \sigma_{Sb} \rangle (x_j - x_b)} \quad \text{(buried layer)} \qquad (6.5\text{-}9)$$

The curves show data for silicon at 300 K; hence all the following results are for that temperature.

A numerical example will illustrate how the curves are used. Say a boron predep diffusion is performed on a silicon wafer doped at 10^{15} cm^{-3} with arsenic. The diffusion conditions are

$$T_1 = 950°\text{C}, \qquad D_1 = 6.6 \times 10^{-15}\,\text{cm}^2/\text{s}$$
$$t_1 = 900\,\text{s}, \qquad N_{01} = 3.9 \times 10^{20}\,\text{cm}^{-3}$$

Calculate R_{S1}, the diffused layer sheet resistance.

To select the correct Irvin curve we note the following: The layer dopant is boron, so the *layer* will be p-type. Since this is a predep diffusion, the profile is erfc u, and $N_b = 10^{15}$ cm^{-3}. Since the sheet resistance of the entire surface layer is desired, $x_b = 0$ and $x_b/x_j = 0$. Thus we use Appendix F.6.5(c) and read $\langle \sigma_{S1} \rangle = 550$ S/cm. (Note that S stands for Siemens, the conductivity unit that replaced the deprecated unit mho.)

We next calculate x_{j1}. Using the SOA

$$\Delta_1 = D_1 t_1 = 5.94 \times 10^{-12}\,\text{cm}^2$$

$$\mathscr{P} = \text{erfc}^{-1}\left(\frac{N_b}{N_{01}}\right) = 3.32$$

$$x_{j1} = \sqrt{4\Delta_1}\,\mathscr{P} = 0.162\,\mu\text{m}$$

Then

$$R_{S1} = \frac{1}{\langle \sigma_{S1} \rangle x_{j1}} = \frac{1}{550(0.162 \times 10^{-4})} = 112\ \Omega/\text{sq}$$

Say we double t_1 to 1800 s, all else remaining unchanged. Then N_{01} and x_b/x_{j1} are both unchanged. Referring to Appendix F.6.5(c), we observe that $\langle \sigma_{S1} \rangle$ remains unchanged, but x_{j1} which is proportional to $\sqrt{t_1}$, will increase by factor $\sqrt{2}$. Therefore R_{S1} will decrease by factor $\sqrt{2}$ so

$$R_{S1} = 112/\sqrt{2} = 79\,\Omega/\text{sq}$$

There are some interesting things to observe about these calculations over and above their use of the Irvin R_S curves. As the predep time was increased, the sheet resistance decreased, a change easily related to physical concepts. The average conductivity remains the same, but with increasing predep time more dopant is diffused into the layer, which is getting thicker. The net area under the profile increases, so a decrease in sheet resistance is not surprising.

Later in the chapter we shall find that the desired value of predep sheet resistance often is used as a benchmark for predep testing. Also, it is easy to show that given an erfc profile, R_{S1}, N_{01}, and x_{j1} are interrelated for a given predep temperature; hence any two may be specified and the third calculated for designing the predep diffusion schedule.

The Irvin predep R_S curves are valid only for the erfc profile. If the predepped dopant diffusivity is concentration dependent, the profile is non-erfc and other methods for R_S calculation are required (Sections 7.11.3 and 7.12.3).

6.6 Oxide Masking

We saw in Chapter 1 that *oxide masking* makes possible selective doping of specific regions of the wafer *during predep*. The basic idea of masking is illustrated in Figure 1–8. The dopant may enter the silicon where there is no oxide, but where the oxide is present, it should prevent the dopant from reaching the underlying silicon. Masking will work if, for the particular dopant, $D_{ox} < D_1$ and if the oxide is thicker or equal to some $(x_{ox})_{min}$, the lowest value that will provide *adequate masking*. A number of definitions of this term can be set up, each corresponding to a different predep dopant level that may be tolerated in the underlying silicon. We address this problem later.

A second aspect of masking has to do with *lateral* diffusion of the predep dopant. Dopant that has diffused vertically through the mask windows into the Si serves as a source for *lateral* diffusion under the mask edges. As a result, the regions of doped silicon are not defined precisely by the mask windows. These aspects of masking are considered in the next two sections.

6.6.1 Indiffusion Through Oxide (Case III)

Consider predep through an *existing* oxide layer of some thickness x_{ox}. Assume that both host wafer and oxide are initially free of the indiffusing dopant. Consider the initial and boundary conditions for solving FSL.

The full descriptive title for this case would be "indiffusion from an infinite source, through an undoped oxide of thickness x_{ox}, into an undoped host, for a limited time." By using earlier methods, we translate the title into these initial and boundary conditions:

$$\left.\begin{aligned} &N(x,0) = 0, \qquad -x_{ox} \le x \le \infty \\ &N(-x_{ox},t) = N_{01}U(t) \\ &N(\infty,t) = 0 \end{aligned}\right\} \tag{6.6-1}$$

where $x = 0$ lies at the oxide/silicon interface. There are two other conditions that must be satisfied. First, the interface is formed by two different solids, so we invoke the *segregation coefficient m*, which is considered in some detail in the next chapter. For the moment we shall define it by

$$m = \frac{\text{equilibrium concentration of the dopant in Si}}{\text{equilibrium concentration of the dopant in SiO}_2}$$

For a given dopant m is a constant, less than one for boron and greater than one for phosphorus.[4] In the present instance m will be the ratio of the dopant concentration on the two sides of the interface so that we write

$$N(0^+,t) = m\,N(0^-,t) \tag{6.6-2}$$

[4] Note that m is roughly analogous to k_d of Chapter 4 except that k_d refers to the distribution of a dopant between the liquid and solid phases of a *single* material, while m refers to the distribution of a dopant at the two sides of an interface between two *different solids*.

Second, the dopant flux density must be continuous across the interface so that

$$\mathscr{F}(0^+, t) = \mathscr{F}(0^-, t) \tag{6.6-3}$$

The solution to FSL, subject to these conditions, has been found to be two infinite series [7].

For $-x_{ox} \le x \le 0$ (i.e., in the *oxide*)

$$N(x,t) = N_{01}\left\{\sum_{n=0}^{\infty} \alpha^n \operatorname{erfc}\left[\frac{(2n+1)\,x_{ox} + x}{\sqrt{4D_{ox}t}}\right] - \alpha \operatorname{erfc}\left[\frac{(2n+1)x_{ox} - x}{\sqrt{4D_{ox}t}}\right]\right\} \tag{6.6-4}$$

where

$$D_{ox} = \text{dopant diffusivity in } SiO_2$$

$$D = \text{dopant diffusivity in Si}$$

$$m = \text{segregation coefficient}$$

$$\alpha = \frac{m-r}{m+r} \quad \text{and} \quad r = \sqrt{\frac{D_{ox}}{D}}$$

If, however, $x_{ox} > 0.7\sqrt{4D_{ox}t}$, only the first pair of terms of the infinite series is necessary, viz:

$$N(x,t) = N_{01}\left[\operatorname{erfc}\left(\frac{x_{ox} + x}{\sqrt{4D_{ox}t}}\right) - \left(\frac{m-r}{m+r}\right)\operatorname{erfc}\left(\frac{x_{ox} - x}{\sqrt{4D_{ox}t}}\right)\right]$$

$$\text{for } -x_{ox} \le x \le 0 \quad \text{and} \quad x_{ox} > 0.7\sqrt{4D_{ox}t} \tag{6.6-5}$$

The general solution for $x > 0$ (i.e., in the *silicon*) is the infinite series

$$N(x,t) = m(1-\alpha)\,N_{01}\sum_{n=0}^{\infty} \alpha^n \operatorname{erfc}\left[\frac{(2n+1)\,x_{ox} + rx}{\sqrt{4D_{ox}t}}\right] \tag{6.6-6}$$

If, however, x is restricted to values that are small relative to $x_{ox}\sqrt{D/D_{ox}}$, the first term alone is adequate, the other terms being negligible. This first term can be written as

$$N(x,t) = 2N_{01}\frac{mr}{m+r}\operatorname{erfc}\left(\frac{x_{ox}}{\sqrt{4D_{ox}t}} + \frac{x}{\sqrt{4Dt}}\right) \quad \text{for } 0 < x \ll x_{ox}\sqrt{D/D_{ox}} \tag{6.6-7}$$

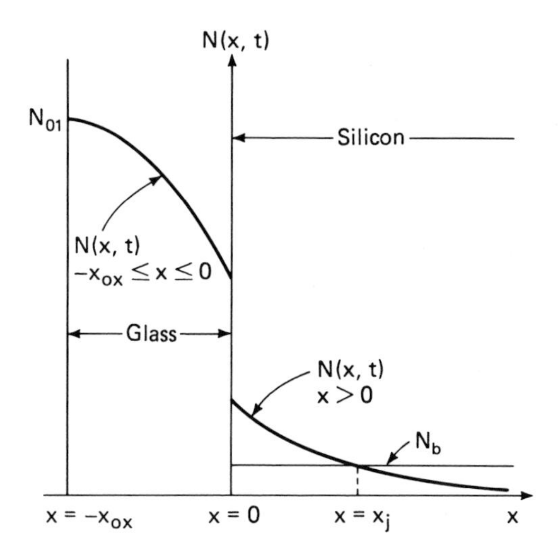

Figure 6–10. Diffusion of phosphorus through an oxide film on silicon. Phosphorus profiles in both materials are shown [7]. (Reprinted with permission from *J. Phys. & Chem. Solids*, Sah, Sello, and Tremere, "Diffusion of Phosphorus in Silicon Oxide Film," Copyright 1959. Courtesy of Pergamon Press PLC.)

The corresponding concentration profile for phosphorus has the form shown in Figure 6–10. It is clear from the figure that the segregation coefficient is less than unity for the condition shown; in fact, Sah and his coauthors argue from experimental results that $m < 0.7$ [7]. This points out the common problem in using the equations: as we have seen before, it is difficult to get "good" values for m and D_{ox}.

In this regard remember that the silicon here is not in contact with *pure* SiO_2 at the interface, but with a phosphosilicate (PSG) glass of some unknown composition $P_xSi_yO_z$. This glass develops as the phosphorus starts to diffuse through and react with the SiO_2 mask. By the time the phosphorus reaches the silicon, the entire oxide layer of thickness x_{ox} has been converted to PSG, but not necessarily of uniform composition, accounting for the PSG/Si interface at $x = 0$.

We now determine the typical range of x values for which the single-term solution of Eq. (6.6-7) is valid.

At 1200°C the diffusivities of phosphorus are $D_{ox} = 2.5 \times 10^{-15}$ and $D = 2.5 \times 10^{-12}\ \text{cm}^2/\text{s}$. Say $x_{ox} = 2000$ Å (0.2 μm). Then Eq. (6.6-7) is valid for

$$x \ll \sqrt{\frac{D}{D_{ox}}}\,x_{ox} = \sqrt{\frac{2.5 \times 10^{-12}}{2.5 \times 10^{-15}}}(0.2) = 6.3\ \mu\text{m}$$

Thus, the single-term solution is valid for x up to roughly 0.05 μm. We shall see later that for adequate masking, x will be much less than this, in fact zero.

With this fact established we now consider a criterion for *adequate masking.*

Say the predep and wafer dopants are of opposite types and form a junction under the masking oxide at x_j *in the silicon.* The value of $(x_{ox})_{min}$ to give a specified x_j is determined from Eq. (6.6-7) to be

$$x_{ox} = \sqrt{4D_{ox}t}\left(¥ - \frac{x_j}{\sqrt{4Dt}}\right) \tag{6.6-8}$$

where ¥ ("Yen") is just a shorthand symbol:

$$¥ = \text{erfc}^{-1}\left[\frac{N_b}{N_{01}}\left(\frac{m+r}{2mr}\right)\right] \tag{6.6-9}$$

A commonly used criterion for adequate masking is that $x_j = 0$, which implies that the predep dopant will never exceed N_b in the silicon (i.e., the silicon type, p or n, will not be changed). For this condition we have

$$(x_{ox})_{min} = \sqrt{4D_{ox}t}\,¥ \tag{6.6-10}$$

Note that D has dropped out of the equation, except in r, and that D_{ox} becomes the governing diffusivity.

Curves of $(x_{ox})_{min}$ are plotted as a function of the predep time t_1 on semilog coordinates. Recall that

$$D_{ox} = D_\infty \exp(-b/T)$$

Then, rearranging Eq. (6.6-10) and taking logs, we get

$$\log(x_{ox})_{min} = \log\sqrt{4D_\infty} - \frac{0.4343b}{2T} + \log ¥ + \frac{1}{2}\log t_1 \tag{6.6-11}$$

This plots as a straight line on semilog coordinates of $(x_{ox})_{min}$ and t_1 with T as the parameter. Typical curves for boron and phosphorus as the predep dopant are given in Appendices F.6.6 and F.6.7.[5] No information is given on N_b/N_{01}, or what is practically the same thing, ¥, for these curves. The reasons for this are (1) log ¥ is a small term wrt the others, and (2) $\log(\text{erfc}^{-1}y)$ is a

[5] Notice that different curves are required for boron and phosphorus, even though they share the same intrinsic value of D in *silicon.* Remember that D_{ox}, which is the governing diffusivity in masking, is different for B and P.

very slowly varying function of y and hence does not change much even for large variations in y. To verify this, consider the following example.

Boron is the predep dopant. The segregation coefficient is $m = 0.3$. The predep is carried out for 100 min (6000 s) at 1000°C. From Appendix F.6.1, $D = 2.6 \times 10^{-14}$ cm^2/s. For D_{ox}, $D_\infty = 2.8 \times 10^{-4}$ cm^2/s and $b = 3.55 \times 10^4$ K, so

$$D_{ox} = 2.8 \times 10^{-4} \exp\left(\frac{-3.55 \times 10^4}{1273}\right) = 2.17 \times 10^{-16}\,\text{cm}^2/\text{s}$$

$$r = \sqrt{\frac{D_{ox}}{D}} = \sqrt{\frac{2.17 \times 10^{-16}}{2.60 \times 10^{-14}}} = 0.0914$$

Substituting into Eq. (6.6-11), we have

$$\log(x_{ox})_{\min} = \log\sqrt{4(2.8 \times 10^{-4})} - \frac{0.4343(3.55 \times 10^4)}{2(1273)} + \log \text{¥} + \frac{1}{2}\log 6000$$

$$= -1.48 - 6.06 + \log \text{¥} + 1.89 = -5.65 + \log \text{¥}$$

Next, consider a typical magnitude for ¥ and how much it might vary in practice.

$$\frac{m + r}{2mr} = \frac{0.3 + 0.0914}{2(0.3)(0.0914)} = 7.14$$

We can expect N_b/N_{01} to vary from 10^{-5} to 10^{-2}, so the limits on log ¥ are $\log[\text{erfc}^{-1}(7.14 \times 10^{-5})] = 0.45$ and $\log[\text{erfc}^{-1}(7.14 \times 10^{-2})] = 0.11$. These numbers verify that log ¥ is indeed small relative to the other terms, and a change of 1000 in N_b/N_{01} causes only a change of 4 in the already small value of log ¥.

The curves of Appendices F.6.6 and F.6.7 are based on empirical data, and follow the form predicted by Eq. (6.6-11). It should be stressed that the oxide thickness read from the curves represents a *minimum* value for masking. In practice the value is increased, even doubled, for safety.

6.6.2 Under-Mask Lateral Diffusion

Ideally, the doped areas in the silicon wafer should be defined precisely by the windows in the oxide mask. This condition is not met during the predep diffusion because dopant atoms, once they pass through a mask window and enter the silicon, may diffuse *laterally* under the mask edges. To compensate for this, photomask windows must be *smaller* than the region to be doped in

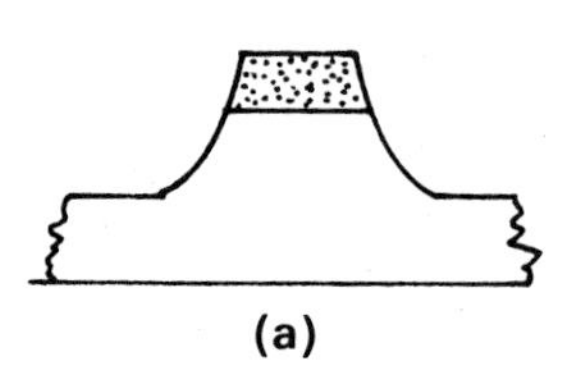

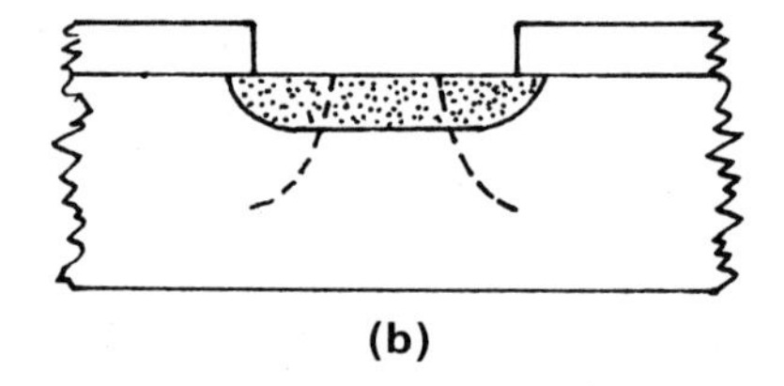

Figure 6–11. Plane and planar junctions. (a) A mesa diode has a flat junction. (b) A planar junction curves under the mask edges. Dashed lines show how a mesa junction can result from etching.

the silicon. The results of this section show that the compensation depends on the depth of the junction to be diffused and on the diffusion type—predep or drive.[6]

A rather unusual bit of terminology is associated with this problem. The term *planar junction* refers to a junction made by the planar process in which all doping steps are made through the upper surface or plane of the substrate. The resulting *diffused* profile of the junction, however, is not a plane; as we shall see shortly, it is curved under the mask edges. A flat or *plane* junction is associated with the mesa type of structure shown in Figure 6–11(a). One way of fabricating such a junction is to etch a planar junction along the dashed lines shown in Figure 6–11(b). Other methods for making the mesa structure are available, however. Remember that *planar* junction refers to the fabricating process rather than to the junction profile normal to the x-direction.

Kennedy and O'Brien have analyzed the problem of lateral diffusion under an oxide for two conditions [8]. The first assumes an infinite source or constant N_{01} and is for the predep diffusion. Because of mixed boundary conditions, they solve a two-dimensional form of Fick's first law (FFL). This solution is applicable to a circular mask window or to a rectangular mask window at points well away from the window corners. The solution involves hypergeometric functions and can be written in the general form

$$N(r, \theta, t) = N_{01}\eta(r, \theta, t) \tag{6.6-12}$$

Their results are plotted in normalized form in Figure 6–12(a), where the parameter is given by the last equation.

Notice in Figure 6–12(a) how the contours of constant concentration are indeed curved near and under the mask edge, and that these contours give the junction shape, if the wafer has uniform doping of the opposite type from the predep dopant. To the right in the figure (i.e., well away from the window

[6] This is in contrast to the situation where the dopant is introduced by ion implantation. As discussed in Chapter 8, implanted areas are well defined by the mask windows because the temperature is too low to support diffusion.

edge), the variation of $N(x)$ follows an erfc$[u(x)]$ form as might be expected. The variation of $N(0, y)$ along the substrate surface can be *approximated* by an erfc$[u(y)]$ form minus a constant. Observe that the curvature near the mask edge is not circular and that the ratio of the lateral penetration, just under the mask, to the vertical penetration well into the window—namely, $N(0, y, \eta)/N(x, \infty, \eta)$—varies roughly from 0.2 to 0.9.

The second condition is for the drive diffusion, assuming an instantaneous source of dose Q (to be discussed in Chapter 7). The three-dimensional solution of FSL is in terms of Green's function; it has the general form

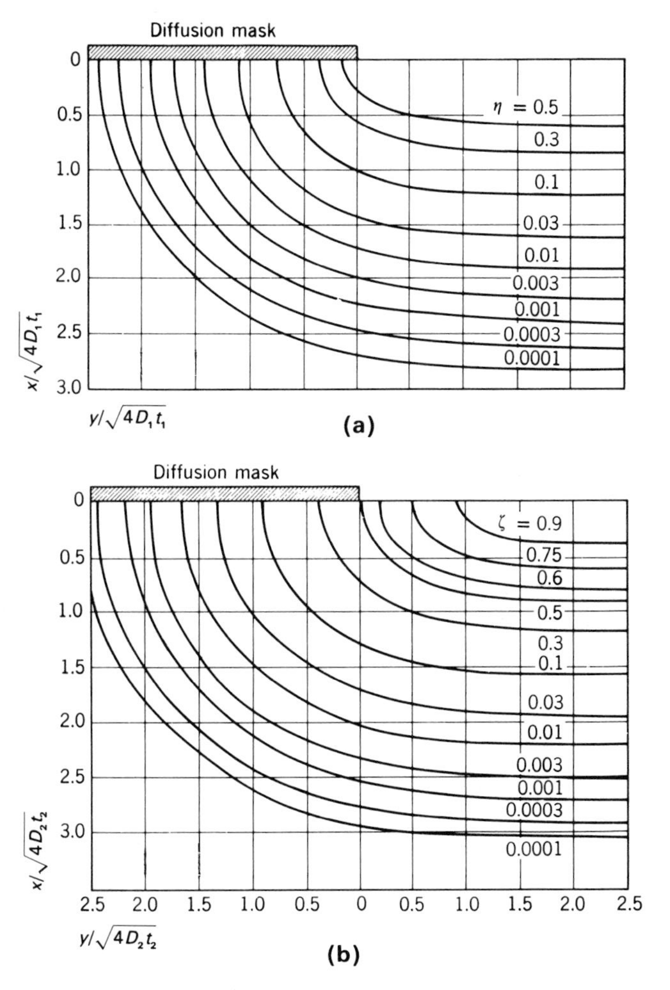

Figure 6–12. Lateral diffusion under a mask edge. (a) Predep diffusion. (b) Drive diffusion. Kennedy and O'Brien [8]. (Copyright 1965 by International Business Machines Corporation; reprinted with permission.)

$$N(x, y, z, t) = \frac{Q}{\sqrt{\pi D_2 t_2}} \zeta(x, y, z, t) \tag{6.6-13}$$

and is plotted in Figure 6–12(b), where the parameter is of the normalized form

$$\frac{N(x, y, z, t)\sqrt{\pi D_2 t_2}}{Q} = \zeta(x, y, \infty, t) \tag{6.6-14}$$

In this case the noncircular curvature of the junction near the mask edge has $N(0, y, \zeta)/N(x, \infty, \zeta)$ varying roughly from 0.4 to 0.8.

These profile shapes are important in terms of lateral under-mask diffusion, but they also affect the breakdown voltage of reverse biased p–n junctions.

6.7 Processing Parameters and x_j

Consider the relative importance of the several predep processing parameters on some predep dependent quantity such as the junction depth x_{j1}. From Section 6.4.1, if the background dopant is a slow diffuser, x_{j1} is given by the slow outdiffuser approximation, viz

$$x_{j1} = \sqrt{4D_1 t_1 \mathscr{P}} \tag{6.7-1}$$

where

$$\mathscr{P} = \text{erfc}^{-1}\left(\frac{N_b}{N_{01}}\right) \tag{6.7-2}$$

The equations show four parameters that affect x_{j1}: time t_1, D_1 (which contains predep temperature T_1 implicitly), N_b, and N_{01}. The last two only appear in a ratio, so we consider them as a unit given by $\mathscr{P}$.

We wish to force in temperature explicitly, and so substitute $D_1 = D_\infty \times \exp(-b/T_1)$. Then, we may rewrite Eq. (6.7-1) as

$$x_{j1} = \sqrt{4D_\infty[\exp(-b/T_1)]t_1 \mathscr{P}} \tag{6.7-3}$$

Note that T_1, t_1, and $\mathscr{P}$ are all subject to control by the processing engineer.

To consider the relative effect of *changes* in the parameters, we shall use the logarithmic derivative of x_{j1}. First take natural logarithms:

$$\ln x_{j1} = \ln\sqrt{4D_\infty} - \frac{b}{2T_1} + \frac{1}{2}\ln t_1 + \ln \mathscr{P} \tag{6.7-4}$$

Then the logarithmic derivative will be

$$\frac{dx_{j1}}{x_{j1}} = \frac{b}{2T_1}\frac{dT_1}{T_1} + \frac{1}{2}\frac{dt_1}{t_1} + \frac{d\mathscr{P}}{\mathscr{P}} \tag{6.7-5}$$

Mathematicians' protests notwithstanding, we shall interpret the logarithmic derivative as the finite fractional change. (See Problem 6–13 to ease your conscience.) Since the equation is linear, we may consider the changes independently, term by term.

6.7.1 Temperature Effect

For t_1 and $\mathscr{P}$ constant we have

$$\left.\frac{dx_{j1}}{x_{j1}}\right|_{t_1,\mathscr{P}} = \frac{b}{2T_1}\frac{dT_1}{T_1}, \qquad T_1 \text{ in Kelvins} \tag{6.7-6}$$

Some typical values are considered in this example. Consider a predep of B (or P) under SOA conditions with t_1 and $\mathscr{P}$ constant. From Appendix F.6.1, $b = 4.28 \times 10^4$ K for both boron and phosphorus at 1000°C. Then, by Eq. (6.7-6)

$$\left.\frac{dx_{j1}}{x_{j1}}\right|_{t_1,\mathscr{P}} = \frac{4.28 \times 10^4}{2(273 + T_C)}\frac{dT_C}{T_C}\frac{1}{(1 + 273/T_C)}$$

where T_C is the temperature in degrees Celsius.

Say $T_C = 1000°\text{C}$ and dT_C is 10°C, so the *fractional change* in T_C is $10/1000 = 0.01$ or 1%. Then

$$\left.\frac{dx_{j1}}{x_{j1}}\right|_{t_1,\mathscr{P}} = \frac{4.28 \times 10^4}{2(1273)}\frac{1}{1.273}1\% = 13.2\%$$

These numbers show that the fractional change in x_{j1} is 13 times larger than a small fractional change in predep temperature! This has some consequences in processing and in furnace design: tight control of the diffusion furnace temperature, not only of its mean value, but also of its variations over the entire wafer batch is essential (Appendix H). Observe that the fractional change in x_{j1} is greater than the fractional change in T_C because the forefactor $(b/2T_1)$ in Eq. (6.7-6) is greater than one.

6.7.2 Time Effect

For constant T_1 and $\mathscr{P}$, Eq. (6.7-5) reduces to

$$\left.\frac{dx_{j1}}{x_{j1}}\right|_{T_1,\mathscr{P}} = \frac{1}{2}\frac{dt_1}{t_1} \tag{6.7-7}$$

In this situation the forefactor, being constant at 1/2, is out of the process engineer's control. Thus, willy-nilly any percentage change in t_1 reflects to one-half that value in x_{j1}.

Regarding control, t_1 should be long enough so that only a small part of the total t_1 is used for inserting wafers into the furnace and removing them after predep. Recall that wafer requirements make necessary a slow push into—and a slow pull out of—the furnace to avoid too rapid changes in the *wafer* temperature. (See Section 7.3 for calculating this effect.) As a general rule a minimum of 15 min would be required for t_1, that value increasing with the length of the wafer boat.

If the boat is moved into—and out of—a furnace whose temperature remains high (say $\geq 900°C$) and constant, the classic "FILO" (first in, last out) problem arises: the wafers at the front end of the boat go into the furnace first, and are the last out, and so experience a longer effective predep time than those at the other end. Recall that this problem may be eliminated by ramping the furnace temperature with the wafers in the furnace (Section 4.11.2).

6.7.3 Doping Effect

For changes in the doping ratio N_b/N_{01} with T_1 and t_1 constant, Eq. (6.7-5) gives

$$\left.\frac{dx_{j1}}{x_{j1}}\right|_{T_1,t_1} = \frac{d\mathscr{P}}{\mathscr{P}}, \qquad \mathscr{P} = \operatorname{erfc}^{-1}\left(\frac{N_b}{N_{01}}\right) \tag{6.7-8}$$

$$= -\lambda\frac{d(N_b/N_{01})}{(N_b/N_{01})} \tag{6.7-9}$$

where

$$\lambda = \frac{\sqrt{\pi}}{2}\frac{(N_b/N_{01})}{\mathscr{P}}\exp\mathscr{P}^2 \tag{6.7-10}$$

The magnitude of the forefactor λ determines if this is an important effect. Its value is plotted as a function of N_{01}/N_b in Figure 6–13. Consider a typical numerical example.

Say $N_{01} = 10^{19}$ cm^{-3} and $N_b = 10^{16}$ cm^{-3}. Then $N_b/N_{01} = 10^{-3}$. From Appendix F.6.3, $\mathscr{P} = \operatorname{erfc}^{-1}(0.001) = 2.32$. Reading from Figure 6–13 or by calculation, $\lambda = 0.083$. Thus a 1% change in the concentration ratio results in only an 0.083% change in x_{j1}.

Figure 6–13 shows that λ is well under 0.5 for normal values of the N_b/N_{01} ratio; therefore, it is clear that the concentration ratio has the smallest effect on x_{j1} of the three parameters. Nevertheless, we must consider what causes variations in that ratio.

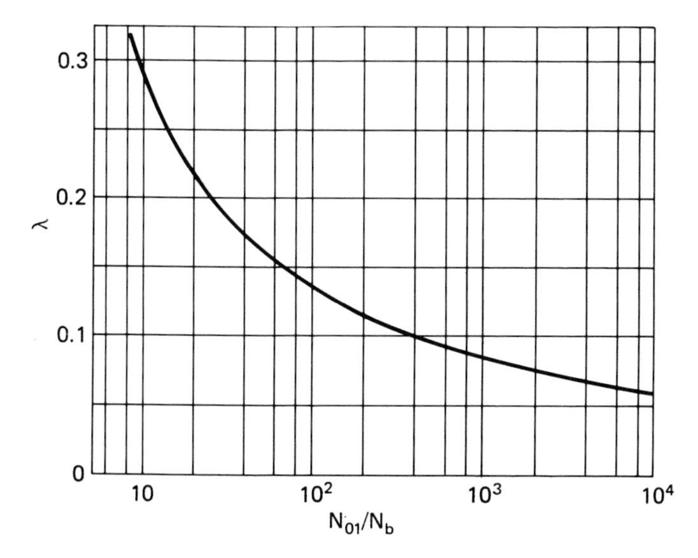

Figure 6–13. The parameter λ as a function of N_{01}/N_b.

The value of N_b is fixed within a certain tolerance when wafers are purchased. Checks of N_b by bulk resistivity measurements can keep the values for wafers to be batch-processed together within a close range. The value of N_{01}, however, is a true process variable, being affected by the predep temperature and the method by which the dopant is introduced into the furnace, a subject considered later. Certainly, in batch processing the dopant should be supplied such that N_{01} is the same for all the wafers in the batch.

6.7.4 Summary

This section has shown that of the three processing parameters, temperature has the greatest effect on x_{j1}. By the same token N_{01} and x_{j1} have the greatest effect on R_{S1}, and N_{01} depends on T_1; hence, temperature has the greatest effect on sheet resistance. It follows that furnace temperature must be under tight control. The design of—and control of temperature in—diffusion (and oxidation furnaces) is of great importance.

6.8 Furnaces

The temperature requirements on diffusion, oxidation, and annealing (Section 8.13) furnaces are the same: all the wafers in a batch, located in the *flat zone*, must be at the same temperature to within, say, $\pm 0.5°C$. Typical processing temperatures run from 850°C, for annealing, to 1100°C, for deep well diffusions. Then, if the nominal temperature is, say, 1000°C, dx_{j1}/x_{j1} will be less than 0.07% for either boron or phosphorus as the dopant. These values are quite adequate, and will be even lower at higher temperatures.

Most furnaces have horizontal tubes, as indicated schematically in Figure 1–5. Wafers are stacked in a line, with faces vertical, on a boat that carries them into the furnace.

Recently, a different furnace configuration has come into use. These employ a vertical heater housing, but the boat may be either vertical with horizontally stacked wafers or horizontal with vertically stacked wafers. Details vary from one manufacturer to another. One prototype form is sketched in cross section in Figure 6–14 [9].

Some of the advantages claimed for this form over the horizontal type are that (1) there is no FILO problem because the boat remains fixed while a large bell jar and heater unit move vertically relative to the boat; (2) wafers up to 10″ in diameter may be accommodated easily; (3) the boat may be rotated in the bell jar to provide more uniform temperature and gas distributions; and (4) the whole unit, for a given wafer batch size, occupies less clean-room floor space, a significant factor in view of high fabrication-facility costs.

Furnace construction details are discussed in Appendix H.

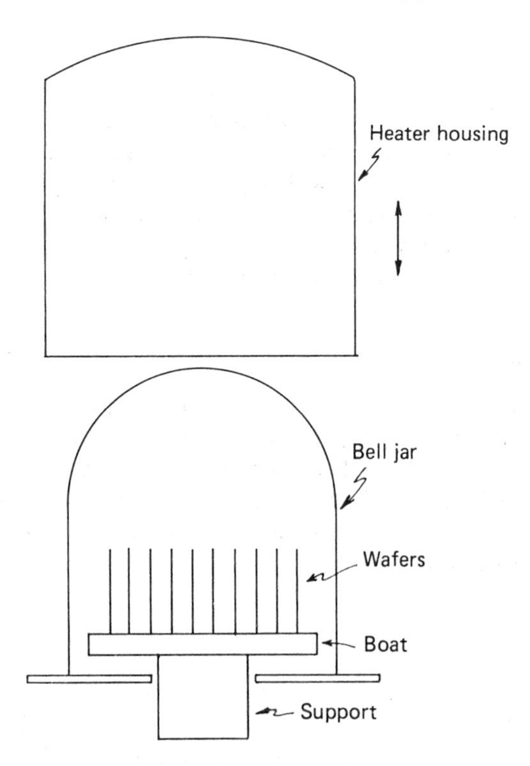

Figure 6–14. Cross section of a prototype vertical diffusion furnace. A radiant heater is located within the heater housing.

6.9 Predep Sources

Recall that the predeposition diffusion introduces dopant atoms into the host wafer. This section considers how this is done. Dopant transfer directly from the gas phase into the wafer is not satisfactory; solid-to-solid transfer works much better.

Predep actually consists of two, sometimes separate, processes: (1) deposition of the desired dopant into a solid layer on the wafer surface to provide a *local* source (i.e., on the wafer), a process called *local source deposition* (LSD), or *sourcing*, and (2) transfer of the dopant from the local source into the wafer by diffusion, which is the actual predep diffusion or *soak*. All the predep variations to be discussed here use the soak step, but they differ in the manner of sourcing.

Sourcing may take place either (1) before, or (2) after the wafers are in the diffusion furnace. In (1) the dopant material is applied directly to the wafer at room temperature as an *on-wafer* source. In (2) the dopant from a *primary* or *off-wafer* source is transferred at furnace temperature to the wafer surface to become the local source. In either event, the solid local-source-to-wafer transfer takes place during the *soak* at the desired temperature T_1.

Since batch processing of many wafers is used, predep should satisfy the following conditions for all wafers, independent of their location and orientation in the furnace.

1. Unmasked parts of all wafers must be doped equally.
2. The surface concentration N_{01} and junction depth x_{j1} must be uniform.
3. Wafers must be oriented for maximum use of the tube volume within the flat zone.
4. Hazardous source and by-product materials must be minimized. Effluents must be disposed of properly.
5. Dopant material deposition onto the furnace wall must be minimized.

In addition, for off-wafer sources:

6. The partial pressure of the dopant vapor should be (a) independent of the predep time t_1, and (b) not a strong function of the *wafer* temperature T_1.

6.10 On-Wafer Sources

When a metal such as gold is to be indiffused, a local wafer source may be prepared by depositing a solid layer of the metal onto the wafer by physical vapor deposition, PVD (Section 1.8.6 and Chapter 12), or electroplating. Another alternate form of on-wafer source uses a liquid compound of the dopant that is applied to the wafer by painting with a brush, spraying, or spinning. Whatever the actual method of application, this type of material usually is referred to as a *spin-on* source, since spinning is the most common application method.

In spinning the wafer is held by vacuum on a motor-driven chuck, and a few drops of the source liquid are dispensed onto the wafer surface. Typically, 50 μl per cm^2 of wafer surface to be covered is used [10]. When the wafer is spun at a few thousand r/min, the liquid is spread by centrifugal force into a uniform, thin layer on the wafer surface. This is dried at about 200°C (not in the predep furnace) to drive off volatile components and to free the dopant from its compound form. The resulting solid layer is the on-wafer source. Instructions for applying and spinning proprietary sources should be followed carefully. Subsequently, the wafer is placed into the furnace for the high-temperature soak.[7]

6.10.1 Plated Sources

A typical deposited metal, local source has a thickness of about 5000 Å. This is very thin indeed, yet this thin layer can act as an *infinite* dopant source—i.e., it shows no signs of depletion during the soak. This comes about primarily because, as illustrated by the following numbers, predep requires such a relatively small dose.

Consider a predep of Au, of density $\rho = 19.32$ and molecular weight M = 197. The Au layer on the wafer has a thickness $\Theta = 5000$ Å $= 5 \times 10^{-5}$ cm. Let n_a = Avogadro's number $= 6.02 \times 10^{23}$ atoms/mole. Then

$$N = \text{concentration of Au} = \frac{\rho n_A}{M}$$

This is easy to verify by simple dimensional analysis, viz

$$\frac{(\rho)(n_A)}{(M)} = \left[\frac{(\text{g/cm}^3)(\text{atoms/mole})}{(\text{g/mole})}\right] = \left[\frac{\text{atoms}}{\text{cm}^3}\right] = (N)$$

Substituting, we have in the Au local source

$$N_{\text{Au}} = \frac{(19.32)(6.02 \times 10^{23})}{197} = 5.9 \times 10^{22}\ \text{atoms/cm}^3$$

[7] An early and unusual application of this dopant type was used by Delco Electronics to fabricate p–i–n diodes used with car alternators for battery charging. A very lightly doped Si wafer was painted on both sides—one with a p-type source liquid, the other with an n-type liquid. After drying, the wafer was placed in the soak furnace. Boron and phosphorus have essentially identical diffusivities, so the doped regions approached the midline of the wafer cross section at the same rate, thus forming the structure with a single soak.

(Compare this with the concentration of Si, namely $N_{Si} = 5 \times 10^{22}$ atoms/cm^3.) So the number of atoms in 1 cm^2 of the 5000-Å-thick gold layer is

$$Q_{Au} = N_{Au}\Theta = (5.9 \times 10^{22})(5 \times 10^{-5}) = 2.95 \times 10^{18} \text{ atoms/cm}^2$$

We saw earlier in the chapter that the typical predep dose will run about $Q_1 \approx 10^{14} - 10^{15}$ atoms/cm^2. Thus, we may conclude that the source layer has at least 1000 times the required number of atoms for the predep, so there is no problem of source depletion.

These numbers show that $N_{source} > N_{SL}$, the limit of Au solid solubility in the silicon wafer, so we expect that N_{01}, the dopant concentration just inside the wafer surface, will be temperature dependent. Also, diffusivity is exponentially dependent on temperature, so the process is temperature limited during soak.

6.10.2 Spin-On Sources

Liquid proprietary solutions are used to prepare spin-on dopant sources. These usually comprise an organic solvent, viscosity-adjusting materials, and a compound of the dopant. During heat treatment the solvents evaporate and the compounds form a doped *oxide*, which then serves as the local source. What remains of this oxide layer after the soak can be removed, by etching in a buffered HF etch, without affecting the underlying Si.

A model for the situation resembles Figure 5–9, but with gas replaced by the local source, oxide by a region of the local source, and flux density across the source/Si interface given by Eq. (6.3-10). Notation would differ also. It can be shown, however, that N_{01}, just inside the Si surface, is directly proportional to N_S, the initial, uniform dopant concentration in the local source.

When the wafer is placed into the soak furnace, $N(0^+, t)$ in the Si, is initially zero. As the wafer temperature rises, $N(0^+, t)$ rises very rapidly to its steady state value N_{01}. Indications are that steady state is reached by the time the wafer is at T_1.

The Irvin R_S curves show that for an erfc profile the predep sheet resistance R_{S1} is determined by N_{01} and x_{j1}. Thus, for a fixed local source formulation (N_S fixed), temperature, and time, R_{S1} is fixed. For this reason, manufacturers of spin-on dopants often specify these parameters for the user to obtain a desired value of R_{S1}.

The range of spin-on formulations to give specified N_{01} values is amazingly large. Some values are listed in Table 6–1 for one manufacturer's product line, but they are typical for other manufacturers also.

Dopants other than those listed in the foregoing table are available in spin-on form. Typical of these are platinum, gold, and zinc, the latter for doping III-V semiconductor compounds. The manufacturers provide a wealth

Table 6–1. Range of N_{01} for Emulsitone Spin-On Sources

Dopant	Name	N_{01} Range, cm^{-3}	n Range
B	Borosilicafilm	1, 3, 5 × 10^n	16–20
P	Phosphorosilicafilm	5 × 10^n	16–20
		3 × 10^n	19–20
		1 × 10^n	17–21
As	Arsenosilicafilm	5 × 10^n	16–18
Sb	Antimonysilicafilm	5 × 10^n	16–19

Courtesy of Emulsitone Company.

of information in terms of temperature and time schedules to obtain specific values of R_{S1} and x_{j1}. Such data also are available in the literature [10].

Low N_{01} values, say in the 10^{16} cm^{-3} range, are relatively easy to obtain with the spin-ons, but not with other types. Undoped spin-on solutions for forming different sorts of glasses are available, too. These can serve as alternatives to thermal oxidation or CVD for forming a glass layer.

6.10.3 Practical Aspects

The use of on-wafer sources allows the transfer of dopant from solid to solid; the dopant does not go into vapor phase in the transfer process. Critical gas mixtures are not required in the furnace tube, in fact only an inert gas, such as nitrogen, at slightly greater than atmospheric pressure is needed to keep gases from back-flowing into the tube from the ambient and causing possible contamination. Wafers may be stacked very close together and normal to the tube axis, as in Figure 5–17, to provide better utilization of both the tube volume and the flat zone than does the arrangement shown in Figure 1–5. Since on-wafer sources are so thin, effectively only wafers occupy space in the boat; no other predep system can provide better space utilization. In fact, comparison to the ideal predep properties in Section 6.9 shows that the solid, on-wafer sources come off very well.

Of the common dopants only arsenic is toxic, especially when it gives off vapor on heating; hence when arsenic-doped spin-ons are being applied and heat-treated, it is essential that adequate venting be provided.

Despite the advantages of on-wafer sources, many wafer processors prefer to use off-wafer sources so that sourcing and soak both take place in the furnace tube.

6.11 Off-Wafer Sources

As an alternative to the on-wafer sources, a dopant, usually a gas-phase compound, is transferred from an *off-wafer* or *primary source* to the wafer surface, where the local deposition source (LDS), usually a doped oxide, is formed. Once this sourcing is completed, soak proceeds as it did in the last section.

Since B and P are the two most common silicon dopants, emphasis will be on them. We saw in Chapter 1 that the corresponding gas-phase compounds normally used are either B_2O_3 or P_2O_5. When these gases reach the wafer they react with the silicon to form a glassy layer of borosilicate or phosphosilicate glass that serves as the local source.

Off-wafer sources are categorized by their states at *room temperature*: gas, liquid, and solid. In all three cases, actual transfer from primary source to wafer surface for sourcing is in the gas phase. When the primary source is to be located *outside* the furnace tube at or near room temperature, the very low vapor pressures of B_2O_3 and P_2O_5 preclude their direct use as transfer gases; hence other boron- and phosphorus-bearing compounds are mixed with O_2 to generate those oxides in the hot tube.

These transfers and soaks in silicon technology are of the *open tube* type: the tubes are not sealed off, rather their effluent ends vent to the atmosphere. An end cap is provided that has an outlet tube of much smaller diameter than the furnace tube proper. This permits the carrier and dopant gases within the tube to be at a positive pressure relative to the atmosphere, to prevent backflow. The greater pressure in the tube is supported by the pressure drop in the constriction (Appendix D.2).

6.12 Gas Primary Sources

The generic gas-supply predep system is shown in Figure 6–15. Note provisions for delivering the source gas, N_2, and O_2 to the furnace. The vapor pressure of B is less than 10^{-8} torr at less than 1275°C, a range that extends above normal B predep temperature [11]. Therefore, a boron compound B_2H_6 (diborane, boron hydride, or boroethane), which is a gas at room

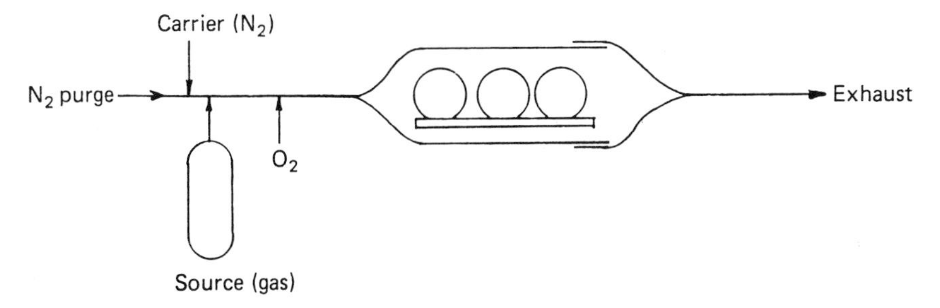

Figure 6–15. Generic gaseous source predep system. Regulators and flow meters are not shown.

temperature, is used as the primary gas source to generate B_2O_3. For satisfactory local source deposition, this must react with oxygen before reaching the silicon wafers. The B_2H_6/O_2 mixture reacts in the hot furnace tube to give:

$$2\ B_2H_6\uparrow + 6\ O_2\uparrow \rightarrow 2\ B_2O_3\uparrow + 6\ H_2O\uparrow \tag{6.12-1}$$

The B_2O_3 also will be in gas phase because it is at elevated temperature. The two end products then react with the wafer silicon to give further reactions.

$$2\ B_2O_3\uparrow + 3\ Si\downarrow \rightarrow 4\ B\downarrow + 3\ SiO_2\downarrow \tag{6.12-2}$$

$$6\ H_2O\uparrow + 3\ Si\downarrow \rightarrow 3\ SiO_2\downarrow + 6\ H_2\uparrow \tag{6.12-3}$$

The net effect of these reactions is that for each two molecules of diborane there are six molecules of silicon dioxide doped with four boron atoms. Six molecules of hydrogen are not used and pass out as effluent. Excess oxygen also can generate additional SiO_2.

The boron-doped oxide is a borosilicate glass and serves as the local source on the silicon wafers. It acts much like the spin-on boron sources, except that the local source here continues to form throughout the soak interval, replenishing itself.

With boron, N_{01} should not exceed $5 \times 10^{19}\ cm^{-3}$ or severe pitting of the wafer surfaces takes place, and the excess boron will locate interstitially and be inactive electronically.

The system shown in Figure 6–15 also is used for phosphorus sourcing from a gaseous primary source. Despite the fact that it is very toxic and explosive, phosphine (PH_3) is the gas of choice. Safety precautions are considered in the next section.

The system normally is operated with oxygen added to the gas flow to generate P_2O_5, and the probable sequence of reactions that takes place is

$$4\ PH_3\uparrow + 8\ O_2\uparrow \rightarrow 2\ P_2O_5\uparrow + 6\ H_2O\uparrow \tag{6.12-4}$$

and the two end products react with wafer silicon:

$$2\ P_2O_5\uparrow + 5\ Si\downarrow \rightarrow 4\ P\downarrow + 5\ SiO_2\downarrow \tag{6.12-5}$$

$$6\ H_2O\uparrow + 3\ Si\downarrow \rightarrow 3\ SiO_2\downarrow + 6\ H_2\uparrow \tag{6.12-6}$$

With excess oxygen present, the silicon also can be oxidized directly as given by Eq. (5.4-1). Thus, in this system the on-wafer local source is a phosphosilicate glass comprised of silicon dioxide doped with elemental phosphorus. The proportions are such that four molecules of PH_3 go over into four phosphorus atoms in $(8 + x)$ molecules of SiO_2. By adjusting the relative flow rates of PH_3 and O_2, one may adjust the phosphorus concentration in the local SiO_2 source. A marked similarity to the spin-on dopants may be noted again, but soak temperature has a very strong effect on N_{01} in the silicon.

It is essential that ample B_2H_6 or P_2O_5 reach every wafer of the batch in the furnace tube. Wafers often are aligned parallel to the axis of the tube as shown in Figure 1–5 when the primary source is a gas. If they are normal to the tube axis, they must be spaced far enough apart to allow gas flow to reach their surfaces without turbulence, and in a uniform manner. This normal stacking can waste much of the available tube volume and flat zone.

6.12.1 Safety Precautions

Gas primary sourcing for both B and P involves hazardous materials and proper safety precautions must be observed. Diborane is poisonous and explosive. To minimize its hazards, it generally is used in highly diluted form, say 0.1% B_2H_6 in an inert gas such as argon. Diborane is flammable *in air* in the percentage range from 0.8 to 98% diborane [12]. Thus, one simple means of reducing the possibility of fire is to keep the B_2H_6/air ratio less than 0.8%.

Phosphine (PH_3) is very toxic and explosive, so it, too, requires the same precautions as were outlined for the B_2H_6 system, particularly in having the supply tank filled with only a fraction of a percent by volume of PH_3 in argon. Phosphine is so flammable that it is considered to be *pyrophoric*, i.e., it will ignite spontaneously in air at or below room temperature in the absence of additional heat, shock, or friction (Section 9.4).

Toxicity is rated by a *threshold limit value*, or TLV—i.e., the average concentration of the gas in air to which a worker averaging five 8-hour days a week of exposure can withstand without adverse effect. The TLV for phosphine is 0.3 ppm in air, and 0.1 ppm for diborane [12].

Hydrogen is flammable, having a flammability range in air of 4 to 75% [12]. It, too, can be flamed off safely if its concentration in the gas mixture lies in the safe range. Flame-off consists of feeding the gas past a burning pilot light, so controlled burning occurs if the safe concentration range is exceeded.

Another important safety practice is *purging*—clearing air out of the system by thorough flushing with an inert gas such as nitrogen (Appendix D.4). If the predep system has been standing idle for some time with no gases flowing, its furnace tube and all the connecting plumbing will be back-filled with air. If a flammable gas, such as B_2H_6 or SiH_4, is then introduced into the system, an explosion may take place as the gas comes into contact with air. To prevent this from happening, N_2 should be turned on first and allowed to flow for some time to insure that all the air is flushed from the system. Then, and only then, is it safe to turn on the flammable gas source. By the same token, a similar procedure is used in shutdown. The nitrogen purge source is opened so that the system is flooded with nitrogen. With this still flowing, the flammable gas supply is shut off. This procedure insures that air cannot flow back into the tube while the flammable gas is still present. Diffusion furnaces usually are held at reduced temperature during standby intervals, and it is wise to keep a modest, but continuous, flow of nitrogen through them.

The foregoing safety precautions are not limited to diffusion systems. They are used wherever flammable gases are encountered. This is particularly true in CVD systems that use silane as a source gas (Chapter 9).

6.12.2 Summary

The gaseous primary source predep systems may be compared as a group to the ideal system.

1. Uniform dopant supply to all the wafers may present difficulties; turbulence between adjacent wafers must be avoided to prevent nonuniform doping on the surface of any wafer.
2. Some control of N_{01} is provided by adjusting relative gas flow rates and T_1.
3. Furnace space utilization is poorer than for solid sources.
4. Safety features must be provided to handle hazardous materials.
5. There will be some dopant deposition on the furnace walls.
6. Critical gas mixtures are required, so sophisticated flow-rate controllers are needed.
7. The dopant partial pressure may be made independent of predep duration t_1 with proper controllers and is not a strong function of furnace temperature T_1.

Despite some disadvantages, these sources are used widely. Relatively simple system hardware and the ability to deposit the local source and perform the soak simultaneously in the furnace tube are desirable. Gaseous primary-source systems have been around for a long time and if they are performing satisfactorily with high yields, managers are reluctant to change them unless great improvement is guaranteed.

6.13 Liquid Primary Sources

The gas primary sources may be replaced by liquids at room temperature, but the basic operation remains the same: vapor from the source liquid is oxidized to B_2O_3 or P_2O_5, which provide the actual local source deposition on the wafer surface.

As a group the liquid primary sources tend to involve materials less flammable than those of the gas sources, while still giving good control of N_{01}.

They provide an additional process variable in that the vapor pressure of the liquid source is temperature dependent.

The liquid-source system resembles the generic diagram shown in Figure D–3(d). An inert carrier gas, typically nitrogen, is bubbled through the source liquid, mixes with the source vapor, and *carries* it to the rest of the system. The liquid-source vessel (bubbler) is immersed in a liquid bath of constant temperature T_L. The source has significant vapor pressure at temperature T_L, and its partial pressure, p_1 here, is regulated by the bath temperature, while p_2 is controlled by a pressure regulator on the carrier gas supply line.

As derived in Appendix D.3.5 the *molecular* ratio, r, of the primary source vapor (subscript 1) to the carrier gas (subscript 2) is

$$r = \frac{p_1}{p_2} \tag{6.13-1}$$

and so is subject to control by T_L and p_2. A U-tube and petcock assembly comprising a liquid trap and drain for any condensates that form in the system, and a nitrogen purge line and an oxygen supply are provided as in Figure 5–13.

The liquid primary source of choice for boron doping is boron tribromide (BBr_3); its room-temperature vapor pressure is 55 torr, and it boils at 90.1°C. Appendix F.6.8 shows that its vapor pressure can be controlled over more than a decade by varying T_L over a 30°C range. Boron tribromide can be used either with or without added oxygen. The second case is spelled out in detail in several references [13]. More commonly the BBr_3 vapor is oxidized *before* the gases reach the silicon wafers. The basic gas-phase reaction is

$$4\ BBr_3\uparrow + 3\ O_2\uparrow \rightarrow 2\ B_2O_3\uparrow + 6\ Br_2\uparrow \tag{6.13-2}$$

The gas-phase B_2O_3 then reacts with the wafer silicon as in Eq. (6.12-2), while excess oxygen also reacts with the silicon as given by Eq. (5.4-1). Thus, 4 molecules of the tribromide result in $(3 + x)$ molecules of SiO_2 doped with 4 boron molecules in the local source. Control of N_{01} in the silicon is provided by adjusting the BBr_3(vapor)/O_2 ratio for a fixed soak temperature.

The liquid primary source of choice for phosphorus doping is phosphorus oxychloride, $POCl_3$ (or *pockle* as it is commonly called in the industry). Its room-temperature vapor pressure is 28 torr; it boils at 105.3°C. Appendix F.6.8 shows that a variation of about 54°C around room temperature changes its vapor pressure by a factor of ten.

Pockle vapor from the bubbler is mixed with oxygen in the system of Figure 5–13; the gas-phase reaction is

$$4\ POCl_3\uparrow + 3\ O_2\uparrow \rightarrow 2\ P_2O_5\uparrow + 6\ Cl_2\uparrow \tag{6.13-3}$$

The two reaction gases move on through the plumbing toward the furnace tube and wafers. If the input end of the tube is cool enough, some of the P_2O_5 will condense out into small particles. If these reach the hot wafers they may stick there and cause trouble in the subsequent deposition and LDS removal steps; hence, a wad of alumina wool is placed at the gas-input end of the furnace tube to act as a mechanical filter for the solid P_2O_5 particles.

Once P_2O_5 is formed, it reacts with Si to form the phosphosilicate glass local source. The overall set of reactions gives: four molecules of $POCl_3$ generate $(5 + x)$ molecules of SiO_2 doped with four molecules of elemental phosphorus; hence the concentration of phosphorus in the local source may be controlled by the $POCl_3/O_2$ ratio. This is quite similar to the situation for the BBr_3 system.

6.13.1 Safety Precautions

While none of the primary source materials involved in these reactions is flammable or explosive, the by-products are toxic. Bromine, Br, in Eq. (6.13-2), is highly toxic, having a TLV of 0.1 ppm in air. The primary source BBr_3 also requires care in handling. Air contains oxygen; hence, if the tribromide vapor reaches air, the reaction of Eq. (6.13-2) takes place and free bromine is released. Safe operation requires that the effluent be highly diluted with nitrogen (or air) before being released into the atmosphere.

There is one principal hazardous material resulting from pockle—the by-product chlorine, Cl_2. It is not flammable but will support combustion, since it is an oxidizing agent. Chlorine is extremely toxic, however, and contact with the gas-phase causes burns on the skin, eyes, and mucous membrane. It has a TLV of 1 ppm in air, and should be diluted thoroughly before being discharged into the atmosphere [12]. Pockle itself requires special handling because it generates Cl_2 on contact with air, Eq. (6.13-3).

6.13.2 Summary

In comparison with the ideal predep system, the liquid- and gas-primary-source systems are quite similar, differing only in a few details. We note the following points.

1. The transfer of dopant to the wafers takes place in the gas-phase, so supplying dopant uniformly to all the wafers in a batch has problems.
2. Control of N_{01} is similar in both systems—namely, through control of dopant concentration in the local source and through soak temperature T_1. Bath temperature T_L gives additional control with liquid sources.
3. Both types have the same volume utilization in the furnace tube by virtue of 1.

4. Both involve hazardous by-products.
5. Deposition on the furnace walls is the same for both types by virtue of 1.
6. Both types require critical gas mixtures, but methods of controlling dopant partial pressure differ.
7. The partial pressure dependence on t_1 and T_1 is the same for both types. With liquid sources the partial pressure is very dependent on the source-bath temperature T_L.

Several semiconducdor fabricators have preferred liquid $POCl_3$ over the gas PH_3 system since the toxicity and flammability hazards are less severe. In balance, the B_2H_6 gas system seems to find greater use than the liquid BBr_3 system.

6.14 Solid Primary Sources

Solid off-wafer primary sources consist of specially formulated wafers that contain either the desired dopant oxide itself or materials that can be converted to it. This type of solid source wafer is called a *planar diffusion* (or *dopant*) *source* (PDS), and is used *inside* the predep furnace.

The basic setup for predep with PDS wafers is shown in Figure 6–16. Prior to a predep run, silicon and *oxidized* PDS wafers are stacked into a boat in the pattern shown and then inserted into the furnace. The PDS and silicon

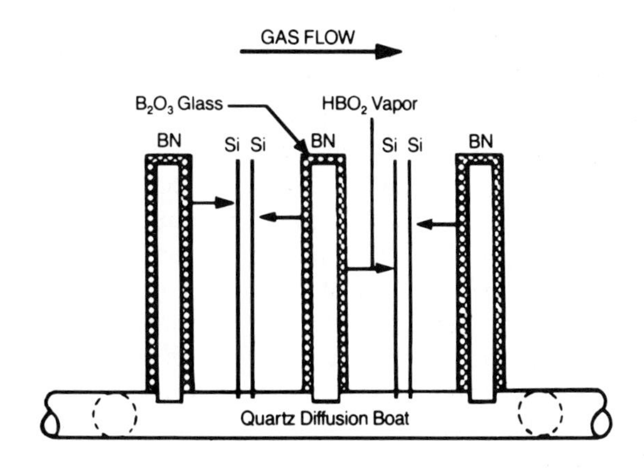

Figure 6–16. The PDS solid source system. Silicon and source wafers are stacked normal to the furnace tube axis [16]. (Courtesy of The Carborundum Company, Electronic Ceramics Division.)

wafers are at the *same* temperature. The desired dopant is transferred in *gas-phase* from a PDS to the adjacent silicon wafer, the transfer gas being B_2O_3 or P_2O_5, as the case may be. (An exception is discussed in Section 6.14.2.)

The front side of each silicon wafer sees a PDS of the same size very close by, a configuration that makes for uniformity of dopant transfer. The vapor reacts with the silicon surface to form either a borosilicate or phosphosilicate glass, which acts as the on-wafer local source. Overall uniformity of dopant transfer contributes to uniformity of local sources and final doping in the silicon wafers.

The wafers are spaced on centers of 0.08 to 0.125 in., are concentric to the furnace tube for uniform heating across each wafer, and the batch is centered in the furnace's flat zone. Generally the only *supplied* gas is dry N_2 [Figure 6–16], which flows along the length of the furnace tube—i.e., normal to the wafers as shown. This is not a *carrier* gas in the usual sense, but serves to prevent backflow of contaminants from the open end of the tube. The flow rate should be low enough to avoid turbulence around the wafer edges where it would militate against doping uniformity. Typical flow rates run in the range of 0.5 to 3 slpm (standard liters per minute), the larger value being for wafers 125 mm in diameter [15].

Planar dopant source wafers are available for boron, phosphorus, and arsenic doping.

6.14.1 Boron Sources

As made by one manufacturer, the composition of the PDSs are adjusted for the predep temperature range in which they are to be used. For $T_1 \leq 975°C$ the boron PDS contains polycrystal boron nitride (BN) with 5 to 8% of B_2O_3 distributed throughout the wafer [16]. This B_2O_3 is *not* the source material, but serves as a binder for the small, platelike crystals or *platelets* of BN. There are trace materials, too, typically carbon, calcium, and various chlorides, whose vapor pressures are much lower than that of B_2O_3 and do not affect the sourcing process [17]. These materials are hot-pressed to form a boule from which PDS wafers are sawed.

The formulation for higher predep temperatures is changed to BN and SiO_2 with the relative percentages changing from 60 BN/40 SiO_2 (for use up to 1100°C) to 40 BN/60 SiO_2 (for operation up to 1250°C). The SiO_2 is the binder material, and may require some etching after use to expose BN at the surface again.

All types require three treatment steps before use: cleaning, activation by oxidation, and stabilization by annealing. Detailed instructions are given by the wafer vendors so we consider only the high points.

The activation step is crucial since it converts BN to the B_2O_3 source material. The PDSs are placed in a furnace with an oxygen ambient and oxidized for about 30 min at 900 to 975°C. The following reaction takes place

$$4\ BN\downarrow + 3\ O_2\uparrow \rightarrow 2\ B_2O_3\downarrow + 2\ N_2\uparrow \qquad (6.14\text{-}1)$$

with the right-hand solid component forming a glassy oxide layer on the surface that is the source material as shown in the wafer cross sections of Figure 6–16. There are no toxic by-products.

After oxidation, the source wafers must be stabilized by annealing in nitrogen at furnace temperatures. Both oxidation and anneal may be carried out in the soak furnace but *with no silicon wafers present*; all the materials and temperatures are perfectly compatible.

The storage of PDSs is of concern; because B_2O_3 is hygroscopic, they will pick up water vapor from the air. The wafers may be stored in a sealed glass vessel with a desiccant present, or they may be left in the predep furnace, in the wafer boat, and at the lower standby temperature.

Another type of PDS wafer of quite different composition is the BORON$^+$ Planar Dopant Source, available from Owens-Illinois, Inc. Essentially, this is a glass wafer with a large amount of B_2O_3 distributed throughout it. Other components of the glass include BaO, MgO, Al_2O_3, and SiO_2, all of which have very low vapor pressures and so do not interfere with the transfer of gas-phase B_2O_3 [18].

The glass PDS has the advantage of not requiring activation by oxidation; the large amount of B_2O_3 in the wafer acts as the source material and at soak temperature diffuses to the surface from the wafer interior. Since the wafer is not covered entirely with a heavy B_2O_3 layer, water vapor absorption is not severe, making storage less of a problem. On the other hand, the glass type gives values of predep sheet resistance R_{S1} of two to three times higher than do the BN types for comparable times and temperatures.

Both the glass and BN types of wafers use the same boat-stacking pattern, that shown in Figure 6–16. In both types the PDS surface B_2O_3 has nonzero vapor pressure (Appendix F.6.9), that will transfer in vapor phase over to the adjacent silicon wafer. Typical R_{S1} figures cited for the PDS predep process are $\pm 2\%$ over an individual wafer, $\pm 3\%$ among wafers in the furnace flat zone, and $\pm 4\%$ overall from batch to batch [16]. These are excellent results.

Numerous curves of sheet resistance R_{S1} versus time and temperature for various PDS types are available in the literature and maufacturers' data. A typical curve set is given in Appendix F.6.10 [19].

The *silicon* wafers must be *deglazed* after soak, i.e., the borosilicate glass layer must be removed by a dip in HF solution. The PDS wafers must be stored as mentioned earlier and be reactivated periodically by oxidation. If the batch-to-batch R_{S1} values are monitored by four-point-probe (FPP) measurements, the need for reactivation is signaled by R_{S1} rising above acceptable values.

6.14.2 Hydrogen Injection

Over the years, as silicon wafers got larger, predep uniformity decreased, due in part to exacerbation of the FILO (first in, last out) problem by the greater thermal mass of the Si wafers. Practically speaking, all the wafers did not have the same $D_1 t_1$ value.

A solution was to add a pre-predep temperature equalization step and a modified sourcing method [20–22]. Furnace temperature initially is set much lower than normal T_1, say around 725 to 800°C, with a 1 : 1 N_2/O_2 mix by volume flowing, and the wafer-loaded boat is inserted. During this temperature equalization or *recovery* step, which lasts for something over 15 min, all the wafers reach the same temperature. The recovery time will depend on the wafer size and the number of wafers in the boat.

Negligible sourcing takes place during recovery because the vapor pressure of the B_2O_3 from the PDS is too low for effective transfer. Appendix F.6.9 shows that the vapor pressure will be less than 10^{-8} torr for a recovery temperature in the 700s°C range.

At the end of recovery time, *hydrogen injection* (HI) is used to initiate sourcing. With no change in temperature, hydrogen is added to the gas flowing through the furnace tube. Recall that H_2/O_2 ratios between 4 and 75% are flammable mixtures; hence the hydrogen is introduced at a rate to keep the ratio below 4%. Water vapor forms from the hydrogen and oxygen and this reacts with the B_2O_3 on the oxidized PDS wafers:

$$B_2O_3\downarrow + H_2O\uparrow \rightarrow 2\ HBO_2\uparrow \tag{6.14-2}$$

The product is *metaboric acid*, whose vapor pressure is orders of magnitude higher than that of B_2O_3 (Appendix F.6.9). Hence HBO_2 serves as the transferring species for local source deposition.

The reaction between HBO_2 and the Si wafer surface is

$$2\ HBO_2\uparrow + 2\ Si\downarrow \rightarrow 2\ B\downarrow + 2\ SiO_2\downarrow + H_2\uparrow \tag{6.14-3}$$

and as usual the excess oxygen or water vapor will oxidize the silicon directly so

$$x\ Si\downarrow + 2x\ H_2O\uparrow \rightarrow x\ SiO_2\downarrow + 2x\ H_2\uparrow \tag{6.14-4}$$

We see, then, that the local source on the Si wafers is a borosilicate glass whose doping level is two molecules of boron in $(2 + x)$ molecules of SiO_2. Soak temperature T_1 has stronger control on R_{S1} than the doping level of the borosilicate layer, however, so R_{S1} control by adjusting x of the oxygen flow rate is not used.

The transfer of HBO_2 and the buildup of the glassy local source layer on the silicon is very rapid, taking about 1 to 2 min. Note that HI provides sourcing at a temperature significantly *lower* than the soak T_1. This means that the soak can be optimized for R_{S1} adjustment.

Following sourcing, both H_2 and O_2 sources are turned off (in that order) and the system is purged with dry N_2 for 10 min. This is followed by a ramp-up in temperature to the desired T_1 soak value.

After the soak, furnace temperature is ramped down to the original recovery value, still with nitrogen flowing. Then the loaded boat may be removed from the furnace tube.

Note that sourcing and soak take place at different temperatures, and that good gas flow control is required. The advantage is that excellent uniformity in R_{S1} values is obtained on each wafer and from wafer to wafer in the batch. A disadvantage of the high vapor pressure of HBO_2 is that the inner surface of the furnace tube becomes coated with a boron-doped glassy layer during sourcing.

Hydrogen injection significantly reduces silicon surface pitting associated with low R_{S1} and high N_{01} values. To a large extent this is due to the ability of HBO_2 to source at a lower temperature than is possible with B_2O_3 [23].

Stach and Turley observed that at high soak temperature, say from 1100 to 1200°C (used for getting low R_{S1} without hydrogen injection), the sheet resistance value became dependent on the ambient gas [21]. They used He, Ar, and N_2 for their tests and found that the R_{S1} values decreased in that order. They also observed that the predep doping profile departed from the erfc u model usually assumed. Hydrogen injection, however, allowed lower T_1 values to be used for a given R_{S1}, so these problems were reduced. The higher vapor pressure of HBO_2 than B_2O_3 is the reason for this improvement.

As a side effect of hydrogen injection, a solid Si/B phase is formed at the silicon surface, of undetermined composition but probably containing SiB_4, SiB_6, and amorphous boron [23]. (Silicon is trivalent, having valences of +2, +4, and −4. We are used to the +4 valence in semiconductor work.) Apparently the solid layer serves two purposes. First, it tends to conserve Si by slowing down the reactions there and limiting the thickness of the local source. Second, it serves as a *gettering layer* that draws defects in the silicon to the wafer surface where they are gettered (Section 2.10.3). The Si/B layer is removed during the final etch phase after predeposition is completed.

Use of the HI technique is confined to the BN type of source wafers. It cannot be used with the glass type because the B_2O_3 is distributed throughout them and diffusion to the PDS surface would be too limited by a low sourcing temperature.

The literature indicates that PDSs have a distinct cost advantage over other B predep systems [24, 25].

6.14.3 Phosphorus Sources

Phosphorus-doped PDS sources also are available [26, 27]. These are sliced from a porous matrix of inert refractory materials that incorporates silicon pyrophosphate (SiP_2O_7), a solid at room temperature. No pre-predep oxidation or etching is required; only a 20- to 30-min anneal at T_1 in nitrogen is used. These wafers must *not* be stored at the mouth of the furnace tube.

The in-boat stacking patern of Figure 6–16 is used, and nitrogen flows in the furnace tube with a rate of 0.5 to 1 slpm to establish a back pressure against the atmosphere, without creating turbulence around the wafer edges. Sourcing and soak take place simultaneously at the same temperature T_1.

At predep temperatures of at least 900°C the SiP_2O_7 volatilizes to give the reaction

$$SiP_2O_7\uparrow \rightarrow P_2O_5\uparrow + SiO_2\downarrow \tag{6.14-5}$$

The gas moves from the interior of the PDS wafer by passing through the porous matrix structure. Also note that no oxygen is required in wafer preparation, either before or during predep, and that no hazardous materials are involved.

The P_2O_5 vapor reacts at the silicon wafer surface to produce the phosphorus-doped-oxide local source. Predep values of T_1 and t_1 may be determined from sheet resistance curves available in vendor literature. Data also are available for comparing the PDS with liquid- and gas-source methods. They present the planar sources in very favorable light—in terms of less complicated equipment and lower cost per silicon wafer [28].

6.14.4 Arsenic Source

An n-type PDS source doped with arsenic (As) also is available. The source material here is aluminum arsenate ($AlAsO_4$), which volatilizes at the sourcing temperature of 800°C where the reaction is [29]:

$$2\,AlAsO_4\downarrow \rightarrow Al_2O_3\downarrow + As_2O_3\uparrow + O_2\uparrow \tag{6.14-6}$$

with the arsenic trioxide (As_2O_3) acting as the transfer material. The trioxide is toxic with a TLV of 0.5 mg/m^3 and hence requires special handling. Note that the TLV unit here differ from those stated earlier for other materials.

The reactions at the silicon wafer surface during the sourcing are:

$$2\,As_2O_3\uparrow + 3\,Si\downarrow \rightarrow 4\,As\downarrow + 3\,SiO_2\downarrow \tag{6.14-7}$$

$$O_2\uparrow + Si\downarrow \rightarrow SiO_2\downarrow \tag{6.14-8}$$

so an arsenic-doped glassy layer is deposited and serves as the local source during soak.

Oxygen may be furnished to adjust the As concentration in the local source, but this is not a common procedure, since the soak T_1 and t_1 may be used for adjusting R_{S1}.

6.14.5 Summary

Of the three types of off-wafer sources, the solid PDS form is the most recently developed and has found widespread acceptance in the semiconductor industry. The following points may be noted for comparison with the criteria of the ideal predep system.

1. The equality of dopant supplied to all wafers in the boat is excellent, since the front surface of each silicon wafer sees a very uniform, nearby source of dopant (a PDS wafer) for sourcing.
2. Control of N_{01} (and R_{S1}) is excellent mainly through the choice of soak time and temperature.
3. Volume utilization in the furnace tube is very good but less so than for on-wafer sources, which occupy virtually no space.
4. Toxic materials are not used or produced as by-products except with the arsenic source. With H_2 injection, hydrogen can be flammable in some concentrations, so proper safety facilities must be provided.
5. There is little dopant deposited on the furnace tube walls except with hydrogen injection.
6. Critical gas mixtures are not needed except with hydrogen injection.
7. The dopant partial pressure is independent of t_1 and is not a strong function of T_1.

In general the PDS predep system utilizes simpler plumbing than the liquid and gas source systems, and offers a considerable cost advantage over the other two.

6.15 Predep Evaluation

After predep has been completed and any LDS residue has been etched from the silicon, the wafers should be checked to verify dose and doping distribution. A four-point-probe measurement of sheet resistance R_{S1} may be used under certain circumstances. Measurements made at different points on a wafer's surface give checks of predep uniformity over that wafer. Measurements on several wafers at, say, their centers will give a check for uniformity over an entire wafer batch.

In addition, measurements of this kind give an overall check of the primary source condition. For example, if R_{S1} readings are consistently high, with T_1 and t_1 known to be correct, primary source depletion is probably the problem.

Checking of surface concentration N_{01} requires measurement of the junction depth x_{j1} (Section 3.14). These values are quite small, say less than 0.5 μm as shown in the illustrative example in Section 6.4.1, and so are difficult to measure. But given R_{S1}, x_{j1}, and an assumed erfc profile, N_{01} may be read from the Irvin R_S curves. Two other profiles are considered in Sections 7.11 and 7.12.

6.16 Thermomigration

In the late 1970s H. E. Cline and T. R. Anthony of the General Electric Research and Development Center developed a process, *thermomigration*, for doping silicon to very deep levels in times very short in comparison to diffusion [30].

The wafer to be doped is subjected to a temperature difference between its two faces, and a liquid-phase dopant at the cooler surface moves into the silicon toward the hotter side. Thermomigration has been used primarily with aluminum as the dopant.

In practice an aluminum film is deposited on one surface of the wafer by vacuum evaporation. This is etched into the desired doping pattern by conventional photolithographic techniques. The wafer is placed horizontally in a furnace with the Al-coated side down and facing a water-cooled heat sink. The upper surface is brought to a temperature between 800°C and the silicon melting point, typically 1000°C, by a radiant heater. (There are problems in maintaining this temperature difference across a thin wafer.) Melting takes place at the 577°C eutectic at Al/Si interfaces on the lower surface, and the melted aluminum migrates rapidly toward the hot upper surface. The process is so rapid, it may be carried out in air. By way of example, these figures are cited: Thermomigration will dope an 11-mil-thick wafer in 2 min as opposed to 124 hours using diffusion from a gaseous primary source. The process has been used for making power transistors, SCRs (silicon controlled rectifiers), and thyristors.

Problems

6–1 Using Figure 6–5 as an aid, derive FSL, Eq. (6.2-2).

6–2 Write a computer program to solve Eq. (6.4-2). Use the algorithm of Appendix F.6.4 and the Newton-Raphson method.

6–3 Given $N_1(x_1, t_1)$ for a predep diffusion, derive equations for

(a) $\mathscr{F}_1(0, t_1)$, the indiffusing dopant flux density at the wafer surface. Begin your derivation with Fick's first law and assume D_1 is constant.

(b) Q_1, the predep dose, using your results from part (a).
(c) Check your results against those in the text.

Note: The next four problems use only a predep diffusion. For each case calculate x_{j1} and R_{S1}. Tabulate your results. Compare results and note trends. Use computer solutions if available. Indicate if the SOA is valid.

Units are N, cm^{-3}; t, s; T, °C. In all cases $N_b = 5 \times 10^{15}$ cm^{-3}, $T_1 = 1050$°C.

Problem	Background	Indiffuser	N_{01}	t_1
6–4(a)	As	B	2×10^{20}	900
	As	B	2×10^{20}	1800
	As	B	2×10^{20}	2700
(b)	B	As	2×10^{20}	1800
6–5(a)	P	B	2×10^{20}	900
	P	B	2×10^{20}	1800
	P	B	2×10^{20}	2700
(b)	P	P	2×10^{20}	1800
6–6	Al	P	2×10^{20}	900
	Al	P	2×10^{20}	1800
	Al	P	2×10^{20}	2700
6–7	B	P	10^{20}	1800
	B	P	10^{19}	1800
	B	P	10^{18}	1800

6–8 (a) By calculation, verify the computer solutions for Problems 6–4(a) and (b), and 6–5(a) and (b).
(b) Calculate the dose Q_1 for these conditions. Tabulate your results.

6–9 A Si wafer is to be uniformly doped with both Al and P. The P concentration is to be 5×10^{18} cm^{-3}. A single outdiffusion is to be performed for 4 h at 1200°C. The desired junction depth is 0.95 μm. Assume infinite sinks for both diffusants.

(a) What N_{Al} is required? Do not assume the SOA. Note that it is useful to evaluate erf u here by the first three terms of the series expansion.
(b) Why must the SOA not be used?
(c) Make a rough sketch of the Al and P profiles at the end of the 4-h diffusion.

6–10 (a) The grade constant G for a junction is defined as the net profile slope magnitude at the junction.

$$G = |dN(x)/dx| \quad \text{at } x_j$$

Derive an expression for G for an erfc profile with a slow diffuser background. Your expression should involve only N_{01}, x_{j1}, and $\mathscr{P}$.
(b) Evaluate G for $N_b = 10^{16}$ cm^{-3}, $N_{01} = 10^{20}$ cm^{-3}, and $x_{j1} = 5$ μm.

6–11 You have a Si wafer 18 mils thick, doped at 10^{15} cm^{-3} with As.

(a) Calculate ρ and R_S.
(b) The wafer is given a 15-min B predep at 950°C with $N_{01} = 3.9 \times 10^{20}$ cm^{-3}. Calculate x_{j1}, $\langle \sigma_{S1} \rangle$, and R_{S1}.

6–12 Consider the masking provided by SiO_2 against indiffusion of B into Si during predep. Calculate and tabulate $(x_{ox})_{min}$ for these conditions: $T_1 = 1200$°C, $t_1 =$ 40 min, $D_{ox} = 5 \times 10^{-15}$ cm^2/s, $D_1 = 2.5 \times 10^{-12}$ cm^2/s, $N_{01} = 5 \times 10^{20}$ cm^{-3}, $m = 0.3$, and $N_b = 10^{15}, 10^{16}, 10^{17}$ cm^{-3}.

6–13 Calculate $\Delta x_{j1}/x_{j1}$ directly using finite increments and compare your results with the logarithmic derivative, Eq. (6.7-5). Comment.

6–14 Hydrogen and oxygen are supplied separately to a torch. Which single gas should be turned on before being ignited, or may both be on before ignition? Explain.

6–15 Consider an Si wafer doped with As at 10^{17} cm^{-3}. A boron predep is run for 30 min at 1100°C.

(a) Read R_{S1} from the Goldsmith curves of Appendix F.6.10.
(b) Assume $N_{01} = N_{SL}$, and calculate x_{j1} and $\langle \sigma_{S1} \rangle$. Determine, say, N'_{01} from the appropriate Irvin curve.
(c) Compare N_{01} and N'_{01} and comment on your results.
(d) To get a closer check should you assume a larger or smaller value of N_{01}? Why? What practical problem is involved?

6–16 As part of the Motorola MOSAIC process the As diffusion schedule shown in Figure 7–4 is used. We assume that T varies linearly in the ramp regions, with m symbolizing slope—e.g., in ramp 1, $T = 800 + mt + 273$, K. Use $E_a = 3.22$ eV and $D_\infty = 0.026$ cm^2/s.

(a) Using 5-min intervals calculate Δ_{ramp} for ramp 1 by the trapezoidal rule.
(b) Calculate Δ for the entire diffusion.
(c) Need the contributions of the ramps to the total Δ be considered? Explain briefly.

References

[1] H. Wolf, *Semiconductors.* New York: Wiley, 1971. Chapter 2.
[2] B. G. Streetman, *Solid State Electronic Devices, 2nd Ed.* Englewood Cliffs: Prentice-Hall, 1980.
[3] A classic paper on diffusion in forming semiconductor devices is F. M. Smits, "Formation of Junction Structures by Solid State Diffusion," *Proc. IRE,* **46,** (6), 205–233, Jan. 1958.
[4] Many references are available for solving the partial differential equation of FSL subject to different initial and boundary conditions—for example, S. K. Ghandhi, *The Theory and Practice of Microelectronics.* New York: Wiley, 1968. Ghandhi gives solutions in familiar solid state terminology.
[5] N. Holonyak, Jr., private communication.
[6] J. C. Irvin, "Resistivity of Bulk Silicon and of Diffused Layers in Silicon," *Bell Syst. Tech. J.,* **41,** (2), 387–410, March 1962. This paper should be consulted

for a discussion of the approximations used by the author. Here we shall assume that *all* the dopant atoms are ionized. Notation is changed to conform to our usage here.

[7] C. T. Sah, H. Sello, and D. A. Tremere, "Diffusion of Phosphorus in Silicon Oxide Film," *J. Phys. & Chem. Solids,* **11,** (3), 288–298, March 1959.

[8] D. P. Kennedy and R. R. O'Brien, "Analysis of the Impurity Atom Distribution Near the Diffusion Mask for a Planar p–n Junction," *IBM J. Res. Dev.,* **9,** (3), 179–186, May 1965. Notation has been changed to be consistent with that used in this book. In particular, C for concentration is replaced by our N, and the dose has been identified as such and replaced by our Q.

[9] P. H. Singer, "Trends in Vertical Diffusion Furnaces," *Solid State Technol.,* **9,** (4), 57–60, April 1986. See also p. 216.

[10] B. H. Justice, D. F. Harnish, and H. F. Jones, "Diffusion Processing of Arsenic Spin-On Diffusion Sources," *Solid State Technol.,* **21,** (7), 39–42, July 1978.

[11] *Handbook of Thin Film Materials.* El Segundo: Sloan Technology Corp., Sloan Materials Division, 1971.

[12] *Safety Precautions and Emergency Procedures for Specialty Gases, Form No. L12-237B of 11/82.* Somerset: Union Carbide Corporation, Linde Division, 1982.

[13] For example, see D. Rupprecht and D. B. Ott, "Boron Nitride as a Diffusion Source," *Solid State Technol.,* **15,** (1), 6, Jan. 1972. This paper compares liquid- and solid-primary-source systems.

[14] F. A. Trumbore, "Solid Solubilities of Impurity Elements in Germanium and Silicon," *Bell Syst. Tech. J.,* **39,** (61), 205–233, Jan. 1960. Fig. 1.

[15] J. Monkowski and J. Stach, "System Characterization of Planar Source Diffusion," *Solid State Technol.,* **19,** (11), 38–43, Nov. 1976. Has some interesting illustrations of the gas flow patterns around and between the wafers in the boat as a function of the gas volumetric flow rate.

[16] *Transtar™ Boron Nitride Low Temperature Planar Diffusion Source (Grade BN-975), Form A-14,005.* Niagara Falls: The Carborundum Company, Electronic Ceramics Division, 1983.

[17] N. H. Ditrick and M. S. Bae, "An Improved Boron Nitride Glass Transfer Process," *Solid State Technol.,* **23,** (7), 69–73, July 1980.

[18] J. J. Steslow, J. E. Rapp, and P. L. White, "Advances in Solid Planar Dopant Sources for Silicon," *Solid State Technol.,* **18,** (1), 31–34, Jan. 1975.

[19] N. Goldsmith, J. Olmstead, and J. Scott, "Boron Nitride as a Diffusion Source for Silicon," *RCA Rev.,* **28,** (2), 344–350, June 1967.

[20] *Low Defect Boron Diffusion Process Using Hydrogen Injection, Form A-14,004, 2/86.* Niagara Falls: The Carborundum Company, Electronic Ceramics Division, 1986.

[21] J. Stach and A. Turley, "Anomalous Boron Diffusion in Silicon from Planar Boron Nitride Sources," *J. Electrochem. Soc.,* **121,** (5), 722–724, May 1974.

[22] D. Rupprecht and J. Stach, "Oxidized Boron Nitride Wafers as *In-Situ* Boron Dopant for Silicon Diffusions," *J. Electrochem. Soc.,* **120,** (9), 1266–1271, Sept. 1973.

[23] T. R. Facey, J. Stach, and R. E. Tressler, "The Use of Boron Rich Layer Formation During Boron Predeposition as a Method of Crystal Defect Elimination in

Silicon," *Proceedings of the 30th Electronic Components Conference, San Francisco, CA, 28–30 April 1980.* New York: IEEE, 1980. pp. 28–30.

[24] D. B. Ott, *Economic and Technical Advantages of Planar Diffusion Sources Form C-910, 5/79.* Niagara Falls: The Carborundum Company, Electronic Ceramics Division, 1979.

[25] *PDS^R Brand Planar Diffusion Sources for Boron and Phosphorus, Form C-909, 5/79.* Niagara Falls: The Carborundum Company, Electronic Ceramics Division, 1979. Comparisons are made of BN sources relative to BBr_3, BCl_3, and B_2H_6.

[26] *PDS^R Phosphorus PH-950* n-*Type Planar Diffusion Source, Form A-14,028, 12/85.* Niagara Falls: The Carborundum Company, Electronic Ceramics Division, 1985.

[27] *PDS^R Phosphorus PH-1000* n-*Type Planar Diffusion Source, A-14,041, 11/85.* Niagara Falls: The Carborundum Company, Electronic Ceramics Division, 1985.

[28] See, for example, *Solid Source Phosphorus Doping, A Report from a Large MOS Manufacturer using 150-mm PH-990 Wafers.* An unsourced and undated report in product information furnished by The Carborundum Company, Electronic Ceramics Division, Niagara Falls, NY.

[29] *PDS^R Arsenic 1000-L* n-*Type Planar Diffusion Source, Form A-14,042, 12/85.* Niagara Falls: The Carborundum Company, Electronic Ceramics Division, 1985.

[30] "Dope Wafers in Minutes with Thermomigration," *Circuits Manuf.*, **17,** (9), 14, 16, Sept. 1977.

Chapter 7
Diffusion; Redistribution

Once the dopant atoms have been predepped into the host wafer and the initial source is removed, they may be redistributed by diffusing from high- to low-concentration regions, if the temperature is high enough. The *drive* diffusion, in which dopant atoms are driven deeper into the host, is an example of redistribution. Predep and drive often are used in sequence as a two-step process.

Much of this chapter is concerned with the drive diffusion and topics related to it. Near the chapter's end, some important examples of anomalous diffusions (in which the dopant profiles differ considerably from the more commonly assumed forms of the complementary error function, or gaussian) are considered.

7.1 Drive Diffusion (Case IV)

Consider the two-step diffusion process. The first step, or predep diffusion, introduces a relatively thin layer of dopant just within the host surface, with a large value of surface concentration, N_{01}. If the predep diffusivity D_1 is concentration independent, the profile has the complementary error function form.

The second step, known as the *drive* or *redistribution* diffusion, then moves the predepped dopant farther into the host, with no additional dopant furnished externally. The surface concentration drops below N_{01} during this process. Initially we assume that the predepped dopant does not outdiffuse during drive so that the total amount of dopant introduced into the host during predep (or dose Q_1) remains unchanged but assumes a different profile shape.

We begin by considering conditions for a solution of Fick's second law (FSL) for the drive following an erfc predep. Then, two approximate forms which are much easier to evaluate and are adequate for most situations will be considered.

7.1.1 Exact Solution

The descriptive phrase for this case is "diffusion within a host from an initial erfc predep distribution for a limited time." The words "diffusion within" imply that no dopant atoms outdiffuse from the host during the drive. The initial condition for this case is sketched in Figure 7–1(a).

Consider the initial (IC) and boundary (BC) conditions. To save writing, let

$$\Delta_1 = D_1 t_1 \tag{7.1-1}$$

The IC is given by

$$N_2(x, 0) = N_1(x, t_1) = N_{01}\,\mathrm{erfc}\frac{x}{\sqrt{4\Delta_1}} \tag{7.1-2}$$

The BC *for a limited time* is handled as we did for the predep in Chapter 6, by assuming an infinitely thick host:

$$N_2(\infty, t_2) = 0 \tag{7.1-3}$$

The second BC, that of no outdiffusion, can be stated as

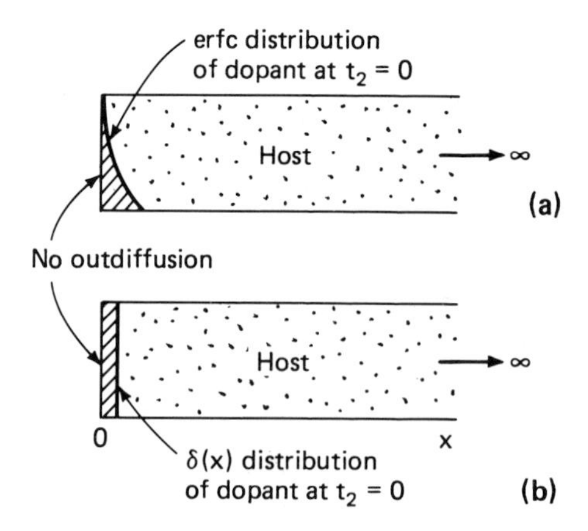

Figure 7–1. Boundary conditions for the drive diffusion. (a) Exact case. (b) Gaussian approximation.

$$\left.\frac{\partial N_2(x, t_2)}{\partial x}\right|_{x=0} = 0 \tag{7.1-4}$$

Note that, if this is true, Fick's first law (FFL) assures that the dopant flux density across the host surface at $x = 0$ is zero; there can be no outdiffusion (and, for that matter, no indiffusion from the outside, either).

Smith has given a heat-diffusion solution for these conditions that applies in this case and has the same IC and BCs [1]. His results require the use of a tabulated function. This so-called "exact solution" is troublesome to use: the solution requires interpolation of two parameters in a table. It also depends upon the assumption of Eq. (7.1-2) and usually is not used. A simpler initial condition has greater applicability and provides ease of use.

7.1.2 Gaussian Approximation (Case IVa)

Assume that a *shallow* predep of any reasonable shape may be approximated by a Dirac δ function, if the dose Q is kept at the predep value Q_1.[1] This means that we may model the thin $\text{erfc}(x/\sqrt{4D_1 t_1})$ layer by an infinitesimally thin layer [Figure 7–1(b)]. Keeping Q_1 unchanged guarantees that the total amount of dopant introduced into the host during predep is preserved during drive without gain or loss. This assumption changes only the initial condition which now becomes

$$N_2(x, 0) = Q_1 \delta(x) \tag{7.1-5}$$

where

$$Q_1 = 2N_{01}\sqrt{\frac{\Delta_1}{\pi}} \tag{7.1-6}$$

is still the same as given by Eq. (6.3-11). The BCs remain as before.

The solution to FSL for these three conditions is

$$\begin{aligned} N_2(x, t_2) &= \frac{Q_1}{\sqrt{\pi \Delta_2}} \exp\left(\frac{-x^2}{4\Delta_2}\right) \\ &= \frac{2N_{01}}{\pi}\sqrt{\frac{\Delta_1}{\Delta_2}} \exp\left(\frac{-x^2}{4\Delta_2}\right) \end{aligned} \tag{7.1-7}$$

This often is referred to as the *gaussian approximation* for the drive diffusion.

---

[1] The Dirac function $\delta(x)$ is rectangular in shape, of infinite height, and infinitesimal width, yet its area is unity.

The surface concentration may be obtained by setting $x = 0$ in the last equation; thus

$$\left.\begin{aligned} N_2(0, t_2) = N_{02} &= \frac{2N_{01}}{\pi}\sqrt{\frac{\Delta_1}{\Delta_2}} = \frac{2N_{01}}{\pi}\alpha \\ &= \frac{Q_1}{\sqrt{\pi \Delta_2}} \end{aligned}\right\} \tag{7.1-8}$$

where

$$\alpha = \sqrt{\frac{\Delta_1}{\Delta_2}}, \quad \text{and} \quad \Delta_2 = D_2 t_2$$

Remember that $\Delta_i = D_i t_i$.

The gaussian approximation of Eq. (7.1-7) is easy to use because it may be evaluated on a calculator without resort to tables. However, the question is, under what conditions may it be used? We can argue physically that as the drive becomes deeper wrt the erfc predep, the latter "looks" more like the Dirac δ form. It may be shown from Smith's solution that the gaussian approaches the exact form as $\alpha = \sqrt{\Delta_1/\Delta_2}$ and if $\beta = x^2/4(\Delta_1 + \Delta_2)$ become small (e.g., for $\alpha = 0.2$ and $\beta \leq 2.5$ the gaussian approximation checks to within 5%. If β increases to 5, the difference remains less than 8%.

Consider an illustrative example. It is common practice to perform drives at higher temperatures and for longer times than predep, two facts that make it easy to keep $\alpha \leq 0.25$. For example, using the diffusivity numbers for boron and phosphorus from Appendix F.6.1 we have

Diffusion	T, °C	D, cm^2/s
Predep	900	1.5×10^{-15}
Drive	1000	2.6×10^{-14}

Thus

$$\alpha = \sqrt{\frac{\Delta_1}{\Delta_2}} = \sqrt{\frac{D_1 t_1}{D_2 t_2}} = \sqrt{\frac{1.5 \times 10^{-15} t_1}{2.6 \times 10^{-14} t_2}} = 0.24\sqrt{\frac{t_1}{t_2}}$$

Then even for equal predep and drive times ($t_1 = t_2$) the inequality would be satisfied by these two temperatures.

7.1.3 S/M Approximation (Case IVb)

Shrivastava and Marshak have presented a slight modification of the gaussian approximation that does away with the need for checking α. An empirical corrective term is added to Δ_2 to provide adequate agreement with the exact form [2]. The modified form of this SMA approximation is

$$N_2(x, t_2) = N_{02} \exp\left(\frac{-x^2}{4\Delta_2'}\right), \quad \text{SMA} \tag{7.1-9}$$

where

$$N_{02} = \frac{2N_{01}}{\pi} \sqrt{\frac{\Delta_1}{\Delta_2'}} \tag{7.1-10}$$

and

$$\Delta_2' = \frac{\Delta_1}{1.3} + \Delta_2 \tag{7.1-11}$$

It is clear that as Δ_2 increases relative to Δ_1, the correction term drops out, leaving the original gaussian solution. The S/M approximation (SMA) may be used in appropriate problems in this book except where another method of solution is prescribed for pedagogic reasons.

7.1.4 Drive Profiles

Both of the approximate solutions for the drive diffusions are of the general gaussian form $\exp(-u^2)$ and plot as shown in Figure 7–2. In both sets of curves the surface concentration is time dependent, as indicated in the equations for N_{02}. Notice that Figure 7–2(a) uses linear scales, and the areas under the several curves are all equal. This is easy to reconcile physically: the area under a curve is proportional to all the redistributing dopant in the host, and that remains constant.

If the foregoing equations are to be used for the two-step diffusion, we must insure that the IC and BCs for the second diffusion are satisfied. No dopant atoms are to be added to or removed from the host wafer during the drive. There are two aspects of this problem that must be considered in practice.

The first has to do with adequate cleanup of the wafer after predep. Recall that wafers must be *deglazed*, i.e., stripped of the local-source borosilicate or phosphosilicate glass with an HF etch (Section 1.7.1), and that separate furnaces are used for predep and drive. If these glasses are not removed completely, they will continue to act as a *predep* source when the wafer is placed in the drive furnace. This extraneous source will dope other wafers in the batch and the furnace tube as well. Tube-wall doping can carry over to subsequent wafer batches and cause departures from drive conditions. It is therefore essential to maintain the integrity of the drive-furnace tube, making sure that it does not harbor any dopants.

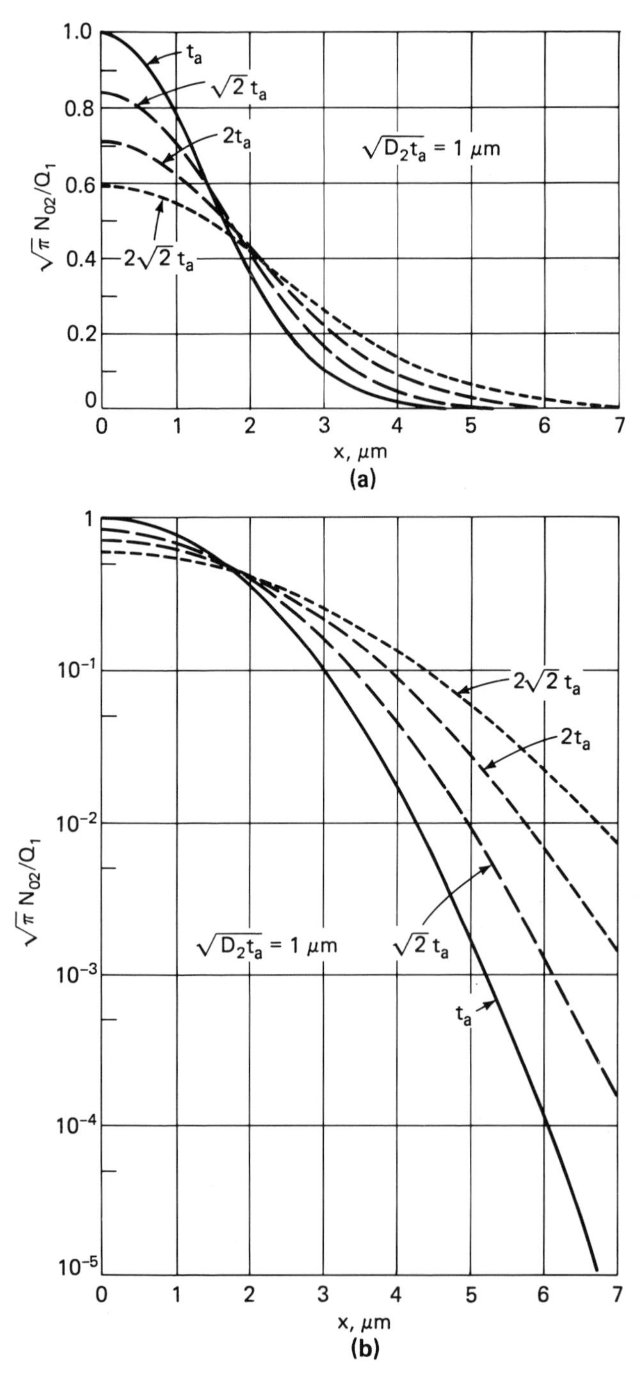

Figure 7–2. Gaussian solutions for the drive diffusion. The x scales are calibrated for $\sqrt{D_2 t_a} = 1\ \mu m$. (a) Linear coordinates. (b) Semilogarithmic coordinates.

The second aspect involves inadvertent removal of atoms from the wafer during drive with a corresponding decrease in dose. This is extremely difficult to overcome in practice. The requirements for PR processing demand that drives be carried out, at least in part, in an oxidizing atmosphere so that SiO_2 grows on the wafer surface as the drive progresses. We shall see in later sections that some of the dopant redistributes itself into the growing oxide during drive and affects the profile of the diffusing atoms, having the greatest effect near the surface and hence on N_{02}.

The equations derived thus far are based on assumptions that are not always realized in practice; also, second-order effects come into play. Even when elaborate computer programs, which can account for these effects and interactions among successive processes, are used, inevitably some empirical fine-tuning must follow. Our equations do give first-order approximations to the real world, however, and can give an understanding of the processes involved while serving as a first step in developing process specifications.

7.2 Drive Junction Formation

If a p–n junction was formed during predep, it will move to x_{j2}, deeper in the wafer, during drive. We consider the worst case, where the *background* dopant outdiffuses into an infinite sink (as in Case II of Chapter 6). Remember that the dopant will outdiffuse during both the predep and drive cycles of the indiffuser, and that its diffusivity may differ from that of the indiffuser; hence, we follow the notation of the last chapter and use the first subscript o to identify its values. Borrowing results from the next section, we shall assert that if, as assumed, the boundary condition at the wafer surface remains the same for the outdiffuser during both cycles, its profile will continue in the error function form.

The indiffuser, on the other hand, will convert from erfc u to a gaussian, which we shall model by Case IVb—i.e., the SMA form. At the junction x_{j2} the *net* concentration, i.e., the algebraic sum of the indiffuser and outdiffuser concentrations, will be zero. (Sign conventions are defined in Section 6.4.) Thus, we write

$$\pm N_{02} \exp\left(\frac{-x_{j2}{}^2}{4\Delta_2{}'}\right) \mp N_b \operatorname{erf}\left(\frac{x_{j2}}{\sqrt{4(\Delta_{o1} + \Delta_{o2})}}\right) = 0 \qquad (7.2\text{-}1)$$

Again, we shall assume a slow outdiffuser approximation (SOA), which is based on the inequalities

$$\operatorname{erf}\left(\frac{x_{j2}}{\sqrt{4(\Delta_{o1} + \Delta_{o2})}}\right) \geq 0.995 \approx 1$$

or

$$\frac{x_{j2}}{\sqrt{4(\Delta_{o1} + \Delta_{o2})}} \geq 2 \qquad (7.2\text{-}2)$$

Subject to this condition, Eq. (7.2-1) can be written as

$$\left.\begin{aligned} x_{j2} &= \sqrt{4\Delta_2' \ln\left(\frac{N_{02}}{N_b}\right)} \\ \text{where} \qquad & \\ N_{02} &= \frac{2N_{01}}{\pi}\sqrt{\frac{\Delta_1}{\Delta_2'}} \end{aligned}\right\} \text{Drive, SOA} \qquad (7.2\text{-}3)$$

Consider some typical calculations. A silicon wafter doped with antimony at $N_b = 10^{15}\ \text{cm}^{-3}$ undergoes a two-step boron diffusion. Conditions are $T_1 = 950°\text{C}$, $t_1 = 15$ min, $T_2 = 1150°\text{C}$, $t_2 = 60$ min. During predep N_{01} is at the limit of solid solubility. Calculate N_{02} and x_{j2}.

The diffusivities and limit of solid solubility for boron can be read directly from Appendix F.6.1:

$$D_1 = 6.6 \times 10^{-15}\ \text{cm}^2/\text{s}, \qquad D_2 = 9.1 \times 10^{-13}\ \text{cm}^2/\text{s},$$
$$N_{01} = N_{SL} = 3.9 \times 10^{20}\ \text{cm}^{-3}$$

The antimony diffusivities must be calculated from the D_∞ and b values in Appendix F.6.1. Remember that temperature must be in Kelvins. Thus

$$\begin{aligned} D_{o1} &= \text{diffusivity of Sb at } 950°\text{C} \\ &= 3.94 \exp\left(\frac{-4.49 \times 10^4}{1.223 \times 10^3}\right) = 4.48 \times 10^{-16}\ \text{cm}^2/\text{s} \\ D_{o2} &= \text{diffusivity of Sb at } 1150°\text{C} \\ &= 3.94 \exp\left(\frac{-4.49 \times 10^4}{1.423 \times 10^3}\right) = 7.80 \times 10^{-14}\ \text{cm}^2/\text{s} \end{aligned}$$

Calculating the several Δs we have

$$\begin{aligned} \Delta_1 &= D_1 t_1 = (6.6 \times 10^{-15})(15)(60) = 5.94 \times 10^{-12}\ \text{cm}^2 \\ \Delta_2 &= D_2 t_2 = (9.1 \times 10^{-13})(60)(60) = 3.28 \times 10^{-9}\ \text{cm}^2 \\ \Delta_{o1} &= D_{o1} t_1 = (4.48 \times 10^{-16})(15)(60) = 4.03 \times 10^{-13}\ \text{cm}^2 \\ \Delta_{o2} &= D_{o2} t_2 = (7.8 \times 10^{-14})(60)(60) = 2.81 \times 10^{-10}\ \text{cm}^2 \end{aligned}$$

and

$$\Delta_2' = \frac{\Delta_1}{1.3} + \Delta_2 = \frac{5.94 \times 10^{-12}}{1.3} + 3.28 \times 10^{-9}$$

$$\approx 3.28 \times 10^{-9}\ \text{cm}^2 \qquad \text{(note that the first term is negligible)}$$

$$\Delta_{o1} + \Delta_{o2} = (4.03 \times 10^{-13}) + (2.81 \times 10^{-10}) \approx 2.81 \times 10^{-10}\ \text{cm}^2$$

Then

$$N_{02} = \frac{2N_{01}}{\pi}\sqrt{\frac{\Delta_1}{\Delta_2}} = \frac{2(3.9 \times 10^{20})}{\pi}\sqrt{\frac{5.94 \times 10^{-12}}{3.28 \times 10^{-9}}}$$

$$= 1.1 \times 10^{19}\ \text{cm}^{-3}$$

and

$$x_{j2} = \sqrt{4\Delta_2' \ln\left(\frac{N_{02}}{N_b}\right)} = \sqrt{4(3.28 \times 10^{-9}) \ln\left(\frac{1.1 \times 10^{19}}{10^{15}}\right)}$$

$$= 3.49 \times 10^{-4}\ \text{cm} = 3.5\ \mu\text{m}$$

We now must check that the SOA criterion is satisfied. From Eq. (7.2-2)

$$\frac{x_{j2}}{\sqrt{4(\Delta_{o1} + \Delta_{o2})}} = \frac{3.49 \times 10^{-4}}{\sqrt{4(2.8 \times 10^{-10})}} = 10.4$$

This is greater than 2, the criterion is satisfied, and use of the SOA was legitimate.

The foregoing calculations are straightforward. If, however, we keep all the conditions the same except for t_2 and x_{j2} and change the problem to read, "What t_2 is required to give an x_{j2} of, say, 3 μm?" the calculations become more troublesome. The reason for this is that one must solve Eqs. (7.2-3) simultaneously for N_{02} and Δ_2'. This may be done by trial and error. Then t_2 follows from the definition of Δ_2'.

On the other hand, if both t_1 and t_2 are not given, it is possible to specify both N_{02} and x_{j2}. Then the calculation for the two diffusion times is quite straightforward again. The problems at the end of the chapter consider both of these conditions.

The effect of changing processing parameters on x_{j1} during predep was considered in Section 6.7. Similar effects arise during the drive diffusion also, and may be investigated mathematically by the same techniques. Once again, it is true that drive temperature has the greatest effect. If we assume that $\Delta_2' = \Delta_2 = D_2 t_2$, and that all *predep* conditions such as Δ_1, T_1, and N_{01} are fixed, we can show that

$$\left.\frac{dx_{j2}}{x_{j2}}\right|_{\substack{\text{predep} \\ N_b, t_2}} = \frac{b}{2T_2}\left[1 - \frac{1}{2\ln\left(\dfrac{2N_{01}}{N_b}\sqrt{\dfrac{\Delta_1}{\Delta_2}}\right)}\right]\frac{dT_2}{T_2} \tag{7.2-4}$$

where $b = E_{a2}/k$.

Comparison with Eq. (6.7-6) shows that the percentage change in x_j is smaller during drive than it was during predep by the bracketed factor in Eq. (7.2-4).

7.3 Time Dependent Diffusivity

A temperature change during diffusion will affect the diffusivities of all the dopants present, and corresponding $\Delta = Dt$ products will change, too.

As an example, consider the diffusion history of a bipolar transistor *base*. Say the *base* predep is run at 900°C. Following the deglaze step at room temperature, the *base* drive is run at 1050°C. Subsequently, after other intermediate room-temperature processing, the *emitter* predep is performed at 1120°C. It is clear that the *base* dopant will diffuse some more during this third high-temperature process. The question is, will it continue with a gaussian profile or assume some other shape? Notice that (1) at each new temperature the *base* dopant will have a different diffusivity value, so in effect the base dopant D changes with time, a fact we shall stress by replacing D by $D(t)$, and (2) the diffusion will be negligible during any room-temperature steps.

Recall Fick's second law (FSL) for *concentration* independent $D(t)$, which is the basis for these diffusion profile solutions

$$D(t)\frac{\partial^2 N(x,t)}{\partial x^2} = \frac{\partial N(x,t)}{\partial t} \tag{7.3-1}$$

Define a function of t, say $\Delta(t)$, such that

$$N(x,t) = N[x, \Delta(t)] \tag{7.3-2}$$

and

$$\frac{d}{dt}\Delta(t) = D(t) \tag{7.3-3}$$

or

$$\Delta(t) = \int_0^t D(t)\,dt \tag{7.3-4}$$

Differentiating Eq. (7.3-2) twice wrt x gives

$$\frac{\partial^2 N(x,t)}{\partial x^2} = \frac{\partial^2 N[x,\Delta(t)]}{\partial x^2} \tag{7.3-5}$$

and differentiating Eq. (7.3-2) wrt t gives

$$\frac{\partial N(x,t)}{\partial t} = \frac{\partial N[x,\Delta(t)]}{\partial \Delta(t)}\frac{d\Delta(t)}{dt} \equiv \frac{\partial N[x,\Delta(t)]}{\partial \Delta(t)} D(t) \tag{7.3-6}$$

where the right-hand identity is obtained from Eq. (7.3-3). Substituting Eqs. (7.3-5) and (7.3-6) into (7.3-3) we obtain

$$D(t)\frac{\partial^2 N[x,\Delta(t)]}{\partial x^2} = \frac{\partial N[x,\Delta(t)]}{\partial \Delta(t)} D(t) \tag{7.3-7}$$

These steps show that $D(t)$ cancels out and the equation returns to its initial form except for a variable change from t to $\Delta(t)$, which is still time dependent. Therefore, *the* x*-variation of N is not affected by the changes of D in time.* The conclusions are:

1. Given a fixed set of initial and boundary conditions, the *spatial variation* (variation in x) or profile shape in the solution to FSL will not be affected if D changes with time. This implies for the given conditions, that an erfc u profile remains an erfc u profile, a gaussian remains a gaussian, and so on.
2. All of the time domain solutions that we have (and will have) apply equally well to the $\Delta(t)$ domain if we
   (a) replace D by 1 and t by $\Delta(t)$ so that $Dt = \Delta$
   (b) evaluate $\Delta(t)$ by Eq. (7.3-4).

The definition of $\Delta(t)$ here is quite consistent with what we have been using right along. The time dependent diffusivity concept is illustrated in the following example.

Consider a BJT base that is subjected to a double-drive diffusion cycle. All the following calculations deal with the *base*, and the subscript significance is: 1 refers to base predep; 2 refers to first base drive; and 3 refers to second base drive (same time and temperature as the emitter predep). We shall assume that the temperature remains constant through each of the base drive diffusions so the drive $D(t)$ versus t variation has the form shown in Figure 7–3.

This is a Case IVb situation, so we use the SMA form

$$N(x,t_2) = \frac{2N_{01}}{\pi}\sqrt{\frac{\Delta_1}{\Delta_2'}}\exp\frac{-x^2}{4\Delta_2'}$$

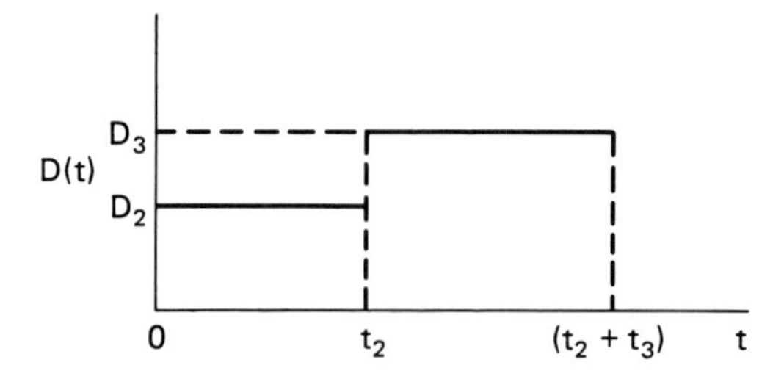

Figure 7–3. Assumed diffusivity versus time for a double drive diffusion cycle.

where

$$\Delta_2' = \frac{\Delta_1}{1.3} + D_2 t_2$$

Converting to the Δ [we drop the (t) for convenience] domain we write

$$N(x, \Delta) = \frac{2N_{01}}{\pi} \sqrt{\frac{\Delta_1}{(\Delta_1/1.3) + \Delta}} \exp \left\{ \frac{-x^2}{4[(\Delta_1/1.3) + \Delta]} \right\}$$

But

$$\Delta = \int_0^{(t_2+t_3)} D(t)\, dt = \int_0^{t_2} D_2\, dt + \int_{t_2}^{t_2+t_3} D_3\, dt$$

$$= D_2 t_2 + D_3 t_3 \equiv \Delta_2 + \Delta_3$$

Converting back to the time domain we have

$$N(x, t_3) = \frac{2N_{01}}{\pi} \sqrt{\frac{\Delta_1}{\Delta_B}} \exp \frac{-x^2}{4\Delta_B}$$

where

$$\Delta_B = \frac{\Delta_1}{1.3} + \Delta_2 + \Delta_3$$

This exercise may seem a waste of time because the result is "obvious." But is it really? Consider a slightly more complicated situation: whenever a boat of wafers is inserted into and later removed from a furnace at diffusion temperature, the wafers see a comparatively slow change in temperature as they go from room- to furnace-temperature and back down again, as shown in Figure 7–4. The same sort of situation prevails if the boat is inserted into

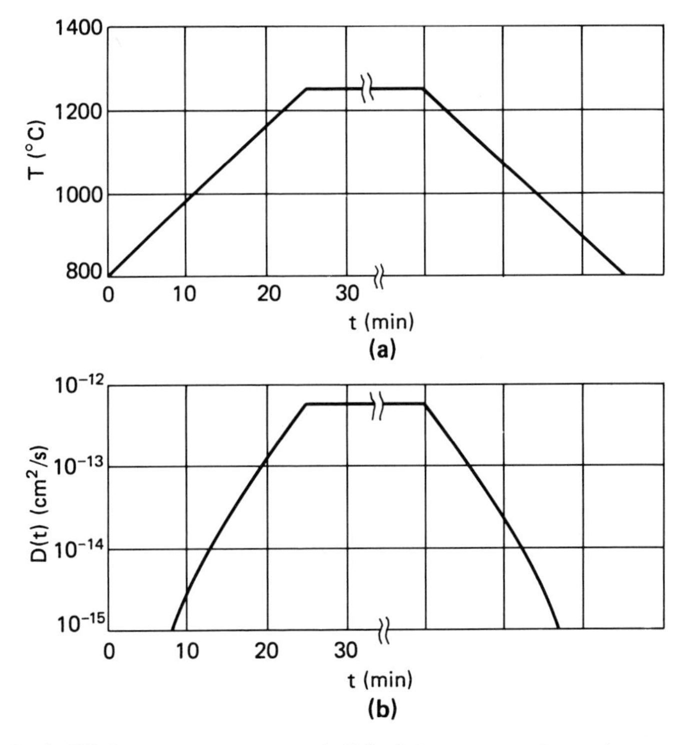

Figure 7–4. Wafer temperature and diffusivity versus time when wafers are inserted into and removed from a furnace. Nominal furnace temperature is 1250°C.

a cool furnace whose temperature is ramped up and then down again before the wafers are removed. Since $D(T) = D_\infty e^{-b/T}$ and T varies with time, we clearly have examples here of the time dependent diffusivity. The value of Δ is not obvious and must be found by integrating as in Eq. (7.3-4). It will be found that this integration cannot be performed analytically, so resort must be made to some form of numerical integration such as the trapezoidal rule.

These results assume no change in boundary conditions during the temperature changes. If they do, profile changes take place, as discussed in Section 7.6.

7.4 BJT Processing Equations

The results of the foregoing sections may be arranged into a procedure to calculate, on a first-order basis, the diffusion cycles for fabricating a discrete double-diffused transistor. They neglect many interactive effects, but do show the basic ideas involved. Computer programs such as SUPREM consider

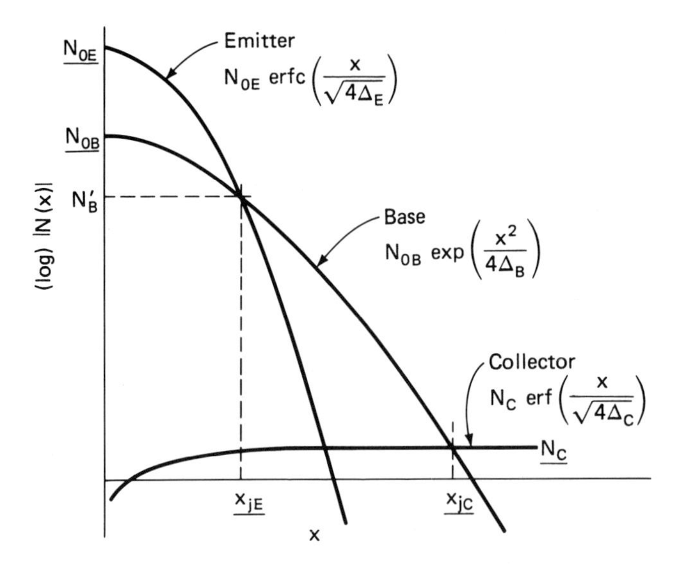

Figure 7–5. BJT profiles after all diffusions are completed.

many of the effects but obscure the ideas for a student [19]. The word "discrete" here implies that isolation diffusion and buried layer diffusion are not used because each transistor will be mounted in its own individual header.

The term "double-diffused" implies that both the base and emitter are produced by diffusion. We have seen, however, that the actual base is subjected to three diffusions: base predep, first base drive, and finally the second base drive during the emitter predep. We designate these three base diffusions by the subscripts 1, 2, and 3, respectively. The emitter predep will be designated by the subscript E. An *emitter* drive is seldom used, and for good reasons: Transistor β increases with the buried base-to-emitter sheet resistance ratio. Since x_{jE} will be small, N_{0E} must be large to reduce emitter sheet resistance. If the emitter were subjected to a drive, inevitably its surface concentration N_{0E} would drop. High N_{0E} also enhances the formation of an emitter ohmic contact with aluminum after metallization has been applied to the wafer.

In a BJT the wafer substrate (or epitaxial layer if one is present) serves as the transistor collector. For this reason we shall replace the symbol N_b by N_C, with subscript C indicating collector.

The individual dopant profile magnitudes of the transistor *after all diffusions are completed* are shown in Figure 7–5. Particular attention should be paid to the notation since it will be used throughout this section. The corresponding notation for the several diffusion cycles is shown in Table 7–1.

Consider the pertinent equations. The base surface concentration *after all diffusions are completed* will be

$$N_{OB} = \frac{2N_{01}}{\pi}\sqrt{\frac{\Delta_1}{\Delta_B}}, \qquad \Delta_B = \frac{\Delta_1}{1.3} + \Delta_2 + \Delta_3 \tag{7.4-1}$$

At both junctions, x_{jE} and x_{jC}, the algebraic sum of the emitter, base, and collector profiles, must equal zero. Reference to Figure 7–5 will show that this condition may be satisfied at either x_{jE} or x_{jC} by

$$N(x_j, t_3) = \pm N_{OE}\,\text{erfc}\left(\frac{x_j}{\sqrt{4\Delta_E}}\right) \mp N_{OB}\exp\left(\frac{-x_j^2}{4\Delta_B}\right) \pm N_C\,\text{erfc}\left(\frac{x_j}{\sqrt{4\Delta_C}}\right)$$
$$\equiv 0 \tag{7.4-2}$$

This is difficult to solve analytically, so use is made of reasonable approximations.

First, in typical transistors the emitter and base profiles are each at least two orders of magnitude greater than N_C *at the emitter junction.* Furthermore, erf u cannot exceed 1 for positive u; therefore, the third term in Eq. (7.4-2) is negligible wrt the two other terms at x_{jE}. Thus

$$\pm N_{OE}\,\text{erfc}\left(\frac{x_{jE}}{\sqrt{4\Delta_E}}\right) \mp N_{OB}\exp\left(\frac{-x_{jE}^2}{4\Delta_B}\right) = 0 \tag{7.4-3}$$

Notice from Figure 7–5 that each term in Eq. (7.4-3) is the quantity N_B', so

$$N_B' = N_{OB}\exp\left(\frac{-x_{jE}^2}{4\Delta_B}\right) = N_{OE}\,\text{erfc}\left(\frac{x_{jE}}{\sqrt{4\Delta_E}}\right) \tag{7.4-4}$$

Second, we assume the slow outdiffuser approximation (SOA) for the *collector* dopant at the collector junction (i.e., at x_{jC} the collector doping is at

Table 7–1. Diffusion Notation for Discrete Double-Diffused BJT

Name	Source	N_0	Temp	Time	(Diffusivity × time)
Base Predep	Infinite	N_{01}	T_1	t_1	$\Delta_1 = D_1 t_1$
1st Base Drive	Planar	[a]	T_2	t_2	$\Delta_2 = D_2 t_2$
2nd Base Drive	Planar	[b]	T_3	t_3	$\Delta_3 = D_3 t_3$
Emitter Predep	Infinite	N_{OE}	T_3	t_3	$\Delta_E = D_E t_3$[c]
1st Coll. Out	Constant	0	T_1	t_1	$\Delta_{o1} = D_{o1} t_1$
2nd Coll. Out	erf	0	T_2	t_2	$\Delta_{o2} = D_{o2} t_2$
3rd Coll. Out	erf	0	T_3	t_3	$\Delta_{o3} = D_{o3} t_3$

Notes: $\Delta_B = (\Delta_1/1.3) + \Delta_2 + \Delta_3$, $\Delta_C = \Delta_{o1} + \Delta_{o2} + \Delta_{o3}$.

[a] Varies during t_2.

[b] Varies during t_3. Final value is N_{OB}. See Figure 7.5.

[c] If boron and phosphorus are the base and emitter dopants, $D_E = D_3$ so $\Delta_E = \Delta_3$.

the background level N_C), so

$$\operatorname{erf}\left(\frac{x_{jC}}{\sqrt{4\Delta_C}}\right) \approx 1 \tag{7.4-5}$$

and the third term in Eq. (7.4-2) becomes simply N_C.

Third, in Figure 7–5 the emitter profile is several orders of magnitude below the base profile and N_C at x_{jC}; hence, at x_{jC} we can write

$$\pm N_B(x_{jC}) \mp N_C = 0$$

or

$$\pm N_{0B} \exp\left(\frac{-x_{jC}^2}{4\Delta_B}\right) \mp N_C = 0 \tag{7.4-6}$$

Another useful equation can be derived by combining the left-hand member of Eq. (7.4-4) with Eq. (7.4-6), namely

$$\frac{N_B'}{N_C} = \left[\frac{N_{0B}}{N_C}\right]^{[1-(x_{jE}/x_{jC})^2]} \tag{7.4-7}$$

The foregoing equations may be arranged into a useful design procedure. Typically, a device designer will specify the underlined quantities in Figure 7–5 to the process engineer, namely the three concentrations, N_{0E}, N_{0B}, and N_C, and the two junction depths, x_{jE} and x_{jC}. Remember that these are values *after all diffusions are completed*. We shall consider simultaneously a typical design procedure and an illustrative example based on these five values as the given quantities.

Calculate the diffusion schedules for an n–p–n transistor to meet these specifications:

$$N_C = 10^{16}\,\text{cm}^{-3} \quad \text{(As doped)}, \qquad N_{0B} = 10^{18}\,\text{cm}^{-3} \quad \text{(B doped)}$$

$$N_{0E} = 5 \times 10^{19}\,\text{cm}^{-3} \quad \text{(P doped)}$$

$$x_{jE} = 1.5\,\mu\text{m} \qquad x_{jC} = 2.5\,\mu\text{m}$$

Since boron and phosphorus are the base and emitter dopants, $\Delta_E = \Delta_3$ for this example.

Calculate N_B' from Eq. (7.4-7)

$$N_B' = N_C\left[\frac{N_{0B}}{N_C}\right]^{[1-(x_{jE}/x_{jC})^2]} = 10^{16}\left(\frac{10^{18}}{10^{16}}\right)^{[1-(1.5/2.5)^2]}$$

$$= 1.91 \times 10^{17}\,\text{cm}^{-3} \tag{7.4-8}$$

Solve the right-hand member of Eq. (7.4-4) for Δ_E

$$\Delta_E = \frac{1}{4}\left[\frac{x_{jE}}{\operatorname{erfc}^{-1}(N_B'/N_{0E})}\right]^2 \tag{7.4-9}$$

$$= \frac{1}{4}\left\{\frac{1.5 \times 10^{-4}}{\operatorname{erfc}^{-1}[(1.91 \times 10^{17})/(5 \times 10^{19})]}\right\}^2$$

$$= 1.34 \times 10^{-9}\,\text{cm}^2$$

In this example, $\Delta_3 = \Delta_E$, because boron and phosphorus have the same diffusivity.

Guess a value of emitter predep temperature T_3 as 1150°C. Then, from Appendix F.6.1,

$$D_3 = 9.1 \times 10^{-13}\,\text{cm}^2/\text{s}$$

and

$$t_3 = \frac{\Delta_3}{D_3} = \frac{1.34 \times 10^{-9}}{9.1 \times 10^{-13}} = 1473\,\text{s} \approx 25\,\text{min} \tag{7.4-10}$$

This value is satisfactory. From practical considerations t_3 must be at least 15 min; if it is not, choose a lower value for T_3 and reevaluate t_3 and D_3 for the *base* dopant.

Solve Eq. (7.4-6) and evaluate Δ_B

$$\Delta_B = \frac{x_{jC}^2}{4\ln(N_{0B}/N_C)} = \frac{(2.5 \times 10^{-4})^2}{4\ln(10^{18}/10^{16})}$$

$$= 3.39 \times 10^{-9}\,\text{cm}^2 \tag{7.4-11}$$

Solve Eq. (7.4-1) for Δ_1

$$\Delta_1 = \Delta_B\left(\frac{\pi N_{0B}}{2N_{01}}\right)^2 \tag{7.4-12}$$

All quantities in this equation are known except Δ_1 and N_{01}, one of which must be chosen. At this point the easiest choice is to make N_{01} the limit of solid solubility of the *base dopant* in silicon at the *base predep temperature* T_1. (Note that lower values may be chosen if they can be realized with the intended predep source.) With this choice made, we evaluate Δ_1.

Say we choose T_1 as 900°C and let N_{01} be the limit of solid solubility of boron in silicon at that temperature. Reading from Appendix F.6.1, we have

$$N_{01} = N_{SL} = 3.7 \times 10^{20}\,\text{cm}^{-3}$$

Then we can calculate

$$\Delta_1 = 3.39 \times 10^{-9}\left[\frac{\pi \times 10^{18}}{2(3.7 \times 10^{20})}\right]^2 = 6.11 \times 10^{-14}\,\text{cm}^2$$

From Appendix F.6.1 the boron diffusivity at 900°C is

$$D_1 = 1.5 \times 10^{-15}\,\text{cm}^2/\text{s}$$

so

$$t_1 = \frac{\Delta_1}{D_1} = \frac{6.11 \times 10^{-14}}{1.5 \times 10^{-15}} = 41\,\text{s}$$

Clearly this is too short for a predep, so another choice must be made. Try $N_{01} = 5 \times 10^{19}\,\text{cm}^{-3}$. Then

$$\Delta_1 = 3.39 \times 10^{-9}\left[\frac{\pi \times 10^{18}}{2(5 \times 10^{19})}\right]^2 = 3.35 \times 10^{-12}\,\text{cm}^2$$

and

$$t_1 = \frac{\Delta_1}{D_1} = \frac{3.35 \times 10^{-12}}{1.5 \times 10^{-15}} = 37\,\text{min} \tag{7.4-13}$$

This value is satisfactory, but may be raised or lowered by suitable changes in T_1 or N_{01}.

From the definition of Δ_B we have

$$\begin{aligned}\Delta_2 &= \Delta_B - \left(\frac{\Delta_1}{1.3} + \Delta_3\right)\\ &= 3.39 \times 10^{-9} - \left(\frac{3.35 \times 10^{-12}}{1.3} + 1.34 \times 10^{-9}\right)\\ &= 2.05 \times 10^{-9}\,\text{cm}^2\end{aligned} \tag{7.4-14}$$

(Note that the Shrivastava/Marshak correction term is negligible.)

Choose a first base drive temperature, say $T_2 = 1100$°C. From Appendix F.6.1, $D_2 = 3 \times 10^{-13}\,\text{cm}^2/\text{s}$, so

$$t_2 = \frac{\Delta_2}{D_2} = \frac{2.05 \times 10^{-9}}{(3 \times 10^{-13})60} = 114\,\text{min}$$

This value is satisfactory, so the design process is complete; all of the temperatures, diffusion times, and the base predep surface concentration have been determined to meet the specifications. Both N_{0E} and N_{01} are below the corresponding limits of solid solubility. This can be brought about by appropriate predep conditions.

Notice that times for the wafer to enter and leave the furnace have been neglected. With Δ values calculated, the methods of the last section allow suitable adjustments to be made.

7.5 Sheet Resistance

The use of the Irvin R_S curves for calculating sheet resistances for surface and buried layers was covered in Section 6.5.2. Emphasis was placed on the predep case. We now look more closely at drive diffusion R_S calculations.

7.5.1 Drive Sheet Resistances

The *method* of using the Irvin R_S curves is the same for predep and drive profiles, but here the curves for the *gaussian*, rather than the erfc u, profile must be used. Furthermore junction depths and surface concentrations must be those after the *drive* is completed. Use of the curves is best illustrated by some numerical examples.

For the first case, say that the predep diffusion of the example in Section 6.5.2 has been completed. Calculations showed that the original predep for 900 s @ 950°C gave $\Delta_1 = 5.94 \times 10^{-12}\,\text{cm}^2$ and $R_{S1} = 112\,\Omega/\text{sq}$. Say the drive parameters are

$$T_2 = 1200°\text{C}, \qquad D_2 = 2.5 \times 10^{-12}\,\text{cm}^2/\text{s}$$

and we wish to calculate R_{S2} after several different drive intervals to observe trends; hence, we prepare the equations for repeated calculation. Minimum t_2 will be 300 s. Thus

$$\Delta_2 = D_2 t_2 \geq (2.5 \times 10^{-12})(300) = 7.5 \times 10^{-10}\,\text{cm}^2$$

This is two orders of magnitude greater than Δ_1; hence, by Eq. (7.1-11), we conclude that $\Delta_2' = \Delta_2$ and

$$N_{02} = \frac{2N_{01}}{\pi}\sqrt{\frac{\Delta_1}{D_2 t_2}} = \frac{3.83 \times 10^{20}}{\sqrt{t_2}}$$

$$x_{j2} = \sqrt{4D_2 t_2 \ln\frac{N_{02}}{N_b}} = 3.16 \times 10^{-6}\sqrt{t_2 \ln\frac{N_{02}}{10^{15}}}$$

$$R_{S2} = \frac{1}{\langle\sigma_{S2}\rangle x_{j2}}$$

The calculated results are tabulated below. Note that $\langle \sigma_{S2} \rangle$ is read for $N_b = 10^{15}$, a p-type *layer*, $x_b/x_j = 0$, and a *gaussian* profile, so Appendix F.6.5(d) is used.

t_2, s	N_{02}, cm^{-3}	x_{j2}, μm	$\langle \sigma_{S2} \rangle$, S/cm	R_{S2}, Ω/sq
300	2.21 E19	1.73	70	83
900	1.28 E19	2.92	40	87
1200	1.11 E19	3.34	35	86
1500	9.89 E18	3.71	33	82
1800	9.03 E18	4.05	30	82

Interpolation is required to read the average conductivities from the Irvin curves, so even two significant figures are questionable. Note that R_{S2} remains essentially constant!

The second case illustrates a buried layer calculation. Consider the layer extending from $x_b = 1.5\ \mu$m to x_{j2}, and use a drive time of 1200 s under the previous conditions.

N_{02} and x_{j2} are already shown in the table. But now we have a different parameter value for the curve, namely

$$\frac{x_b}{x_{j2}} = \frac{1.5}{3.34} \approx 0.45$$

Reading from the same family of curves for $N_{02} = 1.1 \times 10^{19}\ \text{cm}^{-3}$, and interpolating by eye midway between the curves for X_b/x_{j2} values of 0.4 and 0.5, we read $\langle \sigma_{Sb2} \rangle \approx 7$ S/cm. Then

$$R_{Sb2} = \frac{1}{\langle \sigma_{Sb2} \rangle (x_{j2} - x_b)} = \frac{1}{7(3.34 - 1.5)\,10^{-4}} = 780\ \Omega/\text{sq}$$

where the b subscript indicates "buried layer."

Consider some observations regarding these results. After drive the surface-layer sheet resistance is lower than after predep. This is surprising: there is no increase in the amount of dopant in the layer but the profile shape has changed from erfc u, with constant N_{01}, to the gaussian, with much lower and decreasing values of N_{02}. Also, the sheet resistance remains essentially constant even as the drive time triples.

If the drive is carried out under oxidizing conditions, and part of the drive must be, redistribution will take place with considerable change in the profile from the idealized gaussian shape. (This is considered later.) In practice it turns out that R_{S2} is usually *greater* than R_{S1} for a boron layer, in direct contrast to the idealized, calculated results! This may be handled *empirically* by writing

$$R_{S2} = nR_{S1}, \qquad 2 \leq n \leq 4 \tag{7.5-1}$$

The value of n must be determined empirically for the particular process that is being used.

Compare the calculations at $t_2 = 1200$ s for the buried and surface layers. The buried layer has the larger sheet resistance as is expected from Figures 6–9(b) and (c), which show the comparative shaded areas under the gaussian profile for the two layer types.

The sheet resistance and diffusion equations may be combined into a design procedure to give specified values of drive R_{S2} and x_{j2}, subject to first-order assumptions. Say these values are given for a two-step diffusion process. From Eq. (6.5-8), with appropriate second subscripts to refer to the drive diffusion

$$\langle \sigma_{S2} \rangle = \frac{1}{R_{S2} x_{j2}} \tag{7.5-2}$$

Then, N_{02} can be determined from the appropriate Irvin curve, using $x_b/x_{j2} = 0$.

Assuming that $\Delta_1/1.3 \ll \Delta_2$, Δ_2 may be calculated by Eq. (7.2-3):

$$\Delta_2 = \frac{x_{j2}{}^2}{4 \ln(N_{02}/N_b)} \tag{7.5-3}$$

Then a good engineering choice among (T_2, t_2) value pairs can be made to give the required Δ_2 value. This procedure has been illustrated in previous examples.

Initially, if redistribution effects during the drive/oxidation cycle are neglected, Δ_1 is given by Eq. (7.4-12). As in previous examples Δ_1 and N_{01} must be determined by trial and error subject to these constraints: (1) $\Delta_1/1.3 \ll \Delta_2$, (2) $t_1 \geq 15$ min, and (3) $N_{01} \leq N_{SL}$ at the chosen value of T_1.

To take drive/oxidation into account roughly, an empirical approach in evaluating N_{01} and Δ_1 is used. Say experience with a given production line shows that a boron diffused layer gives an n value of 3 in Eq. (7.5-1). Then we would design for an R_{S1} value of $R_{S2}/3$. The appropriate Irvin curve and predep diffusion equations then can be manipulated to give values of x_{j1}, N_{01}, T_1, and t_1 as illustrated in earlier numerical examples.

The curves of Appendix F.6.10 (or other appropriate PDS curves) allow an alternate method of determining the (T_1, t_1) value pair to give a specified R_{S1}, provided the proper type of planar diffusion source is used for the predep.

A disadvantage of the Irvin curves is their large number. If only *surface-* and not *buried-layer* sheet resistances need be determined, x_b/x_j is fixed at zero and the total number of required R_S curve sets is reduced to four: for each profile type, namely erfc u or gaussian, one set of N_b curves will be for p-type *layers* and one for n-type *layers*. Curves of this type are encountered in industry and some of them antedate the Irvin curves [3, 4]. But a *buried-layer* sheet resistance cannot be determined by taking two readings from these curves, one from the surface to x_j, and subtracting a second, from the surface to x_b.

Despite the large number of Irvin curves, they cannot cover all buried-layer sheet resistance calculations unless additional simplifying approximations are made. This will be discussed in the next section.

7.5.2 BJT Sheet Resistances

It can be shown from elementary considerations that the common emitter, short-circuit, forward current gain of a BJT may be approximated by

$$\beta \approx \frac{R_{BB}}{R_{SE}} \tag{7.5-4}$$

where

$$R_{BB} = \text{base } \textit{buried}\text{-layer sheet resistance}$$

$$R_{SE} = \text{emitter } \textit{surface}\text{-layer sheet resistance}$$

at (1) signal frequencies well below the α cutoff frequency, and (2) mid-current levels small enough so emitter crowding is not present, and yet large enough that recombination in the emitter–base transition region has negligible effect. We wish to evaluate these important sheet resistances.

Figure 7–6 shows the BJT doping profiles to identify R_{SE} and R_{BB} and to aid in estimating their values from the Irvin R_S curves. Since N_{0E} and x_b/x_{jE} ($= 0$) are known, we need the equivalent background dopant level for the emitter. It is roughly N_B'. A practical consideration narrows the choice: Irvin curves are plotted only for background levels that are powers of ten, so choose the one nearest N_B', assume an erfc u profile, and read $\langle\sigma_{SE}\rangle$.[2]

For the base buried layer we assume that the area for R_{BE} is negligible wrt the R_{BB} area in Figure 7–6. Then N_{0B}, N_C, x_{jE}, and x_{jC} are all known, so $\langle\sigma_{BB}\rangle$ may be read from the appropriate gaussian profile Irvin curve. Then the sheet resistances are:

$$R_{SE} = \frac{1}{\langle\sigma_{SE}\rangle x_{jE}}, \qquad R_{BB} = \frac{1}{\langle\sigma_{BB}\rangle (x_{jC} - x_{jE})} \tag{7.5-5}$$

[2] In Sections 7.11.3 and 7.12.3, we shall consider some cases with anomalous doping profiles that are neither erfc u nor gaussian. The Irvin R_S curves are not applicable at all in such situations, so other techniques are used.

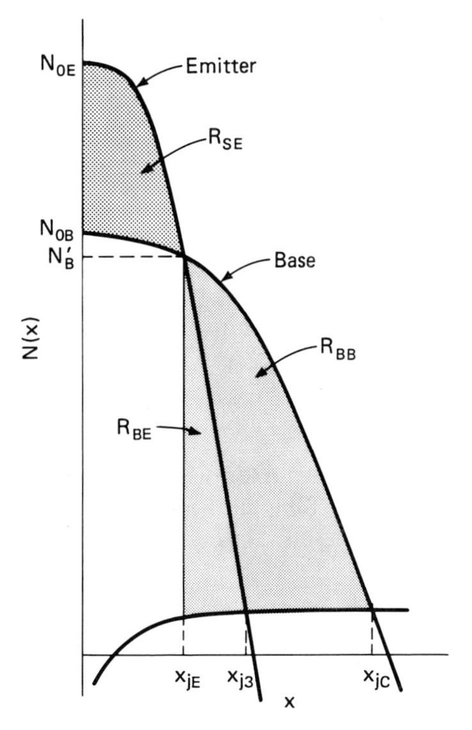

Figure 7–6. BJT profiles identifying R_{SE} and R_{BB}. A first subscript "B" indicates a buried layer. Note: Area R_{BB} includes R_{SE}.

Remember that several approximations are involved in these "calculated" values.

7.6 Outdiffusion into Growing Epitaxial Layers (Case V), Autodoping

Section 6.3.3 covered the outdiffusion of background dopant into an infinite sink, a worst-case situation. We now consider dopant outdiffusion into a solid layer *growing* on the wafer surface. Three cases will be studied.

A typical chemical vapor deposition (CVD) process for depositing homoepi silicon on a silicon substrate involves the pyrolysis of silane (Chapter 9):

$$SiH_4\uparrow \xrightarrow{\text{heat}} Si\downarrow + 2\,H_2\uparrow \tag{7.6-1}$$

The reaction requires externally supplied heat and is a high-temperature operation during which any dopant in the host can diffuse into the epi layer

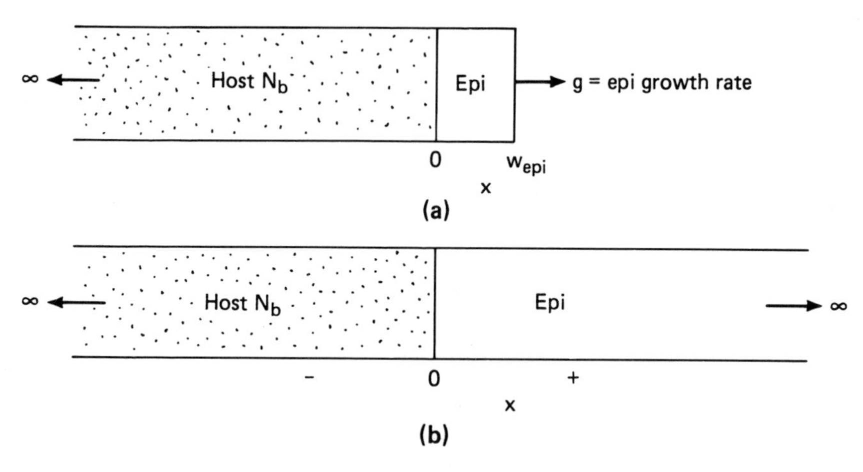

Figure 7–7. Outdiffusion from a uniformly doped host into a growing epitaxial layer. (a) Initial model. (b) Simplified model if the epi layer grows much faster than the dopant diffuses into the epi layer.

at significant rates. Silicon for the epi layer is furnished by externally supplied SiH_4.

A model for this case is shown in Figure 7–7(a). The host lies to the left of the original gas/host interface at $x = 0$, so negative values of x correspond to the host, and positive values to the growing epi layer. Assume that the host is initially doped at a uniform value N_b.

A simpler solution will result if we make another assumption that usually is satisfied in practice. Say that $g = w_{epi}/t$ is the epi layer growth rate, where t is the time (or duration) of epi deposition (and dopant diffusion), and w_{epi} is the final epi width. Say further that g is large relative to the rate of dopant diffusion from host to epi. If this is true, the dopant layer never "sees" the right-hand edge of the epi layer and so views the epi layer as being infinitely thick. This leads to the model of Figure 7–7(b) that is easier to describe mathematically. Observe that D_o, the diffusivity of the outdiffusing dopant, has the dimension cm^2/s. Then dimensionally $\sqrt{D_o/t}$ is cm/s or velocity, so the *infinitely thick epi* criterion is that

$$g = \frac{w_{epi}}{t} \gg \sqrt{\frac{D_o}{t}} \quad \text{or} \quad \frac{w_{epi}}{\sqrt{D_o t}} \gg 1 \qquad (7.6\text{-}2)$$

Let us check some typical numbers to see if this is reasonable. The pyrolysis of SiH_4 typically gives a growth of 25 μm of homoepitaxial silicon in 50 min (3000 s) at 1200°C. Consider three host dopants, B/P, As, and Sb, whose diffusivities can be calculated from data in Appendix F.6.1. Then for the specified time and temperature:

Dopant	D_o, cm^2/s	$w_{epi}/\sqrt{D_o t}$
B/P	2.50 E−12	28.9
As	2.96 E−13	83.9
Sb	2.24 E−13	96.4

The last column shows that the inequality of Eq. (7.6-2) is well satisfied by As and Sb, and probably by B and P, so we shall assume that the criterion is satisfied for typical epi growth conditions.

We conclude that Figure 7–7(b) shows a satisfactory model for this case and write the corresponding initial and boundary conditions for solving Fick's second law (FSL).

The IC for uniform initial doping of the host is stated by

$$N(-x,0) = N_b \tag{7.6-3}$$

and since there is no epi layer initially, it must be true that

$$N(+x,0) = 0 \tag{7.6-4}$$

We again assume a limited time of epi growth and diffusion such that the far side of the wafer is unaware of outdiffusion taking place; hence the BCs

$$N(-\infty,t) = N_b \tag{7.6-5}$$

and

$$N(+\infty,t) = 0 \tag{7.6-6}$$

The solution to FSL for these conditions is [5, 6]

$$N(x,t) = \frac{N_b}{2}\,\text{erfc}\left(\frac{x}{\sqrt{4D_o t}}\right), \qquad -\infty \le x \le +\infty \tag{7.6-7}$$

Erf u is tabulated in Appendix F.6.3 only for positive values of u. To overcome this limitation recall the identity

$$\text{erfc}\; u = 1 - \text{erf}\; u \tag{7.6-8}$$

Second, the series expansion for erf u (Appendix F.6.2) contains only odd powers of u, so erf u is an odd function of u and we may write

$$\text{erf}(-u) = -\text{erf}\,u, \quad \text{and} \quad \text{erfc}(-u) = 1 + \text{erf}\,u \tag{7.6-9}$$

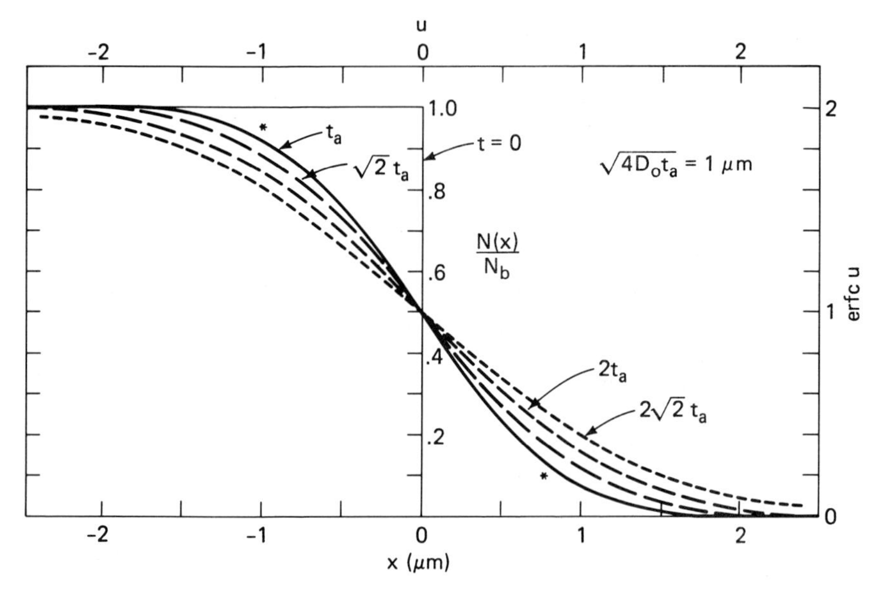

Figure 7–8. Complementary error function profiles for positive and negative values of the independent variable. The curve labeled t_a with the top and right-hand coordinates shows erfc u versus u. All the curves with the center and bottom coordinates show outdiffusion from a uniformly doped host into a growing homoepi layer, with values based on $\sqrt{4D_o t_a} = 1\ \mu\text{m}$. Note the $t = 0$ curve.

Thus the erfc u plot for both positive and negative u appears as in Figure 7–8. The basic erfc u curve is marked with the asterisk and its coordinates are at the top and right-hand side of the plot. Note that the crossover value at $u = 0$ is unity, and that the curve is skew-symmetric about the point (0, 1), as would be expected from Eqs. (7.6-8) and (7.6-9).

Corresponding plots for $N(x)/N_b$ versus x for different times also are shown in the same figure. The crossover value now is 0.5 because of the factor 1/2 in Eq. (7.6-7). These results show that as time goes on, the dopant gradually diffuses from the host into the epi layer, with the greatest effect being observed near the substrate/epi interface.

The principal qualitative lesson to be learned from these results is that epitaxial growth of Si requires high temperatures in the 900 to 1200°C range in which diffusion is significant for the usual silicon dopants. Consequently if a homoepi layer is grown on *doped* silicon, it inevitably will be doped partially by the substrate dopant, as will the substrate by the epi dopant (if there is one). The effects are greatest near the epi/substrate interface. The shape of the profiles may be predicted by Eq. (7.6-7), if the substrate was uniformly doped initially.

7.7 Outdiffusion into Growing Oxide Layer (Case VI)

An oxide layer, needed for the next photolithographic sequence, is required at diffusion termination; hence, diffusion may be started in an inert ambient and finished off in the presence of oxygen or water vapor. Or, as is the case in modern CMOS processes and especially with P as dopant, diffusion may be started in O_2 and finished in an inert ambient. As the oxide layer grows, the substrate dopant will outdiffuse into it. This is referred to as *redistribution during oxidation.*

The situation here is more complicated than Case V for two principal reasons: (1) the host and growing layers are *different* materials, and (2) while the oxygen for the oxide is furnished externally, the silicon is robbed from the host. Therefore the dopant's *segregation coefficient m* is very important. We have used m in Chapter 6, but consider it again more carefully. Its definition is

$$m = \text{segregation coefficient}$$
$$= \left[\frac{N_{\text{dopant}} \text{ in silicon}}{N_{\text{dopant}} \text{ in silicon dioxide}}\right]_{\text{at equilibrium}} \tag{7.7-1}$$

In order to see the importance of this quantity, consider Figure 7–9. The original host, with initial uniform doping assumed, is in (a). In part (b) an oxide layer has been grown at the expense of wafer silicon being consumed. Whereas the silicon surface was originally located at $x = 0$, it is now at the nonzero value x_1. The total oxide width is $(x_1 + x_2)$. Recall that the Pilling-Bedworth ratio (PBR) is defined in Section 5.57 as

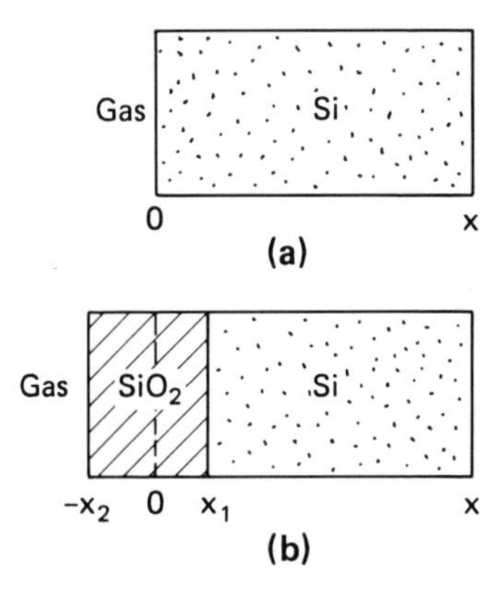

Figure 7–9. Sketches showing the loss of silicon to oxide growth. (a) The original unoxidized silicon surface lies at $x = 0$. (b) After oxide growth the silicon surface lies at $x = x_1$. The Pilling-Bedworth ratio is $(x_1 + x_2)/x_1$.

$$\text{PBR} = \frac{\text{thickness of oxide grown}}{\text{thickness of Si consumed in oxide growth}} \tag{7.7-2}$$

so that in our particular case here

$$\text{PBR} = \frac{x_1 + x_2}{x_1} = 2.22 \tag{7.7-3}$$

Thus if x_1 is one unit, x_2 is 1.22 units, giving a total oxide thickness of 2.22 units as required by the PBR value.

The segregation coefficient in effect tells us which way that dopant prefers to go. For example, if $m < 1$, as it is for boron, Eq. (7.7-1) teaches that boron prefers the oxide, so that the dopant initially in the x_1 region will remain in the oxide. On the other hand, if $m > 1$, as it is with phosphorus, the dopant originally in x_1 tends to be rejected by the oxide and pushes over to the right into the silicon at $x > x_1$. As time increases, the phosphorus doping at x_1 (the oxide/silicon interface location) tends to increase above its initial N_b value. Remember that x_1 increases with time during oxidation. Because of these facts, the doping profiles are affected greatly by the segregation coefficient of the particular dopant, and by its relative diffusivities in the oxide and silicon.

Consider the hypothetical case of $m = 1$ and assume that no dopant can escape at the gas/oxide interface. We can argue that the dopant concentration in the oxide must be lower than it was in the x_1 region of the silicon, if for no reason other than the fixed amount of dopant is spread out over a greater region $(x_1 + x_2)$. It should be clear, then, that the oxidation of doped silicon inevitably will be accompanied by redistribution that will affect the doping profile in the silicon.

7.7.1 Uniformly Doped Substrate

Grove, Leistiko, and Sah have analyzed the situation, and experimentally verified their results for the case with constant initial background doping N_b, and where the dopant can escape from the oxide into the gas [7]. Their principal results are reviewed here, with notation changed to what we have been using.

Four initial and boundary conditions are specified.

IC: Uniform initial doping of the silicon at N_b, or

$$N_{\text{Si}}(x, 0) = N_b, \qquad x > 0 \tag{7.7-4}$$

BC: Surface concentration at the gas/oxide interface is constant at N_0, or

$$N_{\text{ox}}(x_2, t) = N_0 U(t) \tag{7.7-5}$$

Note that N_{Si} and N_{ox} here are dopant concentrations in the silicon and oxide, respectively.

BC: The far side of the silicon is unaffected by redistribution, or

$$N_{Si}(\infty, t) = N_b \tag{7.7-6}$$

BC: The ratio of the concentrations at the oxide/silicon interface equals the segregation coefficient, or

$$\frac{N_{Si}(x_1^+, t)}{N_{ox}(x_1^-, t)} = m \tag{7.7-7}$$

The results are reflected in the profiles of Figure 7–10. In each case the effect of the boundary condition of Eq. (7.7-7) is evident. A surprising additional result, regarding the two interface values of concentration, is that the *magnitudes* of $N_{Si}(x_1^+, t)$ and $N_{ox}(x_1^-, t)$, and not just their ratio, remain constant during the oxidation and distribution processes. This means that

$$N_{Si}(x_1^+, t) = \text{constant} \quad \text{and} \quad N_{ox}(x_1^-, t) = \text{constant} \tag{7.7-8}$$

Figure 7–10(a) shows the profiles predicted by the theory for $m < 1$ and D_{ox} small wrt D, as is typical of boron. The small D_{ox} value limits the rate at which the dopant can escape from the oxide to the gas. But remember that for any given case Eqs. (7.7-8) prevail. The effect on $N_{Si}(x_1^+, t)$ of **b**oron being **b**orrowed at the interface is clearly evident.

Figure 7–10(b) is drawn for $m < 1$ but D_{ox} large, the situation for both In and Ga. Even though both Eqs (7.7-7) and (7.7-2) must be satisfied, the large value of D_{ox} keeps the interface concentrations much smaller than in the prior case because the dopant can escape from the gas/oxide interface at a greater rate.

Experimental results show that the dopant concentration is depressed below the N_b value at the silicon surface for the three acceptors B, In, and Ga.[3] Remember the **b**oron mnemonic, **b**oron is **b**orrowed at the surface; it applies to all three acceptors.

Figure 7–10(c) gives the profiles for $m > 1$ and $D_{ox} < D$, as typical of phosphorous. Note again that $N_{ox}(x_1^-, t)$ can remain comparatively high, since the rate of dopant outdiffusion is limited by D_{ox}. The experimental results show

[3] Experimental results like this can be difficult to verify theoretically because values of m and D_{ox} are highly variable, depending upon the conditions under which they are measured. For example, Donovan shows data for the diffusivity of P in SiO_2 that cover some three orders of magnitude at 1200°C, depending upon the method of dopant introduction [8]! Values of m also show wide variations.

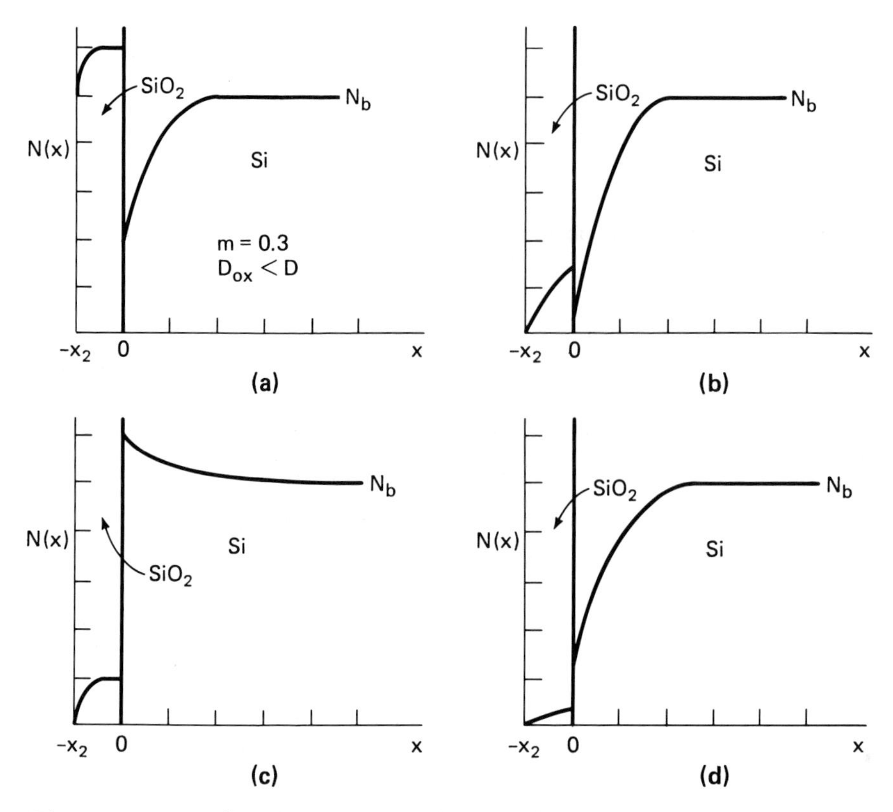

Figure 7–10. Profiles showing redistribution of host dopant during oxidation. Initial host dopant concentration is N_b. (a) Profile for boron: $m = 0.3, D_{ox} < D$. Note that **b**oron is **b**orrowed at the silicon surface. (b) Profile for boron in a hydrogen ambient: $m = 0.3, D_{ox} < D$. (c) Profile for $m > 1$ and $D_{ox} < D$ as with phosphorus. Note that **p**hosphorus is **p**ushed up at the silicon surface. (d) Profile for $m > 1$ and $D_{ox} \approx D$. After Grove, Leistiko, and Sah [7]. (Courtesy of the *Journal of Applied Physics*.)

that the three donors P, As, and Sb all push up the $N_{Si}(x_1{}^+, t)$ value above N_b. A good mnemonic is: **p**hosphorus **p**iles up at the surface.

Figure 7–10(d) shows a theoretical case for $m > 1$ and D_{ox} large so that the dopant can outdiffuse at a high rate. Equations (7.7-7) and (7.7-8) still must be satisfied but $N_{Si}(x_1{}^+, t)$ is depressed below N_b; the high D_{ox} value seems to dominate the effects on the profile.

Grove has calculated the curves of Figure 7–11 [5]. Again for an assumed uniform initial doping N_b, both boron and phosphorus show that the "borrow" or "pile up" effect, as the case may be, is minimized ($N_S/N_b \rightarrow 1$) by using dry—rather than wet—oxidation at high rather than low temperature. This is a good rule of thumb to remember for a uniformly doped substrate.

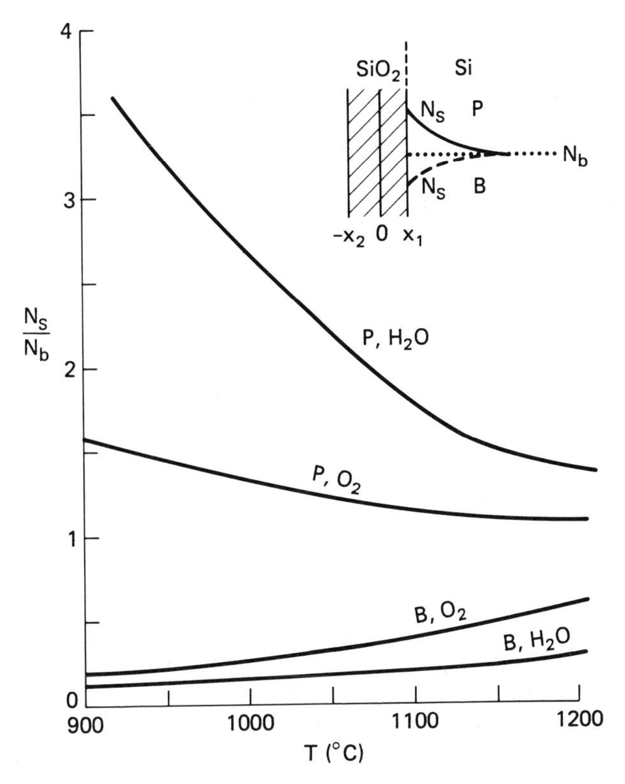

Figure 7–11. Showing the effect on redistribution of boron and phosphorus as a function of oxidizing temperature for both wet and dry oxidizing ambients. A host of initial uniform doping concentration is assumed. After Grove [5]. (*Physics and Technology of Semiconductor Devices*, A. S. Grove. Copyright 1967 by John Wiley & Sons, Inc. Used with permission.)

7.7.2 Diffused Layer

Huang and Welliver have studied a more complicated case—redistribution in a diffused boron-doped BJT *base* region during wet oxidation—and have developed a closed solution for the appropriate differential equations and boundary conditions [9].

In contrast to the previous case, where the substrate has a limitless supply of dopant that can support the statement of Eq. (7.7-8), here the base layer has a finite dose Q_1 of dopant deposited during the base predep. In the subsequent first base drive, which is carried out in an oxidizing atmosphere, silicon oxidation and redistribution of the dopant take place simultaneously. Some of this dose is lost from the base layer to the growing oxide with, say, only Q_2 left in the layer. Experimental results show that the fractional loss of dopant, namely $(Q_1 - Q_2)/Q_1$, is typically over 60%, so the loss here is truly significant, so much so that Eqs. (7.7-8) are no longer even approximately correct.

The basic equation for this profile may be reduced to this form in *our* notation:

$$N(x', t) = N_{02}\left\{\exp\left[-\left(\frac{x'}{\sqrt{4Dt}} + \alpha\right)^2\right] - W\alpha\,\mathrm{erfc}\left(\frac{x'}{\sqrt{4Dt}} + \alpha\right)\right\} \quad (7.7\text{-}9)$$

where

$$\alpha = \frac{1}{2P}\sqrt{\frac{B}{D}}, \qquad F = \frac{P}{m} - 1$$

$$W = \frac{1}{1/(\sqrt{\pi}\,F) + \alpha\exp(\alpha^2)\,\mathrm{erfc}(\alpha)}$$

$$N_{02} = \frac{Q_1}{\sqrt{\pi Dt}} = \frac{2N_{01}}{\pi}\sqrt{\frac{\Delta_1}{\Delta_2}}$$

and P = Pilling-Bedworth ratio and m = segregation coefficient.

The subscripts on D_2 and t_2 have been dropped because we are considering only the drive/oxidation/redistribution part-cycle. All the calculated values to be presented here use Huang and Welliver's data for B and D, with $m = 0.111$ and $P = 2.63$; B is the oxide parabolic growth rate constant of Eq. (5.5-19), and D is the diffusivity of boron in silicon. The typical profile in silicon that reflects this loss is shown in Figure 7–12.

The results show that $N(x', t)$ is the difference between a gaussian and a weighted erfc u function. The former is quite similar in form to what we expect for a drive without oxidation and redistribution. The new variable x' here is measured from the oxide/silicon interface, that moves with time; thus $x' = x - x_1$. To simplify notation here we replace $N_{Si}(x_1{}^+, t)$ by N_S, S designating silicon *surface*. In Figure 7–12 note how the profile (1) peaks with value N_{max} at $x' = x'_{max}$, and (2) remains below the basic gaussian of a drive alone.

The authors give Arrhenius plots of B for both wet and dry oxidation and also of D, and appropriate values may be obtained from these curves for plotting purposes. Since it is the *B/D ratio* which is important, it is plotted versus °C in Figure 7–13 for both types of oxidation.

Given Huang and Welliver's equations one can find the expressions for N_S, x'_{max}, and N_{max}. Algebraic manipulation shows that x'_{max} is proportional to $\sqrt{Dt}$, as might be expected, with a weighting factor that depends on P, m, and $\sqrt{B/D}$. A rather surprising result is that the ratio N_{max}/N_S is time independent, as is the ratio Q_2/Q_1, where Q_1 is the original predep dose and Q_2 is the dose in the silicon after drive/oxidation/redistribution. To get a simple expression

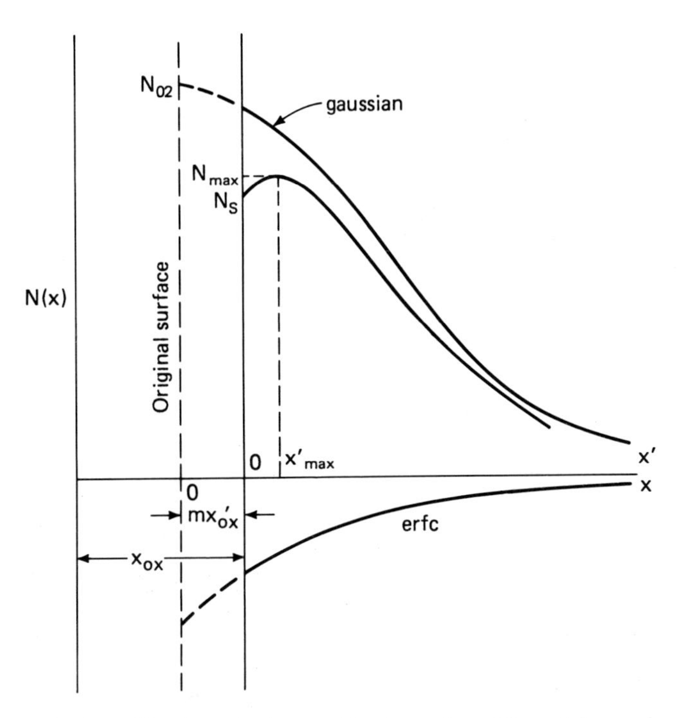

Figure 7–12. Outdiffusion during oxidation of a diffused layer having an initial gaussian doping profile. Note that the concentration peak after oxidation is not at the silicon/oxide interface. After Huang and Welliver [9]. (Reprinted with permission of the publisher, The Electrochemical Society, Inc.)

for this ratio, the authors neglect any junction formed by the base and substrate dopants and define

$$Q_2 = \int_0^\infty N(x', t)\, dx' \tag{7.7-10}$$

where $N(x', t)$ is given by Eq. (7.7-9). Other curves shown in Figure 7–13, for both wet and dry oxidation, should be considered carefully. Note that to minimize the *borrow* effect, Q_2/Q_1 and N_S/N_{02} should be as near to unity as possible. Consider some observations on these curves.

For wet oxidation: For most of the oxidation temperature range, $B > D$. At these temperatures, the oxide grows faster than the boron can drive deeper into the silicon, so boron cannot move away from the oxide/silicon interface as fast as that interface moves into the silicon. Thus, the larger the B/D ratio, the greater will be the loss of dose in the silicon. This is reflected by the smaller Q_2/Q_1 and N_S/N_{02} ratios at lower temperatures.

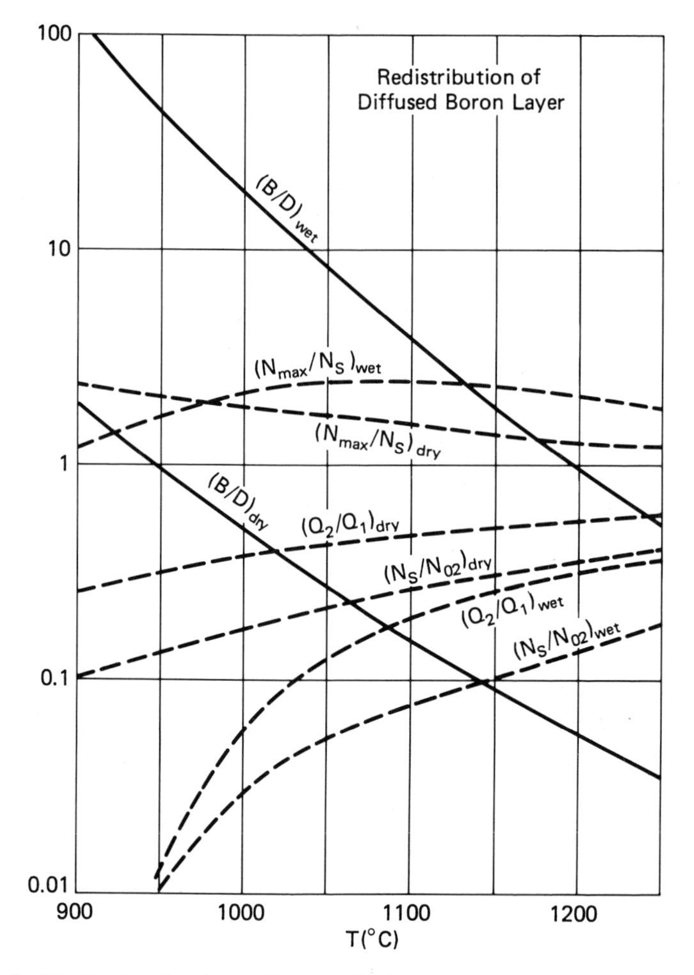

Figure 7–13. Redistribution of boron during oxidation of a gaussian profile layer. B = oxidation parabolic growth rate constant. Calculated from Huang and Welliver [9].

For dry oxidation: Over most of the oxidation temperature range, $B < D$, so diffusion drives the boron into the silicon faster than the interface moves. Thus, we would expect N_S/N_{02} and Q_2/Q_1 to be higher for dry than wet oxidation at a given temperature. This is confirmed by the curves and checks our rule of thumb for the uniformly doped, infinitely thick boron layer (Section 7.7.1). Dry oxidation shows less "borrow" effect than does wet.

To see the dominant effect of the B/D ratio, compare N_S/N_{02} and Q_2/Q_1 for wet and dry conditions at given ratio values—e.g., at B/D of 1 or 2, which values are readily observable in Figure 7–13.

The authors confined their measurements to wafers that were wet-oxidized. One interesting sidelight of this work has to do with the value of segregation coefficient for boron [Eq. (7.7-1)], which usually is taken to be 0.3 for silicon (uniformly doped with boron) in contact with SiO_2. On comparing their theoretical and experimental results, they found closer correlation if m were taken to be 0.111. In this regard, remember that boron moves into the oxide so, rather than SiO_2, borosilicate glass interfaces with the Si.

It should be apparent that if the actual boron profile, as shown in Figure 7–12, deviates from the gaussian, the calculation of junction depth will be affected. The two software programs mentioned in Section 5.7.3 take this effect into account. We already know that the sheet resistance of a diffused layer is affected by the dopant profile. This makes evaluation difficult because the Irvin R_S curves are valid for only erfc u or gaussian profiles.

The oxidation/redistribution effect can be large, and one would like to minimize it. The simplest method for doing this is to carry out the bulk of the drive in an inert atmosphere of nitrogen or argon so that no oxidation takes place initially, and then finish the drive cycle in wet or dry oxygen to get the required oxide thickness. If this procedure is not feasible, the drive can be started in dry oxygen at high temperature to minimize the redistribution effects and then finished in wet oxygen at lower temperature. This could be followed by a short anneal cycle in dry oxygen to densify the oxide without changing its thickness significantly.

The results of this section amply demonstrate how complex the interactions among processing steps are, and why computer programs, such as SUPREM, are used so commonly.

7.8 Redistribution During Oxide CVD

An oxide layer also may be deposited onto a silicon wafer by CVD, a typical reaction being

$$SiH_4\uparrow + O_2\uparrow \rightarrow SiO_2\downarrow + 2\,H_2\uparrow$$

Both the silane and the oxygen are furnished by external sources, so the silicon wafer acts only as a substrate on which the oxide is deposited.

This silane oxidizing process runs at about 500°C, a temperature so low that any diffusion of dopant from substrate into oxide, or vice versa, is negligible. A deposited oxide will not reduce silicon surface states as does a thermally grown oxide, however.

7.9 Junction Formation by Outdiffusion (Case VII)

A special case of outdiffusion, which has been used commercially, produces a junction in a *counterdoped* host (i.e., one initially doped with two dopants of

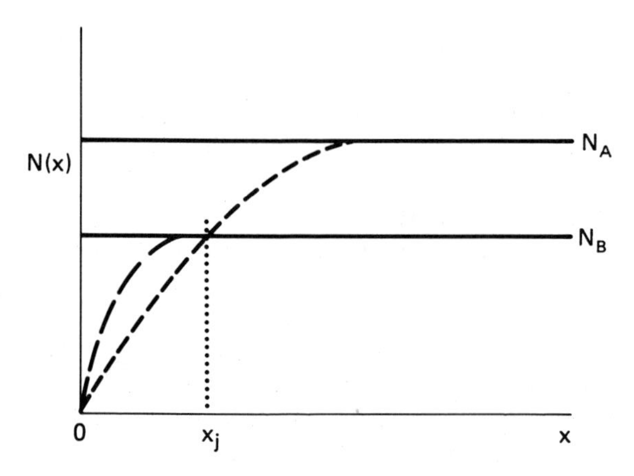

Figure 7–14. Junction formation by outdiffusion from a counterdoped host.

opposite types). Consider such a silicon wafer with dopants A and B with

$$N_A > N_B \quad \text{and} \quad D_A > D_B \tag{7.9-1}$$

The solid lines in Figure 7–14 show the initial uniform concentrations for the two dopants.

Assume that when the wafer is in the diffusion furnace, both dopants react with ambient gas so that their surface concentrations drop several orders of magnitude below their intial values. That is, we assume that both N_{A0} and N_{B0} are zero once outdiffusion begins, so that Case II of Section 6.3.3 is valid. Since dopant A is the faster diffuser, after some time the profiles, both of erf u shape, will have the forms shown by the dashed lines in Figure 7–14. Notice that while the two concentrations are at their initial values with $N_A > N_B$ well in from the surface, near the surface $N_B > N_A$; hence a p–n junction has been formed at x_j where the two curves cross. The value of x_j will depend upon the duration of diffusion. Remember that both dopants are outdiffusing and have erf u profiles. This effect has been used in fabricating semiconductor controlled rectifiers (SCRs).

Junction formation also may take place if the counterdoped substrate is oxidized. If the dopants have different m values, redistribution can produce even two junctions [10].

7.10 Anomalous Profiles

The erf u, erfc u, or gaussian profiles often serve as first-order approximations to the real world. This is not true for shallow arsenic or phosphorus emitters with high N_{0E} values such as are found in small, very-high-frequency BJTs.

We shall consider three main subjects regarding these anomalous effects: (1) Why arsenic with high N_{0E} values is used as the emitter dopant in high-frequency transistors. (2) Why the associated high doping concentrations produce anomalous profiles. (3) What the profile shapes are. This will lead to determination of the corresponding emitter sheet resistance. We finish the chapter by considering the effects when phosphorus with high N_{0E} is used as the emitter dopant.

7.10.1 High-Frequency Considerations

The α cutoff frequency of the *intrinsic* (or internal device) BJT is approximated by

$$(f_T)_i \propto \frac{D_{mb}}{{w_b}^2} \tag{7.10-1}$$

where D_{mb} is the base minority carrier diffusivity and w_b is the transistor's physical base width $(x_{jC} - x_{jE})$. It follows that we expect a BJT to be designed with small w_b if high $(f_T)_i$ is to be obtained.

The f_T of the *extrinsic* device, which includes parasitic resistances and capacitances of the device and its package, will be less than the intrinsic value, because of parasitic capacitance shunting effects. Since these capacitances are directly proportional to device area, high-frequency BJTs are designed with small horizontal geometries. Further, in Section 6.6.2 we saw that deeper diffusions lead to greater lateral diffusion under mask window edges. This leads to larger device areas, so we also expect high-f_T bipolars to have shallow diffusions.

By way of example, the following parameters have been cited for a 3.2-GHz microwave BJT: $x_{jE} = 34$ μin., $x_{jC} = 40$ μin., and $w_b = x_{jC} - x_{jE} = 6$ μin. $= 0.154$ μm [11, 12]. These devices used arsenic, rather than phosphorus, as the emitter dopant. In devices with the same dimensions but with phosphorus emitters, the cutoff dropped to 680 MHz!

The improved performance with arsenic emitters is due to the following properties of As:

1. Lower diffusivity in silicon than P. This allows better control of shallow emitter diffusions and so of x_{jE} and w_b.
2. Arsenic has a higher limit of solid solubility in silicon; for any given diffusion temperature, a higher surface concentration can be obtained. This reflects in a modified profile shape and a lower value of R_{SE}. Arsenic also has a better lattice match in silicon as compared to phosphorus, so even with very high N_{0E} values fewer defects occur, and yields go up. The effective atomic radii in silicon of P, Si, and As, respectively, are 1.10, 1.17, and 1.18 Å; note the close Si/As match [13].

3. Arsenic with high N_{0E} gives an anomalous (not erfc u, not gaussian, and not Dirac δ), nearly triangular emitter profile. For given x_{jE} and N_{0E} this means lower R_{SE}, which leads to higher β [Eq. (7.5-4)].

7.10.2 High Doping Effects

In earlier diffusion studies we have assumed that the diffusivity of the indiffusing dopant species was constant, independent of any dopant concentrations that might be present in the host material. The result was a *linear* second-order differential equation, Fick's second law. The erf u, erfc u, and gaussian profiles are solutions of that linear equation.

As a practical matter the constant-D linear equation assumptions are valid only if $N(x)$ remains less than $n_i(T)$, the intrinsic carrier concentration in silicon at the *diffusion* temperature T. Values of this parameter are plotted versus temperature by Streetman [14]. For typical diffusion temperatures around 1000°C, $n_i \approx 10^{19}\ \text{cm}^{-3}$, and emitter N_{0E} values may exceed this substantially. Then D no longer can be considered concentration- and x-independent at the so-called *intrinsic* value that was determined from Appendix F.6.1 and used in earlier numerical examples. The modified or *extrinsic* D_e and intrinsic D values differ because of different mechanisms which come into play with different dopants (such as arsenic and phosphorus).[4] More important, at high concentration values D_e becomes a function of the doping concentration.

Consider the effects of position-dependent diffusivity on Fick's laws. The first law will become

$$\mathscr{F}(x,t) = -D_e(N)\frac{\partial N(x,t)}{\partial x} \tag{7.10-2}$$

The bookkeeping statement for the small volume $1 \cdot \Delta x$ of Figure 6–5 becomes

$$\left[\mathscr{F}(x,t) - \mathscr{F}(x+\Delta x,t)\right]\cdot 1 = (1\cdot\Delta x)\frac{\partial N(x,t)}{\partial t}$$

or

$$-\frac{\Delta\mathscr{F}(x,t)}{\Delta x} = \frac{\partial N(x)}{\partial t}$$

Then, taking the limit and substituting the derivative of Eq. (7.10-2), we get

---

[4] Most authors use D_i for the intrinsic value in discussing this subject and D for the extrinsic value. In order to maintain consistent notation here, we shall retain the unsubscripted D as the intrinsic value as used earlier; D_e will be used for the extrinsic value.

$$\frac{\partial}{\partial x}\left[D_e(N)\frac{\partial N(x,t)}{\partial x}\right] = \frac{\partial N(x,t)}{\partial t} \tag{7.10-3}$$

This is the general nonlinear form of Fick's second law (FSL) that allows for changes of D_e in N. Since N also is x-dependent, we could replace $D_e(N)$ by $D_e(x)$. Note that if D_e is independent of x and N, it may be factored out of the brackets in Eq. (7.10-3), which then reduces to the linear form.

For convenience, in the rest of this section, we shall drop the parenthetical quantities and write simply D_e and N. The subscript e on D will be retained, however, as a reminder that the extrinsic value is used. The absence of the subscript will indicate the intrinsic value.

When dealing with the highly doped As and P emitters here, we must use the nonlinear form, Eq. (7.10-3), and different profile shapes will result.

7.11 Arsenic Emitter

In this section we consider the expression for D_e and the solution for FSL for the highly doped As predep profile. Apparently there are two principal causes of the enhanced value of D_e over D. (1) At diffusion temperatures all the dopants in the host are ionized and set up electric fields. The indiffusing As, which also is ionized, is subject to *drift* as well as diffusion—both in the same direction—and *field-aided diffusion* results. When $N(x)$ exceeds about 10^{19} cm^{-3}, this effect becomes significant and the extrinsic values of D_e must be used. (2) At the high doping levels, interaction between the incoming species and point defects in the silicon becomes significant and causes further increases above the low-level, intrinsic D values. Since doping levels vary with position, then, so will D_e.

Fair has shown that for a dopant such as arsenic and for $N \geq n_i$, the extrinsic diffusivity depends linearly on N and is given by [15]

$$D_e = D\left(\frac{2N}{n_i}\right) \tag{7.11-1}$$

where n_i is the intrinsic carrier concentration at the *diffusion* temperature. For this form of D_e Eq. (7.10-3) reduces to

$$\frac{2D}{n_i}\frac{\partial}{\partial x}\left(N\frac{\partial N}{\partial x}\right) = \frac{\partial N}{\partial t}$$

Multiplication by N_0/N_0 reduces this to the normalized form

$$\left(\frac{2DN_0}{n_i}\right)\frac{\partial}{\partial x}\left(\frac{N}{N_0}\frac{\partial N}{\partial x}\right) = \frac{\partial N}{\partial t} \tag{7.11-2}$$

This is the equation to be solved for predep conditions in order to get the arsenic emitter profile.

7.11.1 Profile

Fair used Chebychev polynomials to solve the nonlinear equation, and the solution in normalized form for $N \geq n_i$ is

$$\frac{N}{N_0} = 1.00 - 0.87Y - 0.45Y^2 \tag{7.11-3}$$

where

$$Y = \frac{x}{\sqrt{\frac{2N_0}{n_i}(4Dt)}} \tag{7.11-4}$$

This curve of the arsenic profile is plotted on *linear* coordinates in Figure 7–15(a), and should be compared with the erfc u curves of Figure 6–6(a).

To see how the profile looks as a function of x, it must be corrected by Eq. (7.11-4)

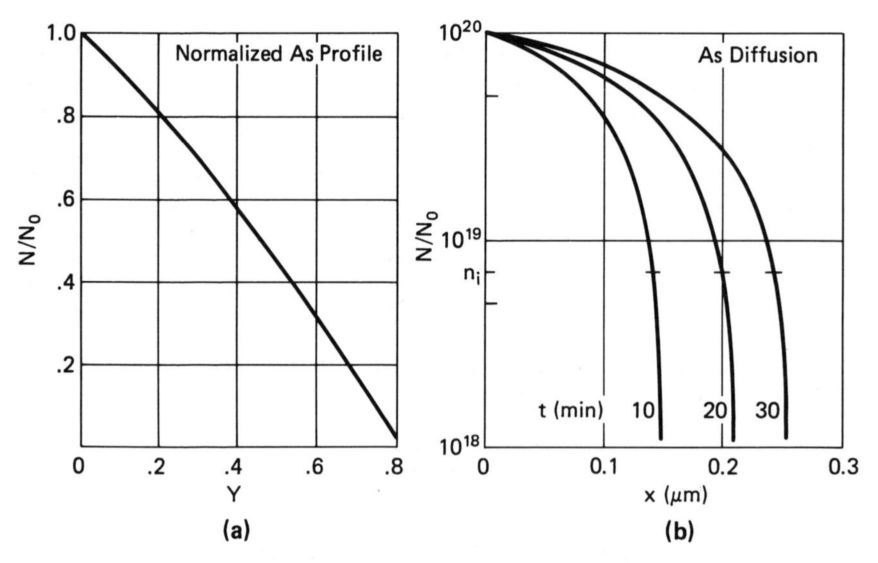

Figure 7–15. Normalized arsenic profile. (a) Plotted on linear coordinates. (b) Plotted on semilogarithmic coordinates to show time variation. Calculated from Eq. (7.11-3).

$$x = \sqrt{\frac{8N_0}{n_i}Dt}\;Y \qquad (7.11\text{-}5)$$

so the shape is unchanged. A typical set of profiles for $T = 1000°C$, with $N_0 = 1 \times 10^{20}$ cm^{-3} is shown on *semilog* coordinates for three values of diffusion time in Figure 7–15(b). The way a curve actually looks for a given set of parameters depends markedly on the value of the forefactor of Y in Eq. (7.11-5). As the ratio N_0/n_i increases, the curve, plotted on semilog coordinates, gets a much sharper knee and a steeper cutoff, but the junction depth, which will be evaluated in short order, increases for a given value of Dt. This may lead to problems of control in shallow junction devices.

The curves of Figure 7–15(b) also should be compared with those for a conventional predep erfc u diffusion as in Figure 6–6(b). It is quite evident that the As profile appears more rectangular on semilog coordinates than does the erfc u form, although we actually are more concerned with a comparison of the linear plots. The following points may be noted in this regard:

1. The As profile has a smaller slope magnitude near $x = 0$. This may be verified mathematically by evaluating the slopes at the surface for both profiles.

For the erfc u case,

$$\frac{\partial}{\partial x}\left[N_0 \operatorname{erfc}\left(\frac{x}{\sqrt{4Dt}}\right)\right]_{x=0} = -\frac{N_0}{\sqrt{\pi Dt}} \qquad (7.11\text{-}6)$$

while for the As case

$$\left.\frac{\partial N}{\partial x}\right|_{x=0} = -\frac{N_0}{\sqrt{\pi Dt}}\left(\frac{1}{\sqrt{3.36N_0/n_i}}\right) \qquad (7.11\text{-}7)$$

In the range of interest where $N_0 > n_i$ the initial slope for As is smaller by the bracketed factor; hence if $N_0 = 10n_i$, that factor leads to an 83% reduction of the initial slope.

2. The profile is very steep in the falloff region. This is verified in the next section.

7.11.2 Junction Depth

Consider equations for the junction depth. If we use a common criterion for Y_j (or x_j), it will be the value at which $N(x) = N_B'$. Substituting this criterion into Eq. (7.11-3) yields the quadratic

Table 7–2. Junction Depth versus N_B'/N_0 for the As Emitter

N_B'/N_0	10^{-1}	10^{-2}	10^{-3}	10^{-4}	10^{-5}
Y_j	0.7455	0.8029	0.8085	0.8091	0.8091

Calculated from Eq. (7.11-8).

$$Y_j^2 + 1.93Y_j - 2.22\left(1 - \frac{N_B'}{N_0}\right) = 0$$

whose solution is

$$Y_j = 0.965\left[-1 + \sqrt{1 + 2.38\left(1 - \frac{N_B'}{N_0}\right)}\right] \tag{7.11-8}$$

Table 7–2 shows corresponding values of Y_j and N_B'/N_0 over a typical range of the latter parameter. Values of Y_j are given to four significant figures, simply to show that Y_j is slightly over 0.8, say 0.81, essentially independent of N_B'/N_0. From Eq. (7.11-5) and rounding off to two significant figures we get[5]

$$x_j = Y_j\sqrt{\frac{8N_0}{n_i}Dt} = \left[2.3\sqrt{\frac{N_0}{n_i}}\right]\sqrt{Dt} \tag{7.11-9}$$

By Eq. (6.4-7) the corresponding result for the erfc u profile, assuming the slow outdiffuser approximation, is

$$x_j = \sqrt{4Dt}\,\mathrm{erfc}^{-1}\left(\frac{N_B'}{N_0}\right) \tag{7.11-10}$$

Since N_B'/N_0 cannot be eliminated as it was in the As case, a simpler approximation cannot be obtained. We can, however, evaluate x_j for a specific N_B'/N_0 ratio, say 10^{-2}:

$$x_j = 3.64\sqrt{Dt} \quad \text{for } N_B'/N_0 = 0.01 \tag{7.11-11}$$

Since $N_0 > n_i$ in the range of interest, the *expression* for x_j shows that the As profile has the larger value for a given Dt product. There are mitigating circumstances, however. For example, if phosphorus is used to generate the

[5] Fair arrives at the value of $Y_j = 0.81$ by a slightly different criterion for defining the junction location [15]. Notice that the numerical value of 2.3 is *not* ln 10 rounded off to two significant figures; rather it is the two-significant-figure value of $0.81\sqrt{8}$.

erfc u profile, we can see from Figure 6–3 that the intrinsic diffusivity for arsenic is about one tenth that of phosphorus. In the next section we shall find that D_e for P can exceed that for As by a factor greater than 100!

The As profile falls off very rapidly near the junction, as may be seen in the semilog plot of Figure 7–15(b). It does not show in the linear plot at (a) in the figure because the cutoff range is too small to be seen on the linear ordinate scale. A mathematical check to compare the slope magnitudes G at x_j (*grade constants*) of profiles is left as an exercise (Problem 7–15). A principal advantage of the sharp falloff of the arsenic curve at the junction is that arsenic does not, to any significant extent, carry over into the base region and decrease R_{BB}.

7.11.3 Sheet Resistance

It is evident that the Irvin R_S curves should not be used to evaluate highly doped arsenic sheet resistances: consequently we shall derive an approximate expression. Note that the second subscript E has been dropped as a matter of brevity and convenience. We begin with the basic equation for the average sheet conductivity of a doped layer, viz

$$\langle \sigma_S \rangle = \frac{1}{Y_j} \int_0^{Y_j} q\,\mu_M\{N\}\,N(x)\,dY$$

The old problem of how to handle $\mu_M\{N\}$, which presumably is position dependent, is simplified here: if $N \geq 6 \times 10^{19}$ cm^{-3}, $\mu_M = \mu_n$ (remember that electrons are the majority carriers here) is constant at the degenerate value of about 75 cm^2/Vs, and can be factored out of the integral as $\langle \mu \rangle$,

$$\langle \sigma_S \rangle = \frac{q\langle \mu \rangle}{Y_j} \int_0^{Y_j} N(x)\,dY \tag{7.11-12}$$

Since Eq. (7.11-3) gives $N(x)$ as a polynomial the integral can be obtained easily. Many approximations are being used here, so we might as well note that the integral is simply the nearly triangular area under the curve in Figure 7–15(a). Thus we may write

$$\langle \sigma_S \rangle \approx \left[\frac{q\langle \mu \rangle}{Y_j}\right] \frac{N_0 Y_j}{2} = \frac{q\langle \mu \rangle N_0}{2} \tag{7.11-13}$$

and

$$R_S \approx \frac{1}{\langle \sigma_S \rangle x_j} = \frac{1}{0.5q\langle \mu \rangle N_0 x_j} = \frac{1}{(6 \times 10^{-18}) N_0 x_j}\ \Omega/\text{sq} \tag{7.11-14}$$

for the values $q = 1.6 \times 10^{-19}$ C, and $\langle \mu \rangle = 75$ cm^2/Vs.[6] The units of x_j are cm.

7.12 Phosphorus Emitter

Phosphorus at high doping comcentrations also exhibits concentration-dependent diffusivity, and in a much more complicated fashion than does arsenic. In these circumstances, P also obeys the nonlinear form of Fick's second law (FSL) and an anomalous concentration profile is inevitable. The measured shape of such a P profile is shown in Figure 7–16(a) [16]. The two profiles result from measurements made by different methods. The open data points were obtained by *secondary ion mass spectroscopy* (SIMS) and give the *total* phosphorus concentration. The closed points are based on *differential conductivity* (DC) measurements and give the concentration of *active* phosphorus atoms that act as donors. These latter phosphorous ions, whose concentration is the same as that of the electrons n, determine the electronic properties of the P-doped layer, and their profile is the one of concern.

The difference between the two curves is the number per cm^3 of *inactive* P atoms (i.e., those that have not entered a vacancy and donated their fifth electron to the host). There is strong physical evidence that for very high surface concentrations a chemical reaction takes place in the nearly constant-concentration surface region of width x_0, whereby a phosphorus silicide phase is formed. Thus, up to x_0 the silicon is saturated with phosphorus and the excess phosphorus goes into the silicide phase. This may account for the nearly constant active phosphorus concentration in this surface region.

The *active* P profile in Figure 7–16(a) shows three regions:

1. A surface region, for $N \approx 4 \times 10^{20}$ cm^{-3}, that is nearly flat (constant), and where $D \propto n^2$.
2. A transition region for $4 \times 10^{20} \geq N \geq 6 \times 10^{19}$ cm^{-3}, where $D \propto n^{-2}$.
3. A tail region for $n < 6 \times 10^{19}$ cm^{-3}, where $D \approx$ constant.

Notice how the dependence of *phosphorus* diffusivity on *electron* (or *active* phosphorus) concentration n varies [16].

The actual profile is difficult to analyze so we shall use Tsai's simplified model, shown in Figure 7–16(b), to make computation easier [17]. Notation is changed to our usage.

[6] Fair uses $0.55 N_0 Y_j$ as the integrated area under the profile as over against our value of $0.5 N_0 Y_j$. His numerical constant in Eq. (7.11-14) is 6.67×10^{-18}. The difference is trivial considering the approximations involved in evaluating $\langle \mu \rangle$. Not the least of these is neglecting the effect of the base profile, which is of the opposite doping type, in calculating the area.

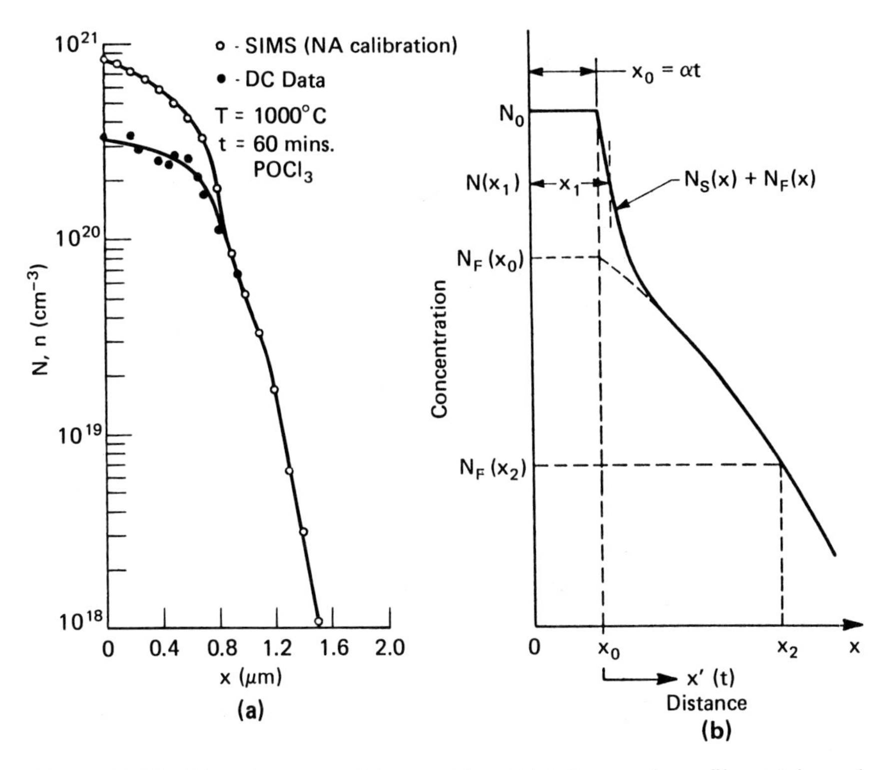

Figure 7–16. Phosphorus emitter profiles. (a) Measured profiles. Fair and Tsai [16]. (Reprinted with permission of the publisher, The Electrochemical Society, Inc.) (b) Idealized profile showing the coordinate system. Tsai [17]. (Courtesy of The Institute of Electrical and Electronic Engineers. © 1969 IEEE.)

7.12.1 Profile Calculation

A chief feature of the simplified model is the assumption that the active concentration is constant at value N_0 in the surface region. To the right of x_0, phosphorus diffusion is characterized by two different values of diffusivity: a slow or low value that we designate D_S, and a fast or higher value, D_F. The effects of both values may be observed in the transition region, while D_F dominates the tail. The boundary between the constant and changing portions of the profile, x_0, is assumed to increase *linearly* with time; hence one basic equation is

$$x_0 = \alpha t \tag{7.12-1}$$

Table 7–3. Arrhenius Parameters for Shallow Phosphorus Diffusion Profile

Parameter (Λ)	Λ_∞	E_a, eV
α, cm^2/s	0.18	1.75
D_S, cm^2/s	49.3	3.77
D_F, cm^2/s	2.49 E−5	2.0
$N_F(x_0)$, cm^{-3}	3.95 E23	0.9

$\Lambda = \Lambda_\infty \exp(-E_a/kT)$; T in Kelvins
E$m = 10^m$
After Tsai [17]. Courtesy of The Institute of Electrical and Electronic Engineers. © 1969 IEEE.

and we note that

$$N(x,t) = N_0, \qquad x \le x_0 \tag{7.12-2}$$

The boundary rate constant, α, is an exponential function of absolute temperature in the usual Arrhenius form.

For $x > x_0$ the two species are apparent, each with its characteristic diffusivity, so

$$D_S \frac{\partial^2 N_S}{\partial x^2} = \frac{\partial N_S}{\partial t}, \qquad x > x_0 \tag{7.12-3}$$

$$D_F \frac{\partial^2 N_F}{\partial x^2} = \frac{\partial N_F}{\partial t}, \qquad x' > 0 \tag{7.12-4}$$

where

$$x' = x - x_0 = x - \alpha t \tag{7.12-5}$$

$$N_S(x,0) = 0, \qquad N_F(x,0) = 0, \qquad N_F(x_0,t) = (1 - v)N_0$$

$$v = \frac{N_F(x_0)}{N_0} \tag{7.12-6}$$

$$N_S(\infty,t) = 0, \qquad N_F(\infty,t) = 0$$

The Arrhenius parameters for α, D_S, D_F, and $N_F(x_0)$ [see Figure 7–16(b)] are given in Table 7–3. Calculations show that for $T < 1100°C$, $D_F > D_S$; hence the choice of subscripts.

The solutions for $x > x_0$ are:

$$N(x,t) = N_S(x,t) + N_F(x,t) \tag{7.12-7}$$

$$N_S(x,t) = \left(\frac{1-v}{2}\right) N_0 \exp\left[\frac{-\alpha}{2D_S}(x-x_0)\right] F_S(x,t) \tag{7.12-8}$$

$$F_S(x,t) = \operatorname{erfc}\left(\frac{x+x_0}{\sqrt{4D_S t}}\right) + \operatorname{erfc}\left(\frac{x-3x_0}{\sqrt{4D_S t}}\right) \tag{7.12-9}$$

$$N_F(x,t) = \frac{v}{2} N_0 \exp\left[\frac{-\alpha}{2D_F}(x-x_0)\right] F_F(x,t) \tag{7.12-10}$$

$$F_F(x,t) = \operatorname{erfc}\left(\frac{x+x_0}{\sqrt{4D_F t}}\right) + \operatorname{erfc}\left(\frac{x-3x_0}{\sqrt{4D_F t}}\right) \tag{7.12-11}$$

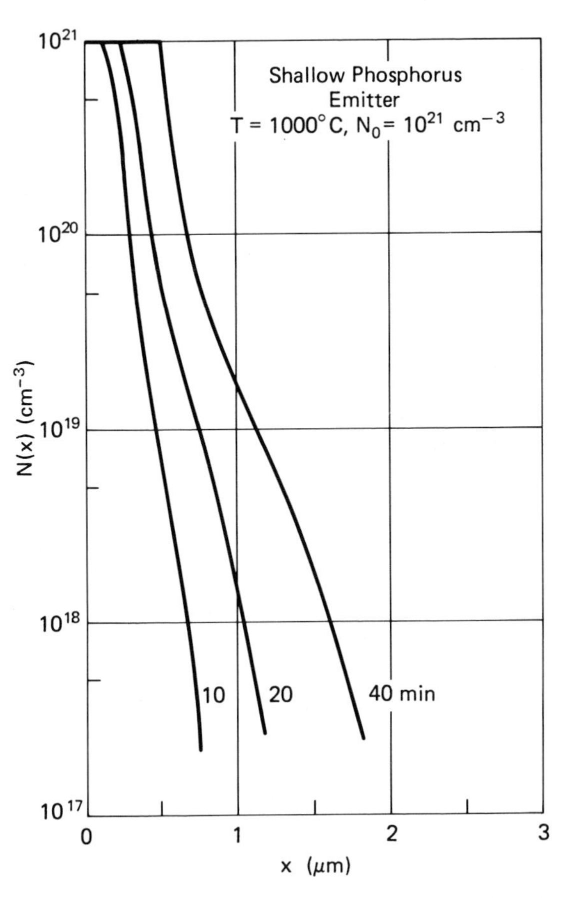

Figure 7–17. Calculated shallow phosphorus emitter profiles based on Tsai's model.

Some simplification results if $\sqrt{4D_S t} \gg (x + x_0)$ and $x < 3x_0$ so that the erfc terms are each nearly unity; then $F_S(x,t) \approx 2$, and

$$N_S(x,t) \approx (1 - \nu) N_0 \exp\left[\frac{-\alpha}{2D_S}(x - x_0)\right] \tag{7.12-12}$$

These equations may be used for profile calculation but the help of a computer is advised. Note in the foregoing equations, if $(x - 3x_0) < 0$, the erfc of a negative argument is required. Recall that $\text{erfc}(-u) = 1 + \text{erf}(u)$. Typical calculated profiles are plotted in Figure 7–17.

The foregoing treatment is based on physical observations, and is not concerned with the mechanisms that cause the changes in diffusivity. Fair and Tsai have reported work relating phosphorus diffusion mechanisms to *charged* vacancies in the Si and their interactions with diffusing P ions [16]. Some of the results illustrate the profound effect of these mechanisms—e.g., the 1000°C diffusivity of P in silicon is increased by factors of roughly 50 and 200 over D_i in the surface and tail regions, respectively. Intermediate values apply in the transition range.

Methods of numerical analysis have been used for solving the nonlinear differential equations relating to the anomalous P profiles. One uses finite difference operators in place of derivatives [18]. These methods are becoming more common.

7.12.2 Junction Depth

In principle, the junction depth for the high-concentration, shallow phosphorus emitter profile may be found simply by noting that at $x = x_j$

$$N(x_j, t) = N_B' \tag{7.12-13}$$

where, as before, N_B' is the common value of the emitter and base profiles where they cross each other at x_j. Direct solution of the foregoing equations for x_j is quite cumbersome, even with a computer; hence we seek some simpler approximation.

For shallow emitter profiles, diffusion temperatures will be kept comparatively low, say not over 1000°C, and diffusion times will be kept comparatively short. Assume that the junction lies on the profile tail, where $N_F(x_j, t)$ dominates, such that

$$N(x_j, t) \approx N_F(x_j, t) = N_B'$$

Now consider the two terms in $F_F(x,t)$: for $x > x_0$, $(x + x_0) > (x - 3x_0)$. Then since erfc u increases rapidly as u decreases

$$\operatorname{erfc}\left(\frac{x + x_0}{\sqrt{4D_F t}}\right) = L \ll R = \operatorname{erfc}\left(\frac{x - 3x_0}{\sqrt{4D_F t}}\right)$$

This effect will be enhanced if x_j is less than $3x_0$ because R will exceed one. Consider some typical numbers to check the inequality. From the data of Table 7–3, at $T = 1000°\text{C} = 1273$ K

$$\alpha = 2.13 \times 10^{-8} \text{ cm/s} \quad \text{and} \quad D_F = 3.02 \times 10^{-13} \text{ cm}^2\text{/s}$$

For a 30-min diffusion

$$x_0 = \alpha t = (2.13 \times 10^{-8})(1800) = 3.83 \times 10^{-5} \text{ cm}$$

and

$$\sqrt{4D_F t} = \sqrt{4(3.02 \times 10^{-13})(1800)} = 4.66 \times 10^{-5} \text{ cm}$$

Say $x = 2.5x_0$; then

$$L = \operatorname{erfc}\left[\frac{3.5(3.83 \times 10^{-5})}{4.66 \times 10^{-5}}\right] = \operatorname{erfc}(2.88) = 4.62 \times 10^{-5}$$

$$R = \operatorname{erfc}\left[\frac{-0.5(3.83 \times 10^{-5})}{4.66 \times 10^{-5}}\right] = \operatorname{erfc}(-0.411) = 1.44$$

These results show the validity of the assumptions in the range for which the calculations were made.

If these approximations are used, $F_F(x, t) \approx R$ and Eq. (7.12-13) reduces to

$$\frac{vN_0}{2} \exp\left[\frac{-\alpha}{2D_F}(x_j - x_0)\right] \operatorname{erfc}\left(\frac{x_j - 3x_0}{\sqrt{4D_F t}}\right) = N_B' \qquad (7.12\text{-}14)$$

In essence the solution will be by trial and error. The task is made much simpler if we define a new variable p by $x_j = px_0$, so

$$\left.\begin{aligned}(x_j - x_0) &= (p - 1)x_0 \\ (x_j - 3x_0) &= (p - 3)x_0\end{aligned}\right\} \qquad (7.12\text{-}15)$$

Use of these quantities and the definition of v yields

$$\exp\left[\frac{\alpha x_0(p - 1)}{2D_F}\right] \operatorname{erfc}\left[\frac{x_0(p - 3)}{\sqrt{4D_F t}}\right] - \frac{2N_B'}{N_F(x_0)} = 0 \qquad (7.12\text{-}16)$$

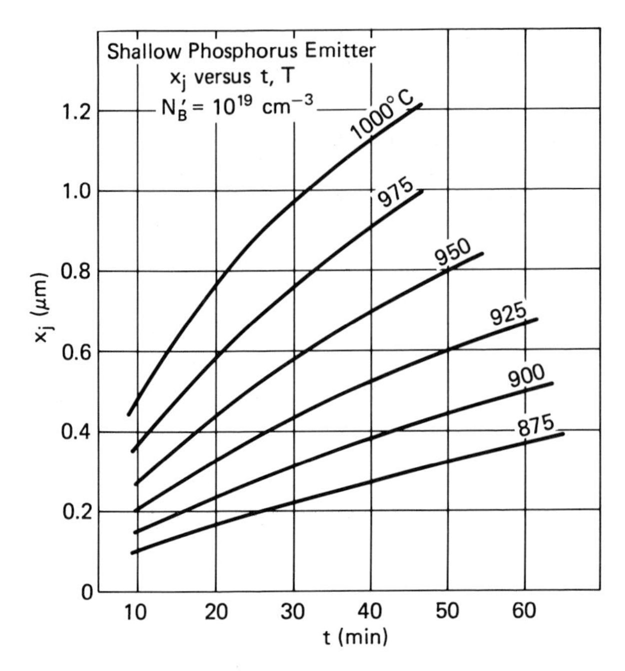

Figure 7–18. Calculated curves of junction depth with time and diffusion temperature as variables, based on Tsai's model.

Given a computer and an assumed value of N_B', this may be solved quite rapidly by using the Newton-Raphson method. Typical results are shown in Figure 7–18 for $N_B' = 10^{19}$ cm^{-3}. The profiles of Figure 7–17 show that x_j is quite sensitive to the value of N_B'.

7.12.3 Sheet Resistance

An approximate expression for the sheet resistance of the high-concentration, shallow *phosphorus* emitter is based on Tsai's model of the profile. Assume that the doping level exceeds 6×10^{19} cm^{-3} throughout the entire layer, so the electron mobility in the emitter region will be constant at $<\mu> = 75$ cm^2/Vs (Section 7.11.3).

The average conductivity of the emitter layer is

$$\langle \sigma_S \rangle = \frac{1}{x_j} \int_0^{x_j} q \langle \mu \rangle [N(x) - N_B'] \, dx \tag{7.12-17}$$

Since q and $\langle \mu \rangle$ are constant, they may be factored out of the integral, which is simply the area under the emitter profile from $x = 0$ to x_j, minus N_B'. On linear coordinates the area is appoximately that of a trapezoid: $(N_0 - N_B')$ to x_0, and dropping nearly linearly to 0 at x_j, or

$$\text{area} = \tfrac{1}{2}(N_0 - N_B')(x_0 + x_j) \approx \tfrac{1}{2}N_0(x_0 + x_j) \qquad (7.12\text{-}18)$$

The last approximation comes about because invariably N_B' is at least a decade below N_0.

It follows that

$$R_S = \frac{2}{q\langle\mu\rangle N_0(x_0 + x_j)} = \frac{1}{(6 \times 10^{-18})N_0(x_0 + x_j)} \qquad (7.12\text{-}19)$$

The values of x_0 and x_j may be determined from earlier calculations.

A numerical example gives a feel for typical orders of magnitudes. A shallow phosphorus diffusion is run at 925°C for 38 minutes. $N_0 = 10^{21}$ and $N_B' = 10^{19}$ cm^{-3}. Calculate R_S. Computer evaluation gives $x_0 = 0.179$ μm, and $x_j = 0.501$ μm. Then

$$R_S = \frac{1}{(6 \times 10^{-18})(10^{21})(1.79 + 5.01)10^{-5}} \approx 2.5\ \Omega/\text{sq}$$

Note that the value of $(x_0 + x_j)$ must be entered in cm in order to maintain consistent units.

7.12.4 Emitter Dip

It has been known for many years that shallow n–p–n structures with highly doped phosphorus emitters exhibit *emitter dip* or *emitter push* as illustrated in Figure 7–19, where the base/collector junction is deeper under the emitter than elsewhere. It appears as if the emitter *pushed* the B/C junction ahead of it, causing that junction to *dip* underneath the emitter. The effect comes about, however, because the phosphorus emitter tail extends into the base, causing

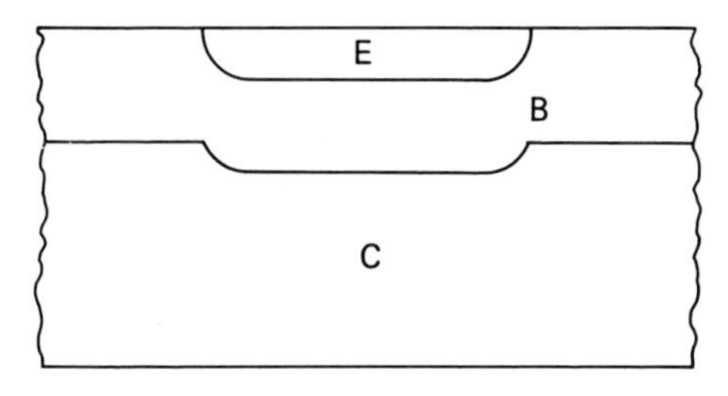

Figure 7–19. Emitter dip.

the diffusion of *boron* there to be enhanced by the presence of the same vacancy species that increased phosphorus diffusivity in the emitter tail. For this reason Fair and Tsai proposed that the diffusivity of boron is enhanced by the same factor as phosphorus in the emitter tail. During emitter diffusion, the base region underneath the emitter undergoes a second drive cycle with the enhanced diffusivity value, while base regions not under the emitter continue to drive with the normal value. This accounts for the dip in the junction under the emitter. A corresponding emitter dip has not been observed with arsenic emitters.

7.13 Correction for ρ versus N Curve

Calculation of room-temperature resistivities normally assumes all dopant atoms to be ionized and active electronically. With this assumption the Irvin resistivity curves of Appendix F.2.1 are used to convert differential resistivity measurements on a sample to the doping profile. Figure 7–16(a), however, shows that the assumption breaks down for *phosphorus* doping where N exceeds about 10^{20} cm^{-3}.

Based on measured data, Fair and Tsai relate the *total* phosphorus concentration, N, to the electron (or electronically active phosphorus) concentra-

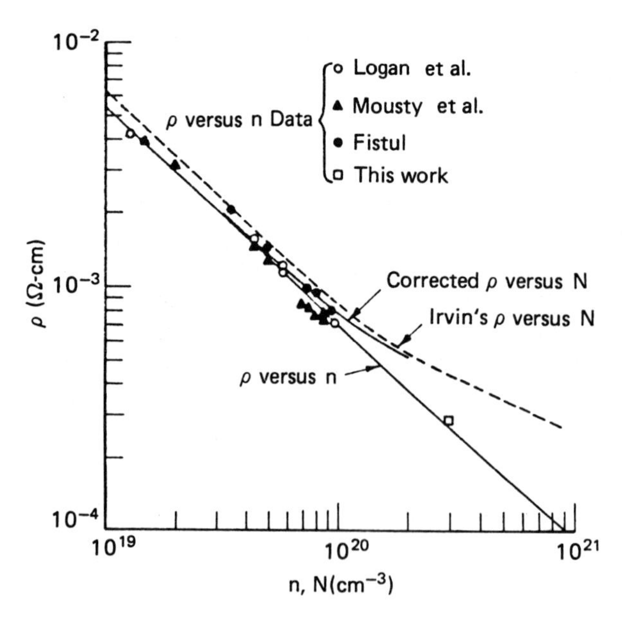

Figure 7–20. Resistivity versus phosphorus and electron concentrations for phosphorus doping in silicon. Fair and Tsai [16]. (Reprinted with permission of the publisher, The Electrochemical Society, Inc.)

tion, n, with the following equation

$$N = n + (2.04 \times 10^{-41})n^3 \tag{7.13-1}$$

Resistivity is determined by n, so the second term on the RHS gives the concentration of electronically *inactive* P atoms.

Figure 7–20 shows how this affects the ρ versus N curves. Remember that this correction applies to *phosphorus* only.

Problems

7–1 A boron predep and drive are performed on a Si wafer doped at 5×10^{15} cm^{-3} with As. $T_1 = 1000°C$, $t_1 = 900$ s, $N_{01} = 2 \times 10^{20}$ cm^{-3}, $T_2 = 1100°C$.

(a) Using a computer program, determine and tabulate N_{02}, R_{S2}, and x_{j2} at the following t_2 values; 800, 1800, and 2700 s.
(b) Comment on the trends.
(c) Compare the trends with those of Problem 6–4(a).

7–2 By using a computer program you are to compare the results of a B predep alone to those of a B predep and drive sequence, both of which give roughly the same junction depth. In both cases the wafer is doped uniformly with As at 5×10^{15} cm^{-3}. Round off results to 3 significant figures.

(a) Predep alone: $T_1 = 1000°C$, $N_{01} = 10^{20}$ cm^{-3}, $t_1 = 2410$ s. Determine N_{01}, R_{S1}, x_{j1}, and if the SOA is valid.
(b) Predep plus drive: $T_1 = 900°C$, $t_1 = 900$ s, $N_{01} = 10^{20}$ cm^{-3}, $T_2 = 1000°C$, and $t_2 = 2700$ s. Determine N_{02}, R_{S2}, x_{j2}, and if the SOA is valid.
(c) Compare the results of (a) and (b), and comment.
(d) Check the values of N_{02} and x_{j2} by hand calculation for part (b). Assume SOA and SMA.

7–3 You are to do a B predep and drive into an n-type Si wafer doped at 3×10^{15} cm^{-3}. Use $T_1 = 950°C$ and $N_{01} = N_{SL}$. After the drive you want $N_{02} = 10^{19}$ cm^{-3} and $x_{j2} = 2$ μm. Assume SOA and SMA and hand calculate.

(a) Calculate the required values of Δ_2', Δ_1, and Δ_2.
(b) Calculate t_1 in min.
(c) Calculate t_2 for $T_2 = 1050$, 1100, and 1150°C. Which combination would you use? Why?

7–4 Consider a double-diffused n–p–n BJT. The substrate is doped with Sb at 6×10^{15} cm^{-3}.

Base diffusions—B predep: 1800 s @ 950°C, $N_{01} = 10^{20}$ cm^{-3}
drive: 3600 s @ 1100°
Emitter diffusion—P predep: 3600 s @ 1100°C, $N_{0E} = 10^{21}$ cm^{-3}

Remember that the B continues to drive during the emitter predep. Assume SOA and MSA.

Calculate N_{OB}, x_{jC}, x_{jE}, and w_b (base width). Use trial and error for finding x_{jE}. [*Hint:* It lies between 1.7 and 1.8 μm.]

7–5 Design the diffusion process for a p–n–p BJT, using B and P to fit these parameters: Assume SOA.

$$N_{OE} = 5 \times 10^{20}\ \text{cm}^{-3}, \qquad x_{jE} = 2.04\ \mu\text{m} \qquad N_C = 10^{15}\ \text{cm}^{-3},$$

$$x_{jC} = 4.817\ \mu\text{m}, \qquad N_B' = 1.179 \times 10^{18}\ \text{cm}^{-3}$$

7–6 The wafer of Problem 6–11(b) is given a 1-hour drive @ 1200°C. Calculate N_{02}, x_{j2}, $\langle \sigma_{S2} \rangle$, and R_{S2}. Save your results for use in Problem 7–7.

7–7 (a) Repeat Problem 7–6 but use a 2-hour drive time.
(b) Repeat Problems 6–11(b) and 7–6 but use a 30-min predep.
(c) Compare your results with those problems.

7–8 Consider a two-step diffusion process in a Si wafer doped at 10^{15} cm^{-3}. The indiffusant is B.

(a) The process specs are: $T_1 = 900$°C, $N_{01} = 0.5\, N_{SL}$, $T_2 = 1150$°C, $x_{j2} = 4\ \mu$m, $R_{S2} = 125\ \Omega$/sq. Calculate t_1 and t_2 in min using SOA and SMA. Show all your calculations. If you extract data from the Irvin curves, state which one(s) you use.
(b) What, if any, practical problems would arise if R_{S2} were changed to 357 Ω/sq, other values remaining unchanged? Show your calculations.

7–9 An n–p–n discrete BJT has these parameters: $R_{SE} = 10\ \Omega$/sq, $N_C = 10^{15}$ cm^{-3}, $N_{OB} = 10^{18}$ cm^{-3}, $x_{jE} = 1.8\ \mu$m, $w_b = 1.2\ \mu$m. Calculate R_{SB}, R_{BB}, and the max possible value of β. Show all your work.

7–10 A Si wafer is doped with Sb at 6×10^{15} cm^{-3}. A boron predep and drive sequence is performed under the following conditions: $t_1 = 20$ min, $T_1 = 1000$°C, $N_{01} = 2 \times 10^{18}$ cm^{-3}, $t_2 = 30$ min, $T_2 = 1000$°C.

(a) Calculate and tabulate N_{02} and x_{j2} by the following methods. (1) Assuming a gaussian drive profile with the predep considered as a Dirac δ function. (2) SMA.
(b) Compare Δ_1 and Δ_2. Would you expect (1) to be a good approximation?
(c) Compare the results of (1) and (2). Comment.

7–11 A boron predep and drive are to be performed on a Sb doped wafer at 10^{16} cm^{-3}. After both diffusions the sheet resistance is to be $R_{S2} = 26\ \Omega$/sq. Experience in the lab shows that R_S changes by a factor of 4 due to redistribution into the growing oxide during drive.

(a) What whould be the value of R_{S1} after the B predep? State all the assumptions and approximations you use.
(b) If the predep is run at 1100°C with an oxidized BN source wafer, what t_1 in min is required?

7–12 Derive an expression for the grade constant G (see Problem 6–10) for a gaussian profile with a slow diffuser background. Your expression should involve only N_b, N_{02}, x_j, natural logs, and constants.

7–13 A shallow As emitter is diffused for 15 min @ 1000°C. Calculate and plot the profile, x_j, and R_S.

7–14 Repeat Problem 7–13 but use P. Assume $N_B' = 10^{19}\ \text{cm}^{-3}$.

7–15 Derive equations for and compare the values of grade constant G for the erfc u, As, and P profiles.

References

[1] R. C. T. Smith, "Conduction of Heat in the Semi-infinite Solid with a Short Table of an Important Integral," *Australian J. Phys.*, **6**, (2), 127–130, June 1953. Table 1.

[2] R. Shrivastava and A. H. Marshak, "A Simple Approximation for Calculating Impurity Profiles for a Two-Step Diffusion Process," *Proc. IEEE*, **65**, (11), 1614–1615, Nov. 1977.

[3] G. Backenstoss, "Evaluation of Diffused Layers in Silicon," *Bell Syst. Tech. J*, **37**, (3), 699–710, May 1958.

[4] R. A. Colclaser, *Microelectronics: Processing and Device Design.* New York: Wiley, 1980.

[5] A. S. Grove, *Physics and Technology of Semiconductor Devices.* New York: Wiley, 1967.

[6] D. J. Hamilton and W. G. Howard, *Basic Integrated Circuit Engineering.* New York: McGraw-Hill, 1975.

[7] A. S. Grove, O. Leistiko, Jr., and C. T. Sah, "Redistribution of Acceptor and Donor Impurities During Thermal Oxidation of Silicon," *J. Appl. Phys.*, **35**, (9), 2695–2701, Sept. 1964.

[8] R. P. Donovan, "Oxidation." In *Fundamentals of Silicon Integrated Device Technology, Vol. I*, ed. R. M. Burger and R. P. Donovan, p. 157. Englewood Cliffs: Prentice-Hall, 1967.

[9] J. S. T. Huang and L. C. Welliver, "On the Redistribution of Boron in the Diffused Layer during Thermal Oxidation," *J. Electrochem. Soc.*, **117**, (2), 1577–1580, Dec. 1970. Their notation is nonstandard: Their $m = 1/P$, and their "segregation coefficient," designated by k, is the reciprocal of the conventional m as defined by our Eq. (7.7-1).

[10] M. M. Atalla and E. Tannenbaum, "Impurity Redistribution and Junction Formation in Silicon by Thermal Oxidation," *Bell Syst. Tech. J*, **39**, (4), 933–946, July 1960. Use care regarding notation. They use a segregation coefficient k which is the reciprocal of our m. They use m as the exponent relating oxide growth and time, viz., $x_{ox}{}^m = Kt$.

[11] M. B. Vora and H. N. Ghosh, "Arsenic A Better Emitter Impurity Than Phosphorus," *Circuits Manuf.*, **11**, (1), 16, Jan. 1971.

[12] "Arsenic emitter doping ups cutoff frequency," *Microwaves*, **9**, (12), 16, Dec. 1970.

[13] See for example, H. Wolf, *Semiconductors.* New York: Wiley-Interscience, 1971. p. 199.

[14] B. G. Streetman, *Solid State Electronic Devices, 2nd Ed.* Englewood Cliffs: Prentice-Hall, 1980. Figure 3.17.

[15] R. B. Fair, "Profile Estimation of High-Concentration Arsenic Diffusions in Silicon," *J. Appl. Phys.*, **43**, (3), 1278–1280, March 1972.

[16] R. B. Fair, and J. C. C. Tsai, "A Quantitative Model for the Diffusion of

Phosphorus in Silicon and the Emitter Dip Effect," *J. Electrochem. Soc.*, **124**, (7), 1107–1117, July 1977. See also: Discussion, op cit, **125**, (6), 995–998, June 1978.

[17] J. C. C. Tsai, "Shallow Phosphorus Diffusion Profiles in Silicon," *Proc. IEEE*, **57**, (9), 1499–1506, Sept. 1969.

[18] V. Arandjelovic, Lj. Milikovic, and D. Tjapkin, "The Numerical Analysis of Anomalous Doping Profiles of Phosphorus in Silicon," *Solid State Electron.*, **22**, (4), 355–359, Apr. 1979.

[19] D. T. Antoniadis, S. E. Hansen, and R. W. Hutton, "SUPREM II—A Program for IC Modeling and Simulation," *Technical Report No. 5019-2. Integrated Circuits Laboratory*. Stanford: Stanford University. June 1978.

Chapter 8
Ion Implantation

Ion implantation (I^2) provides an alternative to the predep diffusion for doping a host material. In contrast to diffusion which is a thermal process, I^2 is controlled electrically. Voltages can be switched on or off very rapidly, whereas changing temperatures is very slow. Thus, ion implantation provides greater flexibility, particularly if shallow doping depths and low doping densities are needed. The price to be paid is more complicated and expensive apparatus.

8.1 Basic Mechanism

The basic mechanism of doping a semiconductor by means of ion implantation is this: the dopant atoms first are ionized so that each has a positive charge. Through the use of electric and magnetic fields the ions are focused into a well-defined beam and accelerated to very high energies *in vacuo.* (Vacuum operation minimizes collisions with ambient molecules.) Then the beam is scanned in a *raster* on the surface of the wafer (substrate, target, or host), much like the pattern of horizontal lines on a television screen. By this means, the distribution of ions over the wafer surface is made very uniform.

Due to the high accelerating voltage, the ions have sufficient energy to penetrate into the wafer, giving up energy by colliding with lattice atoms or their electrons, until finally they come to rest. Note that when an ion stops moving, it unites with an electron in the host and again becomes a dopant *atom.* We then say that the dopant atom has been *implanted* into the wafer. An ion may be deflected by each collision, so the actual path and number of collisions a particular ion makes before stopping is random and unpredictable. Hence, the depth of penetration and the distribution of the stopped dopant ions must be calculated statistically. Once these implanted atoms are moved to lattice sites by annealing (Section 8.13), they will behave as donors or acceptors, as the case may be, just as do dopants introduced by diffusion.

When the charged dopant ions are in flight, they constitute a flow of electric current and their number can be "counted" by integrating a sample of the beam current over the duration of implant. The implant may be terminated abruptly, simply by deflecting the beam off the wafer (a very rapid operation indeed). This leads to very precise control of the dose Q, the number of dopant atoms implanted per unit area of the wafer surface.[1]

Materials are available that can mask the ion beam from penetrating the underlying wafer. Thus, ion implantation can provide selective doping of specific wafer areas.

A side effect of ion implantation is the *radiation damage* that results from very energetic dopant ions colliding with lattice atoms as they lose energy before coming to a halt within the wafer. The host atoms may be dislodged from their normal lattice sites, and the implanted atoms may be prevented from locating substitutionally for host atoms on lattice sites. In that event, they are not activated electronically to serve as donors or acceptors as the case may be. Radiation damage is removed after implant by a heating process known as *annealing*. Three variations are considered in Section 8.13.

Several aspects of ion implantation are considered in this chapter. We begin by studying a generic form of implanter so that we can relate details to the basic apparatus in subsequent sections.

8.2 Generic Ion Implanter

The diagram of a basic ion implanter is shown in Figure 8–1. In order to perform its function properly it must satisfy at least these basic requirements [1]:

1. Furnish ions of the desired dopant species. [ion source]
2. Insure that only ions of the desired species reach the wafer. [analyzer and neutral beam trap]
3. Focus the ions into a beam of desired cross section at the target surface. [focus]
4. Accelerate the ions to the desired energy level to control the penetration depth in the wafer. [accelerator]

---

[1] We use the symbol Q for dose in both diffusion and I^2 to make comparisons of the two processes easier. In a sense the notation is unfortunate here where electrical quantities are involved, because one expects Q to symbolize charge. The usage is widespread, though not universal, in the semiconductor industry, so be alert to avoid confusion.

Some authors use Φ or ϕ as the dose. Others use $Q_\square$ as the *charge* dose in coulombs/cm^2 so $Q_\square = q\Phi$, where q is the electronic charge magnitude. Still other writers use Φ as the ion flux density in ions/cm$^2 \cdot$ s, so $Q = \Phi t$, t being the implant duration. Even the word *fluence* sometimes replaces *dose*. In short, when reading the literature, carefully check the symbols and terms being used.

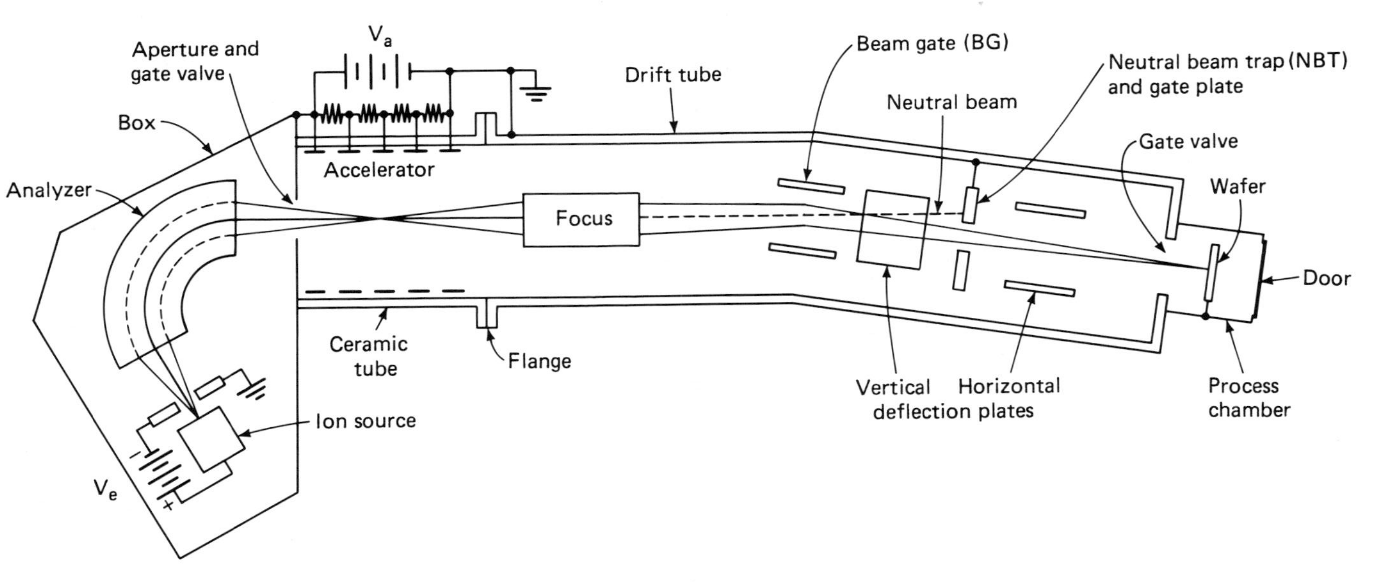

Figure 8–1. Layout of a generic ion implanter.

5. Scan the beam over the wafer surface to provide uniform doping in unmasked areas. [deflection plates]
6. Control the beam current and dose. [ion source and monitor]
7. Provide means for changing wafers rapidly without losing vacuum in the entire system. [process chamber]
8. Provide vacuum level pressures in the range of 10^{-7} to 10^{-5} torr throughout the entire system. [not shown in figure]

The part of the implanter that serves each function is given in square brackets in the foregoing list and may be identified in Figure 8–1. The relative placement of the parts may be changed, e.g., the analyzer may be located downstream from the accelerator. Some of these variations will be discussed in subsequent sections.

8.3 Electric and Magnetic Deflection

Many of the functions provided by the implanter are based on the interaction of a moving ion beam with an electric and/or a magnetic field. Several functions depend on the former: ionization and extraction in the ion source, acceleration in the accelerator, and deflection and focusing within the drift tube. Interaction between the beam and a magnetic field is the basis for mass separation in the analyzer. These interactions are basic to the implanter's operation, and are reviewed in Appendix B. Applications of these principles to the implanter are considered in later sections.

8.4 Implanter Sections

Having considered the overall operation of the implanter, we now look in more detail at the different sections identified in Figure 8–1.

8.4.1. Ion Sources

The function of an ion source is to furnish source material, ionize it by collisions with high-energy electrons, and extract the ions at 10 to 20 kV.

The source material for a dopant species, usually a gas or solid at room temperature, contains the desired species as one if its components. Typical gas sources are: boron trifluoride (BF_3) for boron doping; and phosphine (PH_3) or phosphorus pentafluoride (PF_5) for phosphorus doping; and arsine (AsH_3) or arsenic pentafluoride (AsF_5) for arsenic doping. Boron trioxide (B_2O_3) and phosphorus pentoxide (P_2O_5), or elemental boron (B), phosphorus (P), and arsenic (As) are solids at room temperature and may be used as source materials also. Consider one type of ion source used with gases [2].

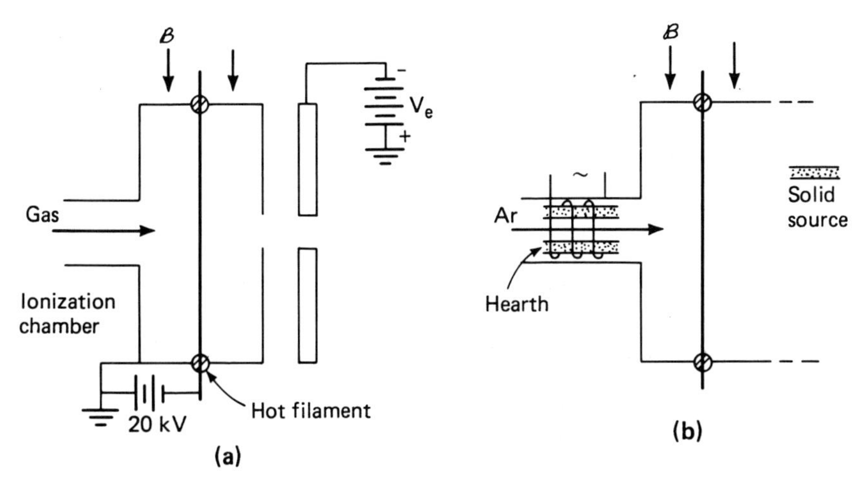

Figure 8–2. Harwell ion sources. (a) Gaseous source type. (b) Solid source type.

Figure 8–2(a) illustrates the Harwell type of ion source. The chamber is evacuated to about 10^{-5} torr and is held at about 20 kV wrt the filament, whose temperature is high enough to provide a copious supply of electrons. The electrons accelerate toward the chamber walls, across the $\mathscr{B}$ field that is supplied by a magnet, and so follow spiral paths whose lengths are much greater than the chamber radius. The source gas (e.g., BF_3) is let into the chamber, where the long electron paths greatly enhance the chance of collision between electrons and the gas molecules.

Such collisions can ionize the gas if one or more electrons are removed from the BF_3 molecule. Dissociation also may take place. As a matter of fact, a large number of different ion species will be produced in the ion chamber. Boron has two isotopes, ^{10}B and ^{11}B, which occur naturally in the ratio of 19/81. Both of these will be ionized, as will other species. Typically, these singly charged species will appear in the chamber: $^{10}B^+$, $^{10}BF^+$, $^{10}BF_2{}^+$, $^{10}BF_3{}^+$, F^+, $F_2{}^+$, ..., $^{11}B^+$, $^{11}BF^+$, $^{11}BF_2{}^+$, $^{11}BF_3{}^+$, ..., electrons, and contaminants.

Collisions in the ionization chamber also may result in two electrons being stripped off from a gas molecule, in which case a doubly charged ion results ($\nu = 2$ in the notation of Appendix B.3.1). Thus, still other ions may appear in the chamber, such as: $^{10}B^{++}$, $^{10}BF^{++}$, ..., $^{11}B^{++}$, $^{11}BF^{++}$, ...

Since for this example only boron is to be implanted, no radicals should reach the wafer. Also the dose is determined by integrating the beam current during the implant. The current due to a charge in motion depends upon the charge magnitude; if the dose-measuring equipment is calibrated to count singly charged $^{11}B^+$ ions, then $^{11}B^{++}$ or even neutral ^{11}B atoms in the beam will cause a miscount of the implanted *atoms* and the dose reading will be incorrect. Therefore means for filtering out all but one boron species must be provided.

Whether singly or doubly charged, all the ion types that are present have different charge-to-mass ratios q_i/m_i, and behave differently in electric and/or magnetic fields, so a single boron species may be selected, usually the singly ionized, heavier, and more abundant isotope $^{11}B^+$. The separation is carried out in the analyzer section of the implanter.[2]

A Harwell source may be modified to sublime (or sublimate) solid source materials to vapor, so they can be ionized like the gas sources. Even though the electron-source filament can furnish heat to sublime a solid, the solid source scheme of Figure 8–2(b) shows a separate hearth heater [3]. The solid source materials tend to pick up moisture from air; hence they require special storage facilities, usually under vacuum.

Figure 8–2(b) shows additional components that are required to handle solids: a hearth that holds, heats, and sublimates the solid sources, and a neutral carrier gas, such as argon, that carries the dopant vapor into the ionization chamber.

Solid elements such as boron, phosphorus, and arsenic may be used instead of compounds to reduce the number of species in the chamber. Solid B is seldom used, however, because of its very low vapor pressure, while As and P are quite satisfactory—e.g., the temperatures required for them to yield a vapor pressure of 10^{-4} torr are 1700, 204, and 129°C, respectively, for B, As and P [4]. In contrast to B, B_2O_3 requires only 600°C for the same vapor pressure [5]. In balance, the solids produce fewer ion species than the gas sources.

In both versions of Figure 8–2, ions "see" a negative extraction voltage of $V_e = 20$ kV through the exit hole in the ion source chamber; hence, they will be accelerated through the hole or aperture and reach a velocity given by Eq. (B.1-6) in Appendix B (note $V_a = V_e$ here). Some beam narrowing will take place because of the small size of the aperture and the positive voltage on the chamber wall.

8.4.2 Analyzer

The analyzer section of the implanter insures that only one species from the source output enters the remainder of the implanter system. Species separation

---

[2] In some special instances $^{11}B^{++}$ will be the ion of choice. Say an ion of charge q_i moves through a potential difference of V volts. Its energy will be $q_iV = vqV$ electron volts. Thus for a given accelerating voltage in the implanter if the doubly ionized species is chosen with no mass change—i.e., if $v = 2$, its members will attain twice the energy of the singly ionized species, effectively doubling the voltage insofar as energy is concerned.

is based on the bending angle's dependence on q_i/m_i in a magnetic field, Appendix B.3.

The expression for r, the radius of the circular path, is given by Eq. (B.3-4) and we already have calculated in the appendix that for $\mathscr{B}^* = 5$ kgauss and $V^* = 150$ kV the values for $^{10}B^+$ and $^{11}B^+$ ions are, respectively, $r_{10} = 12.8$ cm and $r_{11} = 13.5$ cm. If the exit aperture from the analyzer is small enough and located properly, only the mass 11 u ions will pass on through to the accelerator.

Figure 8–3(a) assumes a 90° bending angle and shows paths of different radii for ions with different charge/mass ratios arriving on a common path and entering the magnetic field at point A. Notice how the aperture allows only one path of exit into the accelerator region, and so selects only one q_i/m_i species.

Figure 8–3(b) assumes a 120° bending angle for the selected species. Note that the spreading is greater for the larger angle, so in choosing the value of magnetic flux density $\mathscr{B}$, one can trade off the radius of curvature r against the bending angle θ.

Figure 8–3(c) shows a slightly different situation. Since the beam is diverging slightly, ions of a single species coming from the ionizer section may arrive at the edge of the magnetic field on slightly different paths. All three paths in the magnetic field have the same value of r but the centers of curvature differ. The net effect is that the diverging paths are forced together again, giving a focusing action.

If the analyzer is located ahead of the accelerator as in Figure 8–1, the system is of the *pre-accelerator* analysis (often abbreviated *pre-analysis*) type. Notice that the magnet is at V_a (the accelerating voltage) wrt ground, which can bring up insulation problems on the magnet winding. On the other hand v_o depends on V_e rather than V_a; Eq. (B.3-4) shows that $r \propto \sqrt{V_e}$. Furthermore, since analysis occurs before acceleration, the beam contains only one species of ion during acceleration; it is therefore *clean*, and all its ions will be accelerated uniformly [6].

An alternative configuration places the analyzer downstream from the accelerator, giving what is called a *post-acceleration* analyzer or *post-analysis* type of system. The magnet then will be at ground potential, which solves the magnet winding insulation problem. On the other hand, v_o is determined by V_a rather than the lower-valued V_e, making it harder to bend the ion path in the analyzer. Also, the beam in the accelerator is not *clean*, so the different ion species have different velocities when they enter the magnetic field. This complicates species separation.

It is one thing to separate species of different q_i/m_i values, but it is another to ensure that only the desired species exits through the analyzer aperture; hence, instrumentation is required for identifying the several ion-type locations in the beam.

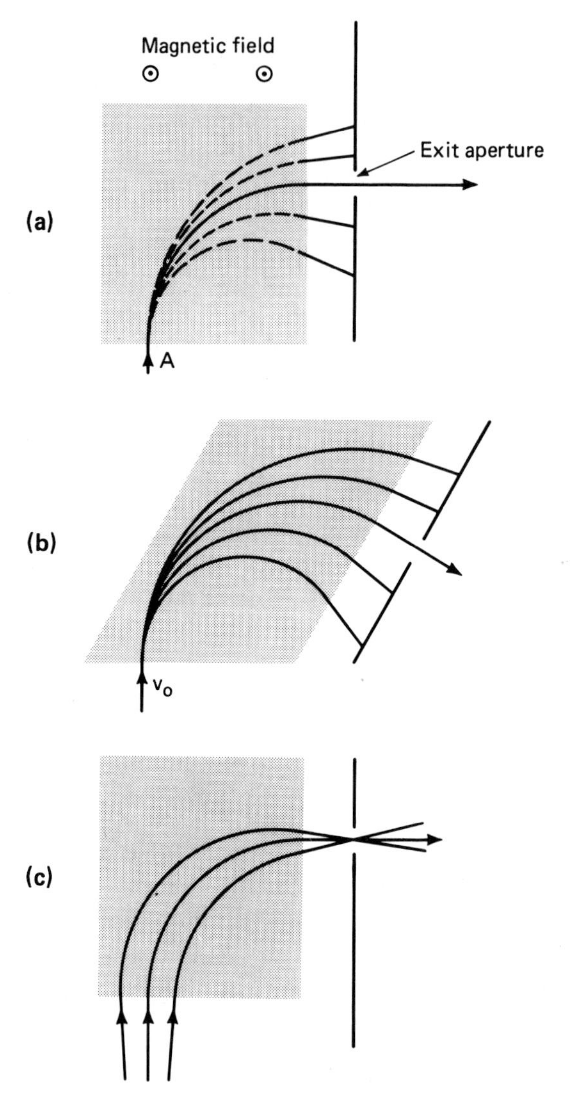

Figure 8–3. Operation of the implanter analyzer section. (a) Ions of different charge/mass ratios have different paths in the magnetic field. Only one type can pass through the exit aperture. (b) A greater bending angle provides greater separation among ion species in the plane of the exit aperture. (c) Ions of the same charge/mass ratio, but arriving at slightly different angles to the analyzer field, may be focused at the exit aperture.

8.4.3 Accelerator; HV Supplies

Reference to Figure 8–1 shows that the ions entering and emerging from the analyzer have energies of $q_i V_e = 20$ keV. The accelerator section functions to accelerate these ions to higher energy for implanting. Consider some of the problems involved here.

First the source and analyzer metal *box* or housing are connected to the high-voltage end of the accelerator column and so is at $+V_a$ volts wrt to the equipment ground. The other end of the accelerator contacts the metal drift tube, and is grounded for safety. Source box, accelerator tube, drift tube, and eventually the process chamber, comprise an evacuated enclosure for the ions to move in; hence, these component parts must be joined with vacuum-tight seals. Since the box and drift tube differ by V_a volts, the accelerator tube must provide the necessary insulation. Typically it will be of ceramic or a glass, Pryex being a common example. The high-voltage (HV) electrodes are cylindrical and are located within the glass tube, with their leads passing through the tube wall via vacuum-tight seals to the HV supply.

Breakdown problems due to leakage paths in and on the tube may be minimized by using several electrodes, say n in number, spaced uniformly along the length of the accelerator tube as in Figure 8–4(a). Connections to these electrodes are tapped off a high resistance voltage divider connected across the V_a supply. Consequently the voltage between adjacent electrodes is reduced by factor $1/(n - 1)$ and the possibility of breakdown is reduced.

The isolation of power supplies from ground is of great concern in the implanter. For the generic system of Figure 8–1 everything within the ion source "box" is at $+V_a$ wrt the equipment ground. Thus all the power supplies needed for the ion source, the extraction voltage source, the analyzer magnet current, and all the instrumentation associated with the functions inside the box must be insulated sufficiently from ground to withstand rated V_a s of 50 kV or more, plus a safety margin.

Large-current ion implanters most commonly use the insulation provided by a 60-Hz, iron-core transformer potted in oil. An alternate approach uses a motor/alternator set with a long, insulator coupling shaft between the two machines [3]. The a-c motor can be power-line grounded, while the alternator can float well above ground with insulation provided by the shaft.

Under proper operating conditions few, if any, ions reach the accelerator electrodes so the current requirements on the accelerator d-c supply (and the focusing and deflection supplies also) are small. This means that high-voltage, 60-Hz, iron-core transformers are not essential for voltage step-up. An alternative is to use a transistorized oscillator running at an ultrasonic frequency, say around 50 kHz, making the use of easy-to-insulate, air-core transformers possible [3]. Feedback can be used around the oscillator circuit to provide excellent voltage regulation.

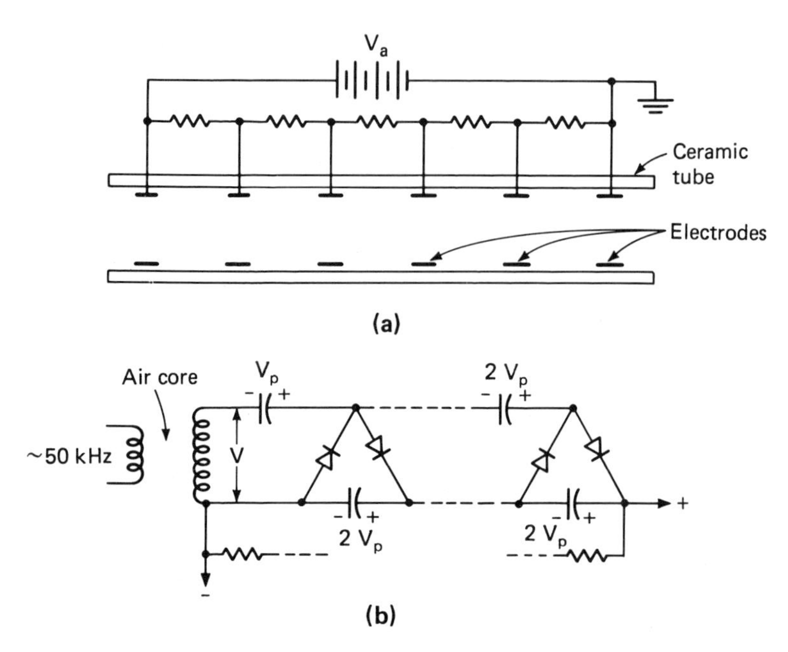

Figure 8–4. The accelerator and high-voltage supply. (a) The accelerator tube is ceramic to provide electrical insulation. Breakdown problems are reduced by distributing the acceleration voltage in equal increments among several equispaced electrodes. (b) The high-voltage supply uses cascaded voltage doublers to boost the input voltage V and to provide rectification.

Neither type of transformer, iron- or air-core, need step up the voltage to the full V_a value since voltage multiplier/rectifier circuits can be used to boost the d-c voltage. A typical circuit is shown in Figure 8–4(b). For n sections of the basic two-diode, two-capacitor voltage doubler, the no-load d-c output voltage will be $2nV_p$, where V_p is the peak value of the transformer secondary voltage.

In order to minimize collisions between the selected ions and any air molecules present in the accelerator and drift tubes, the pressure should be in the 10^{-6} to 10^{-7} torr range, while the pressure in the source box may be about 10^{-4} to 10^{-5} torr. Remember, these two regions are connected together, but the small orifice between the source box and accelerator, through which the ion beam passes, can support such a pressure differential (Appendix D.2). In some implanters another orifice may be provided farther downstream to help keep the process-chamber end of the system well evacuated. Usually, separate vacuum pumps are used for the sections on either side of an orifice. Also, valves usually are provided so that the separate parts of the system may be isolated from each other. This is of particular importance, say, when a filament in the ion source needs to be replaced while keeping the other parts of the system under vacuum.

8.4.4 Focus

Beam ions moving *in vacuo* are subjected to two forces because of their own presence. First, since the ions are of the same sign, they repel each other—an action that tends to make the beam diverge as it moves toward the target. This effect is usually small unless the beam current is quite large, which implies high ion density.

Second, since the beam comprises a flow of current, it creates a magnetic field that encircles the beam. This field tends to crowd the ions together, and works to counterbalance the diverging tendency.

The beam also may be diverging as it leaves the accelerator as shown in Figure 8–1. Ideally beams should arrive at the target with parallel sides and be of relatively small diameter so that focusing would not be necessary. Typically a beam diameter will be around 1 cm or less, with the diameter increasing with beam current. The function of the focus unit, then, is to bend ions traveling off-axis so the beam becomes convergent and focused at the target.

Focusing heavy ions with $\mathscr{B}$ fields is not practicable. But we see in Appendix B.2 that electrostatic deflection is independent of m_i; hence focusing in ion implanters is based on bending by an $\mathscr{E}$ field. Different electrode geometries may be used, but they should provide focus-bending action over a rather long distance.

One such geometry is the *quadrupole* assembly shown in Figure 8–5(a). A double unit with cylindrical electrodes is shown, but half-cylinders are used frequently. The positive and negative voltages are relative to ground, so their average value is zero and hence does not affect the x-directed ion velocity.

A rough sketch of the $\mathscr{E}$ fields among the four left-hand electrodes is shown in Figure 8–5(b). Mathematical analysis is difficult because the fields are not

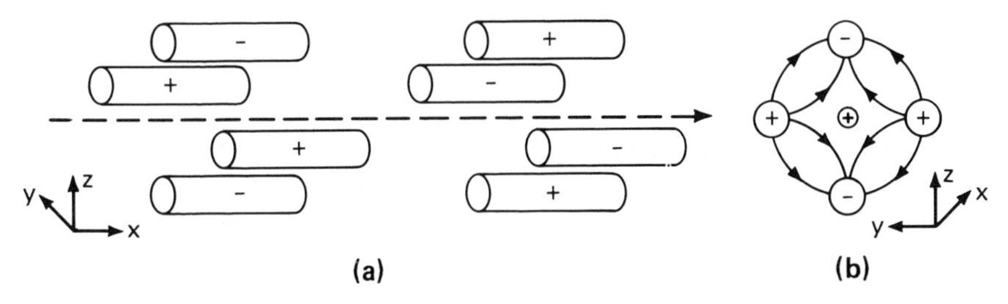

Figure 8–5. A double quadrapole lens for focusing the ion beam. (a) The eight focusing electrodes. (b) Fields seen by a positive ion centered among one set of four electrodes.

uniform, but the general effect on a divergent beam of ions can be seen qualitatively. Positive ions moving directly along the x-axis in the center of the electrodes remain unaffected because the net effect of the fields there is zero. Other ions, however, are repelled by the positive electrodes and attracted toward the negative electrodes. Hence, in passing through the left-hand quadrupole lens the ions are "squeezed up" along the z-axis, narrowing the beam in the y-direction. A similar narrowing action, but in the z-direction, takes place in the right-hand assembly due to the interchanged electrode polarities. The narrowing effects do not match exactly, so the combined action produces a beam of slightly elliptical cross section. Focus onto the plane of the implant target is brought about by adjusting the quadrupole voltages.

8.4.5 Neutral Beam Trap and Gate

Reference to Figure 8–1 shows the beam gate (BG), a pair of electrodes located between the focus and scanner (or deflection) assemblies. Consider its action.

As the beam moves through the system, collisions among ions may produce *neutral atoms* of the implanting species. For example an ion of choice, say $^{11}B^+$, may collide with a molecule of residual gas, causing an electron to be transferred to the boron ion, which results in a neutral boron atom ^{11}B. If neutral atoms reach the wafer, they would cause excessive doping at its center, and an incorrect reading of implanted dose (because dose-measuring equipment integrates *current*, neutral atoms cannot be counted). The neutral atoms are not deflected by electric fields and continue on with unchanged speed and direction, whatever the voltages on the BG plates.

The physical layout in Figure 8–1 is such that the neutral atoms pass through the BG and vertical deflection plates with no change in direction (i.e., along the dashed line), until intercepted by the neutral beam trap and gate electrode NBT. Therefore they are *trapped* in that they cannot reach the wafer.

What is the fate of the *ions* as they pass through the BG electrodes? If these electrodes have *equal d-c* voltages relative to ground, there is zero voltage between them and the ion beam is undeflected by them, follows the dashed path, also is intercepted by the NBT, and cannot reach the deflection plates and the wafer.

If the ion beam is to reach the wafer, it must be deflected downward in the figure until it passes through the hole in the NBT as shown by the solid lines. This *gating on* action results if *unequal d-c* voltages are applied to NBT electrodes with the lower plate negative and of proper magnitude wrt the upper. This has no effect on the neutral atom beam, and if the average of the NTG voltages is ground potential, the bending voltages will not affect the toward-wafer ion speed. Once gated on, the ion beam may be scanned across the wafer by suitable voltages of zero average value, and applied to the deflection plates.

The implant is terminated by the reverse action: by reducing the voltage *between* the NTG plates to zero, the ion beam is switched away from the target to be intercepted by the NBT. If the voltage difference is restored to its initial value, the ion beam reaches the target again so implant may be resumed.

8.5 Scanning

The ion beam diameter is typically much smaller than the wafer, so some relative motion between the two is required. This is known as *scanning*. Even if the two diameters were identical, motion still would be required because the radial distribution of ions across the beam is not uniform, but is rather like a gaussian. Thus, even during scanning, the paths of the ion beam on the target surface should overlap to insure uniform doping by the implant [6].

Generally relative motions in the y- and z-directions, across the face of the wafer, are used. Three types of scanning systems are used to do this:

(a) Electronic, in which the ion beam is scanned across the fixed target in two directions by $\mathscr{E}$ fields.
(b) Hybrid, where both target and ion beam are moved, one axis being scanned electronically and the other mechanically by moving the wafer.
(c) Mechanical, in which the ion beam remains in a fixed position and the target is moved across it in both directions.

In the hybrid and mechanical systems the target motion takes place in the process chamber of the implanter, while electronic deflection is performed in the scanner section.

8.5.1 Electronic Scanning

In the all-electronic scanning systems, symmetrical triangular voltages are applied to both the y- and z-direction scanner plates. This results in the beam scanning linearly across the wafer in both directions, a desirable result since it contributes to implant uniformity. If sinusoidal scanning voltages were used, changes in the instantaneous beam scanning velocity would make the beam move faster at the center of the scan causing lighter doping there.

Both horizontal and vertical scanning frequencies are usually in the range of 100 to 1200 Hz, with values around 900 Hz quite common. They must not be equal or integrally related, however, since that would give a scanned *raster* (or pattern) of fixed position on the wafer surface. Greater uniformity results when the raster lines move about on the wafer so that the effects of nonuniform ion density within the beam tend to cancel out. The raster is made larger than the wafer, since *overscanning* tends to equalize the doping near the edges and the center of the wafer. Also, some current measurement systems require the overscan to intercept ions for counting (Section 8.7). Basically, all-electronic scanning handles only one wafer at a time, and can hold doping uniformity to within 1% over a single wafer, and within 2% from water to wafer. Batch processing is discussed later.

Scanner plate-to-plate voltages are in the few kilovolt range, the actual values being determined by the required scan amplitude, the scanner plate dimensions, and V_d, as discussed in Appendix B.2.

As V_a is raised to increase implant depth, ion kinetic energy increases too, and increased wafer heating may become a problem, as shown by these numbers.

A wafer 9 cm in diameter is scanned by a 10-cm × 10-cm square raster. The dose is $Q = 10^{16}\ \text{cm}^{-2}$, $V_a = 150$ kV, and the beam current is $I = 1$ mA. This is a large current for an implanter and the dose is typical of that used for a BJT emitter.

The power delivered to the raster area is

$$P_r = V_a I = (1.5 \times 10^5)10^{-3} = 150 \text{ W}$$

Because of overscanning, not all this power is delivered to the wafer. The wafer power P_w is proportional to the wafer area, so

$$A_w = \frac{\pi d^2}{4} = \frac{\pi(9)^2}{4} = 63.6 \text{ cm}^2$$

and

$$P_w = \frac{A_w}{A_r} P_r = \frac{63.6}{100}(150) = 95.4 \text{ W}$$

These numbers show that the wafer absorbs a significant amount of power with a corresponding temperature rise as the ions come to rest and give up their kinetic energy. This can cause troubles, particularly if photoresist is used as the masking material. Wafer temperature may be controlled by water cooling the wafer holder, or by other scanning schemes that allow a cooling interval between successive scans of the beam. Usually they do not give as uniform an implant as does the all-electronic scan.

8.5.2 Hybrid Scanning

One system of hybrid scanning is illustrated in Figure 8–6(a). The wafers are mounted on a turntable in the process chamber and the turntable is rotated at a constant speed. Note that several wafers are scanned in a batch. The ion beam is scanned radially to the turntable by a triangular scanning voltage on

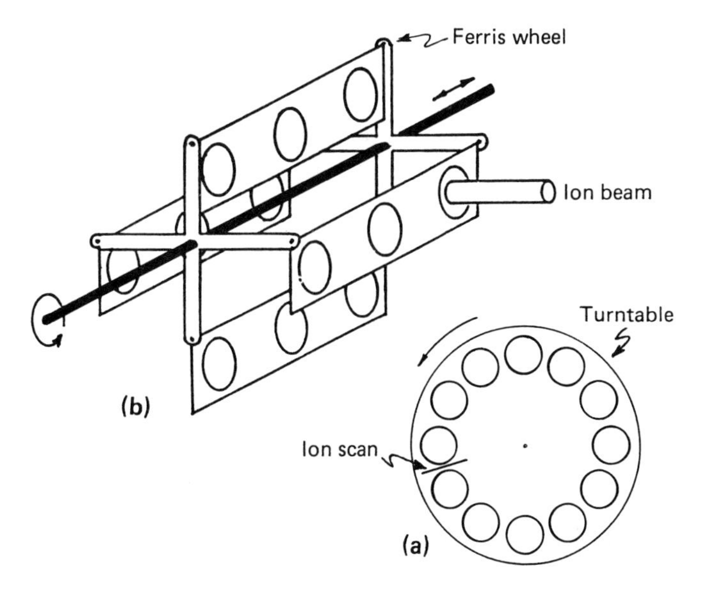

Figure 8–6. Scanning systems. (a) A hybrid system where the ion beam is scanned electronically along a radius as the turntable rotates. (b) A mechanical system where the beam itself is not scanned. The wafers are moved across the fixed beam by combined rotation of the Ferris wheel and linear movement along the wheel's axis.

the scanner plates. The two scanning directions are not orthogonal, but this effect is lowered as the turntable diameter is increased.

Typical rotational speeds run about 1000 r/min, and 20 wafers on the turntable are common, although the number will get smaller for larger wafer sizes. For these numbers the rotation period is 60 ms and a wafer is scanned once in one-twentieth of the period or in 3 ms. Several turntable revolutions are required for a complete implant on each wafer; the exact number will depend upon the required dose. Note that a wafer cools for 57 ms between successive scans.

8.5.3 Mechanical Scanning

Mechanical scanning in both directions is usually reserved for implants of large dose at high energies, such as those required for MOS device sources and drains, and for emitters and buried layers of bipolar devices. Wafer heating can be severe in such situations. The beam remains undeflected and the wafer is moved across the beam.

One system uses the Ferris wheel structure shown in Figure 8–6(b). More wafer support platens and more than three wafers per platen may be used.

The two directions of scan are provided by rotating the Ferris wheel about, and moving it along, its axis. Usually, a complete implant involves only one pass along the wheel axis so that multiple *complete* scans over any wafer are not used. In this event beam overlap on successive revolutions must be adjusted for doping uniformity. Water cooling is not practicable; cooling relies on the fact that only one line is scanned across a given wafer in one rotation of the Ferris wheel, and the wafer cools for the rest of that rotation, another adjacent line is scanned, and so on. Contrast this with the previous system where a wafer surface is scanned completely for every rotation of the turntable.

8.6 Process Chamber

The primary function of the process chamber is to provide a vacuum-tight enclosure for the wafer, current monitoring equipment, and related fixtures. The chamber must also be able to be opened to atmosphere, without breaking vacuum in the remainder of the system, so wafers may be changed between implants. The last feature is provided by vacuum valves between the process chamber and the scanner system and the ambient.

The chamber system is designed to minimize the time required to break vacuum, change wafers, and pump down again, in order to maximize the implanter's *throughput* (the number of wafers that can be implanted per hour). One basic principle is to keep the volume of the chamber as small as feasible to minimize the pumpdown time. A variety of designs is used to increase throughput; we shall consider three.

A feature used in some systems provides multiple chambers that can be alternated between the implant and load modes. While one chamber is under vacuum with implantation going on, the other is isolated from the system and at atmospheric pressure so wafers may be changed. Switching the ion beam between chambers can be accomplished by adding a suitable d-c voltage to the triangular sweep voltage on the horizontal scanner electrodes (Figure 8–1). Two chambers can more than double the throughput of a similar single-chamber implanter, because downtime for loading between implants is effectively eliminated.

Consider some fixturing to increase the throughput when all-electronic scanning is used. A *carousel*, or turntable, not unlike that shown in Figure 8–6(a) can be used to expose one wafer at a time in succession to the electronically scanned beam. Instead of rotating at a constant speed, the carousel is *indexed* after each wafer is *completely* implanted. If two chambers are used, each with a carousel, a good throughput may be obtained, the actual value depending upon the time required for the implant (which depends on the dose and ion beam current) and the wafer capacity of the carousel.

Recent designs utilize a small process chamber, capable of rapid evacuation and sealed off from the main implanter vacuum system by a valve. Wafers are

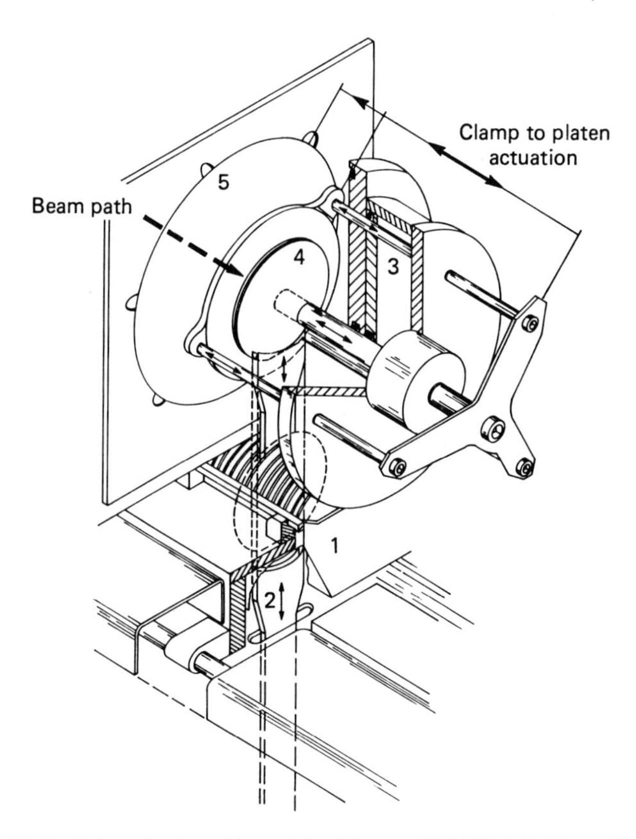

Figure 8–7. Serial wafer handler of the Varian 300 XP Medium Current Ion Implanter. (Courtesy of Varian/Extrion Division.)

shuffled (one at a time) into and out of the single process chamber in succession; to avoid damage, they must not slide against any fixturing. The wafer handler shown in Figure 8–7 does this.

At atmospheric pressure a single wafer is lifted from a cartridge (1) by the lifter (2) into the chamber, (3) which then is closed to atmosphere and evacuated. The platen (4) moves the wafer toward the lock face, and the valve (5) opens to the main system, which houses the ion beam. After implant the valve is closed and the chamber vented to atmosphere. The lifter returns the wafer to the cartridge, which then advances one wafer position, and the process is repeated for all the wafers in the cartridge. The entire operation is automatic.

Later in the chapter we shall see that it is sometimes necessary to have the ion beam arrive a few degrees off-normal to the wafer surface to avoid a condition known as *channeling*. This can be accomplished by mechanically adjusting the wafer-holding fixture relative to the ion beam axis.

8.7 Beam Current Measurement; Faraday Cups

If the implanted dose is to be monitored, the ion beam current must be measured. The setup of Figure 8–8(a) might seem adequate, but there are problems. The measured currents usually are quite small and electronic instruments for low current measurement have high input impedances. This means that the wafer and its fixture cannot be hard-wired to ground, a "must" for safety reasons.

The grounding problem can be solved by *sampling* the current in the overscanned raster regions. Wafers are circular, rasters are rectangular, so the current samplers may be placed at the raster corners as shown in Figure 8–8(b).

The samplers should not disturb the raster by introducing extraneous fields, a condition that is met by the *Faraday cup* of Figure 8–8(c). The cylindrical electrode is held at −55 V and attracts any ions that enter the outer, grounded cup. These pass on to the inner collector, which feeds the current measuring and integrating circuits. Electric field lines from the negative electrode terminate on the inner surface of the grounded outer cup, so, as viewed by the ion beam, the whole cup assembly is at target potential (ground) and scanning is not upset. The negative electrode also suppresses secondary electrons caused by ion impact on the inner collector.

Notice from Figure 8–8(b) that the wafer does not receive the same current throughout the entire raster scan. In fact, when the beam is at the four raster corners where the Faraday cups are located, the *wafer* current may be zero if the beam is of small enough cross section. Also, when the beam is centered on the target the Faraday cups receive zero current. It is clear, then, that geometric factors involving the cup, wafer, and raster areas, must be included in the calibration of the current measuring circuit. Integration of the current for dose determination is performed by conventional electronic circuits.

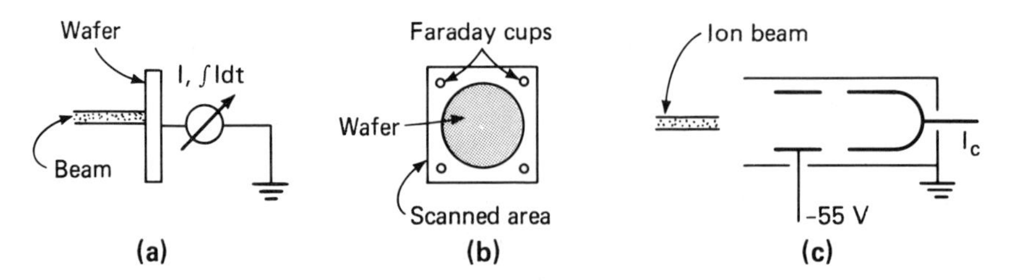

Figure 8–8. Beam current measurement. (a) A series current measuring device creates a safety hazard by removing the device or the wafer from ground. (b) The beam is sampled at the corners of the scanned area. Note that the beam is intercepted by only one Faraday cup at a time. (c) Cross section of a Faraday cup.

8.8 Implant Profiles

Mention was made in Section 8.1 of the motion of an incoming ion as it collides with target nuclei and electrons until it finally comes to rest. If the target presents an irregular pattern of atoms, as in amorphous material, the motion is random and *average* results for the depth of penetration can be computed on a statistical basis. But single-crystal wafers are used normally for device fabrication; hence if the statistical results are to be carried over from amorphous targets, some artifice must be used so that the ion beam "sees" what appears to be an irregular atomic structure. Two general methods are used: *amorphizing* and *dechanneling*. These methods, if used properly, will allow amorphous material statistics to be carried over to single-crystal wafers.

8.8.1 Target Amorphization

The very act of ion implantation disturbs the crystalline structure of a target by knocking some *target* atoms out of their rest positions, producing *radiation damage*, as has been mentioned earlier. Therefore, a predoping implant of Si, or neutral atoms such as argon, into a Si target can produce an amorphous surface layer there. The implant dopant then sees an amorphous target. A single post-implant anneal will remove the damage from both implants and restore the single-crystal structure.

Another amorphizing method uses a thermally grown oxide layer about 300 Å thick on the wafer. This presents an amorphous surface layer to the ion beam, which gives the desired result. Thickness control is important; if the SiO_2 layer becomes too thick, it acts as a mask and prevents the ion beam from reaching the underlying silicon.

8.8.2 Dechanneling; Off-Axis Implantation

If the incident ion beam is directed along one of the principal axes of the single-crystal wafer, it "sees" open channels of the shapes shown in Figures 2–6, 2–7, and 2–10. As the ions move through these relatively open channels the chances of them undergoing direction-changing collisions is small compared to those in an amorphous target. They tend to be "nudged along" the channel by low-energy collisions with lattice-atom electrons; hence they travel much farther in from the surface before losing all their energy and coming to rest. This is known as *channeling*. Some of the ions, however, may suffer collisions with sufficient energy to be deflected out of a channel, in which event they come to rest nearer to the surface than those remaining within the channel. If this condition is present, the implanted profile exhibits two peaks—one for deflected, one for channeled ions [7]. Certainly, amorphous statistics will not be valid here.

The number of ions deflected out of the channels, or *dechannelized*, increases as the target temperature during implant is raised. The reason is that higher temperature causes the target atoms to vibrate more vigorously and with greater amplitudes from their normal positions; therefore, the probability of ion/lattice-atom collisions within the channels nearer the surface increases.

Dechannelization is readily accomplished by inclining the target so that the ion beam does not align with a major crystal axis. Gibbons shows photographs of silicon crystal models viewed under these conditions: 7° off a $\langle 100 \rangle$ axis and 10° off a $\langle 110 \rangle$ axis [2]. Both pictures show no evidence of a channel. This concept is utilized in implanters by arranging the wafer-holding fixtures to provide an off-beam-axis tilt in the range of 7 to 10°. The roughly 3° off-normal from a $\langle 111 \rangle$ plane wafer orientation that was discussed in Section 2.5 is not sufficient for dechanneled implantation; additional offset should be provided to bring it up into the range of 7 to 10°.

If either dechannelization or amorphization is used during implant, the resulting profile of the implanted species will closely resemble that predicted for an amorphous substrate, and greatly simplifies implanted profile calculations.

8.8.3 Projected Range Statistics

The penetration of ions into the target is principally normal to the target surface in the x-direction. The lateral effects common to diffusion are very small, and the concentration peak is not at the surface and is not limited by the thermal limit of solid solubility N_{SL}.

The basic theory of calculating the projected range R_p, the *average* depth of the implanted atoms in the x-direction, was developed by Lindhard, Scharff, and Schiøtt and is known as the LSS theory [2, 8, 9]. Their approach consisted of modeling the mechanism of ion stopping in the amorphous target, and then calculating the range and straggle. To a first-order approximation the distribution is taken to be gaussian. This curve then can be constructed from the known effective range and standard deviation.

The incoming ions proceed on a three-dimensional path through the amorphous target, bumping into target atoms until they come to a halt. The *range* R is the average *total* distance traveled by the ions before they stop. The projected range R_p is the projection of this distance onto the x-axis and is a quantity of practical interest as is the *projected standard deviation* σ_p, again in the x-direction. According to the gaussian model, the bulk of the implanted ions will locate between $(R_p \pm \sigma_p)$, with their average location being R_p in from the wafer surface.

The symbol ΔR_p often is used in implantation work for the *straggle* or *standard deviation*. This notation is cumbersome, especially when it is used with subscripts; we shall use σ since it is a common symbol for the standard deviation, a familiar concept to engineering students.

Since the actual paths of the ions in the wafer are three dimensional, they also will show a gaussian variation in directions normal to the x-axis. The standard deviation in those lateral directions may be designated the *lateral standard deviation* σ_l. In those directions the bulk of the implanted ions will lie within $\pm\,\sigma_l$ of the x-axis. As viewed in three dimensions, then, most of the ions will lie in a cylinder centered on R_p, of length $2\sigma_p$ in the x-direction, and of diameter $2\sigma_l$ normal to the x-direction.

The quantities R_p, σ_p, and σ_l have been determined as a function of three variables: beam energy E, and the species of the implanted ions and the wafer. These will be considered later. At this point we need know only that in general σ_p is small wrt R_p, and that σ_l is of the same order of magnitude as σ_p, but is larger than σ_p for B implanted in Si and smaller than σ_p with both As and P implanted in Si. This has important implication in masking: there will be very little lateral movement under a mask edge as compared to that which results when a predep diffusion is used.

The model envisions two types of collisions: between ions and target-nuclei and between ions and target-atom electrons. The stopping mechanisms due to these collisions are assumed to be independent, but nuclear stopping dominates at low energies where the range is given by

$$\begin{aligned} R &\approx 0.7\left[\frac{(Z_1{}^{2/3} + Z_2{}^{2/3})^{1/2}}{Z_1 Z_2}\,\frac{M_1 + M_2}{M_1}\right]E, \quad \text{Å} \\ &= 0.7E/S_n \end{aligned} \tag{8.8-1}$$

where Z = atomic number, M = atomic mass, and S_n is the reciprocal of the bracketed factor. The subscript a has been dropped on the accelerating energy, so E_a is replaced by E, which is in electron volts. This is quite reasonable since there is no chance for confusion. The subscript "1" designates the ion, and "2" the target.

The significant thing to notice here for *low-energy* implants is that R is proportional to E and decreases as the ion mass increases for a given target material.

At high energies electronic stopping dominates and

$$R \approx 20\sqrt{E} \quad \text{Å} \tag{8.8-2}$$

where again E is in electron volts. The energy crossover point at which the two mechanisms have equal effect is given by

$$E_c = \left(\frac{S_n}{0.2 \times 10^{-15}} \right)^2 \qquad (8.8\text{-}3)$$

Tables of *effective* range R_p, the *projected* straggle or standard deviation σ_p, and the *lateral* straggle σ_l are available as a function of E and the materials involved [10]. Portions of these tables are reproduced in Appendix F.8.1.

These values are used with the assumed gaussian shape for a first-order approximation to the implanted profile. In some instances a single gaussian is not adequate, but may be handled by using two values of σ_p, one for each side of the peak [10].

Inspection of Appendix F.8.1 will show that in general R_p and σ_p both increase as E is raised and that for a given E both are greater for B than P in both Si and SiO_2.

8.8.4 Shifted Gaussian

The peak of the implanted profile is at $x = R_p$, not at the wafer surface as it was in diffusion; hence the gaussian is displaced as shown in Figure 8–9(a). By definition σ_p is half the distance between the $N_p e^{-1/2}$ (or $0.607N_p$) points either side of R_p. The equation for this displaced gaussian is

$$N(x) = \frac{Q}{\sqrt{2\pi}\sigma_p} \exp\left[-\frac{1}{2}\left(\frac{x - R_p}{\sigma_p} \right)^2 \right] \qquad (8.8\text{-}4)$$

where Q is still the dose in atoms/cm^2.

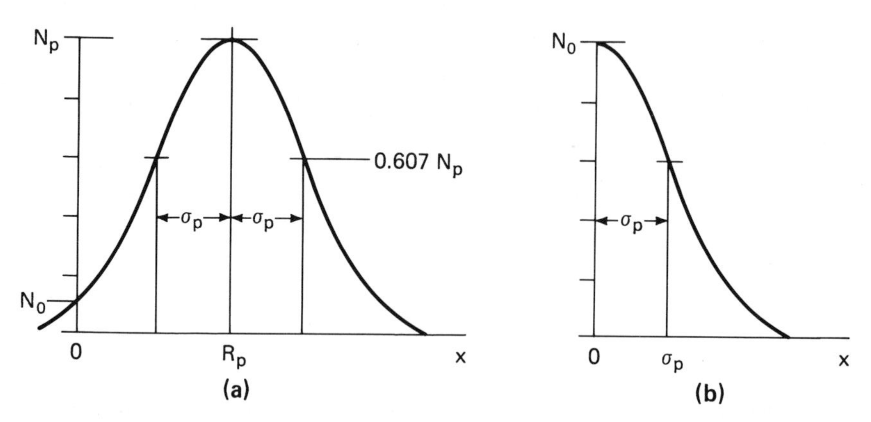

Figure 8–9. Gaussian curves. (a) Shifted, as with ion implantation. (b) Unshifted, as with diffusion.

At $x = R_p$, $N(x) = N_p$, the peak value, so

$$N_p = N(R_p) = \frac{Q}{\sqrt{2\pi}\sigma_p} = \frac{0.399Q}{\sigma_p} \approx \frac{0.4Q}{\sigma_p} \tag{8.8-5}$$

The two foregoing equations combine to give

$$N(x) = N_p \exp\left[-\frac{1}{2}\left(\frac{x - R_p}{\sigma_p}\right)^2\right] \tag{8.8-6}$$

Mathematically the gaussian of Eq. (8.8-6) displays *arithmetic symmetry* about $x = R_p$, i.e., the values of $N(x)$ are the same for equal values of $|x - R_p|$ on either side of R_p. The profile is not valid for negative x values. The normalized surface concentration at $x = 0$ is

$$\frac{N_0}{N_p} = \exp\left[-\frac{1}{2}\left(\frac{R_p}{\sigma_p}\right)^2\right] \tag{8.8-7}$$

Now R_p/σ_p is a function of E alone for a given ion species and target material. Inspection of these quantities in Appendix F.8.1 shows that as E increases, R_p and R_p/σ_p increase, but N_0/N_p decreases. Thus, shallow implants appear less symmetrical beacuse N_0 is a larger fraction of the peak value N_p. If the peak is near the surface, some authors call the curve a *truncated* gaussian. The term *shifted* gaussian (SG) is more apt to distinguish it from the diffused gaussian, which peaks at $x = 0$, and so is a better general term.

It is interesting to compare the gaussian profiles for an ion implantation and a diffusion in Figure 8–9(a) and (b), respectively. Linear scales are used in both sketches. Consider these diffusion equations: We have from Eq. (7.1-7) but replacing Δ_2 by Δ_2'

$$\left.\begin{aligned} N(x) &= \frac{Q_1}{\sqrt{\pi\Delta_2'}}\exp\left(\frac{-x^2}{4\Delta_2'}\right) \\ N(x) &= \frac{2Q_1}{\sqrt{2\pi}\sqrt{2\Delta_2'}}\exp\left[-\frac{1}{2}\left(\frac{x}{\sqrt{2\Delta_2'}}\right)^2\right] \end{aligned}\right\} \tag{8.8-8}$$

Comparison with Eq. (8.8-4) gives the analogs listed in Table 8–1.

Consider some typical numbers for a diffused *predep* and an ion implant. The profiles will differ as will the peak and surface concentrations. In the numerical example of Section 6.3.2 we found that a boron predep into silicon for 15 min at 900°C and a surface concentration of 3.7×10^{20} cm^{-3} gave a dose of 4.85×10^{14} cm^{-2}. Notice that these numbers are independent of the wafer size.

Table 8–1. Comparison of Gaussian Profiles for Implant and Diffusion

	Implant		Diffusion
Standard deviation	σ_p	$\leftrightarrow$	$\sqrt{2\Delta_2'}$
Dose	Q	$\leftrightarrow$	$2Q_1$
Peak value	$N_p = \dfrac{Q}{\sqrt{2\pi}\sigma_p}$	$\leftrightarrow$	$N_0 = \dfrac{Q_1}{\sqrt{\pi\Delta_2'}}$
Peak at	$x = R_p$	$\leftrightarrow$	$x = 0$

Note: $\leftrightarrow$ = "analogous to."

The same dose is to be implanted into a 3 in.-diameter wafer by a small-current implanter with a beam current of 40 μA. The value of E required is independent of the dose. To illustrate, say that R_p for the foregoing implant is to be 0.1 μm. What E is required? From Appendix F.8.1 for boron in silicon we read

$$E = 30 \text{ keV}, \qquad R_p = 0.0987\ \mu\text{m}, \qquad \sigma_p = 0.0371\ \mu\text{m}$$

The value of R_p is close enough to specifications, so we choose $E = 30$ keV. We find the peak concentration, which is located at R_p, by using Eq. (8.8-5)

$$N_p = \frac{0.4Q}{\sigma_p} = \frac{0.4(4.85 \times 10^{14})}{3.71 \times 10^{-6}} = 5.23 \times 10^{19} \text{ cm}^{-3}$$

Note that σ_p was converted from μm to cm to preserve dimensional consistency.

By Eq. (8.8-7) the surface concentration will be

$$N_0 = N_p \exp\left[-\frac{1}{2}\left(\frac{R_p}{\sigma_p}\right)^2\right] = (5.23 \times 10^{19}) \exp\left[-\frac{1}{2}\left(\frac{0.0987}{0.0371}\right)^2\right]$$

$$= (5.23 \times 10^{19}) \exp -3.53 = 1.53 \times 10^{18} \text{ cm}^{-3}$$

Consider another case with a considerably larger value of R_p, say 0.5 μm. $Q = 10^{16}$ cm^{-2} and the $^{11}B^+$ ion is implanted into silicon. The wafer and raster sizes are unchanged from the foregoing example. From [10]

$$E = 190 \text{ keV}, \qquad R_p = 0.5086\ \mu\text{m}, \qquad \sigma_p = 0.0906\ \mu\text{m}$$

This value of R_p is near enough to the specification. Then calculations give $N_p = 4.42 \times 10^{20}$ cm^{-3} and $N_0 = 6.34 \times 10^{13}$ cm^{-3}. Calculation of the beam current and implant time are considered in Section 8.9.

8.8.5 Junction Formation

It is a fortunate fact that successive implants are independent of each other because implantation is a low-temperature process and diffusion is negligible. Furthermore, post-implant anneals usually are performed at temperatures low enough that diffusion is still insignificant, so the as-implanted profiles remain unchanged by subsequent annealing.[3] This simplifies the calculation of junction depths.

The principle of junction formation is the same for ion implantation and diffusion: if incoming and background dopants are of opposite types, a p–n junction will be formed where their concentrations are equal. Hence

$$N(x_j) = N_p \exp\left[-\frac{1}{2}\left(\frac{x_j - R_p}{\sigma_p}\right)^2\right] = N_b \tag{8.8-9}$$

Solving for x_j gives

$$x_j = R_p + \sigma_p \sqrt{2 \ln\left(\frac{N_p}{N_b}\right)} \tag{8.8-10}$$

It is easy to remember that N_p/N_b, and not its reciprocal, appears in the equation for we take the natural logarithm of a number greater than one.

Consider some typical numbers. A p-type substrate is doped at 10^{16} cm^{-3}. Singly charged phosphorus ions are implanted at $E = 100$ keV with a dose of 10^{15} cm^{-2}. Calculate the junction depth.

From Appendix F.8.1 we read for P in Si:

$$E = 100 \text{ keV}, \qquad R_p = 0.1238\ \mu\text{m}, \qquad \sigma_p = 0.0456\ \mu\text{m}$$

By Eq. (8.8-5)

$$N_p = \frac{Q}{\sqrt{2\pi}\sigma_p} = \frac{10^{15}}{\sqrt{2\pi}(4.56 \times 10^{-6})} = 8.75 \times 10^{19} \text{ cm}^{-3}$$

By Eq. (8.8-10)

$$\begin{aligned} x_j &= R_p + \sigma_p \sqrt{2 \ln\left(\frac{N_p}{N_b}\right)} \\ &= 1.238 \times 10^{-5} + (4.56 \times 10^{-6}) \sqrt{2 \ln\left(\frac{8.75 \times 10^{19}}{10^{16}}\right)} \\ &= 3.18 \times 10^{-5} \text{ cm} = 0.318\ \mu\text{m} \end{aligned}$$

[3] In certain special cases post-implant diffusion may be desired, in which event the anneal is run at diffusion range temperatures (Section 8.13).

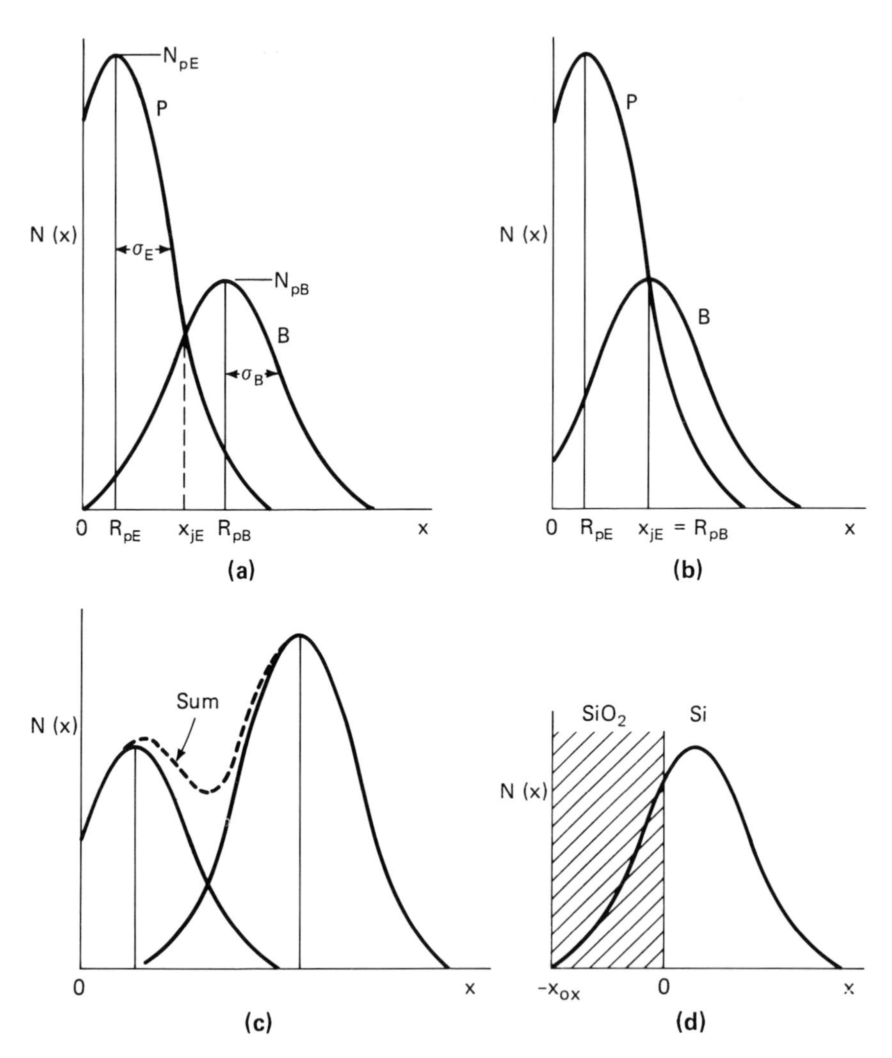

Figure 8–10. Combinations of implants. (a) The two profiles are of opposite dopant types so a junction is formed at the crossover point x_{jE}. (b) A special case where $x_{jE} = R_{pB}$. (c) A tailored profile which is the sum of two profiles of the same dopant type. (d) Tailoring an effective R_p value by implanting through an oxide layer.

If the junction is formed by two *implanted* profiles, the solution becomes more messy. For example, consider the situation of Figure 8–10(a) where both emitter and base of an n–p–n BJT have been implanted. The collector doping is not shown and is assumed to be small wrt the base and emitter concentrations for $x < R_{pB}$ of the base. At the emitter-base junction x_{jE} the two profiles will have the same value so

$$N_{pE} \exp\left[-\frac{1}{2}\left(\frac{x_{jE} - R_{pE}}{\sigma_{pE}}\right)^2\right] = N_{pB} \exp\left[-\frac{1}{2}\left(\frac{R_{pB} - x_{jE}}{\sigma_{pB}}\right)^2\right] \quad (8.8\text{-}11)$$

This may be reduced to the form

$$\left(\frac{x_{jE} - R_{pE}}{\sigma_{pE}}\right)^2 - \left(\frac{x_{jE} - R_{pB}}{\sigma_{pB}}\right)^2 = 2 \ln\left(\frac{N_{pE}}{N_{pB}}\right) \quad (8.8\text{-}12)$$

which may be solved by the quadratic formula, probably on a computer.

Typical numbers for Figure 8–10(a) are:
Phosphorus emitter: $E_E = 80$ keV

$$R_{pE} = 0.0981\ \mu\text{m}, \qquad \sigma_{pE} = 0.0380\ \mu\text{m}, \qquad N_{pE} = 10^{21}\ \text{cm}^{-3}$$

Boron base: $E_B = 190$ keV

$$R_{pB} = 0.5086\ \mu\text{m}, \qquad \sigma_{pB} = 0.0906\ \mu\text{m}, \qquad N_{pB} = 10^{18}\ \text{cm}^{-3}$$

Computer solution gives $x_{jE} = 0.271\ \mu$m.

We have no analog here for the slow outdiffuser approximation of diffusion. One special case does exist, however, which is shown in Figure 8–10(b). The solution for this case is left as an exercise for the student.

8.8.6 Tailored Profiles

Multiple implants of the same dopant species can provide nongaussian profiles. Such profiles may be realized as the sum of a number of gaussians implanted under different conditions. An example illustrating the idea is sketched in Figure 8–10(c). Take note that the profiles in the figure are plotted on linear scales of $N(x)$.

Another type of profile tailoring is possible. Normally the peak lies at R_p (determined by E) in from the surface. By implanting through an oxide layer as illustrated in Figure 8–10(d), we move the peak nearer the surface without changing E or σ_p. This process is related to oxide masking (Section 8.11), and means for calculating the required oxide thickness to place the peak at a desired location are considered there. A number of implanted profiles are given by Gruber [11].

8.9 Dose Calculation

Recall that we have defined the dose Q as the number of ions implanted per cm^2 of the target surface. We wish to derive an equation for this quantity in terms of the ion beam current, implant time, and other related parameters.

Assume that the beam current is constant over the entire scanned raster of area A_r for t, the duration of the implant. Then the number of ions delivered by the beam is It/q_i and

$$Q = \text{number of ions delivered by the beam per cm}^2$$

$$= \frac{It}{q_i A_r} \qquad (8.9\text{-}1)$$

If the current varies during t, I is its *average* value and may be obtained by integration. If the ions are singly charged, $q_i = q$, the electron charge magnitude. This is the basic equation for the implanted dose and it is important to observe that the *scanned* raster area A_r, rather than the *wafer* area A_w, is involved. The reason is that It is the charge delivered by the beam in scanning the whole raster and not the wafer alone. Remember that Q is the dose in ions/cm^2; it is not charge.

The I in the equation is the *total* beam current. If the beam current is sampled by Faraday cups as mentioned in Section 8.7, suitable scaling must be used to convert from sampled to total beam current.

Consider some numbers to compare a diffused predep with an ion implant. This continues the examples of Section 8.8.4. The boron predep diffusion required 15 min at 900°C for a dose of 4.85×10^{14} cm^{-2}. These numbers are independent of the wafer size.

For the 40-μA implanter with an 8-cm × 8-cm raster using singly charged $^{11}B^+$ ions, the implant time is

$$t = \frac{qA_rQ}{I} = \frac{(1.6 \times 10^{-19})64(4.85 \times 10^{14})}{4 \times 10^{-5}} = 124 \text{ s} = 2 \text{ min } 4 \text{ s}$$

Observe that the time is much shorter than for diffusion and that it is inversely proportional to the beam current, so if the beam current were raised by a factor of 10, the implant time would drop to 12.4 s, much too short for diffusion but quite satisfactory for implant. Also observe that these calculations are independent of the beam energy E which is usually chosen to give a specific value of projected range R_p or straggle σ_p.

Consider another example with a considerably larger value of Q, namely 10^{16} cm^{-2}, still with the $^{11}B^+$ ion used. The wafer and raster sizes are unchanged.

If the beam current is 100 μA, the implant time will be

$$t = \frac{(1.6 \times 10^{-19})(64)10^{16}}{10^{-4}} = 1024 \text{ s} = 17 \text{ min } 4 \text{ s}$$

It is clear that for this large dose to be feasible on a production line, an implanter with a larger beam current would be desirable.

8.10 Sheet Resistance

In diffusion the peak of the dopant profile is always at $x = 0$, while for ion implantation it is at $x = R_p$, a function of the implant energy. The straggle also depends on the implant energy, so every E value shifts the gaussian by a different amount and changes the area under the curve; we cannot expect a set of Irvin-like R_S curves for ion implantation. Generally, approximations are used to surmount this difficulty. First, all the implanted ions are assumed to be electronically active after the post-implant anneal, which also is assumed to have negligible effect on the profile.

An *average* majority carrier density $\langle N_M \rangle$ is calculated, although authors differ on what assumptions to use for this. An average value of majority carrrier mobility $\langle \mu_M \rangle$ also is assumed. Then by Eq. (2.7-4)

$$\langle \rho \rangle = [q \langle \mu_M \rangle \langle N_M \rangle]^{-1} \tag{8.10-1}$$

or this equation may be bypassed by the use of curves. Then by analogy to Eq. (3.8-1)

$$R_S = \frac{\langle \rho \rangle}{\Theta} \approx \frac{\langle \rho \rangle}{2\sigma_p} \tag{8.10-2}$$

where Θ is some equivalent thickness of the implanted layer, usually assumed to be $2\sigma_p$ as indicated in the denominator. This assumption considers that the significant portion of implanted atoms is confined to the region $2\sigma_p$ deep and centered on R_p. The total number of atoms implanted in the volume $(A \times 2\sigma_p)$ is QA as seen from Figure 8–11(a). So the average majority carrier concentration in the volume is

$$\langle N_M \rangle = \frac{QA}{2\sigma_p A} = \frac{Q}{2\sigma_p} \tag{8.10-3}$$

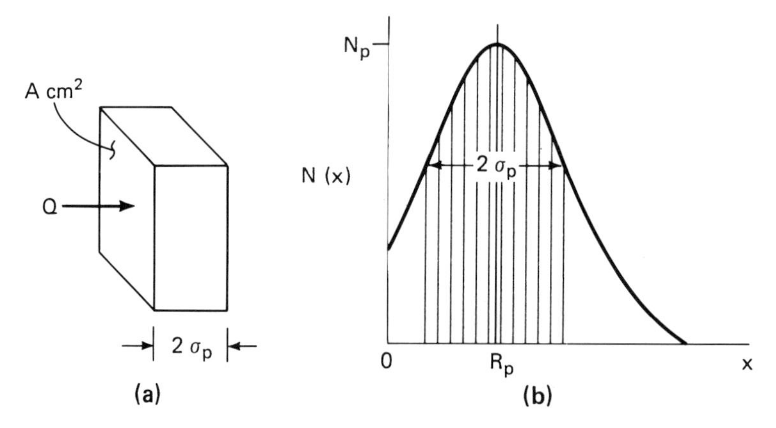

Figure 8–11. Model for sheet resistance of an implanted layer. (a) Model for calculating the number of implanted ions. (b) The ions are assumed to lie in a region $2\sigma_p$ thick.

This result may be checked within 5% by integrating the gaussian profile from $x = 0$ to ∞ to get the *total* number of atoms implanted and then averaging them over $2\sigma_p$. Remember, however, that an approximation still is involved in assuming that all the ions lie within a region $2\sigma_p$ deep.

Some authors favor other approximations such as $2.5\sigma_p$ in the denominator of Eq. (8.10-3) [12]. Then given $\langle N_M \rangle$, some reasonable value of average majority carrier mobility $\langle \mu_M \rangle$ must be used for substitution into Eq. (8.10-1). Usually the value is chosen to correspond to $\langle N_M \rangle$ for the appropriate p or n doping, which may be obtained from curves or from Irvin's data, but a simpler approach uses the Irvin bulk resistivity curves of Appendix F.2.1. From these one can read off $\langle \rho \rangle$ for a given $\langle N_M \rangle$, thereby bypassing Eq. (8.10-1), and R_S can be calculated from Eq. (8.10-2).

Consider a typical numerical example. Calculate the sheet resistance of the implanted layer in the example of Section 8.8.4 where $Q = 4.85 \times 10^{14}$ cm^{-2} and $E = 30$ keV for an $^{11}B^+$ implant in silicon. We found that $\sigma_p = 0.0371$ μm. Then

$$\langle N_M \rangle = \frac{Q}{2\sigma_p} = \frac{4.85 \times 10^{14}}{2(3.71 \times 10^{-6})} = 6.54 \times 10^{19} \text{ cm}^{-3}$$

Note that the units of σ_p where converted from μm to cm to maintain dimensional consistency.

Implanted boron gives a p-type layer, so from Appendix F.2.1 we read that $\langle \rho \rangle \approx 4.5 \times 10^{-3}$ Ω-cm. Then the sheet resistance is

$$R_S = \frac{\langle \rho \rangle}{2\sigma_p} = \frac{4.5 \times 10^{-3}}{2(3.71 \times 10^{-6})} = 606\ \Omega/\text{sq}$$

8.11 Masking

As in diffusion, the wafer areas are selectively doped in ion implantation by using a mask as illustrated in Figure 1–8. As a general rule, if the window edges in the mask are vertical, the dopant will be implanted in the direction of beam incidence with negligible lateral doping effects. Sloping window edges, however, allow some sidewise doping, not due to lateral diffusion or lateral implantation, but to imperfect masking by the thinner mask parts.

Both insulators and metals may be used as mask material, if they: (1) have good stopping properties against the incident ions, (2) are easy to apply to the wafer and easy to remove after implant, (3) are easy to etch into the required window pattern, and (4) are not harmful to the substrate.

Commonly used insulators are: photoresist (PR), SiO_2, Si_3N_4 (silicon nitride), and rarely Al_2O_3 (aluminum oxide). From a practical point of view, PR is used for patterning all the others; hence if PR can be used alone, it eliminates steps in preparation of the on-wafer mask. Also if PR is used alone, the post-bake after development may be eliminated, since no following etch is required. The minimum thickness of PR required for adequate masking, x_m, is highly dependent upon the type of photoresist used. Tables and curves are available in proprietary publications. A typical curve for KTFR resist for 0.01% transmission may be approximated by

$$x_m = 0.12(E/10)^{0.708} \qquad (8.11\text{-}1)$$

where x_m is in μm and E in keV.

The heat generated by implantation tends to polymerize PR, a process that toughens the resist and makes it extremely difficult to remove from the wafer. Heating increases with implant dose and beam energy; thus, the use of PR masking is limited to low and medium doses at moderate E values. Where heating is a problem, SiO_2 or Si_3N_4 are usually used for the mask, but they must be patterned by the usual photolithographic process, which involves PR.

For very deep implants that require high energies, a reasonable thickness of either oxide or nitride cannot provide adequate masking. In such cases they may be augmented by a top layer of a heavy metal such as Au, Pt, W, Ta, or Mo, all of which have greater stopping power. These, too, must be patterned before implant along with the underlying insulator.

Consider means for calculating x_m, the minimum mask thickness for adequate masking. Remember that E will be set by the R_p requirements in the *substrate*, which we take to be silicon. To avoid confusion between mask and

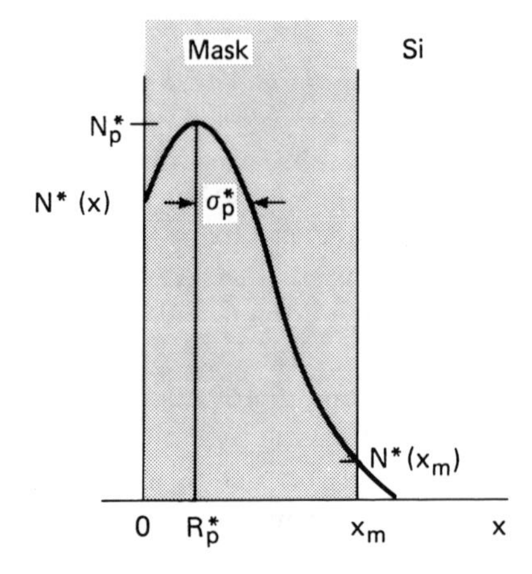

Figure 8–12. Model for calculating minimum masking thickness during implant.

silicon values of projected range, straggle, and concentration, we shall mark those in the mask with a superscript asterisk, *. For convenience in solving the masking criterion we set $x = 0$ at the surface of the *mask* as shown in Figure 8–12.

We need a criterion for "adequate masking." If the wafer is uniformly doped at N_b, we may use the same criterion that was proposed in Section 6.6.1, namely that $N^*(x_m) = N_b$. Using this with Eq. (8.8-6) we get

$$N^*(x_m) = N_p^* \exp\left[-\frac{1}{2}\left(\frac{x_m - R_p^*}{\sigma_p^*}\right)^2\right] = N_b \qquad (8.11\text{-}2)$$

whence

$$x_m = R_p^* + \sigma_p^* \sqrt{2 \ln\left(\frac{N_p^*}{N_b}\right)} \qquad \text{assuming } N^*(x_m) = N_b \quad (8.11\text{-}3)$$

Remember that E is set by the implant conditions in the *silicon.* An illustrative example follows.

Consider the first example in Section 8.8.4 where B was implanted in a 3-in. Si wafer with a dose of $Q = 4.85 \times 10^{14}$ cm^{-2} and $R_p = 0.1$ μm. It was found from Appendix F.8.1 that the required energy was $E = 30$ keV and $\sigma_p = 0.0371$ μm. What minimum masking thickness is required?

From Appendix F.8.1 we read that for B in *silicon dioxide*

$$E = 30 \text{ keV}, \qquad R_p^* = 0.0954\ \mu\text{m}, \qquad \sigma_p^* = 0.0342\ \mu\text{m}$$

The dose is the same for both the oxide mask and the silicon, so by Eq. (8.8-5) we calculate

$$N_p^* = \frac{0.4Q}{\sigma_p^*} = \frac{0.4(4.85 \times 10^{14})}{3.42 \times 10^{-6}} = 5.67 \times 10^{19}\ \text{cm}^{-3}$$

and by Eq. (8.11-3)

$$x_m = R_p^* + \sigma_p^* \sqrt{2 \ln\left(\frac{N_p^*}{N_b}\right)}$$

$$= 0.0954 + 0.0342 \sqrt{2 \ln\left(\frac{5.67 \times 10^{19}}{10^{16}}\right)} = 0.238\ \mu\text{m}$$

In the foregoing calculation of x_m the segregation coefficient of the dopant in SiO_2 and Si does not appear (Section 6.6.1), because implantation is not a thermal process and the segregation coefficient is of little concern unless high temperatures are used subsequent to the implant. To avoid any consideration of this, the tendency in ion implantation is to design for an absolute value of *the masking* such that $N_p^* = 10^4\ N^*(x_m)$, i.e., for 0.01% transmission. *The masking* then is $[N_p^* - N^*(x_m)]/N_p^* = 1 - 10^{-4} = 99.99\%$. The corresponding value of x_m to bring this about is obtained from Eq. (8.11-3) but with N_b replaced by $10^{-4}\ N_p^*$:

$$x_m = R_p^* + \sigma_p^* \sqrt{2 \ln(10^4)} \approx R_p^* + 4.3\sigma_p^* \qquad \text{assuming } N_p^* = 10^4\ N^*(x_m) \tag{8.11-4}$$

This value of x_m is widely used in the industry, and would give $x_m = 0.0954 + 4.3(0.0342) = 0.243\ \mu\text{m}$ for the numbers of the previous example. Note that this is close to the previous value simply because the N_b and $N^*(x_m)$ values were almost identical.

The methods of this section can be used to design an oxide thickness to make the peak value of implant concentration lie at *the silicon surface*, as mentioned in Section 8.8.6. Details are left to the student as an exercise.

8.12 Radiation Damage

Ion implantation causes *radiation damage* to the substrate. This damage must be repaired by annealing, so that carrier mobility and lifetime are restored to their normal values and the implanted atoms are activated electronically.

Incoming dopant ions can interact with host atom nuclei or their electrons. Collisions with nuclei result in loss of ion energy and change in ion direction in the wafer. Interaction with electrons, however, results in energy loss only, without direction change [13].

Ion collisions with lattice atom nuclei can result in the latter being dislodged from their normal lattice sites, which causes the damaged region. We saw in Section 8.8.3 that ion/nucleus collisions dominate at lower implant energies and with heavier ions. Heavy ions penetrate less deeply into the wafer; hence, for a given E their overlapping paths and resulting damage pile up over a shallower region near the surface.

Damage also increases with implant dose. A single ion zigzags through the wafer, leaving a path of damage, until it comes to rest. As the number of incoming ions increases with increased dose, there are more such overlapping damaged paths, so the number of dislodged lattice atoms increases. We expect, then, that the degree of damage increases with high ion dose, with decreasing implant energy, and with the molecular weight of the dopant.

The damage can be so great as to render the layer amorphous such that all long-range order disappears. By way of example, arsenic does this even at low doping levels because it has the largest molecular weight of the three common silicon dopants: B, P, and As.

8.13 Annealing

Basically, in annealing sufficient thermal energy is supplied to the damaged layer atoms so they can rearrange themselves. Since the underlying wafer is single-crystal, it serves as a *seed* for the epitaxial regrowth of the layer being treated. Generally, the temperature required to complete epitaxial regrowth is between 450°C and something less than the melting point of silicon. Thus, the regrowth usually takes place in the solid phase. An exception to this is considered in Section 8.13.2.

It is easier to restore monocrystalinity by annealing if the layer is amorphous rather than only partially damaged or polycrystalline. This results from the higher diffusivity of the dopant atoms in the amorphous layer. Ghez et al. cite their measured value of diffusivity of P implanted at $E = 50$ keV and $Q = 2 \times 10^{15}$ cm^{-2} as $D \approx 3 \times 10^{-12}$ cm^2/s at an anneal temperature of 1100°C [14]. This is ten times the usual value of 3×10^{-13} cm^2/s in single crystal silicon. Amorphization occurs at significantly lower values of dose if the wafer is implanted with liquid nitrogen cooling [7].

The temperatures required for making the dopant atoms electronically active lie in the same range, but are higher than those for epi regrowth, because the dopant and Si atoms are of different size, and distortion of the Si lattice is required for the dopants to enter lattice sites.

The heat energy may be supplied to the damaged region in three ways: (1) by heating the entire wafer in a furnace, (2) by irradiating the wafer *surface* with a laser beam, or (3) by scanning the wafer *surface* with a beam of electrons.

The last two methods have the distinct advantage of confining the heat to the thin layer near the wafer surface so that anneal times are so short that little diffusion takes place.

8.13.1 Furnace Anneal

As the name implies, furnace anneal requires that the implanted wafers be heated in a furnace, similar to those used for diffusion or oxidation. The temperature is raised to the desired value and held there for the specified anneal time. Many wafers may be batch processed simultaneously. As a practical matter a separate post-implant anneal is not necessary if, after implant, the wafer is to be subjected to a high temperature process such as diffusion or oxidation.

Since the entire wafer must be heated, a typical furnace anneal cycle will take about 30 min at 450 to 1000°C. If the implanted layer is amorphous, essentially complete recrystallization and activation of the dopants can be obtained at about 550°C [15]. If the layer is not amorphous, at least 950°C is needed [16]. With a 30-min anneal time, this temperature is high enough to allow some diffusion to take place, a situation that can be handled mathematically by considering the implant as a gaussian predep followed by a drive diffusion. This diffusion effect can make furnace anneal unsatisfactory if the devices in the wafer are of small geometry, say of 1 μm or less.

8.13.2 Laser Anneal

In laser anneal the wafer surface is scanned with radiation from a laser [17]. The radiation energy E depends on its wavelength:

$$E^* = \frac{h^* c^*}{\lambda^*} = \frac{1.242}{\lambda^*}, \quad \text{eV} \tag{8.13-1}$$

where

$$h^* = \text{Planck's constant} = 4.14 \times 10^{-15} \text{ eV-s}$$

$$c^* = \text{velocity of light} = 3 \times 10^{14} \ \mu\text{m/s}$$

$$\lambda^* = \text{wavelength}, \ \mu\text{m}$$

The superscript * calls attention to the useful, though nonstandard, units being used.

Some of the incident energy is reflected from the surface and the rest passes into the wafer, so the wafer's surface reflectivity is important. Depending upon the wafer's optical properties, it either absorbs or transmits radiation, so those are important also.

Table 8–2. Properties of Three Lasers Used for Annealing

Type	State	Mode	λ, μm	Band
Ruby	solid	pulsed	0.6943	infrared red
Nd : YAG	solid	pulsed	0.5330	visible
Ar	gas	cw	0.4880	visible

Note: cw = continuous wave.
Abstracted from Ready, McLure, and Larson [17]. Courtesy of the *Scientific Honeyweller*, copyright 1981.

If the radiation photon energy is *greater* than the bandgap energy E_g, the radiation will be absorbed by the semiconductor, if less than E_g, the radiation will pass on through. Let $\lambda_g{}^*$ be the corresponding wavelength of $E_g{}^*$; so,

$$\lambda_g{}^* = \frac{1.242}{E_g{}^*} \tag{8.13-2}$$

and for silicon where $E_g{}^* = 1.1$ eV

$$\lambda_g{}^* = \frac{1.242}{1.1} = 1.13\ \mu\text{m}, \quad \text{for silicon} \tag{8.13-3}$$

Since wavelength is *inversely* proportional to energy, wavelengths *less than* λ_g will be absorbed by the substrate surface layer, and annealing may take place.

Three lasers whose wavelengths are shorter than 1.13 μm and may be used for Si annealing are listed in Table 8–2.

If the radiation is absorbed, an important question is: how deep does it penetrate the semiconductor before it has negligible intensity? [Recall that the energy *per photon* depends on the wavelength as in Eq. (8.13-1), while the intensity depends upon the *number of photons* and hence upon the total energy.] The basic equation for the intensity of radiation as a function of distance into the wafer is

$$I(x) = I_0 \exp(-\alpha x) \tag{8.13-4}$$

where:

I_i = incident intensity at the wafer surface

I_0 = component of intensity not reflected at the wafer surface $= I_i(1 - R)$

R = surface reflectivity, numeric

α = absorption coefficient, cm^{-1}

The nature of the decaying exponential in Eq. (8.13-4) is such that most of the energy is absorbed within a layer of the semiconductor $1/\alpha$ thick. Thus the thickness of the annealed layer depends strongly on the value of α in the semiconductor.

For a given semiconductor, R is a function of λ and the surface condition: rough, smooth, mirrorlike, and so on. The absorption coefficient α also is wavelength dependent and varies with the degree of the semiconductor's crystallinity. Ready et al. cite these data for a pulsed ruby laser irradiating silicon: single-crystal, $\alpha \approx 500\ \text{cm}^{-1}$; amorphous, $\alpha \approx 5000\ \text{cm}^{-1}$ [17]. Here the heated layer is roughly $(1/500)\ \text{cm} = 20\ \mu\text{m}$ thick for single crystal and about $(1/5000)\ \text{cm} = 2\ \mu\text{m}$ for amorphous silicon.

Since the beam is scanned across the wafer surface, the duration of energy absorption through any elemental area is short, and since the thermal conductivity of silicon is low, the bulk of the wafer remains cold. Furthermore, even if the layer temperature rises to the diffusion range, the amount of diffusion is negligible since the time at high temperature is small. The exposure time can be even smaller with a pulsed laser, where exposure is controlled by the pulse duration between beam movement positions.

Lasers of the continuous wave (cw) type usually have lower intensity capabilities than those of the pulsed type. For these cw lasers the layer temperature is well below the silicon melting point, so epitaxial regrowth is from the solid phase, with the underlying, undamaged silicon acting as the seed. With the more intense pulsed lasers, however, it is believed that actual melting occurs in the damaged layer, so that regrowth is from the liquid phase, still with seeding provided by the underlying silicon. Both types of laser anneal yield equally good results and are superior to furnace anneal in that negligible diffusion occurs during the anneal process. Laser anneal is not readily amenable to batch processing; however, the anneal time per wafer is so short compared to that for furnace anneal that this is no severe handicap.

8.13.3 Electron Beam Anneal

A beam of electrons also can transfer energy to a layer near the surface of a semiconductor for annealing damage there [17]. One principal advantage over laser anneal is that the electron beam (E-beam) process is independent of semiconductor optical properties. To balance this, there are some complications.

The E-beam process is roughly analogous to ion implantation in that charged particles, namely electrons, are accelerated *in vacuo* by an electric field. Scanning or deflection of the beam over the wafer surface can be accomplished easily by either electric or magnetic fields since the electron mass is so much smaller than that of, say, a boron ion. Also, all electrons have the same charge/mass ratio, which means that a magnetic field bends all the electrons through the same angle (Appendix B.3.3). The small electron mass also means that electrons are not scattered as much in moving through the semiconductor as are ions in implantation, so the range statistics are not the same. The plot of absorbed energy versus distance, however, is still essentially a shifted gaussian, so the peak absorption is below the surface at some *effective range* that can be controlled by the beam energy. Contrast this to the laser case, where the peak lies at the surface. In both, however, heating is confined to a thin layer near the surface.

The wafer must be grounded to complete the electrical circuit. If it were insulated, the arriving electrons would build up negative charge that would prevent subsequent electron arrival.

As E-beam technology becomes more common, as in E-beam lithography and other applications, its use for annealing certainly will increase. Like laser annealing it is not adaptable to batch processing, but the scan time per wafer is so short that this is not a limitation, and diffusion during the anneal is negligible.

8.14 Applications

Initially, ion implantation was used in fabricating MOS (metal-oxide-semiconductor) devices; only small doses implanted at comparatively low energies were required. Shortly thereafter, high-R_S resistors were implanted, still with low Q and E.

Bipolar devices require much larger doses and higher energies so more sophisticated implanters were needed before these applications could be realized. Implantation is so widely used at present that we can consider only a few applications to show why the technique is superior to diffusion in certain circumstances.

8.14.1 MOS Threshold Adjustment

An accumulation mode MOS field effect transistor (MOST) is normally OFF (nonconducting), and requires a *threshold voltage* V_T to be applied between gate and source to turn it ON. Complete equations are available in the literature [18], but for our purposes an abbreviated form is adequate:

$$V_T = V_o + \frac{1}{C_i}(\mp \mathcal{Q}_d - \mathcal{Q}_i) \begin{cases} \text{n-substrate, p-channel (PMOS)} \\ \text{p-substrate, n-channel (NMOS)} \end{cases} \tag{8.14-1}$$

where

$\mathcal{Q}_d$ = specific charge (i.e., charge per unit area) in the semiconductor depletion layer under the gate, C/cm^2

$\mathcal{Q}_i$ = specific charge at the gate-insulator/substrate surface, C/cm^2

c_i = specific capacitance of the gate insulator, F/cm^2

$= \varepsilon_i / t_i$

ε_i = gate insulator permittivity, F/cm

t_i = gate insulator thickness, cm

Be aware of notation here: most texts use Q for specific charge in this context, but we use script capital $\mathcal{Q}$ to avoid confusion with dose Q.

The voltage V_o depends on substrate doping and type as does $\mathcal{Q}_d$. Once the gate oxide is in place, all quantities in the RHS of Eq. (8.14-1) are fixed, and so is V_o.

Typical magnitudes of these quantities are such that for *accumulation* mode devices V_T has the following signs: negative for PMOS (= p-channel, = n-substrate) and positive for NMOS (= n-channel, = p-substrate). The quantity $\mathcal{Q}_i$ is positive, irrespective of the semiconductor type; hence, if it is large, it may invert the channel in NMOS so that a depletion mode, or normally ON, device results. This charge may be controlled to some extent by proper processing procedures.

If diffusion alone or substrate background doping is used to dope the channel region under the gate, it is difficult to realize a well-controlled value of V_T, so it often is necessary to tweak V_T to a desired value even if the gate oxide is in place. This may be done by implanting a small dose, *through the gate oxide*, to create a *thin* charge layer of specific charge qQ C/cm^2 that adds algebraically to $\mathcal{Q}_i$. This changes the threshold voltage by the term

$$\Delta V_T = -\frac{qQ}{c_i} \qquad (8.14\text{-}2)$$

The dose Q must be signed properly: "+" for added donors, and "−" for added acceptors.

In principle a diffusion predep could place the charge where it is desired and in a thin layer—i.e., with the peak at the substrate surface, and with a narrow erfc u distribution, but only *before* the gate oxide is in place. But the required dose is so small that diffusion cannot handle the job adequately. Therefore, ion implantation, which can implant small doses very well, is used for *threshold adjust*.

Proper location of the dose is critical. Since the implant is usually made through an oxide in the gate region E is chosen so that the peak of the gaussian distribution lies at or just below the semiconductor surface. Some of these ideas are illustrated in the following example.

An NMOST is to be made using an aluminum gate, and an SiO_2 gate insulator. The silicon substrate is doped at 5×10^{16} cm^{-3}. Calculate the threshold voltage.

Calculations (not shown) give $V_o = 0.012$ V, and $\mathcal{Q}_d = 1.14 \times 10^{-7}$ C/cm^2. A typical value of $\mathcal{Q}_i$ for usual processing procedures is

$$\mathcal{Q}_i = 10^{11}q = 10^{11}(1.6 \times 10^{-19}) = 1.6 \times 10^{-8} \text{ C/cm}^2$$

For the SiO_2 gate oxide, $\varepsilon_i = 3.45 \times 10^{-13}$ F/cm, and say $t_i = 1000$ Å $= 10^{-5}$ cm. Then

$$c_i = \frac{\varepsilon_i}{t_i} = \frac{3.45 \times 10^{-13}}{10^{-5}} = 3.45 \times 10^{-8} \text{ F/cm}^2$$

Then from Eq. (8.14-1)

$$V_T = V_o + \left(\frac{\mathcal{Q}_d - \mathcal{Q}_i}{c_i}\right)$$

$$= 0.012 + \left(\frac{11.4 - 1.6}{3.45 \times 10^{-8}}\right) 10^{-8} = 2.85 \text{ V}$$

Consider adjusting V_T to a lower value of 1.5 V by using a shallow implant. Thus

$$\Delta V_T = 1.5 - V_T = 1.5 - 2.85 = -1.35 \text{ V}$$

From Eq. (8.14-2) the required dose will be

$$Q = \frac{-\Delta V_T c_i}{q} = \frac{-(-1.35)(3.45 \times 10^{-8})}{1.6 \times 10^{-19}} = 2.91 \times 10^{11} \text{ cm}^{-2}$$

This is a small dose and is a positive number, so the implanted ions must be donors; phosphorus is a good choice.

Say that the wafers have a diameter of 100 mm, and that the raster shape is square with a side of 10.1 cm to allow for overscan. The implanter is set for a beam current of 1 μA. Then duration of implant will be

$$t = \frac{qA_rQ}{I} = \frac{(1.6 \times 10^{-19})(10.1)^2(2.91 \times 10^{11})}{10^{-6}} = 4.75 \text{ s}$$

Note should be made of the very short implant time and the very small current.

Now consider what energy is needed for the ion beam. The gate oxide is 1000 Å or 0.1 μm thick. The implant will take place through this so we want R_p* (in the oxide) to be t_i. This will place the charge sheet at the oxide/silicon interface. Reading from Appendix F.8.1 for P in *silicon dioxide* we see that

$E = 100$ keV. We also read for P in *silicon* at this energy that $\sigma_p = 0.0456$ μm = 456 Å, so the charge is indeed confined to a very thin layer near the interface. It must be realized that some fine adjustment of the implant energy probably will be required before a production run can begin. An E value slightly greater than 100 keV might be desirable; this would place the implant profile peak about σ_p below the silicon surface so the bulk of the implant would lie between the two σ_p points in the silicon.

Remember that for the foregoing equations to be valid for threshold adjust, the implant must be narrow and near the oxide/substrate interface so that the charge sheet model is a good one. If the implant range and straggle are such that the implant is wide and not concentrated near the interface in the MOST channel, the charge sheet concept is not valid; the implant must be handled as a modification of the semiconductor doping profile which will not be constant. Then the required ΔV_T is handled empirically.

8.14.2 Depletion Mode MOSTs

It is easy to argue from the preceding example that if the dose is made greater and greater, V_T will decrease, and finally become zero. For this condition the channel is still inverted even if the gate–source voltage $V_{GS} = 0$, and drain current can flow; the device is ON. It is therefore in the *depletion* mode, and more negative V_{GS} is required to turn it OFF.

Even larger doses, still with the same sign, allow the device to remain conducting up to some negative value of V_{GS} when the OFF condition is reached. This will be the new V_T value. Positive or less negative V_{GS} values give conduction; larger negative values give cutoff. These depletion mode devices are often used as *active loads* for active accumulation devices.

We see, then, that ion implantation allows a change in V_T and a change between accumulation and depletion modes. This frees the designer from the tyranny of a fixed N_b value and effects of processing on V_T.

8.14.3 Parasitic C Reduction

Lateral gaps between the gate (G) edges and the edges of the source (S) and drain (D) cannot be tolerated for proper operation of MOSTs. If such gaps do exist, the gate does not control charge over the entire channel length and the device becomes inoperative.

When the S and D regions of a MOST are diffused into the substrate there is poor lateral control of their edges at the channel ends. This places a severe burden on mask alignment operators. To simplify their task, masks are designed so that the gates *overlap* the S and D. These overlaps are labeled "o" in Figure 8–13(a). Note that the *thin* gate oxide is the insulator for the resulting G/S and G/D capacitances; hence c_i and these capacitances can be relatively

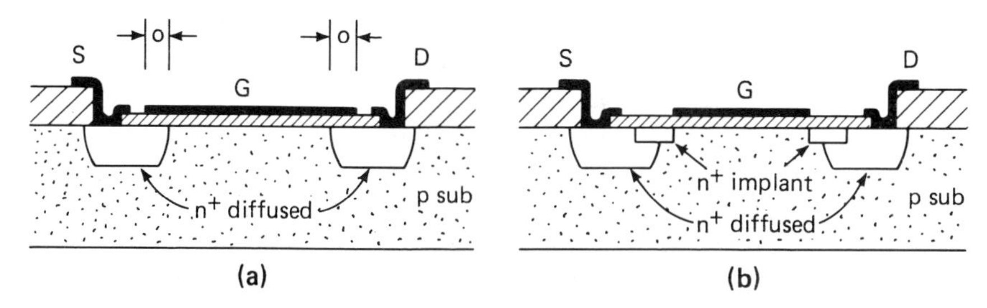

Figure 8–13. Elimination of overlap "o" in MOSTs. (a) Overlap with diffused-alone source and drain. (b) Overlap is eliminated by implanting two supplementary S and D regions, with the gate edges serving as mask edges for the implant.

large. Their presence reduces the switching speeds of MOSTs, so it is desirable to reduce the overlaps as much as possible.

An excellent solution is afforded by ion implantation in the *self-aligned gate* technique, which gives the structure of Figure 8–13(b). The gate is etched so that it does not overlap the *diffused* S and D regions. Subsequent to gate etching, the lateral gap is eliminated by performing an ion implant that uses the gate itself to mask one edge (adjacent to the channel) of each implanted region. The implants are now parts of the source and drain, and there is virtually no overlap because (1) the lateral movement of the implant is negligible, and (2) alignment of the gate with the S and D edges is independent of operator alignment error, in fact it is *self-aligned.* The technique minimizes overlap capacitance, and experience has shown that self-alignment can speed up MOST operation by a factor of 2 to 4.

8.14.4 Implanted Resistors

High-R, thin film resistors are sometimes used *on* IC chips. If stable, high-R_S values are attainable, *in-chip* resistors are a viable alternative.

Figure 8–14(a) shows a sketch of a typical IC resistor where the enlarged end sections are necessitated by minimum line width restrictions and the need for contact holes that extend up to the top surface. Notice that the resistor comprises a p-layer in an n-epi tub for isolation purposes.

Neglecting minor corrections for the two end sections, we have for the resistance between the two contacts

$$R = \frac{\ell}{w} R_S = n_{\square} R_S \qquad (8.14\text{-}3)$$

where

$$n_{\square} = \text{number of squares} = \frac{\ell}{w} \qquad (8.14\text{-}4)$$

Since R is directly proportional to R_S, the need for large resistance values, while keeping the area small (chip real estate is expensive), creates the need for large sheet resistance values.

Two readily available values on a bipolar chip are R_{SE} (emitter sheet resistance), which ranges from 2 to 10 Ω/sq, and R_{SB} (base sheet resistance) which is around 50 to 200 Ω/sq. A third value is R_{BB} (sheet resistance of the base layer buried under the emitter), which will reach around 2 to 3 kΩ/sq, but is difficult to hold to close tolerance. A resistor that is configured as shown

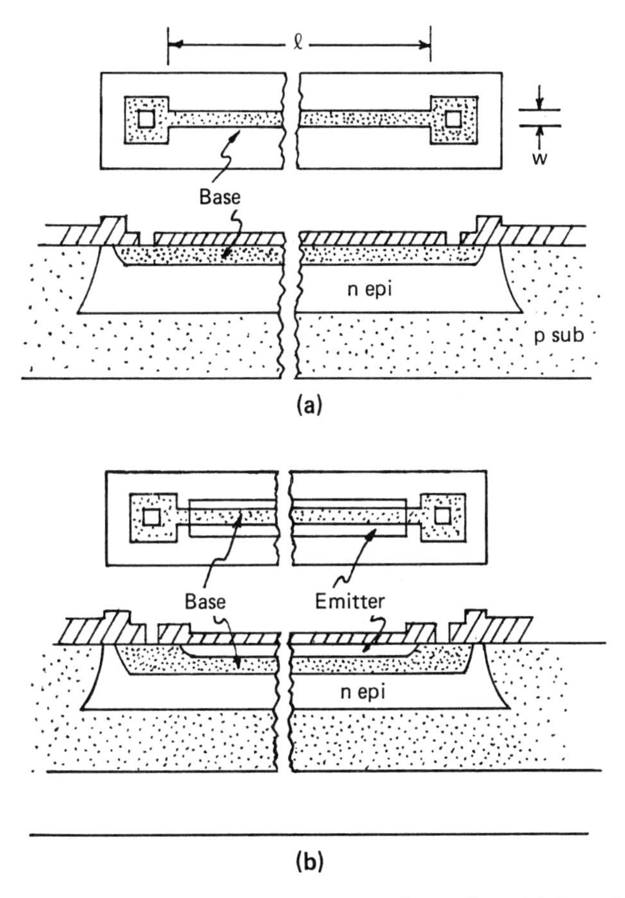

Figure 8–14. Implanted/diffused resistor configuration. (a) Base-layer resistor. (b) Pinch resistor. A portion of the base layer is buried beneath an emitter layer, raising the sheet resistance.

Table 8–3. Implanted Boron Resistors

	Parameter Range for 400–950°C Anneal		Anneal T for Zero TCR,	TCR for for 950°C Anneal
Q, cm^{-2}	R_S, Ω/sq	TCR, %/°C	°C	%/°C
10^{12}	100k–25k	~+0.7	none	~+0.7
10^{13}	13k–2500	−0.1 to +0.45	475	+0.45
10^{14}	3000–400	−0.15 to +0.3	475	+0.15
10^{15}	2000–50	−0.2 to +0.2	540	+0.10

Sansbury [16]. Courtesy of *Solid State Technology*, PennWell Publishing Company, Copyright 1976.

in Figure 8–14(b) is called a *pinch resistor*. Transistors of the n–p–n type dominate on typical bipolar IC chips; hence these three R_S values are tailored to optimize BJT behavior, so control for passive resistors is lost except for $n_{\square}$.

Implantation, however, allows the fabrication of resistors *independent* of the active devices, and vice versa. Since it affords excellent control of low doses, R_S values ranging from 50 Ω/sq to 100 kΩ/sq are feasible. The temperature coefficient of resistance (TCR) may be controlled by use of a proper annealing temperature. Sansbury has summarized the results for implanted boron resistors as shown in Table 8–3 [16]. Once again we see an application where I^2 provides much better control than diffusion and a much larger range of values.

Consider a numerical example. A 100-kΩ resistor is to be fabricated by implanting $^{11}B^+$ ions into an n-epi layer doped at 10^{16} cm^{-3} with arsenic. The minimum line width in the design is 2 μm and space limitations set the resistor length, exclusive of the end sections, at 20 μm. What dose is required? Beam energy E is a variable of choice.

From Eq. (8.14-3)

$$R_S = \frac{Rw}{\ell} = \frac{10^5(2)}{20} = 10^4\ \Omega/\text{sq}$$

Since we want a shallow implant, choose $E = 20$ keV. Then we read in Appendix F.8.1 for boron in silicon: $R_p = 0.0662\ \mu$m and $\sigma_p = 0.0283\ \mu$m.

From Eq. (8.10-2)

$$\langle \rho \rangle = 2\sigma_p R_S = 2(2.83 \times 10^{-6})10^4 = 5.66 \times 10^{-2}\ \Omega\text{-cm}$$

We read from Appendix F.2.1 that the value of $\langle N_M \rangle$ required to give this average resistivity in p-type silicon (boron is an acceptor) is $\langle N_M \rangle = 1.01 \times 10^{18}$ cm^{-3}.

Now the boron implant is in an n-type epi layer doped at 10^{16} cm^{-3} with donors. To override this background doping, the actual equivalent $\langle N_M \rangle$ that must be implanted is

$$\text{implanted } \langle N_M \rangle = 1.01 \times 10^{18} + 10^{16} = 1.02 \times 10^{18} \text{ cm}^{-3}$$

Note that the correction for counterdoping the epi layer is only 1% in this case. From Eq. (8.10-3)

$$Q = 2\sigma_p \langle N_M \rangle = 2(2.83 \times 10^{-6})(1.02 \times 10^{18})$$
$$= 5.77 \times 10^{12} \text{ cm}^{-2}$$

Once again we note that this value should be verified empirically before any production run is started.

8.14.5 Bipolar Devices

Reference to Figure 1–3(f) shows that an n–p–n BJT on an IC chip involves a number of different semiconductor layers in the substrate. These along with their typical doping levels, type, and sheet resistance or resistivity are:

$$\text{Buried layer (BL), n}^+, R_S \sim 10\ \Omega/\text{sq}$$
$$\text{Epi layer (epi), n, } N_d \sim 10^{16} \text{ cm}^{-3}, \rho_{\text{epi}} = 0.5\ \Omega\text{-cm}$$

The emitter, active, and inactive base layers are given in the last section. Notice the wide range of doping levels that are involved. Any or all of these layers may be introduced by ion implantation. GIMIC is an acronym for such an all-implanted BJT process developed at Bell Labs [16].

Another fairly common process involves two implants, one of the emitter *followed* by another for the active base *through* the emitter. A cross section of the device and its processing schedule are shown in Figure 8–15. A particularly interesting feature of this process is that it provides a means of adjusting the transistor β through control of the active base dose. This is analogous to the threshold adjust procedure for MOSTs.

The basis for β adjust may be seen from the first-order equation: $\beta \approx R_{BB}/R_{SE}$, which shows that a change in R_{BB}, say by a dose adjustment, produces a proportionate change in β. If the dose Q_{BB} is *lowered*, R_{BB} and β will both *increase*.

The implanted n$^+$ buried layer (BL) is an interesting application of the combined effects of implantation, oxidation, and diffusion. As is usual, the BL is formed in the substrate *before* the epi layer is deposited. An oxide mask layer is grown over the entire substrate surface and windows are etched to define the BL locations. A thin predep implant of As with Q in the range of

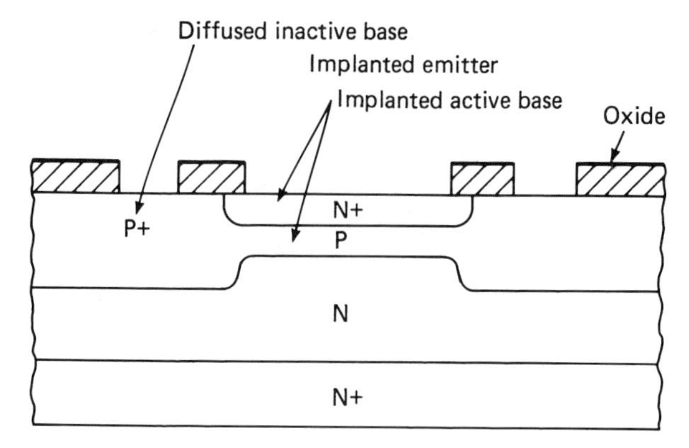

Process Steps:
1. Form inactive base by boron predep and diffusion
2. As implant for emitter: dose = $10^{15} - 10^{16}/cm^2$
3. 1000°C anneal 15-30 min.
4. B implant for active base: dose = $2 \times 10^{12} - 10^{13}/cm^2$
5. 850°C anneal 20 min.

Figure 8–15. A doubly implanted n–p–n BJT. Sansbury [16]. (Courtesy of *Solid State Technology*, PennWell Publishing Company, Copyright 1976.)

10^{15} to 10^{16} cm^{-2} is formed very close to the wafer surface, i.e., with low ion beam energy.

This is followed by a reoxidation step at high temperature, which simultaneously provides a drive diffusion for the implanted As. Remember that As is a slow diffuser, so not very much drive takes place. However, the oxidation (and drive) is made to continue until the entire radiation-damaged surface layer has been consumed in oxide growth. Thus, after the oxide layer has been etched away, a good, single-crystal substrate surface is exposed. This acts to seed the growth of the epi layer that is deposited by CVD. Note that the As-doped regions will be "buried" under the epi layer as desired.

8.15 Implantation in Compound Semiconductors

This book is concerned primarily with planar processing in silicon, an elementary semiconductor. A word must be added here, however, regarding the processing of gallium arsenide (GaAs) and other *compound* semiconductors—which are becoming very important now, because they provide much higher switching speeds than silicon and more desirable optical effects.

As a group, the compounds present a basic problem in high-temperature fabrication: in an ambient pressure near room value and at elevated temperatures they tend to dissociate (break down) into their component elements. For example, gallium (Ga) has a melting point of about 30°C and As has a vapor pressure of one atmosphere at around 600°C. Consider the implications. If GaAs is placed in an open-tube diffusion furnace and the temperature is raised, in short order the As will sublime away and a small pool of molten Ga will be left behind. One way to prevent this is to carry out the diffusion in a sealed ampoule containing the substrate, the dopant, and some excess As (the last to provide a large positive As pressure to prevent dissociation).

It is clear, then, that ion implantation, which operates effectively at room temperature where these compounds are stable, provides an attractive alternative to thermal diffusion as a means of doping GaAs and other compound semiconductors [19].

Problems

8–1 In Figure B–2(a) the ion enters the plate region at $x = 0$, $z = 0$, and $t = 0$. Show that the actual trajectory $z = f(x)$ is parabolic.

8–2 In Figure 8–8(b) the wafer diameter is 4 in. and each Faraday cup diameter is 3/8 in. A 4.5-in. × 4.5-in. raster is used. The scanning voltages are symmetrical triangular. The sum of the four cup currents is integrated.

(a) Evalute S, the scale factor of wafer/cup integrated currents.
(b) Is the location of the cups in the overscan corner regions critical? Explain.

8–3 A Si wafer 4 in. in diameter doped with As at 10^{16} cm^{-3} is given a shallow B implant with a dose of 10^{15} cm^{-2}. A square raster, $4\frac{1}{4} \times 4\frac{1}{4}$ in., is used and the beam energy is 50 keV.

(a) If the beam current is 50 μA, how many minutes of implant time are required for a single wafer?
(b) Calculate x_j.
(c) What thickness of SiO_2 is required for adequate masking?
(d) If photoresist (PR) is about 0.75 times as effective as Si in masking at this beam energy, what PR thickness is required for adequate masking?
(e) If the implant is replaced by a predep diffusion at 1000°C with $N_{01} = N_{SL}$, what predep time in minutes is required for the same dose? Comment on the feasibility.

8–4 Consider the accuracy of Eq. (8.10-3) if averaging over $2\sigma_p$ is acceptable.

(a) Find $\langle N_M \rangle$ by averaging the implanted profile from $x' = -R_p$ to ∞, where $x' = x - R_p$.
(b) For B and P in Si, show that Eq. (8.10-3) is correct to within 5% if $E \geq 20$ keV.

8–5 A p-substrate is doped at 5×10^{16} cm^{-3} with $c_i = 3.45 \times 10^{-8}$ F/cm^2, and an n-channel is formed when the gate-source voltage $V_{GS} \geq 2.85$ V. We wish to convert this to depletion mode operation with current cut off at $V_{GS} = -1.5$ V. What dose magnitude and sign is required?

8–6 A mask provides a window, whose size is 8.5 squares, for an implanted resistor. Other on-chip devices require an implant dose of 5×10^{12} cm^{-2} at 20 keV. What resistance would be implanted? Substrate doping is p-type at 7×10^{15} cm^{-3}.

References

[1] See, for example, P. E. Gise and R. Blanchard, *Semiconductor and Integrated Circuit Fabrication Techniques.* Reston: Reston Publishing, 1979.

[2] A number of different ion sources are described by J. F. Gibbons, "Ion Implantation in Semiconductors— Part I Range Distribution Theory and Experiment," *Proc. IEEE*, **56**, (3), 296–319, Mar. 1968.

[3] Danfysik AS, Jyllinge, DK 4000 Roskilde, Denmark.

[4] R. E. Honig, "Vapor Pressure Data for the Solid and Liquid Elements," *RCA Rev.*, **23**, (4), 567–586, Dec. 1962

[5] *Handbook of Thin Film Materials.* El Segundo: Sloan Materials Division, 1971.

[6] J. L. Stone and J. C. Plunkett, "Recent Advances in Ion Implantation— State of the Art Review," *Solid State Technol.*, **9**, (6), 35–44, June 1976.

[7] Excellent diagrams showing this action are given by F. F. Morehead and B. L. Crowder, "Ion Implantation," *Sci. Am.*, **228**, (4), 55–71, April 1973.

[8] J. Lindhard, M. Scharff, and H. Schiøtt, "Range Concepts and Heavy Ion Ranges," Kgl. Danske Vid. Selskab, *Mat. Fys. Medd.*, **33**, 1963.

[9] A brief summary of the principal ideas is given by A. B. Glaser and G. E. Subak-Sharpe, *Integrated Circuit Engineering, Design, Fabrication and Applications.* Reading: Addison-Wesley, 1979.

[10] J. F. Gibbons, W. S. Johnson, and S. W. Mylroie, *Projected Range Statistics, Semiconductors and Related Materials, 2nd Ed.* Stroudsburg: Dowden, Hutchinson, and Ross, 1975. Renewed © by John Wiley & Sons.

[11] G. A. Gruber, "Ion Implant Testing for Production Control," *Solid State Technol.*, **26**, (8), 159–167, Aug. 1983.

[12] R. A. Colclaser, *Microelectronics: Processing and Device Design.* New York: Wiley, 1980.

[13] An excellent general reference is J. F. Gibbons, "Ion Implantation in Semiconductors—II: Damage Production and Annealing," *Proc. IEEE*, **60**, (9), 1062–1096, Sept. 1972.

[14] R. Ghez, G. S. Oehrlein, T. O. Sedgwick, F. F. Morehead, and Y. H. Lee, "Exact description and data fitting of ion-implanted dopant profile evolution during annealing," *App. Phys. Lett.*, **45**, (8), 881–883, Oct. 15, 1984. This paper also shows the change in dopant profile from the implanted gaussian as the result of annealing at high temperatures in the diffusion range.

[15] W. C. Till and J. T. Luxon, *Integrated Circuits: Materials, Devices, and Fabrication.* Englewood Cliffs: Prentice-Hall, 1982.

[16] J. Sansbury, "Ion Implantation in Semiconductor Processing," *Solid State Technol.*, **19**, (11), 31–37, Nov. 1976. Table I of this reference lists the degree of achieved anneal for different temperatures.

[17] An excellent summary is given by: J. F. Ready, B. T. McClure, and W. L. Larson, "Laser Annealing," *Scientific Honeyweller*, **2**, (3), 37–47, Sept. 1981. Note that this publication is an internal journal of the Honeywell Corporation.

[18] See, for example, B. G. Streetman, *Solid State Electronic Devices, 2nd. Ed.* Englewood Cliffs: Prentice-Hall, 1980.

[19] See, for example, T. Hara and T. Inada, "Ion Implantation in Gallium Arsenide," *Solid State Technol.*, **22**, (11), 69–74, Nov. 1979.

Chapter 9
Chemical Vapor Deposition; Epitaxy

The fabrication of semiconductor devices and integrated circuits requires the laying down of relatively thin layers or films of semiconductors, insulators, and metals. One method of accomplishing this is *chemical vapor deposition* (CVD), in which appropriate source materials in gas phase are furnished near the substrate. With the application of energy by heat, plasma, or radiation, chemical bonds are broken and the source materials dissociate into free radicals. In some cases these form new compounds, in others a component of one source material deposits out. The component or new compound molecules must move around on the substrate surface until they arrive at locations favorable for film formation. Excess energy is given up and film growth proceeds. For successful deposition, by-products must be removed (as shown conceptually in Figure 9–1) or they will interfere with film formation.

Note two important points: (1) chemical action is involved, by materials decomposing and possibly reacting to form a new compound, and (2) all material for the deposited film must be supplied from external sources. For example, in CVD of silicon dioxide, both Si and O_2 must be furnished. Contrast this with the thermal oxidation of Si, where the Si comes from the substrate itself.

9.1 Basic Chemical Processes

There are five basic chemical reactions that may be involved in CVD:

1. *Pyrolysis*—a compound dissociates with the application of heat.
2. *Photolysis*—a compound dissociates with the application of radiant energy that breaks bonds.
3. *Reduction*—a component of a compound is freed by reacting with another component to form a new compound, with a lowering of valence.

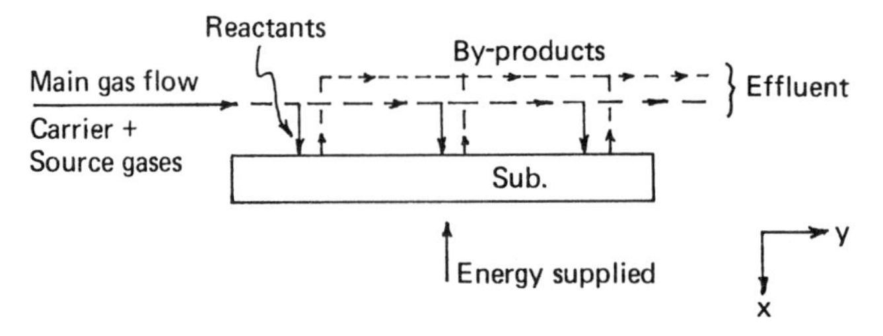

Figure 9–1. The basic mechanisms of a CVD system.

Table 9–1. Typical CVD Processes and Reactions

Pyrolysis:

$$SiH_4\uparrow \rightarrow Si\downarrow + 2\,H_2\uparrow \tag{a}$$

$$SiH_2Cl_2\uparrow \rightarrow Si\downarrow + 2\,HCl\uparrow \tag{a'}$$

Photolysis:

$$SiH_4\uparrow + 2\,N_2O \rightarrow SiO_2\downarrow + 2\,H_2\uparrow + 2\,N_2\uparrow \tag{b}$$

Reduction:

$$SiCl_4\uparrow + H_2\uparrow \rightarrow Si\downarrow + 4\,HCl\uparrow \tag{c}$$

$$SiHCl_3\uparrow + H_2\uparrow \rightarrow Si\downarrow + 3\,HCl\uparrow \tag{c'}$$

Oxidation:

$$SiH_4\uparrow + O_2\uparrow \rightarrow SiO_2\downarrow + 2\,H_2\uparrow \tag{d}$$

Reduction-Oxidation:

$$3\,SiH_4\uparrow + 4\,NH_3\uparrow \rightarrow Si_3N_4\downarrow + 12\,H_2\uparrow \tag{e}$$

Doping Reactions:

$$n\,SiH_4\uparrow + B_2H_6\uparrow \rightarrow n\,Si\downarrow + 2\,B_2\downarrow + (2n + 3)H_2\uparrow \tag{f}$$

$$n\,SiH_4\uparrow + 2\,PH_3\uparrow + n\,O_2\uparrow \rightarrow n\,SiO_2 + 2\,P\downarrow + (2n + 3)H_2\uparrow \tag{g}$$

4. *Oxidation*—a component of a compound is freed by reaction with another component, with a rise in valence, to form a new compound.
5. *Reduction–Oxidation (Redox)*—a combination of reactions 3 and 4 with the formation of two new compounds.

Typical examples of CVD reactions used in the silicon industry are shown in Table 9–1.

Dopant-bearing compounds are available that will pyrolize along with source gases to produce doped layers in the CVD deposition process. Two typical reactions of this type are shown at the bottom of Table 9–1. Notice that two components come out of the reaction as solids that are intermixed, so a doping action occurs. For example, in (f) the two solids give a boron-doped silicon layer. In (g) a phosphorus-doped glass results. The degree of doping can be controlled through the factor n. Other examples of CVD reactions will be considered later.

Two general types of reactions can take place between *gas-phase* chemicals. A *homogeneous reaction* takes place throughout the body of the gas, so a solid product can be produced above, rather than on, the substrate. Resulting particulate matter can fall onto the substrate in a random manner and interfere with the orderly growth process.

The second type of reaction is called *heterogeneous* and requires the presence of a third body. Three-body collisions do not happen readily in the gas phase, so in CVD they take place at the hot surface of the substrate, a more desirable event for epitaxial growth. It is fortunate that the pyrolysis and reduction reactions of silane (SiH_4) and the chlorosilanes ($SiH_xCl_{(4-x)}$) listed in Table 9–1 are of the heterogeneous type.

9.2 Deposited Film Structure

The microstructure of films deposited by CVD can be amorphous, poly-, or monocrystalline. Insulator or metal films generally will be amorphous or polycrystalline. Semiconductor films are used in all three forms, but single-crystallinity is essential if reliable semiconductor properties are required. Amorphous silicon (aSi) is used for special applications such as solar cells; doped polycrystal Si (polySi) often is used as a conductor on ICs.

A single-crystal substrate is necessary if the deposited film is to be monocrystalline. What is desired is for the crystal lattice of the substrate to be propagated into the growing film with both having the same crystal orientation. If this prevails, the film is said to be *epitaxial*. Thus, epitaxial deposition, or simply *epitaxy*, is a special case of CVD that requires more stringent conditions than amorphous or polycrystal films.

Film and substrate are of the *same* material in *homoepitaxy* (also, but rarely, called *isoepitaxy*), e.g., Si film on Si substrate. A less common type of epitaxy occurs when a single-crystal film is deposited on a single-crystal substrate of *different* material. In this event we speak of *heteroepitaxy*. Silicon-on-sapphire (SOS) is an example with single-crystal Si grown on a single-crystal sapphire (Al_2O_3) substrate. Silicon may be grown epitaxially on spinel and cubic zirconia also, as may GaAs on Si.

In the SOS combination the two materials have completely different crystal structures, Si being cubic while Al_2O_3 is hexagonal. This can cause a severe problem of lattice mismatch. Certain orientations of the sapphire, however, will provide the necessary match to support epitaxial growth. It has been determined experimentally that (100) Si can be grown epitaxially on $(1\bar{1}02)$ sapphire and (111) Si on (0001) sapphire [1]. (Note that three Miller indices suffice for specifying crystal orientation of cubic silicon, but four are required for hexagonal sapphire.)

In contrast to sapphire, spinel ($MgAl_2O_4$) has a cubic structure and so has a better lattice match to Si. It has been reported that (111) Si will grow epitaxially on (111) spinel. Also, (100) and (110) orientations have been grown [2]. There is less migration of substrate atoms to the Si epi from spinel than from sapphire. Single crystals of sapphire, spinel, cubic zirconia, and other insulating materials are now grown commercially in substrate sizes.

Sapphire, spinel or similar insulating, crystalline IC substrates provide better isolation between devices and resistance to radiation damage than does silicon. Also, some of those materials have higher thermal conductivity than silicon does and so have greater power-handling capability. For special applications, such as military or space hardware, the additional cost of the heteroepitaxial structure may be warranted.

9.3 Silicon Epitaxial System

A principal application of CVD is the deposition of epitaxial silicon on silicon substrates [3]. A common epi system, shown in Figure 9–2(a), uses heat to reduce or pyrolize the source gas(es). A horizontal reactor is illustrated but other forms, as in (d), may be used [4].

The tube or reactor vessel cross section may be circular or rectangular with rounded corners. The latter form permits side-by-side placement of wafers for a larger load in a tube of given length and of less height. The wafers lie on a tilted *susceptor*, often of SiC-coated graphite, which in addition to being a support for the wafers, enters into the heating process in a manner to be described. The susceptor should not *outgas* (emit vapors) at operating temperature, hence the SiC coating, and is tilted at about 5° from horizontal, high end downstream, to compensate for the depletion of the source gas as it flows past the wafers. The correct tilt angle depends on the gas velocity over the susceptor (Section 9.5).

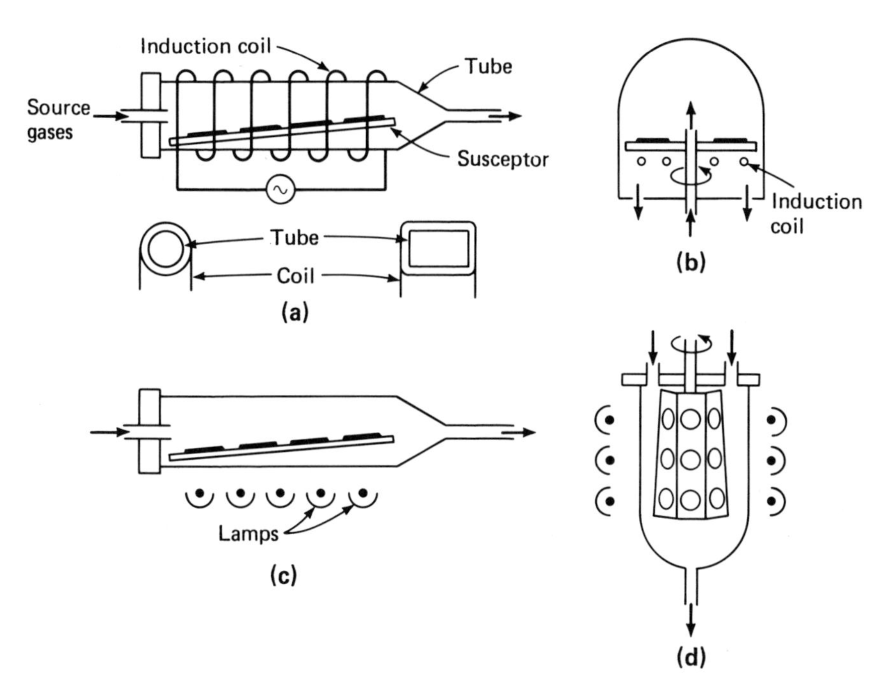

Figure 9–2. Four forms of cold-wall CVD reactors. (a) Horizontal with induction heating. (b) Pancake type with induction heating. (c) Horizontal with lamp heating. (d) Vertical with lamp heating.

Silicon tends to deposit at the hottest parts of the system; hence the wafers should be at higher temperature than the tube walls. This rules out the conventional *hot-wall* furnace of the oxidation/diffusion type except under special conditions. A *cold-wall* system is desired because even at wall temperatures well below 500°C, which is lower than normal epi susceptor temperatures, amorphous Si (*a*Si) may deposit on the walls and then possibly drop off onto the wafers. Forced air or water cooling is used to keep the tube walls cold.

Four cold wall configurations are shown in Figure 9–2. In (a) voltages are induced in the susceptor by r-f fields from the induction coil. The susceptor is heated by bulk I^2R loss when resulting currents flow. Typically the r-f energy supplied to the coil is in the range of 150 to 400 kHz. The induction coils are made from copper or aluminum tubing so cooling water can be circulated through them to keep the nearby tube walls cool.

Heat is transferred by thermal conduction from the susceptor to the wafers lying on it. Consequently, the wafers do not heat uniformly, but from back to front. The back side may be as much as 20°C hotter than the front side, and this may result in warping where large, thick wafers are used. The induction coil turns should be spaced so that the susceptor is heated uniformly. The rectangular tube cross section provides better heating than the cylindrical, as is apparent from the insets in Figure 9–2(a). An alternate form of an induction heated reactor is the pancake type shown in (b).

The systems shown in (c) and (d) in Figure 9–2 also are of the cold-wall type but use radiant heating from lamps to supply energy. These lamp systems provide ease in setting uniform heating along the susceptor length by adjustment of voltages on the several lamps.

Notice that all the heating systems in Figure 9–2 allow forced-air cooling of the reactor tube walls, so Si deposition there is minimized. The first permits water cooling as well.

9.4 Chemicals for Silicon Epitaxy

The silane (SiH_4), dichlorosilane (SiH_2Cl_2) [5], and silicon tetrachloride ($SiCl_4$) reactions of Table 9–1 are all heterogeneous. Other properties influence the choice among these gases for silicon epi. For the time being, we assume that the gas pressure within the reactor is near atmospheric value.

The by-products of the epi reaction should not be toxic, corrosive, or an etchant for silicon. Hydrochloric acid (HCl), which results from $SiCl_4$ reduction, etches Si readily at high temperatures, even in the gas phase, and is highly corrosive. Hydrogen is involved in silane pyrolysis and also can etch silicon at high temperatures, but at a lower etch rate; it is noncorrosive, but flammable. On this basis, the pyrolysis of SiH_4 is preferable to the reduction of $SiCl_4$. Also, silane is noncorrosive and stable at room temperature, but it is *pyrophoric* (it ignites spontaneously on contact with air). It burns with a cold flame, however (Section 6.12.1)

Another factor for consideration is the temperature required for the reaction, and for good growth of the epi layer. Reaction (a) of Table 9–1 requires about 1000°C for good epi while (c) requires 1200°C for a similar deposition rate of about 0.5 to 1 μm/min, the actual value depending upon the gas flow rate, reactor geometry, and other factors. The 200°C lower operating temperature of SiH_4 is desirable because it reduces autodoping and diffusion effects (which are discussed in Section 9.7). Both reactions usually use H_2 as the carrier gas. The reactions will proceed at about 50°C lower if He is used as the carrier. Helium is not used commercially, however, because of its high cost.

As for the silicon source, $SiCl_4$ is a liquid [6] while SiH_4 is a gas at room temperature. The latter has the edge here in available purity and ease of handling. The vapor phase of $SiCl_4$ is used in the epi reactor (Appendix D.3.5) and its reduction is a two-way reaction, so it is possible for Si, already deposited in the layer, to be etched off and converted back to the gas-phase chloride. The pyrolysis of SiH_4 is almost entirely one-way in the decomposing direction, however.

Armirotto argues convincingly that silane is safer than hydrogen since it burns with a very cold flame and so does not cause equipment meltdown [7]. A small silane leak in a gas supply system will usually seal itself by forming solid silicon as it burns. If a large leak occurs, the supply may be shut off at the tank. If, however, hydrogen ignites on contacting air, it burns at a very high temperature and meltdown can occur. Also, hydrogen/air mixtures can explode at the usual susceptor temperatures used in epitaxy. If the induction coil source frequency is lowered, the H_2 carrier can be replaced by an inert gas such as nitrogen, so this danger may be eliminated.

Despite these advantages of SiH_4, $SiCl_4$ was used almost exclusively in the early days of silicon epitaxy because of silane's pyrophoric property. Techniques to overcome the danger by proper handling, however, have been developed [8–11].

In Section 9.8.2 we shall see that a gas phase etch is used just before epitaxial deposition begins. Silane pyrolysis does not give a strong by-product etchant such as HCl, so one must be supplied. The natural choice is HCl, but it is counterproductive to introduce a corrosive agent when it is not already in the system. To this end, a noncorrosive, nontoxic silicon etchant was developed by joint effort of Motorola and Matheson Gas Company personnel. It is a special form of sulfur hexafluoride (SF_6), dubbed "E-Gas," which has the additional advantages of etching isotropically independent of crystal orientation, and etching Si at 1000°C rather than 1200°C used with HCl. Also, if a small amount of E-Gas is supplied with the silane during epi deposition, the process will proceed at a temperature as much as 200°C lower without a change in deposition rate [11].

The gases $SiCl_4$, SiH_4, and SiH_2Cl_2, as well as the common dopant sources phosphine (PH_3), arsine (AsH_3), and diborane (B_2H_6) all are oxidized readily by water vapor, which is present in most environments; hence, sealed gas systems to supply the tube are necessary, and to keep a positive pressure wrt atmosphere, a carrier such as H_2, or an inert purge gas such as N_2, must be kept flowing. This prevents backflow of the ambient into the tube. Furthermore, several of the gases react violently with air, so care is necessary to prevent them from reaching the atmosphere without sufficient dilution (Section 6.12.1).

9.5 Gas-Flow Dynamics

A brief study of the gas-flow equations shows how the epi growth process is affected by system parameters. A factor of principal importance is the flow of reactant gases from the main gas stream to the wafers (or mass transfer), and the chemical reaction rate at the wafer surface also is important. We consider mass transfer first and follow the Deal/Grove model (Section 5.5.1) [12]. We also assume that the dominant transfer mechanism is diffusion, even though some convection usually will be present. We also neglect the effect of reaction by-products that flow in the opposite direction—from wafer to mainstream, where they are carried away to the exhaust system.

Figures 9–1 and 5–8(b) show the close similarity of the off-wafer gas regions in the epi growth and thermal oxidation setups; hence we can borrow results from the oxidation model.

Consider a boundary layer between the wafer surface at $x = 0$ and $x = -x_G$, the value of x at which the main gas stream is unaware of bleed-off for the pyrolytic reaction. Then, from Eq. (5.5-2) and the definition

$$h_G = \text{gas-phase mass transfer constant}$$

$$= \frac{D_G}{x_G} \tag{9.5-1}$$

we have for the flux density of the reactant in the x-direction (normal to the wafer surface)

$$\mathscr{F}_G = h_G(N_G - N_S) \tag{9.5-2}$$

and following Eq. (5.5-5) to recast $\mathscr{F}_G$ in terms of pressures,

$$\mathscr{F}_G = \frac{D_G}{x_G kT}(p_G - p_S) \tag{9.5-3}$$

Numerous approximations have been used for the gas-phase reactant diffusivity D_G. We use a form from Hammond [13]

$$D_G = cT^{3/2}\frac{p_G}{p_T} \tag{9.5-4}$$

where

c = proportionality constant

p_G = partial pressure of the diffusing species at $-x_G$

p_T = total gas pressure at $-x_G$

Then, substituting Eq. (9.5-2), we get

$$\mathscr{F}_G = \frac{c}{kx_G} T^{1/2}\frac{p_G}{p_T}(p_G - p_S) \tag{9.5-5}$$

The temperature dependence here is small since T typically will range from 1000 to 1300°C, or from 1273 to 1573 K. This range gives a $T^{1/2}$ ratio of only 1.11. Thus, raising the temperature at the wafer has little effect on $\mathscr{F}_G$ but, as we shall see later, it has a large effect on the chemical reaction rate.

Grove uses an alternate form, namely

$$h_G = \frac{3}{2}\frac{D_G}{\ell}\sqrt{R_e}$$

where

$$R_e = \text{the Reynold's number of the system}$$
$$= \frac{\rho v_y \ell}{\mu}$$

and ρ = gas density, v_y = mainstream velocity, ℓ = susceptor length, and μ = gas viscosity. Use of this form changes a numerical forefactor slightly in the equations, but has little effect on the overall qualitative discussion.

The flux density will vary in the y-direction, i.e., along the susceptor, because the gas stream depletes as it moves along furnishing reactant, and less reactant is available for the wafers downstream. In effect, the boundary layer in the gas becomes thicker, or x_G gets larger, as function of y, and Hammond has shown this to be [13]

$$x_G(y) \propto \frac{\sqrt{y}}{\langle v_y \rangle} \tag{9.5-6}$$

where $\langle v_y \rangle$ = average gas flow velocity in the y-direction. Then the variation of $\mathscr{F}_G$ along the susceptor will be

$$\mathscr{F}_{G_y} \propto \frac{T^{1/2}}{k}\frac{p_G}{p_T}(p_G - p_S)\frac{\langle v_y \rangle}{\sqrt{y}} \tag{9.5-7}$$

This variation becomes important in batch processing because it causes nonuniform doping within the batch. We correct for this effect empirically by tilting the susceptor at about 5° to cancel out the variation in y of Eq. (9.5-7). More details on system parameter adjustment to give uniform layer thickness are given later in this section. Notice that the last equation shows $\mathscr{F}_{G_y}$ proportional to $\langle v_y \rangle$. This means that an increase in gas velocity will raise the deposition rate. There is a practical limit to this effect, however, since the free silicon atoms must incorporate into the epi lattice.

The second factor to be considered in the Deal/Grove model is the flux density of the reactant incorporated into the epi layer at the wafer surface. Following their model again, we assume the flux density is proportional to N_S, the reactant gas-phase concentration at the wafer surface, viz

$$\mathscr{F}_S = k_S N_S \tag{9.5-8}$$

where k_S = chemical reaction rate constant.

Now, under steady-state conditions $\mathscr{F}_G$ of Eq. (9.5-2) must be equal to $\mathscr{F}_S$. Equating and solving we get

$$\mathscr{F}_S = \frac{k_S h_G}{k_S + h_G} N_G \tag{9.5-9}$$

Following the method of Figure 5–11 we see that the epi layer growth rate g is given by

$$g = \frac{\mathscr{F}_S}{N_{\mathrm{Si}}} \tag{9.5-10}$$

where N_{Si} = number of atoms /cm^3 of Si = 5×10^{22} cm^{-3} or

$$g = \frac{k_S h_G}{k_S + h_G} \frac{N_G}{N_{\mathrm{Si}}} \tag{9.5-11}$$

Since both h_G and k_S are present, this equation shows that the epi growth rate depends on both mass transport and chemical reaction rate [14].

Note the two limiting cases:

$$g \approx k_S \frac{N_G}{N_{\mathrm{Si}}} \qquad \text{if } h_G \gg k_S \tag{9.5-12}$$

and the growth is said to be surface reaction limited.

$$g \approx h_G \frac{N_G}{N_{\mathrm{Si}}} \qquad \text{if } k_S \gg h_G \tag{9.5-13}$$

and growth is mass transport limited. Later we shall see the ranges over which these extremes tend to govern the growth process.

Consider how temperature affects the epi growth rate by using the logarithmic derivative method of Section 6.7. We recast Eq. (9.5-11) in terms of pressure

$$g = \frac{k_S h_G}{k_S + h_G} \frac{p_G}{kTN_{\text{Si}}} \tag{9.5-14}$$

Taking natural logs and the derivative, we get

$$\frac{dg}{g} = \frac{1}{k_S + h_G}\left[k_S \frac{dh_G}{h_G} + h_G \frac{dk_S}{k_S}\right] - \frac{dT}{T} \tag{9.5-15}$$

Combining Eqs. (9.5-1) and (9.5-5)

$$h_G = \frac{c}{x_G} T^{3/2} \frac{p_G}{p_T} \tag{9.5-16}$$

Then taking natural logs and the derivative again

$$\frac{dh_G}{h_G} = \frac{3}{2}\frac{dT}{T} \tag{9.5-17}$$

In order to continue we need an expression for k_S as a function of temperature. This takes on the usual Arrhenius form

$$k_S = k_\infty \exp\left(\frac{-E_a}{kT}\right) \tag{9.5-18}$$

whence

$$\frac{dk_S}{k_S} = \frac{E_a}{kT}\frac{dT}{T} \tag{9.5-19}$$

The last two equations both show that in contrast to h_G, k_S is a very strong function of temperature. Substituting Eqs. (9.5-17) and (9.5-19) into (9.5-15) yields

$$\frac{dg}{g} = \left(\frac{3}{2}\frac{k_S}{k_S + h_G} + \frac{h_G}{k_S + h_G}\frac{E_a}{kT} - 1\right)\frac{dT}{T} \tag{9.5-20}$$

Typical numbers show that the middle term, which depends on the temperature variation of k_S, dominates at low temperatures in most cases. Therefore the deposition process tends to be chemical rate limited there. The behavior is summarized in Figure 9–3. If h_G is large, k_S dominates [Eq. (9.5-12)], and the semilog plot is linear as predicted by Eq. (9.5-18). If h_G is small, it dominates [Eq. (9.5-13)], and curves are no longer of Arrhenius's form.

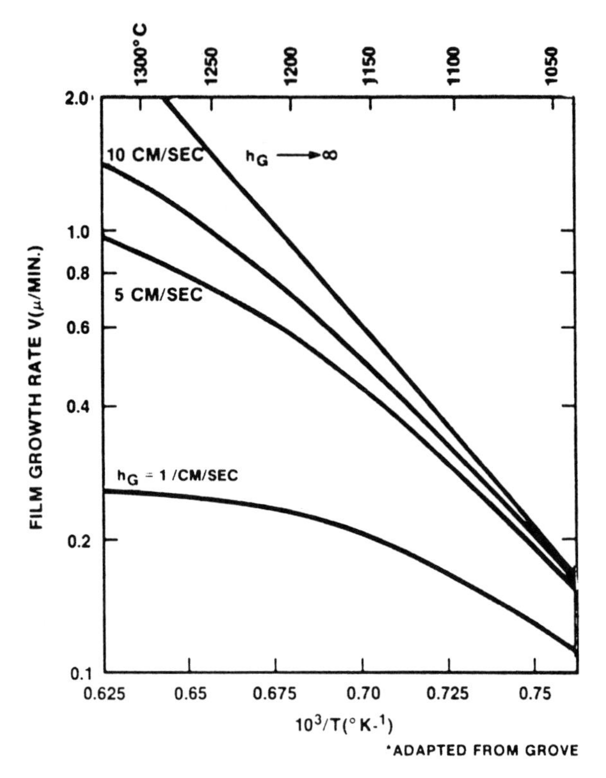

Figure 9–3. Temperature range over which mass transport and reaction rate control the film deposition rate. Atherton [14]. (Courtesy of *Semiconductor International Magazine.*)

We have considered some of the theoretical factors that are involved in silicon epi deposition. Other practical considerations must be attended to if the deposited film thickness δ is to be uniform on all wafers along the length ℓ of the susceptor. Hammond has found that nonuniformity problems may be grouped into upstream and downstream categories and has summarized these as in Figure 9–4 [13]. Some explanation of the figure is necessary.

Distance along the susceptor is y, with the dashed line indicating the midpoint at $\ell/2$. The direction of gas flow is from left to right. The profiles of δ versus y indicate the nonuniformity of the deposited thickness. The notations at the right indicate how the deposition parameters, such as temperature, temperature profile, susceptor tilt angle α, main gas-flow rate, and silicon-bearing reactant gas concentration cause the profile shape shown. These diagrams are of great help in experimentally adjusting a system, and apply to CVD systems generally, not just to epitaxy alone.

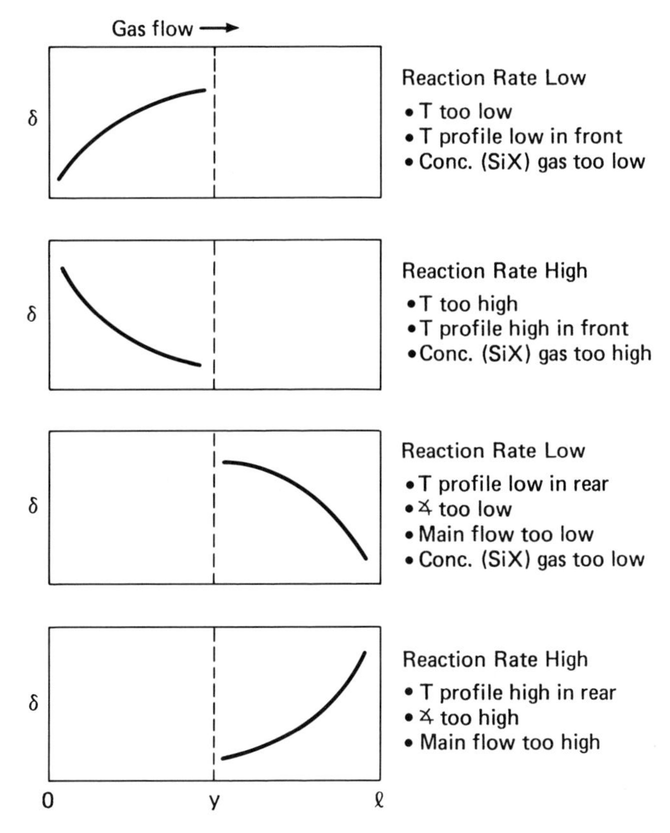

Figure 9–4. The effect of deposition parameters on film thickness variations. Hammond [13]. (Courtesy of *Solid State Technology*, PennWell Publishing Company, Copyright 1979.)

9.6 Doping Considerations

Various dopant sources, both liquid and gas, have been discussed relative to diffusion in Sections 6.12 and 6.13. These are equally applicable to the epitaxial deposition process. Usually, however, the liquid sources—e.g., BBr_3, PCl_3, and $POCl_3$—are not available in as high a purity as gas sources are, so the latter are used exclusively in epi work. Also, $POCl_3$ in the presence of SiH_4 can form SiO_2, a by-product that is definitely not wanted during the epitaxial deposition of silicon.

The commonly used gas dopant sources are hydrides of the doping elements, namely diborane (B_2H_6), phosphine, and arsine (PH_3 and AsH_3, respectively). These gases, while toxic, are noncorrosive and form no etchants when pyrolized; hence they are quite compatible with the silane epi system.

For doped epi layers, both silane and the appropriate doping hydride must be furnished simultaneously, and in the proper proportions, to produce the desired doping level. Consider the typical reaction with both gases present given at (f) in Table 9–1. The ratio of gas *molecular* flow rates of silane to the dopant gas is designated n.

Gas *volumetric* flow rates are easier to measure directly than molecular flow rates, so we consider how they are related.

The number density or number of molecules/cm^3 of a gas is $N = n_a \rho / M$, where n_a = Avogadro's number, ρ = density, and M = molecular weight. This may be checked by simple dimensional analysis. The number of molecules in volume V is $n_v = NV$ and the molecular flow rate will be n_v/t. Then for silane (1) and dopant (2) the volumetric flow rate ratio will be

$$F_v = \frac{V_1 t}{V_2 t} = \frac{n_{v1} M_1}{t \rho_1} \frac{t \rho_2}{n_{v2} M_2} = n \frac{M_1 \rho_2}{M_2 \rho_1} \tag{9.6-1}$$

An alternate formulation is given in Section D.1.

Let r_a be the gas-phase *atomic* ratio of dopant to silicon. Since there are two atoms of boron per molecule of B_2H_6, but only one atom of silicon per molecule of silane, $r_a = 2/n$, and the reaction leads us to believe that the boron *concentration* in the solid phase would be

$$N_B = r_a N_{Si} = \frac{2N_{Si}}{n} \tag{9.6-2}$$

where N_{Si} = number of silicon atoms per cm^3 = 5×10^{22} cm^{-3}. Thus, if n = ratio of gas molecular flow rates = 10^5,

$$N_B = \frac{2(5 \times 10^{22})}{10^5} = 10^{18}\ \text{cm}^{-3}$$

The corresponding reaction for phosphorus doping is

$$2n\ \text{SiH}_4\!\uparrow + 2\ \text{PH}_3\!\uparrow \rightarrow 2n\ \text{Si}\!\downarrow + 2\ \text{P}\!\downarrow + (4n + 3)\,\text{H}_2\!\uparrow \tag{9.6-3}$$

Here there is only one P atom per molecule of PH_3, and still one atom of silicon per silane molecule, so $r_a = 1/n$ and the corresponding solid phase P concentration would be

$$N_P = r_a N_{Si} = \frac{N_{Si}}{n} \tag{9.6-4}$$

As a matter of fact Eqs. (9.6-2) and (9.6-4) do not give the correct doping levels *in the epi layer*. Many factors—such as reaction temperature, gas flow rates, wafer surface conditions, and lattice mismatch between the dopant and substrate atoms—affect the actual incorporation level. Typical curves, such as those in Figure 9–5, give the correct values, but it must be stressed that they apply to only a single set of deposition conditions and are not of general use [15]. They do show general trends, however. Each epitaxial system should be calibrated for its particular set of conditions.

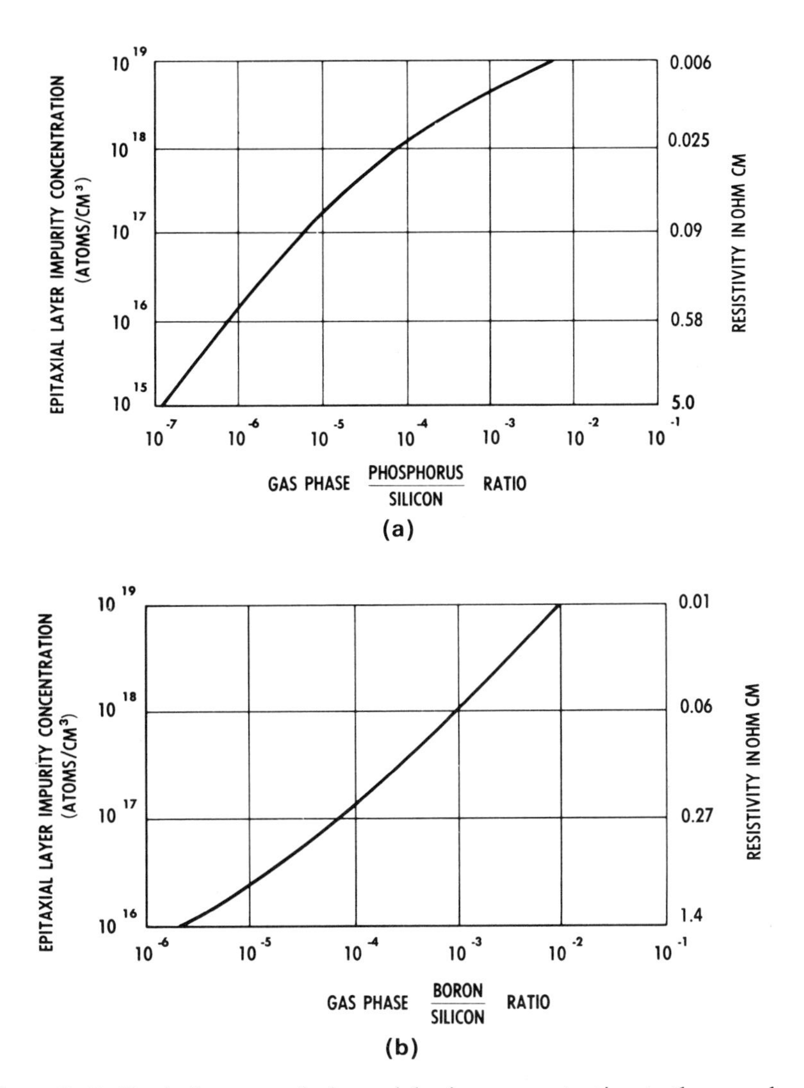

Figure 9–5. Typical curves relating epi doping concentration to the gas-phase dopant/silicon ratio for (a) phosphorus (b) boron. Warner [15]. (Courtesy of McGraw-Hill Book Company.)

A question arises—how can we account for the difference between the actual and calculated values of the doping concentration in the epi layer? What happens to the excess solid dopant atoms? The pyrolytic reactions actually can go both ways. Thus, if some of the dopant atoms cannot fit into the growing epitaxial structure, they may recombine with hydrogen and convert into the gas phase again.

A word should be said about flow rates. The common dopants are dangerous gases such as silane, and so are diluted with another gas. For example, silane might come in a cylinder in the ratio $2\% SiH_4 : 98\% N_2$ or Ar.

Another example, $SiCl_4$, a liquid at room temperature, has H_2 bubbled through it so the combination of hydrogen and $SiCl_4$ vapor might have the ratio of $800H_2 : 1SiCl_4$ in the gas phase (Section D.3.5). This is to be doped with PH_3 to produce an n-type layer. A typical gas ratio might be $(5 \times 10^8) H_2 : 1PH_3$. Then the molecular flow rate ratio would be

$$n = \frac{SiCl_4}{PH_3} = \frac{1}{800} \times \frac{5 \times 10^8}{1} = 6.25 \times 10^5$$

The actual *volumetric* flow rates would depend on the size of the reactor, the number of wafers, and other factors.

It is possible for additional doping of the epi layer to result from outdiffusion from the substrate and from dopant converted to the gas phase from the substrate. These forms of spurious doping are considered next.

9.7 Spurious Doping

Intentional doping during the epitaxial growth process results from the supply of dopant-bearing gas. Spurious doping results from other dopant sources that might be present. Two typical examples are considered.

9.7.1 Autodoping

In Section 6.3.3 we considered how dopants in a substrate, interfaced with a suitable ambient gas, may outdiffuse into the gas phase at high temperature. A typical reaction is the reverse of the pyrolization of B_2H_6, namely

$$2\,B\downarrow + 3\,H_2\uparrow \rightarrow B_2H_6\uparrow$$

Thus, for example, a boron-doped wafer in an H_2 ambient at typical epi deposition temperature may act as a source of diborane. Consider a situation where an epi layer is being deposited on a B-doped substrate. The intended dopant is PH_3. Boron, outdiffusing from the substrate, reacts to form B_2H_6, which serves as a dopant source, competing with the PH_3. If a wafer furnishes the spurious dopant for its own epi layer, this is termed *auto-* or *self-doping*. As temperature and ambient gas pressure within the reactor are lowered, the degree of autodoping is decreased. The mechanism behind this effect is discussed in Section 9.11. If B_2H_6 from one wafer dopes other wafers on the susceptor, *macro-autodoping* is said to be present.

Recall from Chapter 1 that buried-layer regions (BLs), if they are created by diffusion, must be diffused into the substrate *before* the epi layer is deposited. They also can serve as autodoping sources during epi deposition. Typically an n-epi layer should be doped in the 10^{15} to 10^{16} cm^{-3} range, while the BL should be doped as high as possible. Usually As is used as the BL dopant because it is a slow diffuser and it has a good lattice match with Si. This permits its incorporation near the limit of solid solubility without damaging the substrate crystalline structure; hence concentrations in the BL may approach 10^{21} cm^{-3}. These buried regions also can serve as autodoping sources when the epi layer initially starts to deposit. The reason for this is that the reverse reaction to AsH_3 pyrolization is

$$2\ As\downarrow + 3\ H_2\uparrow \rightarrow 2\ AsH_3\uparrow$$

The resulting arsine can act with the intended dopant PH_3. As deposition progresses, the BL will no longer be in contact with hydrogen, so the generation of AsH_3 is shut off and the diffusion effect takes over. This is discussed in the next section.

Autodoping from BLs can be minimized by using low deposition temperatures and gas pressures. Also recall from Chapter 8 that it is possible to *implant* buried layers through an existing epi layer. This procedure eliminates the BL autodoping problem.

9.7.2 Outdiffusion

The growing epi layer also may be doped spuriously by outdiffusion from the substrate and buried regions. In this situation the dopant remains in the solid phase and redistribution diffusion takes place. In particular diffusion Case V of Section 7.6 is applicable. Low epi temperature and high deposition rates are desirable: the diffusion effect is minimized if the Dt product is kept as small as feasible. Recall that Si epi grows from SiH_4 at lower temperature than from $SiCl_4$ and the other chlorosilanes, so it tends to give the lowest diffusivity value.

Autodoping from the gas phase and from diffusion give different profiles. The former yields relatively uniform doping from the *substrate* as the epi layer grows; diffusion gives the typical complementary error function profile.

9.8 Substrate Preparation

In order to get good epitaxial growth without false indication of the BL locations, the substrate lattice must be oriented (and its surface prepared) properly. We consider these questions next.

9.8.1 Pattern Shift and Distortion

It is a physical fact that the location, shape, and size of the buried layer regions may be distorted or even lost on the upper surface of the finished epi layer. Remember that undecorated p- and n-type silicon look the same to the eye, so if the condition is present, it becomes difficult or impossible to complete subsequent mask alignments successfully. Consider a typical example: if the BL region appears shifted from its correct position on the top epi surface, the next masking operation cannot align the base of a BJT with the buried region. To understand this, first consider how one finds the location of a buried layer before the epi layer is deposited.

Figure 9–6 illustrates a number of pertinent fabrication steps. At (a) an n^+ BL region is shown after it has been predepped into a p-substrate through its defining oxide window. Chamfers replace curvature of the diffused region edges to simplify the drawing. The result of a drive/reoxidation step is shown at (b). Note that a slight *dimple* or depressed region appears on the upper surface of the BL, the inevitable result of silicon consumption during thermal oxidation.[1] For a typical reoxidation-layer thickness of 1000 Å, the dimple depth would be $1000/2.2 = 455$ Å (Section 5.5.7). The location of the BL region is discernible by the color difference between the dimple and field oxides, which are of different thicknesses. After epi deposition (following oxide removal) as at (d), the top surface dimple may be observed under proper illumination.

The lower parts of the figure show four different results after an epi layer has been deposited. The ideal situation is shown at (d) where the dimple on the top epi surface lies directly above and has the same size and shape as the original dimple on the BL surface.

The condition at (e), where the top dimple is displaced sideways from the bottom one, is called *shift*. At (f) the pattern has changed size and/or shape, a condition known as *distortion*. The final condition, at (g), shows no top surface dimple at all and is known as *washout*: all evidence of the buried layer is lost on the top surface.

It is clear that for proper operation of the finished device all of these conditions should be eliminated, or at least minimized. To this end it has been

---

[1] Recall that dimple regions usually are omitted in the cross sections illustrating a fabrication process because their depths are small compared to other dimensions, and their omission simplifies the drawings.

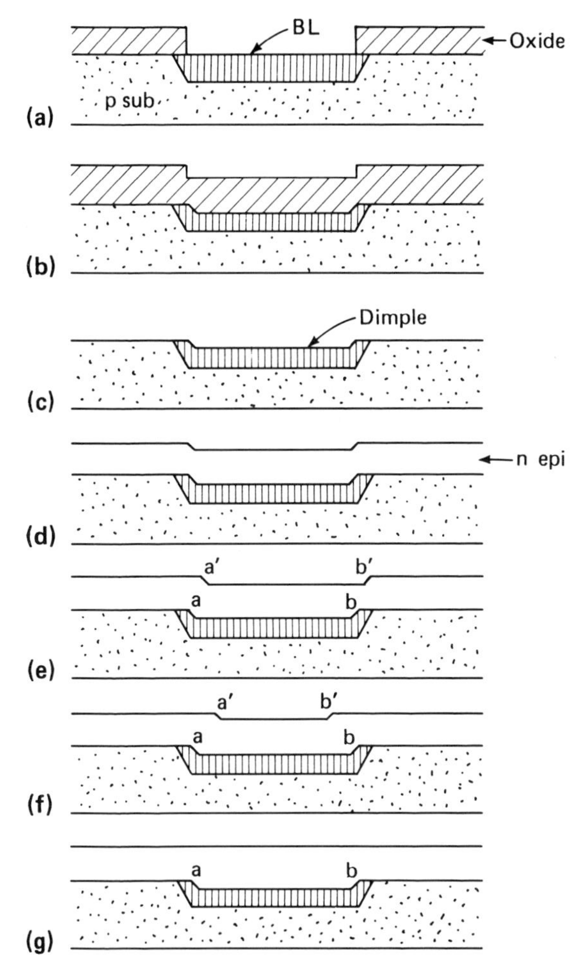

Figure 9–6. Pattern shift and distortion. (a) A buried layer (BL) is predepped through an oxide window. (b) Reoxidation during BL drive. (c) Oxide is removed. A dimple is left on the epi surface. (d) An epi layer is deposited with no distortion or drift. (e) The top layer dimple is shifted wrt the BL dimple. (f) Distortion: the upper dimple is not the same size as the lower. (g) Wipeout. No evidence of the dimple is on the upper surface.

determined experimentally that ⟨100⟩ wafers work quite satisfactorily. On the other hand, ⟨111⟩ wafers can exhibit considerable shift and distortion, but these effects can be minimized if the substrates are tipped 2 to 5° off the ⟨111⟩ axis toward the ⟨110⟩ axis [3, 16]. A value of 3° is fairly standard now.

In general, the pattern shift and distortion may be decreased by raising deposition temperature and lowering the deposition rate. These conditions may conflict with others already considered, so the off-axis orientation of $\langle 111 \rangle$ wafers is particularly important.

9.8.2 Surface Preparation

Wafers ready for epi deposition have been polished to a mirror finish, and in some cases have been oxidized, windowed, and subjected to BL diffusions. Any oxide left on the wafer surface cannot support epi growth. Furthermore, in either case the surface is not atomically clean enough to support true epitaxial propagation of the wafer crystal structure into the growing layer. Therefore, all oxide must be removed and the wafer surface etched to expose an atomically clean region. The oxide may be removed by a usual HF dip. A silicon dry etch should be performed in the epi reactor immediately prior to the start of deposition. As we have seen, gas-phase HCl and E-Gas are commonly used for this, though other etchants are available. Gupta recommends the following "best conditions" for these two gases [17]. HCl: 1175°C, 10^{-2} torr partial pressure, 0.17 μm/min etch rate. For SF_6: 1000°C, 2×10^{-4} torr, 0.21 μm/min. Early practice removed up to 5 μm of substrate, but less is removed in current technologies.

9.9 Deposition Cycle

The proper sequence of gases and temperature cycling must be followed for epitaxial deposition. In addition, safety precautions must be observed to prevent ignition or explosion of hazardous gases that are involved. A typical safe sequence, derived from Armirotto, is listed below for a silane based process [7]. A similar sequence for $SiCl_4$ is given by Warner [15]. Typical values are given by numbers in the square brackets, and "ON" means that the gas is supplied to the reactor vessel. The steps in the sequence may be keyed to Figure D–5 which shows the gas supply system.

1. Load wafers onto the susceptor and place it into the reaction vessel. Close loading door.
2. N_2 (or Ar) ON to purge air from the system [2 min].
3. H_2 carrier ON. [30 l/min; value depends on reactor size.] Heat ON. After 2 min purge, raise susceptor T to etch value [1000°C]. Allow T to stabilize. Note that hot H_2 is a reducing agent, cleans up residual oxides, and etches the wafer slightly.
4. E-Gas ON. Etch off 3 μm of wafer ($t = 3/0.2 = 15$ min).
5. E-Gas OFF. H_2 carrier still ON. Reset susceptor T to deposition temperature if necessary; allow T to stabilize.
6. SiH_4 and PH_3 (or B_2H_6) ON. (800 $H_2 : 1\ SiH_4 + 5 \times 10^8\ H_2 : 1\ PH_3$) Possibly small percent of E-Gas ON (Section 9.4). For 5-μm thickness, $t = 5/0.5 = 10$ min.

7. SiH_4 and PH_3 OFF. H_2 carrier still ON. Heating power OFF.
8. When susceptor is cool, N_2 ON, H_2 OFF ($t = 5$ min to purge system).
9. N_2 OFF. Remove susceptor from reactor.

9.10 Nonplanar Epi

We usually think of epi as a layer of uniform thickness deposited over the entire nonoxized wafer. There are applications, however, where isolated islands of epi material, projecting upward from the main wafer surface, are desired. These may be obtained with photolithographic methods to etch away undesired, between-island portions of an already deposited, uniformly thick layer. An alternative process called *nonplanar epi* is available, too.

In a *pre-epi-deposition* process the substrate is oxidized and windows are etched at the desired island locations by use of photolithography. Then silicon is deposited over the entire surface with the result shown in Figure 9–7: epitaxial Si may form within the windows because it deposits on a single crystal substrate. Since epi Si cannot grow on amorphous silicon dioxide, however, either amorphous (*a*Si) or polycrystal (polySi) silicon will deposit there. A subsequent HF etch, possibly aided by gentle abrasion, will remove the oxide and its overlayer of non-single-crystal material, leaving the epi islands at locations defined by the windows. This is analogous to the *lift-off* technique sometimes used to delineate patterns in metal layers deposited by physical vapor deposition (Section 10.9). A similar nonplanar deposition may be used to deposit epi into grooves etched in the silicon substrate [7].

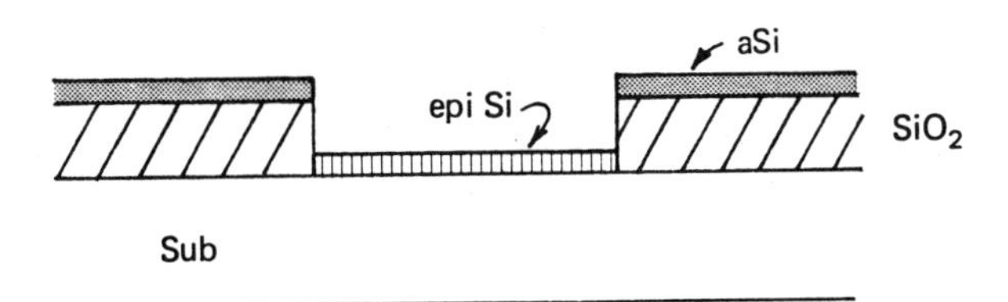

Figure 9–7. Nonplanar epi. The epi layer is deposited through an oxide window. The *a*Si or polySi regions are removed when the underlying oxide is etched away.

9.11 Low-Pressure Epitaxy (LPE)

The discussion thus far has assumed that the total gas pressure in the reactor vessel is at or near the atmospheric value of 760 torr. Experimental work has shown that if the pressure is dropped below this value, giving low pressure epi (LPE), a number of advantages will result [18]. Some of these are:

1. Lowering of deposition temperature without great reduction in the deposition rate. For example, with SiH_4 a drop from 760 to 40 torr allows a drop from 1000 to 850°C, still with good epi formation and deposition rates in the 0.2 to 0.3 μm/min range. If the temperature gets too low, however, the layer will not be epitaxial.
2. Greatly reduced autodoping effects so that higher resistivity epi layers can be deposited. Herring cites these figures for SiH_2Cl_2: with $T =$ 1050°C and reactor pressure in the range 20 to 120 torr, autodoping is reduced by a factor of 5 to 10 with a deposition rate lying in the range 0.1 to 1 μm/min [18].
3. Pattern shift and distortion are reduced but no simple explanation is available.
4. Gas consumption is reduced with a corresponding cost reduction.

The physical setup for LPE is similar to that for atmospheric epi except that a vacuum pump is required at the exhaust end of the reactor vessel. It serves to maintain the low pressure even while gases are being supplied at the input. Some of the effluents that are evacuated by the pump may be corrosive, so special measures and procedures are required to protect the pump (Section D.5).

Consider a probable mechanism behind the LPE effects, one concerned with mass transport. From Section 9.5 the gas-phase, mass transfer flux density of the reactant across the boundary layer is given by

$$\mathscr{F}_G = \frac{c}{kx_G} T^{1/2} \frac{p_G}{p_T} (p_G - p_S) \tag{9.11.1}$$

Note that p_S will change with temperature since its value depends on the chemical reaction rate at the wafer surface. Qualitatively, at least, we note that as the total pressure p_T is lowered, so is the temperature for a given $\mathscr{F}_G$ value. This checks observed facts.

A similar expression holds for dopants, originating from the heated wafer surface, that flow across the boundary layer in the reverse direction, that is from the wafer to the main gas stream. These are responsible for autodoping (Section 9.7.1). We denote these partial pressures for the dopants: p_S' at the wafer surface, and p_G' at $-x_G$. Then we have for the reverse direction dopant flux density

$$\mathscr{F}_G' = \frac{c'}{kx_G} T^{1/2} \frac{p_S'}{p_T'} (p_S' - p_G') \tag{9.11.2}$$

Thus if p_G' is lowered, $\mathscr{F}_G'$ will increase making it easier for the surface-generated dopants to leave the wafer region and pass out into the main gas

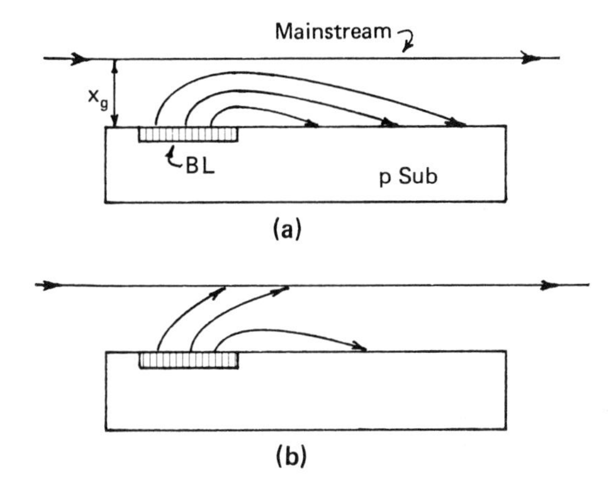

Figure 9–8. Simple model to show reduction of autodoping with LPCVD. (a) Atmospheric pressure: surface-generated dopant returns to the wafer. (b) Low pressure: the surface-generated dopant can join the mainstream.

stream. This concept underlies autodoping reduction. A simple model is shown in Figure 9–8. At (a) the mainstream pressure is large so the surface-produced dopants tend to return to the wafer(s), causing autodoping. In contrast, at (b), some of the dopants pass into the lower pressure mainstream, so the autodoping effects are reduced.

Low-pressure epi is becoming more common for the reasons already discussed. We shall see later in this chapter that low-pressure deposition, sometimes enhanced by plasma formation or by incident illumination, is used in the CVD of oxides and nitrides.

9.12 Epi Layer Evaluation

The principal production line electrical tests used for checking epi layers are performed with the hot point and four point probes as described in Chapter 3, but certain additional information is useful for conditions peculiar to epi layers. One such special case is that of FPP measurements on a lightly doped n-epi layer atop a heavily doped n-substrate. Matare has listed some of these guidelines, which follow [19].

For an n-epi layer on an n^+ substrate the layer thickness δ must satisfy these criteria for valid probe readings:

$$\text{If} \quad \frac{\rho_n}{\rho_{n^+}} \approx 10^3, \quad \delta \geq 0.1s \quad \text{where } s = \text{probe spacing} \tag{9.12-1}$$

$$\text{If} \quad \frac{\rho_n}{\rho_{n^+}} \leq 10^3, \quad \delta = \frac{4 \times 10^3}{\rho_n/\rho_{n^+}} \quad \mu\text{m} \tag{9.12-2}$$

Matare also describes another four-point-probe configuration where two probe tips contact the upper surface of the epi layer and two the bottom surface of the substrate. This configuration eases the requirements on δ given above.

The cases where epi layer and substrate are of opposite types have been considered already in Chapter 3. Recall that the side contacted by the measuring probe is effectively isolated by virtue of the existence of the p–n junction. This junction isolation effect may be enhanced by applying a reverse bias across the layers, separate and independent of the four point probe and its associated circuitry. Remember that such a reverse bias changes the width of the junction depletion region: it will increase from the zero-reverse-bias value, and will extend a greater distance into the more lightly doped side. If this is a very thin epi layer, the effective layer thickness, say δ', may be significantly less than the actual layer thickness δ and should be compensated in calculations. The actual thickness may be measured by methods described in Chapter 3.

9.13 CVD of Polycrystal and Amorphous Films

Recall that the deposition of epitaxial layers puts the most stringent requirements on the CVD process. Take away the monocrystal substrate and the deposition of an epitaxial film becomes impossible; the film will be either polycrystal or amorphous. As a general rule, if these are adequate for the application in hand, the deposition temperature may be lowered by a few hundred degrees since the deposited atoms or molecules no longer need energy for scurrying over the substrate surface to "fit into" a monocrystal structure.

The boundary layer ideas that were presented earlier still apply, but greater flexibility is permissible in the choice of deposition parameters. The data in Table 9–2 summarize several categories of silicon CVD by temperature and pressure [13]. The low-pressure and plasma-enhanced categories are discussed later. We next consider the deposition of a few different, but commonly used, materials.

9.13.1 Polysilicon

Highly doped polysilicon often is used as a conductor in layered, many-conductor-level IC structures and as the gate material in certain MOS devices and circuits. In such applications the silicon is deposited on an amorphous oxide or nitride film and so will be polycrystalline. If the polySi is applied after an aluminum deposition (not a current practice), its deposition temperature must be lower than the 577° eutectic of Al/Si, or any other eutectic of concern. Reference to Table 9–2 shows that low- or mid-T ranges in a cold-wall reactor operating with gases in the atmospheric pressure range are favored. The

Table 9–2. CVD Categories by Temperature and Pressure

	At. P		LP		LP-PE	
	HW	CW	HW	CW	HW	CW
Lo T (200–500°C) Polycrystal (or amorphous)	0	2↓	1↑	0	0	2↑
Mid T (500–1000°C) Polycrystal	0	2	1↓	0	0	0
Hi T (900–1300°C) Monocrystal	0	2	0	1↑	0	0

At. P = atmospheric pressure, LP = low pressure, PE = plasma-enhanced. HW = hot-wall reactor, CW = cold-wall reactor. Commercial use: 0 = not significant; 1 = limited; 2 = dominant. ↑↓ = increasing or decreasing use.
After Hammond [13]. Courtesy of *Solid State Technology*, PennWell Publishing Company, Copyright 1979.

pyrolysis of silane, reaction (a) in Table 9–1, could be used. If, for example, phosphorus doping is required, the reaction of Eq. (9.6-3) may be used.

9.13.2 Silicate Glasses

Silicon dioxide grown thermally on silicon is often referred to as a *primary* oxide or glass. Glasses not in contact with silicon are designated as *secondary* and are used in several applications—for example, as an interlevel dielectric between the polySi MOST gates and the interconnect metal layer. SiO_2 getters contamination (Section 2.10.3), conforms to the surface topography, and reflows well. Second, in topside passivation, SiO_2 (or PSG) often is used alone or combined with Si_3N_4 as a protective coating over the entire chip (Sections 5.3.4 and 9.13.3). Their deposition is referred to as *glassing* or *glass passivation* [20]. Windows subsequently are etched through to the Al bonding pads so connections to the outside world may be made. Passivation oxides are deposited by CVD. Deposition temperature must be low enough to be compatible with all the materials on the chip and lower than eutectic temperatures. These glass coatings should at least form a barrier to water vapor and provide mechanical protection, particularly to the soft, top-level aluminum interconnects.

Plain SiO_2 may be deposited for this purpose by the oxidation reaction of (d) in Table 9–1 in the temperature range of 200 to 500°C, and a cold-wall reactor is preferred as may be seen from Table 9–2. Recall that SiH_4 will burn in O_2 if its concentration is sufficiently high; hence, the silane is diluted with N_2 or Ar in the supply bottle and an excess of N_2 is used as the carrier to avoid this problem.

Passivation layers typically are 0.5 to 2 μm thick, and pure SiO_2 this thick does not cover the top-layer Al interconnects well: it tends to crack in crossing them because the thermal expansion coefficients of Si, Al, and SiO_2 do not match well. This problem is more severe in small geometry configurations where narrow interconnect lines are spaced close together. Beck states that a 1-μm-thick SiO_2 layer needs at least 2% by weight of phosphorus to eliminate the problem [21]. For this reason a binary glass, such as phosphosilicate glass (PSG), comprising phosphorus-doped SiO_2, is preferred since it is more flexible and has less tendency to crack.[2] Also PSG is a gettering agent for Na^+ ions and immobilizes them.

To deposit PSG, PH_3 is added to SiH_4, O_2, and N_2 and the reaction basically is that of (g) in Table 9–1. Usually, however, a nonstoichiometric glass, $P_xSi_yO_z$, is formed; hence the simpler designation PSG. A typical doping level would be 2.5 wt % of P in SiO_2.

Another doping reaction can take place, namely

$$2\,PH_3\uparrow + 4\,O_2\uparrow \rightarrow P_2O_5\downarrow + 3\,H_2O\uparrow \tag{9.13-1}$$

Recall that P_2O_5 is a glass. If this reaction does occur, the typical doping level would be 2 to 5 wt % of P_2O_5 in SiO_2.

Usually the reactions would be run in a cold-wall reactor in the 300 to 500°C range. Film quality and doping of the PSG are determined by these deposition parameters: temperature, O_2/hydride ratio ("hydride" here is $SiH_4 + PH_3$) and hydride flow rate. Figure 9–9, due to Kern, shows the effect of these parameters on the deposition rate [20]. Control of the PH_3/SiH_4 ratio during deposition can create multilayers of glasses ranging from SiO_2 to PSGs of different phosphorus content.

Sheet resistance measurements may be used to give an indirect indication of the percent of phosphorus in a PSG sample. Four-point-probe (FPP) measurements are made to obtain the V/I ratio, uncorrected for geometry, thickness, and so on. Corresponding values of P percent may be obtained from curves given by Beck that are based on direct types of measurements [21]. The FPP method provides a simple means of checking batch-to-batch film uniformity.

Borosilicate glasses (BSG) are another type of binary glass that may be deposited by CVD [22]. The gases supplied are SiH_4, O_2, and N_2 carrier, with B_2H_6 replacing PH_3. Once again, sufficient nitrogen must be present to prevent the SiH_4 and O_2 components from flaming. If the hydride/N_2 ratio exceeds 1.6%, flaming occurs when these gases are mixed with the excess O_2 required to form the glass.

---

[2] Other notation is encountered for CVD oxide films. In this regard, the name SILOX does not refer to a type of glass but rather to a proprietary line of CVD reactors manufactured by Applied Materials, Inc., and used for depositing glasses of various types. Another term, VAPOX, is sometimes used to indicate "vapor deposited oxide," i.e., one laid down by CVD.

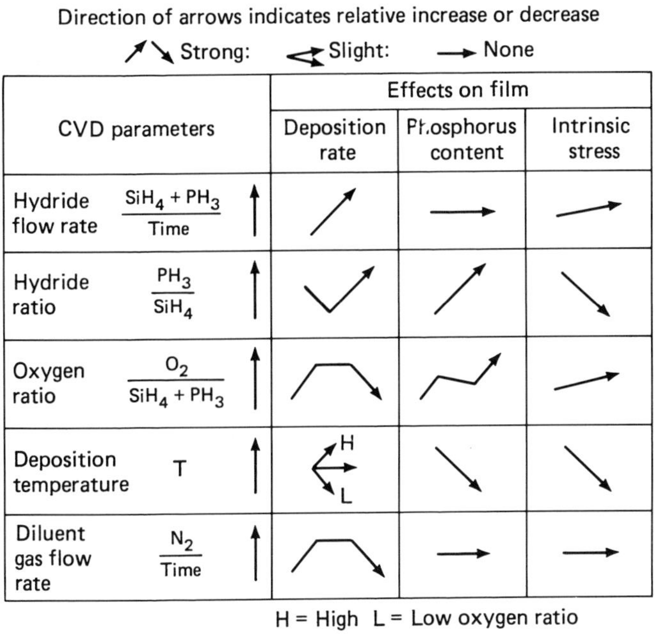

Direction of arrows indicates relative increase or decrease

Strong: Slight: None

CVD parameters			Effects on film		
			Deposition rate	Phosphorus content	Intrinsic stress
Hydride flow rate	$\frac{SiH_4 + PH_3}{Time}$				
Hydride ratio	$\frac{PH_3}{SiH_4}$				
Oxygen ratio	$\frac{O_2}{SiH_4 + PH_3}$				
Deposition temperature	T		H L		
Diluent gas flow rate	$\frac{N_2}{Time}$				

H = High L = Low oxygen ratio

Figure 9–9. Effect of CVD parameters on the deposition of PSG. Kern [20]. (Courtesy of *Solid State Technology*, PennWell Publishing Company, Copyright 1975.)

The oxide/hydride ratio is 2.25 in stoichiometric BSG. If, however, that ratio of gases is used, the films are poor insulators (due to oxygen deficiency) and are not soluble in HF, which causes severe etching problems. The basic deposition reactions for BSG glasses are

$$\left.\begin{aligned} SiH_4\uparrow + 2\,O_2\uparrow &\rightarrow SiO_2\downarrow + 2\,H_2O\uparrow \\ B_2H_6\uparrow + 3\,O_2\uparrow &\rightarrow B_2O_3\downarrow + 3\,H_2O\uparrow \end{aligned}\right\} \qquad (9.13\text{-}2)$$

In addition to the N_2 carrier, excess oxygen is supplied. It might appear that bottling premixed gases in the proper proportions would increase the repeatability of film properties. Kern and Heim report, however, that premixture leads to instability of the gases and is not recommended [22].

Figure 9.10 shows the effect of temperature on the BSG deposition rate. Temperature also affects film composition, with the variation rounding off rapidly around 400°C, and essentially leveling off at a composition of roughly 22.5 mol % of B_2O_3 in the glass at 450°C. Figure 9–10 shows that the deposition rate is low at 450°C, so a compromise value of 400°C is used. To

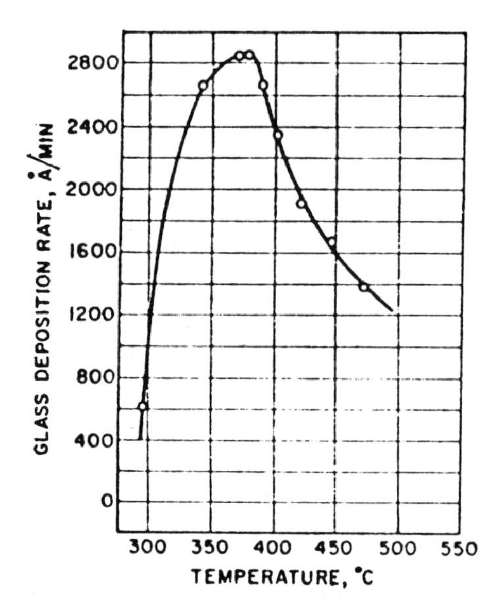

Figure 9–10. Deposition rate of BSG as a function of substrate temperature. Kern and Heim [22]. (Reprinted with permission of the publisher, The Electrochemical Society, Inc.)

improve the glassy properties of the film, thermal densification usually follows the deposition.

If PH_3 and B_2H_6 are supplied simultaneously with SiH_4, O_2, and the nitrogen carrier, phosphoborosilicate glass results. Properties lie between those of PSG and BSG.

9.13.3 Nitrides

Silicon nitride is an alternative passivation agent and dielectric. In fact, its relative dielectric constant, $\varepsilon_r = 3.9$, is roughly twice the value for SiO_2; further, the nitride prevents migration of sodium: hence, it is a better barrier than the glasses and it still getters the Na^+ ion.

Nitride cannot be used as the top layer for EPROMS (electrically programmable read only memories) because it absorbs the incident ultraviolet radiation that is used to erase, or deprogram, the transistors before reprogramming. For these units, an oxynitride layer, which combines the properties of oxide and nitride, may be used.

The thermal growth of nitride on silicon by the reaction

$$3\,Si\downarrow + 4\,NH_3\uparrow \rightarrow Si_3N_4\downarrow + 6\,H_2\uparrow \tag{9.13-3}$$

is too slow to be practicable; even at 1400°C it takes many hours to grow a layer of useful thickness. This set of large *T* and *t* values would wreak havoc with the locations of junctions already in place, and usually the nitride is desired on a material other than silicon; hence, CVD is used. Reaction (e) in Table 9–1, at 850°C and atmospheric pressure in a cold-wall reactor, gives a deposition rate of some 200 Å/min [23]. Nitride CVD films usually are nonstoichiometric, however, being of the form Si_xN_y and often are designated by SiN. Multilayers of SiO_x and SiN can be grown in sequence in one reactor simply by changing the reactant gases and the temperature (900°C being used for the oxide layers) or combined so that oxynitride may be produced.

Hot-wall reactors working at 875°C also may be used for nitride deposition. Low NH_3/SiH_4 ratios give high growth rates while a wide range of medium values gives fairly constant growth rate. Actual gas flow rates depend upon the reactor size, but a typical set of ratios is: $1SiH_4 : 260N_2 : 299NH_3$ by volume. The growth rate may be raised within limits by raising the total volumetric flow rate while maintaining the same ratios [24].

Low-pressure CVD is more common at present for the deposition of silicon nitride. This is discussed in Section 9.14.

9.13.4 Continuous Production Reactors

The CVD systems described thus far employ batch processing. The extensive use of PSG in the semiconductor industry for protecting wafers and chips led to the development of continuous, conveyor-belt deposition systems to provide much greater wafer throughputs. Such systems cannot be sealed off from the atmosphere, so the reactant gases must be released in very close proximity to the wafer surfaces [20, 25].

A basic setup of the continuous type is illustrated in Figure 9–11(a). The wafers are carried past the bottom surface of the water-cooled dispersion head to which the reactant, carrier, and purge gases are supplied and effluents are removed. The bottom of the dispersion head is close to the wafers, so mass transport is confined to a short region. A shroud is used to help confine the gases to the deposition region and to keep out ambient atmosphere. The wafers are brought to deposition temperature by radiation from a heater, located below the conveyor belt, which must be able to withstand the high temperatures involved. Typical belt materials are Inconel (a nickel-based alloy with small amounts of several other components) or other special nickel/chromium alloys.

The figure shows that each wafer in turn is exposed to the same deposition conditions in the same location, so uniformity of product can be maintained very well. Note that the heater extends well beyond the disperser and shroud at the input end in order to preheat the wafers before they reach the deposition zone. A typical throughput is 200 to 400 wafers per hour.

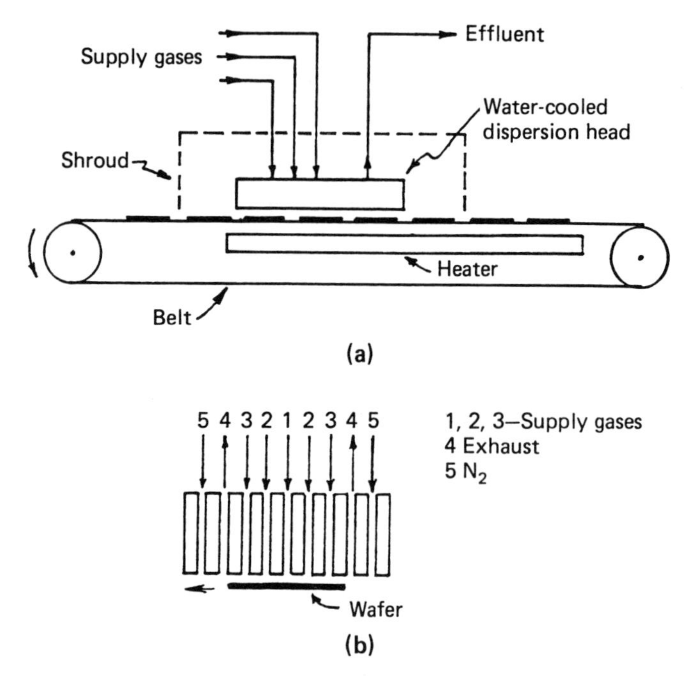

Figure 9–11. Continuous production CVD reactor. (a) Basic form. (b) Detail of a gas disperser where gases are mixed near the wafer surface. Water cooling is not shown.

In one type of unit the hydride, O_2, and N_2 are premixed before being fed to the dispersion head, whence they emerge from bottom slots located about 1 cm above the wafers. With proper adjustment the mixture flows in laminar fashion over the hot wafers, where a heterogeneous reaction takes place. Water cooling prevents the reaction taking place on the disperser.

In another design the separate gases are fed through jets or slots in the disperser as shown in cross section in Figure 9–11(b). Note that the gas flows vertically through the jets to the wafers, but horizontally over the wafers. The jets are very close to the wafers (about 2.5 mm away). The gases mix in this thin space and react on the wafer surface. Note from the figure how exhaust jets are located on the outsides of the reactant supplies, and jets for N_2 are at the outsides of the disperser. This nitrogen acts as an inert blanket to isolate the reaction region from the ambient atmosphere within the shroud, an action that is aided by the positive pressure of the gases wrt atmosphere.

9.14 Low-Pressure CVD (LPCVD)

The effects of using pressures lower than atmospheric on silicon epitaxy were discussed in Section 9.11. In particular it was seen that pressure reduction allowed deposition at lower temperatures. The same principles apply to nonepitaxial CVD as well.

Reference to Table 9–2 shows that the trends in LPCVD are toward the use of hot-wall reactors that are similar to diffusion furnaces. When such a configuration is used, it is no longer necessary to lay wafers flat against a susceptor. On the contrary, they may be stacked vertically and at right angles to the furnace axis and direction of main supply gas flow. This gives a much higher stacking density with a greatly increased throughput. Recall that gas-phase diffusivity increases with lowered pressure, which solves the problem of getting reactants to the wafer surfaces by paths normal to the main gas flow. Another advantage of vertical stacking is that any solid particles, formed as the result of homogeneous reactions in the gas or by heterogeneous reactions on the walls, are less likely to fall onto the wafer surfaces. The reactions used for LPCVD are the same as for deposition at atmospheric pressure. Typical LPCVD deposition conditions are described in the literature [26].

Chemical vapor deposition has not been confined to polySi, SiO_2, and Si_3N_4; demands of VLSI structures involving many layers of add-on materials have seen use of tungsten (W) (both blanket and in selected areas) and tungsten silicide (WSi) among others. Efforts have been made to increase deposition rates for all materials at even lower temperatures. One system for doing this is discussed in the next section.

9.15 Plasma-Enhanced CVD (PECVD)

Thus far, we have considered CVD systems in which thermal energy is used to break bonds in the reactants. We also have seen that below-atmospheric pressures allow reduction of substrate temperature. This lowers the thermal energy available to break bonds and the deposition rate will generally go down. Consider these numbers: thermal energy at 1000°C = 1273 K is $kT = (8.62 \times 10^{-5})\ (1273) \approx 0.11$ eV. Typical activation energies for CVD reactions lie in the range of a few eV. Furthermore the chemical reaction rate at the substrate surface is proportional to $\exp(-E_a/kT) \approx 1.1 \times 10^{-4}$ for 1000°C and a 1 eV activation energy. This is a small number indeed, and will be even smaller at the temperatures used for LPCVD—e.g., 3.03×10^{-7} for $E_a = 1$ eV and $T = 500$°C! An additional energy source is needed to raise the reaction rate, one that does not depend on the substrate temperature.

One such source is the kinetic energy of free electrons in a plasma, established, say, by a glow discharge. The energy distribution of the free electrons in a plasma is of the Maxwell-Boltzmann type with a broad peak at roughly 2 eV [27]. The equivalent temperature of this energy is $T = 2/(8.62 \times 10^{-5}) = 2.32 \times 10^4$ K; therefore, these free electrons can greatly enhance the chemical reaction rate while keeping the substrate temperature in the few-hundred-°C range. Highly reactive, free radical species that can enter into the CVD deposition process are the result of the dissociation process in the reactant gases.

Plasma-enhanced CVD (PECVD) is not without problems. (1) It can cause radiation damage by bombarding the substrate with high-energy particles. (2) Ions can bombard the reactor vessel walls and fixtures, releasing contaminants. The choice of fixture materials is important in this regard. Aluminum is good since it *sputters* (i.e., dislodges atoms by momentum transfer from bombarding ions) at a very low rate. Silicon dioxide is excellent with d-c plasmas, too. (3) The broad energy spectrum of the free electrons allows undesirable reactions to take place, e.g., reactions other than, say, the breaking down of SiH_4 into Si and H_2. This problem and its solution are discussed in Section 9.16.

The PECVD process depends upon the formation of a glow discharge in a low-pressure gas. These discharges are discussed in Appendix C. Two forms of PECVD reactors that employ the plane, parallel electrode configuration are shown in Figure 9–12 [28].

Some typical PECVD deposition conditions for some materials are listed in the remainder of this section [29]. Data are for a plasma generation frequency of 410 kHz.

Amorphous and Polysilicon: SiH_4/Ar, 1 torr, r-f power = 80–90 W, 300 Å/min in the hot-wall reactor of Figure 9–12(b). PolySi at 620°C, *a*Si at 450°C. PH_3 can be added for doping.

Recently, the deposition of *a*Si by PECVD has become very important for at least two applications: (1) *a*Si solar cells, and (2) *a*Si thin-film transistors (TFT) as integral parts of liquid crystal television displays. The application in low-cost solar cells is of particular commercial interest. As we have seen, light is absorbed by only a few micrometers of thickness of Si; hence, it is wasteful and economically unsound to use single-crystal wafers, which must be mils thick. Furthermore, the production of monocrystal Si is highly energy- and cost-intensive. On the other hand, very thin (a few micrometers) *a*Si layers can be deposited with either p- or n-type doping on insulating substrates, and plasma deposition assemblies have been built that can deposit *a*Si on substrates up to 40 cm by several meters in size.

Such layers for solar cells usually contain up to 10% of H_2, so the film is actually an alloy called *hydrogenated aSi.* Deposition temperatures can run as low as 200°C, in which case glass may be used as the substrate, an important cost consideration in some applications.

$SiO_2 : SiH_4/N_2O/O_2$. $1SiH_4 : 24O_2$, 380°C, 1.2 torr, 31–37 W, deposition rate ≈ 400 Å/min. PSG may be deposited by adding PH_3, but the deposition rate is lowered.

$Si_3N_4 : SiH_4/NH_3$. Hot-wall reactor, 360°C, 2 torr. Deposition rate is practically independent of the silane/ammonia ratio and overall flow rates, and is linear with r-f power, a typical value being 350 Å/min at 100 W.

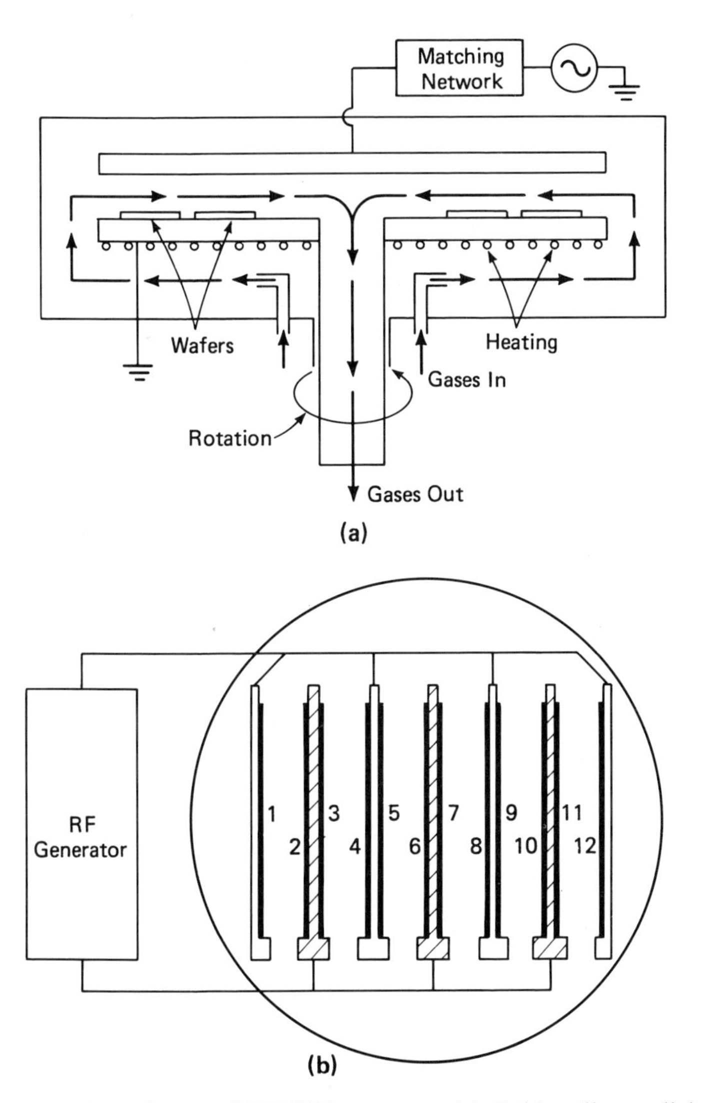

Figure 9–12. Two forms of PECVD reactors. (a) Cold-wall, parallel plate. (b) Hot-wall, where the wafers are vertical and parallel to the reactor tube axis. Gorowitz, Gorcyzca, and Saia [28]. (Courtesy of *Solid State Technology*, PennWell Publishing Company, Copyright 1985.)

Many other sets of deposition parameters may be used, several of which covered in the literature. Conditions also may be found for depositing materials other than those mentioned here, for example tungsten (W) and tungsten silicide, with SiH_4 and tungsten hexafluoride (WF_6) as the reactants [28].

9.16 Photopyrolysis

Radiant-energy photons also may supply energy to drive a CVD reaction [27], either by supplying heat to the wafers or susceptor directly or by interacting with the reactant gas molecules to dissociate them and produce free radicals. We next consider the mechanisms involved.

When radiation is used to produce heat for CVD reactions, the process is called *pyrolysis.* Ultraviolet (UV), visible, or infrared radiation may be used. Usually, this is directed at the susceptor, where it is absorbed, as in Figure 9–2(c), and transferred to the wafers by conduction. The reaction is run at or below atmospheric pressure.

If the radiation is directed at the wafers rather than the susceptor, two types of heating can take place. (1) If $\lambda \leq 1.13\ \mu m$, the radiation is absorbed by the silicon (Section 8.13.2). Note that this covers the UV, visible, and part of the infared spectrum, though the absorption is not uniform over this entire range of wavelengths and also is confined to a thin layer (about 1 to 10 μm) near the silicon surface. Since silicon wafers are typically around 20 mils ($\approx 500\ \mu m$) thick, wafer heating is not uniform from front to back. (2) If $\lambda > 1.13\ \mu m$, a portion of the infrared range, it will pass through the silicon and be absorbed in the susceptor, which will heat up. Lasers may be used instead of noncoherent sources.

9.17 Photolysis

In *photolysis,* the photons interact with the reactant gas molecules directly, causing dissociation with free radicals as the final result. These radicals then can form deposits on the substrate.

Theoretically, photolysis can produce CVD films at any substrate temperature, but if T is too low, the films are of poor quality, because the molecules depositing on the substrate lack sufficient surface mobilities for good film formation. Thus, substrates usually are held to at least 100 to 200°C.

If sufficient radiation reaches the substrate directly or if the radiation has components with wavelengths longer than the silicon bandgap value (so it can pass through the wafers to the susceptor), it will be absorbed and some heating will take place. Otherwise, a separate susceptor heater is required.

Photolysis may be considered in two subcategories, depending upon the radiation source used: broad- or narrow-spectrum.

9.17.1 Broad-Spectrum UV Excitation

Quartz/Hg lamps are typical inexpensive, broad-spectrum UV sources [30, 31]. They have energy peaks outside the usual range of CVD activation energies of a few electron volts, but the peaks are broad and outputs large, so reactions do take place. Special techniques may be used to enhance the energy transfer to the reactants. Ultraviolet radiation has two distinct advantages—the sources are relatively inexpensive, and they do not cause radiation damage.

Recall that in photolysis the photon must transfer energy to the reactants. Consider the deposition of Si_3N_4. Silane and ammonia, which were used in the processes described earlier, efficiently absorb only radiation with $\lambda < 0.22$ μm, while the quartz envelopes of the Hg sources typically cut off at $\lambda \approx 0.20$ μm, so the reaction is driven by a relatively narrow and not very intense source. For better coupling, one option is to replace NH_3 by hydrazine (N_2H_4) which absorbs UV better and causes this reaction:

$$3\ SiH_4\uparrow + 2\ N_2H_4\uparrow \xrightarrow[200^\circ C]{UV} Si_3N_4\downarrow + 10\ H_2\uparrow \qquad (9.17\text{-}1)$$

Some companies prefer not to use hydrazine because it presents additional safety problems, so another approach is used, one that involves the use of mercury vapor as a *photosensitizer* in the basic SiH_4/NH_3 reaction:

$$3\ SiH_4\uparrow + 4\ NH_3\uparrow \xrightarrow[100\text{–}200^\circ C]{UV+Hg\uparrow} Si_3N_4\downarrow + 12\ H_2\uparrow \qquad (9.17\text{-}2)$$

Mercury vapor, which has a strong resonance at 0.2357 μm (= 5.27 eV), is maintained in the CVD system at roughly 10^{-3} torr. The UV from the quartz/Hg lamp excites the Hg vapor molecules into a very high energy state due to the resonance. They are not ionized and so cannot enter into the chemical reaction, but the excited Hg, acting as a photosensitizer, drops back to its ground state and gives up the excess energy to the reactants. By this we mean that the process of Eq. (9.17-2) proceeds efficiently and still allows the use of NH_3 and the inexpensive UV source with the Hg vapor acting as a catalyst. The presence of Hg vapor raises problems, however, because it is hazardous and must be recaptured from the pump exhaust.

9.17.2 Laser Excitation (LCVD)

In contrast to the broad-spectrum UV lamp, lasers have very narrow, high-amplitude spectra [27]. There are advantages to this. We have a choice of radiation wavelengths, which extend from 12.42 to 0.177 μm (E = 0.1 to 7 eV.) Excimer lasers extend this range from 0.351 to 0.157 μm (E = 3.54 to 7.91 eV). These lasers use inert gases to form halides that are stable only in an excited state. Sometimes they are designated by a superscripted asterisk following the halogen, e.g., XeCl*; we shall omit the asterisk. Typical excimer lasers use xenon fluoride (XeF), $\lambda = 0.351$ μm; xenon chloride (XeCl), $\lambda = 0.308$ μm, and argon fluoride (ArF), $\lambda = 0.193$ μm. This wide range of available sources allows choice of a wavelength to which particular reactants are very sensitive [32].

The disadvantages are high equipment cost and possibly the need to substitute a reactant that can absorb energy efficiently from the radiation of the chosen laser. This will be illustrated shortly.

The mechanism of energy transfer should be reviewed briefly to aid in choosing a laser for a particular reaction. If the radiation energy is *less* than (wavelength *greater* than) the dissociation energy (wavelength) of one or more of the reactants, such as SiH_4 or N_2O, dissociation of that reactant will not take place. Excitation is possible, but not the production of free radicals.

As a specific example consider SiH_4, which does not absorb well for $\lambda < 0.150$ μm, i.e. at the short wavelength end of the available laser frequencies. Disilane (Si_2H_6) may be substituted, however, since it absorbs well for $\lambda < 0.22$ μm, which falls nicely in the range of available lasers. This is analogous to the substitution of N_2H_4 for NH_3 in the reaction of Eq. (9.17-1). A reaction based on disilane for silicon deposition is

$$Si_2H_6\uparrow \xrightarrow[630^\circ C]{Xe/Hg} 2\,Si\downarrow + 3\,H_2\uparrow \tag{9.17-3}$$

This reaction with the Xe/Hg source has given deposition rates in the hundreds of Å/min.

Films of SiO_2 have been deposited with the SiH_4/N_2O overall reaction

$$SiH_4\uparrow + 2\,N_2O\uparrow \xrightarrow{ArF} SiO_2\downarrow + 2\,N_2\uparrow + 2\,H_2\uparrow \tag{9.17-4}$$

driven by the ArF excimer laser of $\lambda = 0.193$ μm. Actually, the equation obscures the true action, which involves two steps, first

$$N_2O\uparrow \xrightarrow{ArF} O\uparrow + N_2\uparrow \tag{9.17-5}$$

where the laser radiation dissociates the N_2O into nitrogen and *atomic* oxygen, which is highly reactive; second, this oxidizes the silane

$$SiH_4\uparrow + 2\,O\uparrow \rightarrow SiO_2\uparrow + 2\,H_2\uparrow \tag{9.17-6}$$

Aluminum oxide (Sapphire, Al_2O_3) has been deposited from trimethylaluminum [TMA, $Al(CH_2)_3$] and N_2O. This is a three-step reaction where the laser radiation separately dissociates the TMA and the N_2O. The Al_2O_3 then is formed by reaction of the appropriate free radicals.

9.18 Compound Semiconductors; MOCVD [33, 34]

Since the 1960s, interest has increased in the compound semiconductors, particularly in those made up of one element from column III_A and one from column V_A of the Periodic Table, and called the III-Vs (Figure 2–12). These compounds crystallize in the same diamond lattice as do Si and Ge, except that atoms of the two elements alternate in the lattice structure. The III-V semiconductors do not occur naturally, and yet are of great interest because of their large band gaps (InP, 1.27; GaAs, 1.38; AlAs, 2.16 eV); their optical properties (InP, GaAs, and InAs have direct band gaps); and their large electron mobilities (GaAs, 8500; GaSb, 5000; InAs, 22,600 $cm^2/V\text{-}s$). These properties have allowed the fabrication of much faster devices, semiconductor lasers, and micro-optoelectronic integrated circuits [35].

Since these man-made materials are compounds, their chemistry is quite different from that of elements such as Si and Ge. While they have been deposited in monocrystal form from the liquid phase, and by physical vapor deposition (Chapter 12), CVD provides a useful and rapid method.

Two different types of deposition chemistry have been used. (1) those in which both components are halides such as diethylgallium chloride (DEGaCl) and arsenic trichloride ($AsCl_3$), and (2) those based on column III_A alkyls and column V_A hydrides. Typical alkyls for this use are trimethylgallium [TMGa, $Ga(CH_3)_3$], or triethylgallium [TEGa, $Ga(C_2H_5)_3$]. These types of metal and organic compounds are called *organometallics, metallo-organics,* or simply *metalorganics*; when they are used in chemical vapor deposition the MOCVD nomenclature is used commonly. Since these chemicals are deposited from the vapor phase to give epitaxial semiconductor layers, an alternate name, MOVPE, sometimes is used. The halides used here are inorganics of the types we have considered as n-type dopants for silicon, such as PH_3 and AsH_3. On pyrolysis these alkyl/hydride combinations produce no etchants, a property that provides great flexibility in the choice of substrates for the III-V deposition.

Consider a typical MOCVD deposition of GaAs (gallium arsenide) by the pyrolysis of TMGa and AsH_3. The former is liquid at room temperature and atmospheric pressure, so H_2 is bubbled through it as a carrier (Section D.3.5). The resulting vapor is mixed with the AsH_3 before reaching hot parts of the reactor system. Induction heating may be used with a cold-wall reactor. Pyrolysis occurs in the 650 to 775°C range and the reaction is

$$Ga(CH_3)_3\uparrow + AsH_3\uparrow \rightarrow GaAs\downarrow + 3\ CH_4\uparrow \qquad (9.18\text{-}1)$$

The effluent is methane (marsh gas) diluted with the hydrogen carrier and does not affect adversely any of the reactions.

Different gallium compounds may be formed by replacing AsH_3 by appropriate hydrides—for example, PH_3 for gallium phosphide (GaP), SbH_3

(antimony hydride) for gallium antimonide (GaSb), and so on. The column III component of the compound may be changed by choice of a suitable alkyl. Table 9–3 summarizes some of the possibilities.

The table also shows some *ternary* (three-element) compounds such as gallium arsenide phosphide ($GaAs_{1-x}P_x$), where x is the mole fraction. This may be considered as a pseudobinary compound of GaAs and GaP (Section 4.14). The value of x may be controlled through the relative flow rates of the two alkyls during deposition. It has been found that properties such as band gap may be varied between that of the limiting binaries corresponding to $x = 0$ and $x = 1$, subject, however, to the limitations listed in the right-hand column of the table.

The addition of other suitable metalorganics allows doping of the III-V semiconductor. For example, DEZn and DMCd will dope with zinc or cadmium (column II_B) as an acceptor and DETe (diethyltelluride) with tellurium (column VI_A) as a donor.

As an alternative to the induction heated susceptor, cold-wall reactor setups have been used successfully with lamp heating of the susceptor to 620°C [36]. Another variation combines induction and laser heating to allow lower substrate temperatures during deposition [37]. In this setup the laser radiation interacts with the gas just above the substrate to provide the additional required heat. The laser is chosen to produce a maximum of a vibrational or rotational mode of the desired reactant species; it does not produce dissociation directly, only heat.

Table 9–3. Some Reactants for MOCVD Deposition of Binary and Ternary III–V Compounds

Type	Compound	Reactants	Allowed x values
Binary	GaAs	$TMGa/AsH_3$	
		$DEGaCl/AsH_3$	
	GaP	$TMGa/PH_3$	
	InAs	$TEIn/AsH_3$	
	InP	$TEIn/PH_3$	
	InSb	$TMIn/SbH_3$	
Ternary	$GaAs_{1-x}P_x$	$TMGa/AsH_3/Ph_3$	0.1–0.6
	$GaAs_{1-x}Sb_x$	$TMGa/AsH_3/SbH_3$	0.1–0.3
	$Ga_{1-x}Al_xAs$	$TMGa/TMAl/AsH_3$	0.2–0.9
	$Ga_{1-x}In_xAs$	$TMGa/TEIn/AsH_3$	0.0–1.0
	$InAs_{1-x}P_x$	$TEIn/AsH_3/PH_3$	0.2–0.6

After Dupuis [34]. Courtesy of Electronic Packaging and Production, Cahners Publishing Company, Inc.

A group of high-tech applications of the III-Vs involves the fabrication of a *superlattice*, a multilayer stack of very thin layers of alternating materials. Such fabrication is handled successfully with MOCVD using computer control of the heating, and relative as well as total gas flow rates. Molecular beam epitaxy (MBE) is also used for laying down an epi layer of GaAs on Si and for forming superlattice structures (Section 12.5).

Problems

9–1 An epi system uses a tank supply of ($PH_3 + H_2$), and a bubbler for $SiCl_4$, Figure D–3(d), with H_2 as the carrier gas. Cite all the references you use.

(a) Derive a literal equation for n_P/n_S, the molar ratio of phosphorus to silicon laid down in the growing epi layer. Assume that this ratio is the same as in the gas phase.
(b) Derive an equation for the P concentration in the epi layer. Note there are 5×10^{22} atoms/cm^3 in Si.
(c) Given the following conditions, calculate F_{vT} in l/min: $N_P = 10^{16}$ cm^{-3}, $T_S = 20°C$ with $F_{vH} = 10$ cm^3/min, $T_T = 25°C$ with $P_T = 10$ psi and $r = 3 \times 10^8$.

Note: 1 psi = 51.6 torr.

9–2 For the data of Problem 9–1 do not assume that the molar ratio of P to Si is the same in both solid epi and gas phases. What is the P concentration in the epi layer? Cite any references you use.

9–3 Reduce Eq. (9.5-20) to the two limiting cases of $k_S \leq h_G$, and $h_G \leq k_S$. Reconcile your results with Figure 9–3.

9–4 Using the data of Sections 9.16 and 9.17,

(a) Explain why Si_2H_6 is better suited than SiH_4 for photolysis with an ArF excimer laser, and yet SiH_4 is quite satisfactory for photopyrolysis at 200°C.
(b) What are the activation energies in eV of the two gases?

References

[1] G. W. Cullen, "The Preparation and Properties of Chemically Vapor Deposited Silicon on Sapphire and Spinel," *Proceedings of First International Conference on Crystal Growth from the Vapour Phase*, Zurich, Switzerland, Sept. 1970. pp. 107–125.

[2] G. A. Keig, "Single Crystal Oxide Substrates," *Solid State Technol.*, **15**, (9), 53–58, Sept. 1972.

[3] M. L. Hammond, "Silicon Epitaxy," *Solid State Technol.*, **21**, (11), 65–75, Nov. 1978.

[4] M. L. Hammond, "Radiant vs. Induction Heating for Silicon Epitaxy," *Circuits Manuf.*, **18**, (9), 42, 44, and 45, Sept. 1978.

[5] M. A. Drews et al., "Physical Properties of Semiconductor Industry Chlorosilanes," *Solid State Technol.*, **16**, (1), 39–43, Jan. 1973.

[6] C. L. Yaws et al., "Physical and Thermodynamic Properties of Silicon Tetrachloride," *Solid State Technol.*, **22,** (2), 65–70, Feb. 1979.

[7] A. L. Armirotto, "Silane: Review and Applications," *Solid State Technol.*, **11,** (10), 43–47, Oct. 1968. A good review of the precautions for handling and storing silane is given here. Similar information is available from specialty gas vendors.

[8] M. L. Hammond, "Safety in Chemical Vapor Deposition," *Solid State Technol.*, **23,** (12), 104–109, Dec. 1980.

[9] R. A. Bolman, Jr., "Hazardous Production Gases, Part 1: Storage and Control," *Semicond. Int.*, **9,** (4), 156–159, Apr. 1986.

[10] C. Murray, "Improving Gas Handling Safety," *Semicond. Int.*, **9,** (8), 60–65, Aug. 1986.

[11] S. Cygelman, "Gases for the Electronic Industry, A Changing Technology," *Solid State Technol.*, **14,** (1), 45–48, Jan. 1971.

[12] A. S. Grove, *Physics and Technology of Semiconductor Devices.* New York: Wiley, 1967.

[13] M. L. Hammond, "Introduction to Chemical Vapor Deposition," *Solid State Technol.*, **22,** (12), 61–65, Dec. 1979.

[14] R. W. Atherton, "Fundamentals of Silicon Epitaxy," *Semicond. Int.*, **4,** (11), 117–130, Nov. 1981.

[15] R. M. Warner, Jr., Ed., *Integrated Circuits, Design Principles and Fabrication.* New York: McGraw-Hill, 1965.

[16] S. P. Weeks, "Pattern Shift and Pattern Distortion During CVD Epitaxy on (111) and (100) Silicon," *Solid State Technol.*, **24,** (11), 111–117, Nov. 1981.

[17] D. C. Gupta, "Improved Methods of Depositing Vapor-Phase Homoepitaxial Silicon," *Solid State Technol.*, **14,** (10), 33–39, Oct. 1971.

[18] R. B. Herring, "Advances in Reduced Pressure Silicon Epitaxy," *Solid State Technol.*, **22,** (11), 75–80, Nov. 1979.

[19] H. F. Matare, "The Electronic Properties of Epitaxial Layers," *Solid State Technol.*: Part 1, **19,** (1), 25–34, Jan. 1976. Part 2, **19,** (2), 36–44, Feb. 1976. Part 3. **19,** (3), 38–43, Mar. 1976.

[20] W. Kern, "Chemical Vapor Deposition Systems for Glass Passivation of Integrated Circuits," *Solid State Technol.*, **18,** (12), 25–33, Dec. 1975.

[21] C. Beck, "Phosphorus Concentration in Low Temperature Vapor Deposited Oxide," *Solid State Technol.*, **20,** (12), 58–60, Dec. 1977.

[22] W. Kern and R. C. Heim, "Chemical Vapor Deposition of Silicate Glasses for Use with Silicon Devices. I. Deposition Techniques," *J. Electrochem. Soc., ELECTROCHEMICAL TECHNOLOGY,* **117,** (4), 562–568, April 1970.

[23] R. E. Caffrey and V. E. Hauser, Jr., "Using Chemical Vapor Deposition to Make Dielectric Film," *Bell Lab. Rec.*, **49,** (2), 38–42, Feb. 1971.

[24] R. Ginsburgh, D. L. Heald, and R. C. Neville, "Silicon Nitride in a Hot-Wall Diffusion System," *J. Electrochem. Soc.*, **125,** (9), 1557–1559, Sept. 1978.

[25] W. C. Benzing, R. S. Rosler, and R. W. East, "A Production Reactor for Continuous Deposition of Silicon Dioxide," *Solid State Technol.*, **16,** (11), 37–42, Nov. 1973.

[26] W. Kern and G. L. Schnable, "LPCVD for VLSI Processing—A Review," *IEEE Trans. Electron Devices,* **ED-26,** (4), 647–657, Apr. 1979.

[27] R. Solanki, C. A. Moore, and G. J. Collins, "Laser-Induced Chemical Vapor Deposition," *Solid State Technol.*, **28,** (6), 220–227, June 1985.

[28] B. Gorowitz, T. B. Gorczyca, and R. J. Saia, "Applications of Plasma Enhanced Chemical Vapor Deposition in VLSI," *Solid State Technol.*, **28,** (6), 197–203, June 1985.

[29] R. S. Rosler and G. M. Engle, "Plasma Enhanced CVD in a Novel LPCVD-type System," *Solid State Technol.*, **24,** (4), 172–177, April 1981. This paper has many curves interrelating deposition parameters. The system that is of the form shown in Figure 9–12(b) handles seventy 100-mm wafers per load.

[30] J. W. Peters, F. L. Gebhart, and T. C. Hall, "Low-Temperature Photo-CVD Silicon Nitride: Properties and Applications," *Solid State Technol.*, **23,** (9), 121–126, Sept. 1980.

[31] S. C. Su, "Low-Temperature Silicon Processing for VLSI Fabrication," *Solid State Technol.*, **24,** (3), 72–81, March 1981.

[32] R. A. Lawes, "Use of Excimer Lasers in Photolithography," *Semicond. Int.*, **9,** (7), 76–77, July 1986.

[33] J. J. Coleman and P. D. Dapkus, "Metalorganic Chemical Vapor Deposition". In *Gallium Arsenide Technology*, ed.-in-chief D. K. Ferry, Chapter 3. Indianapolis: Sams, 1985.

[34] R. D. Dupuis, "Chemical Vapor Deposition for New Materials Applications." *Electron. Packag. & Prod.*, **18,** (6), 140–143, June 1978.

[35] N. Holonyak, Jr., et al., "Method for Producing Integrated Semiconductor Light Emitter," U.S. Patent 4,378,255, March 1983.

[36] "MOVPE Process Grows GaAs Layers on GaInAs," *Semicond. Int.*, **9,** (7), 20, July 1986.

[37] K. A. Jones, "Laser Assisted MOCVD Growth," *Solid State Technol.*, **28,** (10), 151–156, Oct. 1985.

Chapter 10
Etching

Chapter 1 showed the need for selective etching, whereby an etchant attacks one material, such as SiO_2, while leaving relatively unaffected adjacent materials, such as Si and Al. Etching processes are discussed in greater detail in this chapter.

10.1 Types; Processes

Etches may be classified by their physical state: liquid, dry plasma, or vapor. The E-gas used in chemical vapor deposition (Section 9.4) is an example of a vapor phase etchant.

Another classification is based on the manner in which the etch attacks monocrystal materials: in *isotropic* etching the action is independent of the crystal orientation, and hence of direction; in *anisotropic* etching different crystal planes etch at different rates. We shall see some practical examples of anisotropic etching (e.g., in fabricating V-shaped grooves for VMOS devices).

All etching processes, irrespective of the physical state in which they take place, involve three basic events: (1) movement of the etching species to the surface to be etched, (2) chemical reaction to form a compound that is soluble in the surrounding medium, and (3) movement of the by-products away from the etched region, allowing fresh etchant to reach the surface. Both (1) and (3) usually are referred to as *diffusion*, although convection may be present, too. The slowest of these processes primarily determines the etch rate, which thus may be diffusion- or chemical- reaction–limited. Most liquid etching processes used commercially in silicon technology are reaction-limited.

Different chemistries are required for selectively etching the materials used in silicon technology. Many sources are available in the literature for specific applications [1, 2, 3].

10.2 Wet Etching

In wet etching the masked substrate contacts the actual etching agents in liquid solution. Two of the most common components in *electronic grade* etch come diluted with water: hydrofluoric acid (HF) is 49% and nitric acid (HNO_3) is 70% in H_2O. These give etch rates too high for good control in many applications, so further dilution, to moderate the etching action, is necessary. Water may serve as the *diluent* or diluting agent, but sometimes acetic acid (CH_3COOH) will be used with HNO_3 because it buffers against HNO_3 depletion.

In order to remove depleted solution and by-products from the substrate surface and replenish the etchant, the system is agitated gently, often in a circular swirling action, or N_2 may be bubbled through the mixture. Machines that do this automatically are available commercially. Alternatively, the etchant may be sprayed onto the wafer.

Most Si technology etches are acid-based, so the etching tank effluent is acidic, and requires special disposal precautions. For the usual acids—such as nitric, acetic, and sulfuric—neutralization in a catch basin before disposal is adequate. On the other hand, HF requires special treatment and must not be discharged into a sewer system.

10.3 Oxide Etching

To make windows in SiO_2 without affecting the underlying Si, a selective etch is needed. Hydrofluoric acid (HF) attacks SiO_2 but leaves Si unaffected at room temperature, so HF is the basis of SiO_2 etches. The overall etching reaction is [4]

$$SiO_2 + 6\,HF + 6\,H_2O \rightarrow SiO_2 + 6\,H_3O^+ + 6\,F^- \qquad (10.3\text{-}1)$$

$$\rightarrow H_2SiF_6 + 8\,H_2O \qquad (10.3\text{-}2)$$

The by-product H_2SiF_6 (fluosilicic acid) is water-soluble and so can move away from the region of chemical reaction. Equation (10.3-1) shows production of the *hydronium ion* (H_3O^+), which really is the basis for the acid action [4, 5]. The etch rate may be moderated by adding more water because it lowers the hydronium ion concentration.

During etching, HF tends to deplete rapidly so ammonium fluoride (NH_4F) is added to form a *buffered* solution, typical proportions being $10NH_4F:1HF$. Since NH_4F is a solid, it is used in a water solution; a typical mix would be $6(40\%NH_4F):1(49\%HF)$. The etch rate for this solution at room temperature is around 1600 Å/min, too fast to control etch depth precisely by adjusting the etch time. But oxide thickness versus oxidation time is well controlled; hence the thin gate oxide of an MOS transistor is produced by etching away a thick oxide completely and thermally regrowing a new oxide of the desired thickness, rather than by partially etching away the thick oxide.

A *surfactant* or *wetting agent* may be added to the solution to obtain a more uniform etch over the entire area. Typical are Dow Chemical Company Triton-XR or 3M Company FG-95. The usual concentration is very small: 1 part of surfactant to roughly 500 parts of etchant solution.

Fast-acting etches tend to remove photoresist (PR) from the oxide. This effect can be reduced by spinning a layer of HMDS (hexamethyldisilizane) onto the oxide and drying it *before* the PR is applied (Section 11.8).

Even though HF stops etching when it reaches Si, there is no visible indication that the SiO_2 etch is completed. One can make a good estimate by knowing the etch rate, oxide thickness, and etch time, but a direct visual check can be made after the wafer is removed from the solution, and rinsed in water. Silicon is *hydrophobic* and is wetted by neither water nor HF, while both wet SiO_2 and PR. Thus the etch is completed if the windows are water-free. Usually the wafer is returned to the etch for another half-minute or so (to insure clean window edges), rinsed, and dried with dry N_2.

> **Caution:** Remember—HF attacks the human body and, in contrast to other common acids, does not cause pain on contact. Proper gloves should be worn, HF fumes should be avoided, and proper eye protection should be in place. All operations involving HF must be performed in a fume hood. Remember that HF attacks glass; use plastic or Teflon beakers.

The etch rate is affected by strength of the etchant, temperature, and SiO_2 doping. Boron-doped oxide etches slowly, phosphorus-doped, very rapidly. The latter fact must be kept in mind when contact windows are etched in the oxide after a phosphorus diffusion. A $5H_2O:1$ buffered HF etch is typical for B-doped oxides, while more water is added for P-doped—up to $20H_2O:1$ buffered HF for heavy doping, such as in emitters.

Steam-grown oxides are less dense than dry-grown oxides and etch faster. The etch rate for B-doped oxide may be increased by a pre-etch immersion in HBF_4 (fluoboric acid). Apparently this treatment modifies the glass structure [6].

10.4 Isotropic Si Etch

The isotropic etching of silicon is based on a two-step process. First, the Si is oxidized to SiO_2 by a strong oxidizer, usually HNO_3:

$$3\,Si + 4\,HNO_3 \rightarrow 3\,SiO_2 + (\text{by-products}) \qquad (10.4\text{-}1)$$

Second, SiO_2 is dissolved by HF into a water-soluble compound, as described in the last section.

$$3\ SiO_2 + 18\ HF + \text{(by-products)} \rightarrow 3\ H_2SiF_6 + 8\ H_2O + 4\ NO\uparrow \quad (10.4\text{-}2)$$

While we think of the reaction as a two-step sequence, actually the two reactions take place simultaneously, giving the overall reaction

$$3\ Si + 4\ HNO_3 + 18\ HF \rightarrow 3\ H_2SiF_6 + 8\ H_2O + 4\ NO\uparrow \quad (10.4\text{-}3)$$

The etch rates of the HNO_3/HF combination can be high—the maximum at room temperature being roughly 800 μm/min, for the ratio $4.5HNO_3:5.5HF$ [6]. This value is too high for good control. Changes in the proportions in either direction slow the rate. For example, $7HNO_3:3HF$ drops the rate to roughly 120 μm/min because the reaction tends toward HF limitation [2].

The etch rate is moderated more often by adding water or CH_3COOH (acetic acid), which act primarily as diluents (although the acid also provides buffering against loss of HNO_3, and in sufficient quantity may be used to stop the etching action completely). Typical component ratios run in the range $5HNO_3:3CH_3COOH:3HF$ to $7HNO_3:7CH_3COOH:1HF$. Curves of iso-etch-rate versus content are available in many references [4, 7].

PolySi often is used for interconnects and gates in MOS devices. Since polySi presents all possible crystal orientations to the etchant, an isotropic etch should be used. A typical formulation is $10HNO_3:1HF:10H_2O$.

10.5 Anisotropic Si Etches

The three principal silicon lattice planes have different combinations of atomic density and number of bonds normal to the plane (Table 2–1). Some etches sense these differences and attack the planes at different rates, giving anisotropic etching. Theoretical reasons for this selective action are meager, but a plane that combines high atomic density and a small number of bonds etches more slowly. On this basis we expect etch rates to decrease from $\langle 100 \rangle$ to $\langle 110 \rangle$ to $\langle 111 \rangle$ planes.

Anisotropic Si etches usually are not based on HNO_3/HF mixtures but rather on NaOH (sodium hydroxide), KOH (potassium hydroxide), N_2H_4 (hydrazine), or exotic organics, all complexed with water, alcohol, or other organics such as catechol [$C_6H_4(OH)_2$ = pyrocatechol]. Many formulations are described in the literature [8, 9].

A quasi-anisotropic silicon etch (Canadian patent No. 903650) is based on As_2O_3 (arsenic trioxide) in reflexed orthophosphoric acid (H_3PO_4), and yields vertical side walls irrespective of the wafer orientation [10]. The bottom profile may be controlled, however. Figure 10–1 shows a deep well etched in

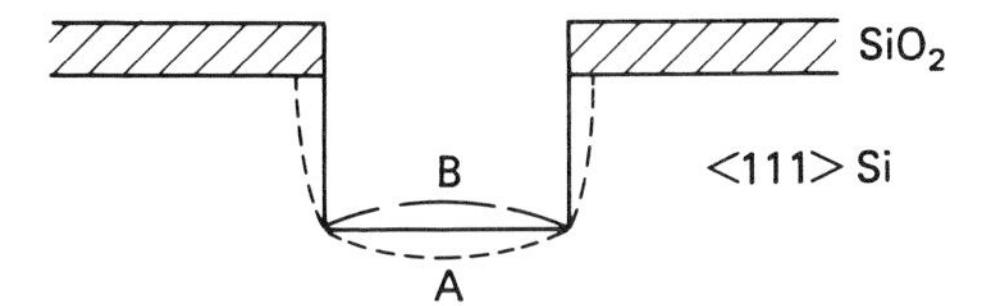

Figure 10–1. An ideal well formed with arsenic trioxide etch is shown by the solid lines. Profiles "a" and "b" are typical of conventional etches.

⟨111⟩ Si as used frequently for dielectric isolation in IC fabrication. Increasing the fraction of As_2O_3 can change the bottom profile from A to B, so a flat bottom may be obtained. The dashed lines at the sides show the effect of undercutting beneath the mask when conventional etch is used.

10.6 Anisotropic Si Etch: Applications

Anisotropic etching is used in several silicon processing applications. We shall consider only three of these, briefly, by way of illustration. These techniques must be used judiciously, however, to avoid unwanted crystal-orientation effects. For example in etching mesa structures, ⟨331⟩ and ⟨211⟩ planes may be exposed, with a resultant cutting away of the mesa corners.

10.6.1 V-Groove Etching

A whole technology, VMOS, has been developed that is based on the ability to etch a V-shaped groove in silicon. A ⟨100⟩ wafer is used as the substrate, as shown in Figure 10–2(a). A window of width w is cut in the masking medium, nominally SiO_2, with one pair of window sides parallel to the ⟨110⟩ directions, as shown in Figure 10–2(c); this will be parallel to the normal wafer flat. Then the wafer is immersed in an etch that attacks the ⟨100⟩ planes faster than the ⟨111⟩ planes. The dashed line α shows the groove after a short etch interval. The etch proceeds at a high rate vertically by acting on ⟨100⟩ planes. The ⟨111⟩ faces, however, are exposed on the sides, faces that etch very slowly; hence, as shown in (b), the etch effectively "stops" at the two exposed ⟨111⟩ planes that are inclined at 54.7° wrt the ⟨100⟩ planes.

The dashed line β shows the groove at a later time, and finally the end result of the etch is shown by the solid line where no ⟨100⟩ plane is left exposed. The etch action effectively stops when this limit is reached, because only the slow-etching ⟨111⟩ faces are exposed.

An interesting result comes about because of the self-limitation at the bottom of the groove. If the etch is allowed to proceed until the V is formed, the vertical depth d of the groove is controlled by the widow width w. Reading from Figure 10–2(b) we note that

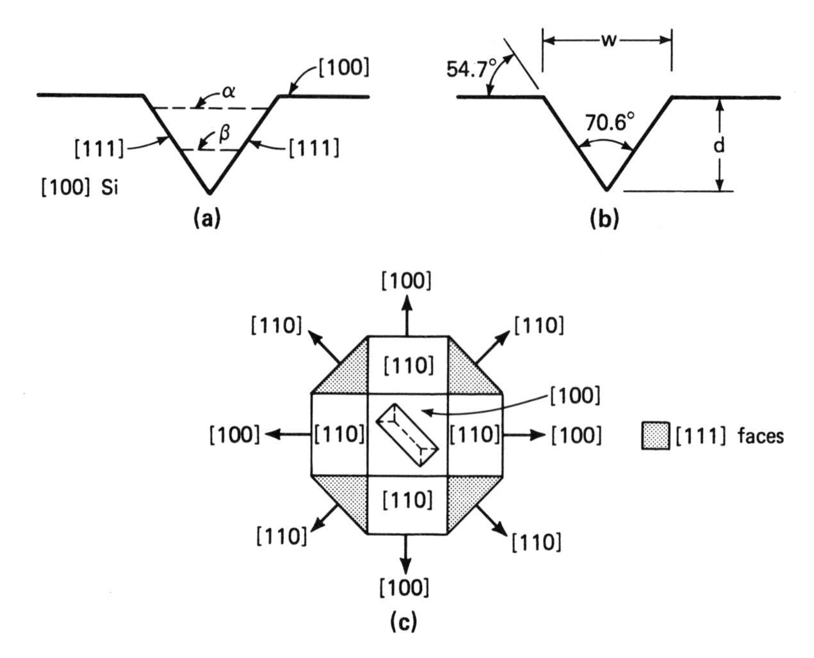

Figure 10–2. A V groove is formed with an anisotropic etch. (a) Pertinent crystal planes. (b) Angles of the V groove. (c) Top view of Figure 2–4 showing the proper groove orientation.

$$\tan\frac{70.6}{2} = \frac{w}{2d}$$

whence

$$d = 0.706w \tag{10.6-1}$$

If a square mask window is aligned as shown in (c), the two ends also will be ⟨111⟩ planes, so a V is formed in the second direction also. On the other hand, if the mask is rectangular, the V will appear normal to the shorter dimension w as in Figure 10–2(c). The groove will be flat-bottomed normal to the longer dimension, as shown by one of the dashed lines in Figure 10–2(a).

If the mask is not aligned exactly, there will be some undercutting beneath the window edges. The usual practice is to use a 5-min dip in 10% HF, after the anisotropic etch is completed, to remove any SiO_2 overhanging the edges [11].

Wafers of ⟨110⟩ orientation also may be used for V-groove etching, but ⟨100⟩ wafers are readily available so they are preferred. Many references discuss how the V-grooves are used in MOS technology [9, 12].

10.6.2 Narrow-Groove Etching

An anisotropic etch of (44 wt% KOH):H_2O at 85°C can etch ⟨110⟩ planes up to 400 times faster than ⟨111⟩ planes. Hence, if rectangular mask windows are aligned parallel to ⟨111⟩ planes on a ⟨110⟩ wafer, it is possible to get very narrow grooves with minimum undercut. Kendall has shown the importance of proper window alignment and has etched grooves only 0.4 μm wide [13].

High-C capacitors and vertical, multijunction solar cells are applications that use several deep, parallel, narrow grooves. The capacitors are fabricated by oxidizing the sides of the narrow grooves and then coating the oxide with deposited metal to give several capacitor segments in parallel.

10.6.3 Plane and Defect Revealing Etches

Certain anisotropic etches may be used to reveal wafer crystal orientation and crystalline defects. The former application has been illustrated in Figure 2–11(a) where a ⟨111⟩ triangle has been revealed by application of Sirtl etch (Section 2.6). Other etches, such as Secco and Wright, will pit a wafer to reveal defects in the crystal structure.

10.7 Other Insulator Etches

After SiO_2 (or silicate glasses), the two most commonly used insulators in silicon technology are Si_3N_4 (silicon nitride), and polyimide (PI).

Si_3N_4: A typical nitride etch comprises an 85% solution of H_3PO_4 (orthophosphoric acid) in water. This formulation etches both Si_3N_4 and SiO_2 at rates of 100 and 25 Å/min, respectively, at 180°C. Unfortunately, it also attacks PR, so pattern etching can be a problem. This may be circumvented by depositing an SiO_2 layer by CVD on top of the nitride, and covering this with PR. (This idea has been considered in Section 5.9.2 in connection with nitride masking.) Dry plasma etching, which is discussed later in this chapter, is preferred, because it provides greater selectivity in etching the two layers.

Polyimide: Polyimide (PI) is a generic name for a family of organic polymers that contains the *imide* group shown in Figure 10–3(a). It is mixed with solvents in liquid form and is spun on the wafer as photoresists are (Section 11.8). Typical solvents for the polyimides are NMP (methyl pyrrolidinone) and DMF (dimethyl formamide). Sometimes these are used in mixture with acetone or methanol.

The thickness for a given formulation is determined by the spin rate and time. The layer is dried and cured at 300°C; subsequently, it is stable up to 500°C. The material is a good between-layer insulator, and is used for wafer passivation and especially for *planarization* [14, 15]. With proper additives it also may serve as a photoresist.

In multilayer structures, the top wafer surface becomes less flat as more windows are etched and additional layers are applied. Eventually, nonplanar-

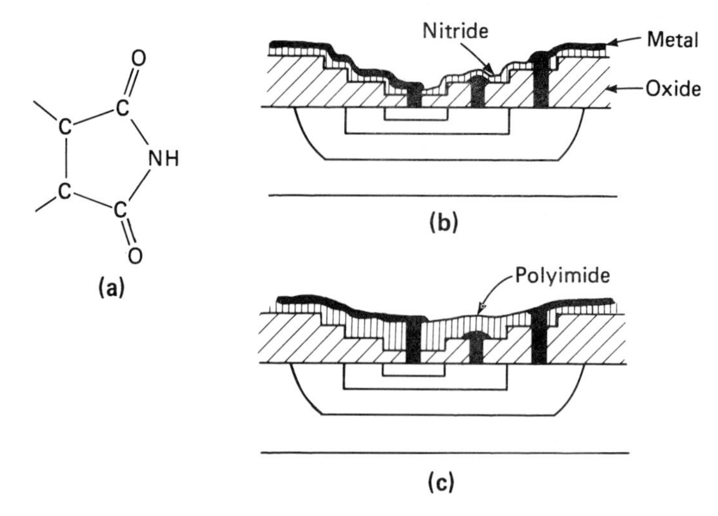

Figure 10–3. Polyimide planarization. (a) The "imide" group. (b) Sketch of the profile with nitride insulation. (c) Replacement of the nitride with polyimide provides a flatter top surface.

ity reaches the point where metallization cannot cover edges, because of their excess steepness. Planarization involves applying an insulator to fill up the valleys so that the top surface becomes flatter. If, as in Figure 10–3(b), a CVD layer of, say, SiO_2 or Si_3N_4 is deposited, the layer tends to have nearly uniform thickness and conforms to the underlying features. A spun-on PI layer, however, tends to provide a more nearly flat surface, with the thickness varying from point to point to compensate for the underlying topography, as in Figure 10–3(c). Rothman states that with the polyimide, flattening can be up to 80% [14]. Furthermore, the layer does not tend to crack as do correspondingly thick glass layers, even those of phosphosilicate glass (PSG). Some sort of PI etch is necessary since windows must be cut in the polyimide to allow contact to the underlying metal regions.

Typical PI wet etches involve solvents in water, and are masked by conventional photoresists. An attractive alternative exists, however, because polyimides may be photosensitized. This means that the layer to be etched can act as its own PR. It can be exposed to UV through a photomask in contact with the layer, and then developed and etched simultaneously. Processing is described by Iscoff [15].

10.8 Metal Etches

Metal etches abound, but care must be used to ensure that a proper selective type is chosen so that other materials on the wafer are not attacked. Typical formulations are given in handbooks. Conventional PR techniques are used for masking, and special photoresists for metal etching, such as Kodak KTFR and KMER, are available.

10.9 Lift-Off Technique

In most photolithographic processes, such as the etching of windows in SiO_2, the PR is applied on top of the existing layer that is to be etched. The PR is exposed and developed before the wafer is immersed in the etchant. For thin metal films an alternative process is available: lift-off. A similar process was described in Section 9.10, but a different masking material is used here.

Lift-off is illustrated in Figure 10–4, where the PR is applied, exposed, and developed *before* the metal deposition. A postbake is not used (Section 11.8). The metal layer is deposited by CVD or PVD over the entire surface, windows included, as in Figure 10–4(b). The wafer then is immersed in PR stripper that can act laterally under the metal located atop the PR. The stripping action removes the PR and overlying metal with it. Metal deposited through the PR windows remains in place as shown in (c). Positive PR is preferred, because it can support finer details and no acids are involved in its use. Note that in lift-off the metal itself is not actually etched.

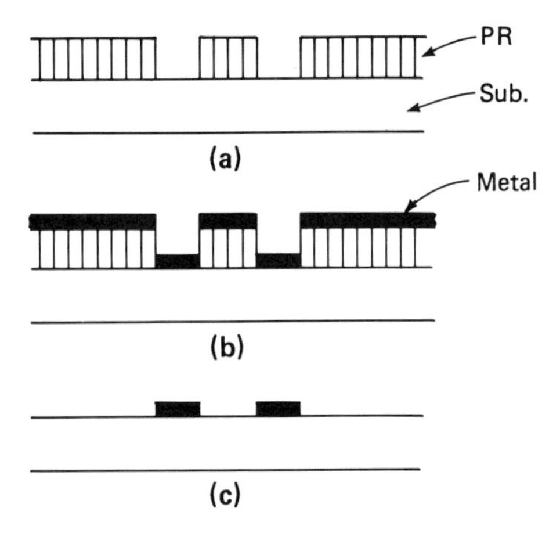

Figure 10–4. The lift-off technique. (a) Patterned photoresist. (b) Metal is deposited over the entire surface. (c) The PR is dissolved and removed leaving the desired metal pattern.

10.10 Wafer Cleaning

Technically, wafer cleaning is not etching, but the process is included here because it may involve etching, and some steps are related chemically to etching.

The processing steps for a given batch of wafers may be spread over some two months or even longer. Most of that time the wafers are in storage between processing steps. Clean wafers can become oxidized and pick up dust and other contaminants within four hours or so of exposure to air. Hence, on return to the production line they should be cleaned just prior to each processing step. In the next sections we consider some of the common wafer cleaning methods.

Four general types of contaminants are found on wafers:

1. Dirt and dust. This type is the result of wafer sawing, polishing, and dicing, all of which produce particles of various sizes. Dirt (actually *detritus*) may fall from plastic storage boxes, while hair and minute skin particles come from personnel. Chemicals also can contribute particulate contamination, but are available with low particle count.
2. Unwanted oxides. Silicon on exposure to air at room temperature can grow a very thin (≤ 100 Å) native oxide in just a few minutes. Also, some oxidizing agents that are used in processing can produce a thin oxide on exposed Si. As discussed in Section 10.3 they may be removed by HF-based etches.
3. Organics. These result from certain processing chemicals, and particularly from fatty acids deposited if skin contacts wafers. Lange gives a summary of contamination sources and discusses contamination by human skin [16].
4. Metallic residues and ions. These come mainly from processing chemicals. For example, analytical reagent grade HNO_3 assays out the heavy metals as Pb at 0.05, and Fe at 0.06 ppm (parts per million). The same grade of acetone is rated at < 0.1 ppm for Cu, Ni, and Fe, with the heavy metals at < 0.5 ppm. Electronic grade chemicals have smaller, but nonzero, amounts of trace metals. If these remain on the wafer surface, they may be introduced into the silicon during high-temperature operations such as oxidation, diffusion, and chemical vapor deposition. Metal ions can lower breakdown voltages and change MOS threshold voltages, hence they should be removed before high-temperature operations are started.

Generally the dirt and dust contaminants can be removed by water solutions and some sort of mechanical scrubbing action; chemistry is unnecessary. The three remaining types of contaminants require chemical reactions with oxidizers, acids, organics, or alkalis.

10.10.1 Scrubbing

Mechanical wafer scrubbing is performed by automatic machines that involve rotating brushes, high-pressure water jets, or sonic agitation [17]. Not only does it remove dust and dirt, scrubbing also improves adhesion of photoresist to the wafer.

In operation the brush bristles are forced down toward the wafer, so they are bent to make a line contact with it. The pressure along that contact line can be very high. If dry operation were permitted, the bristles would damage the wafer if they were harder than it. If the wafer were harder, the bristles would break and contribute more particles to be cleaned away. Moreover, such particles tend to lodge in etched windows or other low spots, making removal more difficult; thus it is essential that water be present so bristle and wafer never make contact. Preferably, a surfactant (detergent) should be added to lower surface tension and to improve wetting of both wafer surface and bristles. The detergent also acts as a degreasing agent. Unfortunately, bacteria thrive on detergent, so frequently NH_4OH is added to kill them. To further aid the wetting process, the bristles are made from *hydrophilic* (water-liking) material such as nylon or propylene. The high pressure between the wafer surface and the water layer produces a rapid cleaning action, a few seconds being enough to bring about thorough cleaning [18]. After being scrubbed and rinsed, the wafers are spun dry.

The problems associated with brushes may be bypassed by replacing them with high-pressure spray jets. The water solution is forced from the jets at 300 to 3000 psi, with the jets located not over half an inch from the wafer surface. As the jets move relative to the wafer, a vigorous scrubbing action is supplied by the liquid itself. An added advantage is that the jet spray can remove very small particles that brush bristles would miss.

Ultraclean water has very high resistivity. As the water flows past the metal jets at high velocity, friction can build up a static electric charge (an example of triboelectricity). Additives to the bath help dissipate this charge; sometimes CO_2 (carbon dioxide) is bubbled through the liquid for this purpose.

Vigorous liquid scrubbing action also can be produced by *cavitation*—the formation of bubbles when shock waves, launched into the liquid from a sonic generator, hit the wafer surface. These bubbles do the actual scrubbing. Generators for this purpose operate around 0.8 MHz. Lower frequencies, in the 20 to 80 kHz range, which are more typical of conventional ultrasonic cleaning, cannot remove the submicron particles encountered in semiconductor processing. A mix of $1NH_4OH : 1H_2O_2 : 5H_2O$ often is added to aid the bubbles in wetting the wafer surface [17, Part 3].

10.10.2 Degreasing

Contaminants of the three other types require chemical reactions to dislodge them from the wafer surface. As mentioned in the last section, greases, such as fingerprints, lightly bound to the surface sometimes can be removed by scrubbing with detergent solutions. Organic solvents can be used, too. Simple dip-and-dry procedures do not work very well, as the solvent rapidly becomes

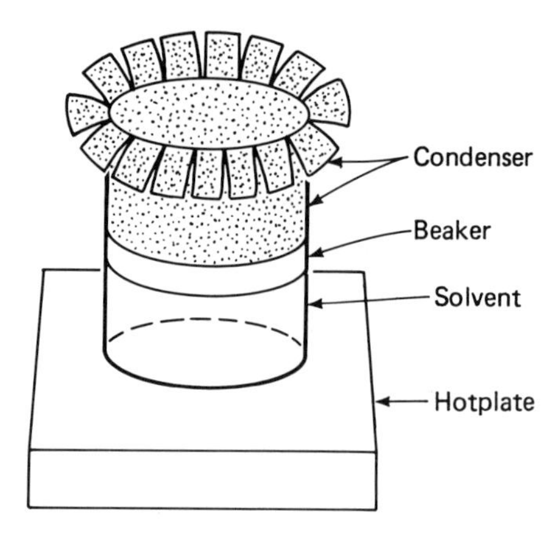

Figure 10–5. A small vapor degreaser. Solvent fumes are condensed in the condenser region.

polluted as successive contaminated wafers are dipped into it. An alternative is *vapor degreasing*.

A simple form of vapor degreaser is shown in Figure 10–5. An organic solvent such as TCE (trichloroethylene)—or preferably TCA (trichloroethane) for safety reasons—is placed in the beaker, whose top is surrounded by a finned, aluminum cooling sleeve. The solvent is boiled, the vapors are cooled enough in the sleeve region to condense, and the liquid returns for boiling again. This is a *reflux process* because the solvent flows back into the sump.

A wafer to be cleaned is held in the solvent vapor, which dissolves the grease and condenses. The liquid falls and carries the dissolved materials back into the sump. The wafer is not dipped into the liquid, so it does not pick up dissolved, carry-over contaminants. Rather, it contacts only clean solvent vapor while the sump collects more and more of the contaminants. Another advantage is that the reflux action is efficient in preventing vapor loss by condensing the vapor before it can rise above the top of the beaker and cooling sleeve. This extends the useful life of a given batch of solvent, with a corresponding cost reduction.

After vapor degreasing with TCE or TCA, the wafer surface must be made water-compatible again, so it is rinsed in acetone, in methyl or isopropyl alcohol, and finally in water before drying. The reasons for this sequence are given in Section 3.14.1.

10.10.3 Organic/Inorganic Removal

General cleaning to remove organics and inorganics usually involves a strong oxidizer such as hydrogen peroxide (H_2O_2). This reagent must be handled carefully because the 30% concentration used for semiconductor work is much stronger than the dilute 3% form available at the local drugstore. The peroxide usually is combined with either an acid or a base. One example is given here. (Standard references, such as those given earlier for etching, should be consulted for more details, as the actual, multistep procedures used for cleaning can be rather long and involved.)

Piranha etch, 7(conc H_2SO_4):3(30%H_2O_2), attacks organics vigorously. The wafer is immersed for 10 min at 125°C, and then rinsed in H_2O_2. Organic traces are removed by a subsequent dip in $NH_4OH:H_2O_2$ or $HCl:H_2O_2$. A stop is available commercially for protecting equipment from piranha.

10.11 Wet versus Dry Etching

Dry etching provides an alternative to the methods already discussed; it involves gases, rather than liquids. Consider the pros and cons of the two etching types. The equipment required for wet etching is less expensive, its cost running only some 10% of that for dry plasma etching [19]. Equipment for both types is easy to automate, and both are able to etch several different materials with only a change in the etchant being used. The cost factor is important, so wet etching is used where it is adequate; even in 1986 over 85% of chips manufactured in the United States were processed by wet etching.

Wet etching also provides better selectivity. In plasma etching, a single etch medium that attacks, say, SiO_2, also attacks Si, but with different etch rates. Dry etching can give smaller line width (i.e., the size of the windows and interconnect lines). Wet etching undercuts more, so it is more isotropic, and minimum line width is limited; control is more difficult, and fewer devices can fit on a chip of a given size. Undercutting also raises interconnect resistance. The finer line structure available with anisotropic plasma etch thus allows much greater packing density on wafers and becomes predominant for widths of 2 μm or less.

Both input and output of the dry etching system are in gas form; this provides some distinct advantages over wet etch systems: Many of the gases used are safe, nontoxic, and easy to handle in vendor-supplied cylinders, and the effluents raise fewer pollution problems. We have seen earlier that dry etching provides better resolution, or smaller line widths, and a clean/etch/strip sequence can be handled more easily in a single reactor system than in wet etching. Temperature effects, which are inevitable since chemistry is involved, are much smaller with dry etch than with wet—where only a 1 to 2% change in etch rate per degree Celsius is the norm. Finally, wet etching presents more critical problems in effluent disposal since it usually involves liquid acids as against gases in the other type (Section 10.2).

We tend to think in terms of high-tech applications of silicon technology, VLSI and the like, which involve sub-micron geometries. Yet the geometries of the vast majority of chips are not this small. Wet etching will be around for a long time. Murray gives an interesting table comparing the two etching types [19].

10.12 Dry Etching

In contrast to wet etching, dry etching is based on the use of one or more gases as the vehicle for the etching species. Gases are neutral; hence ionization, usually in a plasma (Appendix C), is required to release the etching species. A vacuum system is required for initial pumpdown before the gas is introduced, as is the plasma generating equipment. By-products are gaseous and can be removed by pumping.

There are two categories of dry etching, based on the properties of the chosen gas. (1) In *reactive* etching the ionized gas has radicals that react chemically with the material being etched. Gases can be chosen for different chemical reactions, so reactive etching tends to be selective but not necessarily anisotropic. (2) In *nonreactive* etching an inert gas such as argon (Ar) is used. The ionized argon atoms, Ar^+, are accelerated by a large electric field toward the material to be etched and dislodge the atoms there by momentum transfer. This is a physical/mechanical process called *sputtering*. Ionized inert gases support no chemistry, so nonreactive etching tends to be nonselective, but may be anisotropic. There are other intermediate types of dry etching that combine physical and chemical processes, RIE (reactive ion etching) being an example. This form combines the properties of chemical etching and sputtering.

All types of dry etching require the neutral gas to be broken down into radicals or ions. In semiconductor work the conversion usually is accomplished by establishing a plasma that gives a copious supply of the desired species. Since a plasma is used, these types are called *plasma etching*.

Another type of reactive etching, not based on a plasma, utilizes a photochemical process to produce the reactive species by exciting the appropriate gas with a laser beam [20]. Because choice of chemistry may be made, this type tends to be selective, but it is not anisotropic.

10.13 Plasma Etching Reactors

In this section we consider some of the basic forms of plasma etching reactors and the reasons why each is suited to a particular type of etching. Other configurations are described in the literature. Appendix C discusses the formation of plasmas and related information.

10.13.1 Basic Reactor Types

Three typical reactor configurations used for plasma etching are shown in Figure 10–6. All of the configurations are r-f driven and require a vacuum

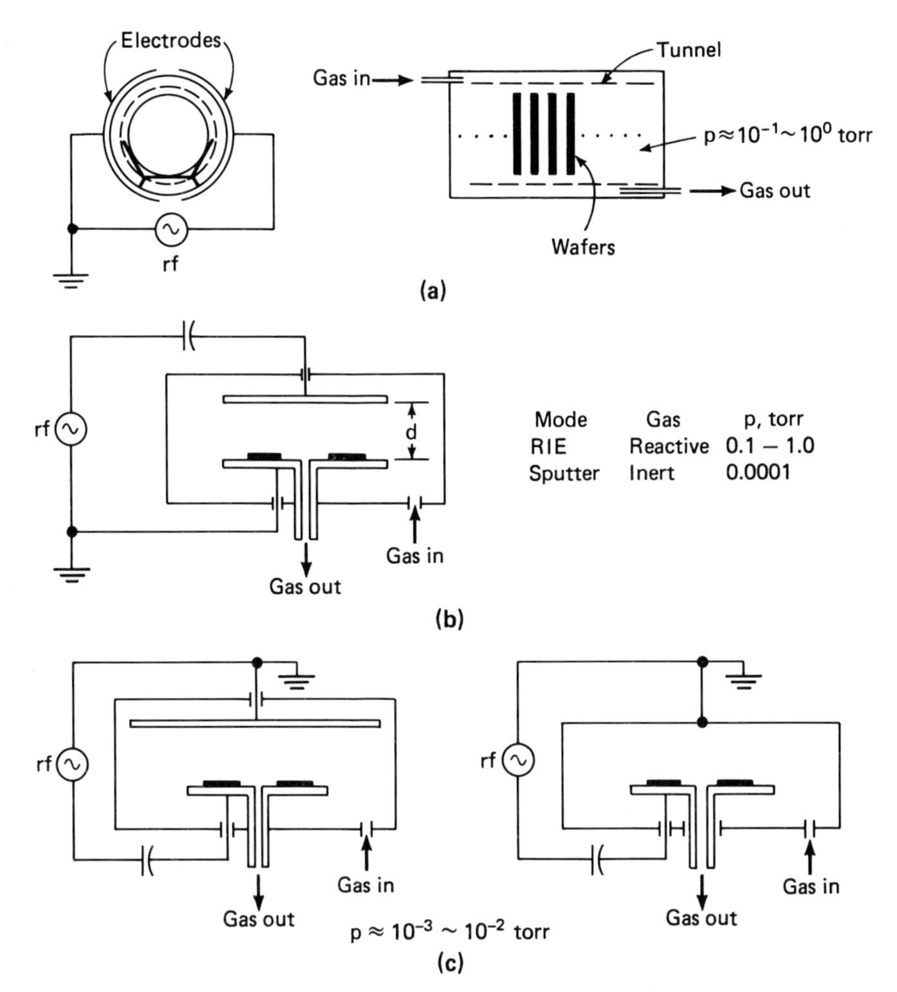

Figure 10–6. Four plasma etching reactors. (a) Barrel. (b) Parallel plate used for PPE and sputter etching. (c) Two forms used for RIE, the wafers are on the smaller, ungrounded electrode.

chamber or vessel, fed by a suitable gas supply and exhausted with a vacuum pump to remove effluent gases and maintain the proper pressure within the chamber. The gas system details, which include means for regulating the gas flow at very low values, are not shown. The barrel reactor shown in (a) was the first type to find commercial use in the semiconductor industry. It is unique in that its electrodes are outside the vacuum vessel and so are not in contact with the discharge gas. The barrel reactor also may be driven by an inductance coil surrounding the vessel.

In some applications a perforated aluminum cylinder, called a *tunnel*, is placed concentrically between the vessel walls and the wafers. It is shown dashed in Figure 10–6(a), and acts to confine the plasma outside the tunnel so that ions do not reach the wafers.

The reasons for having different reactor configurations may be explained in terms of the d-c voltage or *self-bias* that appears across the electrodes. A description of how this voltage is developed is given in Appendix C.3.

10.13.2 Batch Processing

In any commercial process, such as plasma etching or ion implantation, that requires processing in an evacuated chamber, one must reckon with the time required for pumpdown. It is directly related to chamber volume. Shall a large chamber that can process a large batch of wafers for each pumpdown cycle be used, or will a small chamber which processes one wafer at a time but has a short pumpdown time be better? This problem also was encountered in Chapter 8. In the early days of plasma etching the large-batch/large-chamber method was preferred. The trend now is toward a one-wafer-at-a-time approach.

An important factor here has to do with the uniformity of etch. If many wafers are exposed simultaneously to a single plasma between plane parallel electrodes, it is difficult to have every wafer in exactly the same environment, so etch uniformity suffers. The hex reactor of shape similar to that of Figure 9–2(d) or the cylindrical configuration of Section 12.8.2 overcomes this problem very well. Another method uses several wafers on a rotatable carousel, so that only one wafer at a time is exposed to the plasma, even though several are in the chamber.

The present trend toward wafers of larger diameter tips the scales in favor of processing one wafer at a time in the chamber. With this design approach, the electrodes and chamber may be shaped to provide excellent uniformity of etch over the entire wafer surface. The wafers may be shuffled into and out of the small chamber without breaking vacuum by using a lock-and-shuffle mechanism in principle not unlike that shown in Figure 8–7. These systems often are referred to as the *load-lock* type. Another advantage they have is that the chamber and gas are not exposed to air during the wafer shuffle cycle, thereby eliminating the problems associated with chloride gases contacting air (Section 10.15.2).

The principal exceptions to this trend of processing wafers one at a time are the hex and barrel reactors; however, the barrel now is used only for PR ashing (the plasma removal of photoresist).

10.14 Endpoint Detection

A single plasma may etch both mask and underlying material, but not necessarily at the same rate. If the mask etch rate is lower, endpoint detection is essential so the etch can be stopped when the mask windows are completed. Endpoint detection methods monitor some system parameter that changes significantly when the mask etch stops or the underlying etch begins. We briefly consider five typical methods [21].

10.14.1 Optical Spectroscopy

The plasma-generating gas glow discharges are characterized by a color that depends upon the gas being used. For example N_2 gives a pinkish glow, while O_2 has a bluish color. It turns out that atoms and free radicals such as F and $CF_3{}^+$, respectively, also can become excited and subsequently emit radiation of characteristic wavelength. Some typical values are: F*, 7040 Å; CO*, 2977, 4835, and 5198 Å; and N*, 6740 Å. The superscripted asterisk here represents an excited species [22]. One such emission may be monitored with a typical setup shown in Figure 10–7(a), where the monochromator serves as an adjustable filter. A simpler system uses a fixed optical filter to pass the radiation from a single species. The intensity in both cases is then monitored by an appropriate photodiode or photocell.

If an active species such as F* is monitored, its concentration drops during etch, causing a decrease in the monitored radiation. On the other hand, if a by-product such as CO (carbon monoxide) is monitored, its emission intensity will increase during etch. Thus, the endpoint can be detected as a change in radiation intensity at the particular wavelength being monitored.

If the complete spectrum is plotted by a spectrophotometer (recorder in the figure), the method also may be used as a leak detector. For example if a peak occurs at the hydrogen value of 6563 Å, the odds are large that water is getting into the system, unless H_2 is a supplied or by-product gas. The presence of N_2, indicated by a peak at 6740 Å, usually signals an air leak in the system, unless N_2 is being supplied.

10.14.2 Mass Spectroscopy

Another similar means of endpoint detection uses a mass spectrograph, as described in Section 8.4.2, to separate out one species by its mass. The monitored quantity essentially gives a count of the selected species. The spectrometer generally is located between the reactor chamber and the exhaust pump (i.e., at some distance from the discharge so it cannot sense short-lifetime species). Usually an etching by-product is measured [23].

10.14.3 Laser Interferometry/Reflection

The reflecting properties of the wafer surface can change as a result of etching. This change may be sensed for endpoint detection by focusing a laser beam

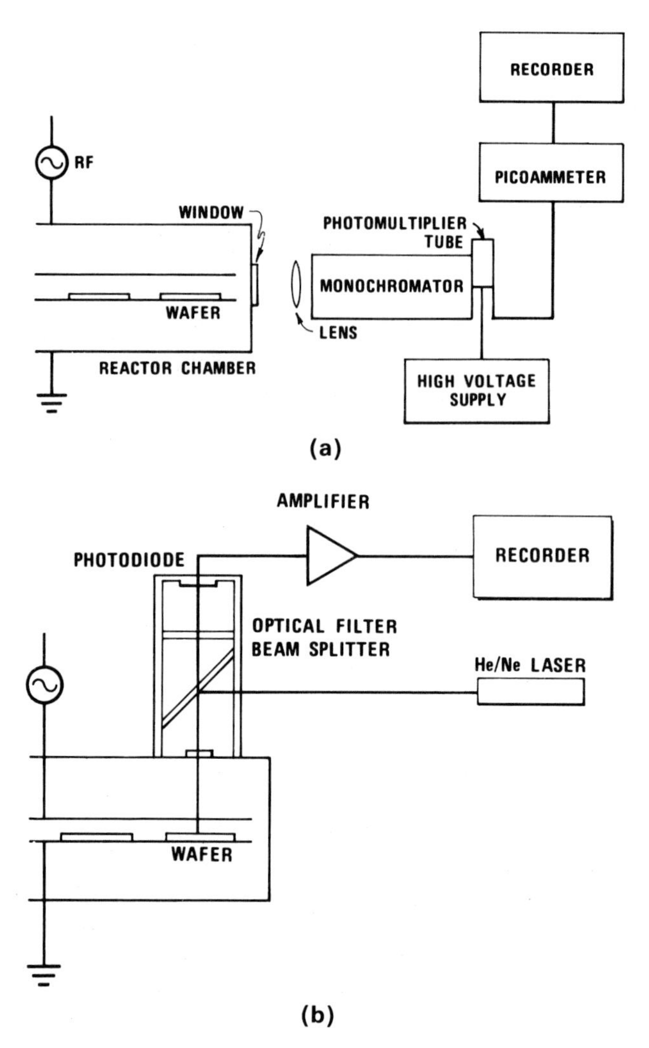

Figure 10–7. Optical endpoint detection apparatus. (a) Spectroscopy. (b) Interferometry. Marcoux and Pang [23]. (Courtesy of *Solid State Technology*, PennWell Publishing Company, Copyright 1981.)

on a spot on the wafer and monitoring the intensity of the reflected beam. If the etched surface is a thin film, interferometry is preferred. A typical setup is shown in Figure 10–7(b). Due to its high reflectance, aluminum works well with this method.

The reflecting method is inadequate for batch processing because only a small spot on one wafer is monitored. Also, if very fine lines are being etched, the change in reflection from the test spot will be very small. This can be circumvented by using masks that provide for a fairly large test spot. Careful orientation of the wafer in the reactor is needed so that the wafer is reached by the laser beam.

10.14.4 Self-Bias Detection

This is a very simple system, which requires only a high impedance d-c voltmeter connected across the electrodes. The magnitude of the self-bias voltage (Section C.3) is influenced by changes in the discharge, so a change in the concentration of any species will cause a variation in the voltage reading. The variation may be small, however, and give less sensitivity than methods that monitor a single species in the discharge.

10.14.5 Pressure Sensing

The total pressure in a discharge is affected by a change in the partial pressure of any of its components; hence, the pressure in the discharge or effluent may be monitored with a vacuum gage to sense the endpoint. This is a simple method to realize, but precautions are necessary. Small variations in the gas flow rate into the reactor will cause small pressure variations, resulting in noise in the gage's electrical output. This may be filtered out with a simple RC filter having a time constant of about 5 s. Chlorine-based etches can contaminate the sensing element in certain gage types, causing false readings. This can be remedied by heating the gage periodically with an air gun.

10.15 Reactive Plasma Etching

In reactive plasma etching the source gas is broken down by the plasma into excited elements, here represented by a superscripted asterisk, and into charged and/or neutral free radicals (groups of atoms). These, for example F^* or CF_3^+, are then available to do the actual chemical etching. Reaction by-products are gases at the operating temperature, and their partial pressures are kept high enough to be removed by the system's vacuum pump. Gases are replenished by *leaking* them (introducing them at low flow rates) into the system to balance the pump action. Two type of reactors are used for reactive plasma etching: the barrel and the parallel plate electrode (PPE) forms, shown in Figure 10–6(b) and (c), respectively.

The choice of gases for use in reactive dry etching has been largely empirical. A partial list of these gases is given in Table 10–1, which shows that CF_4 (Freon-14) etches a wide range of materials. It is one of the fluorocarbons that were developed originally as nontoxic, noncorrosive, and nonflammable refrigerants, and was one of the first gases used for plasma etching. Their

Table 10–1. Partial List of Plasma Etching Gases

	Nonmetals						
	Si	polySi	SiO_2	Si_3N_4	Al_2O_3	GaAs	PR
CF_4	√	√	√	√			
CF_4/O_2	√	√	√	√			
CF_4/Ar	(√)		√	√			
CF_4/H_2			√				
CF_4/N_2		√					
CCl_2F_2	√		√			√	
CCl_4	√	√	(√)				
CCl_4/Ar					√		
C_2F_6		√					
C_3F_8	(√)		√	√			
C_4F_8	(√)		√				
HF			√				
O_2						√	√
BCl_3					√	√	
$CClF_3$	√						
CF_3Br						√	
CHF_3	(√)		√				
	Metals						
	Mo	W	Au	Cr	Al	Ti	Ta
CF_4	√	√				√	√
CF_4/O_2	√	√					
$C_2Cl_2F_4$			√				
$C_2Cl_3F_3$	√						
Cl_2				√			
Cl_2/O_2				√			
CCl_4	√			√	√		
CCl_4/Ar				√	√		
BCl_3					√		
CHF_3			√				

Note: (√) indicates low etch rate. Compiled from several sources.

properties enhance the fluorocarbons' value as etchant source gases, but, as we shall see, they sometimes lack selectivity. Unfortunately, recent experience indicates that the related chlorofluorocarbons (CFCs) affect the earth's ozone layer.

10.15.1 Fluorocarbons

Consider some basic reactions resulting from a CF_4 plasma. A copious supply of free electrons (e^-) is present in a plasma, and some of them enter into the fundamental reaction that gives free radicals, namely

$$CF_4\uparrow + e^- \rightarrow F^* + CF_3^+ + 2\,e^- \tag{10.15-1}$$

It has been determined that the F* species has a long lifetime (τ) on the order of 0.1 to 1 s, depending upon the partial pressure of CF_4. This means that atomic F* can exist outside the plasma proper. In contrast, the CF_3^+ radical, which is a strong reducing agent, has a relatively short τ of roughly 10 μs, and so cannot exist outside the plasma.

These facts have strong implications for the barrel etcher. The charged species CF_3^+ is prevented from reaching the wafers by lifetime and tunnel-shielding effects; hence etching by CF_4 in the barrel reactor must be by the excited atomic form F*.

Consider some reactions of the two radicals F* and CF_3^+ with various materials to be etched. It is atomic fluorine that provides the etching of silicon:

$$Si + 4\,F^* \rightarrow SiF_4\uparrow \tag{10.15-2}$$

The resulting compound SiF_4 is a gas at etch temperatures and pressures, and so may be removed by the system pump. From τ considerations the silicon need not be located directly in the plasma for etching to take place, so a barrel etcher can be used.

Silicon dioxide also is etched by a CF_4 plasma. There is disagreement on the etching action, but the consensus holds that the active radical is CF_3^+. The overall reaction may be written in the somewhat ambiguous form

$$SiO_2\downarrow + CF_3^+\uparrow \rightarrow SiF_m\uparrow + CO/CO_2\uparrow \tag{10.15-3}$$

where the CF_3^+ comes from the basic plasma reaction of Eq. (10.15-1).

In typical processing an SiO_2 mask on Si is frequently encountered. Since CF_4 etches both materials, the problem of relative etch rates arises. If mask windows are to be etched in the oxide, the etch will not stop automatically on reaching the silicon, as it did with liquid HF; hence it is desirable to have the SiO_2:Si etch rate ratio, R_{OS}, as large as possible. On the other hand, if silicon is to be etched through oxide mask windows, we desire R_{OS} to be as small as possible (or $R_{SO} = 1/R_{OS}$ as large as possible). What can be done to satisfy these contradictory conditions?

If the barrel etcher is used, the wafers are shielded from the charged CF_3^+ species, but they can be reached by the long-lifetime, excited F* atoms; hence the silicon etches faster, with a typical R_{OS} value of 1/10.

If a PPE (parallel-plane-electrode) reactor is used, the wafers are in the plasma, so both species can etch; experience shows an R_{OS} value of roughly one, so for etching SiO_2 the etch ratio must be enhanced in some manner. One approach adds a scavenger gas that consumes the F* species before it can

Table 10–2. Etching Gas Nomenclature

BCl_3	boron trichloride	
$CBrF_3$	bromotrifluoromethane	
$CClF_3$	chlorotrifluoromethane	Freon-13
CCl_2F_2	dichlorodifluoromethane	Freon-12
CCl_2O	carbonyl chloride, phosgene†	
CCl_4	carbon tetrachloride	
CF_4	tetrafluoromethane	Freon-14
CHF_3	fluoroform	
CH_3F	fluoromethane, methyl fluoride	
$C_2Cl_2F_4$	dichlorotetrafluoroethane	Freon-114
$C_2Cl_3F_3$	trichlorotrifluoroethane	Freon-113
C_2F_6	perfluoroethane	
C_3F_8	perfluoropropane	
C_4F_8	perfluorocyclobutane	
HF	hydrogen fluoride	
NF_3	nitrogen trifluoride	
SF_6	sulfur hexafluoride	
$SiCl_4$	silicon tetrachloride	
SiF_4	silicon tetrafluoride	

†extremely dangerous

attack the Si, leaving the CF_3^+ free to etch the oxide. Hydrogen is one such gas since it reacts with F* to form HF in gas phase. Principal etching of the oxide will be by CF_3^+, however.[1]

Another approach is to change the source gas. The C/F ratio in the gas affects the etch rate ratio. Singer cites the following data: for CF_4, $R_{OS} = 1$; C_2F_6, 3; C_3F_8, 5; CHF_3, 10. This is an example of how an empirical choice of gas can solve an etching problem. Singer also discusses the etching chemistry of these gases and the variation of etch rates with process parameters [25].

If the scavenging gases such as H_2 or CHF_3 deplete the F* too much, a condition known as *polymerization* can occur. This means that molecules of an unsaturated reaction compound at the wafer surface can join to form more complex molecules. These may form solid residues that can retard or even stop the etching action. Singer gives the curves of Figure 10–8, which show the parameter regions where this effect comes into play [25].

On the other hand, if Si is to be etched through an oxide mask in the PPE setup, the etch rate for Si should be significantly higher than that for SiO_2. A

[1] Anhydrous HF has been used to etch SiO_2 selectively in the DryOx process. Even though this does not involve a plasma, it uses HF at temperatures and pressures common to plasma etching. Bersin and Reichelderfer state that HF in the 150 to 190°C temperature range and at pressures from 0.1 to 30 torr does not etch Si [24]. Etching under a mask also is possible. DryOx is a trademark of the International Plasma Corporation.

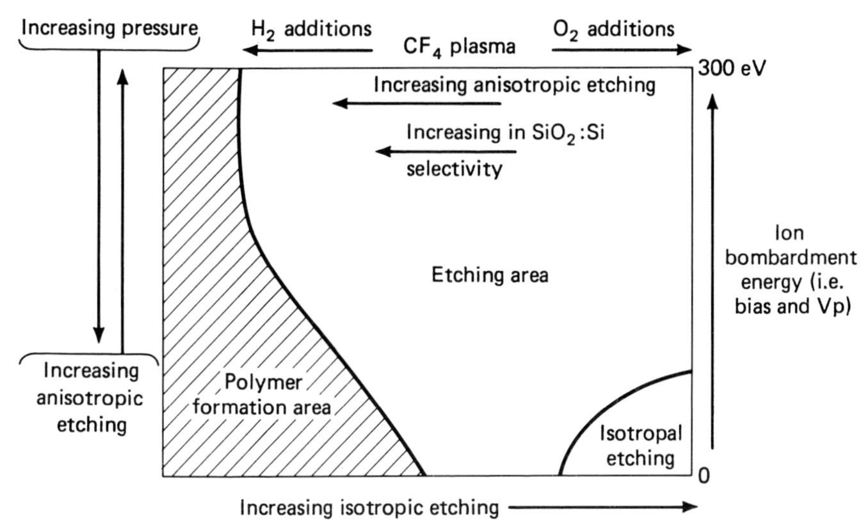

Figure 10–8. Effect of process variables on polymer formation and in producing etching anisotropy. Isotropal and isotropic are synonyms. Singer [25]. (Courtesy of *Semiconductor International Magazine.*)

mixture of CF_4 with about 8% oxygen added is used for this, since it gives an etch rate ratio of (111)Si to SiO_2 of about 17, and of polysilicon to the oxide of about 25 [25, 26]. One of the mechanisms postulated for this effect is that free electrons in the plasma collide with molecular O_2 to form atomic O*. This is not a chemical reaction, but then intermediate chemical steps form fluorinated radicals such as COF*. These react at the surface to give an overall result of

$$CF_4\uparrow + O_2\uparrow \rightarrow 4\,F^*\uparrow + CO_2\uparrow \qquad (10.15\text{-}4)$$

or possibly

$$2\,CF_4\uparrow + 2\,O^*\uparrow \rightarrow 8\,F^*\uparrow + 2\,CO\uparrow \qquad (10.15\text{-}5)$$

In either event, the production of the silicon etching species F* is enhanced, resulting in the higher Si etch rate. No such increase is caused in the production of $CF_3{}^+$ that etches the oxide. In all of the foregoing reactions it is easy to identify an active or by-product species suitable for endpoint detection as discussed in Section 10.14.

Silicon nitride (Si_3N_4) also may be etched in a CF_4 based plasma with the F* species acting as the actual etchant. Following the reaction of Eq. (10.15-1) the result is:

$$Si_3N_4\downarrow + 12\ F^*\uparrow \rightarrow 3\ SiF_4\uparrow + 2\ N_2\uparrow \qquad (10.15\text{-}6)$$

There is a problem in that F*, the active agent for Si_3N_4, also etches Si, and at a faster rate. Typically, the etch rate in a barrel reactor will be less than that for Si but about four times that of SiO_2 [27]. Hence, an oxide may be used as an etching mask for the nitride. Since the same species etches both Si and its nitride, it follows that steps taken to raise (or lower) the etch rate of Si relative to SiO_2 will have the same effect on the nitride. The substitution of SiF_4 improves the nitride/silicon selectivity, however. With 4% oxygen added, the etch rates of Si and SiO_2 are reduced wrt Si_3N_4, that of Si by a factor of about 6. This combination also will etch polyimide [28].

Consider metal etching with CF_4. Both tungsten (W) and molybdenum (Mo) react with atomic F* to give hexafluorides that are volatile at etching temperatures and pressures.

$$W\downarrow + 6\,F^*\uparrow \rightarrow WF_6\uparrow \qquad (10.15\text{-}7)$$

$$Mo\downarrow + 6\,F^*\uparrow \rightarrow MoF_6\uparrow \qquad (10.15\text{-}8)$$

Reference to Table 10–1 shows that Al is not etched by fluorocarbon types of chemistry, a fortunate fact because it provides a form of selectivity. The table also shows that chloride chemistry does work with Al.

10.15.2 Chlorides

Aluminum is a commonly used metal in silicon technology and requires a suitable selective etch. Chlorides such as CCl_4, BCl_3, and $SiCl_4$ are used rather than fluorocarbons because of selectivity. These chlorides serve a double function.

When aluminum is exposed to air a thin layer, of about 30 Å, of Al_2O_3 (aluminum oxide) grows very quickly, even at room temperature [29]. This *native oxide* is resistant to a large range of chemicals and must be removed before the aluminum itself can be etched. Even though CF_4 will not attack Al, it is used sometimes for a 2-min oxide removal before the Al etch is started. But chlorine (Cl_2) alone cannot remove the oxide, e.g., this reaction is not possible:

$$2\ Al_2O_3\downarrow + 6\ Cl_2\uparrow \rightarrow 4\ AlCl_3\uparrow + 3\ O_2\uparrow \qquad (10.15\text{-}9)$$

because the free energy of reaction is $\Delta G = +105.5$ kcal/mole.

Fortunately most chlorides can etch both the native oxide and Al. For example, for etching Al_2O_3 with CCl_4:

$$2\ Al_2O_3\downarrow + 3\ CCl_4\uparrow \rightarrow 4\ AlCl_3\uparrow + 3\ CO_2\uparrow \qquad (10.15\text{-}10)$$

where ΔG is -16.8 kcal/mole and the reaction can proceed [30]. The behavior of BCl_3 is similar to that of CCl_4.

The gases CCl_4 and BCl_3 are used most often for etching aluminum. Their etching chemistries are quite similar, so the choice between them is based largely on their other properties. Both are considered to be analogous to CF_4; when they collide with free electrons in the plasma these radicals will result:

$$CCl_4\uparrow + e^- \rightarrow Cl^*\uparrow + CCl_3{}^+\uparrow + 2\,e^- \qquad (10.15\text{-}11)$$

and

$$BCl_3\uparrow + e^- \rightarrow Cl^*\uparrow + BCl_2{}^+\uparrow + 2\,e^- \qquad (10.15.12)$$

In either case it is the atomic Cl* that provides the etching action. This is a comparatively short-lived species and so cannot exist outside the plasma proper. For this reason, chloride-based etching of aluminum is not practicable in a barrel reactor with tunnel; a parallel plate (PPE) configuration, where the wafer is in contact with the discharge, is necessary. The 2550 Å spectral line of Cl can be used for endpoint detection.

The actual etching reaction is

$$Al\downarrow + 3\,Cl^*\uparrow \rightarrow AlCl_3\uparrow \qquad (10.15\text{-}13)$$

The $AlCl_3$ (aluminum chloride) by-product is a solid at room temperature and sublimes at roughly 180°C; so substrate heating is used to ensure that the chloride volatilizes.

The reactions of both Eqs. (10.15-11) and (10.15-12) can run in the reverse direction, causing Cl* to recombine with $CCl_3{}^+$ or $BCl_2{}^+$, as the case may be, to form the original source chloride. These are three-body reactions and require a surface—of the wafer, of the fixtures, or of the walls and base plate of the reactor. Since the reverse reactions will reduce the quantity of Cl* available for etching, they lower the etch rate. Therefore, good scavenging of $Cl_3{}^+$ or $BCl_2{}^+$ by the system pump is important.

10.15.3 PR Ashing

Dry process removal of photoresists, *PR ashing*, is often carried out in the barrel etcher. Oxygen is used to ash conventional organic resists and plasma excitation results in atomic oxygen (O*), which ideally oxidizes the PR into gases such as CO, CO_2, and H_2O that are removed by the system's vacuum pump. Complete removal is essential so that no solid residue remains on the wafers. An argon purge is often used after the stripping process. Typically, an ashing cycle for a 1-μm-thick resist layer may last 15 to 40 min, with the actual time depending upon such factors as the *load* (number of wafers in the batch) and the applied power [31].

An oxidizing agent weaker than O_2 is desired for stripping resists on sensitive films (e.g., the chromium on some photomasks) to avoid damaging them. Wet air (air bubbled through water) may be used for this.

The ashing time may be reduced by using other gases or by preheating the wafers to 200 to 300°C before insertion into the etcher barrel. This heat treatment can cut the ashing time in half. Also, the CF_4/O_2 mixture can speed the process, although care must be used to ensure that the F* species will not damage exposed Si or Si_3N_4. The CF_4/O_2 mixture also will strip polyimide-based resists.

Special care is required if the photoresist has been used to mask aluminum during etching. Conventional negative resists take up free chlorine-bearing radicals. If the resist is exposed to moist air after the etch, HCl will form and possibly etch the aluminum further. Also the resist becomes more resistant to stripping. To avoid this problem, the wafers should be immersed in distilled water for a minimum of 2 min after etching. The strip process follows after drying. Conventional positive resists, due to different chemistry, do not exhibit this effect, but they should be stripped immediately after the etch cycle.

Irving cites these advantages of dry over wet PR stripping [31]:

1. It is a cooler process.
2. Stripping time is independent of resist history, e.g., pre- and postbake times.
3. It requires fewer steps.
4. Less airborne contamination is present since the process is isolated from the atmosphere in a partial vacuum.

10.16 Anisotropic Etching

Plasma etching with reactive gases at relatively high pressures, say $> 10^{-1}$ torr, usually is isotropic: the etch proceeds horizontally as well as vertically relative to the wafer surface. For small geometries, with line widths of 2 μm or less, anisotropic etching normal to the wafer surface is desired. This may be achieved with two forms of plasma etching: *nonreactive* ion or sputter etching with gas in the 10^{-4} torr pressure range, and *reactive* ion etching (RIE) with reactive gas in the 10^{-2} to 10^{-1} torr range.

10.17 Nonreactive Ion Etching: Sputtering

In ion (or sputter) etching a nonreactive gas such as argon (Ar) is ionized by an r-f plasma in a PPE configuration as shown in Figure 10–6(b). A d-c self-bias voltage is present (Section C.3). The Ar^+ ions are accelerated toward the wafers, which are on, and in contact with, the grounded negative electrode. If the d-c self-bias is large enough, the argon ions will hit the wafer with sufficient momentum to dislodge material, giving sputter etching.

The gas pressure can determine if the etching will be anisotropic or not. The mean free path L_m (the average distance the ions travel between collisions) is given by

$$L_m = \frac{5 \times 10^{-3}}{p^*}, \quad \text{cm} \tag{10.17-1}$$

where p^* is the gas pressure in torr (= 1 mm Hg).

Typically d, the interelectrode spacing, will run a few, say 5, cm, and d', the spacing from the cathode to the near edge of the plasma, will be a small fraction of this, say $d' \approx 1$ cm.

Say $p^* = 1$ mtorr (= 1 micron), so that $L_m = 5$ cm. This is large enough wrt d' that the positive ions (i^+) probably will not collide with other ions in moving from the plasma to the negative electrode and so will travel in essentially straight lines normal to the cathode. If the electric field is large enough to cause sputtering, the etch will tend to be anisotropic.

On the other hand, say $L_m < d'$. For example, if $p^* = 20$ mtorr, L_m will be 0.25 cm. Then an ion will suffer several direction-changing collisions in moving to the cathode, so anisotropic etching cannot be expected. The use of r-f, as opposed to d-c, excitation favors operation at lower pressures, so anisotropic etching can be achieved.

Since no chemical action is involved, all materials exposed to the ions are etched: oxide, nitride, aluminum, and silicon, but not at the same rate. Generally, the selectivity among the materials is poor because it depends on their relative sputtering rates alone, and not on chemical reaction. For this reason sputter etching is seldom used in IC processing.

When ions hit the wafer under usual sputtering conditions, they give up some kinetic energy to heat. To prevent an excessive rise in temperature, the wafer-holding electrode is usually water cooled. This also minimizes the formation of an arc discharge (Appendix C.1).

10.18 Reactive Ion Etching (RIE) [32]

In RIE reactive gas in the pressure range of 10^{-2} to 10^{-1} torr is used with the reactor configurations shown in Figure 10–6(c).[2] In brief, the advantages of PPE and sputter etching are combined to give a combination of reactive chemistry and mild sputtering by positive ions. For a midrange pressure of

[2] The effect of comparative electrode size on voltage magnitudes is discussed in Appendix C.3.

5×10^{-2} torr the mean free path is $L_m = 10^{-1}$ cm. At these pressures the distance between the plasma sheath and the wafers is even smaller, so ions arrive primarily in the direction parallel to the $\mathscr{E}$ field. Also, the vertical features on the wafer surface are much smaller than the interelectrode spacing, being at the most a few micrometers, so they do not disturb the direction of the $\mathscr{E}$ field, which remains normal to the wafer surface. As a result, the sputtering action is anisotropic.

The reactive chemistry and ion sputtering are not necessarily by the same species; a case in point is Si. Recall from Section 10.15.1 that F* is the species that etches silicon in a CF_4 plasma. Since this is a charge-neutral species, it cannot gain momentum due to the $\mathscr{E}$ field and cannot sputter etch; some other mechanism must be involved.

While we tend to think of ion enhancement of etching by direct sputtering, some alternate mechanisms are proposed, all of which are based on the formation of *volatile* reaction products. These mechanisms are (1) polymer removal, (2) creation of dangling bonds at the surface that accelerate the formation of free-radical reactions, and (3) localized surface heating by ion impact that can accelerate chemical surface reactions and the volatilization of compounds.

Consider some facts related to these mechanisms. The creation of dangling bonds at the wafer surface facilitates reaction with free radicals there to form oxides, fluorides, or chlorides—depending upon the source gas composition. The reaction rates of neutral species at a surface (e.g., F* or CF_2) have an exponential temperature dependence of the form $\exp(-b/T)$; therefore, wafer temperature can affect rates at the wafer surface.

Production rates of species in the plasma are relatively independent of the wafer surface temperature; rather, they depend on electron impacts with gas molecules, and so on the free electron concentration in the plasma. If the power delivered to the plasma is increased, free electron concentration increases, so positive ion concentration increases, too. This tends to increase any ion-enhanced process.

The formation of polymers on wafer surfaces depends on the wafer material, its surface condition (whether damaged or having dangling bonds) and the source gas composition. For example most polymers are carbon based. Carbon is available in many gases, such as CF_4, CHF_3, or CCL_4, but RIE gases, such as SF_6 and NF_3, have no carbon. Yet any organic present, such as PR, can furnish carbon, and C-polymer films may form. Subsequently these may be converted to volatile compounds such as CO, CO_2, or H_2O. Si-based polymers usually do not convert to volatiles, however; therefore, they may be used as an etch stop.

Polymer formation is enhanced by a large H : F ratio in the source gases. Thus, the polymer formation by three commonly used gases obeys the hierarchy $CF_4 < CHF_3 < CH_3F$. Where polymerization is not wanted, say for sidewall protection of Si, a gas should be chosen, if possible, that forms heavier ions, such as CCL_x^+ rather than the lighter BCL_y^+.

Polymer formation on the RIE reactor electrodes also can affect wafer etch rates and their constancy. Electrode polymers can act as sources of contaminating particles when sputtered. Tailoring of the gas may decrease polymer formation; Egitto et al. show that the addition of CO_2 or O_2 to CHF_3 has this effect [33]. Apparently these added gases scavenge excess CF_2 and CF_3 radicals that are active in polymer formation.

We shall consider some typical etching applications that are based on the foregoing mechanisms. Remember that the choice of reactive and other source gases is largely empirical. In these examples relative, rather than absolute, etch rates are given because the latter depend on too many parameters—such as electrode geometry and size, and the wafer batch size or *load*.

10.18.1 Silicon Dioxide

The etching of silicon dioxide provides a good example of the combined effects of chemical and sputter actions by RIE in CF_4 plasmas. The oxide often is on or near silicon, so relative etch rates are of concern. We saw earlier that R_{OS}, the oxide-to-silicon etch rate ratio, is relatively small and that the etch is isotropic in PPE, being entirely chemical in action. Under RIE conditions, however, R_{OS} increases because the CF_3^+ radical selectively attacks the SiO_2 by both chemical and sputtering actions.

If F*, the species that chemically etches the silicon, were scavenged, R_{OS} would increase because of the lowered silicon etch rate. This effect has been demonstrated by experiments with the gas combination CF_4/H_2 [34]. As the percentage of H_2 is increased, the etch rates of both Si and SiO_2 decrease but that of Si does so much faster. The R_{OS} ratio increases up to a maximum of 35 : 1 at 40% H_2. At higher percentages a C-based polymer film, which does not sputter well, forms on the wafers and causes the chemical etch by CF_3^+ to decrease. Note that the Si rate decreases because of the scavenging effect of H_2 on F*.

Ephrath also reports that AZ1350B, a Shipley optical, positive photoresist, and PMMA (polymethylmethacrylate), a positive resist used with electron and ion beams, also etch more slowly as the H_2% increases and at a lower rate that SiO_2, so the selectivity of silicon dioxide etching wrt these two resists rises with H_2 percentage. Silicon nitride's etch behavior in the CF_4/H_2 gas mixtures parallels that of SiO_2 [34].

Results similar to those with CF_4 are obtained if CHF_3 is used in place of CF_4/H_2 since hydrogen is available to scavenge the F* species. Polymer formation may occur also, in fact CHF_3 is well known as a polymer precursor. More information on the behavior of CHF_3 when mixed with other gases is given later.

Consider an RIE mechanism other than the combined chemical and direct sputter etching of SiO_2. This mechanism assumes that while CF_4 does not adsorb to oxide and silicon surfaces readily, the plasma-produced species F* and CF_x do [35]. One of these latter, CF_2 (difluorocarbene), forms a polymer on Si that does not sputter well, so the chemical etching by F* on silicon is limited. When CF_2 is adsorbed to SiO_2, however, it releases F, which reacts chemically with silicon *in the SiO_2*, and carbon, which reacts chemically with oxygen *in the SiO_2* to form volatile compounds such as CO and CO_2. The mechanism proposed here differs from the earlier one, but the overall result is the same: $R_{OS} > 1$ in CF_4. The sputter action aids in the formation of volatile compounds on the oxide, but is not effective on the polymer formed on the silicon.

This model is consistent with the result that carbon in fluorocarbon gases has less effect on the etching of SiO_2 than on Si. The effects on SiO_2 depend on the radical CF_2 that forms volatile products on the oxide, but C forms a hard-to-sputter polymer film on the Si; hence a higher C : F ratio in the source gas raises R_{OS} as in C_2F_6 and C_3F_8 over that for CF_4 [36].

The foregoing example illustrates that halocarbon plasma species can react with wafer materials to form polymers that inhibit chemical reactions. If some of these polymers are sputtered off by the accelerated ions, etching will proceed on the sputtered surfaces, but not on surfaces where the film remains [37].

Different polymer films may exhibit different effective etch rates. For example, if Si is exposed to oxgyen during RIE, its etch rate r_S drops. On the other hand, with chlorine substituted for O_2, r_S rises. This is attributed to differences in the polymer films on the Si [35].

The effect of different etch rates may be clarified by the example shown in Figure 10–9. A cross section of silicon masked with oxide is shown at (a). The $CF_3{}^+$ ion direction is vertical as shown. The F* species attacks the Si at "xxx" and forms a polymer that is removed by $CF_3{}^+$ sputtering; thus F* continues to etch the Si with the polymer forming on both the sidewalls and bottom of the well as shown at (b). Again, the ions sputter away the polymer at the bottom. Since they arrive vertically, however, they cannot reach and sputter the sidewall polymer film. This action continues with the final result shown at (c). There is no *undercut* and so no lateral etching of Si under the mask; therefore, the Si etching is anisotropic.

Results with CHF_3 can give some insight into the models. Alone, CHF_3 permits RIE of silicon dioxide but at a lower rate than the fluorocarbons. Even though F* is scavenged by the hydrogen, plasma products, most likely CF_2, cause more polymerization on the oxide; sputter rate there decreases, so R_{OS} decreases. Chang has reported results for CHF_3 mixed with O_2 and CO_2 [37]. For 9 CHF_3/1 O_2, O_2 apparently lowers the CF_2 concentration and raises the F* concentration. Direct chemical etch of Si increases, with less chance of silicon polymer growth, so R_{OS} drops compared to the value with CHF_3 alone.

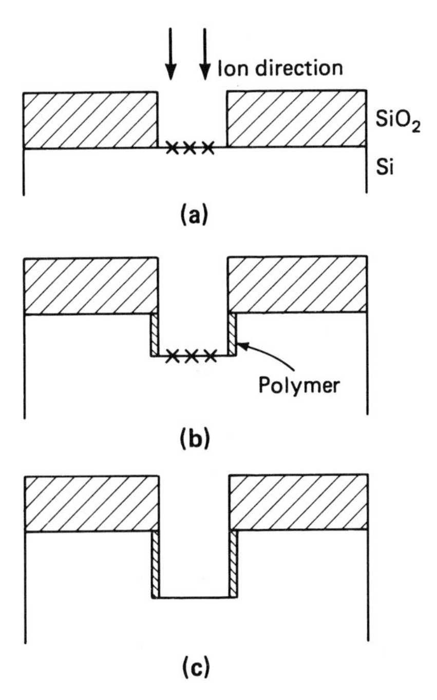

Figure 10–9. Ion direction and polymer formation produce etching anisotropy.

With 9 CHF_3/1 CO_2 the CO_2 aids formation of volatile compounds at the SiO_2 and less polymer film develops there. Chang reports good anisotropy with $R_{OS} > 20:1$.

10.18.2 Silicon

The etching of silicon is of great importance, especially of polysilicon, which in modern MOS ICs is used for gates and some levels of interconnects. In both applications the polySi may be highly doped to lower its sheet resistance. Doping levels affect etch rates, a phenomenon related to the change in Fermi level as doping concentration changes [38].

We already have some data in the preceding section about fluorocarbon etching of Si in the RIE setup. Letting r_S denote the silicon etch rate, we can summarize as follows:

1. In CF_4, r_S is low.
2. Species that scavenge F* lower r_S. Such species are obtained by adding H_2 to the CF_4 or by substituting CHF_3.
3. The addition of O_2 to either CF_4 or CHF_3 enhances the production of F* causing r_S to rise.

4. Polymer formation on Si is enhanced by raising the H : F ratio. This is consistent with (2) and lowers r_S.

We now consider some alternative chemistries for Si etching. Recall that the silicon-to-oxide etch rate ratio is $R_{SO} = 1/R_{OS}$. Sulfur hexafluoride (SF_6) may replace CF_4. Under a photoresist mask r_S is high but undercutting is present so the etch is isotropic. The addition of up to 30% of $CFCl_3$ (fluorotrichloromethane) to the SF_6 makes the etch anisotropic, apparently the effect of the chlorine addition.

If SF_6 is used to etch Si under an SiO_2 mask, r_S remains high but etching is isotropic, a condition that can be remedied by adding one of the chlorides. No carbon is present to combine with the oxygen *in the* SiO_2 to form volatiles. Also Cl_2 does not react appreciably with SiO_2 so the oxide etch rate is low, and R_{SO} is high.

10.18.3 Aluminum

Aluminum and its copper and silicon alloys also may be etched anisotropically under RIE conditions [39, 40].[3] As we have seen, chlorides form the basis for the etch process. Since highly corrosive by-products are formed, they frequently are frozen out from the system exhaust by being passed through a cold trap chilled with liquid nitrogen, or they are converted chemically.

A number of chlorine-bearing gases are used—e.g., Cl_2, BCl_3, CCl_4, and other fluorocarbons—but any one used alone gives poor results, so mixtures, including other additives, are used commonly. Consider the BCl_3/Cl_2 combination. The first gas is needed to remove the native Al_2O_3 oxide. It also acts as a getter for the O_2 released in reduction of the oxide, and so helps to prevent subsequent oxidation, but its aluminum etch rate is low. Chlorine in the 25 to 50% range is added as a remedy, but too much will make the etch isotropic.

While Cl* is the principal active etchant, anisotropy of etch is aided by BCl_x radicals that are formed in the plasma and react at the wafer surface, and by surface bombardment by high-energy positive ions. Since carbon is not present, polymer formation is difficult. Anisotropy also is aided by the addition

[3] Pure aluminum generally is not used for making contact with silicon. The Al/Si phase diagram of Figure 4–7 shows a nonzero solubility for Si in Al at typical processing temperatures; hence Al *leaches* the Si, causing voids in the semiconductor. These may be filled by the aluminum, an effect that deteriorates device performance. If the Al penetration is shallow, the electrical contact is affected and at worst may break. If the penetration extends to an underlying p–n junction, a condition known as *spiking*, the junction may be shorted. With the incorporation of 2 to 4% Si in the metal film, leaching is prevented.

Voltage across an Al/Si contact increases the movement of Si and Al by *electromigration*. The addition of a few per cent of Cu to the alloy can decrease electromigration and minimize spiking. A typical alloy composition is 94Al:4Cu:2Si, but its sheet resistance is higher than that of pure aluminum.

of SiF_4 and reduction of the self-bias voltage. The fluoride also cleans up undesirable residues in this manner. Aluminum chlorides and oxychlorides (remember O_2 is released by Al_2O_3 etching) are usual by-products. These are hygroscopic and form acidic compounds with water vapor on exposure to air, compounds that can severely corrode the vacuum system and fixturing. Usual practice follows the etch cycle with a gas change to a fluorine-bearing type, causing the chlorides to become fluorides which are not hygroscopic. Wafer-shufflers that do not require exposure of the vacuum system to air after each etch cycle further reduce the water vapor problem.

Copper and silicon in the aluminum alloys can lower the aluminum etch rate r_A. Silicon tetrafluoride (SiF_4) or small amounts of O_2 or He can reduce this problem. When CVD SiO_2 is used to mask the Al, raising plasma power lowers the Al/Si-to-SiO_2 etch rate ratio because increased ion bombardment of the oxide raises r_O.

A chlorocarbon, CCl_4, in place of Cl_2 with the BCl_3, gives another chemistry. The carbon permits the formation of CCl_3 radicals that attack the Al_2O_3, and aids in forming sidewall-protecting polymers. The organic residues on the wafer after etch can be oxidized by adding a small amount of oxygen with the chlorides, or by a separate O_2 ashing step.

10.19 Radiation Damage

In both sputter and reactive ion etching high-energy ions strike the wafers, so we expect forms of radiation damage that are common to ion implantation. Ion bombardment in RIE is combined with chemical reaction effects. As a general rule, these effects may be reduced by lowering the self-bias voltage to lower incident ion energy, but this lowers the ion directionality and so affects the degree of anisotropy. Other reactor configurations that involve the use of magnetron action can also help [41], (Section 12.11). Consider some of these effects beginning with ones that are simpler to correct.

Incident ions also can sputter materials from the walls and fixtures of the system, especially from the cathode that supports the wafers. This sputtered material can fall back onto the wafers, causing contamination. The effect is minimized by (1) reducing self-bias to lower ion momentum (as noted above this may be counterproductive for etching), (2) using metals for the fixturing that have low sputter yields, such as aluminum or stainless steel (but Al must not be used with chlorides since it will be etched), and (3) protecting the fixtures with a low-yield, passivating coating.

Combined chemical and sputter effects leave polymeric residues on the wafers. Post-etch ashing in an O_2 plasma can eliminate these by oxidizing them into volatile products. This can be a no-$\mathscr{E}$-field, isotropic process [42].

Bombardment by ions and photons from the plasma can affect the surface density of states, N_{SS}, on silicon. This becomes important in silicon gate MOS

(SIGMOS) technology when the polysilicon gate is etched, because device transconductance and threshold voltage are changed by N_{SS} shifts. Changes in N_{SS} also can affect the barrier height in Schottky diodes, and the behavior of ohmic contacts between Si and a metal such as Al or Au. Thus, all these properties are affected by the choice of plasma gases and the self-bias voltage. In RIE, CF_4 increases N_{SS} more than do $SiCl_4$/Ar or CHF_3.

Ion and photon bombardment of oxides can create traps in the oxide, too. These may be neutral or may become positive by trapping a hole. In either case, drifting MOS threshold voltage can result. For large geometry SIGMOS devices these traps may be removed by a furnace anneal in N_2 at about 1000°C, but small-scale devices need a shorter anneal such as afforded by lasers.

A catastrophic example of radiation damage is the destructive dielectric breakdown of a SIGMOS gate oxide when the anisotropic etching of the polySi gate is terminated [43]. A typical cross section of the materials involved is shown schematically in Figure 10–10(a). The electric charges developed by positive ions and self-bias during RIE are shown on the pertinent layers; the dielectric PR and gate oxide regions behave as capacitances and are shown at (b) as two equivalent capacitors in series with C_E. During RIE the largest voltage will appear across the external capacitor, C_E. Typically, gate oxides are thin, say 1000 Å or less, making them very susceptible to dielectric breakdown. The trend toward ever smaller devices means ever thinner gate oxides, which exacerbates the problem. When the rf is turned off to terminate the etch, a transient readjustment of charge takes place—with positive charge in the gate layer and negative charge from C_E flowing toward the gate oxide.

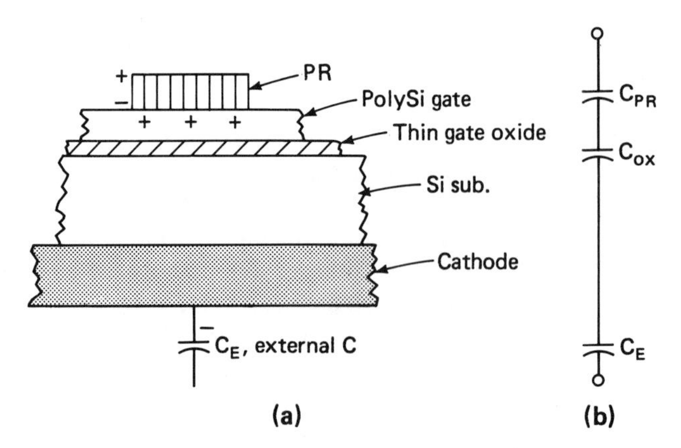

Figure 10–10. Dielectric breakdown in SIGMOS devices. (a) The stack of materials. (b) Equivalent series circuit of three capacitors. After Watanabe and Yoshida [43]. (Courtesy of *Solid State Technology*, PennWell Publishing Company, Copyright 1984.)

As a result the equivalent oxide charge Q_{ox} on the capacitance C_{ox} increases. Since $V_{ox} = Q_{ox}/C_{ox}$, V_{ox} increases with Q_{ox}. Due to the small size of C_{ox}, this voltage can be large, so large, in fact, that the gate oxide can rupture, a condition that becomes more likely as device size decreases. For example when the size of an MOS device is scaled down linearly by a factor of $1/S$, where S is greater than one, all linear dimensions are multiplied by $1/S$. Oxide thickness d scales by $1/S$ and area by $1/S^2$; hence the capacitance, $C_{ox} = \varepsilon A/d$, also scales by $1/S$. Its decrease makes V_{ox} even larger for a given Q_{ox}.

Watanabe and Yoshida suggest these remedies. (1) Use an insulated cathode so the wafer does not contact that electrode. Remember that with r-f excitation dc does not flow in the series circuit; electrical contact between the back side of the wafer and the electrode is not required. This effectively introduces a fourth capacitor in the string, located between C_{ox} and C_E. (2) At the end of the etch cycle, raise the gas pressure *before* turning off the r-f power. This lowers the self-bias, less voltage is on C_E before the transient, and the transient voltage across C_{ox} is lowered.

10.20 Compound Semiconductors

Gallium and indium fluorides tend to be nonvolatile at normal etching temperatures, even though arsenic and phosphorus are highly volatile. As a result, gallium arsenide (GaAs), indium phosphide (InP), and similar compound semiconductors cannot be etched in barrel and PPE configurations under isotropic conditions in fluoride plasmas such as CF_4 and SF_6/O_2. Rather, chlorine-bearing gases are used, such as BCl_3/Cl_2, CF_2Cl_2, CF_2Cl_2/O_2, and CCl_4 for GaAs and Cl_2/O_2, $Cl_2/O_2/Ar$, and CCl_4 for InP, because Ga- and In-chlorides are salts with high vapor pressures at etch temperatures, and As- and P-chlorides are liquids or gases.

Both gallium arsenide and its native oxide, "GaAs-oxide," will etch in $COCl_2$, PCl_3, and HCl. The gas combination 8 Cl_2/2 O_2 will etch both InP and GaInAs.

Under RIE conditions, where additional ion bombardment is present, GaAs will etch in CF_4 and in CHF_3. The availability of C here allows polymer formation, and bombardment generates surface defects. A study of GaAs Schottky diodes shows that RIE produces changes in the barrier heights and breakdown voltages, probably the results of radiation damage. Further changes in device characteristics can come about due to loss of stoichiometry in the compound semiconductor layers under bombardment [41].

An interesting case arises in the fabrication of gallium arsenide phosphide (GaAsP) LEDs (light emitting diodes). Since As and P have very high vapor pressures, encapsulation of the compound semiconductor, usually in Si_3N_4, is necessary to prevent dissociation at diffusion temperatures. The nitride also can serve as a diffusion mask and it is patterned with windows by RIE in

CF_4/O_2. This does not etch GaAsP, so a natural *stop* is provided when the etchant reaches the semiconductor [26].

Problems

10–1 A thin Al interconnect layer is deposited on an oxide-coated Si wafer.

(a) What wet etch can be used to selectively pattern the Al layer? Need it be anisotropic? Explain.
(b) Repeat for the SiO_2 layer.

10–2 You wish to etch a row of holes, 5 μm in diameter, completely through a 20-mil-thick Si wafer.

(a) What wafer orientation should be used? Why?
(b) Recommend three wet etches that may be used.
(c) Explain how this technique may be used to fabricate an inductance "coil." Use sketches.
(d) Do you expect a large or small L value? Why?

10–3 Part of a wafer is to have Au interconnects, the other part Al. Explain the advantages of the lift-off technique for patterning the metals in this situation. Positive resist is used.

10–4 How does a load-lock system help in spot monitoring during plasma etch.

10–5 Explain why N_2 and O_2 have different glow discharge colors.

10–6 A W layer is to be selectively dry etched without attacking some Al interconnects. Specify the gas(es) and reactor type to be used. Explain.

10–7 Explain why the positive ions travel a distance significantly less than d in sputter etching.

10–8 Verify Eq. (C.3-1).

References

[1] H. C. Gatos and M. C. Lavine, "Chemical Behavior of Semiconductors: Etching Characteristics." In *Progress in Semiconductors, Vol. 9*, ed. A. F. Gibson and R. E. Burgess, pp. 1–46. London: Temple, 1965.

[2] D. J. Elliott, *Integrated Circuit Fabrication Technology*. New York: McGraw-Hill, 1982. Table 11–3, pp. 260–263.

[3] W. Kern and C. A. Dickert, "Chemical Etching." In *Thin Film Processes*, ed. J. L. Vossen and W. Kern. New York: Academic Press, 1967. Chapter V-1.

[4] S. K. Ghandhi, *The Theory and Practice of Microelectronics*. New York: Wiley, 1968.

[5] A. Beiser and K. B. Krauskopf, *Introduction to Physics and Chemistry*. New York: McGraw-Hill, 1964.

[6] R. A. Colclaser, *Microelectronics: Processing and Device Design*. New York: Wiley, 1980.

[7] H. Robbins and B. Schwartz, "Chemical Etching of Silicon II, the System HF, HNO_3, H_2O and $HC_2H_3O_2$," *J. Electrochem. Soc.*, **107**, 108, 1960.

[8] H. Wolf, *Semiconductors.* New York: Wiley Interscience, 1971.
[9] D. B. Lee, "Anisotropic Etching of Si," *J. Appl. Phys.*, **40,** (1), 4569–4574, Oct. 1969.
[10] "Etching Solution Keeps IC Walls from Moving," *Circuits Manuf.*, **13**, (1), 18, Jan. 1973.
[11] P. Ou-Yang, "Double Ion Implanted V-MOS Technology," *IEEE J. Solid-State Circuits*, **SC-12**, (1), 3–10, Feb. 1977.
[12] F. E. Holmes and C. A. Salama, "V-Groove M.O.S. Transistor Technology," *Electron. Lett.*, **9**, (19), 457–458, Sept. 1973.
[13] D. L. Kendall, "On etching very narrow grooves in silicon," *App. Phys. Lett.*, **26**, (4), 195–197, 15, Feb. 1975.
[14] L. B. Rothman, "Properties of Thin Polyimide Films," *J. Electrochem. Soc.*, **127**, (10), 2216–2220, Oct. 1980.
[15] R. Iscoff, "Polyimides in Semiconductor Manufacturing," *Semicond. Int.*, **7**, (10), 116–119, Oct. 1984.
[16] J. A. Lange, "Sources of Semiconductor Wafer Contamination," *Semicond Int.*, **6**, (4), 125–128, April 1983.
[17] P. S. Burggraaf, "Wafer Cleaning," *Semicond. Int.*, **4**, (7), July 1981. Part 1: "Brush and High-Pressure Scrubbers," 71–78; Part 2: "State-of-the-Art Chemical Technology," 91–94; Part 3: Sonic Scrubbing," 97–100.
[18] "A Brief Lesson in Wafer Scrubbing Theory," *Semicond. Int.*, **4**, (7), 89, July 1981.
[19] C. Murray, "Wet Etching Update," *Semicond. Int.*, **9**, (5), 80–85, May 1986.
[20] P. D. Brewer, G. M. Reksten, and R. M. Osgood, Jr., "Laser-assisted Dry Etching," *Solid State Technol.*, **28**, (4), 273–278, April 1985.
[21] A. D. Weiss, "Endpoint Monitors," *Semicond. Int.*, **6**, (9), Sept. 1983. "Part I—Endpoint Diagnostics," 98–99. "Part II—Optical Emission Spectroscopy," 100–103.
[22] W. R. Hashbarger and R. A. Porter, "Spectroscopic Analysis of R.F. Plasmas," *Solid State Technol.*, **21**, (4), 99–103, April 1978.
[23] P. J. Marcoux and Pang Dow Foo, "Methods of End Point Detection for Plasma Etching," *Solid State Technol.*, **24**, (4), 115–122, April 1981.
[24] R. L. Bersin and R. F. Reichelderfer, "The DryOx* Process for Etching Silicon Dioxide," *Solid State Technol.*, **20**, (4), 78–80, April 1977.
[25] P. H. Singer, "Dry Etching of SiO_2 and Si_3N_4," *Semicond. Int.*, **9**, (5), 98–103, May 1986.
[26] R. L. Bersin, "A Survey of Plasma-Etching Processes," *Solid State Technol.*, **19**, (5), 31–36, May 1976.
[27] "Dry versus Wet, Plasma Etching/Stripping," *Circuits Manuf.*, **16**, (4), 42–46, April 1976.
[28] G. Turbon and M. Rapeaux, "Dry Etching of Polyimide in O_2-CF_4 and O_2-SF_6 Plasmas," *J. Electrochem. Soc.*, **130**, (11), 2231–2236, Nov. 1983.
[29] D. W. Hess, "Plasma Etching of Aluminum," *Solid State Technol.*, **24**, (4), 189–194, April 1981.
[30] J. E. Spencer, "Management of $AlCl_3$ in Plasma Etching Aluminum and Its Alloys," *Solid State Technol.*, **27**, (4), 203–207, April 1984.
[31] S. M. Irving, "A Plasma Oxidation Process for Removing Photoresist Films," *Solid State Technol.*, **14**, (6), 47–51, June 1971.

[32] B. Bollinger, S. Iida, and O. Matsumoto, "Reactive Ion Etching: Its Basis and Future, Parts I and II," *Solid State Technol.*, **27**, (5), 111–117, May 1984, and **27**, (6), 167–173, June 1984.

[33] F. D. Egitto et al., "Ion Assisted Plasma Etching of Silicon-Oxides in a Multifacet System," *Solid State Technol.*, **24**, (12), 71–75, Dec. 1981.

[34] L. M. Ephrath, "Selective Etching of Silicon Dioxide Using Reactive Ion Etching with CF_4-H_2," *J. Electrochem. Soc.*, **120**, (9), 1419–1421, Aug. 1979.

[35] H. H. Sawin, "A Review of Plasma Processing Fundamentals," *Solid State Technol.*, **28**, (4), 211–216, April 1985.

[36] B. A. Heath, "Etching SiO_2 in a Reactive Ion Beam," *Solid State Technol.*, **24**, (10), 75–78, 85, Oct. 1981.

[37] J. S. Chang, "Selective Reactive Etching of Silicon Dioxide," *Solid State Technol.*, **27**, (4), 214–219, April 1984.

[38] W. Beinvogl and B. Hasler, "Reactive Ion Etching of Polysilicon and Tantalum Silicide," *Solid State Technol.*, **26**, (4), 125–130, April 1983.

[39] D. HG. Choe, C. Knapp, and A. Jacob, "Production IRE-II. Selective Aluminum Alloy Etching," *Solid State Technol.*, **27**, (3), 165–171, March 1985.

[40] A. A. Chambers, "The Application of Reactive Ion Etching to the Definition of Patterns in Al-Si-Cu Alloy Conductor Layers and Thick Silicon Oxide Films," *Solid State Technol.*, **26**, (1), 83–86, Jan. 1983.

[41] S. W. Pang, "Dry Etching Induced Damage in Si and GaAs," *Solid State Technol.*, **27**, (4), 249–256, April 1984.

[42] S. J. Fonash, "Damage Effects in Dry Etching," *Solid State Technol.*, **28**, (4), 201–205, April 1985.

[43] T. Watanabe and Y. Yoshida, "Dielectric Breakdown of Gate Insulator Due to Reactive Ion Etching," *Solid State Technol.*, **27**, (4), 263–266, April 1984.

Chapter 11
Lithography

Etching is used to remove unwanted material and to delineate the regions where material is to be added. The pattern of these regions is specified by the device or circuit designer. The fundamental function of lithography is to transfer the pattern to the wafer itself.

One version of *photo*lithography was outlined in Chapter 1, where the pattern, stored in the emulsion of a glass based photomask, was transferred to the wafer photoresist (PR) by ultraviolet (UV) radiation in a *contact* aligner. Alternative versions are discussed in this chapter.

Lithography is named for the specially prepared stone (Greek: *lithos*) that has been used to transfer a drawing to the printed page for over 200 years. Much of the technology and terminology in the semiconductor industry is derived from the graphic arts or other industries that specify dimensions in British units. Thus we find mixed units in semiconductor industry usage: for example, mils and μin as well as Å, mm, μm, and nm.

11.1 Pattern Transfer Processes

In Chapter 1 we considered use of (PR) as a *stop* to delineate regions to be etched. Our concern now is to transfer the proper pattern to the resist. Figure 11–1 shows several alternative ways to do this, tracing the major steps from the designer's sketch to the patterned resist. The actual transfer during exposure (*print* or *write*) is by radiation. The figure shows two wavelength ranges: Path I uses the ultraviolet (UV), where the whole process is referred to as *photo*lithography. Path II uses a nonoptical range, furnished by x-rays or e-beams, where wavelengths are shorter by orders of magnitude.

The semiconductor industry strives to reproduce ever smaller features on chips, i.e., to increase the *resolution*, or fineness of pattern detail. In this context, we shall use ρ to designate the smallest feature dimension, line width, or space, that can be reproduced in the process. The smallness of ρ is limited

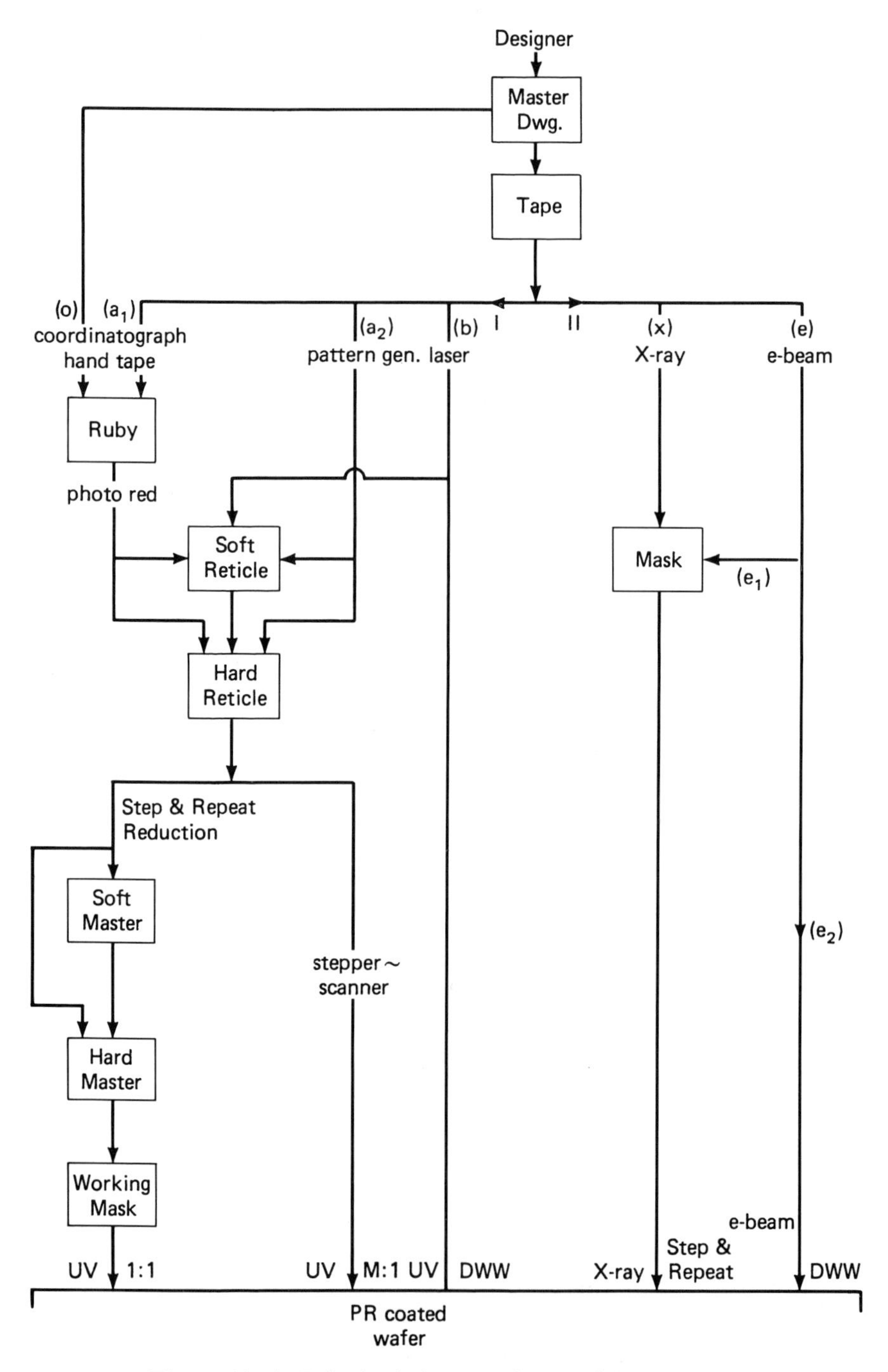

Figure 11–1. Principal alternate forms of lithography.

by several process- or equipment-related parameters, but the transfer radiation wavelength λ also sets a limit because of diffraction, scattering, and other wave-related phenomena. In general, the shorter the wavelength, the smaller the value of ρ (the better the resolution), other things being equal; hence, the drive for better resolution has led to use of shorter wavelengths.

Photolithography uses three ranges: (1) deep ultraviolet (DUV)—from 200 to 260 nm (= 2000 to 2600 Å); (2) ultraviolet (UV)—from 260 to 330 nm; and (3) near ultraviolet (NUV)—from 330 to 450 nm. The notation is unfortunate: as a group these often are called simply ultraviolet (UV). Mercury lamps provide several lines in this range, though other sources, such as excimer lasers, may be used.

By 1989, resolution with UV exposure and production-line process limitations was about 0.5 μm, i.e., good definition of 0.5-μm-wide pattern lines, spaced 0.5 μm between adjacent edges (= 1 μm between centers). Progress has been rapid: in March 1986 0.75-μm features with UV exposure were reported [1]. The 0.5-μm value meets the *optical* lithographic goals of the USA Department of Defense program for very high speed integrated circuits (VHSIC).[1]

Within photolithography are a pair of options: (a_1, a_2) a photomask determines the transferred pattern, or (b) a controlled laser beam is used to *write* the pattern in the resist without a mask, the direct write-on-wafer (DWW) system (see Figure 11–1). Most wafers are processed by mask-based photolithography, which is covered in the early parts of this chapter.

The following notation will be used. The term *IC* will indicate either a device or an integrated circuit that will occupy one *chip* or *field* on both the mask and wafer. The two-dimensional replication of a field into a *matrix* that covers the whole wafer is an *array*. These may be identifed in Figure 1–11. Operations that involve the entire array, rather than one field, are referred to as *global*, as in global exposure. A mask *series* is the set of masks required to complete an IC. As many as 30 may be used, though that number is unusually high. Since each mask of a set controls the etching for only one level on the final IC, all but the first mask be *aligned* to *register* properly with previous levels on the wafer; the first must align properly with the wafer flat (Section 2.6).

11.2 Mask-Based Photolithography

Consider the options for mask-based photolithography. Figure 11–1 shows that all of the methods need a master drawing for each mask (or *level*) in a

[1] Further VHSIC goals are to decrease feature size to below 0.5 μm, so x-rays or electron beams (*e-beams*) are used where the approximate wavelengths are 5 Å and 0.1 Å, respectively. In fact, by late 1987 0.1^+ μm features were realized in certain structures by the use of direct-write-on-wafer PR by e-beam technology [2].

series. These are drawn to large scale so errors will be reduced in subsequent size reductions. In the older technology (o) in Figure 11–1, this drawing was transferred to a *Ruby* (or Rubylith®) master comprised of a two-layer plastic material, the lower layer being of transparent, dimensionally stable polyester.[2] The thin, strippable upper layer is colored. Pattern transfer is accomplished by cutting or *scribing* the pattern edges in the upper layer by hand on a knife-equipped drafting machine (*coordinatograph*), and stripping (removing) the unwanted regions from the support layer. The coordinatograph now is under computer control in the manner to be described shortly. What remains is the field pattern reproduced as dark (opaque) regions on a clear, (transparent) background, or vice versa. Thus, the Ruby serves as an enlarged intermediate storage medium for the pattern. Rubylith pattern transfer is nearly obsolete in IC manufacture, but still is viable in thin and thick film processing.

The remaining alternatives of Figure 11–1 require a control tape prepared by digitizing each master drawing of a set and storing the data. This tape can be checked by having it control an x–y plotter to print out large-scale drawings, each of which is checked against the corresponding master. Also, a composite set can be plotted on a single sheet, using a separate color for each wafer level. Adherence to design rules for overlaps and spacings may be checked from the drawings, but the large majority of rule checking is done with computer software.

After corrections are made, the tape is available for the alternative methods shown in Figure 11–1. The digital tape may be used for *mask making* in three ways: (a_1) It may control a coordinatograph to produce the Ruby artwork from which a *photomask* is prepared by photo reduction (but this technique is obsolete for IC work). (a_2) It may control a *pattern generator* that optically exposes a sensitized photoplate to produce the photomask directly. (b) It may be used to control a suitable laser or electron beam (*e-beam*) to direct-write onto a photoemulsion-coated mask blank. These methods are discussed later.

The mask-to-PR transfer process actually involves two steps: (1) *alignment,* where the mask image is located properly wrt the substrate pattern features (if any) from earlier steps, and (2) *exposure* (or *printing*), where the actual transfer by UV takes place.

A single machine or *tool* is used to provide both steps and commonly is called either an *aligner* or *printer* since these functions always go hand in hand. Because the align/print steps are common to all the mask-based variations shown in Figure 11–1, they impact mask making, photoresist formulation, exposure, etc.

---

[2] Rubylith® is a trademark of Ulano Products Co. Inc. The upper, strippable layer of material is ruby-colored; hence the name. Other colors such as amber, which also is opaque to UV, are available as Amberlith®, and so on.

Manual alignment by an operator requires mask and wafer illumination by *visible* light, a feature not needed for some automatic operations. Once proper registration is achieved, printing (or exposure) with UV radiation, of wavelength appropriate for PR exposure, follows. The UV is shined through the mask whose opaque regions prevent (and transparent regions allow) resist exposure. Subsequently, the resist is developed, and the regions that remain act as a *stop* for etching wafer layers. Since the exposing radiation must pass through the clear portions of the mask, its supporting base must have high transmission of the UV wavelength(s) involved.

11.2.1 Standing Wave Effect

Transfer radiation wavelength also affects standing waves within the resist during exposure. The incident UV passes through the resist and suffers multiple reflections at the surface and at the PR/wafer interface, causing standing waves within the PR; hence, exposure varies across the PR thickness. After development, the edge of a PR line shows roughness in the form of steps because of the exposure variations. The wavelength in the resist λ_{PR} is shorter than λ_{air} in air by a factor of about 3.5 to 1.6 for modern resists. Thus, for the h-line of Hg ($\lambda_{air} = 0.4047$ μm), for example, the shorter wavelength will be $\lambda_{PR} \approx 0.4047/3.5 \approx 0.116$ μm. Then the number of standing wave cycles in a PR layer, say, 1.2 μm thick will be $1.2/0.116 \approx 10$. The developed PR would show about 10 steps or terraces along the edge of a vertical line edge when viewed under a microscope. Ideally, such an edge should be smooth and vertical. If a polychromatic (several wavelengths present) UV source, such as an unfiltered Hg lamp is used, each component causes a different standing wave pattern, so the overall effect is smoothed out. The standing waves are affected by reflective properties of the layer under the PR.

11.2.2 Aligner Types

The type of aligner also affects the choice of mask pattern size and materials. Aligners differ in their spacing between the PR-coated substrate and mask *during exposure.* (1) In *contact* aligners, seldom used now, the wafer is clamped in intimate contact with the mask, PR against the mask's emulsion (or pattern) side, causing mask damage. (2) In *proximity* aligners, the mask and substrate are never in contact but are separated by a very small distance, say ≤ 50 μm. (3) In *projection* aligners, mask and wafer are separated by a relatively large distance, say up to 6 in. (≈ 15 cm).

Types (1) and (2) are 1 : 1 systems in that mask and wafer patterns are the same size and cover the entire wafer. The third or projection type may be 1 : 1 also, but $M : 1$ *reduction* is more common, with M usually 2, 4, 5, or 10. The mask in this latter case usually carries the pattern of one field at M times the wafer field size and is called a *reticle* (Figure 11–1). Relative motion between mask and wafer, called *stepping*, is required after each field exposure until the complete array is exposed over the wafer. Wafer exposure time is longer here than in the 1 : 1 types.

The equipment used for these different types of align/printing must have provisions for both pattern alignment and exposure: *global* for 1 : 1 systems (Section 11.3) and *stepper* for M : 1 systems (Section 11.4).

11.3 Global Aligners

The principles of alignment are the same for all three types of aligner/printers. Figure 1–12 shows a simplified sketch of a contact aligner and the manual alignment steps are described in Section 1.8.5. Ideally the *overlay error*, or departure from perfect registration, should be zero; it can be held to fractions of a micron with current technology. If the mask and existing wafer features are of exactly the same size, as they should be, misalignment can be *translational* or *rotational*. It is relatively easy to correct these, especially if the viewing microscope has *split optics*, whereby each eye of the operator sees a different region of the array. Usually special *alignment*, *register*, or *fiducial* marks are provided on the several masks of a set. During exposure, the mark on each mask is transferred as part of the pattern. Together, these marks form a complete and easily recognizable pattern, so alignment is based on assembling that pattern, mask by mask. Some simple forms of fiducial marks are illustrated in Figure 11–2. Circles and squares of increasing size, one for the fields in each mask of a set are shown at (a) and (b). Successive masks of a set are aligned to bring the marks into concentricity as at (c) and (d). A more elaborate

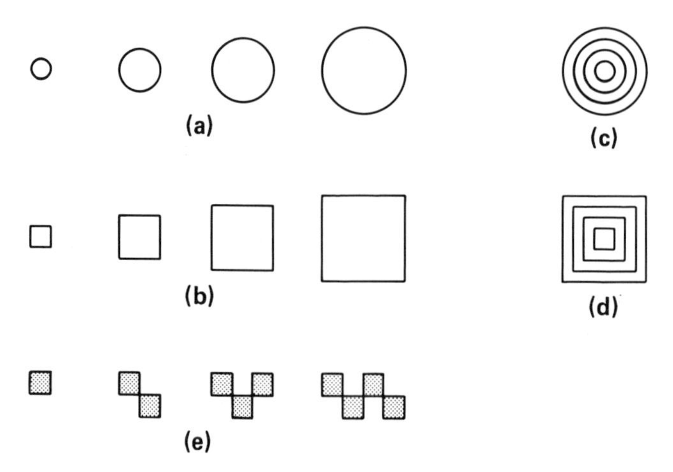

Figure 11–2. Simple forms of fiducial or alignment marks.

pattern of squares is shown at (e). Automatic aligners use fiducial marks, and misalignment error signals are fed to a computer that controls motion of the wafer-supporting table. Overlay error to less than $\pm 0.1\ \mu m$ may be realized.

11.3.1 Contact Aligners

Due to prior deposition and etching sequences, wafer surfaces are not flat; they also may be warped because of earlier high-temperature processes. Contact printers are designed so mask and wafer will be clamped into intimate contact *during PR exposure* to overcome these conditions.

Forcing the mask emulsion and PR into contact during alignment verification and printing damages the mask pattern—due to the resist itself and to any airborne particles on either mask or wafer. Furthermore, if contact printing follows epi deposition, the *epi spikes* that occur sometimes can cause severe damage.

Chapter 1 shows that several contact/separation cycles per align/print are needed with contact aligners, so damage to *soft* emulsion masks is high and their useful life under typical production line conditions is about 20 aligning cycles. Frequent mask replacement adds to fabrication costs, but it is cost effective because emulsion masks have low unit cost and higher yields are obtained by frequent replacement.

Hard masks, where the soft emulsion is replaced by a patterned metal or metal-oxide film, are less susceptible to damage and can extend the useful life by a factor of roughly ten. Overall costs are about the same, however, since hard masks cost about ten times more than the soft variety [3]. Longer mask life reduces equipment down time for mask replacement, however, so hard masks allow greater throughput in a contact aligner. Hard masks are discussed further in Section 11.6.1.

11.3.2 Proximity Aligners

Proximity printers resemble the contact type but do not require mask and wafer to make contact at any time during the complete align/print cycle. Even during printing the two remain separated by a small, constant gap g in the range of 2 to 50 μm. The purpose is to avoid mask damage and increase throughput.

The gap is so small that no optical projection system is needed to focus the mask pattern onto the wafer for image transfer; one relies on parallel rays to minimize distortion. A good collimating system between the illumination source and mask is required, however, and since this is a global system, illumination must be uniform over the entire array.

As the gap is made larger, diffraction-produced distortion increases, and if $g > \rho$, diffraction causes illumination to vary on the wafer surface and nonuniform exposure results. Exposure latitude of the PR can compensate for this to some extent. We have seen that ρ improves (gets smaller) as the illumination wavelength λ decreases. These effects are summarized for proximity printers by the approximation [4]

$$\rho \approx \sqrt{g\lambda} \tag{11.3-1}$$

Thus both g and λ should be as small as feasible for high resolution. For example, for a 10-μm gap and a DUV λ of 250 nm (= 2500 Å = 0.25 μm), $\rho \approx \sqrt{10(0.25)} \approx 1.6\ \mu$m.

With no focusing lens, the mask itself is the only intervening glass between the UV source and wafer; hence its blank must be transparent at the chosen λ. This is discussed in Section 11.6.1, where data show that masks for DUV operation are considerably more expensive than those for longer wavelengths in the overall UV range.

The minimum g value is limited by the height of wafer features. On the other hand, a gap of only a few microns does not leave enough space for *changing wafers*, so, as in the contact aligner, the wafer platen must move up and down. Only one separation/closure per align/print cycle is needed, however, for wafer changing. Stops are provided so that the wafer locks g below the contact position. The gap must be reproducible, so very tight mechanical tolerances and frequent, careful adjustment are needed.

The resolution and construction for the contact and proximity aligners are so similar that models are available commercially that allow for gap adjustment for either mode. Proximity operation yields higher throughput, the same resolution, and virtually no mask damage.

11.3.3 Projection Aligners

Global or 1 : 1 *projection* align/print systems use a mask-to-wafer separation of several inches, and an optical imaging system for focusing the array pattern on the wafer PR. Three advantages over the contact and proximity systems are apparent. (1) Focusing gives better optical imaging. (2) The separation remains fixed because it is large enough to allow wafer changing. (3) Space is available for a thin dust shield, or *pellicle*, over the mask pattern to prevent damage. Dust remaining on the pellicle is out of focus on the wafer and so has little effect on the projected image (Section 11.6.5). Also, there is room for a nitrogen stream to keep the wafer surface or pellicle clean.

Early global projection systems used a lens between the mask and wafer for optical imaging, to give a *refractive* system. When wafer diameters reached three inches and more, however, it was no longer practicable to build 1 : 1 lenses large enough to cover the entire wafer and yet meet the requirements of UV operation, high resolution, and so on. *Reflective* systems that employ mirrors, rather than lenses, offered a solution. In addition, mirrors do not exhibit chromatic aberration, so unfiltered polychromatic sources can be used. The resulting increased intensity lowers exposure time, thereby raising throughput, and also reduces the standing wave effect.

Mirrors require distortion correction, however, and correction costs increase with wafer size. A cost-effective solution lies in using only the mirror's "zone of good correction," shown in Figure 11–3. This is accomplished by placing a slit, shaped as a sector of the annular correction zone, between the UV source and the mask so that only a small portion of the mask pattern is projected at any instant. The slit position is fixed wrt the mirrors, so *both* mask and wafer, which lie in the same focal plane, must *scan* across the slit image to cover the entire array [5]. This system is called a *scanner*.

In the simple system of Figure 11–3 a UV ray is reflected three times in going from mask to wafer, so left and right are interchanged in the image; hence, mask and wafer must travel in opposite directions, making accurate tracking difficult. This is solved commercially by introducing more reflecting

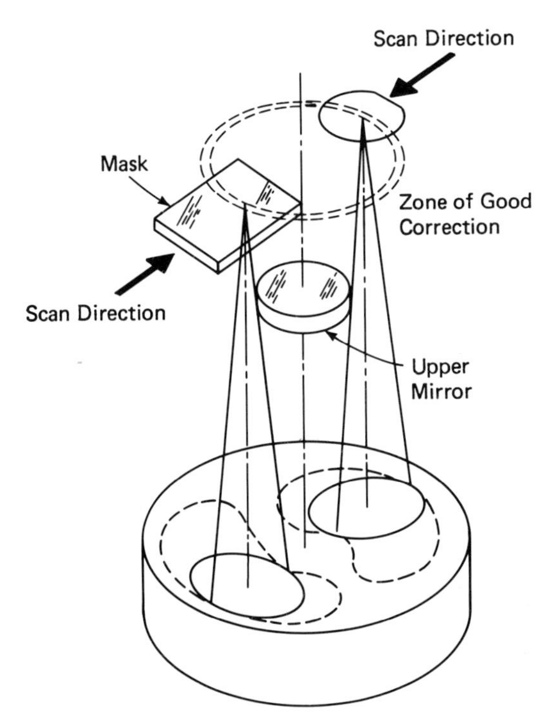

Figure 11–3. A 1 : 1 global projection printer. The slit covers the zone of good correction above the mask. The UV source is located above the mask and slit. Rays to and from the upper mirror are not shown. Markle [5]. (Courtesy of *Solid State Technology*, PennWell Publishing Company, Copyright 1984.)

surfaces, either with mirrors or prisms, so the wafer sees a correct, rather than a reversed, image. Then mask and wafer can be mounted on a single rigid fixture that scans both of them simultaneously in the same direction across the slit image. Markle has described a commercial system of this type [5].

It may be argued that the scanned projection system is not global in the sense that the entire wafer is not exposed at one time. We consider it to be so, however, because the mask is of wafer-array (i.e., global) size—in contrast to the stepper system discussed next.

11.4 Direct Step-on-Wafer System (DSW)

A second type of projection aligner/printer uses field-by-field, rather than global exposure. A pattern for one *field*[3] is repetitively stepped and exposed on the PR-covered wafer to form the wafer-size array. This more complicated type is used to improve resolution, alignment, and focus. The field mask, or reticle, is made M times the wafer field size (*linearly*), so an $M:1$ reduction in the optical projection system is required. On exposure any linear errors on the reticle also are reduced by factor M on the wafer. Since the reticle image must be *stepped* over the wafer to form the array, equipment having this capability is called a *stepper* or DSW (direct step-on-wafer) system.

Steppers provide another advantage: they are designed to align and focus automatically before each field exposure; hence these two parameters are corrected on a field-by-field basis. Contrast this to global systems, where a single align/focus sequence must suffice for the whole wafer array. The stepper takes longer to cover the wafer, however. A table of state-of-the-art steppers is given by Burggraaf; current steppers have throughputs of fifty 6-in. wafers/hr [6].

Reflective reduction systems are becoming common, since mirrors need satisfy less severe specifications than lenses [5, 6]. Some lens relationships are considered in Appendix E.

11.4.1 Stepper Requirements

With a reticle covering only one field, alignment for every field is required. The stepper must perform several functions to do this.

1. Set the wafer parallel to the reticle.
2. Set the wafer to bring a *die-field* to the projected reticle image.

[3] The term "field" is used in several ways in semiconductor work. In this chapter we see it used for a pattern of chip size, the maximum size of a square or rectangular pattern that a lens may reproduce, or the nonfeature region of a mask. The appropriate meaning is apparent from the context.

3. Adjust the wafer-to-lens distance to bring the die-field into proper focus.
4. Align the die-field to the image.
5. Expose the die-field PR.

Fully automated operation with computer control is essential to meet accuracy, precision, and speed requirements.[4] A 100-mm wafer can accommodate about sixty 10 × 10-mm dice, and a throughput of at least 60 wafers per hour is desired. This allows about one second per field alignment and exposure! Automatic operation is essential.

11.4.2 Auto Position and Align

Horizontal autopositioning may be handled in terms of (y, z) coordinates assigned to each die-field and located by counting laser interference fringes from a reference point. The count is delivered to a computer that drives the wafer stage to the correct location. The step size in a typical system is about 880 Å ($= 0.08\ \mu$m) with mercury g-line illumination.

Autoalignment is based on the use of visible light and fiducial marks designed so video/electronic systems can derive error signals from them. Figure 11–4(a) shows a *bright field* microscope system that can be used for visual or auto align. In contrast to contact and proximity aligners, the reticle and die-field images are coplanar, making observation easier. One form of fiducial marks that can be used for autoalign is shown in (b). The four transparent bars (2) are on the reticle, and the dark square (1) is on the die-field from prior lithographic steps. Moving the wafer changes the overlap of the four bars and the square (3). The light from each bar region is sensed by optoelectric devices; unequal signals tell the computer how to move the wafer. Four equal signals result when that field is aligned correctly. Lin gives details on this and other systems [7].

Another autoalign system, used in the TRE 800 SLR Wafer Stepper™ and similar equipments, has a television camera to monitor the microscope image. The camera's electrical output feeds a video monitor for the operator, and the computer input. The basic setup is shown in Figure 11–5(a). The locations of the fiducial mark on the die-field and the window on the reticle are shown in (c) and (b), respectively. The two images are coplanar and merged at the video camera as shown in (d). Note that the cross is not centered in the window, so the wafer is out of alignment. In the television process, the image is scanned in a raster of lines, each of which corresponds to one sweep across the picture at a value of z. The idealized electrical output V of the camera for one scanned line is shown in (e). Computer processing of the whole set of raster-line outputs

[4] *Accuracy* is the deviation from the absolute or conformity to a standard value. *Precision* is the variation in repeatability.

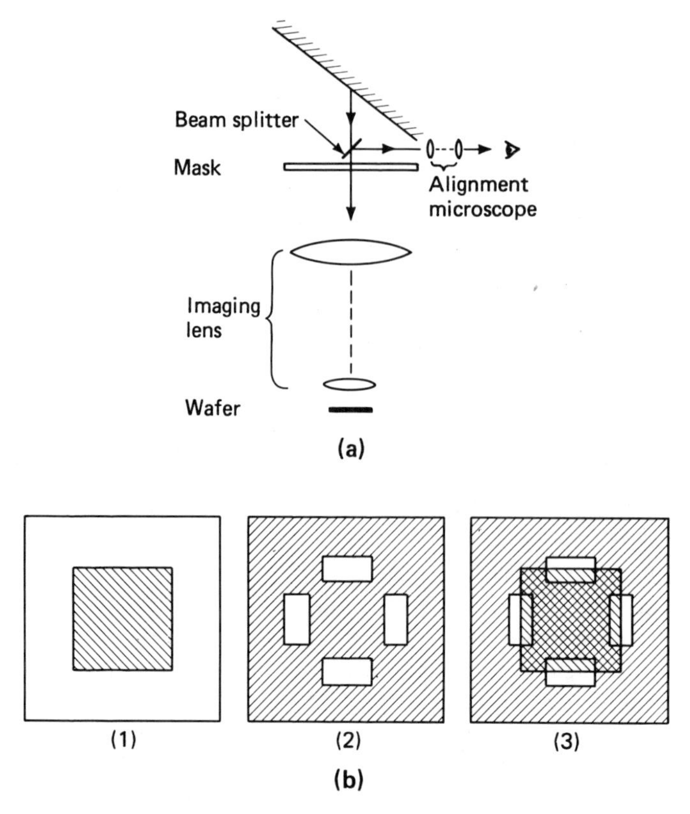

Figure 11–4. Automatic aligner details. (a) Basic bright field microscope system. (b) Fiducial marks. The right-hand sketch shows perfect alignment. After Lin [7]. (Courtesy of North Holland Press, Elsevier Science Publishing Company, Inc.)

locates the cross's center. If the cross is not centered in the window, corrective signals drive the wafer stage until feedback indicates correct alignment. The cross pattern in the 800 SLR is supplemented with bar marks to aid in alignment by an operator. The complete optical system is described in references [8] and [9].

11.4.3 Auto Level and Focus

The autoleveling system of the TRE 800 SLR illustrates the techniques used. It senses back pressures from three air jets, spaced 120° apart and directed at the wafer near its edge, to generate computer input signals. A hemispherical bearing, floating in a socket, supports the wafer table. The computer controls

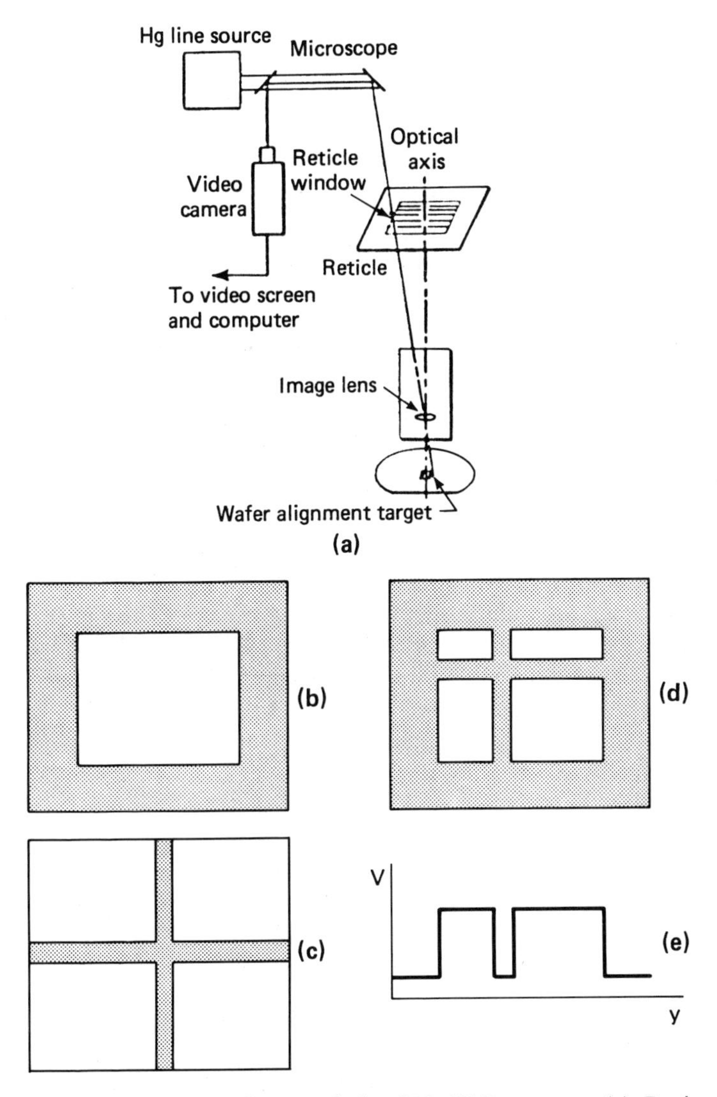

Figure 11–5. Automatic aligner of the 800 SLR system. (a) Basic setup. (b) Reticle window fiducial mark. (c) Wafer cross. (d) TV monitor output for a misaligned condition. (e) Voltage output for one scanned line above or below the cross horizontal arm. After Stover [8]. (Courtesy of *Solid State Technology*, PennWell Publishing Company, Copyright 1981.)

the jet flows until equal back pressures are detected, when the wafer and reticle are parallel. The bearing then is vacuum-clamped in its socket.

11.4.4 Illumination Problems

Field-by-field alignment and exposure limit stepper throughput. Exposure (or print) time with a mercury source represents a large fraction of the total time a wafer is in the stepper. To reduce print time, more powerful illumination sources are preferred. Excimer lasers (Section 9.17.2) are good candidates. For example, an XeCl unit radiates at 308 nm; when driven by a capacitor dump, it can develop pulses 20 ns wide at rates from 10 to 200 pps, each with enough energy to expose a single field adequately.

Potentially this permits *flash-on-the-fly* operation (i.e., the pulse is so short that the stepping action need not be stopped for exposure). This could raise throughput by a factor of 2 to 3 [5].

11.4.5 Step-and-Scan Systems

The 308-nm laser wavelength is shorter than the Hg i-line, so the problem of building a lens for use with it is exacerbated. An alternative uses a mirror system. In order to confine operation to the well-corrected mirror region, scanning is used, but not over the whole wafer. Rather, a strip whose width is that of one field is scanned, Figure 11–6. Stepping occurs from strip to strip, giving a *step-and-scan* mode. Markle describes the system details—including the layout of the 4× mirror reduction system with a numerical aperture (NA) of 0.33 (Appendix E) [5]. Two mirrors are used, so the track length is folded to about 15 cm.

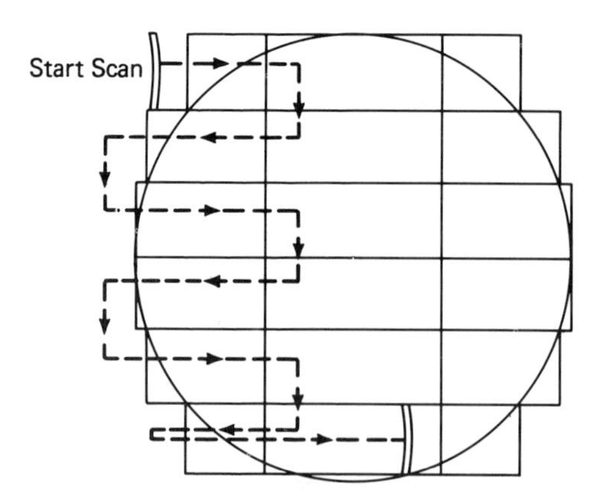

Figure 11–6. Principle of step-and-scan. Markle [5]. (Courtesy of *Solid State Technology*, PennWell Publishing Company, Copyright 1984.)

11.5 Direct Write-on-Wafer (DWW)

The pattern transfer method of path (b) in Figure 11–1 requires no mask; the tape is used to control a direct *write-on-wafer* (DWW) system, where a laser beam *writes* the pattern directly on the wafer PR. Relative motion between the beam and wafer is required to trace out the pattern; common practice keeps the beam position fixed and moves the wafer-supporting table. This type of system is versatile but much slower than a stepper, because the entire wafer is written serially: all detail in a field is written point by point. The system does, however, have the potential for finer resolution. Details of a pattern-generating system are covered in Section 11.6.3.

An alternative system used by IBM involves a pulsed dye laser operating with these conditions: $\lambda = 420$ nm, pulse duration = 10 ns, repetition rate = 400 pps, and 2 kW maximum power. The beam is focused by a lens designed for 420 nm. The laser beam position is fixed, while the support table for the wafer can scan in the x, y, and z directions, and provide rotation about all three axes [10]. This system also may be used for mask generation and repair (Section 11.6.3).

Electron beams can be used in place of laser beams in DWW systems. They potentially have much higher resolution, and are discussed in Section 11.12.3.

11.6 Photomasks

The mask-based photolithographic procedures of Figure 11–1 are certainly the most commonly used, since resolution of 0.5 μm satisfies the requirements for the majority of ICs produced. These procedures use either photomasks at final scale or reticles at enlarged scale. In this section we consider some of the details of masks themselves, and how they are made [11].

11.6.1 Photomask Materials

A photomask comprises a mechanically and thermally stable transparent (to UV) blank, capable of withstanding normal handling on a production line, covered on one side by a thin layer of material where the pattern is stored as an array of clear and opaque regions.

Several glasses may be used as mask blanks. Types and cutoff wavelengths in order of increasing cost are: (1) soda lime, and white soda lime (crown glass), 300 nm; (2) borosilicate, 250 nm; and (3) fused silica, < 200 nm. Borosilicate glass (BSG) has a lower temperature coefficient of expansion (TCE) than the soda limes by a factor of ten but costs twice as much as crown and twenty times more than soda lime. When heating is a problem and exposure wavelength is in the UV range, BSG would be a good choice despite its higher cost [4].

Fused silica must be used for deep UV exposure at 200 nm. It has a still lower TCE by a factor of eight, but its cost is five times greater than BSG. None of these glasses will work for x-rays, whose wavelengths are much shorter.

Vendors furnish mask blanks cut to correct size, with surfaces ground and polished to specified flatness, with edges rounded, and coated with emulsion, or chromium that is covered with photoresist. O'Neill quotes these per-wafer prices for 1981: 5 × 5 in borosilicate, 5 μm flatness over the entire area, $15 to $20. The same size with 2 μm flatness is $50. Note that costs for mask blanks are not trivial. Typical flatness values run from 100 μin/in. (= 2.54 μm/in.) for working masks to half that value for master masks. Ultragrade blanks are flat within 2μ over the entire wafer [3].

There is a recent trend from square to circular masks since the latter are easier to grind and polish. Also, if the mask diameter ≥ 5 in., the circular shape makes for easier handling.

The patterned layer in the common *soft* mask is usually an emulsion photosensitized with silver halide. This layer is fairly soft, even after development, and is damaged easily in contact printing or by rough handling. On *hard masks* the emulsion is replaced by a layer of Cr (chromium, *chrome*) or iron oxide that is less susceptible to damage. The hard layer is patterned by photolithography. A thin, iron oxide layer (about 500 Å), while opaque to UV, is transparent to visible light, a useful property in visual mask aligning. The combination of an r-f sputtered iron oxide layer on a BSG base gives a long-lived mask [12].

Chromium and iron oxide masks reflect light and so may cause light scattering and shadows in projection printing. This results in nonuniform exposure over the mask. The application of an antireflective coating on the chrome minimizes this trouble. The current trend, however, is to use a thin chrome-oxide hard layer that reduces reflection. This hard coating is called *AR-chrome*, or antireflecting chrome [3].

In mask making, PRs for patterning hard coatings have slower photographic speeds than does emulsion; hence, hard masks require longer exposures. Hard coatings etch rapidly because they are thin: about 1000 Å for chrome, and 500 Å for iron oxide. Emulsion runs around 30,000 Å before and 25,000 Å after processing [3]. Photoresist usually is plasma-ashed. Note that mask making is not part of an IC fabrication line.

Relative costs of hard and soft masks are given in Section 11.3.1. Hard masks give better edge acuity of the pattern, which is very thin, so they are preferred now.

11.6.2 Polarity; Correct Reading

The *polarity* (or *sense*) of a photomask describes whether it has clear or opaque images. The polarity is *negative* or *dark field* if the pattern features are clear

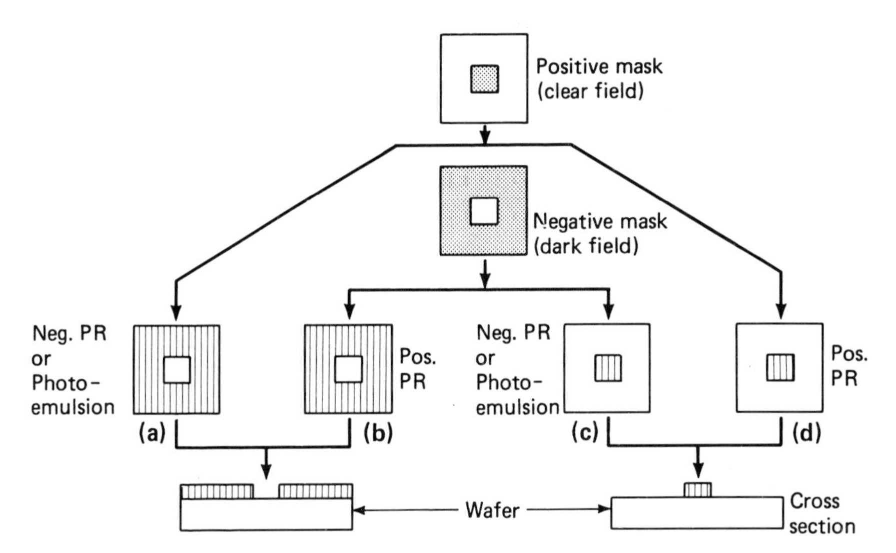

Figure 11–7. Mask nomenclature and effects of PR types and photoemulsion on image polarity.

(transparent) and the between-feature area (or *field*) is dark (opaque). *Clear field* or *positive* polarity masks have dark features on a transparent field. Both types are illustrated at the top of Figure 11–7. Conventional silver halide sensitized emulsion, as used on soft masks, reverses images on development, giving a photographic *negative* as in Figure 11–7; hence, each time an image is transferred photographically to an emulsion layer a polarity reversal takes place.

There are two types of PR, which in effect have opposite polarities. Thus, there are two options in choosing PR type. If positive resist is used, the mask and developed PR have the same polarity [Figure 11–7(b) and (d)], while negative resist reverses the polarity, as shown in (a) and (c). It follows that the number of photographic and PR steps used in producing a mask must be considered so the original artwork will be transferred to the final mask with proper polarity.

All *working* masks should be correct reading. This means that pattern features and lettering should read correctly, not reversed, when the imaged layer is viewed *through* the glass blank [13].

11.6.3 Mask Making; Pattern Generator

Paths (o) and (a_1) of Figure 11–1 require Ruby masters. Path (a_2) bypasses that need and the subsequent photoreduction. The key equipment here is the *pattern generator*, whose basic structure and specifications are shown in Figure 11–8(a) [14, 15]. The direction of light in this figure is downward from

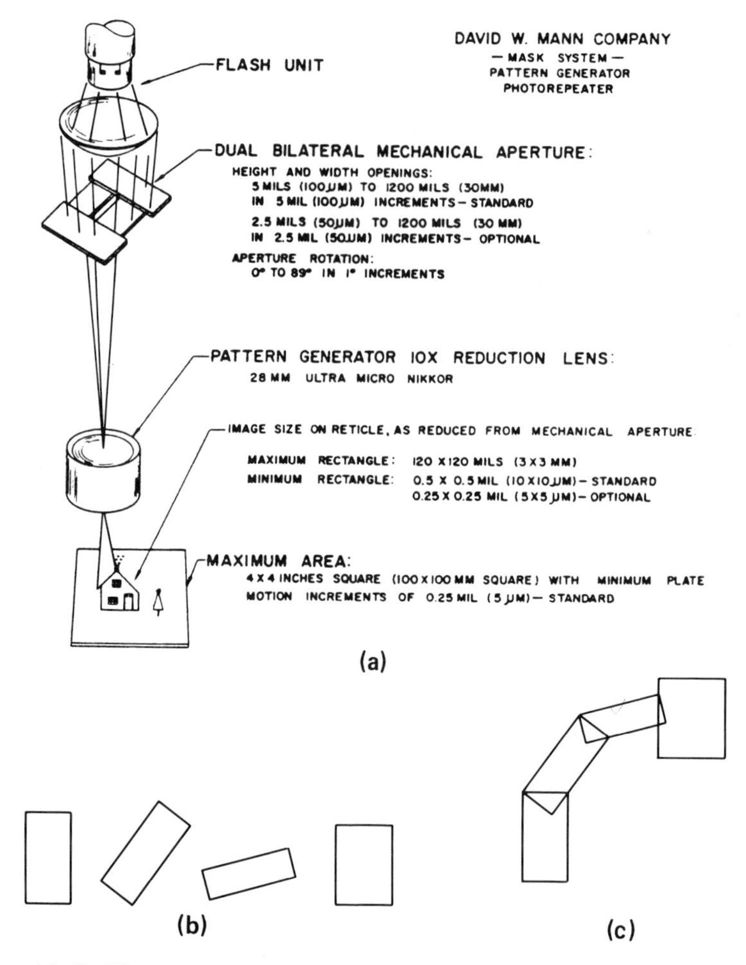

Figure 11–8. The pattern generator. (a) Basic setup. O'Malley [15]. (Courtesy of *Solid State Technology*, PennWell Publishing Company, Copyright 1971.) (b) Four consecutive aperture settings. (c) Superposition of the four settings.

the flashed light source to the reticle. The *dual bilateral mechanical aperture* is made up of two normal pairs of parallel, movable shutter blades. The blades can rotate as a unit through an angle θ. The separation of each pair, and θ are under control of the digital tape. Thus, the unit is capable of forming rectangular apertures of variable size and *aspect ratio* (length-to-width ratio) at any θ. The tape also controls the flashing of the light source and the y–z motions of the reticle support table. This unit also may be used for DWW operation.

Figure 11–8(b) illustrates the method of operation. Four consecutive aperture settings and their angular positions are shown; after each aperture has been set and the table moved to the proper position, the light unit flashes to expose the emulsion-coated reticle plate. Between flashes, the apertures and reticle support table are reset so the next exposure will be located properly wrt earlier ones. Figure 11–8(c) shows the pattern formed for this example.

By extrapolation we see that any pattern on an IC may be synthesized by successive exposures of rectangular elements of proper size, aspect ratio, rotation, and location. Only four are needed in the simple example; over 10^5 rectangular exposures may be required for an IC field mask.

The pattern generator bypasses the coordinatograph, photoreduction equipment, and Ruby master, so changes in a mask may be generated directly from corrected tape to new reticle in one step. This greatly speeds up the process of making mask changes or corrections. A disadvantage of the pattern generator is that if the light source fails to flash even once, the mask will be defective. The generated reticle must be checked carefully for this and other faults. Means for repairing such defects are discussed in the next section. The mechanical specifications given in Figure 11–8 date from 1971 and are quite remarkable for their small values and tight tolerances. Another pattern generator has been described in Section 11.5.

11.6.4 Mask Repair; Cleaning

There are two kinds of mask and reticle defects: a *void* in the pattern, which is transparent where it should be opaque, and a *bridge* in the pattern, which is opaque where it should be transparent. These defects result from faulty mask making—e.g., by a missed pattern generator flash, from mask damage, or from design errors. If the masks are of the hard type, say with chrome coatings, they may be repaired, particularly at the reticle level, which is relatively large in size. Repair frequently offers an economical alternative to replacement.

Consider two methods of patching a void such as that shown at the top of Figure 11–9. This is to be filled with a *patch*. Assume that a clear field (or positive polarity) mask was used to generate the defective reticle, so that the void is positioned in an opaque region of the master.

Say that negative PR is spun onto the defective reticle, reprinted from the master, and developed with the result shown in Figure 11–9(a), line 3. Since *dark develops* away with negative PR, a window appears over the void. A new Cr layer is deposited, the lift-off etching technique is used, and a chrome patch fills the void.

A more recent method of void repair involves bombarding organic photoresist with a beam of high-energy ions, as is illustrated in Figure 11–9(b) [16]. Positive PR is used here so that the opaque region of the master leaves PR in the void after printing and developing as shown in line 3 of Figure 11–9(b). Then the mask is scanned across the ion beam, of energies up to 150 keV, with a scanning time of about 30 min. Bonds are broken in the organic resist material with hydrogen and oxygen being released. The residue is *amorphous*

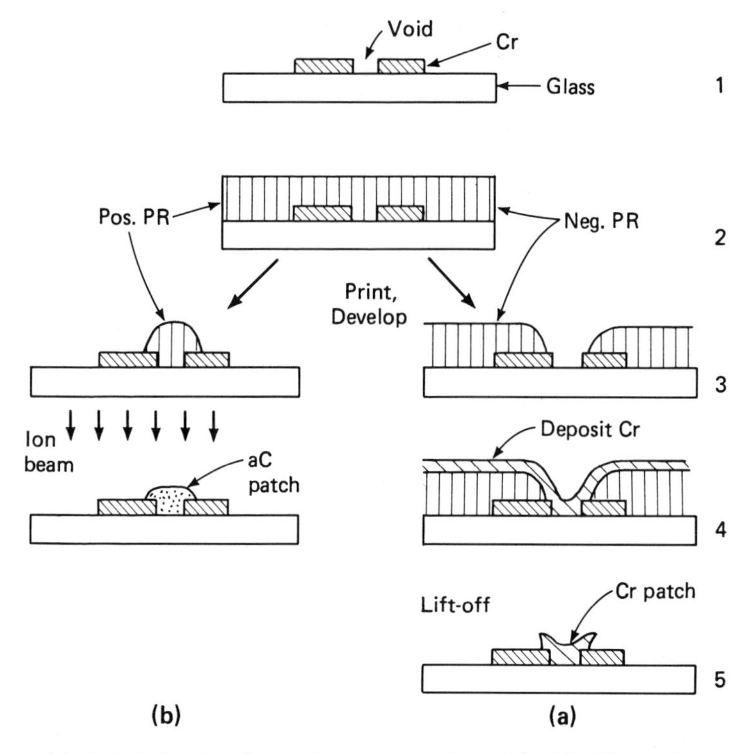

Figure 11–9. Methods of repairing a mask void. (a) Chromium patch. (b) Amorphous carbon patch. After Burggraaf [16]. (Courtesy of *Semiconductor International Magazine.*)

carbon (*a*C), which serves as a patch in the void. The *a*C adheres well, is scratch resistant, and is opaque to UV. This technique may be used for making masks by coating a glass blank with PR, printing it from a master, and then converting it to *a*C by exposure to ions. The amorphous carbon is then the patterned layer.

Ions of lower energy may be used to toughen patterned PR on a chrome-covered blank. This allows the Cr layer to be etched by ion milling, a form of ion etching that gives finer resolution. After ion milling, the hardened resist may be left in place on the Cr as an antireflective coating or it may be removed by ashing in O_2 [16]. Bridges, the second type of defect, also may be removed by spinning on a new layer of PR, exposing and developing it, and etching away the unwanted chrome.

Lasers also may be used for reticle repair. One system used by IBM has been described in Section 11.5. For void repair, the defective reticle is covered with PR that is burned away in the region of the void by the laser beam. A

Cr layer is deposited and lift-off is used to leave a Cr patch. Note the similarity to the process of Figure 11–9(a), except that here the laser burn replaces printing from the master. Bridges may be removed by *zapping* (i.e., by burning away) the metallic bridge with the laser beam.

The need for mask repair or even replacement may be reduced by keeping masks free of airborne particles, fingerprints, and particulate matter such as PR residues. Regular cleaning and/or the use of pellicles (Section 11.6.5) help to do this. Wet cleaning tends to cause pinholes in masks and reticles, so dry methods are preferred.

Blowing off the surfaces with dry air or nitrogen tends to force particles onto the reticle more firmly, so particle removal by suction is recommended. The gap between reticle and suction head should be small (not over 6 to 8 mils), and may be monitored by checking air pressure. The suction head is fixed in position and the reticle scanned under it by a movable table to maintain the small spacing uniformly. Some 95% of particles 5 μm or larger in diameter are removed, which is adequate for 10× reticles. Smaller particles must be removed for 1 : 1 applications. A soft brush vibrating slowly in front of the suction head will do this.

Traces of grease may be removed by a soft, absorbent roller placed before the brush. The rollers clean by absorbing grease, so they must be replaced frequently [17].

11.6.5 Pellicles

The pattern side of a mask or reticle, particularly if it is the soft type, is susceptible to damage by foreign particles. Reticle life can be extended considerably if the patterned surface is protected by a transparent barrier. A film and its mount affixed to the mask for this purpose, called a pellicle, not only seals off the pattern from particles, but also eliminates direct contact. Pellicles are suitable for projection printers only, because contact and proximity types have inadequate gap space. The pellicle film must have minimum effect on the projected image, so it must be thin, in the submicron to 0.5-mil range; it must be transparent to the UV exposure wavelength in use; and it must have a separation from the mask's pattern side (or *standoff*) of 100 mils or less [18].

Typical membrane materials are nitrocellulose and mylar, whose optical transmissions extend to wavelengths of about 350 nm and 250 nm, respectively. Since transmission losses are not zero, a pellicle increases printing time. The very thin membrane requires support, usually a plastic or metal frame that is sealed to the mask. Film and frame may be removed from the mask for cleaning or even replacement, if necessary. When in place, the pellicle, not the mask's pattern side, picks up particles. These lie in the light path that is focused onto a PR-coated wafer and can cast shadows. Since the particles are not on the reticle pattern, however, they are out of focus on the PR, and cause less illumination variation than do particles on the reticle. Considerable data are available relating lens f/number, standoff, and particle size to keep the illumination variations within 10%, a satisfactory value [18].

Pellicles may be cleaned by air blowing or suction, as described in the last section, or in extreme cases may be rinsed in deionized water.

11.7 Photoresists

Photoresists are the medium for transferring a desired pattern to the wafer surface in preparation for etching, and in some cases for mask making and repair. In this section, we consider the common photoresists that are based on organic polymers sensitized to UV, and look briefly at their types, chemistries, and properties. Resists for x-ray and electron beam exposure are considered later.

Most systems of photolithography are based on the selective dissolving of polymers during development. These polymers are linear, synthetic organics of relatively large size and molecular weight M. They must have certain characteristics: (1) absorb radiation, (2) have photosensitive additives that in response to radiation cause a change in solubility in a chosen developer, (3) be liquid for spin-on application, and (4) dry to a thin, tough coating that adheres well to the substrate. In addition, the viscosity of the liquid form must be adjustable by adding suitable solvents to permit proper spin-on application, and their formulation must be adjustable to the wavelength of the exposing radiation, e.g., the formulation for deep UV, where $\lambda < 300$ nm, will differ from that for UV [19].

Positive and negative PRs differ in how exposure to radiation affects their solubility in developer. These differences depend on their dissimilar chemical systems, which we consider separately. Overall behavior rather than chemical details is considered, and their properties are compared in Table 11–1.

11.7.1 Negative Types

Negative resists are mixtures of two components: (1) a rubberlike polymer *vehicle* of large molecular weight ($\leq 150{,}000$) and long chain length (about 5000 Å) that adheres well, forms a good film, is not photosensitive, and is soluble in the chosen organic developer; and (2) a *photopolymerizer* (or *photoinitiator*) that, on proper exposure, causes crosslinking of the polymers to form larger molecules. The affected molecules form a three-dimensional network of linked polymers, a *gel*, that is essentially insoluble in the chosen solvent.

The mask pattern defines regions of exposure and nonexposure. Unexposed regions, being soluble, dissolve away in development, leaving a window. (Recall the mnemonic: *dark develops away*.) Exposed regions, being converted to the insoluble gel, do not dissolve. Even though all exposed regions remain

Table 11–1. Comparison of Negative and Positive Resists

Negative	Positive
	Exposure
• Exposure decreases solubility	• Exposure increases solubility
• Exposed and unexposed attacked by developer, swelling	• Exposed becomes base-soluble, no swelling
• "Dark develops away"; exposed preserves image	• "Clear develops away"; unexposed preserves image
• Shorter exposure, greater latitude	• Longer exposure, smaller latitude
• Larger M, lower resolution	• Smaller M, higher resolution
• Sharper pattern, greater contrast required	• Requires less sharp pattern and contrast
• Overexposure undercuts and narrows windows	• Overexposure erodes resist top shoulders
	Application
• Thickness $<1/3$ min feature, step coverage problem, more pinholes	• Thickness $\approx$ min feature, less step coverage problem, fewer pinholes
• Better adherence	• Poorer adherence; may require adhesion promoter
	Development
• Solvent based, hazardous, effluent problem	• Aqueous alkali solvent, safer, less effluent problem
• Harder to strip, crosslinking present	• Easier to strip, no crosslinking, smaller molecules
	Wet Etch Resistance
• Resists acid or base	• Resists acid only; is base-soluble
	Extraneous Effects
• O_2 sensitivity, degrades image	• No O_2 sensitivity
• Dust and pinholes	(see Figure 11–11)
	Costs
• Less expensive	• More expensive

after development, they swell to some extent in the developer solvent and affect resolution.

Negative PR has a wide *exposure latitude*, i.e., a wide exposure range that still resists development. If an illuminated region is under- or partially exposed, crosslinking is not complete and the PR dissolves only partly on development, causing a *scum* of partially developed material. Scum formation is accelerated by the presence of NO_2 and ozone. Partial development, scum formation, and swelling distort feature edges, limiting the use of negative resists to line widths of 1.5 to 2 μm.

Negative resists are classified as *fast* since they require relatively short exposure times, the value depending on the intensity of the illumination. Some formulations permit dry development in a plasma that eliminates the scum problem.

11.7.2 Positive Types

Positive resists consist of three components:

1. A synthetic, polymeric resin that usually is based on a phenolic such as Novolac [20]. It has lower molecular weight and shorter chain length (about 100 Å) than the negative resists, is *hydrophobic* (water repelling), and has low solubility in alkalis.
2. A photoactive compound such as a diazide that increases the solubility of the resist in alkali, a process known as *photosolubilizing.*
3. A vehicle or solvent system.

In the photosolubilizing reaction, the resist converts to carboxylic acid during exposure; it is soluble in a weak water-based alkali. Unexposed regions remain while exposed dissolve away after exposure and development, just the reverse of negative types. It is significant that the unexposed resist remains insoluble in the developing alkali and so does not swell as negative resists do. Further, the smaller molecule size inherently permits better resolution than the other type.

Positive PRs have less exposure latitude than negative types, so scumming due to partial exposure is less of a problem. Overexposure erodes and rounds the upper shoulders of the resist but does not affect the window size. Positive resists show an aging effect in storage: a diazo dye forms that darkens the liquid with time. While they cost some four to five times as much as negative resists, the difference is insignificant because PR cost is only a small fraction of total processing costs. Positive resists are limited to use with acidic etchants since they dissolve in basic solutions, and they are deemed *slow* because they require longer exposure. The two resist types are compared in Table 11–1.

11.7.3 Optical Implications

The two types of resists—*negative*, which *reverses* pattern polarity, and *positive*, which *maintains* pattern polarity—provide flexibility in the lithographic process (Figure 11–7), but respond differently to anomalous effects, such as light scattering and dust or pinholes on the mask.

In the following discussions remember that in etching oxides it is the small pattern *features* such as windows that normally are developed away in photoresists; the large *field* or *background* surrounding the features remains after development. Conversely, for metal etching the *field* regions of the PR are removed by development, while the *features* remain. We now consider how PR responds to three commonly encountered optical effects.

11.7.4 Scattering and Line Width

Negative resist and positive polarity masks are used with contact and proximity printing for making ICs and devices for which resolution of 2 to 3 μm is satisfactory. Consider the effect of light scattering on line width for this combination as shown in Figure 11–10.

Ideally, the window in the developed PR and the mask feature should have the same width. Light passing the mask feature edges is reflected by the oxide or substrate, however, with some scattering under the mask as shown in (a) and the inset. Consequently PR under, but near, the edge of the pattern, which should not be exposed, receives varying amounts of illumination; hence, full development takes place near the edge, but decreases in from the edge. This has two effects: (1) the windows in the PR will be slightly narrower than defined by the mask, and (2) farther in from the edge where development is incomplete,

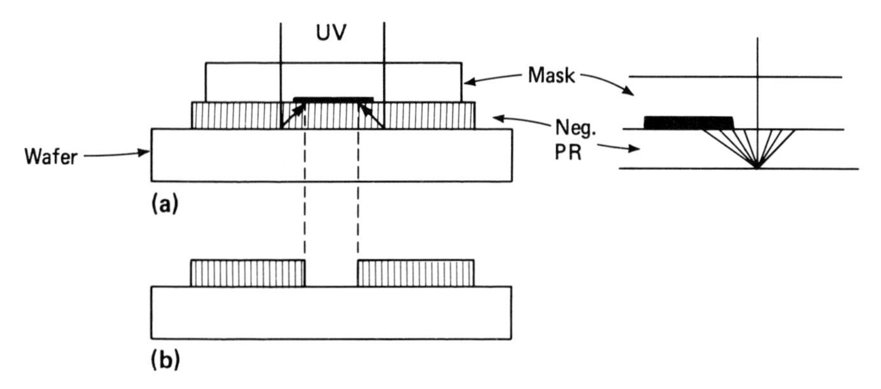

Figure 11–10. Scattering or reflection from the substrate causes a line in developed negative PR to be narrower than the mask line.

a *scum* of partially developed resist will remain, making the window edges ragged. Where line widths are small, these effects are serious. Positive resists are less susceptible to these effects (Section 11.7.2).

11.7.5 Dust and Pinhole Effects

Dust particles on the mask can block illumination from reaching the PR during exposure; their effects differ for the two PR types. Most dust particles will locate on the field regions simply because the fields occupy much more area than do the features. Moreover, the particles affect the PR only if they locate on transparent mask regions. Therefore, dust particles cause fewer defects with dark field masks (when they are used for patterning oxide). What effects they do cause can be serious, however.

Consider the oxide etching effects shown in the top line of Figure 11–11(a). The left-hand side of the figure shows a clear field mask and negative PR. Note the effects: dust on the field creates a void in the PR, but dust straddling a

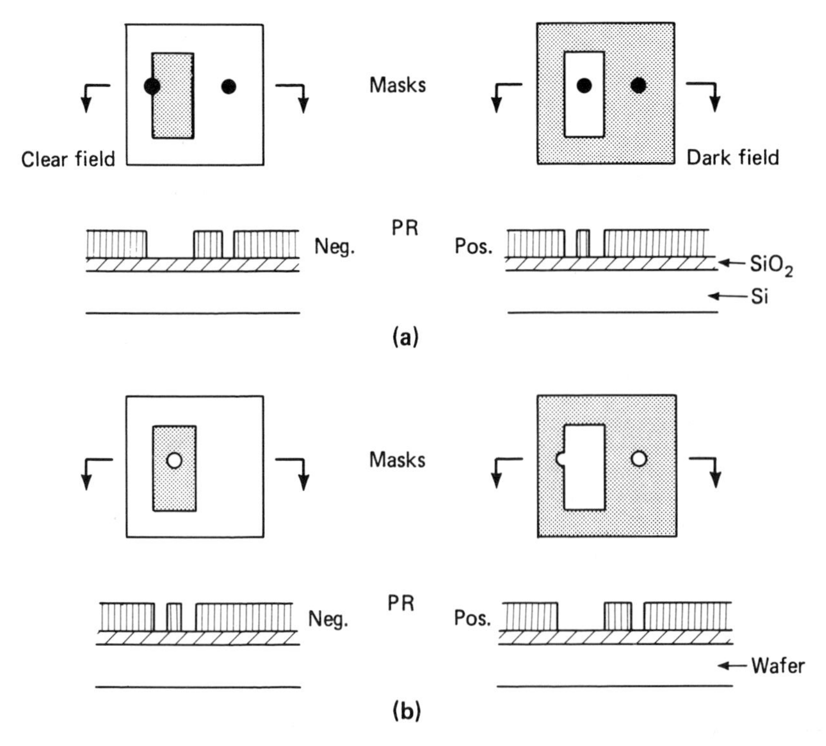

Figure 11–11. Some dust and pinhole effects. (a) Dust on the mask. (b) Pinholes in the mask pattern.

feature edge widens the PR window at the dust location. The dark-field-mask/positive-PR situation is shown at the right.

Where lithography prepares for metal etching, mask fields produce windows, and the combinations are: clear field mask/positive PR, and dark field mask/negative PR. The student can sketch the effects of dust for these cases. The moral is clear: keep masks clean. For oxide window etching, positive resist is less likely to have defects caused by dust because only a small fraction of the mask area is affected.

Pinholes *in the mask pattern* have just the opposite effect of dust on the developed PR in etching windows. Pinholes pass, rather than block, illumination of the PR during printing. The principal effects are shown in Figure 11–11(b). Pinhole patching has been covered in Section 11.6.4. An alternative solution to pinhole problems is afforded by the use of multilayer resists (Section 11.10).

11.7.6 Storage

Long-term storage of PR in opened bottles causes resist deterioration, especially in laboratory use, where the PR consumption rate is low and storage time long. Evaporation raises viscosity. Warm temperature and exposure to light accelerate random crosslinking, causing solid particles to form in negative resists. Positive types tend to darken, causing changes in their light-absorbing characteristics. Storage in tightly capped small bottles in a dark, cool environment minimizes these problems.

Some vendors repackage bulk lots in quart bottles with the resist filtered, say to a particle size of 0.2-μm, and with viscosity adjusted within desired limits. Still, the resist usually is dispensed onto wafers through special filters to remove residual particulate matter. Burggraaf reviews a number of resist-related processes [21].

11.8 PR Application

Spin on is the usual means of applying PR. The wafer is mounted on the chuck of a motor-driven *spinner*. Chucking is critical; it must keep the wafer level. Liquid resist is then dispensed onto the wafer and *spun* at 3000 to 6000 r/min. Centrifugal force spreads the resist into a thin coat over the wafer surface. Most of the solvent is spun off, so the coating is no longer liquid. Thickness τ is fairly uniform over the surface, except for a slightly thicker bead near the circumference.

An empirical relationship between film thickness and *final* rotational speed for certain conditions is

$$\tau = bs^{-a} \tag{11.8-1}$$

where s is the rotational speed in thousands of r/min ($s = n/1000$ where n is in r/min), and b and a are empirical parameters for a specific resist at a specified viscosity, on a substrate of given material and size. Values of a and b are given in Table 11–2 for 3-in. and 4-in. Si wafers coated with KMR (Kodak Micro Resist)-747 with nominal viscosity of 33 centipoises [22]. Values also are given for the KMR on SiO_2 and Si_3N_4. Thickness τ for these values is in μm.

More factors are involved than implied by Eq. (11.8-1); another empirical form is required for the same resist on a 4-in. nitride-covered wafer

$$\tau = 0.55 + 0.5686 \exp[-0.423(s - 3.0)^2] \qquad (11.8\text{-}2)$$

with the same units as before. It is apparent that the τ versus s curve here has a different shape: it maximizes at $s = 3\ \text{min}^{-1}$ whereas for the other cases τ decreases steadily with increasing s. Variations are so great that users should determine their own τ versus s curves.

The desired τ value depends on several factors. For example, to insure proper exposure for *negative* resists, $\tau < \rho/3$ where ρ is the minimum feature size, so for a 2.5 μm line width τ is restricted to roughly 0.8 μm [23]. While small τ is desirable for good resolution, it can lead to problems: pinholing is more likely in thin resists, and good coverage over comparatively large steps on the wafer surface (see Figure 10–3) is difficult to achieve. In contrast, *positive* resists can easily tolerate $\tau \approx \rho$, so pinholing and step coverage are lesser problems. Some authors refer to τ/ρ as the aspect ratio; we have defined aspect ratio in a more conventional manner in Section 11.6.3.

There are two variations in dispensing PR. In *static* dispensing, the liquid is dropped onto a motionless wafer, which then is brought up to speed. In *dynamic* dispensing, PR is placed on the wafer while it rotates slowly at, say, 500 r/min, and then is accelerated to the final speed at which most of the solvent extraction takes place. In both cases, the acceleration rate is important, so automatic control of the spinner motor is used [21, 22].

If PR adherence to the underlying substrate is poor, subsequent operations can lead to undercutting at window edges, edge lifting, and in extreme cases

Table 11–2. Parameters in Eq. (11.8-1) for KMR-747

	3-in. Wafer		4-in. Wafer	
Substrate	b, μm/min[a]	a	b, μm/min[a]	a
Si	1.5260	0.52	1.4689	0.53
SiO_2	1.4725	0.49	1.4754	0.54
Si_3N_4	1.2499	0.41	[See Eq. (11.8-2)]	

O'Hagan and Daughton [22]. Courtesy of *Circuits Manufacturing,* © copyright 1978, Miller Freeman Publications.

to complete loss of small-pattern features. Hence, the wafer surface must be free from contaminants such as residual scum from earlier resist operations, dust particles, and even metallic compounds from earlier etching processes.

Good adherence of resist to substrate depends on chemical bonding, so the number of substrate bonding sites is important. This number varies with substrate material and surface condition (e.g., adherence is better on monocrystal silicon than on polySi).

Amorphous SiO_2 can give troubles because it has comparatively few bonding sites and tends to be hydrophilic (water-attracting). This is particularly bad if the oxide is phosphorus-doped, since phosphorus and water form H_3PO_4 (orthophosphoric acid), which can break bonds. Fresh oxide, grown or deposited directly before PR application, minimizes this difficulty.

The use of an *adhesion promoter* such as hexamethyldisilizane, $(CH_3)_3SiNHSi(CH_3)_3$, on SiO_2 can cause the surface to be hydrophobic: HMDS reacts with SiOH to give $SiOSi(CH_3)_3$ with the trimethyl radicals bonding to the surface and NH_3 being liberated [24]. The HMDS is diluted to a 5% solution in xylene, deposited on the wafer, and spun at about 500 r/min until the wafer appears clear. Resist is then applied in the usual manner and no processing parameters are affected. Xylene can be used for cleanup.

Adhesion also is improved by a heat treatment, before PR application, known as *bakeout*. This usually is carried out in a convection oven with temperature in the 200°C range. Microwave oven heating is an alternative, provided no metal is on the wafer. A *prebake* after PR spin on drives off residual solvent from the PR. Heating should be slow to keep a hard surface coat from forming before evaporation is complete. Since microwave heating proceeds from inside to outside, it is helpful in this regard. Complete loss of solvent at the PR/resist interface aids bonding, and with no trapped solvent remaining, the subsequent development of pinholes is minimized. Complete drying of the resist has another advantage, particularly for contact printing: the PR is no longer tacky and will not stick to the photomask.

The prebake temperature and duration must be restricted to prevent thermal crosslinking that could contribute subsequently to scum and particle formation. The proper conditions depend on the type of resist used. Vendor instructions, which include correct prebake parameters, should be followed carefully.[5]

Since PR adhesion partially depends on the underlying material, manufacturers formulate resists for particular applications, e.g., KTFR for thin films, KMER for metal, and KMR for microimages by Kodak. Other manufacturers have similar product lines.

[5] A *postbake* is used after PR exposure. The prefixes *pre* and *post* imply before and after exposure, while bakeout is used before spin on. Postbake hardens and toughens the resist after development to make it ready for etching operations.

11.9 PR Processing

Figure 11–12 outlines the major steps in processing photoresist when a photomask is used to define the pattern. It is assumed that the wafers have been stored before the sequence begins and that a delay, requiring storage, may be necessary before the wafer goes to the etching station. To prevent contamination, storage should be in airtight containers. The names of major equipment items are enclosed in ellipses in the figure.

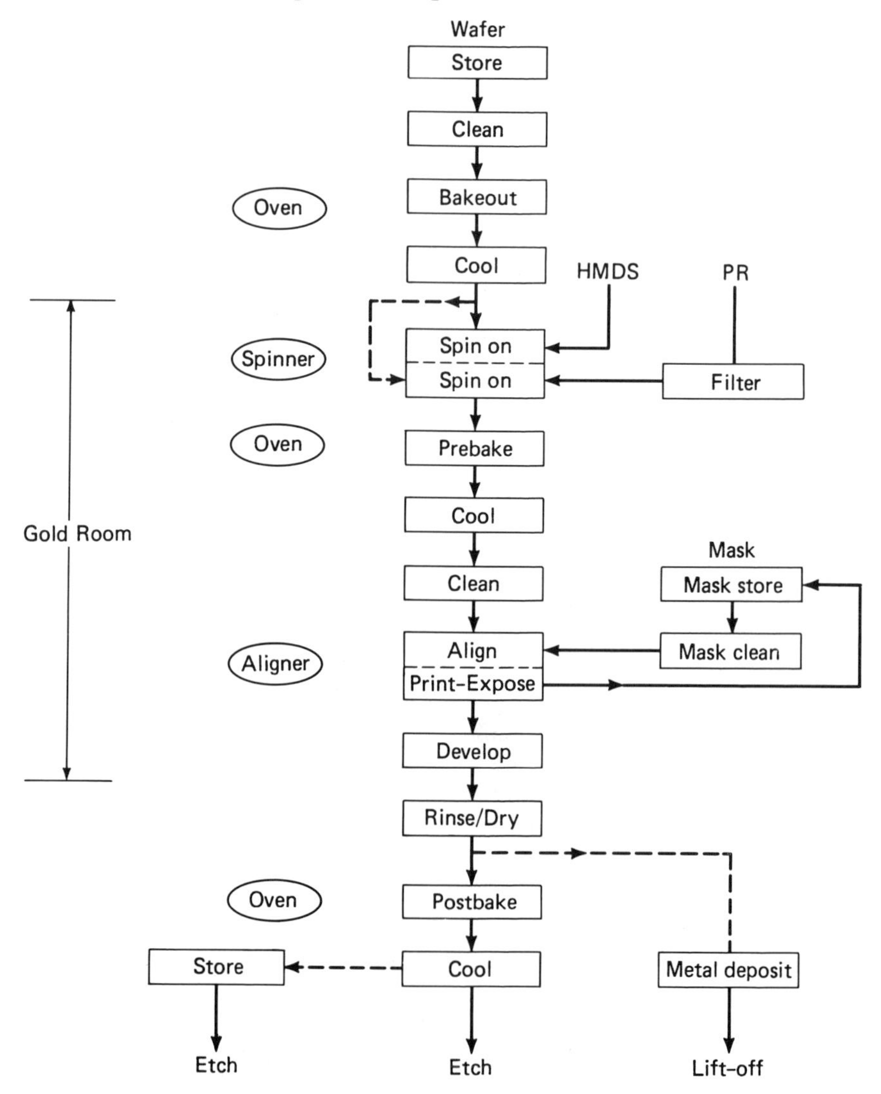

Figure 11–12. The major steps of PR processing

Conventional resists are sensitive to NUV and UV until developed; hence, they must be processed in especially lighted rooms free of those components. Gold-colored fluorescent lamps or gold-filtered daylight lamps are used for this purpose, so the PR processing areas are known as *gold rooms*. The figure shows which steps require this special environment.

The three previously mentioned baking steps are shown. Values on the low side, tailored for use in three-hour teaching laboratory sessions, are: bakeout, 15 min @ 200°C; prebake, 12 min @ 100°C; postbake, 10 min @ 125°C. Manufacturers' recommendations should be followed if feasible. Note that each bake is followed by cooling to return the wafer to room temperature before the next operation begins. Provision is shown for applying HMDS or another adhesion promoter to the wafer before PR spin on.

Only major steps are shown in the figure. Every operating facility should have protocols covering all essential details such as cleaning the spinner of excess resist spun off during use, mounting masks and removing them from the aligner, and so on.

In commercial practice, PR development is handled by automatic machines, which dip, spray, and/or ultrasonically agitate the wafers in the developer in order to promote uniform results. Single wafers can be developed by hand. A wafer is held with clean tweezers and dipped into developer for 40 to 45 s. This may be followed by spraying on more developer. (An artist's air brush serves well here.) Small batches may be handled in a carrier that is dipped into the solutions. Development is continued until visual inspection shows a good, clear pattern. Remember, development must be done in the gold room.

The developer should be of the type recommended by the resist manufacturer. Many solvents are available for use with resists. Resist removal, or ashing, after the etch process, has been discussed in Chapter 10.

11.10 Multilayer Photoresists

A single-layer resist (SLR) must satisfy a number of contradictory conditions: it should be thin for high resolution without causing pinholes and poor step coverage, it must provide good etch resistance without affecting good substrate adhesion, and so on.

Two resist layers exposed in sequence to slightly different masks can solve the pinholing problem. A resist layer is applied to the substrate, printed from a mask, and developed. The process is repeated with the same type of resist, but printed with a mask having slightly larger pattern features. Pinhole reduction results because the chance of pinholes lining up in the two layers is small.

Recent uses of multilayer resists (MLR) are more imaginative and employ different resist types for the two or even three layers. Each resist may be optimized for a particular function. Consider the *portable conformable mask* (PCM) arrangement of Figure 11–13, where a very thin top layer serves as a mask for the lower one [25]. The lower PR's chief function is planarization (i.e., smoothing out steps on the wafer surface). It also is formulated to adhere well to the substrate and to withstand the rigors of etching. In particular, if dry etching is used, the PR of the lower level must withstand the heat produced by incident ions, and possibly the ashing effect of O_2 in the plasma. Resists based on poly(methylmethacrylate) (PMMA) were used for this bottom layer, but they tend to have poor thermal stability. Better replacements are poly(dimethylglutarimide) (PMGI) which can also be developed in an aqueous solution, and GAF PR-514, a polyimide-based positive resist that can hold up to temperatures generated by plasma and sputter etching [26].

Since the top resist is applied to a planarized surface, it may be much thinner than in an SLR application. After development it serves as an *in situ* mask for the bottom layer. Conventional spin-on resists may be used, but their thinness is limited, and interaction between the two resists may form an interlayer that can affect top-layer development. Kodak Micropositive 809, or Shipley Micropositive 2400 or 300 resists prevent such interlayer formation [25].

Typical values for this system are: spun-on PMMA bottom layer with $\tau \approx 2$ μm, prebaked, and sensitive to DUV; spun-on top layer with $\tau \approx 0.23$ to 0.26 μm, and sensitive to near UV. The top layer is printed with UV through a conventional mask, or by DWW with a pattern generator, and developed. A DUV exposure follows that prints the PMMA layer, with the upper layer

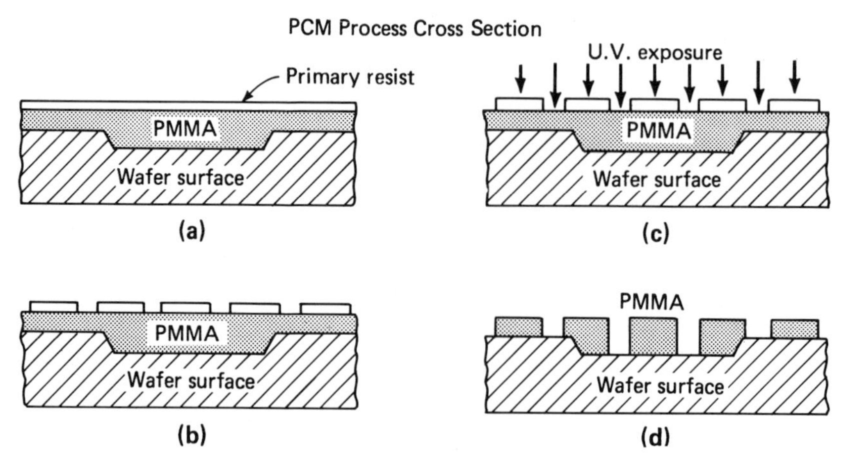

Figure 11–13. An example of two resist layers: the PCM process. After exposure and development the top layer acts as the mask for the lower PMMA layer. Johnson [25]. (Courtesy of *Semiconductor International Magazine*.)

serving as the *in situ* mask. Development then readies the system for etching. The main processing steps are illustrated in Figure 11–13.

The upper layer may use *inorganic* resists, based on the photosensitivity of silver compounds. They can be deposited by physical vapor deposition to thicknesses below 1000 Å, significantly less than those attainable by spin on.

One such system uses a 100-Å top layer of silver selenide (Ag_2Se) on a 600-Å layer of arsenic selenide (As_2Se_3). The bottom layer is polyimide-based and is 3000 Å thick. Silver ions are activated under exposure and migrate into the center layer, which becomes insoluble in the developer while the top layer dissolves. Since the layers are very thin, lateral spreading of the pattern image is negligible. Either wet or dry development may be used. The middle pattern layer can resist O_2 plasma, so dry etching of the lowest layer may be used. Plasma etching without oxygen can transfer the pattern to the substrate. In 1983 these resist systems gave 0.5-μm geometries with available optical systems, but now that value can be achieved with single-layer systems. Several variations of the process that involve organic resists are discussed in reference [27].

11.11 X-Ray Lithography (XRL)

The previous sections have dealt with *optical* lithography, where UV radiation extending down to around 200 nm is used for transferring the pattern to the wafer PR. Ultimate resolution was restricted by the wavelength. We next consider *nonoptical* methods, in which transfer is by x-rays or electron beams (e-beams), and where λ is lower by a factor of at least 100. The lithography paths for these nonoptical methods are grouped under II in Figure 11–1. First consider the x-ray system.

X-ray lithography (XRL) wavelengths lie in the range of 4 to 8 Å (0.4 to 0.8 nm) where a whole new set of rules applies:

1. X-rays cannot be focused by a lens.
2. X-rays are uncharged photons and cannot be deflected by electric or magnetic fields. Exposure relies on straight-line ray paths from mask to wafer in proximity printing, which is chosen to reduce mask wear.
3. X-rays reflect only from a limited number of metals and then poorly.
4. X-rays have high penetrating power, so the *transparency* and *opacity* of materials differ from what they are in the UV range.

For example, consider the Cr/silica optical mask: silica that is transparent through the DUV is opaque to x-rays, while chrome is opaque in the UV but transparent to x-rays [28]! In general, materials of low atomic number (At. No.) are transparent to x-rays while certain metals of high At. No., e.g., Au, are opaque. Hence, x-ray masks require different materials from those for UV use. The resolution of any mask-based system can never exceed that of the masks, so, to provide adequate resolution, XRL masks are generated with e-beams of even shorter wavelength.

11.11.1 Setup

Figure 11–14(a) shows the basic XRL setup with a linear source. The vacuum chamber at the top encloses the electron source, accelerating electrodes (not shown), and the target, often of palladium (Pd), from which the x-rays are emitted. A thin (5-μm) window of beryllium, Be (At. No. = 4), serves to pass the x-rays, and hold the vacuum. The lower chamber is often filled with He at atmospheric pressure because it has negligible absorption for x-rays, helps keep the mask and substrate clean, and improves heat transfer.

Ideally, pattern transfer should be by parallel rays. Even though the rays travel in straight lines from a small source, they are divergent and cause distortion as shown in Figure 11–14. Diverging rays from a point source target, as shown at (b), cause slight magnification of value

$$M = \frac{2w'}{2w} = 1 + \frac{g}{D} \qquad (11.11\text{-}1)$$

where g = mask-to-substrate separation and D = source-to-mask distance as shown in the figure. Magnification is minimized by making $D \gg g$. The value of g is about 20 to 50 μm in typical practice. This provides protection for the mask while it and the substrate are still in the same thermal environment [29].

If the mask field is too large, divergent rays and thick mask features can cause distortion as shown in Figure 11–14(c). The satisfactory field size may be increased by raising D and lowering the mask feature height (thickness) to the minimum required for opacity to the x-rays.

Divergent rays also place a requirement on mask planarity as shown in Figure 11–14(d). Given that $g \ll D$ and $t \ll D$, the deviation in the location of a mask feature edge in the transferred pattern is

$$\Delta y = R\frac{t}{D} \qquad (11.11\text{-}2)$$

where R is the radial distance from the optical axis to the feature edge on the mask, and t is the departure from mask planarity at that edge [30].

If the x-ray source is not a point, but has diameter d, source size can become resolution limiting as shown in Figure 11–14(e). The *smear* or *penumbral blur* is

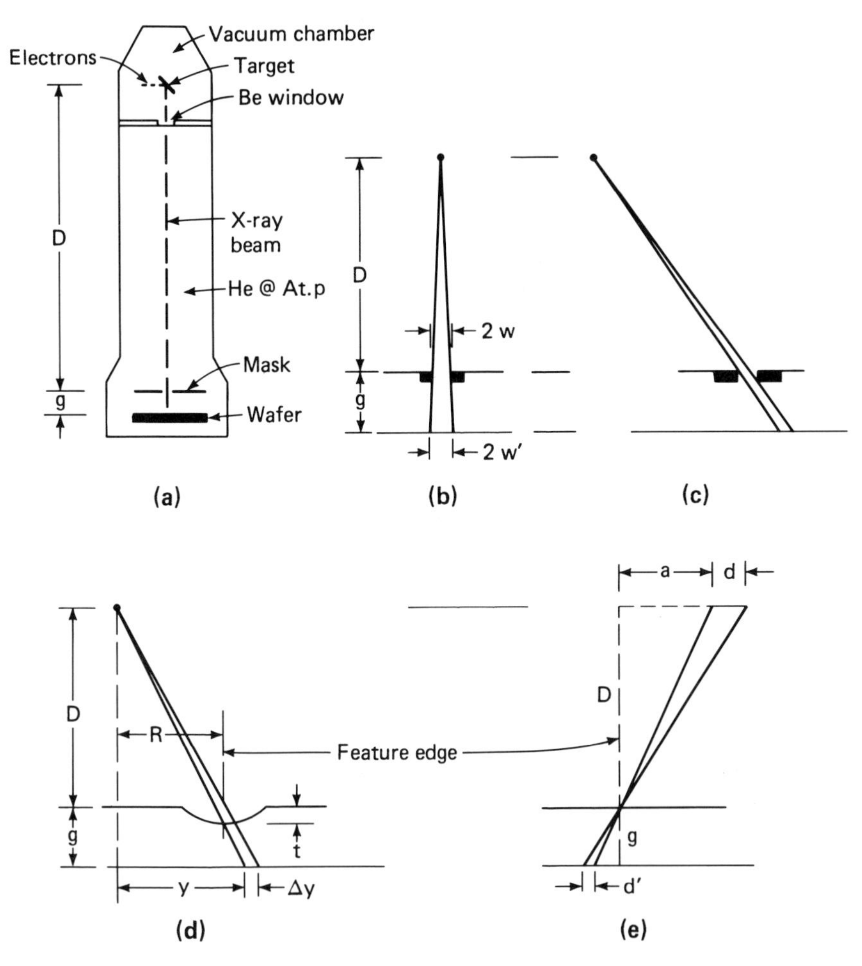

Figure 11–14. X-ray lithography with a linear source. (a) Basic setup. (b)–(e) Various forms of distortion.

$$d' = d\frac{g}{D} \tag{11.11-3}$$

Again, if g is minimized, d' will decrease as the source–mask spacing is made larger.

This is a proximity setup, so Eq. (11.3-1) for diffraction-limited resolution applies. Thus, for $\lambda = 4.37$ Å $(= 4.37 \times 10^{-4}\ \mu\text{m})$ and $g = 10\ \mu\text{m}$, $\rho = \sqrt{(4.37 \times 10^{-4})(10)} = 0.066\ \mu\text{m}$. The realizable value is larger by a factor up to 10 because of other considerations.

Distortion considerations place limits on the wafer size if global masks are used. Since at least 4- or 5-in. wafers are common, the trend is to use step-and-repeat, but with a field considerably larger than chip size—say about 4×4 cm, which is large enough to cover several chips. This keeps the throughput at a reasonable value.

X-rays from a storage source (next section) are well collimated and so are parallel. Consequently, these distortions are essentially eliminated.

11.11.2 X-Ray Sources

The limiting factors with x-rays are absorption requirements for exposing the resist, and lack of mask contrast (i.e., the relative absorptions in the "opaque" and the "transparent" regions of the mask).

There are two basic types of x-ray mechines: *linear*, or nonstorage, and *ring*, synchrotron, or storage. The former has a single electron source and target that generates the x-rays. The electrons travel in straight lines from source to target where the x-rays are generated. The electron beam power typically runs in the 5-kW range at 25 keV, but the transfer of electron energy to x-rays is inefficient and a weak photon beam results, making for a long exposure time. In linear sources the value of λ is determined by the target material, with palladium (Pd), $\lambda = 4.37$ Å, being the most common.

Most machines use water-cooled, stationary anodes of special shapes or rotating conical anodes. Linear machines are available in the same size as an equivalent optical unit and have throughputs of forty to sixty 3-in.-diameter wafers per hour. Resolution is in the 0.5-μm range, and beam divergence is comparatively large.

Photon absorption in x-ray resists is low, running less than 10%. Coupled with low beam intensity, this requires exposure times of around 2 min with $D = 30$ cm [30]. Chlorinated resists about 5000 Å thick absorb about 13% of the incident x-rays.

The second type of source, the *synchrotron* or *ring*, can produce collimated x-ray beams of higher intensity. In newer versions, electrons are injected at high velocity by a *linac* (**lin**ear **ac**celerator) into, perhaps, a square ring with rounded corners. They are accelerated further to velocities near that of light. When their paths are bent by magnetic fields to follow the ring, their change in angular acceleration results in a relativistic effect known as *synchrotron radiation*, whereby electromagnetic radiation is emitted. By proper control of the bending radius and the electron velocity, the emerging radiation is a collimated beam of x-rays about 100 times more intense than can be obtained from existing linear sources. Wavelengths run 4 to 20 Å [31].

X-ray-transparent ports at each ring bend allow the x-rays to be directed to steppers. Progress is being made to develop compact ring sources of about 5 m in diameter with up to 16 ports for driving steppers, but the capital costs are staggering [31, 32]. Exposure costs are comparable to those for optical steppers, but the resolution is significantly better. It is estimated that a ring machine has a potential throughput of about 60 wafers per hour, but that number varies with the number of ports and steppers.

11.11.3 Masks

It is more difficult to make masks for x-ray lithography than for UV. When used with x-rays, chrome is relatively transparent and silica relatively opaque, so a chrome on silica mask would be opaque to x-rays because silica underlies the metal. Satisfactory x-ray masks require two materials with significantly different x-ray absorption values. Layers of boron nitride (BN) <10 μm thick or of single-crystal Si <2 μm thick are transparent relative to gold, so these combinations are suitable for x-ray masks.

In one common method of mask substrate fabrication, a layer of BN : H (a hydrogenated amorphous form of BN) 3 to 4 μm thick, is deposited by CVD on both sides of a sacrificial silicon wafer. Side 1 is bonded to a heavy Pyrex support ring around its edge, and the BN is etched away on that side. The ring has an orientation notch, so the mask can be located precisely in subsequent operations. The BN on side 2 is covered with a support layer of polyimide on which an appropriate metal pattern is formed. All the Si is etched away from side 1 and the mask substrate is complete [28]. Details for making x-ray masks and several optional materials for them are given in references [28] and [32].

Two methods are used for producing the opaque metal regions: back etching and electroplating. In both methods direct write e-beam is used to establish the pattern in the resist. The back-etch process is already familiar, but it must be adapted to e-beam technology. A layer of metal is deposited onto the substrate side 2 window, which is comprised of 50 Å of Cr to promote adhesion, then up to about 6000 Å of Au. A resist is spun on this layer, exposed by e-beam, and developed. Exposed metal is sputter-etched away, and the resist is removed to leave the finished mask [28].

The second method of preparing a metal pattern requires deposition of two thin layers by vacuum evaporation: 50 Å of Cr to promote adhesion plus 300 Å of Au. Resist is applied, and after preparation is patterned by direct write e-beam. After development the unwanted portions are *dry* etched away, and the resist stripped. The remaining Au is too thin to be x-ray opaque; hence, it is electroplated with more Au to the desired thickness (about 6000 Å). The plating-up process yields vertical side walls and so maintains the excellent resolution of the e-beam direct write.

Most dust particles and other particulate matter are transparent to x-rays and so do not present shadowing problems as they do in photolithography. General cleanliness should be maintained, however.

11.11.4 Alignment

X-rays are not visible and cannot be reflected, so a separate visible-light laser source is provided for alignment. Different schemes of alignment marks are used. Several systems use diffraction gratings or Fresnel zone plates. One system uses three circular zone-ring patterns, two on the mask and one etched in the wafer SiO_2. The laser source illuminates the system at the same angle as the x-rays. When the alignment and the mask-to-wafer gap are correct, the three ring patterns appear in a straight line with the image of the wafer zone plate midway between the other two, and with all three in focus [30]. Auto align methods may be based on such optical images.

11.11.5 Resists

X-rays do not scatter in resist and, since the rays have a short wavelength, the resist side walls are quite vertical, a fact that aids resolution. Incident x-ray beams are of low intensity; hence the resist should have high sensitivity to keep exposure times reasonable. The developed resist must withstand the effects of plasma and the reactive ion beam during dry etch.

X-ray resists are based on high-molecular-weight polymers and copolymers. To enhance sensitivity, chlorine atoms, which absorb x-rays, often are added. Both negative and positive x-ray resists are available.

A typical negative type is poly(chloroalkyl acrylate), which is applied 0.7 μm thick and exposed to half-thickness by the incident x-rays. After development, reactive ion etching finishes cutting through the resist and the underlying material. As in photoresists, the negative resists here are resolution-limited by swelling during development. A good negative resist is 2, 3-dichloro-1-propyl acrylate (DCPA). It is usually mixed with copoly(glycidal methacrylate–ethyl acrylate) (COP) to enhance adhesion. The mixture is called DCOPA [33]. These resists have submicron capability.

Poly(methylmethacrylate) (PMMA) is a commonly used positive resist. When applied at a thickness of 300 Å it provides a limiting resolution of about 175 Å (= 0.0175 μm) [29]. Taylor gives a detailed summary of single- and multilayer x-ray resists that are in common use [33].

11.11.6 Appraisal

The principal question regarding x-ray lithography is whether or not its use is justified. Is its equipment cost counterbalanced by its resolution and throughput relative to e-beam technology, which, after all, is used for patterning XRL masks?

When x-ray lithography was first described in 1972 its goal was 1-μm resolution, then unattainable in the UV range, but now achievable. Also, it is tied to e-beam technology for making masks. At present, it is more economical than e-beam for large-scale production, but has inferior resolution in a world that demands ever smaller line widths, with e-beam giving $\leq 0.1\ \mu$m already. Since a mask is used with x-rays, pattern transfer is at least partially parallel, while e-beam DWW is entirely serial, so x-ray has a much greater throughput.

Whether or not x-ray lithography is economically justifiable depends largely on two questions: will x-ray equipment costs be reduced (and e-beam throughput be increased) significantly? The answers are not yet in.

11.12 Electron Beam Lithography (EBL)

Electron beam lithography (EBL) is a second type of nonoptical lithography. Pattern transfer is by a beam of electrons whose wavelength depends upon their energy. The relationship may be derived from de Broglie's equation and the expression for the electron kinetic energy:

$$\lambda = 12.3/\sqrt{E}, \quad \text{Å} \tag{11.12-1}$$

where E is the electron energy in eV and is numerically equal to the voltage through which the electrons are accelerated. For example, for an accelerating voltage of 25 kV, the energy is 25 keV and the corresponding wavelength is $\lambda = 0.078$ Å, a very small value compared to typical x-rays. It is not possible to realize line widths this small, but gold stripes 25 nm (250 Å) wide spaced on 50-nm centers have been produced. This far exceeds the requirements of existing devices [34].

In contrast to x-rays, electrons are charged particles; hence, they may be deflected and focused by electric and/or magnetic fields. The deflecting signals are computer-controlled and so may compensate for errors in the system. For these reasons, EBL gives the finest resolution and lowest distortion of all available types of lithography, but at the expense of low throughput and high capital investment.

The two principal applications of EBL are mask making and direct write-on-wafer (DWW), the former accounting for well over 75% of EBL use. The two applications are listed at (e_2) and (e_1) in Figure 11–1.

11.12.1 Setup

Figure 11–15 shows the column of an EBL system that generates an electron beam of square cross section. A slight change in the aperture plates can produce a circular spot, the type used in the early models. While not shown, y- and z-drive motors for the wafer stage, vacuum system, and the computer are essential parts of the system.

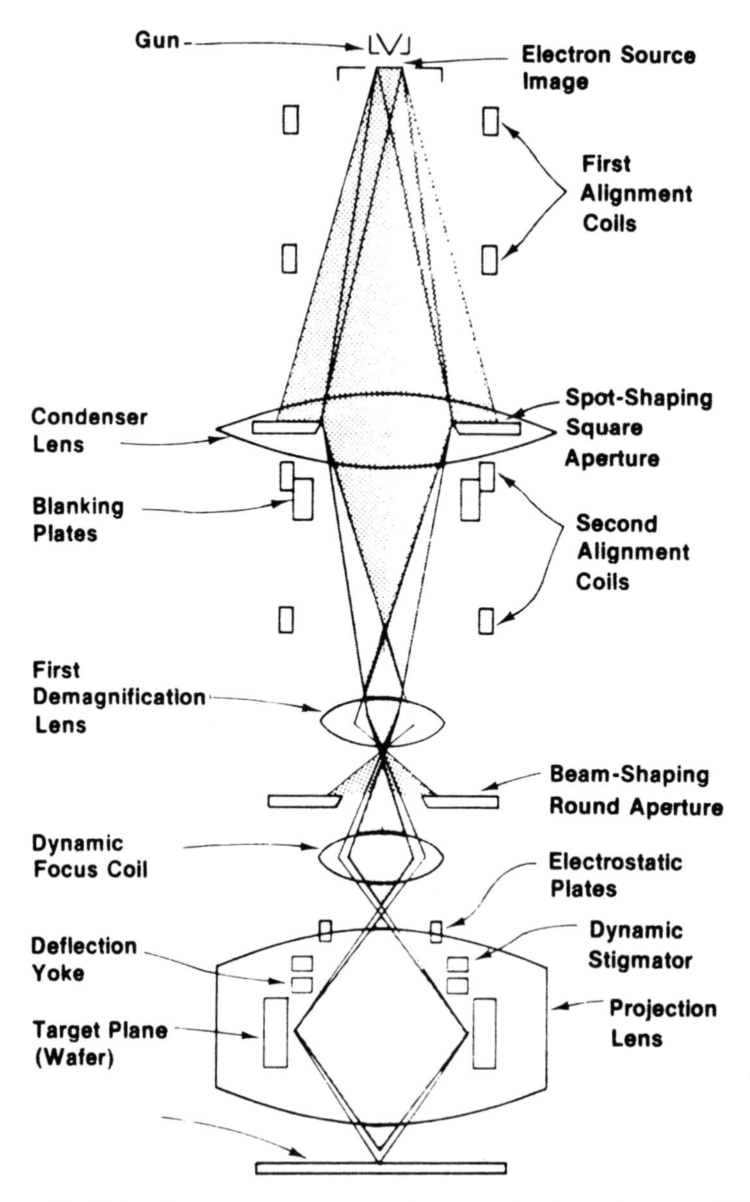

Figure 11–15. Basic square-aperture e-beam tool. Piwczyk and Williams [36]. (Courtesy of *Solid State Technology*, PennWell Publishing Company, Copyright 1982.)

An electron gun, comprised of a cathode and several beam shaping and focusing electrodes, provides the electrons. The cathode is either a directly heated tungsten filament or an indirectly heated sleeve coated with LaB_6 (lanthanum hexaboride) that emits electrons at lower temperature. The beam is accelerated by a high voltage, typically 10 to 100 kV, between the cathode and the electrostatic plates that function as the anode. Various electric and magnetic fields align, shape, focus, blank, and deflect the beam as it moves toward the wafer. Focusing regions in the figure are indicated by an optical lens in cross section. The entire column from electron gun through the wafer stage must be evacuated so that the electron paths will not be affected by collisions with gas molecules.

The drive signals for the several fields are under computer control, with basic scanning instructions furnished from the control tape of Figure 11–1. Through signal processing, the computer can adjust the beam motion to correct distortion and aberration in the several chip fields. The computer also controls the wafer stage motion.

11.12.2 Scattering Effects

When incoming electrons enter the resist, they hit and interact with atoms there and are scattered in all directions, resulting in *forward scattering.* Electrons also are *back scattered* from the substrate into the resist. Scattered electrons can travel up to about 0.5 μm in the resist, so the effect can be large. As a result, the exposed region is larger than the incoming beam, and line width exceeds both electron wavelength and beam diameter. Scattering, rather than diffraction, is the resolution-limiting effect in EBL.

Primary scattering contributes blur to the transferred pattern, while back scattering increases the region of exposure, so much so that the required primary exposure can be reduced by several percent. Also, back scattering contributes to the *proximity effect.* If two adjacent, widened exposure scans are fairly close, their scatter patterns may overlap. The overlap regions require less exposure; this is the proximity effect. Fortunately, signal processing in the controlling computer can compensate for this effect: the computer *knows* the beam's positions in advance and so can adjust the primary beam intensity accordingly.

The magnitude of the effect also depends upon the resist thickness. In mask making the resist is thin and the effect is small. In DWW applications the resist must be thicker (to withstand subsequent ion beam etching for longer times), so the effect is more pronounced. The proximity effect may be reduced by using special resists, especially multilayer types [34]. It also is influenced by the degree of adjacent scan overlap, which depends on the shape and size of the primary beam. This is considered in the next section.

11.12.3 Writing Schemes

Most EBL systems are of the DWW type and write the pattern serially. As it scans, the beam is turned on to write pattern features and blanked off in the no-pattern regions. Several scan-pattern/beam-shape schemes have been devised, always with the goal of increasing writing speed to raise machine throughput. We consider a few.

The first and oldest system was adapted from existing scanning electron microscopes (SEM), where a sharply focused beam of electrons is scanned over a field in a television-like *raster*. The beam has a circular cross section and a gaussian radial current distribution of the form given by Eq. (8.8-6) and shown by the dashed lines in Figure 11–16(a). The beam diameter d is defined here as the spacing between its half-response points. Adjacent line spacing may be adjusted. For example, in Figure 11–16(a), d is adjusted to one fourth of the desired minimum line width ρ and adjacent scans are spaced d on centers, so four adjacent scans are required for a feature line ρ wide. Other spacings may be used as in (b).

The raster scan wastes time: consider the situation of Figure 11–16(c) where the beam writes only when it is in the Γ-shaped feature. Since the raster scans the *entire* field at a constant rate, any time spent with a blanked beam (i.e., outside the Γ) is wasted, since no useful data are being transferred. This type of scan can place severe requirements on the transfer rate if a useful throughput is to be attained. Consider the following numbers.

Say a wafer has an active square area of 8×10^3 mm^2, made up of eighty 10×10 mm fields that are stepped and scanned in sequence. The wafer is to be scanned in 60 s, if feasible.

Say the field is dense with 1×1 μm features, i.e., the *pixel* area is 10^{-6} mm^2, so there are $100/10^{-6} = 10^8$ pixels/field that must be scanned in 60/80 s. Thus the data transfer, or beam writing, rate is $10^8(80/60) \approx 133$ MHz, a rather high value, so the wafer scan time must be raised.

A *vector* scan reduces these requirements by scanning only the feature portions of the field. At the start of the field the computer directs the blanked beam to a feature corner, say point A in Figure 11–16(d). The beam is unblanked and scanning is confined to the Γ-shaped feature. The time saved in changing from raster to vector scan is $t_r(1 - R)$, where t_r is the time to scan the whole field raster, and R is the ratio of the feature to field areas. For usual IC chips R will run about 0.2, so 80% of the field scan time is saved by the vector system, a significant saving indeed.

Examples of the circular-beam/vector-scan machines or *tools* are the Cambridge EBMF-2 and the Texas Instruments EBSP [4]. Descriptions of still other circular beam scan systems are available [35].

In the foregoing scan methods the beam is deflected electronically in two directions over one field at a time, and the wafer is stepped from field to field

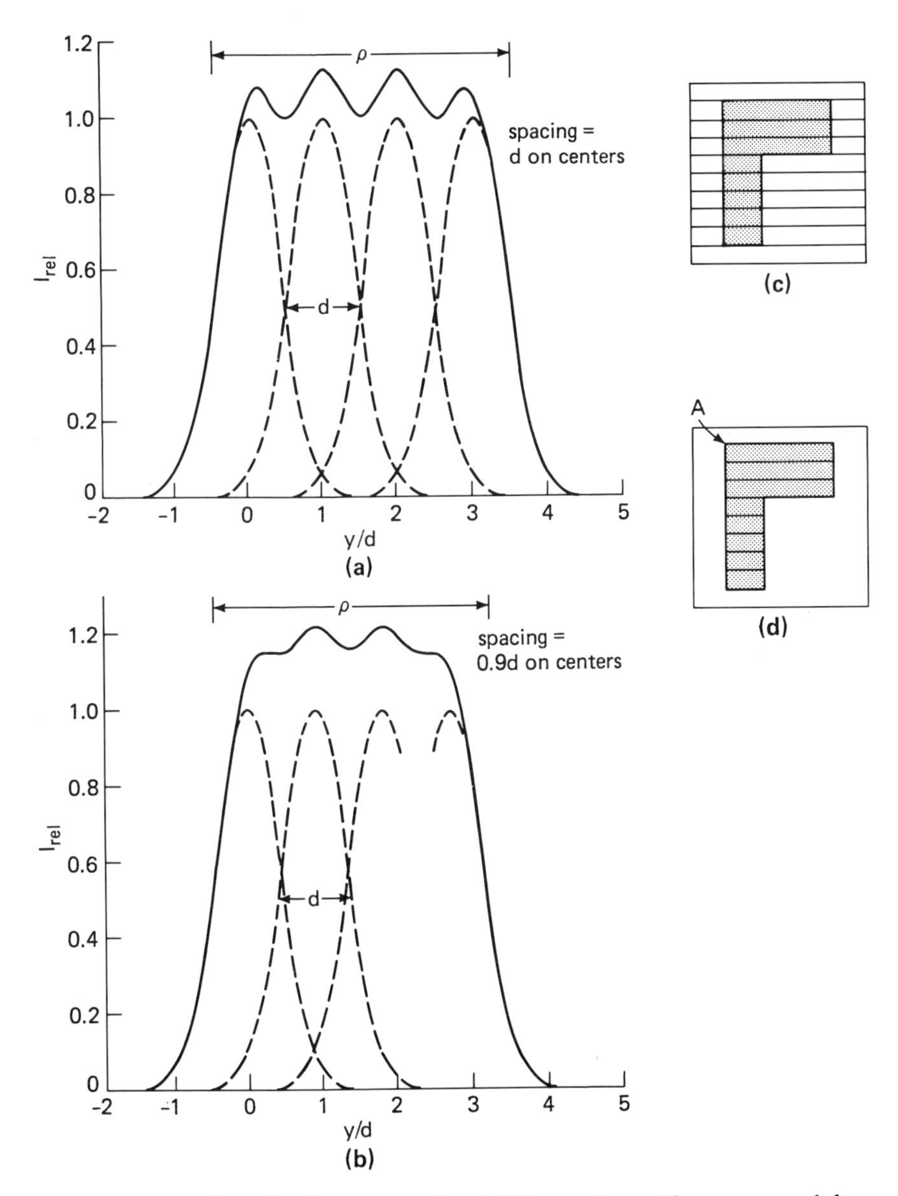

Figure 11–16. Gaussian beam scanning. (a) Four adjacent beams spaced d on centers. The gaussian shape of the individual beams is shown. (b) Four beams with $0.9d$ spacing on centers. (c) Raster scan. (d) Vector scan.

by the motor-driven wafer stage. Scanning can speed up significantly by moving the wafer *continuously* at constant rate across the beam (now scanning in only one direction) as in the electron beam exposure system (EBES) developed by Bell Laboratories [4]. A beam of circular or variable rectangular cross section is deflected at right angles to the direction of wafer movement by electronic means to trace a *stripe* or *ribbon* over the wafer as in Figure 11–17. (The variable beam feature is discussed later.) Each electronic scan line is controlled by 512 addresses in the computer, with each address 0.25 to 0.5 μm wide. Empty stripes in the pattern are not scanned. The data rate is 40 or 20 MHz and the system can write at 5 cm^2/min. For the same numbers as in the preceding example with an active square area of 8×10^3 $mm^2 = 80$ cm^2, the write time is $80/5 = 16$ min/wafer. Some commercial versions of the system

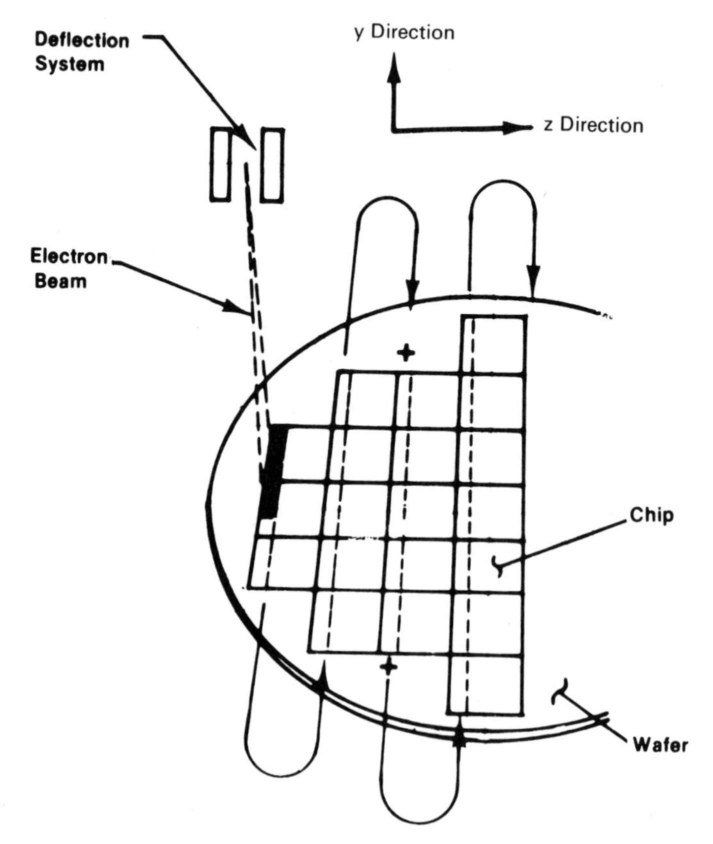

Figure 11–17. EBES scan pattern. The e-beam scans normal to the wafer motion. Piwczyk and Williams [36]. (Courtesy of *Solid State Technology*, PennWell Publishing Company, Copyright 1982.)

are the ETEC Corporation's MEBES (Manufacturing EBES) and the Varian eEBES.

There are advantages in using a beam of square cross section, such as that provided by the dual aperture system of Figure 11–8. The current distribution is uniform across a large part of the beam as shown in Figure 11–18(a). A typical beam shape uses $n \approx 5$. The greater width of the square, relative to the circular, beam allows greater coverage by each pass of the beam as it writes the pattern, so less time is required to scan a given feature in either raster or vector scan modes than with the circular beam.

The *fixed* square aperture cannot write a feature narrower than the aperture, however; for small ρ this limits the effective size of the square in wider

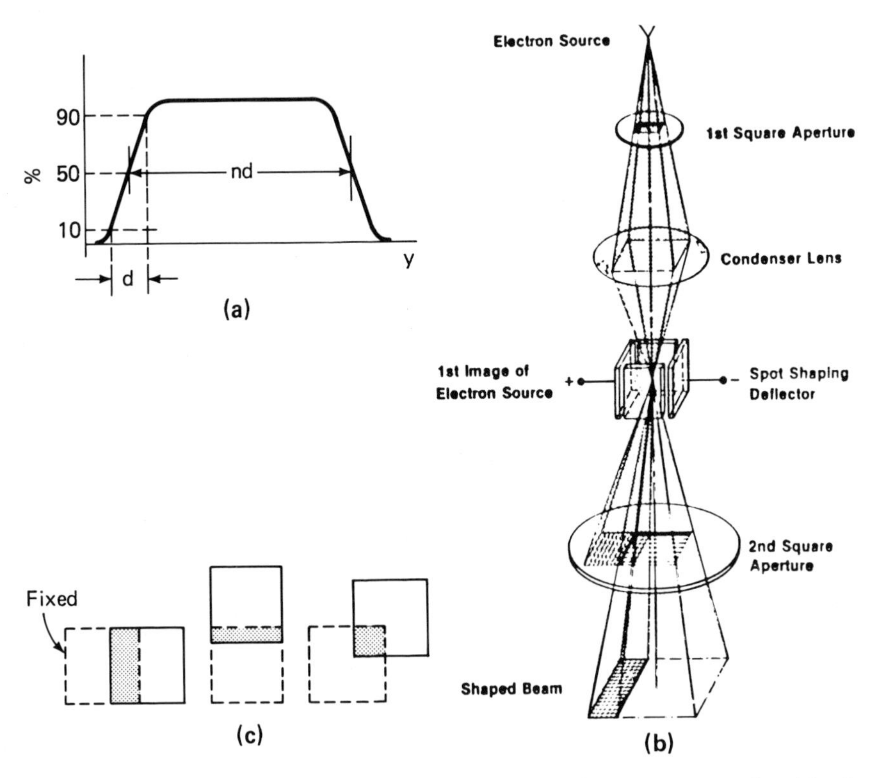

Figure 11–18. Rectangular beam EBL. (a) Square-beam cross section. Typically $n = 5$. (b) Modification of basic e-beam tool for variable width rectangular beam. Piwczyk and Williams [36]. (Courtesy of *Solid State Technology*, PennWell Publishing Company, Copyright 1982.) (c) A fixed and a movable aperture give different rectangular beams, shown shaded. The final beam size may be smaller than the apertures.

pattern features. Much greater flexibility is provided by a variable-width system as in the IBM EL-3 tool. The basic square aperture setup (Figure 11–15) is modified by the addition of two crossed pairs of deflection plates as in Figure 11–18(b) [36]. These can move the electron image of the upper aperture over the lower aperture. Only where the apertures overlap can electrons pass through to the wafer. The final beam shape is rectangular, but its size can be controlled in both the y- and z-directions to give shapes like those shown shaded at (c). Note that the beam may be smaller than the apertures. Demagnification is used to sharpen the edges of the final beam for resolution improvement.

Current densities can reach high values in the beam (e.g., for a total current of 8 μA in a 4 $\times$ 4-μm cross section the current density is 50 A/cm^2). These high values increase coulombic interactions among beam electrons and restrict the maximum allowable beam width. They also greatly affect the main magnetic deflection system that provides the field scan. Modified deflection yokes have been designed to compensate for this.

Heating is a consideration, too. At 25 kV and a beam current density of 50 A/cm^2, the power density is 1250 W/cm^2. This heat spreads through the PR and by thermally breaking bonds can affect unexposed areas as much as scattering.

Beam width is continuously adjusted to feature sizes—narrow for small features, wide for large—so write time is reduced further. The need for computer control is self-evident.

11.12.4 Alignment

Alignment fiducial marks are located on the wafer by using back-scattered electrons as in back-scattered or SEM microscopes. As the primary beam hits the substrate, the back-scattered electron intensity depends upon the material being hit and on the beam angle. The fiducial marks are special physical features on the mask, such as small mesas (Figure 11–19). Their back-scattered beam tells the computer where they are located relative to the scan pattern. The computer then directs the beam to the desired position [36].

11.12.5 Resists [30]

Most resists for masking against ion etching must be at least 0.5 μm thick to withstand the ions; greater thicknesses increase exposure time. Resists for EBL generally have less exposure latitude than do optical resists and suffer covalent bond breakage on exposure just as they would from heat. Whether they are of the positive or negative type depends on how their free radicals and ions then behave. If they cross link, the resists are negative.

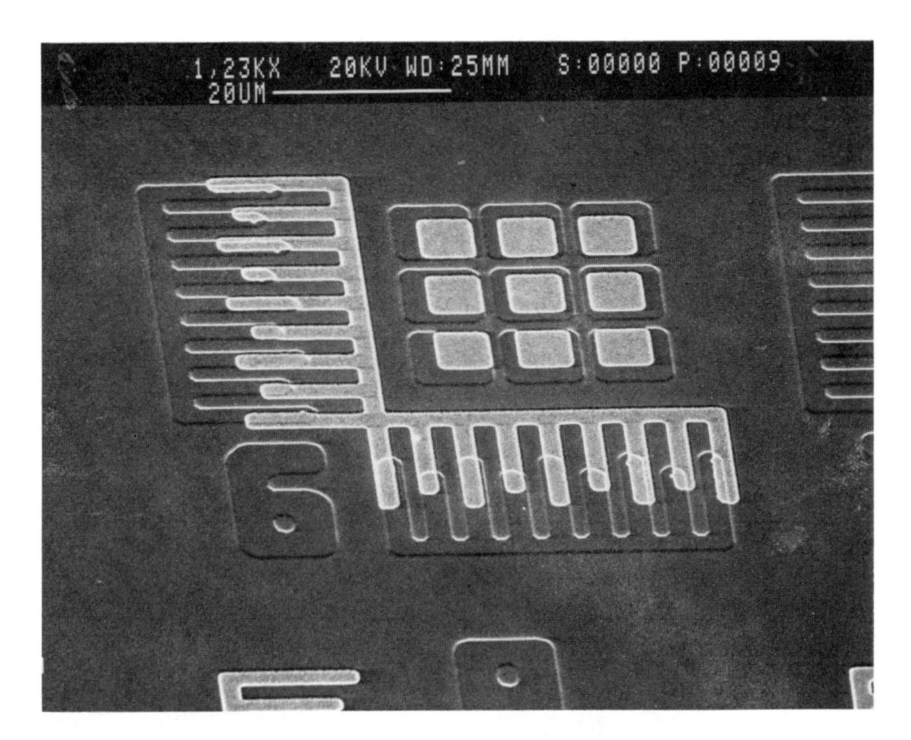

Figure 11–19. SEM photo of wafer features similar to those used for autoaligning. (Courtesy of NSF/ERC Center for Compound Semiconductor Microelectronics, UI-UC.)

Copolymer(glycidyl methacrylate–ethyl acrylate) (COP) is the most commonly used negative resist. The crosslinking process may continue for some time after exposure, so COP-covered wafers are held some 20 to 40 min *in vacuo* after exposure until this process is complete. This argues for a wafer-changing mechanism at the wafer stage that is in the evacuated column enclosure. Another example is PBS, copolymer(polybutene solfone), developed by RCA for chrome mask making. Among positive e-beam resists, PMMA has the best resolution, 10 nm, and reasonable exposure sensitivity.

Dry-developing and positive ablative resists also are available. The latter are self-developing in that they are removed or volatilized when exposed to the e-beam [37].

11.12.6 Mask Projection Systems

E-beam throughput may be increased if the pattern is masked rather than written serially. One tool to do this uses both UV and e-beam. The mask is patterned in titanium dioxide (TiO_2) and is in contact with a photoelectron emitter, as shown in Figure 11–20 [38].

Ultraviolet radiation passes through a UV-transparent window in the source envelope and through the transparent regions of the mask to strike the palladium-covered emitter, giving rise to low-energy photoelectrons (<0.5 eV). These are accelerated and focused in axial fields: a uniform electric field of 10 kV/cm and a magnetic field of 0.1 Tesla ($=1000$ G). The electrons travel in spiral paths in moving to the wafer, but, with proper adjustment of the fields, arrive there in their original relative (y, z) positions, effectively giving a 1 : 1 projection system with electrons serving as the transfer medium between mask and wafer. Details of several systems are in references [36] and [38]. Since the mask limits the ultimate resolution, it should be generated by conventional EBL.

These advantages are claimed: (1) global rather than serial printing, giving higher throughput than DWW systems, and (2) 5 times better resolution and 14 times the depth of field of optical systems; however, the method will have

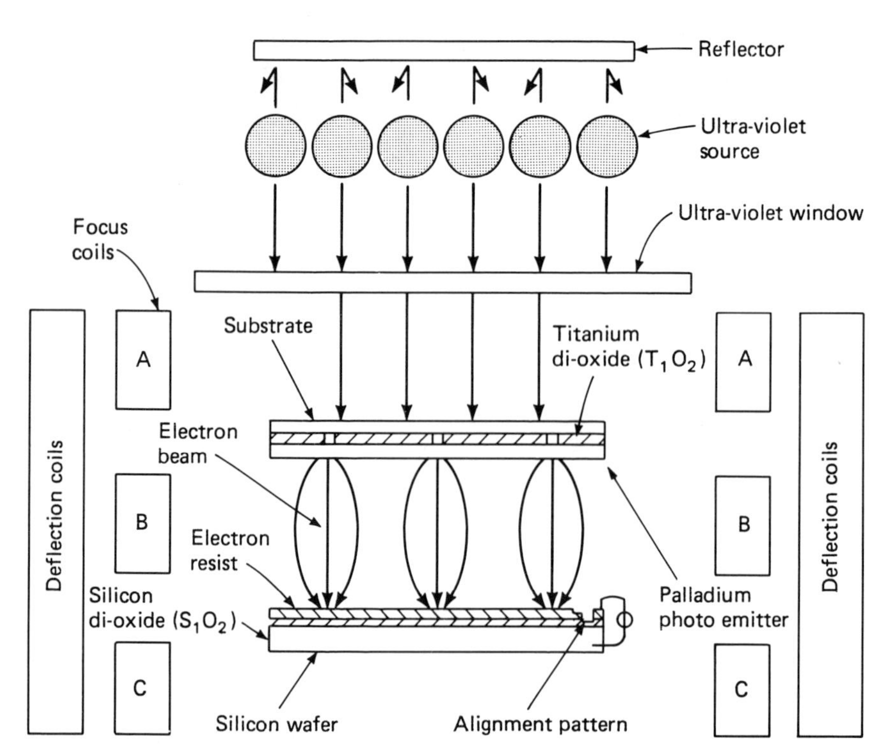

Figure 11–20. One form of electron beam projection system [38]. (Courtesy of *Circuits Manufacturing*, © 1971, Miller Freeman Publications.)

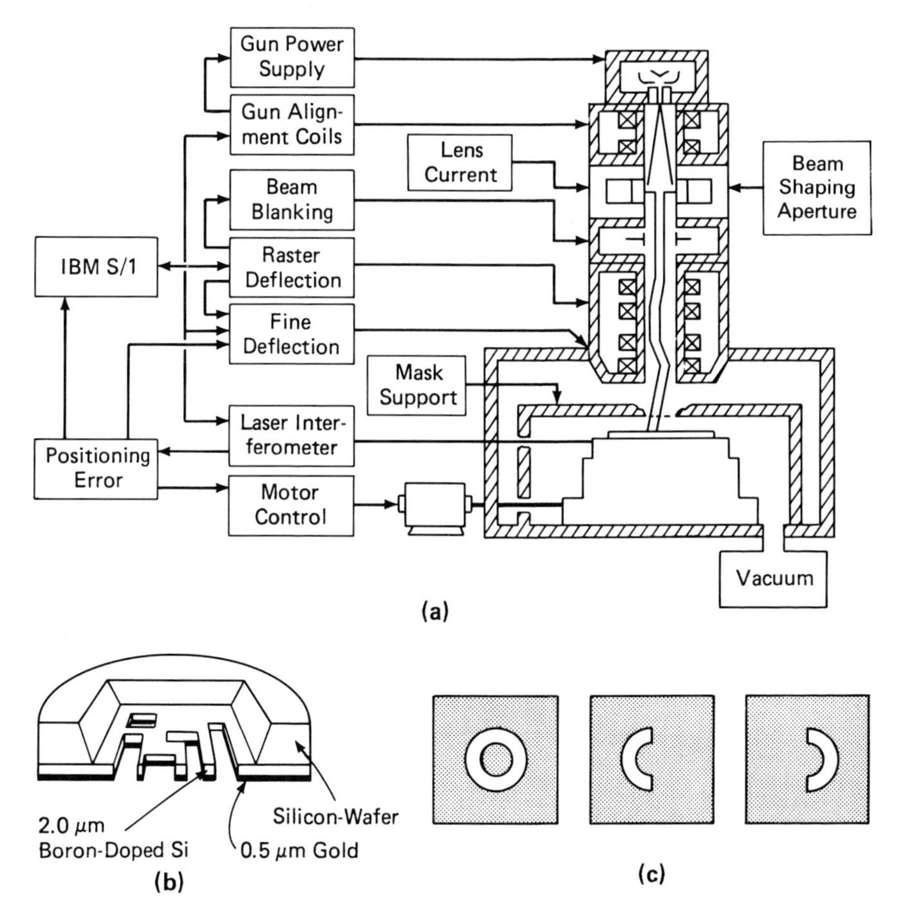

Figure 11–21. Electron beam proximity printing system. (a) Basic tool. (b) Mask construction. Bohlen et al. [39]. (Courtesy of *Solid State Technology*, PennWell Publishing Company, Copyright 1984.) (c) The doughnut problem.

poorer resolution than e-beam DWW systems. Wafer size is limited by the focus coils.

Figure 11–21(a) shows another type of mask-based e-beam printer, developed by IBM Deutschland GMBH, one that uses a *physical transmission* or *shadow* mask [39]. Transparent mask regions are holes through which electrons pass [Figure 11–21(b)]. The familiar *doughnut problem* shown in (c) is present: there is no way to support the center opaque portion in a single mask without blocking part of the transparent annulus. Hence two complementary field *half-masks* are used, side by side and spaced one field width on centers. The electron beam scans both half-masks on the two adjacent fields; after this scan, the wafer carriage is stepped by only one field width. In this manner, each wafer field receives two exposures, one through each half-mask. The time to scan a field is not doubled, except at the ends of each wafer-wide stepping cycle.

The masks are generated by direct write EBL with nonlinearities compensated for by corrections in the computer-generated scanning signals. Throughput is higher than for a serial DWW tool, but the resolution is not as good.

11.12.7 Appraisal

Electron-beam lithography offers dual options: printing with a mask, or the much more common maskless serial DWW. Various scan-pattern/beam-shape systems are available, but vector or ribbon wafer-wide scan of the EBES type with an adjustable-width beam give the highest DWW throughputs. Proximity effect compensation is required in all DWW variations. Compared to optical lithography and x-ray lithography (XRL) EBL has the highest resolution, about 0.1 μm, provides correction for scanning aberrations and distortions, and compensates wafer warpage, but has the lowest throughput.

Column costs for all three are roughly comparable, but of these only the e-beam column and the storage ring x-ray source must be evacuated. E-beam also requires more computer capacity. Relative costs for owning and operating XRL and EBL tools may be calculated [29].

E-beam lithography has two principal applications: mask making and direct writing on wafer. Low throughput in the former is of small consequence since mask making is an off-line procedure. The EBL systems have the greatest potential for pushing the limits of resolution.

Problems

11–1 Explain how and why the first mask of a series is aligned to the wafer's crystallographic orientation.

11–2 Illustrate translational and rotational misalignment errors with simple sketches of a mask and a wafer already showing features.

11–3 Explain:

(a) why hard masks give higher aligner throughput than soft masks.
(b) why proximity aligners have greater throughput than the contact type.

11–4 The following data apply to a lens for UV operation at the Hg i-line: NA = 0.2, $s = 20$ mm, $M = 5$, $f = 100$ mm. Calculate u, v, ℓ, F, F_E, ρ, and δ. Comment on your results. See Appendix E.

11–5 How many 12 × 12-mm dice will fit on a wafer 100 mm in diameter. Show with a sketch.

11–6 Briefly describe the operation of a DWW laser-driven system that does not use a dual bilateral mechanical aperture.

11–7 Draw the analog of Figure 11–11 for these combinations:

(a) Positive PR/Negative Mask.
(b) Positive PR/Positive Mask.
(c) What is the effect on line width in each case?

11–8 Sketch the remaining possibilities for Figure 11–11.

11–9 Make sketches analogous to those of Figure 11–11 but for a metal film replacing the oxide as the medium to be etched.

11–10 (a) Explain why the largest dust particle that can be tolerated on a pellicle is smaller for an f/8 lens than an f/4 lens.
(b) What is the effect on tolerable particle size as g decreases?

11–11 A 3-in. Si wafer is to be coated with PR to $\tau = 6500$ Å.

(a) What terminal rotational speed in r/min is required for a bare, an oxide-coated, and a nitride-coated wafer?
(b) Are your calculations valid for any PR? Explain.

11–12 A two-layer resist is to be exposed to separate masks without development after the first layer is exposed. How can you prevent exposure of both layers simultaneously? Explain.

11–13 Sketch a cross section of the finished mask substrate of the second type described in Section 11.11.3. Label the several regions.

11–14 Derive Eq. (11.11-2) and explain its significance.

11–15 How is scanning over an entire wafer accomplished in an EBES system, given the limited scan capabilities of the e-beam itself? Explain.

References

[1] W. B. Glendinning, "A VHSIC Lithography Review," *Solid State Technol.*, **29**, (3), 97–99, Mar. 1986.
[2] J. S. Hughes, NSF/ERC Center for Compound Semiconductor Microelectronics, UI-UC. Private communication.
[3] T. G. O'Neil, "Photomask Material Review," *Semicond. Int.*, **4**, (8), 81–94, Aug. 1981.
[4] R. K. Watts and J. H. Bruning, "A Review of Fine-Line Lithographic Techniques Present and Future," *Solid State Technol.*, **24**, (5), 95–105, May 1981.
[5] D. A. Markle, "The Future and Potential of Optical Scanning Systems," *Solid State Technol.*, **27**, (9), 159–166, Sept. 1984.
[6] P. Burggraaf, "Wafer Steppers and Lens Options," *Semicond. Int.*, **9**, (3), 56–63, March 1986.
[7] B. J. Lin, "Optical methods for fine line lithography." In *Fine Line Lithography*, ed. R. Newman. Amsterdam: North-Holland, 1980.
[8] H. L. Stover, "Stepping into the 80's with Die-by-Die Alignment," *Solid State Technol.*, **24**, (5), 112–120, May 1981.
[9] R. Carlson, "Manufacturing One Micron," *Solid State Technol.*, **28**, (1), 141–144, Jan. 1985.
[10] J. R. Moulic and W. J. Kleinfelder, "Direct IC Pattern Generation by Laser

Writing," FAM 16.2, *1980 IEEE International Solid-State Circuits Conference*, ISSC 80, Feb. 1980.

[11] D. J. Elliott, *Integrated Circuit Mask Technology*, New York: McGraw-Hill, 1985.

[12] W. R. Pratt and M. P. Risen, "Developments in Semiconductor Microlithography—Photomask Degradation During Contact Printing," *Circuits Manuf.*, **16**, (8), 15–18, Aug. 1976.

[13] P. A. Schulz and D. R. Ciarlo, "Photomask Terminology." In *Technological Advances in Micro and Submicro Photofabrication and Imagery*, ed. J. N. Graf and W. Converse, Proc. SPIE, Vol. 55, Aug. 21–23, 1974, San Diego, CA.

[14] "Pattern-Making Machines: State-of-the-Art Report," *Circuits Manuf.*, **11**, (9), 28–33, Sept. 1971.

[15] A. J. O'Malley, "An Overview of Photomasking Technology in the Past Decade," *Solid State Technol.*, **14**, (6), 57–62, June 1971.

[16] P. Burggraaf, "Ion Beams for Photomask Repair and Production," *Semicond. Int.*, **7**, (10), 31–32, Oct. 1984.

[17] H. R. Rottman, "Dry Cleaning 10× Masks," *Circuits Manuf.*, **16**, (8), 18–21, Aug. 1976.

[18] R. Hershel, "Pellicle Protection of IC Masks," *Semicond. Int.*, **4**, (8), 97–106, Aug. 1971.

[19] E. A. Chandross et al., "Photoresists for Deep UV Lithography," *Solid State Technol.*, **24**, (8), 81–85, Aug. 1981.

[20] T. R. Pampalone, "Novolac Resins Used in Positive Resist Systems," *Solid State Technol.*, **27**, (6), 115–120, June 1984.

[21] P. S. Burggraaf, "Photoresist Processing Systems," *Semicond. Int.*, **4**, (8), 57–76, Aug. 1981.

[22] P. O'Hagan and W. Daughton, "Thickness Variance of Spin-On Photoresist: Dynamic vs. Static Photoresist Dispensing," *Circuits Manuf.*, **18**, (4), 71–72, 74, April 1978.

[23] D. J. Elliott and M. T. Nash, "Image Characteristics of a Positive Photoresist on Semiconductor Surfaces and Their Impact on Device Yield." In *Technological Advances in Micro and Submicro Photofabrication Imagery*, ed. J. M. Graf and W. Converse, Proc. SPIE, Vol. 55, Aug. 21–23, 1974, San Diego, CA., pp. 1–7.

[24] R. H. Collins and F. T. Deverse, U.S. Patent 3,549,368.

[25] D. W. Johnson, "Multilayer Resist Processes and Alternatives," *Semicond. Int.*, **7**, (3), 83–88, March 1984.

[26] D. J. Levinthal, "Photoresist Trends in Semiconductor Processing," *Electron. Packag. and Prod.*, **17**, (11), 49–50, Nov. 1977.

[27] P. S. Burggraaf, "Multilayer-Resist Lithography and Multilayer Resist Research," *Semicond. Int.*, **6**, (6), 48–55, June 1983.

[28] A. R. Shimkunas, "Advances in X-Ray Mask Technology," *Solid State Technol.*, **27**, (9), 192–199, Sept. 1984.

[29] A. D. Wilson, "X-Ray Lithography," *Solid State Technol.*, **29**, (5), 249–255, May 1986.

[30] A. P. Neukermans, "Current Status of X-Ray Lithography," *Solid State Technol.*: *Part 1*, **27**, (9), 185–188, Sept. 1984; *Part 2*, **27**, (11), 213–219, Nov. 1984.

[31] A. Heuberger, "X-Ray Lithography," *Solid State Technol.*, **28**, (2), 93–101, Feb. 1986.

[32] B. Santo, "X-Ray Lithography: The Best Is Yet To Come," *IEEE Spectrum*, **26**, (2), 48–49, Feb. 1989.

[33] G. N. Taylor, "X-Ray Resist Trends," *Solid State Technol.*, **27**, (6), 124–131, June 1984.

[34] A. N. Broers, "The Submicron Lithography Labyrinth," *Solid State Technol.*, **28**, (6), 119–126, June 1985.

[35] N. C. Yew, "Electron Beam—Now a Practical LSI Production Tool," *Solid State Technol.*, **20**, (8), 86–90, Aug. 1977.

[36] B. P. Piwczyk and A. E. Williams, "Electron Beam Lithography for the 80s," *Solid State Technol.*, **25**, (6), 74–82, June 1982.

[37] E. D. Roberts, "Recent Developments in Electron Resists," *Solid State Technol.*, **27**, (6), 135–141, June 1984.

[38] "Projecting Very Accurate Circuit Patterns Onto ICs," *Circuits Manuf.*, **11**, (9), 34–36, Sept. 1971.

[39] H. Bohlen et al., "High Throughput Submicron Lithography with Electron Beam Proximity Printing," *Solid State Technol.*, **27**, (9), 210–217, Sept. 1984.

Chapter 12

Physical Vapor Deposition; Sputtering

It is necessary in wafer processing to add conductive materials to serve as interconnects, and dielectrics as insulators on the wafer. Two commonly used methods for doing this are: physical vapor deposition (PVD), which involves vacuum evaporation and deposition, and sputtering, which takes place in a gas at low pressure. In contrast to chemical vapor deposition (CVD), these methods usually do not involve *chemical reactions* for conductors, but they may for certain nonconducting *compounds*, e.g., silicon dioxide (SiO_2) or silicon nitride (Si_3N_4). We consider these deposition methods in this chapter [1].

12.1 Physical Vapor Deposition (PVD)

The basic idea of PVD is discussed in Section 1.8.6. Molecules (or atoms) from a *supply* of the material to be deposited are evaporated (or sublimed) from a *source* into an evacuated enclosure (bell jar) that also holds the substrates. The molecules move in straight-line paths to the substrate(s) where, given proper conditions, they stick, condense, and coalesce to form a continuous film. The vacuum environment, usually below 5×10^{-5} torr, serves two principal purposes: (1) it prevents a chemical reaction between the evaporated molecules and any residual reactive components such as O_2, and (2) it minimizes intermolecular collisions. For this pressure range the mean free path is $L_m \geq 100$ cm by Eq. (D.2-2). Since this is very large wrt the source–wafer distance, there are few collisions and the evaporant molecules (or atoms) travel in straight lines until they hit a wafer or chamber wall.

It can be seen from Figure 1–13 that PVD involves three basic steps: (1) Evaporation of the material to be deposited, (2) transit of these molecules *in vacuo* to the substrate, and (3) condensation of the molecules on the substrate and film formation. Various postdeposition treatments (such as annealing) may be used to enhance the properties of the film.

12.2 Evaporation

The vapor pressure of an evaporant is highly dependent on its temperature. The latent heat of vaporization ΔH must be supplied to the material by raising the material's temperature to bring about evaporation. Once the temperature is high enough, the material melts and its vapor pressure becomes significant. The Clausius-Clapeyron equation is the basic relationship between the vapor pressure and liquid temperature [1]

$$\frac{dp_v}{dT} = \frac{\Delta H}{T(v_g - v_l)} \tag{12.2-1}$$

where

p_v = vapor pressure of the evaporant

T = absolute temperature of the evaporant

ΔH = latent heat of vaporization

v_g = molar volume of the gas phase

v_l = molar volume of the liquid phase

Usually three simplifying assumptions are made: First, $v_g \gg v_l$. Second, while ΔH is temperature dependent, it may be assumed constant in the range of interest. Third, at pressures much less than atmospheric the general gas law may be assumed valid for the evaporant vapor pressure, viz

$$p_v v_g = R_o T \tag{12.2-2}$$

or

$$\frac{dp_v}{dT} = \frac{\Delta H\, p_v}{R_o T^2} \tag{12.2-3}$$

where R_o is the gas constant. The variables may be separated and the equation integrated to give

$$\ln p_v = A' - \frac{B'}{T} \tag{12.2-4}$$

or

$$p_v = \exp\left(A' - \frac{B'}{T}\right) = \exp_{10}\left(A - \frac{B}{T}\right) \tag{12.2-5}$$

where

$$B' = \Delta H/R_o = 2.303B$$

$$A' = \text{integration constant} = 2.303\ \text{A}$$

Equation (12.2-4) plots linearly on ln p versus $1/T$ coordinates, but usually other coordinates are used. The important thing is that the evaporant vapor pressure rises exponentially with absolute temperature. More accurate results are obtained if ΔH is not assumed constant. For example, Honig uses [2]

$$\log p_v = A + \frac{B}{T} + C \log T + DT \qquad (12.2\text{-}6)$$

where A through D are empirical constants, and Honig's B is a negative quantity. Some typical Honig curves are shown in Figure 12–1. Tabular values are given in several papers and commercial catalogs [2, 3, 4]. Typical listings are given in Table 12–1.[1] Usually it is desirable to have the evaporant vapor pressure in the range of 10^{-4} to 10^{-2} torr (i.e., very much greater than the pre-evaporation residual value). This will be considered in Section 12.3.

It may be shown that molecules at pressure p^* torr arrive at a surface with flux density [1]

$$\mathscr{F} = (3.51 \times 10^{22}) \frac{p^*}{\sqrt{MT}}, \quad \text{cm}^{-2}\text{s}^{-1} \qquad (12.2\text{-}7)$$

where

$$T = \text{temperature, K}$$

$$M = \text{molecular weight}$$

$$p^* = \text{pressure, torr}$$

This equation applies to the evaporant as well, if p_v^*, its *vapor* pressure in torr, is used. If all the molecules stick on hitting a surface, steady state sets in and $\mathscr{F}_v$, the vapor flux density at a surface, equals the evaporation rate.

The evaporant *mass* flux density is

$$G_v = \mathscr{F}_v \frac{M_v}{n_a} = 0.0584 \sqrt{\frac{M_v}{T}}\, p_v^*, \quad \text{g/cm}^2\text{s} \qquad (12.2\text{-}8)$$

where n_a = Avogadro's number [1].

---

[1] Nichrome IV is a proprietary designation of the 80Ni/20Cr alloy and is available also under the trade names Chromel A and Tophet A. Other compositions of Ni/Cr alloys also are used in wafer processing.

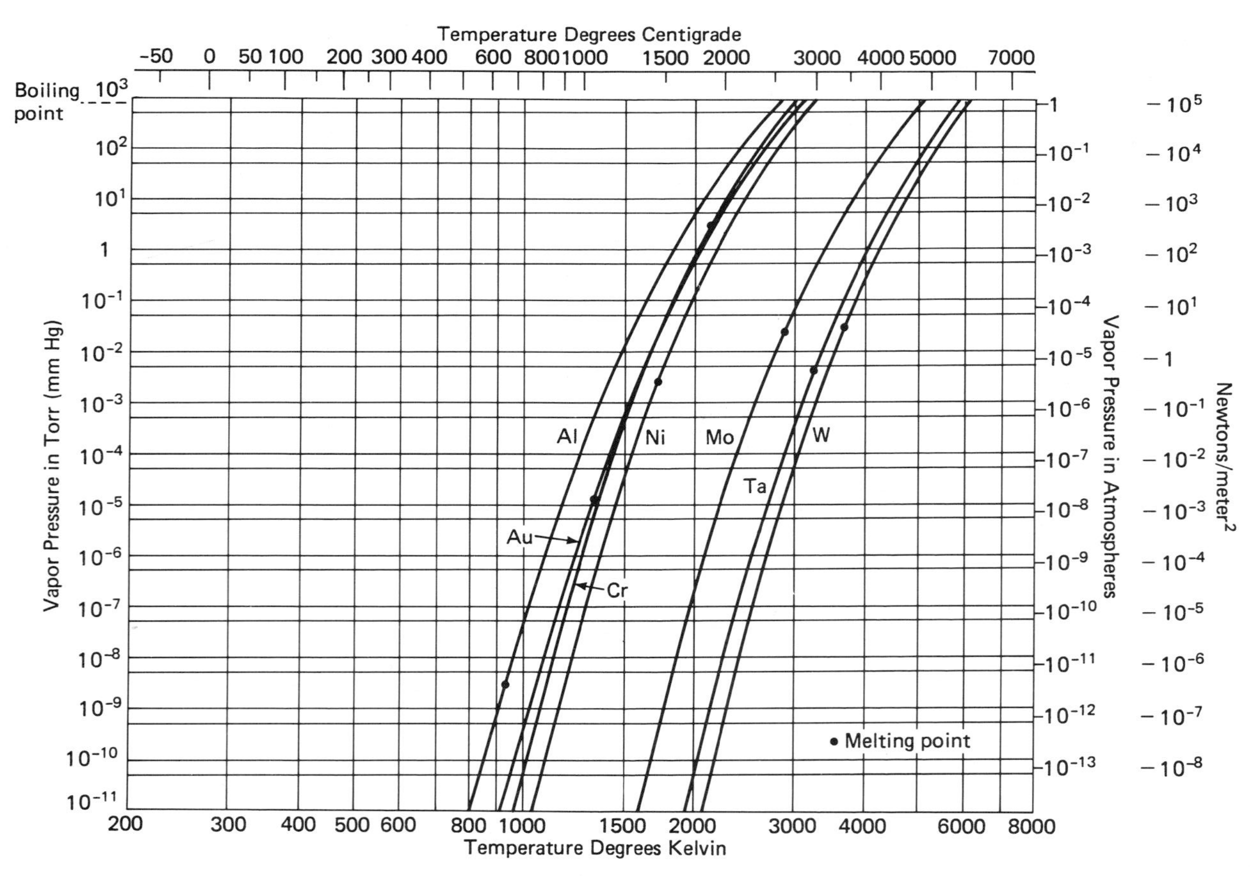

Figure 12–1. Vapor pressure curves for selected materials. After Honig [2]. (Courtesy of General Electric Company.)

Table 12–1. PVD Data

Supply	MP, °C	T, °C at p, torr 10^{-8}	10^{-6}	10^{-4}	E-Beam	Crucible	Coil	Boat	Basket	Sputter	Remarks
Al	660	677	812	1010	Xlnt	TiB_2-BN ZrB_2-BN	W	TiB_2-BN Al_2O_3	W	RF	Alloys w/ and wets stranded W
Au	1062	807	947	1132	Xlnt	Al_2O_3, BN Vit C	W	W, Mo coated Al_2O_3	W	DC RF	
Ni/Cr IV	1350	847	987	1217	Xlnt	Al_2O_3 Vit C BeO	W W	Al_2O_3 coated	W, Ta	DC RF	Alloys w/ refractories
SiO	1702	sublimes		600	Xlnt	Ta	W	Ta	W	RF RF react.	[a]
Cr	1890	837 sublimes	977	1157	Good	Vit. C	W plated	W plated	W	DC RF	Will sublime w/ controlled rate source
Mo	2610	1592	1822	2117	Xlnt	—	—	—	—	DC RF	Fine wire, flash evap.
Ta	2996	1960	2240	2590	Xlnt	—	Ta	—	—	DC RF	Fine wire, flash evap.
W	3410	2117	2407	2757	Good	—	W	—	—	DC RF	Flash evap. Forms volatile oxides

[a] Baffle box source best for resistance evaporation. Low rate suggested.
Abstracted from [4]. Courtesy of Sloan Technology Division of Veeco Instruments Inc.

In both equations p_v* is nearly exponential in temperature; hence the $\sqrt{T}$ factor has little effect. A better idea of the importance of temperature on mass flux density can be obtained by eliminating p^* in terms of Eq. (12.2-5) and taking the logarithmic derivative:

$$\frac{dG}{G} = \left(2.3\frac{B}{T} - \frac{1}{2}\right)\frac{dT}{T} \tag{12.2-9}$$

Typically $2.3B/T \sim 20$–30, so taking the average

$$\frac{dG}{G} \approx 25\frac{dT}{T} \tag{12.2-10}$$

This shows the high degree of G sensitivity to temperature changes. When several wafers are coated at once in a planetary fixture (Section 1.8.6), it is important that G remain constant so all receive the same film thickness. Usually, a thickness monitor (Section 12.2.5) is used as the sensor in a feedback loop that controls the source heating current.

The heated source also radiates thermal energy at a rate $\propto T^4$, while the evaporant flux density increases faster, roughly in an exponential manner, with temperature. Thus high evaporation temperatures favor supply evaporation over radiative thermal loss, giving a more efficient use of the source heater energy. The temperature is limited, since I^2R-heated sources can burn out and evaporants tend to splatter at high temperatures.

Evaporation is inherently an inefficient process because of the high heat of vaporization, high radiation loss, and conductive losses through the leads of the evaporation source that holds the evaporant material. Only a small amount of energy remains with the evaporant in kinetic form. Consider the evaporation of Al at 1200°C (1473 K). At this temperature $\Delta H = 2.77$ eV/atom. By contrast, the average kinetic energy per atom on escape is only $3kT/2$ or 0.19 eV.

12.2.1 Directly Heated Sources

One method of raising the supply material temperature is to use I^2R heating of the evaporant itself, the simplest form being *flash* evaporation. A fine wire of the evaporant material is bent into a loose, self-supporting coil connected between current-supply terminals. A large a-c current is passed through this filament that heats so rapidly that it vaporizes *in a flash.* Table 12–1 indicates some of the materials that are amenable to this method. An obvious drawback is rapid depletion—one flash and all the evaporant is gone, so multiple sources often are necessary. If the evaporant cannot be formed into a wire, a variation uses an electromechanical device to drop small pellets of the material onto an already heated source boat.

Chromium (Cr) does not flash-evaporate well because it *sublimes* at relatively low temperatures and would form a poor film. In this case, a tungsten rod that is plated to a considerable thickness with Cr is attached to the current terminals. The rod serves both as support for the Cr and as the I^2R heater. The uniformity of the Cr plating aids better film deposition [5].

Most conventional sources provide both support and I^2R heating for the evaporant and are fashioned in forms more useful for holding the evaporant material supply. Four fairly common ones are shown in Figure 12–2 [4].[2] Table 12–1 lists how these may be used for a few common evaporants; manufacturers furnish other applications [4, 5, 6].

The several styles of direct-heated sources in Figure 12–2 are adapted to different forms of evaporant. The crucible at (a) will handle any evaporant form, although pellets and powders work especially well. The crucibles proper are fashioned from different materials (e.g., alumina, silica, and vitreous carbon) chosen for minimum reaction with the particular evaporant to be used. They also are available in several shapes to provide different directivity patterns.

The boat-type sources, shown in Figure 12–2(b), are useful for evaporating short lengths of metals, and powders, particularly those that sublime. The boats usually are fashioned from Ta or W, both of which are refractory (high-melting-point) materials and also have low vapor pressures. Coils and baskets, shown in (c) and (d), invariably are made from Ta or W wire. They are useful where short lengths of metal wire evaporants may be placed within the coils or hung on the coil wires in the form of *clips* (short bent pieces). As the source temperature is raised, Al, for example, will first melt (when the latent heat of fusion has been supplied) and be held to the coil by surface tension. (Multiple-strand coils often are used so that *wicking* aids in preventing the melted aluminum from dropping off.) Further heating to supply ΔH results in evaporation.

The evaporants of the first five rows in Table 12–1 have significantly lower melting points and higher vapor pressures than do the refractories of the last three rows. This is fortunate: for example, at 1010°C Al has a vapor pressure at 10^{-4} torr, a high but satisfactory value for PVD, while that of W certainly will be much less than 10^{-8} torr. This means that at 1010°C the Al will

[2] Tungsten has low ductility; to form it into the shapes shown in Figure 12–2 requires special techniques. A small amount of Na improves the ductility, but at 1000°C sodium's vapor pressure is several orders of magnitude greater than that of, say, Al. Hence, Na will coevaporate with the Al and contaminate the deposited film. This is particularly bad for MOS devices since it causes voltage-dependent shifts in the threshold voltage. Sources with low alkali content are available, and other techniques such as *electropolishing* make it easier to bend the tungsten. Heating is still required for bending; W cannot be cold-formed.

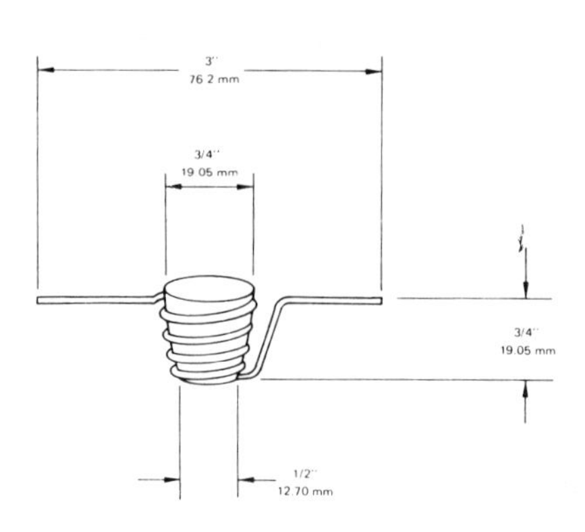

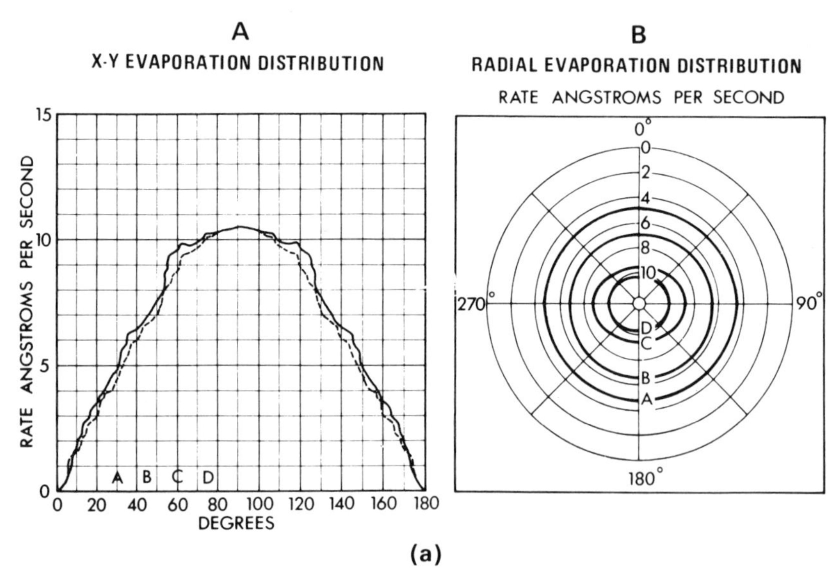

Figure 12–2. Typical directly heated evaporation sources and their directivity patterns. (a) Crucible and heater. (b) Boat. (c) Coil. (d) Basket. Sloan [4]. (Courtesy of Sloan Technology Division of Veeco Instruments Inc.)

evaporate without contamination by tungsten. Also, there is no worry about the source melting or sagging at that temperature.

The directivity pattern in which the evaporated material leaves the source is important when several wafers are processed simultaneously in the bell jar, because the wafers must be mounted at different angles wrt the source. The deposition rate is not uniform in all directions, so wafers in fixed positions

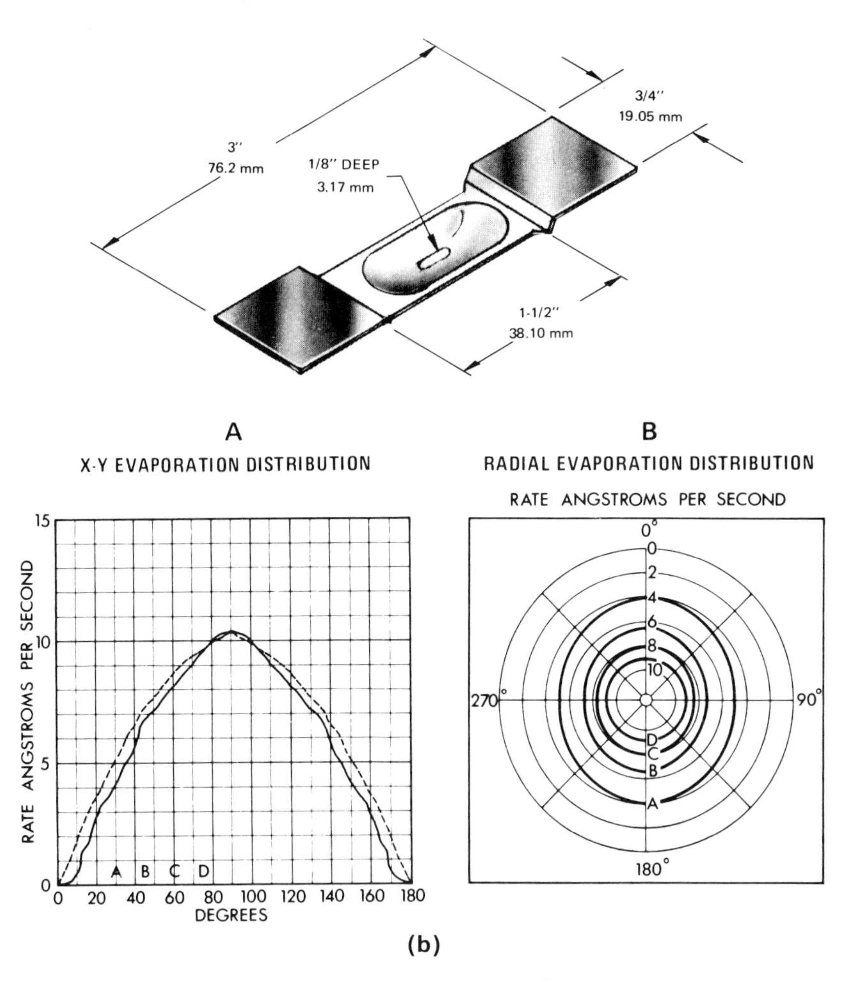

Figure 12–2 (*Continued*)

would not receive equal deposits of the evaporant. A planetary fixture (Section 1.8.6) changes the wafers' positions relative to the source during evaporation to reduce these variations. The directivity patterns of the sources also are shown in Figure 12–2. Figure 12–3 gives the key for these patterns. All data for the directivity patterns in Figure 12–2 were taken with the source temperature at 1000°C and Ag as the evaporant except for SiO [4]. It is easy to argue that the deposition rates to several wafers become more uniform as the source-to-wafer separation increases. Sources deplete rapidly, a problem that

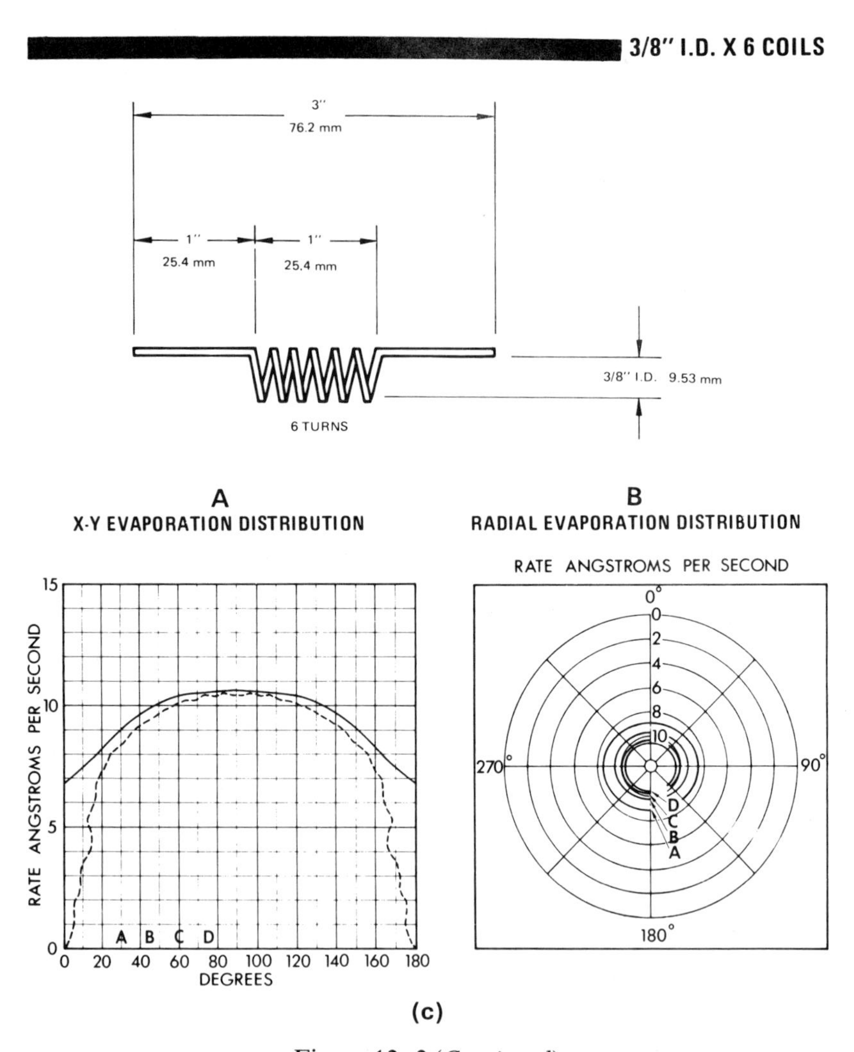

Figure 12–2 (*Continued*)

may be handled with an automatic feed system that unreels material in wire form onto the source.

Direct-heated evaporation sources may be made from nonmetals. Intermetallic rectangular bar stock is available that can be machined into boat shapes as needed. The current and voltage requirements to raise the bars to the desired temperature can be calculated readily [7].

12.2.2 Indirectly Heated Sources

Since direct-heated sources of the types shown in Figure 12–2 can cause sodium contamination and have limited supply capacity, some wafer fab lines

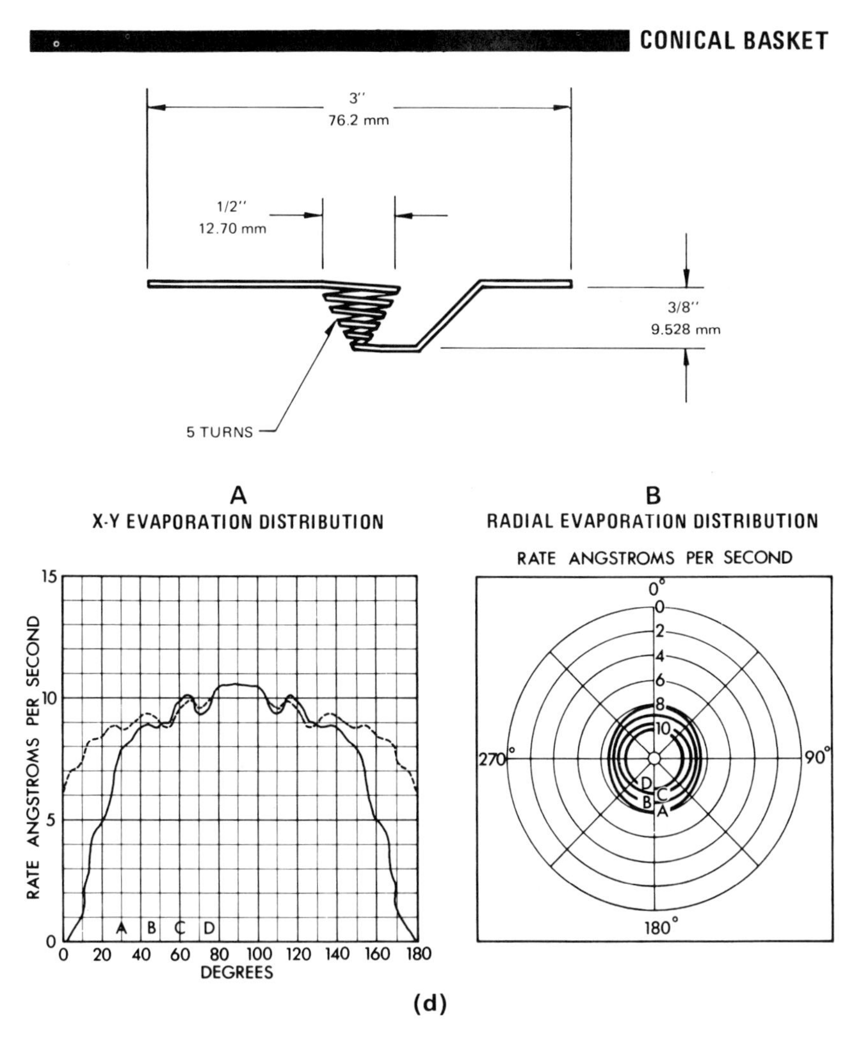

Figure 12–2 (*Continued*)

use alternative methods for heating the evaporant. For example, the evaporant supply or *charge* may be held in a crucible and heated by r-f excitation (Section 9.3). The crucible is held in a coil of vacuum-compatible copper or stainless steel *tubing* through which cooling water is circulated. When r-f energy is applied, the charge is heated by eddy currents, but the coil and crucible remain cool and contamination is eliminated. Vacuum feedthroughs are required to bring the water and rf through the vacuum chamber wall to the coil. The directivity pattern, being determined by the crucible shape rather than by the heating method, will resemble that of Figure 12–2(a).

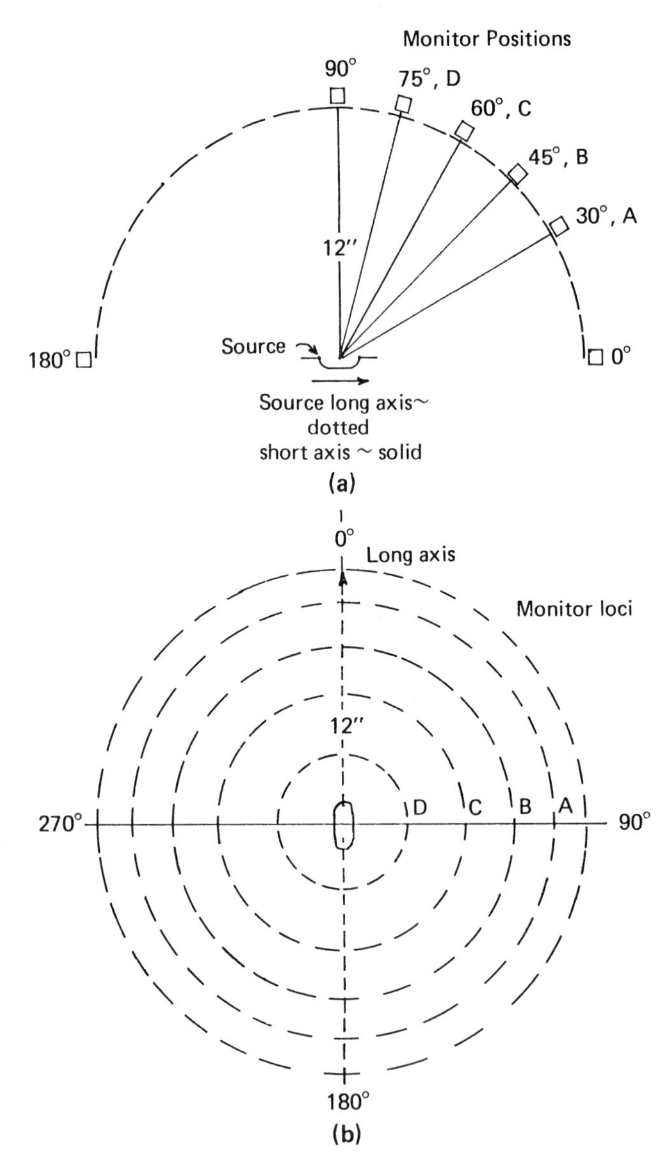

Figure 12–3. Key to source directivity patterns. (a) Deposition monitor measurements are made along the long (x, dotted) and short (y, solid) axes of the source at a fixed distance of 12 in. (b) The monitor is rotated about the source at five angles above the horizontal to measure the polar diagrams. The angles are identified in (a).

Another very common indirectly heated source uses a beam of electrons for heating the evaporant supply, which is held in a water-cooled *hearth*, usually of copper. The cavity in the hearth is shaped like a crucible. The diagram of an *in-line* electron gun system is shown in Figure 12–4(a) [4]. The electron gun is comprised of a robust filament, capable of high electron emission because beam currents up to a few hundred milliamperes are used, and an anode that contains a beam-defining aperture. More elaborate gun structures, such as the Pierce type, may be used to improve the actual beam-

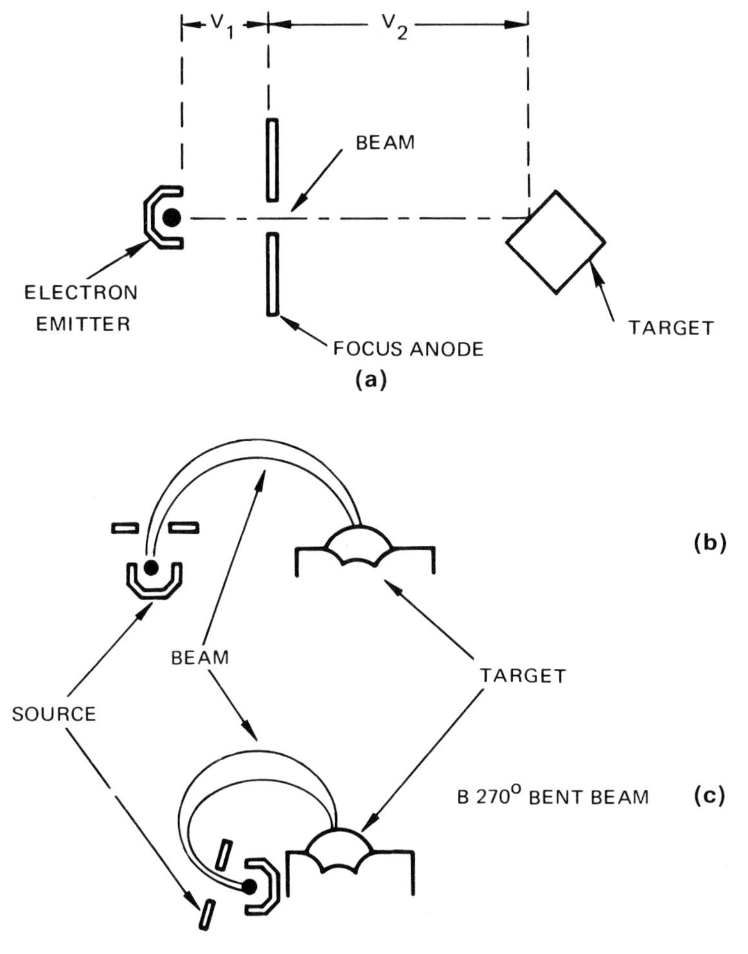

Figure 12–4. Configurations of three electron beam evaporation sources. (a) In-line. (b) 180° bent beam. (c) 270° bent beam. Sloan [4]. (Courtesy of Sloan Technology Division of Veeco Instruments Inc.)

to-cathode current ratio for more efficient operation [8]. The filament is a tungsten coil of the *naked type*, i.e., it is open in the vacuum chamber and so must withstand exposure to atmosphere, though not at high temperature, every time the bell jar is opened for changing wafers. Special oxide-coated cathodes are more efficient electron emitters and can operate at lower temperatures, but are not practicable, since repeated exposure to atmosphere damages them.

The in-line gun is located so that its electron beam has a straight-line path to the evaporant supply. The electron kinetic energy is converted to heat on impact, causing a region of the supply to melt and form a pool of liquid much larger than that given by directly heated sources. This makes for higher evaporation rates. Beam powers up to 6 kW (say, of 0.3 A @ 20 kV) are common. The heating is much larger than with the direct-heated sources and larger charges of the supply are feasible.

If $V_2 = 0$ in Figure 12–4(a), all acceleration of the electrons is furnished by V_1 on the gun anode, and the system is said to be *self-accelerated*. If $V_2 \neq 0$, additional electron acceleration takes place between the gun anode and the evaporant supply (target in the figure), and the designation *work-accelerated* is used. In either case, a return path for the electrons must be provided between the hearth and the gun cathode.

Since the gun must have line-of-sight orientation to the evaporant supply in this configuration, it must be located above the plane of the hearth in the bell jar proper. This means that the filament must be allowed to cool before the jar is opened to atmosphere for wafer changing.

Also, the gun is exposed to evaporated material and so may become coated with the evaporant. Filament contamination by the evaporant is not a severe problem, however, since the gun's operating temperature is about 2300°C, which is high enough to inhibit condensation. If the filament has contaminants, such as sodium, which are evaporated and then pick up negative charge from beam electrons, the contaminants will accelerate to the source and evaporate with the supply material.

It is common practice now to incorporate the electron gun and some beam-bending and focusing permanent magnets in the water-cooled hearth assembly. The magnetic field is arranged wrt the electron gun so the beam is bent in a circular path (Appendix B) before hitting the supply. Two designs are common—giving 180° or 270° bends as shown in Figure 12–4(b) and (c) [4]. With both types, the gun/hearth assembly may be located below a valve or lock that may be closed before the bell jar is opened to atmosphere. The gun is no longer in the line of sight to the supply, so it will not pick up deposited evaporant. Furthermore, since Na ions are very heavy relative to electrons the ions will not bend enough in the magnetic field to arrive at the hearth. The 270° type is better in this regard but does not focus the beam as well as the 180° type. In some 270° units a component of the $\mathscr{B}$ field is furnished by an electromagnet to allow focus adjustment. In addition an a-c component driving the electromagnet can make the beam scan back and forth across the evaporant supply to keep a larger liquid pool [4].

In its basic form the water-cooled copper hearth is in direct contact with the evaporant supply or charge. It is interesting that any liquid supply that reaches the cooled hearth will freeze out quickly and provide an *inner hearth* of the evaporant material itself. This aids in minimizing contamination by the hearth. Nonetheless the cooled copper hearth can lead to severe heat loss, of up to 90%, by conduction from the melted supply, to the hearth, to the cooling water [4]. The loss may be lowered by placing a crucible or hearth *liner* of low thermal conductivity (e.g., vitreous carbon) between the hearth and charge.

Typical deposition rates for common fabrication materials are given in Table 12–2. In commercial practice hearths are quite large, the truncated-cone shape having an average diameter of 19 mm and a 12-mm height. To prevent overflow, the hearth is filled only to about 80% of capacity. An Al charge will amount to about 10 g or $\frac{1}{3}$ oz.

Despite its higher cost, electron beam evaporation offers several advantages over the direct-heated types of sources:

1. Large charge with much longer depletion time.
2. Less contamination.
3. Heating at the evaporant upper, rather than the lower, surface, resulting in less *spitting* of the evaporant.
4. With the bent-gun variations the gun/hearth assembly may be located below a valve to isolate it from the opened bell jar.
5. The gun, hearth, and magnets may be built as a single unit, which makes for compactness and ease of alignment.
6. The composition of a deposited alloy is closer to the bulk value because of localized heating at the charge surface.

Table 12–2. Electron Beam Evaporation Data

Evaporant	Beam Power, kW	Rate at 25 cm Å/min	Charge Form	Notes
Al	6.0	10,000	Wire	melts
Au	6.0	3,000	Wire	melts
Ni	6.0	84,000	Chunk	melts
Cr	1.3	12,000	Hot-pressed	sublimes
Mo	6.0	8,000	Chunk	melts
Ta	6.0	12,000	Chunk	semimelts
W	6.0	5,600	Chunk	semimelts

Abstracted from [4]. Courtesy of Sloan Technology Division of Veeco Instruments Inc.

12.2.3 Alloy Evaporation

Nichrome, a Ni/Cr alloy, is used for making high-resistance, thin-film resistors on certain ICs. Its evaporation from direct-heated sources (except by flash evaporation) is troublesome: the composition of the deposited film may differ significantly from that of the bulk source, which affects the value of thin-film resistance. The cause is related to the change in vapor pressures of the two metals, when they evaporate alone and when they are alloyed. The interaction effect is approximated by Raoult's law for two components, say A and B, namely [1]

$$\frac{p_A'}{p_B'} = \frac{f_A}{f_B} \frac{p_A}{p_B} \tag{12.2-11}$$

where

p = isolated vapor pressure

p' = modified vapor pressure in the alloy

f = mole fraction

or

$$f_A = \frac{W_A/M_A}{\dfrac{W_A}{M_A} + \dfrac{W_B}{M_B}} \quad \text{and} \quad f_B = \frac{W_B/M_B}{\dfrac{W_A}{M_A} + \dfrac{W_B}{M_B}} \tag{12.2-12}$$

where M = molecular weight and W = weight. Therefore,

$$\frac{p_A'}{p_B'} = \frac{W_A}{W_B} \cdot \frac{M_B}{M_A} \cdot \frac{p_A}{p_B} \tag{12.2-13}$$

Then, invoking Eq. (12.2-8) and noting that both metals are at the same temperature in the source, we get

$$\frac{G_A'}{G_B'} = \frac{W_A}{W_B} \sqrt{\frac{M_B}{M_A}} \frac{p_A}{p_B} \tag{12.2-14}$$

Consider some typical numbers for the 80Ni/20Cr alloy.

Let A = Cr and B = Ni. Calculate the G' ratio at a source temperature of 1200°C. From Honig's curves: $p_{Cr} \approx 3 \times 10^{-4}$ torr, $p_{Ni} \approx 3 \times 10^{-5}$ torr. $M_{Cr} = 52.01$ and $M_{Ni} = 58.71$. Then

$$\frac{G_{Cr}'}{G_{Ni}'} = \frac{20}{80}\sqrt{\frac{58.71}{52.01}} \cdot \frac{3 \times 10^{-4}}{3 \times 10^{-5}} \approx 2.7$$

But the ratio $G_{Cr}/G_{Ni} \approx 9.4$, a significant change indeed.

These results show that initially Cr deposits down faster than the Ni. This is advantageous since Cr sticks well to SiO_2 on the wafer, but the Cr depletes at the supply surface and the relative rate of Ni increases while Cr evaporation becomes limited by its rate of diffusion from the bulk supply to the surface. The effect is less important with electron beam evaporation since a large pool of the melt is present.

Flash evaporation of nichrome wire maintains the bulk ratio in the deposited film very well, but source depletion is a limitation. *Coevaporation* may be used where the Ni and Cr components are evaporated simultaneously, but from two separate sources. Raoult's law does not apply then, but excellent temperature control of the two sources is required. The change in composition is not such a problem in sputtering, as we shall see later (Section 12.6).

12.2.4 Oxide Evaporation

Oxides, as well as other compounds, can present unique problems in vacuum deposition since they tend to *dissociate* at the high temperatures required for evaporation. This is particularly true when one of the two components has much higher vapor pressure than the other, a case in point being cadmium sulfide (CdS), where sulfur is the component with higher pressure. For example, from Honig's curves the vapor pressures of S and Cd at 100°C are, respectively, 4×10^{-3} and 2×10^{-7} torr. As a result, when evaporation of CdS is attempted, the sulfur evaporates off very rapidly and Cd is left behind in the boat. Coevaporation from separate Cd and S sources at different temperatures is a solution, but requires special techniques.

Silicon monoxide (Section 5.3.5) sublimes, and has high vapor pressure: 10^{-2} torr at 1020°C and 10^{-1} torr at 1110°C. There is a tendency for loss of stoichiometry in evaporation, however, so the resultant film is a mixture of Si, SiO, and SiO_2 called SiO_x. The following empirical guides are helpful: (1) for $T < 1250$°C, SiO_x is deposited but with poor mechanical properties, (2) for $1250 < T < 1400$°C, SiO_x with optimum mechanical properties is deposited, and (3) $T > 1400$°C favors SiO_2 deposition, but the film tends to crack because of excessive deposition rates due to high vapor pressures [1].

Powdered SiO that serves as the supply tends to pick up moisture and *spits* when the trapped vapor is released on heating. This causes small particles of the charge to be emitted in solid phase; these particles can damage the film.

Use of a source that eliminates line-of-sight paths from charge to substrate solves this problem. A number of these sources, such as the Drumheller [1]

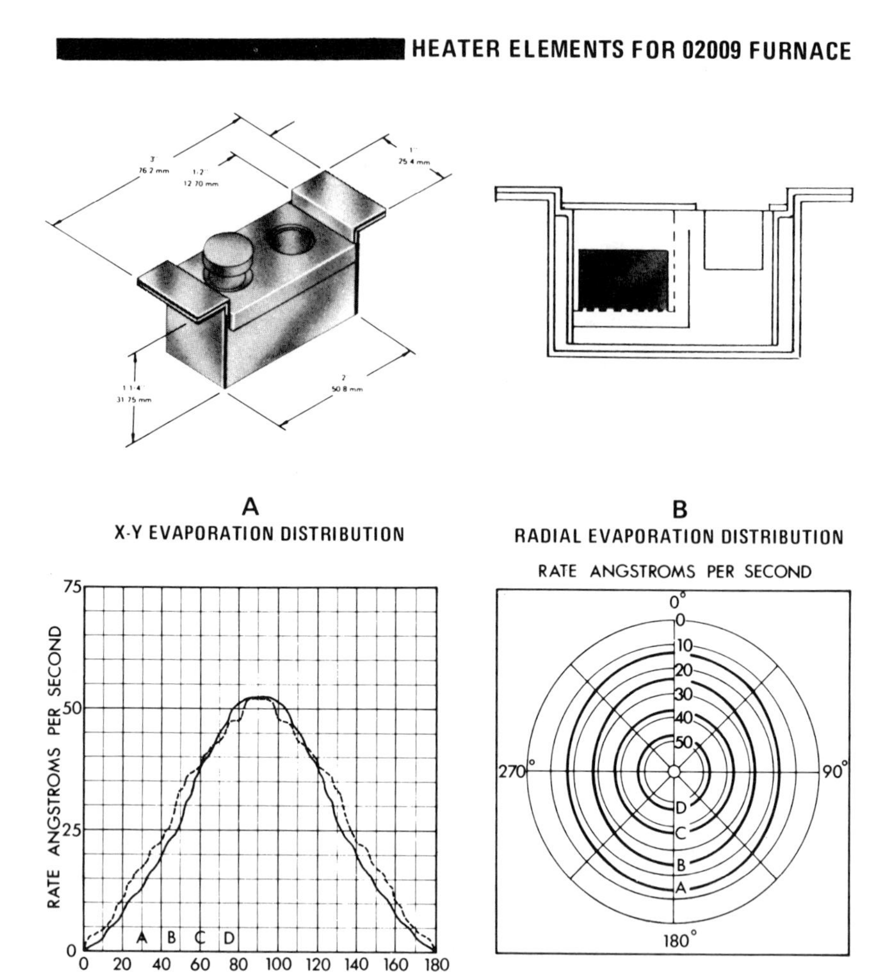

Figure 12–5. A box source for evaporating silicon monoxide. Sloan [4]. (Courtesy of Sloan Technology Division of Veeco Instruments Inc.)

and the box type of Figure 12–5, are available [4]. Note that baffles are arranged to eliminate any direct path for solid particles between the charge and ambient. Spitting also may be reduced by adding roughly 5% mole fraction of boron oxide (B_2O_3) to the SiO supply charge. This has no apparent effect on the properties of the deposited film.

12.2.5 Thickness Monitor

It is necessary to monitor film thickness during deposition if consistent results are to be obtained. A common form of monitor uses the principle that the resonant frequency of an oscillating plate is sensitive to its inertial mass. Thus a quartz crystal, serving as the frequency-determining element in an oscillator circuit, is positioned so that it receives evaporant just as the wafers do. As the thickness τ_x of the film on the crystal increases, added mass loads the crystal and the oscillator frequency drops [9].

The film thickness on the monitor crystal is related approximately to the *change* in oscillator frequency Δf by

$$\tau_x = \Theta \cdot \Delta f \tag{12.2-15}$$

where Θ is an empirically determined, evaporant-dependent constant that depends inversely on the material density. Typical Θ values are 0.6 and 0.3 Å/Hz for Al and SiO, respectively.

When several wafers, mounted in a rotating planetary fixture, are being coated simultaneously, the oscillator unit must be positioned for minimal interference with deposition on the wafers. For this reason, an independent calibration to relate the average film thickness τ on the wafers to Δf or to τ_x is required.

The frequency change may be monitored visually on frequency meters of either the heterodyne/analog or direct-reading digital types. The Δf signal also may be used to terminate deposition when τ reaches the desired value. Deposition should not be terminated by shutting off the heating supply of the evaporant source; the thermal lag in cooling prevents abrupt cut off. A better solution interposes a rotatable shutter between wafers and source.

In time, as τ_x builds up, oscillations will cease and crystal replacement is necessary. Crystal blanks may be recycled many times by selectively etching off the film (e.g., hydrochloric acid will remove aluminum without damaging the quartz).

In practice, the oscillator circuit is enclosed in an electrically shielded, water-cooled housing and mounted in the bell jar. A quartz plate, with a nominal resonant frequency of 5 MHz, is clip-mounted, for easy replacement, on the front of the oscillator housing and facing the evaporating source. The oscillator output is isolated by an emitter follower that feeds the monitoring equipment via a coaxial cable. These precautions isolate the frequency from external capacitance changes.

12.3 Transit

Evaporated molecules must move from the source to the wafers, preferably in straight lines. To this end, the residual gas pressure in the bell jar before the onset of evaporation is held to well below 10^{-5} torr to minimize collisions. If the evaporant vapor pressure rises to 10^{-2} torr or higher, the mean free path of the evaporant atoms will drop to 0.5 cm or less (Eq. D.2-2), and the probability of collisions becomes significant. If such collisions do take place, the molecules are deflected from linear paths and some may return to the source, causing a decrease in the steady state flux density of Eq. (12.2-7) at a given source temperature. For this reason, 10^{-2} torr is usually an upper limit for the evaporant vapor pressure [1].

Based on Eq. (12.2-7) we can write the ratio

$$\frac{\mathscr{F}_v}{\mathscr{F}_r} = \frac{p_v}{p_r}\sqrt{\frac{M_r T_r}{M_v T_v}} \tag{12.3-1}$$

where subscript v refers to the evaporant with vapor pressure p_v (obtainable from Honig's curves) and subscript r is used for the residual gas of partial pressure p_r. This is related to temperature T_r by the general gas law, Eq. (12.2-2), with appropriate subscript changes. Direct evaluation of the equation is difficult because T_r is not known, but certainly T_r will exceed the pre-evaporation value of 300 K because of radiative heating from the evaporant source. It is clear, however, that the ratio p_v/p_r should be large to maximize the flux density ratio, a desirable condition if the deposited film is to be of high purity.

Observation of bell jar vacuum gages during evaporation shows a great rise in pressure from the pre-evaporation residual value of $p_r \leq 10^{-5}$ to the total pressure, once evaporation begins, of nearly $10^{-4} \leq p_v \leq 10^{-2}$ torr.

12.4 Film Formation

After evaporation and transit from source to substrate the evaporant molecules (or atoms) must stick to the substrate surface if they are to build up a film. Sticking is not assured, however, since the atoms may be reflected from the surface or dwell on it for only a small fraction of a second and leave, without giving up the heats of vaporization and fusion as required for reverting to the solid phase. In both these events the atoms retain most of the $3kT/2$ kinetic energy that resulted from evaporation.

For sticking to take place on the surface, a proper balance must exist between surface temperature T_S and the incident evaporant flux density $\mathscr{F}_v$. For a given T_S, $\mathscr{F}_v$ must exceed a specific minimum value. The relationship is determined empirically for any given setup, rather than theoretically.

It is convenient to define the concept of *sticking coefficient* Σ:

$$\Sigma = \frac{\text{no. of atoms sticking/cm}^2\text{s}}{\mathscr{F}_v} \tag{12.4-1}$$

Thus, if T_S increases during the initial stages of film formation, Σ decreases; at too high a surface temperature the atoms cannot condense out and are reflected.

If conditions are right, some atoms will arrive at the surface and will give up the necessary energy to remain attached by going through the process of *thermalization*, whereby the atom skids around on the surface undergoing multiple, energy-losing collisions. Apparently there are favored sites on the substrate where such atoms come to rest. When two such atoms join up, a *nucleus* is formed for further growth. *Nucleation* will occur at many places on the surface. The energy given up to convert to the solid phase also may go to bonding.

As more atoms subsequently arrive, they have a chance of sticking to their own kind as well as forming new nucleation sites, and Σ begins to increase toward unity as the film thickness τ increases. Nuclei increase in size as more atoms link up with them, a process known as *agglomeration*. These islands of film material finally grow enough to *coalesce* and form a more or less continuous film.

Growth of the deposited film is not a perfectly ordered process where a uniform buildup, monolayer by monolayer, takes place. Its initial growth is very random, with grain boundaries, voids, and other defects being incorporated into the film. As the thickness increases, uniformity improves and values of electrical resistivity approach bulk values. References are available that illustrate the film growth process with figures and more detailed analyses [1, 10].

The common metals that may be vacuum-deposited onto SiO_2-covered wafers are Al, Ni/Cr (nichrome), and gold. In general, metals that oxidize readily stick well to SiO_2, so Al and nichrome adhere well. Gold does not, so it usually is preceded by a thin Cr layer. The Cr adheres well to the oxide, and Au to the Cr. Use of this technique was mentioned in Section 11.11.3.

12.5 Molecular Beam Epitaxy (MBE)

Typical flux densities of evaporants are high in conventional PVD methods—e.g., for Al at 1200°C, $p_v{}^* = 4 \times 10^{-3}$ torr and $\mathscr{F}_v \approx 10^{23}$ atoms/cm^{2}s. The corresponding growth rate of the deposited film is roughly 4 μm/s if $\Sigma = 1$ and source depletion does not limit the deposition. This is so high that epitaxial growth is essentially impossible to achieve, as is fine control of doping profiles. Contrast this with conventional CVD vapor-phase epi growth of Si on Si at about 30 μm/hr.

In the early 1960s a modified method of PVD was developed that overcame these difficulties with Si as both evaporant and substrate, viz *molecular beam epitaxy* (MBE). The essential difference lay in the use of modified sources, which permitted operation at much lower temperatures, in an ultrahigh vacuum (UHV) of 10^{-11} to 10^{-9} torr. At source temperatures in the 600 to 900°C range the vapor pressures varied from $< 10^{-11}$ to 10^{-9} torr. With similar low substrate temperatures the sticking coefficient Σ was small and film buildup rates τ/t were around 1 μm/hr (less than 1 atom of thickness per second)! With these incredibly low rates it became possible to control film doping profiles at the atomic level, with nongaussian and nonerfc profiles quite feasible. Even *retrograde* profiles (doping surface concentration lower than within the bulk) could be obtained, as could very abrupt junctions. The low substrate temperatures virtually eliminated outdiffusion from the substrate into the growing film. Since other methods of growing Si epi, although with less control, were available, the method was not used commercially.

In 1968, Cho and Arthur of Bell Labs applied MBE to the deposition of very uniform films of compound semiconductors [11]. Several III-V, II-VI, and IV-VI compounds, as well as several metals, soon were in use. Interest at that time was high in layered structures for optoelectronic applications, which led to concentration on the III-Vs: binary GaAs and ternary aluminum gallium arsenide (AlGaAs or more properly $Al_xGa_{(1-x)}As$). It became possible to grow layered structures of alternating GaAs and AlGaAs in layers only 100 Å thick. These films were of top quality, and competition among groups using MOCVD (Section 9.18) and MBE arose, resulting in ever-improved capabilities. Also, doping of the layers in the direction of film growth (our x-direction) was controllable on the same atomic scale. These results were better by nearly two orders of magnitude than results with conventional liquid-phase or vapor-phase epi. The key to the method, as stated earlier, is the use of very low growth rates in an ultrahigh vacuum; the penalties are very long deposition times and complex equipment.

Many doped semiconductors will not maintain their doping levels and type during conventional PVD[3]; the same is true of the III-V semiconductors, in that they dissociate on evaporation. For this reason, coevaporation is the rule, with a separate source for each component of the compound and for each dopant.

Developments in the late 1980s have made desirable the deposition of an epitaxial layer of GaAs on a Si substrate. Even though both semiconductors are cubic, their lattice constants differ enough to cause severe dislocations, and the Ga and As atoms do not alternate as they should in the crystal layer.

These problems have been resolved with MBE by inclining the substrate 4° off the [100] axis, and preceding the GaAs deposition with a thin layer of As on the Si [12].

12.5.1 Setup

The basic setup for an MBE system is shown in Figure 12–6 [13]. Details of the vacuum system are not shown, although they are of utmost importance. The normal ambient pressure before the onset of deposition is 10^{-11} to 10^{-9}

[3] A striking example of this is seen when degenerate n-type Ge doped with As is evaporated under usual conditions and the deposited film is lightly doped p-type.

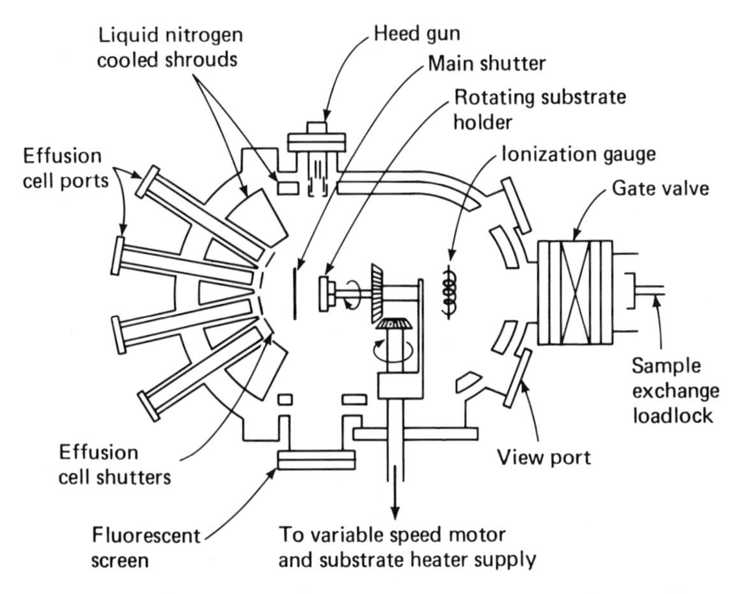

Figure 12–6. A basic molecular beam epitaxy deposition system. Singer [13]. (Courtesy of *Semiconductor International Magazine.*)

torr. The pumpdown time is so long that it is not feasible to vent the entire system to atmosphere for wafer changing; locks must be used. Also, special pumps are used that minimize contamination, and means are provided for cleaning the wafer *in situ* [14].

The figure shows the several sources, one for each component and dopant required for the film structure to be deposited, all aimed at the substrate. In typical equipment, up to eight sources are provided, each with a separate shutter so that any supply beam may be interrupted abruptly. The substrate holder can be rotated to ensure uniform deposition over the entire substrate surface. Uniform substrate temperature is required also for the same reason; the temperature is raised by enough to provide the required heat of reaction at the surface [13]. The figure shows the wafer mounted vertically. A horizontal, face-down orientation also is satisfactory, provided that the source positions are changed. These two wafer orientations prevent particulate matter from hitting the growth surface.

The figure also shows a HEED (high energy electron diffraction) gun and its fluorescent screen. Other commonly used analytical tools are a reflection electron diffraction (RED) mass spectroscope, and a secondary ion mass spectroscope (SIMS). These may be used to monitor film properties *in situ*. The vacuum environment makes their use possible; they cannot be used during LPE or VPE.

The evaporant sources used in MBE differ greatly from those of conventional PVD; they even are named differently (as *effusion cells*, furnaces, or molecular beam sources). The thermal type is most common; Figure 12–7 shows one that is used with a solid-phase supply. Since chamber vacuum must be broken to reload the supply material, the cells must be designed to hold a large supply to lengthen the depletion life.

Furnace temperatures for the column III and V elements are much lower than, say, for Si, but must be high enough for evaporation to take place. The flux density is temperature dependent and may be adjusted with the appropriate shutter closed. Typical temperatures run from 350°C for phosphorus to 1300°C for Al [14]. Computer control is essential to deposit thin layers of desired composition and doping. Metal parts of the furnace usually are of Ta (it is refractory and easier to bend than tungsten). Pyrolytic boron nitride (PBN) is the material of choice for the crucible that holds the evaporant

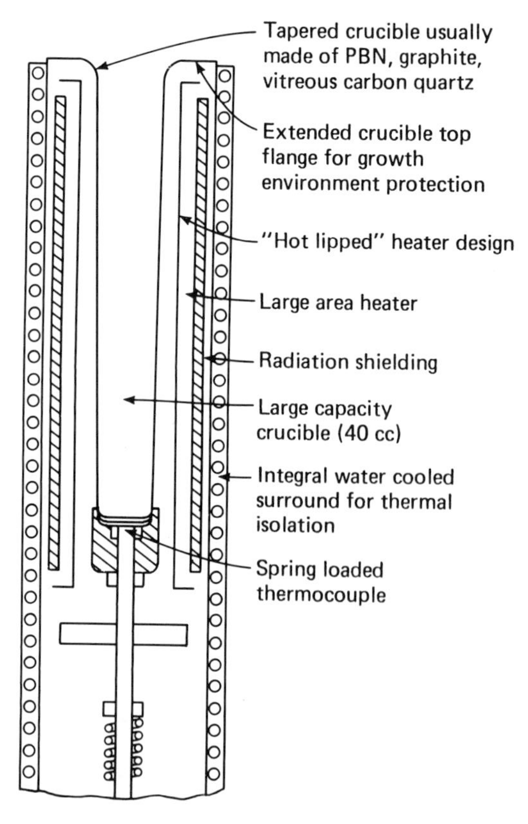

Figure 12–7. An MBE solid supply effusion cell. Singer [13]. (Courtesy of *Semiconductor International Magazine.*)

supply. Note in the figure that multiple heat shields are used in the effusion cell. These reduce radiative heat loss, raising the efficiency and reducing outgassing from nearby walls.

Most evaporant materials used in MBE have sticking coefficients of $\Sigma \approx 1$ at room temperature; hence, water cooling is used to condense out gas-phase materials confined near the furnaces when the shutters are closed. This reduces cross contamination among the effusion cells.

Gas sources may be used in place of the solid-phase type. Since the gas is piped in from the outside of the vacuum chamber, source depletion is no problem. The purity of gas is not as good as that of the solids, and toxicity of gases such as arsine (AsH_3) is another problem. Where chemical reaction between the components is desired, other types of supplies and furnaces may be used—e.g., metalorganic supplies (MOMBE) or materials for chemical beam deposition (CBE). Singer describes some of these variations [13].

When Si is used as a dopant in III-V compound semiconductors or as a film material, an electron beam source is used rather than the thermal effusion cell because the vapor pressure of Si is much lower than for the other materials—e.g., only 6×10^{-6} torr at 1200°C [2]. The wafer temperature is in the 600 to 900°C range for Si on Si, to minimize outdiffusion of dopants into the growing film. For these conditions the sticking coefficient is low, namely $\Sigma \approx 10^{-4}$ [15].

12.5.2 Film Formation

The basic steps of PVD film formation (namely sticking, nucleation, agglomeration, and coalescence) are present in MBE, but the temperature ranges are quite different. For example, substrate temperature T_S for GaAs ranges from 500 to 620°C, so outdiffusion into the growing film is negligible [13]. This is low compared to the 700 to 900°C for LPE, and about 750°C for VPE [14]. Also, the lower temperatures cause less stress due to thermal mismatch between adjacent layers or film and substrate, and fewer vacancies remain in the film. As might be expected, it is observed that the surface smoothness improves as the film thickness τ increases.

With the most commonly used solid-supply effusion cells, the film growth rate τ/t depends on $\mathscr{F}_v$ (which is temperature dependent) on cell-to-substrate separation, and on T_S. As stated earlier, the rate is typically 1 μm/hr.

Stoichiometry is obtained in III-V compound films by virtue of an interesting phenomenon. The group V components apparently adsorb to the surface only in sufficient numbers to satisfy bonding with the group IIIs already on the surface; therefore, the growth rate is determined by the group III $\mathscr{F}_v$ value and Σ [14]. Stoichiometry is obtained by adjusting the group V $\mathscr{F}_v$ to produce an excess at the surface, where only the required number incorporate into the film.

Film contamination is of concern because of the low flux densities used. The substrate is shielded from particulate matter by proper orientation, and cross contamination among sources is kept low by use of chilled baffles. Also, the ambient pressure is held at 10^{-11} to 10^{-9} torr. Reactive residual gases such as O_2, which can react with film components and incorporate into the growing film, must be held to around 10^{-14} torr partial pressure [14]. If, for example, such a gas were at 10^{-6} torr, and if $\Sigma \approx 1$, the gas would incorporate into a growing film at the same $\tau/t = 1$ μm/h rate which is typical of the desired material; hence, its partial pressure is held at orders of magnitude lower.

12.6 Sputter Deposition

Sputtering is the ejection of molecules (or atoms) from a solid at a gas/solid interface by momentum transfer from impacting ions of another material. Film deposition by sputtering is closely related to plasma etching of Chapter 10, but mechanisms taking place at both anode and cathode become important. In Figure 12–8 we see the basic two-electrode or *diode* sputter deposition rig. A d-c voltage is maintained across plane parallel electrodes, spaced d apart. The *target*, made of the material to be deposited onto the substrates, is connected to the negative supply voltage and serves as the cathode. The substrates are on the positive electrode or anode. A plasma is present in the interelectrode region (Appendix C); hence, there is a large supply of positive ions, usually Ar^+. Under the action of the electric field, these accelerate toward the negative target, where they hit and by momentum transfer dislodge and remove *uncharged* molecules or atoms from the target. This requires the breaking of atomic or molecular bonds at the target surface [16]. The sputtered species leave with relatively high energies (from 1 to 10 eV as compared to the $3kT/2 \approx 0.2$ eV value of vacuum evaporation). The action

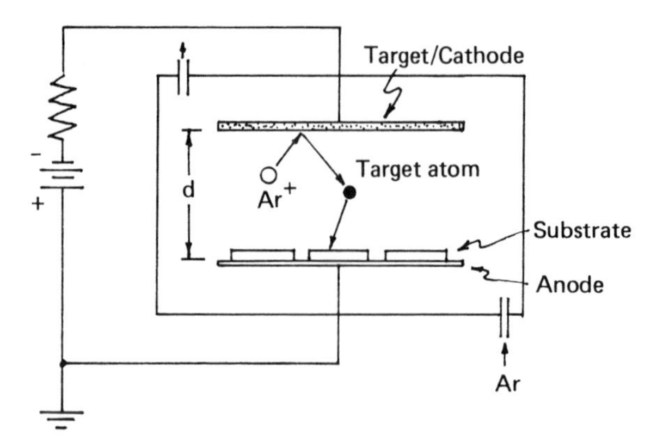

Figure 12–8. A basic parallel plane electrode, diode sputter-deposition system.

described thus far constitutes plasma- or sputter-etching at the target. We should note that some secondary electrons also are dislodged from the target by the positive ions. Due to the high positive anode potential, a large percentage of these electrons reach the substrates.

Because of their high energies, a large fraction of the target molecules also reaches the substrates and forms a film there (usually of the same composition as the target because of the momentum transfer mechanism). It is this film on the anode that is our concern.

Sputter deposition is based on momentum transfer and not on thermal processes, so film growth proceeds at comparatively low temperatures. Among the advantages of sputter deposition are

1. Compounds tend not to dissociate.
2. Refractories sputter easily; evaporation at high temperature is not required.
3. Good coverage is not confined to planar substrate surfaces.
4. The target is a large source and can deposit uniform films over larger areas than PVD.
5. Both conductors and insulators may be deposited.
6. Film adhesion is excellent due to the high particle energies and sputter scrubbing of the substrates.
7. The deposition rate does not depend on target material vapor pressure.

Sputter deposition also may deposit epi layers to form superlattices at low temperatures [17, 18].

12.7 Plasma Environment

Essential to the whole sputter deposition process is a copious supply of positive ions i^+, and their etching or sputtering action on the negatively biased target surface. These ions arise in a gas plasma, as is discussed in Appendix C.

A gas, Ar for example, must be present for the plasma to exist. In usual practice, the equipment is housed in a vacuum chamber that is initially pumped down to $<10^{-5}$ torr to minimize the effects of residual gas. The chamber then is back-filled with Ar to 10^{-2} to 10^{-1} torr $= 10$ to 100 mtorr $= 10$ to 100 μ (*micron*). At these pressures the mean free path $L_m \leq 5/p_m = 5/10 = 0.5$ cm (p_m is the pressure in microns), so with a typical value of $d \approx 4$ cm, a large number of collisions takes place between the positive ions and the sputtered molecules, whose paths between the target and substrates therefore will not be in straight lines.

When dc or rf is applied across the electrodes to strike a glow discharge, the interelectrode spacing is adjusted so that the anode touches the positive column glow. This is the region where the plasma, containing equal numbers of electrons (e^-) and positive ions (i^+) is present. With rf applied at 13.56 MHz or some other ISM (industrial/scientific/medical) frequency, self-bias appears across the electrodes, negative on the target/cathode, so the positive ions move toward the target as required.

With both d-c and r-f types the electrons generated in the plasma, and secondary electrons from the target as well, are attracted to the substrate-bearing anode. While the electrons have insufficient mass to cause sputtering there, their kinetic energy is converted to heat. The resulting temperature rise usually is significant, so the anode (as well as the cathode) is water-cooled.

A serious drawback of d-c sputtering is that it cannot be used with an insulating target. In brief, an insulating target in Figure 12–8 makes a break in the d-c circuit, so dc cannot flow. The positive ions arriving at the target build up positive charge there. Further positive ions are repelled, the voltage *between the electrodes* drops below the breakdown value and the whole sequence shuts down. With r-f excitation, however, the high-mobility electrons from the plasma have half a cycle to reach the target and neutralize the accumulated positive charge from the previous half-cycle, so the process will be sustained [19].

No advantages in the sputtering of metals or other conductive materials are afforded by r-f over d-c sputtering; hence, because of the additional equipment cost and complexity, the r-f type is generally reserved for sputtering insulators [20]. Neither d-c– nor r-f–driven glow discharges are efficient ion sources, so sometimes they are augmented to increase the rate of sputtering-ion production. Two methods are described next.

12.7.1 Electron Injection

The ionizing rate can be improved by providing a separate, independently controlled electron source (namely, a thermionic cathode and associated anode) to initiate and maintain ionization of the gas. This cathode and anode are added to the basic diode form and give the name *triode sputtering.* A typical configuration uses the cathode/anode combination to ionize the gas in the region between the target and the substrate-bearing anode or *collector*. The first anode has a high positive bias relative to the hot cathode; to avoid attracting the thermionic electrons, the collector will be less positive by about 50 V.

Since the thermionic electrons ionize the gas, the latter can be at a pressure of about 0.1 μ, two orders of magnitude less than usual, with a corresponding longer mean free path L_m. Thus, the sputtered atoms/molecules suffer fewer collisions in moving to the substrates and arrive there with higher energies; this makes for better and faster film formation.

In essence, triode or hot-cathode sputtering allows separate parameter adjustments of the ionization and sputter-deposition processes. Ion currents will be higher, at lower voltage, than for the diode case.

12.7.2 Magnetic Field Control

As positive ions strike the target they sputter out secondary electrons as well as target atoms. The electrons are repelled by the negative potential on the target and attracted to the positive anode, where they serve no useful purpose but produce heat on striking the substrates. Many arrive at the substrates with energy corresponding to the full target/anode voltage drop (i.e., they have not lost energy to collision processes).

By the application of proper magnetic fields in the plasma region these electrons can be made to follow curved or spiral paths in moving toward the anode, thereby increasing their chances of ionizing neutral gas atoms by colliding with them [21]. This increases the efficiency of the overall ionization process and the sputter-deposition rate. Some specific applications are considered later.

12.8 Setup

The parallel plate type of diode sputtering rig, shown in basic form in Figure 12–8, is the most commonly used configuration. An alternative coaxial form has been used, but on a smaller scale. We consider these in order.

12.8.1 Parallel Plate Configuration

Figure 12–9 shows the fixturing that is typical for laboratory use with small substrates [22]. A target 10 cm in diameter is used, although diameters up to 50 cm are not unusual in production-line equipment. One unusual feature, however, is the copper tube (10), which serves four functions:

1. It provides a coaxial voltage feed to the cathode assembly, minimizing stray capacitance to ground. This reduces power losses and is satisfactory for both d-c and r-f excitation.
2. It houses the water supply and return hoses (not shown) connected to (11) for cooling the target, thus eliminating the need for separate feedthroughs for them. Note that the water spreads radially over the rear surface of the cathode/target.
3. It supports the entire sputtering assembly so that additional supports are not required for the grounded anode that is connected to the copper tube. Electrical insulation between cathode and anode is provided by the Teflon block (5).
4. The entire assembly is connected to the feedthrough ring (9), so it may be removed from the vacuum system as a unit.

If copper is used for the tube, it should be of the OFHC (oxygen-free, high-conductivity) type to minimize outgassing; stainless steel would be preferable.

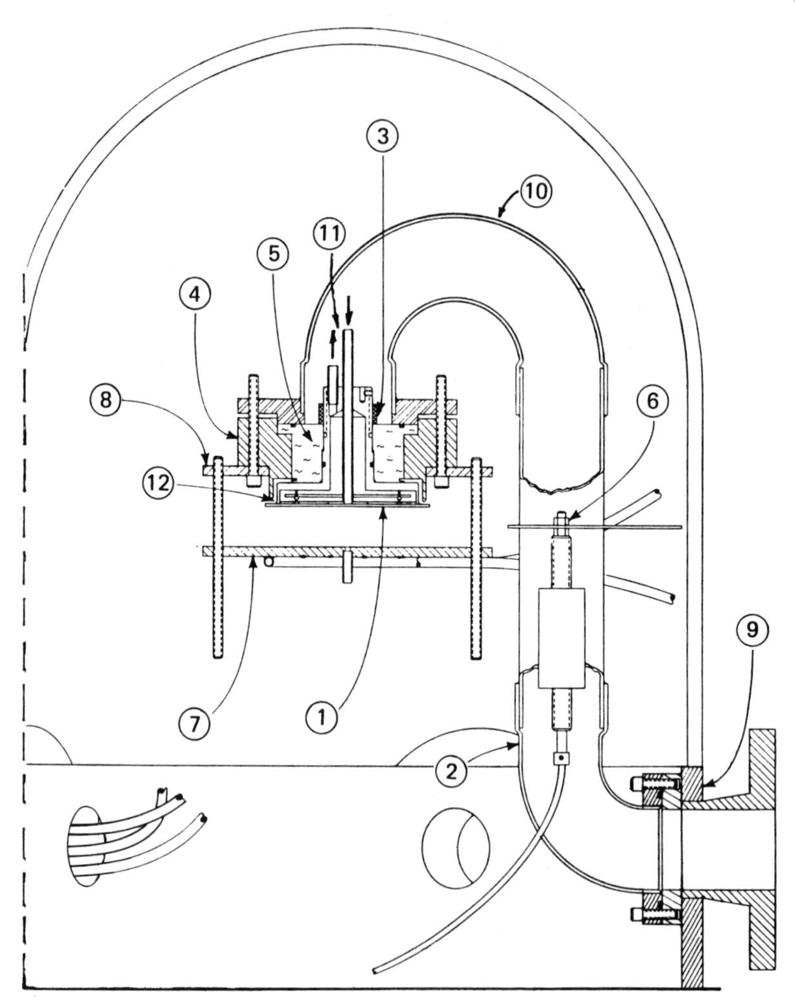

Key: 1. Target and cathode. 2. Copper tube. 3. Retaining nut. 4. Al block. 5. Teflon insulator. 6. Shutter and rotary feedthrough. 7. Anode/substrate holder. Note cooling coil. 8. Support yoke. 9. Feedthrough ring. 10. Copper tube, coax feed. 11. Cooling water supply and return. 12. Gap.

Figure 12–9. Fixturing for a parallel electrode, diode sputter-deposition system. Starnes [22].

The need for water cooling on both electrodes presents an interesting insulation problem, particularly for d-c excitation. A study of Figure 12–9 shows that the anode/substrate holder (7) is connected electrically to the copper tube support (10), which must be grounded for safety. Cooling water passes through a copper tube brazed or hard-soldered to the electrode's lower surface; the water there is therefore at ground potential.

The cathode/target, however, is off ground by the full d-c supply voltage, usually ≥ 1 kV, so cooling water there cannot be grounded or the d-c supply will be shorted. A simple solution is to have the water that flows between the grounded water supply and the cathode run through plastic tubing at least 30 feet long to provide the necessary insulation. Tidy people must be warned against shortening this length.

Note that the grounded Al block (4) and target are separated by a very small gap (12), typically about 2.5 mm for operation at a gas pressure of around 50 mtorr ($= 50\ \mu$). One tends to think that breakdown will occur over such a small gap, but Paschen's Law teaches otherwise (Section C.1).

The shutter (6) in Figure 12–9 plays an important role. It is grounded (at anode potential) and is operated by a rotatable feedthrough in the ring (9). The shutter is between the electrodes on startup. It serves as the anode when the discharge strikes, and intercepts the target molecules. Thus any contamination on the target surface is removed and does not reach the substrates. Note that the target–shutter spacing must be large enough to allow the discharge to strike. When the cleaning is completed, the shutter is rotated out of the way, as in Figure 12–9, and deposition onto the substrates begins. The shutter also can be used to terminate deposition abruptly.

Any fixturing that is located near enough to the glow discharge to be sputtered should be of material with low sputter yield (e.g., stainless steel). This practice reduces the possibility of contaminating the substrate film with fixture material. In d-c systems cylindrical parts may be covered with silica tubing, which sputters poorly. Silica has higher r-f yield, however, so it is not used with r-f excitation. Edges on metal parts should be smoothed or rounded to prevent high localized electric fields.

Impedance matching must be considered for r-f operation. Standing wave measurements on the setup of Figure 12–9 with $d = 3.7$ cm and $p = 10$ mtorr of Ar gave a value of $Z = 78 + j210\ \Omega$ at 13.56 MHz. This must be matched to a 50-Ω coaxial cable from the r-f source (Section C.4). An L section is adequate [23]. Since the impedance is highly dependent on sputtering and gas parameters, automatic adjustment of the matching system is desirable. Several systems have been described [24].

Varian has used a different matching configuration on some of their sputtering equipment, in that adjustable mutual inductance transforms a parallel tuned circuit, one branch of which is the discharge impedance, to a series tuned circuit to match the coax. This allows independent adjustment of the real and imaginary parts of the impedance [25].

The gas supply and vacuum systems are standard; they usually pump against a leak valve to the Ar supply to maintain the desired gas pressure during sputter deposition. A Pirani gage is a good choice for monitoring pressure in the 10 to 100 μ pressure common to sputtering.

Another variation uses *substrate tuning*, in which an adjustable impedance is introduced between the substrate holder and ground. The rf develops a self-bias voltage at the substrates that affects the film properties. Logan has studied its effect on sputter deposition of silica films [26].

12.8.2 Coaxial Configuration

The coaxial sputter rig is an alternative to the parallel electrode form [27]. A unit of this type has been described that employs a cylindrical cathode, about 5 cm in diameter and 50 cm long, coated with target material [28]. The coaxial anode cylinder that bears the substrates is about 40 cm in diameter. An axial magnetic field furnished by a solenoid is used to improve the ionization efficiency, as discussed in Section 12.7.2. The configuration is that of a basic magnetron. The axial magnetic and radial electric fields cause the secondary electrons to move in curved paths in planes normal to the axial direction. The longer curved paths increase the chances for ionizing collisions with the gas. Moreover, with proper adjustment of the two field amplitudes, the electrons do not hit the substrates at all [29]. This eliminates substrate heating by electron impact and can reduce power loss by about 10%. The magnetic field has little effect on the heavy ions and hence does not affect their sputtering action. For the dimensions cited, the anode cylinder can handle about 6000 cm^2 of substrate area.

This rig has sputter-deposited Cu at 110 Å/min with these conditions: $P = 1.5$ kW, $p = 3$ mtorr, $\mathscr{B} = 20$ G. Other features of this system are the same as those described in the last section.

12.9 Film Formation

The formation of sputter-deposited films involves the same steps outlined earlier for PVD. The high energies of the target molecules (or atoms) as they arrive at the substrates cause differences in degree, however, for they can scurry over the surface very well to locate nuclei to join and can cause some scrubbing action on the surface to aid sticking. Some have enough energy to partially embed into the surface, in an incipient implantation action that further aids nucleation.

Deposition rates depend on the arrival rate of the atoms at the substrates, and this must depend on the sputtering rate at the target. Let

$$y = \textit{yield} = \text{no. of target atoms (molecules) ejected per ion striking the target} \tag{12.9-1}$$

In turn, the number of ions striking the target per second is proportional to the ion cathode current I^+, so we can write

$$\text{arrival rate at substrate} \propto yI^+, \quad \text{atoms/s} \tag{12.9-2}$$

This shows a functional dependence, but it is not easy to evaluate because y depends on many variables and I^+ is the ion, not the total, cathode current (electrons do not produce sputtering). Empirical results are preferable for any given setup. Wehner and Anderson give detailed yield data [30, 31].

The yield depends upon the gas pressure, target material, ion species, and energy. The yield peaks when the mass ratio of target molecule (or atom) to incident ion is unity [17]. Values of y for the common metals used in silicon processing and for Si itself range from 0.1 to 5.6. Increasing ion energy raises y while increasing gas pressure lowers it [32].

In general, the various types of materials have decreasing y values in the following order: elemental metals (with Au having the high value for Si processing types); alloys; glasses (with SiO_2 having the highest deposition rate); binary crystalline solids (e.g., Al_2O_3); and ternary crystalline solids such as barium titanate ($BaTiO_3$) [16, 32]. Davidse and Maissel list y values for some of the common dielectric compounds [33].

Compounds, one component of which is a gas, tend to dissociate on being sputtered from the target. For example, silicon nitride (Si_3N_4) dissociates and deposits a Si film if ionized Ar is used, but it sputter-deposits satisfactorily if N_2 replaces Ar. Mixtures of Ar and N_2 produce silicon-rich films, with the gas ratio determining the composition. Typical deposition rates with rf applied at 600 W and 10 μ of N_2 pressure are roughly 90 Å/min [23].

Not all the material sputtered from the target reaches the substrates and incorporates into the growing film—e.g., sputter rates of 500 to 1000 Å/min translate to deposition rates of roughly one-tenth that value [34].

Deposited films often fail to attain their bulk densities and may be amorphous for crystalline target materials. Heat treatment may help, but excess substrate temperature reduces the deposition rate by allowing deposited material to sublime; hence, postdeposition annealing is preferred [16].

Even though the target is physically a large, planar source, it does not give a uniform deposit over a substrate field of the same size. Electric fields are not uniform over the target surface and across the plasma, and positive ion density is largest where the fields are highest—notably, near the electrode edges. In time, more target material is dislodged near the edges, with further adverse effect on the field distribution. Also, the plasma between the electrodes is nonuniform, tending toward a toroidal shape [35].

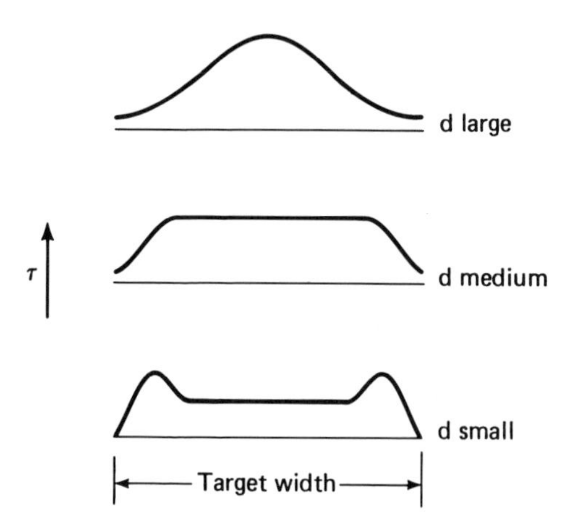

Figure 12–10. Variation of depositeα film thickness τ for three values of target/anode spacing d.

Empirical results are shown in Figure 12–10, where d = target–substrate spacing and s = Crookes' dark space length (Section C.2), with pressure and voltage assumed constant. Chopra has recommended $d \approx 2s$ as the optimum condition, with uniform deposition covering about half the target area [36].

12.10 Alloys, Compounds, Reactive Sputtering

Because of the momentum transfer mechanism, alloys sputter well and tend to maintain their target composition in the deposited film. If a target of the desired alloy is not available, one made of alternating thin strips of the two component metals, supported on a metal plate, can be used to codeposit the alloy. The atomic ratio of the two components, say 1 and 2, in the deposited film is

$$\frac{N_1}{N_2} = \frac{y_1 A_1}{y_2 A_2} \tag{12.10-1}$$

where y is the yield and A is the area on the target. Because of multiple collisions, sputtered atoms do not travel in straight lines in passing through the plasma region, so the strip distribution is not duplicated on the substrates. To keep the field as uniform as possible, the target surface should be kept smooth.

Another method, called multiple target sputtering (MTS), uses simultaneous deposition from two or more targets. The advantage here is that each target is composed of a single component metal; since separate discharges are involved, the *y*s may be tailored by adjusting the two discharge powers independently. To accommodate the multiple targets, interelectrode spacings will have to be larger than in a single-discharge system.

Silicides are alloys of great importance in IC fabrication because they are used for low-resistance interconnects. The silicides of Al, Mo, W, and Ti are in common use. If Al is deposited by PVD as a thin layer on polySi, subsequent high-temperature operations cause the two to react to form aluminum silicide at the interface. Silicide formation also will take place for refractory metals sputter-deposited on polySi, but rough surfaces result.

Sputtering offers two other alternatives. The two materials, refractory and Si, may be codeposited from separate targets to form the silicide as a deposited film, or they may be deposited sequentially from two separate targets (as the wafers rotate between them) so that the film comprises alternate, thin layers of each material. In the latter case subsequent high temperature is required for silicide formation. Kammerdiner and Reeder have presented comparisons of the two cases for $TiSi_2$ [37].

Targets for many compounds or mixtures are formed from powdered material hot-pressed into shape. Purity of the original powder may be high—in the 5-9s (99.999%)—but gases may become incorporated during hot-pressing and cause contamination during sputtering. The *cermets* (ceramic/metals), usually SiO_x and metal, are an example of pressed targets. These are not compounds, but mixtures that are useful for depositing high-resistance, thin-film resistors.

Compounds tend to dissociate or lose stoichiometry in sputtering. In some cases, where neither component of the compound is a gas, the strip target may be used as well as the multiple target system. Both binary and ternary compound semiconductors have been produced by sputtering with the MTS technique [18]. Superlattice structures, made up of several very thin layers of different materials, have been fabricated. These have multiple targets in a ring under which the substrates are rotated to produce the alternating layers. Again, *y* for each target may be tailored by power adjustment.

Silicon nitride has been mentioned in Section 12.9. The nitrides as well as the oxides have a gas as one of the components, so two targets cannot be used; rather, the gaseous component that *reacts* with the solid component can be furnished as all or part of the gas supply to give *reactive sputtering*.

The reactions can take place at the target, in the gas/plasma region, and at the substrates. Because they form an insulator on the target surface, the first type are to be minimized; therefore, high sputter and deposition rates are desirable. A mixture of argon and the reactive gas, rather than the latter alone, normally is used and proper adjustment of the ratio can minimize compound formation at the target surface [38]. Aluminum oxide (Al_2O_3) may be deposited from an Al target with an Ar/O_2 gas mixture. Silicon dioxide also may be deposited from an Si target with Ar/O_2, but experience has shown that an SiO_2 target with a high Ar/O_2 ratio gives better results; only enough oxygen is needed to maintain stoichiometry. Silicon nitride may be deposited with Si and either N_2 or ammonia (NH_3) as the reactive gas [23]. Nitrogen is preferred because NH_3 is corrosive.

12.11 Planar Magnetron Sputtering

The use of a magnetic field in a coaxial diode sputtering rig was mentioned in Section 12.8.2; a $\mathscr{B}$ field also may be used to enhance the ionization rate in the basic parallel electrode array. A simple form that uses permanent magnets has a circular target/magnet assembly as shown in Figure 12–11. The center magnet is cylindrical and the outer annular; both are ferrites. Typical field strengths run 100 to 500 G (=0.01 to 0.05 T). Note that the $\mathscr{B}$ field is parallel—and the $\mathscr{E}$ field normal—to the target face, about midway between the center and ring magnets. Due to the cross fields, the secondary electrons from the target follow spiral paths along the flux lines, thereby increasing the ionization rate. The electrons also are prevented from reaching the target except under the middle of the center magnet. There they see parallel $\mathscr{B}$ and $\mathscr{E}$ fields, and so go to the anode; substrates should be arranged to avoid this region. Typical sputter deposition rates for this system are available [20]. Metals are sputtered with d-c voltage between target and anode.

Magnetron sputtering with applied rf can be used for sputtering insulators. Bar magnets, typically of Alnico, and polepieces may be arranged differently from the form shown in Figure 12–11, but the $\mathscr{B}$ field must be parallel to the target face in at least one region. The following results have been reported for

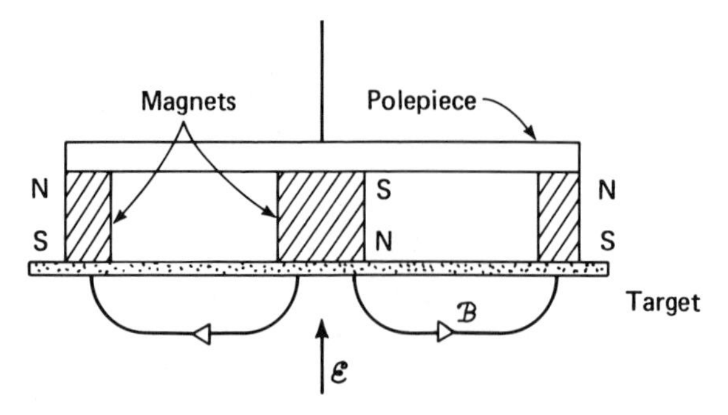

Figure 12–11. Cross section of a simple circular magnet, polepiece, and cathode/target assembly for magnetron sputter-deposition.

SiO_2 : 700 Å/min at 13.56 MHz and 1.8 kW in pure Ar, about twice the deposition rate for comparable conditions without the magnetic field [39].

12.12 Bias Sputtering

The properties of sputter-deposited films can be affected adversely by argon becoming incorporated during deposition. The number of such atoms can be reduced by sputter-etching the films with *bias sputtering*. The basic d-c configuration is shown in Figure 12–12(a) where the *collector*, an auxiliary anode that supports the substrates, is located between the main electrodes and biased negatively wrt the anode, but still positive relative to the target. Because of the negative bias, the collector attracts Ar^+ ions from the discharge. These sputter-etch the growing film; the higher the bias, the greater the sputtering rate—up to a point. If the bias gets too large, positive ions may implant into the substrate film, which defeats the purpose of substrate sputtering.

Bias sputtering may be described as deposition with simultaneous ion bombardment. Since two competing processes are at work at the substrate surfaces, conditions must be adjusted so that deposition predominates. Christensen discusses the mechanisms in more detail and characterizes the properties of the resulting films [40].

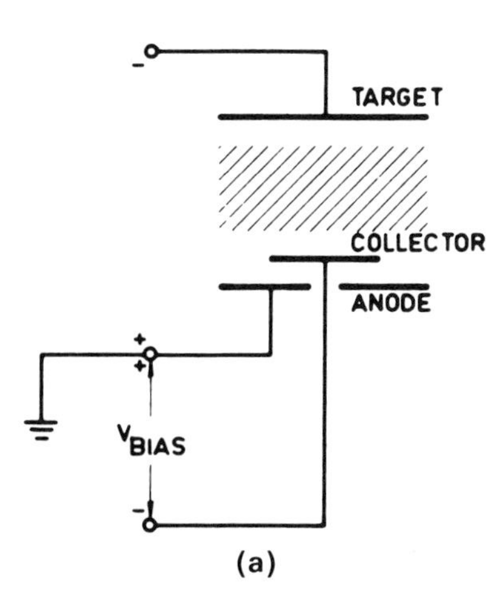

Figure 12–12. Connections for three forms of bias sputtering for metals. (a) Diode with d-c bias supply. (b) Diode with r-f bias supply. (c) Triode with r-f bias supply. Christensen [40]. (Courtesy of *Solid State Technology*, PennWell Publishing Company, Copyright 1970.)

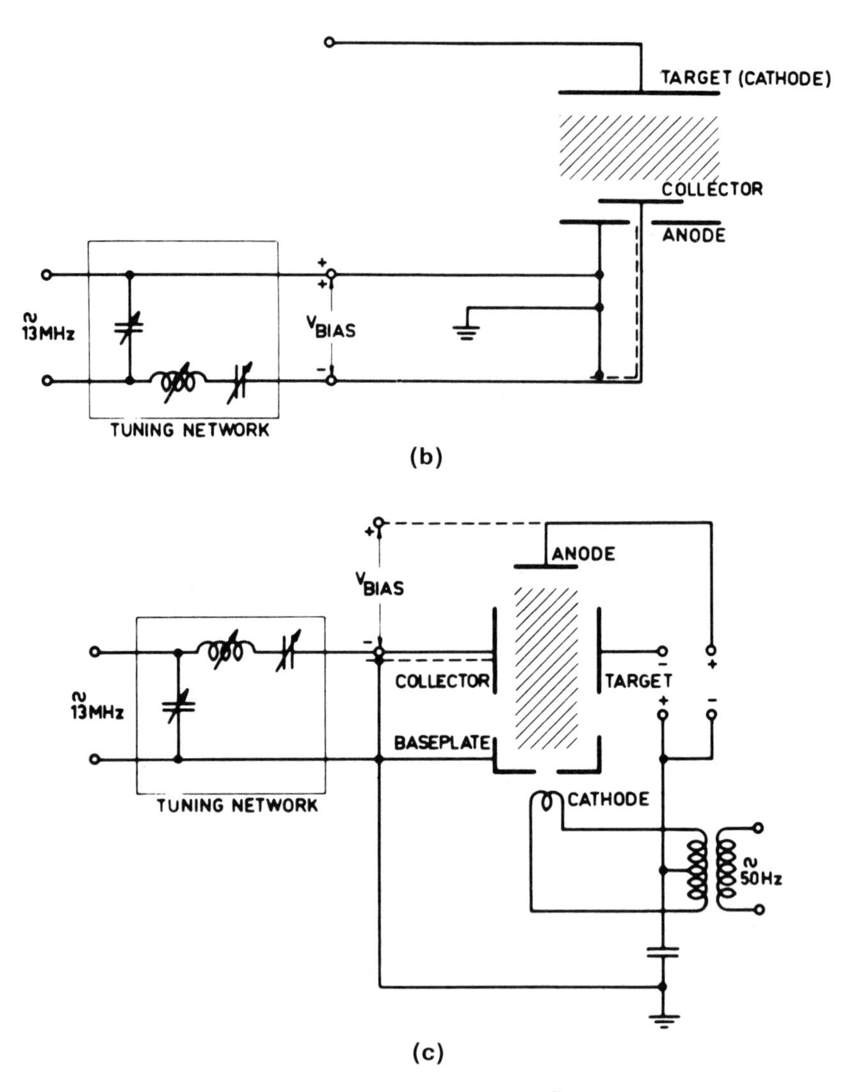

Figure 12–12 (*Continued*)

The bias may be developed from an r-f source, and a triode structure also may be used, as shown in Figure 12–12(b) and (c), respectively. Insulating substrates may be sputter-cleaned before the main supply between target and anode is applied when r-f biasing is used, a significant advantage.

The bias technique provides better step coverage in sputter-depositing Al, an effect that may be due to the increase of the deposited atoms' surface mobility on the substrate [41].

12.13 Ion Plating

Ion plating uses the combined effects of PVD and sputtering to deposit films at a higher rate than sputtering and with better adhesion than PVD. Further, it can coat substrate side walls without rotating them, and with adequate cooling can coat even low-melting-point plastics [42].

The basic setup is shown in Figure 12–13. An evaporation source and shutter are at grounded anode potential, while the substrates are on the cathode, held at 2 to 5 kV negative to ground during operation. Argon is back-filled to 20 to 60 μ after an initial pumpdown to between 5×10^{-7} and 5×10^{-6} torr. The discharge is struck and the substrates cleaned by sputter-etching with Ar^{+} with the shutter closed. The evaporation source is heated, to outgas the evaporant supply, and the shutter is opened, to allow the evaporant to travel to the substrate. The evaporant becomes ionized in passing through the plasma and aids in the sputter-etch, but the lower energy atoms stick to the substrate and form the film. Since competing processes are involved, conditions must be adjusted so that a film *is* deposited [28]. The three-dimensional-coating capability comes about because the substrates are biased negatively and the evaporant ions are positive.

More than one evaporant source may be provided, so that layers of different materials may be deposited, and higher deposition rates may be obtained by using electron beam evaporation instead of a conventional heated evaporation source.

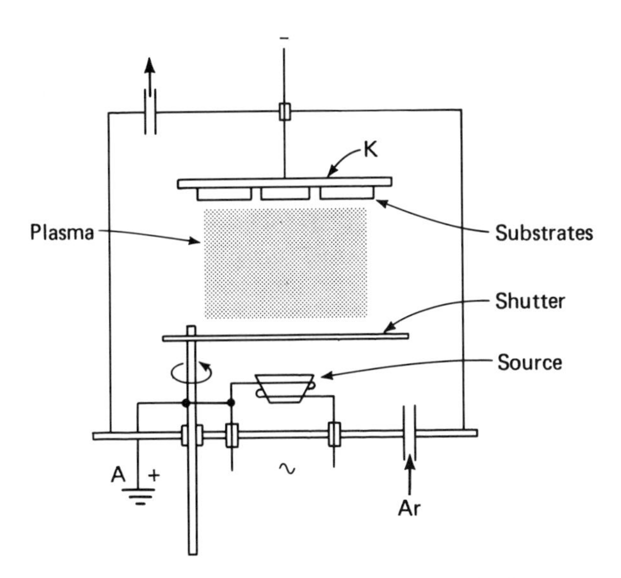

Figure 12–13. Basic setup for ion plating.

Problems

12–1 (a) Explain why Honig's curves differ in shape from the predicted form of Eq. (12.2-1).
(b) Why is that equation valid over a limited temperature range?

12–2 Using equations, show why a bent e-beam system causes no concern about Na atoms, which have picked up electrons, reaching the evaporant supply.

12–3 An evaporant supply in a boat source usually melts completely as evaporation takes place. Explain why a much larger charge can be used with an e-beam system.

12–4 Verify that a deposition rate for Si of 1 μm/hr is equivalent to about 1 atomic layer/s.

12–5 Explain why the component ratio in an alloy is maintained better from a large pool of melt, as in e-beam evaporation.

12–6 Cadmium and sulfur are to be coevaporated from separate sources to give a CdS film. Recommend reasonable temperatures for the two sources, and justify your recommendations.

12–7 Sketch the water cooling system for the rig of Figure 12–9. The anode and cathode cooling coils are to be in series.

(a) Suggest lengths for the vinyl hoses.
(b) Justify your suggestions.
(c) Why is the anode copper cooling coil not soft-soldered to the anode?

12–8 Design an L section to match the plasma impedance of Section 12.8.1 to $50 + j0$ Ω. Since phase shift is unimportant, should the C be in series or shunt? Explain.

References

[1] R. W. Berry, P. H. Hall, and M. T. Harris, *Thin Film Technology*. Princeton: Van Nostrand, 1968.

[2] R. E. Honig, "Vapor Pressure Data for the Solid and Liquid Elements," *RCA Rev.*, **23**, (4), 567–586, Dec. 1962.

[3] R. E. Thun, "Thick Films or Thin," *IEEE Spectrum*, **6**, (10), 73–79, Oct. 1969.

[4] *Sloan Technology Corporation Handbook of Thin Film Materials*. Santa Barbara: Sloan Technology Division of Veeco Instruments Inc., 1970.

[5] *Sylvania Bulletin 235-1*. Exeter, NH: Sylvania Emissive Products.

[6] *Sylvania Bulletin 225 GR-2*. Exeter, NH: Sylvania Emissive Products.

[7] J. B. Hedge and W. A. Bagot, "Power Parameters of Metallizing Boats," *Res./Dev.*, **13**, (12), 32–37, Dec. 1971.

[8] *Veeco Catalog 71*. Plainview, NJ: Veeco, 1971.

[9] A. F. Plant, "A Little Weigh," *Ind. Res.*, **13**, (7), 36–39, July 1971.

[10] S. N. Levine, *Principles of Solid-State Microelectronics*, New York: Holt, Rinehart and Winston, 1963.

[11] A. Y. Cho, "Morphology of Epitaxial Growth of GaAs by a Molecular Beam Method: The Observation of Surface Structures," *J. Appl. Phys.*, **41**, (7), 2780–2786, June 1970.

[12] T. E. Bell, "Innovations: Growing GaAs on silicon," *IEEE Spectrum*, **23**, (4), 25, April 1986.
[13] P. H. Singer, "Molecular Beam Epitaxy," *Semicond. Int.*, **9**, (10), 42–47, Oct. 1986.
[14] P. E. Luscher, "Crystal Growth by Molecular Beam Epitaxy," *Solid State Technol.*, **20**, (12), 43–51, Dec. 1877.
[15] S. Wolf and R. N. Tauber, *Silicon Processing for the VLSI Era, Volume 1—Process Technology*. Sunset Beach: Lattice Press, 1986.
[16] J. L. Vossen and J. J. O'Neil, "R-F Sputtering Processes," *RCA Rev.*, **29**, (2), 149–179, June 1968.
[17] J. E. Greene, "Epitaxial Crystal Growth by Sputter Deposition: Applications to Semiconductors, Part 1." In *CRC Critical Reviews in Solid State and Materials Sciences*, ed. D. E. Schuele and R. W. Hoffman, Vol. 11, Issue 1, pp. 47–97. Boca Raton: CRC Press, 1983.
[18] J. E. Greene, "Part 2," loc. cit. [17] Vol. 11, Issue 3, pp. 189–227, 1984.
[19] H. R. Koenig and L. I. Maissel, "Application of RF Discharges to Sputtering," *IBM J. Res. Dev.*, **14**, (2), 168–171, Mar. 1970.
[20] R. K. Waits, "Planar Magnetron Sputtering." In *Thin Film Processes*, ed. J. L. Vossen and W. Kern, Chap. II-4, New York: Academic Press, 1978.
[21] J. L. Vossen and J. J. Cuomo, "Glow Discharge in Sputter Deposition," loc. cit. [20], Chap. II-1.
[22] R. M. Starnes, *Design and Installation of a Diode Sputtering System*, MS Thesis, EE Dept., UIUC, 1969.
[23] R. M. Starnes, *A Study of Dipolar Polarization in Silicon Nitride Films Using an Adapted Thermally Stimulated Current Technique*, PhD Thesis, EE Dept. and Coordinated Science Lab., UIUC, April 1972.
[24] N. M. Mazza, "Automatic Impedance Matching System for RF Sputtering," *IBM J. Res. Dev.*, **14**, (2), 192–193, Mar. 1970.
[25] F. Turner, "A New Frontier in Sputtering Equipment," *Varian Vacuum News*, pp. 1, 4, Feb. 1973.
[26] J. S. Logan, "Control of RF Sputtered Film Properties Through Substrate Tuning," *IBM J. Res. Dev.*, **14**, (2), 172–175, Mar. 1970. Also in *Solid State Technol.*, **13**, (12), 46–48, 53, Dec. 1970.
[27] J. A. Thornton and A. S. Penfold, "Cylindrical Magnetron Sputtering," loc. cit. [20], Chap. II-2.
[28] V. Hoffman, I. Weissman, and D. Sanservino, "The Thin Film," *Ind. Res.*, **14**, (11), 50–53, Oct. 1972.
[29] T. S. Gray, *Applied Electronics, 2nd Ed.* New York: Wiley, 1954. Chap. 1, Art. 8d.
[30] G. K. Wehner and G. S. Anderson, "The Nature of Physical Sputtering." In *Handbook of Thin Film Technology*, ed. L. I. Maissel and R. Glang, Chap. 3. New York: McGraw-Hill, 1970.
[31] G. K. Wehner and G. S. Anderson, loc. cit. [30], Chap. 4.
[32] I. H. Pratt, "Thin Film Dielectric Properties of RF Sputtered Oxides," *Solid State Technol.*, **12**, (2), 49–57, Dec. 1969.
[33] P. D. Davidse and L. I. Maissel, "Dielectric Films through rf Sputtering," *J. Appl. Phys.*, **37**, (2), 574–579, Feb. 1966.
[34] "Sputter Etching & Deposition, Theory and Applications of Glow Discharges," *Circuits Manuf.*, **217**, (2), 78–81, Feb. 1981.

[35] L. T. Lamont, Jr., "Thin Film Notebook: Chap. VIII, R. F. Sputtering," *Varian Vacuum News*, pp. 2, 4, Nov. 1972; also Chap. IX, p. 2, Feb. 1973.
[36] K. L. Chopra, *Thin Film Phenomena.* New York: McGraw-Hill, 1969.
[37] L. Kammerdiner and M. Reeder, "Codeposition vs Layering of Sputtered Silicide Films," *Semicond. Int.*, **7**, (8), 122–126, Aug. 1984.
[38] P. S. McLeod, "Reactive Sputtering," *Solid State Technol.*, 26, (10), 207–211, Oct. 1983.
[39] K. Urbanek, "Magnetron Sputtering of SiO_2: An Alternative to Chemical Vapor Deposition," *Solid State Technol.*, **20**, (4), 87–90, April 1977.
[40] O. Christensen, "Characteristics and Applications of Bias Sputtering," *Solid State Technol.*, **13**, (12), 39–45, Dec. 1970.
[41] J. F. Smith, "Influence of DC Bias Sputtering During Aluminum Metallization," *Solid State Technol.*, **27**, (1), 135–138, Jan. 1984.
[42] "Ion Plating: The Best of Sputtering and Evaporation," *Circuits Manuf.*, **12**, (1), 10, 12–14, Jan. 1972.

Appendix A

Four-Point-Probe Derivations; Optical Interference

This appendix derives collinear four-point-probe (FPP) equations of Chapter 3 relating measured V/I ratios to sample resistivity ρ or sheet resistance R_S for several geometries, and equations of optical interference.

A.1 Semi-Infinite (S-I) Sample

Equation (3.4-1) for the semi-infinite sample may be derived by exploiting an electrostatic analog that will be considered now. Let a current I be introduced at a point on the surface of an S-I sample as shown in Figure A–1(a). Imagine a hemispherical surface, of radius r and of area $2\pi r^2$, within the sample and centered at the point of current entry.

With the return current collected at infinity, there is no preferred direction for current flow within the sample; hence, the current density is everywhere uniform on, and normal to, the hemispherical surface, so

$$I = \int_{\text{hemisphere}} \sigma_\infty \mathscr{E}\, da = \sigma_\infty \mathscr{E}(2\pi r^2)$$

where σ_∞ = conductivity of the S-I sample = $1/\rho_\infty$. Then

$$\mathscr{E} = \frac{\rho_\infty I}{2\pi r^2} \tag{A.1-1}$$

where $\mathscr{E}$ = the electric field magnitude on the hemisphere.

Next consider a point charge of magnitude Q located in an infinite region of permittivity $\varepsilon = \varepsilon_o \varepsilon_r$. Imagine a spherical surface of radius r centered on the charge as shown in Figure A–1(b). Again there is no preferred direction; hence the electric displacement $\varepsilon\mathscr{E}$ is everywhere uniform on, and normal to, the

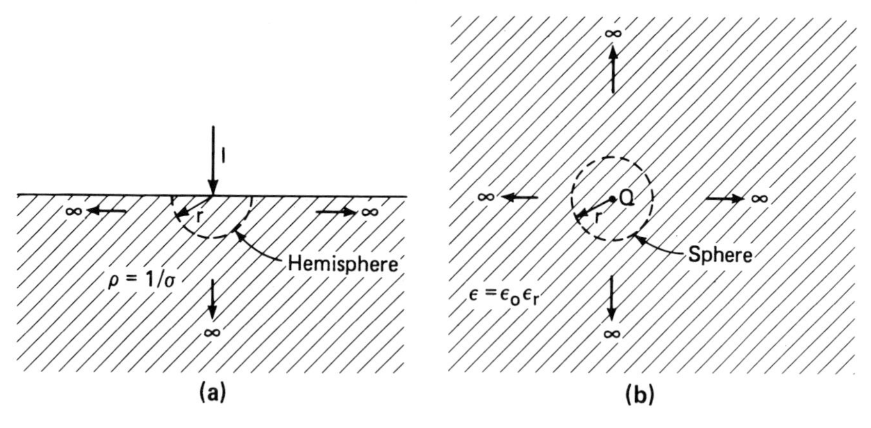

Figure A–1. The electrostatic analog used for deriving the basic FPP equation. (a) Current I introduced into a semi-infinite, conducting medium. (b) Charge Q in an infinite, nonconducting medium.

sphere's surface, and we can write

$$Q = \int_{\text{sphere}} \varepsilon\mathscr{E}\, da = \varepsilon\mathscr{E}(4\pi r^2)$$

or

$$\mathscr{E} = \frac{Q}{4\pi r^2 \varepsilon} \qquad \text{(A.1-2)}$$

Comparison of the two diagrams in Figure A.1 shows that the lines of current flow at (a) would be radially outward from the center and identical in form to the lines of electrostatic flux in (b); hence, we can pass from one picture to the other by using the analog

$$\rho_\infty I \leftrightarrow \frac{Q}{2\varepsilon} \qquad \text{(A.1-3)}$$

Thus, if we know $\mathscr{E}$ at any point in the lower half of the medium in the electrostatic case, we can determine $\mathscr{E}$ for the corresponding point in the medium for the current flow case simply by substituting the last equation.

Furthermore, if the current in Figure A–1(a) is reversed in sign, thereby implying that it is introduced into the sample at infinity and removed at the point on the surface, the current magnitude at every point within the S-I medium will remain unchanged, but its direction will be reversed. The corre-

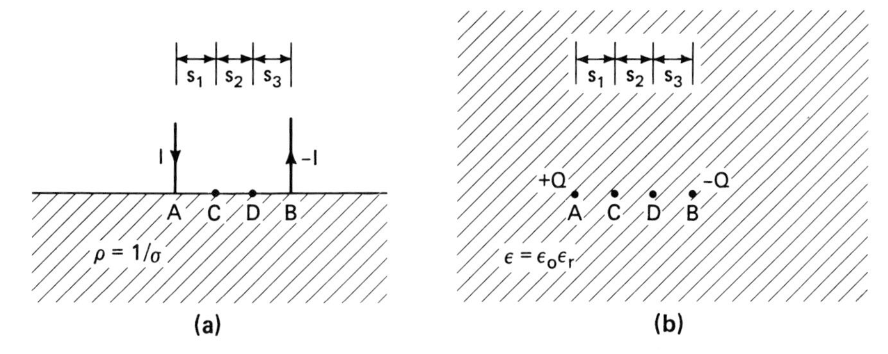

Figure A–2. Extension of the electrostatic analog to apply to the FPP. (a) The FPP on the surface of the semi-infinite sample. (b) The analogous charge distribution in an infinite nonconducting medium.

sponding electrostatic analog in the infinite medium still has the form shown at Figure A–1(b), but the sign of Q would be reversed.

Since $\mathscr{E}$ is linear in both I and Q in Eqs. (A.1-1) and (A.1-2), superposition may be used if more than one current or one charge is present. This means that the FPP problem can be solved by setting up an analogous electrostatics problem involving two charges, as shown in Figure A–2.

Even though the FPP is usually built with uniform intertip spacing s, we shall generalize by allowing variations in tip spacing as indicated in Figure A–2. By so doing, we obtain equations in a form that allows easy error analysis for bent probes (Problems 3–3 and 3–4).

In Figure A–2(a) current enters the S-I sample at point A and leaves at B. This is modeled in the electrostatic analog by charges $+Q$ at A and $-Q$ at B in the infinite medium shown at (b). The potential difference or voltage between the inner points C and D is to be determined. Elementary electrostatics theory teaches that the potential at any point due to a number of charges Q_i in the electrostatic analog is given by

$$\psi = \frac{1}{4\pi\varepsilon}\Sigma\frac{Q_i}{r_i}$$

where r_i is the distance between charge Q_i and the point where the potential is being evaluated. Only two charges are involved, so the potentials at points C and D are

$$\psi_C = \frac{1}{4\pi\varepsilon}\left(\frac{+Q}{s_1} + \frac{-Q}{s_2 + s_3}\right) \quad \text{and} \quad \psi_D = \frac{1}{4\pi\varepsilon}\left(\frac{+Q}{s_1 + s_2} + \frac{-Q}{s_3}\right)$$

The potential *difference* between the two points is

$$V = \psi_C - \psi_D = \frac{Q}{4\pi\varepsilon} K_S \tag{A.1-4}$$

where

$$K_S = \left[\frac{1}{s_1} - \frac{1}{s_2 + s_3} - \frac{1}{s_1 + s_2} + \frac{1}{s_3}\right] \tag{A.1-5}$$

Invoking the analog and replacing $Q/2\varepsilon$ by $\rho_\infty I$ we have

$$V = \frac{\rho_\infty I}{2\pi} K_S \tag{A.1-6}$$

or

$$\rho_\infty = 2\pi \frac{V}{I} \frac{1}{K_S} \tag{A.1-7}$$

Now if, as in the case of a usual commercial FPP, $s_1 = s_2 = s_3 = s$, then $K_S = 1/s$ and

$$\rho_\infty = 2\pi s \frac{V}{I} = 2\pi s R \frac{V}{V_I} \quad \left\{ \begin{array}{l} \text{S-I sample} \\ s_1 = s_2 = s_3 = s \end{array} \right. \tag{A.1-8}$$

where $R = V_I/I$. This is Eq. (3.4-1). Remember that the subscript ∞ is just a reminder that the equations are valid only for an S-I sample.

A.2 Thickness Correction for I-t Samples

Consider Figure A–3(a), where the sample is shown shaded extending to infinity horizontally between the parallel upper and lower surfaces or *planes* p_u and p_l, respectively. The thickness between the planes is t. We shall designate this infinite plane of finite thickness by "I-t."

The surrounding medium, unshaded in the figure, is an insulator, so no current can cross either surface except at points A and B. (Note that this requires the wafer to be resting on an insulator, not a conductor.) Since current density across a surface is given by

$$j_{\text{normal}} = \sigma \mathscr{E}_{\text{normal}}$$

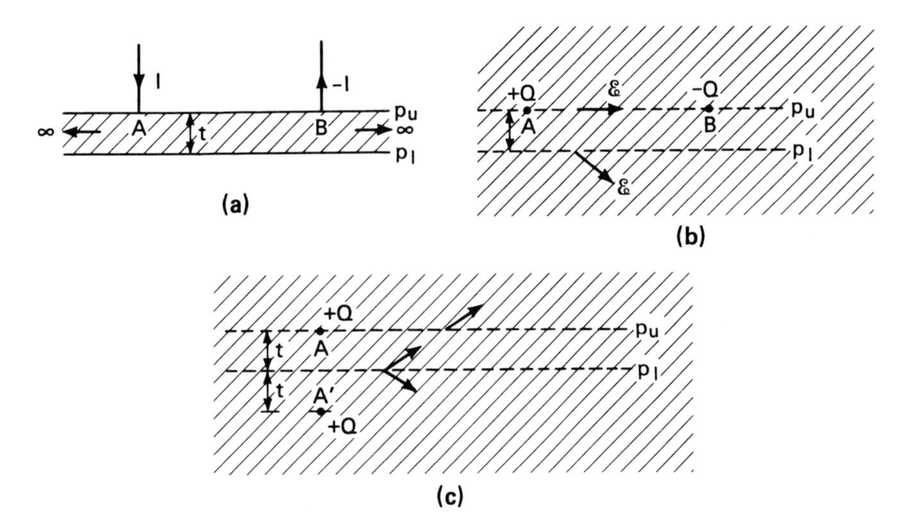

Figure A–3. Setting up the electrostatic analog for a sample of finite thickness, t. (a) The currents due to the FPP. The normal component of the $\mathscr{E}$ field is zero everywhere on the upper and lower sample surfaces (p_u and p_l), except at points A and B. (b) A single charge Q at A in the analog satisfies the condition on p_u, but not on p_l. (c) An image charge Q is added at A′, $2t$ below A.

where σ is the sample conductivity, the condition will be satisfied if the *normal* component of $\mathscr{E}$ on *both surfaces* is identically zero except at A and B. This is a boundary condition. In contrast to the S-I case, the model here must specify that $\mathscr{E}_{\text{normal}}$ on *both* p_u and p_l will be zero except at A and B. The two currents in (a) are modeled by the two charges in (b).

The $\mathscr{E}$ field from a point charge is radial; hence the field along the upper plane, p_u, is wholly tangential, with normal component zero. Therefore, the charge at A satisfies the boundary condition on p_u. The radial field from $+Q$ at A, however, is inclined to the lower plane, p_l, as shown, and has a nonzero normal component along p_l; the boundary condition is not satisfied, so the model must be modified.

An image charge $+Q$, placed a distance $2t$ vertically below the $+Q$ charge at A, will produce a field that can reduce the net normal component on p_l to zero. But note that the image charge has now upset the boundary condition on p_u! What should be done? Well, what has been done once can be done again: by adding image charges of $+Q$, always spaced by intervals of $2t$ along a vertical line through A, we can patch up the model to satisfy the boundary condition—now on the upper surface, then on the lower surface, and so on. In final form, the model will have an infinite line of $+Q$ charges in order

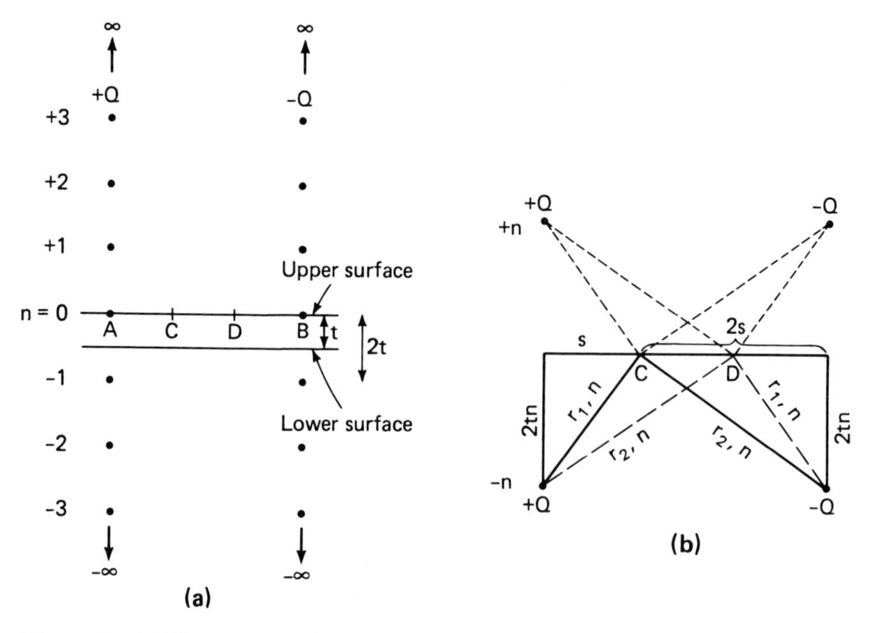

Figure A–4. The corrected analog for a sample of thickness t. (a) Indicating the infinite array of charges to set the normal $\mathscr{E}$ component on both p_u and p_l to zero, except at A and B. (b) Showing the symmetries between top and bottom, and between left and right in the analog.

to satisfy the boundary condition of zero normal component of $\mathscr{E}$ on both surfaces.

A similar argument holds for modeling the current at point B, except that now an infinite string of $-Q$ charges spaced at intervals of $2t$ along a vertical line through B is required. An indication of the complete array of charges needed to model the I-t case of the two current tips of the FPP is shown in Figure A–4(a). (The lines of charges have not been carried out to plus and minus infinity in the figure!)

The array has a twofold symmetry, between top and bottom and between left and right in the diagram, that can be exploited. The appropriate trigonometry is illustrated in Figure A–4(b). Note that in both (b) and (a), the intervals between adjacent charges have been replaced by a counting number n that tells the number of $2t$ units vertically, between any one of the charges and p_u. The voltage V between the two inner points located at C and D may be evaluated by the method used with the S-I case, except that here we have many more charges that contribute to the potentials at C and D. Reading from Figure A–4(b) we have

$$\psi_C = \frac{Q}{4\pi\varepsilon} \sum_{n=-\infty}^{+\infty} \left(\frac{1}{r_1, n} - \frac{1}{r_2, n} \right) \tag{A.2-1}$$

where $r_1, n = \sqrt{s^2 + (2tn)^2}$ and $r_2, n = \sqrt{(2s)^2 + (2tn)^2}$.

This may be simplified, as can the corresponding expression for ψ_D. After use of the analog and considerable algebraic manipulation, $V = \psi_C - \psi_D$ becomes

$$V = \frac{\rho I}{2\pi s}\left[1 + 4\frac{t}{s} \sum_{n=+1}^{+\infty} \left(\frac{1}{\sqrt{(s/t)^2 + (2n)^2}} - \frac{1}{\sqrt{(2s/t)^2 + (2n)^2}} \right)\right]$$

$$= \frac{\rho I}{2\pi s} \frac{1}{a} \tag{A.2-2}$$

or

$$\rho = a\left(2\pi s \frac{V}{I}\right) = a\rho_\infty \tag{A.2-3}$$

This is Eq. (3.5-1). The symbol ρ here is the actual resistivity of the I-t sample; ρ_∞ is a measured value given by Eq. (3.4-1).

The factor a is the reciprocal of the bracketed factor in Eq. (A.2-2) and is the thickness correction factor for the I-t sample. The factor has been tabulated by Valdes and is plotted in Appendix F.3.2 [1]. Notice that the s/t variable of Eq. (A.2-2) has been replaced by its reciprocal in the curve.

A.3 Logarithmic Potential Derivation for Thin Samples

Equation (3.5-2) for thin samples is derived in this appendix through use of the logarithmic potential. Students protest a second derivation, pleading Occam's razor, but the forefactor 4.53 that was obtained from graphical values in Section 3.5.3 is placed on a sound theoretical basis here.

Consider an I-t sample whose thickness t is so small that equipotential surfaces within the sample are normal to the infinite plane (i.e., no voltage drops vertically within the sample). A current is introduced at a point A on the surface as shown in Figure A–5(a) and collected at infinity, so there is no preferred direction and the equipotential planes are concentric cylinders centered on A. Then V_{CD}, the voltage difference between two points C and D located distances c and d from A, respectively, and due to I, is of the logarithmic potential form

$$V_{CD}|_{+I} = \frac{\rho I}{2\pi t} \int_c^d \frac{dr}{r} = \frac{\rho I}{2\pi t} \ln \frac{d}{c} \tag{A.3-1}$$

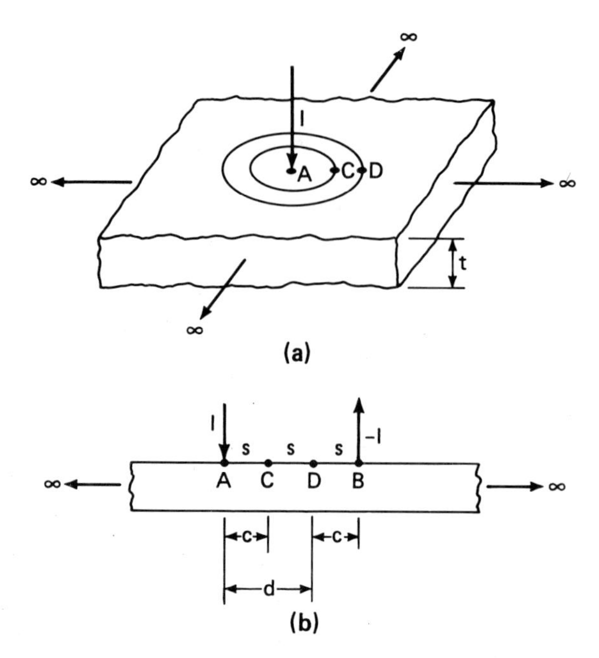

Figure A–5. Calculation of the voltage V_{CD} due to I on the surface of a sample t thick, but of infinite size horizontally. (a) Setup for the logarithmic potential. (b) The FPP on the sample.

Note the order of the integration limits and the subscripts on V. The lower limit corresponds to the first subscript, and the upper limit to the second. If I flows out from the sample at point A, its sign is negative in the equation. The corresponding FPP setup is shown at (b). Assume that $d = 2c = 2s$ as in the FPP.

Since the voltage is linear in I, superposition may be used to find the total voltage between C and D due to the two currents. It may be shown that

$$\rho = \frac{\pi}{\ln 2} t \frac{V}{I} \tag{A.3-2}$$

so the statement regarding the constant 4.53 is correct.

A.4 Optical Interference

Consider the situation shown in Figure A–6(a), where we wish to measure t, the thickness of a very thin air space between the parallel surfaces of a half-silvered optical flat of glass and an optically dense material. In a real-world situation the glass is much thicker than the air space, so we neglect what

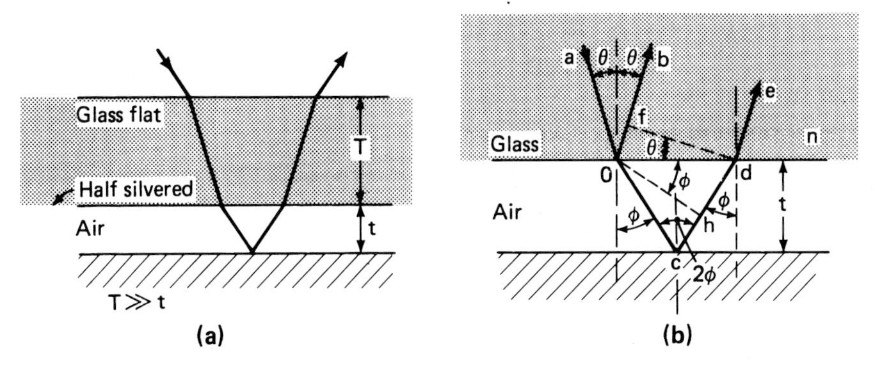

Figure A–6. Determination of the thickness t of a very thin air space. (a) The principal optical path. (b) Detail of the optical paths for calculating destructive interference.

happens at the upper surface of the glass and consider only the lower portion of the system, shown at (b).

Part of the incident light ray *ao* is reflected at *o* as the ray *ob*. The glass is more dense optically than the air, so the reflection takes place with no phase shift. Another portion of the incident ray is refracted at *o* as the ray *oc*, is then reflected at *c*, and emerges as ray *de* after refraction again at *d*. The reflection at *c* takes place with a phase shift of π radians because the air is less dense optically than the reflecting solid below *c*. We seek the phase difference between *ob* and *de*. When this difference is an odd multiple of π radians (i.e., when the difference is $(2m + 1)\pi$ for $m = 0, 1, 2, \ldots$), destructive interference takes place: the two rays tend to cancel and leave a dark region. Let us find this phase difference.

Construct *fd* normal to *ob*, and *oh* normal to *cd*. Then a wavefront at *oh* will be reflected as a wavefront at *fd*, i.e., *of* and *hd* contain the same number of wavelengths and so contribute no phase *difference* between *ob* and *de*. This may be verified quite easily. Reading from (b) in the figure:

$$hd = od \sin \phi \quad \text{and} \quad of = od \sin \theta$$

These two equations with Snell's law of refraction yield

$$\frac{hd}{of} = \frac{\sin \phi}{\sin \theta} = n = \frac{\lambda_a}{\lambda_g} \quad \text{or} \quad \frac{hd}{\lambda_a} = \frac{of}{\lambda_g} \tag{A.4-1}$$

where n = refraction index of the glass, and λ_a and λ_g are the wavelengths of the monochromatic light in air and glass, respectively. The equation shows that the two paths *of* and *hd* do indeed contain the same number of wavelengths and so contribute no phase difference, as stated earlier.

For this reason, any phase difference that does exist between *ob* and *de* is due to the difference in path length $\delta = oc + ch$ that is traversed in air. From Figure A–6(b)

$$oc = \frac{t}{\cos\phi} \quad \text{and} \quad ch = oc\cos 2\phi$$

or

$$\delta = oc(1 + \cos 2\phi) = \frac{t}{\cos\phi} 2\cos^2\phi = 2t\cos\phi$$

The corresponding phase shift in radians due to δ is

$$\psi = 2\pi\delta/\lambda_a$$

and the total phase difference between *ob* and *de* is ψ plus the π radian shift caused by reflection at *c*. Destructive interference occurs when this total shift is an odd multiple of π radians. Thus, the condition for destructive interference is

$$\left(\frac{4t\cos\phi}{\lambda_a} + 1\right)\pi = (2m + 1)\pi, \qquad m = 0, 1, 2, \ldots$$

or

$$2t\cos\phi = m\lambda_a, \qquad m = 1, 2, 3, \ldots \tag{A.4-2}$$

This principle may be used to measure x_j and the thickness of thin films.

Reference

[1] Data from L. B. Valdes, "Resistivity Measurement on Germanium for Transistors," *Proc. IRE*, **42**, (2), 420–427, Feb. 1954. © 1954 IEEE.

Appendix B

Ion/Field Interactions

This appendix considers some basic interactions between ions and electric and/or magnetic fields that are used in ion implanters (Chapter 8).

B.1 Parallel Initial Velocity and $\mathscr{E}$ Field

Consider the idealized situation shown in Figure B–1(a) where an ion of charge $+q_i$ and mass m_i is at rest between two parallel plates in the y–z plane and located at $x = 0$. A negative voltage $-V_1$ produces an $\mathscr{E}$ field that accelerates the ion in the x-direction. It reaches the V_1 electrode with velocity v_{x1} and kinetic energy (KE)

$$\mathrm{KE} = \frac{m_i v_{x1}{}^2}{2} \tag{B.1-1}$$

and potential energy (PE)

$$\mathrm{PE} = q_i V_1 \tag{B.1-2}$$

Conservation of energy demands that these two energies be equal, whence

$$v_{x1} = \sqrt{\frac{2q_i V_1}{m_i}} \tag{B.1-3}$$

The velocity is independent of d.

Consider the two right-hand electrodes. Since they are at the same potential, the right-hand region in Figure B–1(a) will be space-charge-free. If the electrode at x_1 contains a hole through which the ion can pass, the ion (initially at $x = 0$) will move into this right-hand region and continue with constant

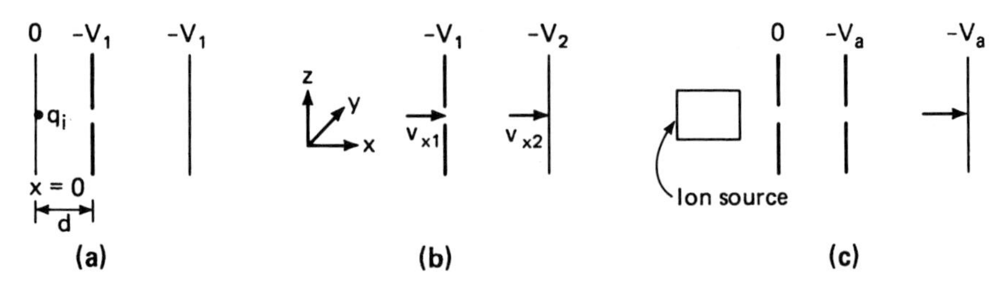

Figure B–1. Acceleration of a charged particle by an electric field. (a) The ion is initially at rest at the left-hand electrode. The two other electrodes are at the same potential. (b) The ion enters a hole in the center electrode with velocity v_{x1} parallel to the electric field. (c) Skeletal form of the ion implanter of Figure 8–1.

velocity v_{x1} in the x-direction. Notice from Eq. (B.1-3) that v_{x1} may be expressed in terms of V_1 for a given q_i/m_i ratio; hence V_1 might be called the *voltage equivalent of velocity* v_{x1}.

Consider Figure B–1(b) where the two electrodes are at different potentials with $|V_2| > |V_1|$. The velocity v_{x1} is the same as before. Conservation of energy requires the final KE to be the sum of the initial KE and the PE *added* by the ion moving from x_1 to x_2, or

$$\frac{m_i v_{x2}^2}{2} = \frac{m_i v_{x1}^2}{2} + q_i(V_2 - V_1) \tag{B.1-4}$$

whence

$$\frac{m_i v_{x2}^2}{2} = q_i[V_1 + (V_2 - V_1)] = q_i V_2 \tag{B.1-5}$$

Thus, the parallel electrodes, normal to the x-direction, can be used to accelerate ions moving in the x-direction. The last equation shows that the final ion energy is determined by the voltage difference between the point where the ion energy was zero and the point where the ion leaves the accelerating field. For application to the ion implanter in Figure 8–1, the notation should be changed so that

$$v_x = \sqrt{\frac{2q_i V_a}{m_i}} \tag{B.1-6}$$

Note that the velocity depends on q_i/m_i, the ion charge-to-mass ratio, and is proportional to $\sqrt{V_a}$. This type of ion/electric field interaction is used to accelerate ions in the implanter and for extraction in the ion source.

B.2 Initial Velocity Normal to $\mathscr{E}$ Field

Consider the configuration shown in Figure B–2(a) where an electric field $\mathscr{E}_z$, normal to the ion initial velocity, is applied by a voltage V_z across two parallel deflecting electrodes separated by distance d. Assume that there is no field fringing near the plate edges. The x-directed velocity v_x remains constant since there is no x-directed component of $\mathscr{E}$ field present, and the time of flight between the plates, whose length in the x direction is ℓ, is

$$t = \frac{\ell}{v_x} \tag{B.2-1}$$

The z-directed velocity v_z, at $x = \ell$, is given by Eq. (B.1-6) with subscript changes from x to z and a to z. The ion's acceleration in the z-direction is given by

$$a_z = \frac{f_z}{m_i} = \frac{q_i \mathscr{E}_z}{m_i} = \frac{q_i V_z}{m_i d} \tag{B.2-2}$$

so that v_z at $x = \ell$ also is given by

$$v_z = a_z t = \frac{q_i \ell V_z}{m_i d v_x} \tag{B.2-3}$$

The angle θ at which the ions emerge from between the plates at $x = \ell$ is

$$\tan\theta = \frac{v_z}{v_x} = \frac{q_i \ell V_z}{m_i d v_x{}^2} = \frac{\ell V_z}{2 d V_x} \tag{B.2-4}$$

where V_x is the accelerating voltage that produced the initial velocity v_x. If fringing of the $\mathscr{E}_z$ field is neglected, v_z will remain constant once the ion emerges

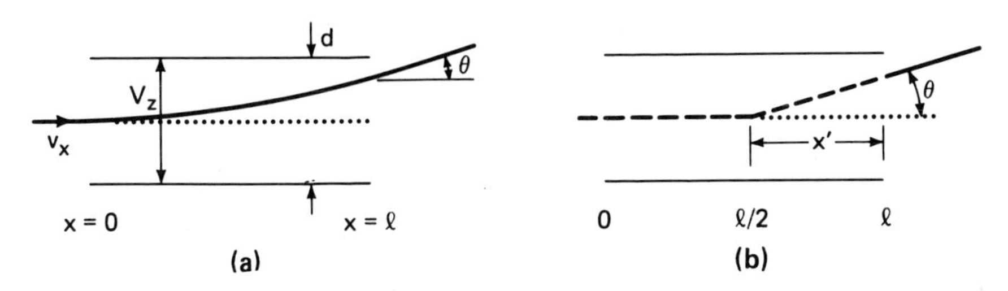

Figure B–2. Acceleration of an ion by an electric field that is normal to the ion's initial velocity at $x = 0$. (a) The ion path is parabolic. (b) The virtual ion path consists of two linear segments meeting at $x = \ell/2$, $z = 0$.

from between the plates, as will v_x; hence, the ion will continue in a straight line at angle θ wrt the x-axis. Note that the bending angle is independent of the ion charge-to-mass ratio, and that $\tan\theta$ is inversely proportional to the accelerating voltage V_x. This bending action by a normal $\mathscr{E}$ field is the basis for focus, scan, and neutral trap functions in the implanter. It cannot be used for separation of ions of different q_i/m_i ratios.

It is easy to show that the total ion deflection in the z-direction at $x = \ell$ is

$$z = \frac{a_z t^2}{2} = \frac{q_i V_z}{2 m_i d}\left(\frac{\ell}{v_x}\right)^2 = \frac{\ell^2 V_z}{4 d V_x} \qquad \text{(B.2-5)}$$

Thus a limit is placed on the maximum value of V_z (and θ), for if the ion enters along the midline of the deflection assembly, and if z exceeds $d/2$, the ions will hit the deflection plate. Consider some illustrative numbers.

For a typical deflection system in an ion implanter:

$$\theta = 10^\circ, \qquad \ell = 6\,\text{in.}, \qquad d = 2\,\text{in.}, \qquad V_x = 150\,\text{kV}$$

Then by Eq. (B.2-4) the required voltage across the deflection plates must be

$$V_z = \frac{2 d V_x}{\ell}\tan\theta = \frac{2(2)(1.5 \times 10^5)\tan 10^\circ}{6} = 17.6\,\text{kV}$$

With the plates separated by 2 in., the z-directed field will be

$$\mathscr{E}_z = \frac{V_z}{d} = \frac{17.6}{2(2.54)} = 3.46\,\text{kV/cm}$$

which is low enough to present no breakdown problem in the evacuated drift tube. For comparison—for the same conditions, an electron would be bent by the same angle even though its q/m ratio is vastly different from that of, say, $^{11}B^+$, a singly charged boron ion of mass 11 u.[1]

As a matter of interest we calculate by Eq. (B.1-6) that the x velocity for an $^{11}B^+$ ion, of mass 11 u and charge magnitude equal to that of an electron, with $V_x = 150$ kV is

$$v_x = \sqrt{\frac{2 q_i V_x}{m_i}} = \sqrt{\frac{2(1.6 \times 10^{-19})(1.5 \times 10^5)}{11(1.66 \times 10^{-27})}} = 1.62 \times 10^6\,\text{m/s}$$

---

[1] Use of the common *atomic mass unit* (amu) is deprecated in favor of the *unified atomic mass unit* (u), which is defined as one-twelfth of the ^{12}C carbon isotope mass. Numerical values of the two units are equal to better than three significant figures, however. Numerically 1 u = 1.66×10^{-27} kg.

On the other hand, an electron would have a velocity in the opposite direction of

$$v_x = \sqrt{\frac{2(1.6 \times 10^{-19})(1.5 \times 10^5)}{9.11 \times 10^{-31}}} = 2.30 \times 10^8 \text{ m/s}$$

From Eq. (B.2-5) we calculate for both ion and electron beams

$$z = \frac{1}{4}\frac{\ell^2}{d}\frac{V_z}{V_x} = \frac{1}{4}\frac{6^2}{2}\frac{17.6}{150} = 0.528 \text{ in.}$$

so neither beam will hit a deflection plate.

It can be shown quite easily that the path of the ion while it is between the deflection plates is parabolic (Problem 8–1). Despite this fact, a useful, *virtual* path that still gives the correct deflection angle may be considered. Thus, in Figure B–2(b) we extrapolate the path of emergence, at angle θ, linearly back to $z = 0$ as shown by the dashed line, and solve for x':

$$x' = \frac{z}{\tan\theta} = \left(\frac{\ell^2 V_z}{4dV_x}\right)\left(\frac{2dV_x}{\ell V_z}\right) = \frac{\ell}{2} \tag{B.2-6}$$

so the *break* in the dashed path is at the midpoint of the deflection array. Therefore, we may think of the ion's actual parabolic path as being replaced by two linear segments: one along the x-axis at $z = 0$, from $x = 0$ to $x = \ell/2$, and a second from $(x = \ell/2, z = 0)$ to $(x = \ell, z = z)$ at θ wrt the x-axis.

B.3 Normal Initial Velocity and $\mathscr{B}$ Field

Consider the situation shown in Figure B–3(a) where a uniform $\mathscr{B}$ field is established normal to the plane of the page (x–z plane) and directed toward the reader ($-y$-direction). We neglect fringing at the field edges. A positive ion enters the field in the z-direction with constant velocity v_o. The force on the ion due to the magnetic field has magnitude

$$f_{\mathscr{B}} = q_i v_o \mathscr{B} \tag{B.3-1}$$

and its direction always is normal to both the velocity and $\mathscr{B}$; hence the ion will move on a circular path of some radius r to be determined. With fringing neglected, the ion will exit the field on a linear path that is tangential to the circular path at the point of exit.

While in the magnetic field and on its circular path the ion is subjected to a centrifugal force

$$f_c = \frac{m_i v_o^2}{r} \tag{B.3-2}$$

The two forces must be equal for the ion to remain on its circular path.

The initial velocity v_o, given by Eq. (B.1-6), is imparted to the ion by accelerating it to a potential of V volts before it enters the $\mathscr{B}$ field. Solving the three appropriate equations yields

$$r = \frac{1}{\mathscr{B}} \sqrt{\frac{2m_i V}{q_i}} \tag{B.3-3}$$

with MKS units used—viz, r, m; V, volts; m_i, kg; q_i, C; and $\mathscr{B}$, Tesla (= Weber/m^2 = 10 kgauss). Conversion to more familiar units gives

$$r^* = \frac{4.56}{\mathscr{B}^*} \sqrt{\frac{V^* m_i^*}{v}} \tag{B.3-4}$$

where the units now are: r^*, cm; $\mathscr{B}^*$, kgauss; V^*, kV; $m_i^* = m_i/u$, and $v = 1$ or 2 for singly or doubly charged ions, respectively.

Consider some typical numbers. For singly ionized $^{10}B^+$ and $^{11}B^+$ ions, m_i^* is 10 and 11, respectively, and v is 1 for both of them. Calculation shows that, for $\mathscr{B}^* = 5$ kgauss and $V^* = 20$ kV, $r_{10} = 12.9$ cm and $r_{11} = 13.5$ cm. For the $^{11}B^{++}$ ion, v would rise from 1 to 2, and r_{11} would be reduced by $1/\sqrt{2}$ to the value of 9.57 cm.

For comparison with the $\mathscr{E}$ field case we calculate the expression for θ, the angle of beam bend. With fringing neglected, the beam stops bending on

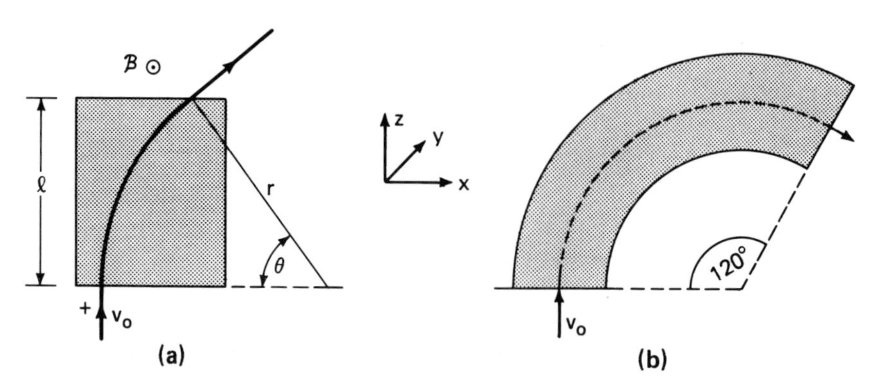

Figure B–3. Deflection of an ion by a magnetic field that is normal to the ion's initial velocity. (a) The ion path is circular. Bending ceases at the edge of the magnetic field. (b) Curved metal polepieces conserve ampere turns in establishing the magnetic field.

leaving the $\mathscr{B}$ field, since the bending force becomes zero. For small angles of bend the bending is limited by the length ℓ of the field in the z-direction (i.e., parallel to the entry direction of v_o). This is illustrated in Figure B–3(a) where we read that

$$\sin\theta = \frac{\ell}{r} = \frac{\ell^*}{r^*} \tag{B.3-5}$$

For the numbers already given for the $^{11}B^+$ ion we had that $r_{11} = 13.5$ cm; hence, for a 10° deflection the required field length would be $\ell = 13.5 \sin 10° = 2.34$ cm.

Since this beam-bending action by a $\mathscr{B}$ field depends on q_i/m_i, it is used for species separation in the analyzer section of the implanter, as discussed in Section 8.4.2. For this use θ is large, usually in the 90 to 135° range, and a little common sense is more useful than trigonometry. Say that θ is chosen to be 120°, then the critical length of the field is in the x-direction (i.e., normal to the entry direction of v_o). It is wasteful of ampere-turns on the electromagnet winding if $\mathscr{B}$ is established in regions where the ions will not be moving. For this reason, metal pole pieces, shaped as sectors of an annulus of mean radius r, are used to confine the magnetic field to a region near the ion's circular path [Figure B–3(b)]. It is apparent, then, that the sectors must have an angular length of the desired θ, 120° in this particular case.

In contrast to $\mathscr{E}$ field deflection, here the angle depends on q_i/m_i and is inversely proportional to the *square root* of the voltage used to produce v_o. These differences form the basis of choice between the two types of deflection systems in different regions of the ion implanter, as we may see in the body of Chapter 8.

Appendix C

The Glow Discharge

The plasma associated with a glow discharge in a gas is a source of free electrons e^-, positive ions i^+, and sometimes free radicals. These species are needed for PECVD (plasma-enhanced chemical vapor deposition, Chapter 9), dry etching (Chapter 10), and sputtering (Chapters 10 and 12). The glow discharge comprises a particular domain of operation in a gas when it is excited by magnetic and/or electric fields. We consider the latter, as it is more common. A good first model is a pair of plane parallel electrodes, surrounded by a gas and connected to a d-c power source as in Figure C–1(a) [1].

C.1 General Gas Discharge

Consider the I–V curve, shown at (b), that is obtained as the d-c *applied* voltage V_a is raised slowly. The current scale extends over many orders of magnitude; note that there are two breaks in the ordinate scale. Actual values are not given because they depend on the gas, electrode separation d, and other factors. A brief description of the processes that take place as V_a is increased follows.

A small fraction of the gas will be ionized even at $V_a = 0$, due to external agencies such as cosmic rays or photoemission from the device electrodes under excitation by ambient light. Free electrons e^- and positive ions i^+ are present in equal but small numbers; no current flows, however, because $V = 0$.

In the range O to A, electrons move to the anode, causing current flow. The i^+ are large, have much lower mobility, and make negligible contribution to the current. As V is raised further, more of the e^-s are pulled to the anode in unit time, so the current increases.

The current saturates from A to B: all electrons produced in unit time by the external ionizing agency move to the anode. Note, though, that as V is increased, the e^-s move faster and so gain more kinetic energy as they move

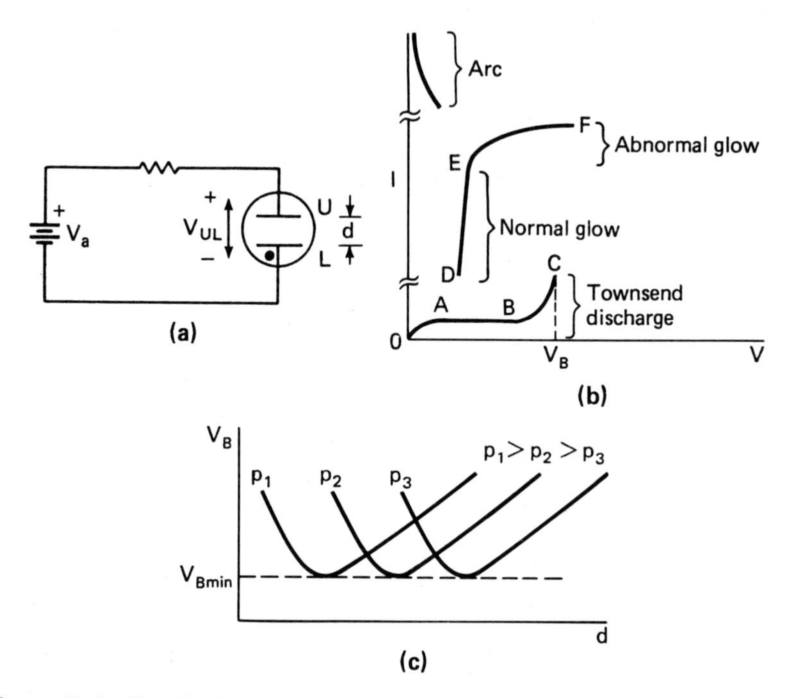

Figure C–1. Gas discharge. (a) Circuit for determining the $I-V$ characteristic. The large dot indicates that gas is present. The independent variable is V_a. (b) The $I-V$ curve. (c) Illustrating Paschen's law.

toward the anode. At B, some electrons have sufficient energy to produce ionization by collision with gas molecules, so the generation rate of e^- and i^+ increases.

Current continues to increase in the B to C range; as V is raised (by increasing V_a) electrons gain more energy. This increases the ionization-by-collision rate, more electrons are produced, and so on; *avalanche* begins to set in, and the current increases very rapidly. The entire discharge from O to C is called a Townsend discharge and is not self-sustaining: if the original source of ionization is removed completely, current will drop to zero. Currents in this range would be much smaller than those in the range above C.

The *breakdown voltage* V_B lies at C, where another mechanism comes into play. Electrons arriving at the anode have sufficient energy to produce *secondary* electrons by dislodging them from the anode. The secondary to primary ratio runs around 0.1 [2].

These secondaries produce more ionizing collisions with the gas molecules. The resulting i^+s strike the cathode, producing more secondaries, and so on. This is the avalanche mechanism. If the secondaries produce enough i^+s to replenish their numbers at the cathode, the discharge becomes self-sustaining: if the original source of ionization is removed completely, current continues to flow.

At breakdown, ionization is taking place so rapidly that a region of essentially equal numbers of e^-s and i^+s, a *plasma*, is formed. This is a charge-free region and has negligible voltage drop across it. In the so-called *normal* glow region most of the voltage drop is confined to the gas region near the cathode, a condition that will be discussed shortly. Note that there is a break in the current scale below the point D in Figure C–1(b).

In the *normal glow* range of D to E, voltage remains nearly constant as current increases, but the curve does have a positive, finite slope. The name *glow* derives from the light that is emitted from portions of the interelectrode region. The mechanism for this is discussed in Section C.2. It is of historical interest that cold-cathode gas diodes operating in the normal glow region once were used as voltage regulators, the zener diodes of the pre-semiconductor era.

At the onset of normal glow, i^+s arrive at a small area of the cathode, probably favored by some surface condition, and a glow is observed there. The glow gradually spreads over the cathode as V increases, with current density remaining constant, until the latter's surface is completely covered by glow. This signals the end of the normal glow domain, E in Figure C–1(b).

The range E to F is called the *abnormal* glow region, where current density at the cathode and intensity of glow increase with rising I. The name abnormal is unfortunate: there is nothing unusual about this region, in fact it is the region of interest for semiconductor processing. The voltage increases rapidly with current because of changes in the regions between the electrodes.

If V is increased beyond another breakdown point F that is larger than V_B, an *arc* occurs with extremely high currents. The number of positive ions hitting the cathode increases rapidly, and they produce heat in addition to dislodging secondary electrons. If the current density increases to about 100 mA/cm^2, the cathode temperature can rise enough to cause thermionic emission of electrons from the cathode as well [2]. Thus, another electron producing mechanism is added. These effects can destroy the device unless current is limited properly. The I–V curve in the arc range has a negative slope, and the arc must be avoided for device-fabricating plasma processes. Adequate water cooling of the cathode can keep the cathode temperature low enough, however, to prevent thermionic emission.

If the initial *applied* voltage V_a is somewhat greater than V_B, operation will jump into the glow region. The current limiting resistor in series with V_a in Figure C–1(a) must be chosen to limit the current to the abnormal glow range.

The value of breakdown voltage V_B depends on the gas type and pressure, and on the electrode geometry and material. For a given gas, and electrodes of a given material in the plane, parallel configuration, Paschen's law states that V_B is a function of the *product* of gas pressure p and electrode spacing

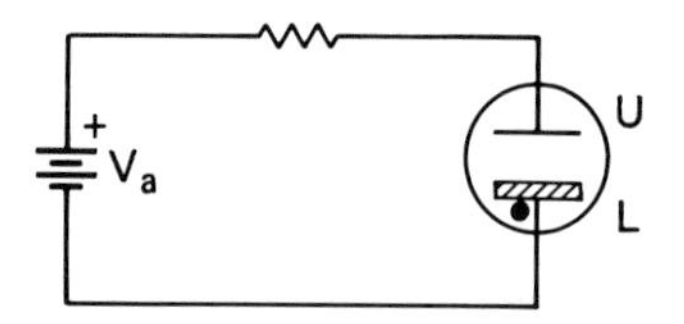

Figure C–2. The lower electrode is insulated.

d—i.e., $V_B = f(pd)$. Figure C–1(c) shows a family of typical curves that illustrate this relationship [1].

Gray cites $V_{B\min}$ as 350 V for air between plane, parallel electrodes at a *pd* product of 0.6 torr-cm. Note that, if an air-glow were desired at minimum V_B, even at 1 torr the plate separation would be only $d = 0.6/1 = 0.6$ cm, which does not leave much working space between the electrodes. The curves show, however, that for a given pressure d may be raised (or lowered for that matter) if a higher V_B is acceptable.

The discussion has assumed d-c excitation, so in the glow domain e^-s move to the positive, upper electrode U. No charge accumulates on either electrode, however; both electrodes are conductors, and current flows as the sum of the electron and positive ion components moving between and reaching the electrodes. These conditions prevail irrespective of which electrode is grounded.

If V_a is large enough, i^+ will have sufficient momentum to dislodge electrode L material. This is the basis of sputter etching (Chapter 10) and sputter deposition (Chapter 12).

Now consider that the lower electrode is covered with an insulator, say SiO_2 or Al_2O_3, as in Figure C–2. When the glow discharge is struck, electrons still move to U and positive ions to L. But positive charge will accumulate on the insulated surface of L; the intervening insulator prevents the charge from being neutralized by electrons moving in from the external circuit. Thus, in short order more i^+s are repelled by the positive charge on L—no steady state dc flows and the discharge ceases, but $V_{UL} \neq V_a$ by virtue of the charge on L. The implication is that insulating materials prevent a glow discharge with d-c excitation. Alternating current excitation, particularly in the r-f range, overcomes this difficulty (Section C.3).

C.2 The Glow

The preceding section looked at the I–V curve of the entire device. Now say that the I–V values are set for operation in the glow domain, and consider how voltage and other parameters vary with x in the interelectrode region. The cathode is at $x = 0$, and the anode at $x = d$.

Potential measurements within the interelectrode region yield the curve of Figure C–3 [1, 2]. The shaded regions indicate dark regions in the discharge.

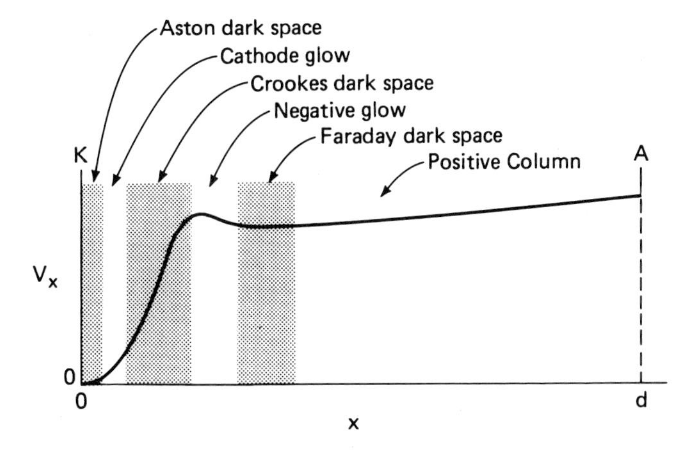

Figure C–3. Voltage variation between the electrodes of a glow discharge. Regions of glow are not shaded.

Positive ions off the cathode (K) serve to terminate most of the electric field lines originating at A, so most of the voltage drop appears across the near-cathode region. Consider the glow mechanism. Secondary electrons resulting from i^+ impact on the cathode start to accelerate. Initially, they have insufficient energy to *excite* gas molecules on collision. (By excitation we mean that an outer orbital electron in a molecule is raised from ground to a higher energy state.) No photons are emitted.

In the region of cathode glow, the rise in voltage accelerates the electrons even more. Some, on colliding with molecules, do produce excitation. Electrons are raised to higher orbits, then decay to ground state again, emitting photons of wavelength $\lambda = \Delta E/h$ as they do so. The quantity h is Planck's constant ($= 4.14 \times 10^{-15}$ eV-s), and ΔE is the energy difference between the two states, which is a function of the gas type. If λ is in the visible range, a glow is seen in this region of the interelectrode space and is called the cathode glow.

Most of the potential difference between the electrodes appears across the Crookes dark space, so the electrons gain most of their final energy there. The energy values, while too great to provide excitation, are insufficient to produce ionization when an electron and a gas molecule collide. Consequently molecular electrons do not move from ground to a higher state, and so cannot fall back to ground state again with the release of a photon—no light is emitted. Since the electrons move rapidly in this region, there is a fairly large number of positive ions in this dark space. These furnish the positive charge for the rise in potential.

In the negative column (or glow region), electrons have attained enough energy to ionize gas molecules on collision, and they do so copiously. The result is the production of the plasma where essentially equal numbers of e^-s and i^+s are present. There is little excess charge of either sign in the plasma, so the voltage drop across it is small. The very large number of free charges in the plasma is conducive to recombination, so photons are emitted with λs corresponding to the ionization potential of the gas.

The remaining regions of Figure C–3 are of little interest to us except that very little voltage drop appears across them; hence they are regions of essentially equal numbers of negative and positive charges. The discharge will be maintained if d is decreased, so long as the anode (A) stays clear of the Crookes dark space. For fabrication conditions, d usually is chosen so that the anode lies in the negative column, which, as we have seen, produces a visible glow due to a plasma there. It is this plasma that is so important.

Thus far, we have assumed an inert gas between the electrodes. If, as in PECVD (Section 9.15) , reactive gases such as SiH_4, O_2, and PH_3 are used, some of these will become ionized, highly reactive, and will participate in the deposition reactions. It is also true that not all the electrons will reach the ionizing energies of the reactants, so not all of the gas will be involved.

For simplicity in the foregoing discussion it was tacitly assumed that there are clear-cut boundaries between the glow and dark regions. This is not strictly true, because there is some light throughout the interelectrode region, due to either excitation or ionization. This comes about because electron energies are distributed statistically, and occasionally an electron at any location will have enough energy to cause a photoemission event subsequent to collision with a molecule. Practically, however, the emitted light from the "dark" areas is so relatively weak that the dark designation is quite reasonable.

C.3 A-C/R-F Glow Discharge

If the d-c source V_a [Figure C–1(a)] is replaced with an a-c source as in Figure C–4(a), other mechanisms enter in starting and maintaining a glow discharge. For low frequencies, say below 50 kHz, the half-cycle time is at least $1/(5 \times 10^4) = 20\ \mu s$. Positive ion mobilities are high enough to establish a glow discharge in that time interval. Thus, at such frequencies the glow discharge behaves as if dc were applied but with the cathode and anode interchanging roles each half-cycle.

At higher frequencies, say in the r-f range, the half-period is too short, and another mechanism enters in: electrons can oscillate in the region of glow due to the alternating field and acquire enough energy to make ionizing collisions. Consequently, secondary electrons produced by i^+s hitting the cathode are not needed to maintain the discharge.

An interesting and important property of the r-f glow discharge is that it can develop d-c voltages if there is a break in the d-c path. Consider the configuration shown in Figure C–4(a), where U and L are *conducting* elec-

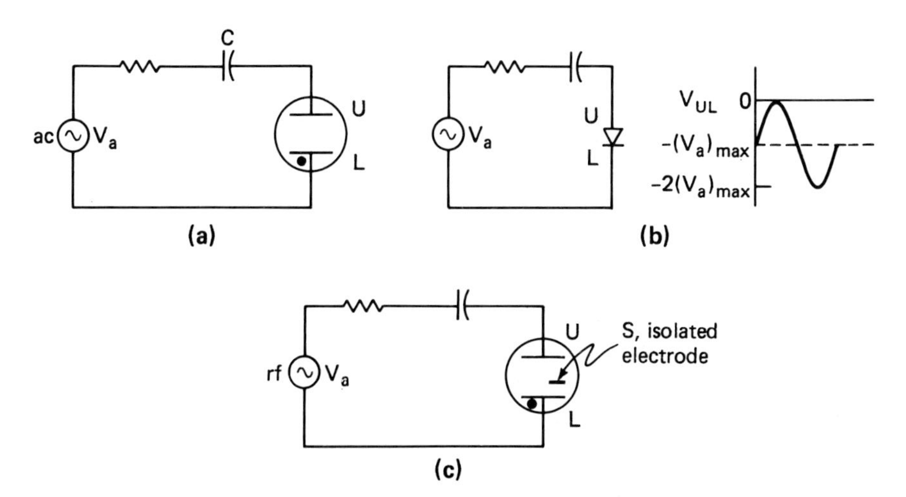

Figure C–4. Development of self-bias in an r-f plasma. (a) Basic configuration. (b) Analogous clamp circuit. (c) An isolated electrode is located between electrodes U and L.

trodes, and a series capacitor is present. When the discharge first strikes, high mobility e^-s move to U during the positive half-cycle of V_a, and i^+s move to L; negative charge starts to build up on U and positive charge on L, causing a small d-c voltage across the external capacitor C. During the negative half-cycle of V_a the highly mobile electrons rush in to neutralize the positive charge on L, but the positive ions cannot move fast enough within the half-cycle interval to neutralize the negative charge on U. In fact, when we speak of an *r-f plasma*, we imply that the frequency of V_a is high enough to satisfy this condition. Typically $f > 100$ kHz, although one of the ISM frequencies (defined later, Section C.4), say 13.56 MHz, is usually used. After a few more cycles of V_a, a steady state condition is reached when the d-c voltage across C equals the d-c voltage V_{UL}, which nearly equals the max value of V_a. Remember that no steady state direct current can flow because of C.

The action here is analogous to that of the clamp circuit shown in Figure C–4(b), where $V_{UL} \approx -(V_a)_{max}$, and the effect of the movement of the high-mobility electrons and the less mobile holes is simulated by the diode's action. The net effect is that the a-c applied voltage is superimposed on a d-c self-bias of magnitude $\sim (V_a)_{max}$ between the electrodes. This condition prevails no matter which electrode, U or L, is grounded.

If one of the electrodes is insulated, the external series capacitor C may be omitted (or shorted out) and the same development of self-bias takes place. The capacitor C is necessary only if neither electrode is insulated, or if it is not part of the impedance matching network between the V_a source and the discharge electrodes (Section C.4).

Next, consider the configuration of Figure C–4(c), where an *isolated* electrode S has been inserted in the region between conducting electrodes U and L. Since both e^- and i^+ species are present, initially both move toward the surface of S, but once again, electrons win the race due to their higher mobility. Negative charge starts to build up on S and is balanced by an excess of i^+ nearby in the plasma. The result is that S becomes self-biased negatively wrt the plasma; since the electrode is isolated, the charge cannot leak off. If U and L are of the same area, this voltage will be comparatively small. Note that the wafers in the barrel reactor of Figure 10–6(a) are just such isolated elements; hence we expect them to be slightly negative wrt the plasma near them.

A silicon wafer that has an insulator coating such as SiO_2 or Si_3N_4 and is resting on an electrode, say L in Figure C–4(c), will have a surface corresponding to S. This will be slightly negative relative to the plasma and so will attract i^+s to provide ion etching. If we think of S as the surface of a wafer to be etched, we see that it is desirable to increase the magnitude of this self-bias to provide greater ion acceleration for RIE (reactive ion etching) and sputter (or ion) etching.

It turns out that this voltage may be increased by using U and L electrodes of different area, as shown in Figure 10–6(c). In the final form, U becomes the reactor vessel itself and so has a very large area compared to L. The electrode-to-plasma voltages V_{UP} and V_{LP} are inversely related to some power n of the area ratio:

$$\frac{V_{LP}}{V_{UP}} = \left(\frac{A_U}{A_L}\right)^n \qquad \text{(C.3-1)}$$

Koenig and Maissel have shown that theoretically $n = 4$ for a simple model that assumes the two voltages to be across two series capacitors [3]. Practically, however, $1 < n < 2$ [4]. Thus if $A_U \gg A_L$, most of the available self-bias will appear between the insulated surface of a wafer on L and the plasma. It is for this reason that RIE uses the configurations of (c) rather than (b) in Figure 10–6.

C.4 R-F Problems

The use of r-f excitation of a glow discharge raises some special problems. First, to prevent radiation beyond the limits prescribed by the Federal Communications Commission requires careful shielding and/or severe power limitation. Where frequency may be held to tight tolerance as with crystal-controlled oscillators, operation is usually at one of the frequencies allocated for industrial/scientific/medical (ISM) use. Unlimited radiation is permitted at these special frequencies. Those most commonly used for plasma processes are 13.56 ± 0.00678 MHz, and its harmonics at 27.12 ± 0.0136 and $40.68 \pm$

0.020 MHz [5]. Self-excited units can operate at any frequency, but shielding against radiation becomes very critical.

Second, the a-c load impedance must be matched to the r-f excitation source. For reproducible results, the load power must remain constant.

The impedance presented by the glow discharge is not just that of a parallel plate capacitor comprised of the two electrodes. The gas itself and the plasma profoundly affect the impedance, which is a function of geometry, current, gas type and pressure, number of wafers being etched, and so on. Any variation in these causes a change in the load impedance, so the matching network should be adjustable, preferably by automatic means. Usually the generator is designed to have a 50-Ω resistive output impedance to match a coaxial cable. The impedance matching network transforms the complex load impedance to this value and can be of the conventional T, Π, or L types, although the circuit element values of the last form may not be as convenient [6].

To handle variations in the load, a feedback loop is provided that senses the reflected power component from the load as measured by a reflectometer and tunes the matching network elements to minimize that component.

In some comparatively rare instances the r-f generator is replaced by a source operating in the 40 to 400 kHz range. This does away with the need for impedance matching [7].

C.5 Modified Techniques

In certain applications the generation of glow discharges is modified by electron injection and by applying magnetic fields to provide magnetron action. These techniques are discussed in Sections 12.7.1 and 12.7.2.

References

[1] A general reference for this section is T. S. Gray, *Applied Electronics, 2nd Ed.* New York: Wiley, 1955.

[2] See, for example, J. L. Vossen and J. J. Cuomo, "Glow Discharge Sputter Deposition." In *Thin Film Processes*, eds. J. L. Vossen and W. Kern, pp. 24–29. New York: Academic Press, 1978.

[3] H. R. Koenig and L. I. Maissel, "Application of RF Discharges to Sputtering," *IBM J. Res. Dev.*, **14**, (2), 168–171, March 1970.

[4] H. H. Sawin, "A Review of Plasma Processing Fundamentals," *Solid State Technol.*, **28**, (4), 211–216, April 1985.

[5] *U.S. Federal Communications Commission Rules and Regulations, Part 18.*

[6] W. L. Everitt and G. E. Anner, *Communication Engineering, Third Ed.* New York: McGraw-Hill, 1956.

[7] "Plasma Etching Silicon and Dielectric Films—Planar vs. Barrel Etchers," *Circuits Manuf.*, **18**, (4), 27–36, April 1978.

Appendix D
Gas Systems

In this appendix we consider some basic ideas of gas systems that are common to the processes of Chapters 5, 6, 7, 9, 10, and 12. These systems distribute gases under controlled pressure and flow rates. Our interests will center on systems working near atmospheric pressure; space limitations prevent a discussion of systems in the vacuum range, where pressures are at least a few decades below atmospheric value.

D.1 Basic Concepts

Consider a volume V in a tube through which gas is flowing at a constant rate and temperature, as in Figure D–1(a). The pressure difference across the volume is $(p_2 - p_1)$, and the average pressure is $\langle p \rangle$. The general gas law teaches that

$$\langle p \rangle V = nkT \qquad \text{(D.1-1)}$$

where

n = number of gas molecules in volume V

k = Boltzmann's constant = 1.38×10^{-23} J/K

$= 8.62 \times 10^{-5}$ eV/K

T = absolute temperature in Kelvins (K)

Unfortunately, pressure units are not standardized. For pressures below atmospheric value the deprecated unit torr is still in common use. It is equal to 1 mm of Hg within two parts in 10^7. The millibar (mbar) and Pascal (Pa = N/m^2) also are common. Pounds/in.^{2} (PSI) and atmosphere (atm.) are

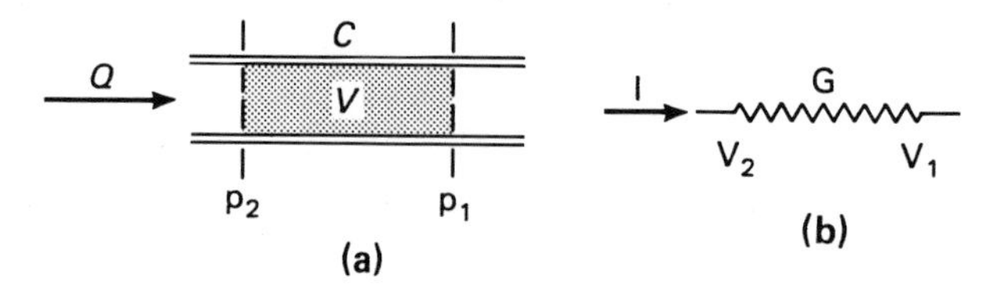

Figure D–1. Conductances. (a) Gas flow in a cylindrical pipe. (b) Electric current flow in a resistor of conductance G.

common near and above atmosphere. Choice is set by the pressure gage in use. Conversion factors are:

$$\left.\begin{aligned} &1\ \text{torr} = 1.33\ \text{mbar} = 133\ \text{Pa} \\ &1\ \text{mtorr} = 1\ \mu\,(\text{micron}) \\ &1\ \text{atm.} = 1.01 \times 10^5\ \text{Pa} = 14.7\ \text{PSI} \end{aligned}\right\} \tag{D.1-2}$$

If the gas flow rate is such that volume V of gas is moved in time t, then

$$\frac{\langle p \rangle V}{t} = \frac{nkT}{t} \tag{D.1-3}$$

and we can define

$$F_v = \text{volumetric flow rate} = \frac{V}{t} \tag{D.1-4}$$

and

$$F_n = \text{molecular flow rate} = \frac{n}{t} \tag{D.1-5}$$

or

$$\langle p \rangle F_v = F_n kT \tag{D.1-6}$$

These flow rates may be measured by *flowmeters* [1–5]. Typical units will be l/min and molecules/min, respectively.

An alternative form of flow rate, the *throughput* Q, also is used, particularly for vacuum systems. It is given by the LHS of Eq. (D.1-3)

$$Q = \langle p \rangle \frac{V}{t} = \langle p \rangle F_v = F_n kT \tag{D.1-7}$$

Thus the throughput may be defined as the molecular flow rate at standard temperature of 25°C (300 K). Typical units are torr-l/s.

The small tube in Figure D–1(a) impedes the free flow of gas, and its effect on the throughput is characterized by a parameter

$$C = \text{conductance} = \frac{Q}{p_1 - p_2} \tag{D.1-8}$$

Note that this equation is analogous to that for the electrical conductance of Figure D.1(b), viz

$$\mathrm{G} = \frac{\mathrm{I}}{\mathrm{V}_1 - \mathrm{V}_2} \tag{D.1-9}$$

The electrical quantities in Eq. (D.1-9) are given here in roman type to avoid confusion with their gas analogs. Note the analogs: C and G, Q and I, and p and V.

D.2 Conductance Calculations

Gas at near atmospheric pressure typically behaves like a fluid in flowing through a tube and is said to be in the *viscous range*. For practical purposes this is defined by [6]

$$L_m/d < 0.01, \qquad \text{Viscous Range} \tag{D.2-1}$$

where

$$\begin{aligned} L_m &= \text{mean free path of the gas} \\ &= \text{average distance traveled by a molecule between collisions with other molecules} \\ &= \frac{5 \times 10^{-3}}{p^*}, \quad \text{cm} \end{aligned} \tag{D.2-2}$$

where p^* is the pressure in torr.

In this viscous domain, molecules in a pipe are more likely to collide with each other than with the pipe walls; hence, they will drift along the pipe under the influence of a pressure difference between the ends. Consider at what pressure a 2-in.-diameter pipe drops out of the viscous range. From the last two equations we may reduce the viscous domain criterion to

$$p^* > \frac{5 \times 10^{-3}}{10^{-2}d} = \frac{5 \times 10^{-3}}{10^{-2}(2)(2.54)} = 0.0984\,\text{torr}$$

The conductance of a cylindrical pipe or tube when it is in the viscous domain can be derived from Poiseuille's equation for viscous flow, and in the commonly used bastard units is [6]

$$C = \frac{180\, d^4 \langle p^* \rangle}{L} \quad \text{l/s} \quad \text{(Viscous Range)} \tag{D.2-3}$$

where

C = conductance in l/s of a cylindrical pipe L cm long and of diameter d in cm, for air at room temperature

and $\langle p^* \rangle$ = average pressure in the pipe in torr.

By virtue of the analog between Eqs. (D.1-8) and (D.1-9) it is apparent that the total conductance of two tubes of different diameters in series is the product over the sum of the individual conductances, with the smaller controlling.

A circular hole or orifice between two parts of a gas system, or a change in effective tube diameter, has a nonzero value of conductance, and hence can support a pressure difference. Use was made of this concept in Section 8.4.3. It also is the basis for valve operation: the lowered conductance can maintain a pressure difference while reducing the flow rate.

In the *transition range*, where

$$0.01 \leq \frac{L_m}{d} \leq 1 \qquad \text{(Transition Range)} \tag{D.2-4}$$

and for a pressure difference of at least one decade between the two sides of the orifice, the conductance is given by [6]

$$C = 15A, \quad \text{l/s} \tag{D.2-5}$$

where A = area of the circular hole in cm^2. For example, a 1-in.-diameter orifice has a conductance of 76 l/s.

At typical *high vacuum* pressures of several decades below atmospheric value, gas behavior changes markedly and is in the *molecular range*, which is defined by [6]

$$\frac{L_m}{d} > 1.0, \qquad \text{(Molecular Range)} \tag{D.2-6}$$

where d in the case of a cylindrical tube is its diameter and L_m is the mean free path of the gas molecules, both expressed in the same units. Because L_m

exceeds the vessel size in this range, a molecule has a greater probability of colliding with the vessel walls than with another molecule. Under this condition, molecules will not *drift* along a pipe under the influence of a pressure difference between the pipe ends, because this mechanism depends on intermolecular collision. Molecular motion here is determined solely by thermal agitation, so the magnitude and direction of the molecular velocities are probabilistic. A pressure difference cannot force molecules along a pipe; they must move through the tube by their own thermal motion.

In this range the cylindrical pipe conductance becomes [6]:

$$C = \frac{12.2d^3}{L} \qquad \text{(Cylindrical Pipe)} \qquad \text{(D.2-7)}$$

For the circular aperture

$$C = 11.7A \qquad \text{(Circular Aperture)} \qquad \text{(D.2-8)}$$

where, as before, C is in l/s, L and d are in cm, and the aperture area A is in cm^2.

D.3 Gas Supply Systems

Consider some supply systems for processing gases and how gases from different sources are mixed. Emphasis will be on those systems where sources are gases or liquids at *room temperature*. Among the former are oxygen, nitrogen, argon, silane (SiH_4), arsine (AsH_3), and phosphine (PH_3). Typical liquid sources at room temperature are phosphorus oxychloride ($POCl_3$ or *pockle*), silicon tetrachloride ($SiCl_4$), boron tribromide (BBr_3), and trichloroethane (TCA). These liquids have nonzero vapor pressures near room temperature and hence will provide a vapor or gas phase when in a confining vessel. This vapor then may be used for processing.

D.3.1 Supply Tanks

Gases of desired composition and purity in high-pressure tanks or cylinders are available from vendors. The tanks come equipped with a *tank valve* that prevents gas escape. The user connects a pressure-regulator/pressure-gage/valve combination to this for metering and flow control (Figure D–2). Tanks and regulators are available with different fittings and threads, some being left-handed, to minimize the chance of interconnecting dangerous and explosive gas combinations. For example, different sizes and threads prevent connection of an oxygen regulator, with some carryover oxygen, to a tank of hydrogen, a gas combination that might explode. Therefore, the user must match threaded regulators and gages to the tanks properly.

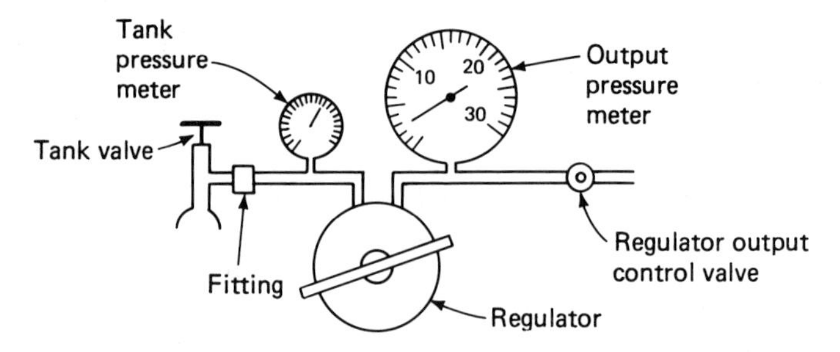

Figure D–2. Diagram of the valves, pressure regulator, and gages associated with a gas tank.

Table D–1. Fitting Numbers for Certain Gases

CGA #	Gas
320	CO_2
350	AsH_3, B_2H_6, H_2, PH_3, SiH_4 (high pressure)
510	SiH_4 (low pressure)
540	O_2
580	Ar, He, N_2
590	Industrial air
660	Freon 12

Courtesy of Matheson Gas Products, Inc., Secaucus, NJ [7].

Fortunately, standards exist for the threaded fittings, which are identified by a CGA (Compressed Gas Association) number. Numbers for a few typical gases are given in Table D–1. More complete lists are available in vendors' literature [7, 8]. Note that gases with the same fitting number are compatible and no danger arises if they are mixed.

Processing tanks come in several sizes, and the pressure gage often has a scale calibrated in cubic feet of gas [7, 8]. This calibration gives the gas volume at standard conditions of 70°F and 1 atmosphere, and will far exceed the physical volume of the tank. The relationship between the two volumes may be calculated with help from the gas law.

Say a full tank of oxygen has an indicated gage pressure of 2000 PSIG, corresponding to a gas volume of 220 ft^3 under standard conditions of 70°F and 14.7 PSI. What is the volume of the tank?

From the gas law we have that

$$P_T V_T = P_S V_S$$

where T refers to the tank and S to standard conditions. Note that P_T is the gage pressure plus atmospheric pressure. Then

$$V_T = \frac{P_S V_S}{P_T} = \frac{14.7(220)}{(2000 + 14.7)} = 1.6\,\text{ft}^3$$

This tank closely matches a Matheson 1-A size of 1.55 ft^3 rated internal volume, but note that tank sizes vary among vendors.

Improper use can damage a tank regulator, so manufacturers' instructions should be followed exactly. Special care also is required in handling high-pressure gas cylinders, since they can act like bombs. They must not fall or collide against each other; they never should be left free-standing without support. Safety requires that they be strapped or chained to a solid support such as a table, wall, or gas cabinet. Also, great care must be exercised in moving cylinders to avoid bumping them. They should be moved on a cart to which they are firmly fastened.

D.3.2 Single Tank

Consider a gas cylinder containing a gas G_1 flowing at pressure p_1 and temperature T_1 as shown in Figure D–3(a). (Symbols are defined in Figure D–4.) The volumetric flow rate can be set manually and read on a Rotameter

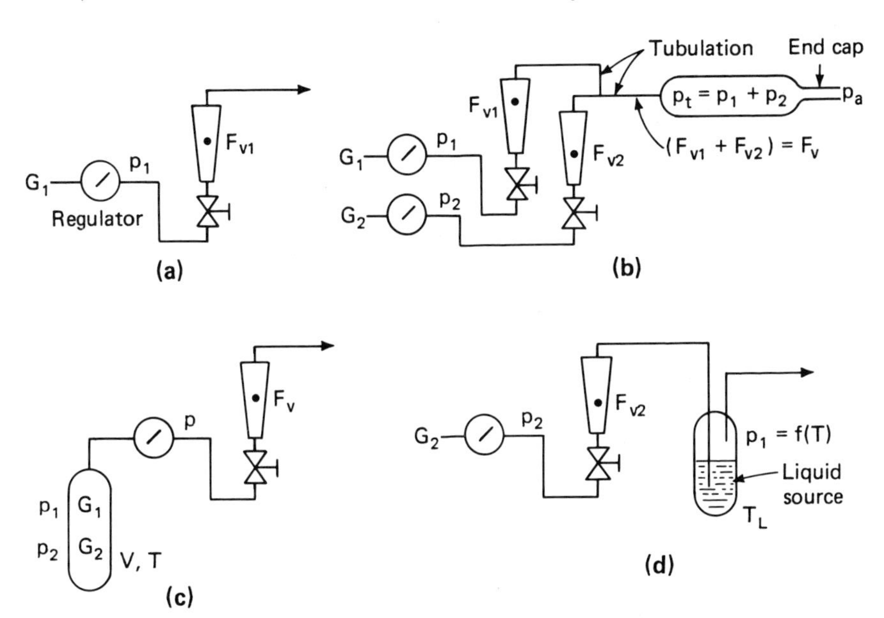

Figure D–3. Representative gas supply systems. (a) Single gas supplied from a tank. (b) Two gases supplied from separate tanks and mixed. (c) Two mixed gases supplied from a single tank. (d) A tank-supplied carrier gas mixed in a bubbler with the vapor from a liquid source.

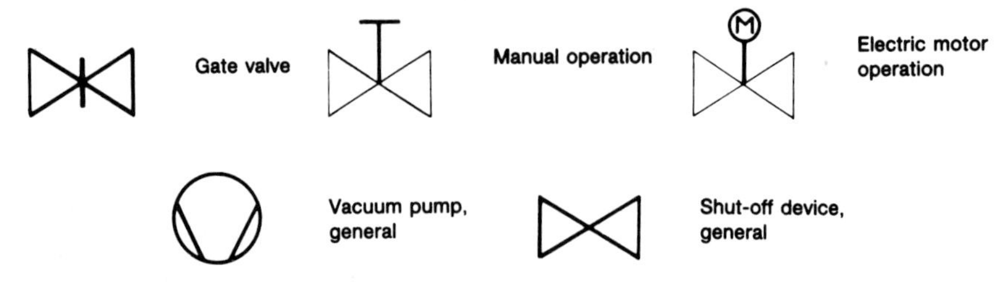

Figure D–4. Some standard symbols used for gas systems [18]. (Courtesy of Leybold Vacuum Products, Inc.)

(flowmeter) as F_{v1}. Then by Eq. (D.1-6) the molecular flow rate will be

$$F_{n1} = \frac{p_1 F_{v1}}{kT_1} \tag{D.3-1}$$

Note that Boltzmann's constant k has these values in commonly used bastard units:

$$k = 6.24 \times 10^4 \frac{\text{torr} \cdot \text{cm}^3}{\text{mole} \cdot \text{K}} = 1.03 \times 10^{-19} \frac{\text{torr} \cdot \text{cm}^3}{\text{molecule} \cdot \text{K}}$$

$$= 1.99 \times 10^{-21} \frac{\text{PSIG cm}^3}{\text{molecule K}}$$

(PSIG is pounds/in.^{2} gage pressure.) Be sure to distinguish between *mole* and *molecule*; Avogadro's number, $n_a = 6.02 \times 10^{23}$, is the number of molecules in a mole.

D.3.3 Two Tanks

Figure D–3(b) shows a situation where two gases G_1 and G_2 from separate tanks are to be mixed. The individual volumetric flow rates can be set and read so that a total volumetric flow rate of the mixed gases can be set as desired. By Eq. (D.3-1) the molecular flow rates will have the ratio

$$\frac{F_{n1}}{F_{n2}} = \frac{p_1 F_{v1} T_2}{p_2 F_{v2} T_1} \tag{D.3-2}$$

If pressures and temperatures are the same for both gases, the molecular and volumetric flow rate ratios are identical.

Consider the rest of the system that uses the mixed gases to feed a diffusion (or oxidation) tube as shown in Figure D–3(b). First note that gas molecules are neither gained nor lost in leak-free tubulation, so F_{v1} and F_{v2} remain unchanged, and the total volumetric flow rate to the diffusion tube is their sum $(F_{v1} + F_{v2})$.

Second, consider the pressures. Dalton's law requires that p_t in the tube be the sum of the partial pressures of the two gases less any pressure drops in the tubulation. If the sum p_t is greater than p_a, the pressure difference must be absorbed by the tube of small diameter in the end cap. The condition that $p_t > p_a$ is desirable because it prevents backflow from atmosphere into the tube.

D.3.4 Single Tank, Mixed Gases

A number of semiconductor processing gases are toxic and/or pyrophoric (can ignite spontaneously on contact with air). Silane, for example, is pyrophoric if it mixes with air in a volume ratio of between 4 and 96% of silane. Hence, for safety reasons it is common practice to purchase tanks of, say, 2% silane in an inert gas such as nitrogen or argon. Then, even if this mixture is inadvertently released into air, the percentage of silane will remain below the 4% danger margin.

Such a mixed-gas system is shown in Figure D–3(c). Dalton teaches that each component gas behaves as if it alone occupies the total volume V. Using p_1 and p_2 as the partial pressures of the two gases, we have

$$p_1 V_1 = n_1 kT \quad \text{and} \quad p_2 V_2 = n_2 kT$$

Let

$$r = \textit{molecular} \text{ ratio of gas } G_1 \text{ to gas } G_2$$

$$= \frac{n_1}{n_2} = \frac{p_1}{p_2} \tag{D.3-3}$$

Adding, we get

$$(p_1 + p_2) V = (n_1 + n_2) kT$$

But the sum of the partial pressures is the total pressure p, which is indicated by gages, so

$$pV = (n_1 + n_2) kT = \left(1 + \frac{1}{r}\right) n_1 kT$$

If volume V is flowing in time t

$$pF_v = \left(1 + \frac{1}{r}\right) F_{n1} kT$$

or the molecular flow rate of the lesser component is

$$F_{n1} = \frac{pF_v}{(1 + 1/r)kT} \tag{D.3-4}$$

Note again that F_{n1} is set by adjusting the total F_v, which can be determined from a flowmeter.

D.3.5 Vapor from a Liquid Source; Bubbler

Figure D–3(d) shows a situation in which the tank-supplied gas serves as a carrier component G_2 for a vapor G_1 coming from a liquid source in a bubbler (Figure 5–15). Gas temperature in the vessel is T; the liquid source temperature is T_L. Then within the volume V of the vessel we can write for the carrier gas G_2

$$p_2 V = n_2 kT$$

and for the liquid source's vapor G_1

$$p_1 V = n_1 kT$$

Note that T is the same for both *gases* (and so is V by virtue of Dalton's law), but p_1 is a function of the liquid-phase source temperature, T_L, and is obtained from vapor pressure tables or curves for the source in question—e.g., Appendix F.6.8. The tank regulator determines p_2.

Dividing the equations, and dividing again by time t, we have for the *molecular* ratio of the source gas to the carrier

$$r = \frac{n_1}{n_2} = \frac{F_{n1}}{F_{n2}} = \frac{p_1}{p_2} \tag{D.3-5}$$

These equations can be combined with those of the earlier sections to solve flow problems in diffusion, chemical vapor deposition, and other semiconductor processes.

Further information on gas supplies are given in appropriate sections throughout the text and references [9].

D.4 Gas Distribution Systems

Some of the gases used in silicon processing are hazardous and impose special conditions on the distribution system that transports them from supply to process chamber. A system typical of atmospheric-pressure epitaxy is illus-

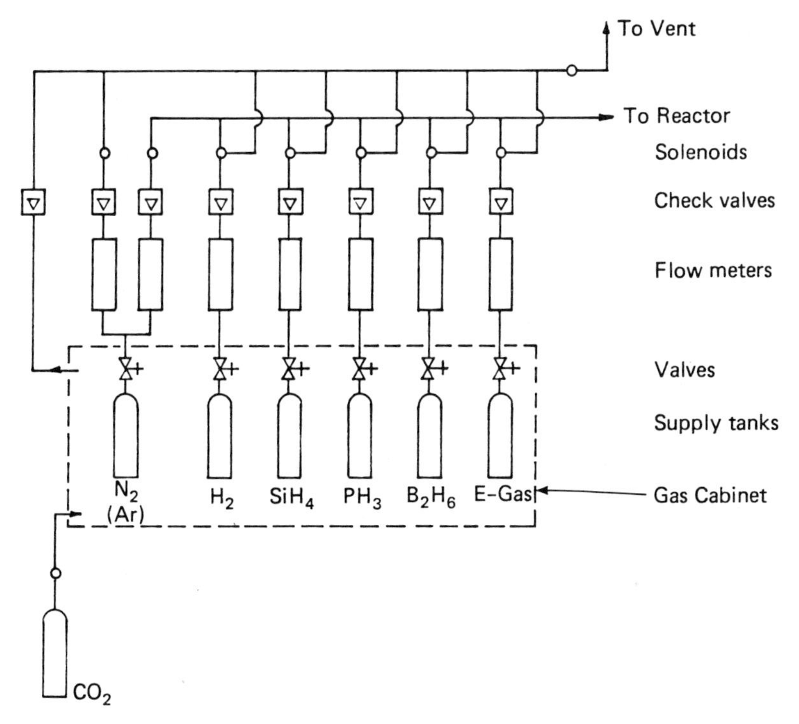

Figure D–5. A gas distribution system for a silane-based epitaxy deposition system.

trated in Figure D–5. The additional components needed for a liquid source such as $SiCl_4$ are shown in Figure D–3(d).

Consider a few of the features in the figure. Each gas is supplied from a tank with a shutoff valve. Both p- and n-type dopants are shown for versatility. Silane and the dopant supplies are diluted with an appropriate gas, either inert or compatible with the system, to keep concentrations down to a safe level. The tanks are enclosed in a gas cabinet that is connected to the vent line to prevent buildup of gas pressure due to leaks in the tanks and valves. As an added safety measure, facilities may be provided to purge the gas cabinet with CO_2 or another fire extinguishing gas (e.g., Halon) in the event of fire. Appropriate temperature and/or smoke sensors would be provided to turn on the extinguisher supply if needed.

Pressure regulators and flow meters are provided for checking and setting the flow rates of the several gases individually. The check valves are one-way devices to prevent backflow of gases from the vent or reactor lines toward the flow meters and supply tanks. This eliminates cross-contamination among the supply lines.

There are two types of solenoid-controlled valves shown: simple on/off and two-position-on. The latter are used to direct gas, from all but the N_2 (or Ar) supplies, to the reactor or to the vent lines. These ensure there is no pressure buildup when a gas supply is disconnected from the reactor line. The solenoid valves may be operated remotely either manually or by computer.

It is easy to envision how the whole system may be adapted to computer control by means of the solenoid valves. Control can be based on the response to sensors or time, or a combination of the two [10–12]. The furnace or reactor heating system also may be under computer control. If gas flow rates are to be controlled also, adjustable solenoid valves, under control of the flowmeters, may be added. The obvious advantage of computer control is that it provides excellent duplication of results from batch to batch.

D.5 Exhaust Pump Considerations

In systems operating at less than atmospheric pressure—e.g., low-pressure epitaxy (Chapter 9) or plasma etching (Chapter 10)—the effluent gases are removed by a vacuum pump. Some of the effluents may be corrosive, so special measures and procedures are required to prevent pump damage [13–15]. If oxygen is removed by an oil-type pump, special pump oil and operating precautions are necessary to prevent explosions [16].

In these low-pressure systems, operating pressure is higher than that afforded by the vacuum pump alone; hence, *throttling* is necessary and is effected by introducing a suitable (and possibly variable) conductance between the operating chamber and the pump [17].

References

[1] *Technical Bulletin T-022, Revision H.* Hatfield, PA: Brooks Instrument Division, Emerson Electric Co., 1979.

[2] *Complete Flow and Level Measurement Handbook and Encyclopedia™.* Stamford, CT: Omega Engineering, Inc., 1987.

[3] *Bulletin MFC 11/82.* Burlington, MA: MKS Instruments, Inc., 1982.

[4] C. Murray, "Mass Flow Controllers: Assuring Precise Process Gas Flows," *Semicond. Int.*, **8**, (10), 72–78, Oct. 1985.

[5] J. J. Sullivan, J. H. Ewing, and R. P. Jacobs, "Calibration Techniques for Thermal-Mass Flowmeters," *Solid State Technol.*, **28**, (4), 345–349, April 1985.

[6] R. W. Berry, P. M. Hall, and M. T. Harris, *Thin Film Technology.* Princeton: Van Nostrand, 1968.

[7] *Gases for Equipment for Semiconductor Manufacturing.* Secaucus, NJ: Matheson Gas Products, Inc., 1984.

[8] *Specialty Gas Regulation Equipment.* Denton, TX: Victor Equipment Company.

[9] "Handling and Safety Information," *Specialty Gases and Equipment.* Chap. 4. Allentown, PA: Air Products and Chemicals, Inc., 1982.

[10] J. Jackson, "Computer Control of Epitaxial Production Systems," *Solid State Technol.*, **15**, (11), 35–39, Nov. 1972.

[11] R. A. Rockhill and E. R. Werych, "Computer Control of Diffusion Systems," *Solid State Technol.*, **16**, (2), 29–32, Feb. 1973. Note that the supply problems for diffusion and epi systems are quite similar.

[12] "Epitaxial Reactors for Silicon Wafer Processing," *Circuits Manuf.*, **10**, (11), 66–67, Oct. 1970. This shows control systems for general CVD with additional supplies for other process gases.

[13] L. F. Dahlstedt, "Vacuum Equipment Considerations for Gases in Plasma and LPCVD Applications," *Semicond. Int.*, **2**, (5), 62–67, June 1979.

[14] M. Baron and J. Zelez, "Vacuum Systems for Plasma Etching, Plasma Deposition, and Low Pressure CVD," *Solid State Technol.*, **21**, (12), 61–65, Dec. 1978.

[15] G. R. Koch, "Vacuum System Considerations for Plasma Etching Equipment," *Solid State Technol.*, **23**, (9), 99–101, Sept. 1980.

[16] D. McKinniss, "Fluorosilicone Oil for Vacuum Pumps in Oxygen Service," *Semicond. Int.*, **9**, (8), 132–133, Aug. 1986.

[17] W. R. CLark and J. J. Sullivan, "Comparison of Pump Speed Control Techniques for Pressure Control in Plasma/LPCVD Systems," *Solid State Technol.*, **25**, (3), 105–107, Mar. 1982.

[18] *Vacuum Technology—Its Foundations Formulae and Tables.* Export, PA: Leybold Vacuum Products, Inc., 1983.

Appendix E
Lens Relationships

Stepper lenses are complex structures: one commercial unit involves 11 separate lens elements, two of which are cemented together, giving 20 air/lens interfaces [1]. They must meet some severe and sometimes conflicting specifications such as: operate in the UV range where usual optical glasses do not behave properly, have a flat field large enough to project a die field, operate with the object greater than the image with both being located near the lens, have high resolution and large depth of focus, and pass sufficient UV energy to keep exposure times reasonable. We shall review some basic relationships of *thin* lenses to gain some insight into these more complicated structures, and to introduce some terms used with steppers.

The *diameter* d of a lens limits the amount of light entering it, and hence the required exposure. The f *number*, designated here by f/# or F, is

$$F = f/d \tag{E-1}$$

where f is the *focal length* [defined after Eq. (E-2)]. If its value is given as f/4.5, the focal length is 4.5 times the diameter. For fixed f, as d increases, F decreases, and exposure time decreases.

With a stepper lens the object (the reticle field pattern) and its image (focused by the lens onto the wafer PR) are located at *conjugate foci*, u and v, as shown in Figure E–1(a). The object of linear size U is located u from the lens center, and the image of linear size V at v from the center on the other side of the lens. Given the object, its location, and the focal length, the image size and location may be determined by ray tracing as shown.

The principal focus and conjugate foci are related by

$$\frac{1}{u} + \frac{1}{v} = \frac{1}{f} \tag{E-2}$$

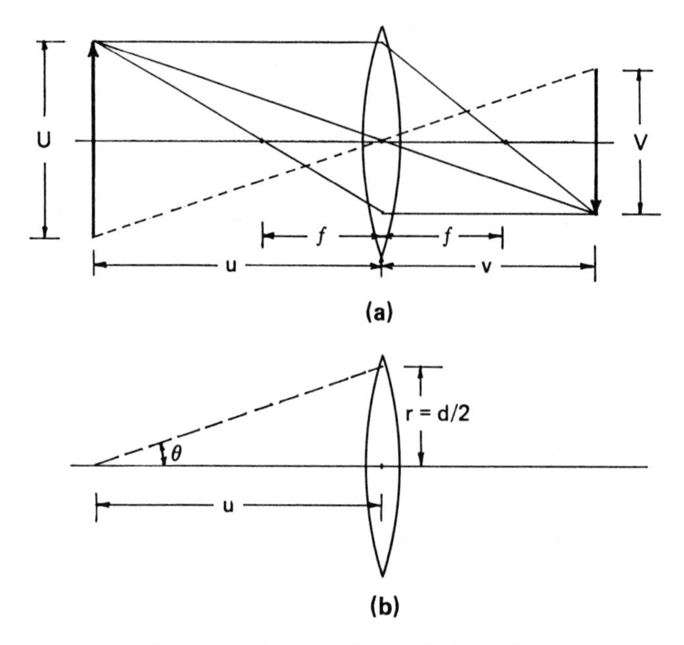

Figure E–1. Basic lens relationships.

For a real image to be focused at v as required in the stepper, u, v, and f are all positive numbers with $u \geq f$ and $v \geq f$. If f is fixed, v increases as u decreases. Note that if $u = \infty, f = v$. This defines the focal length.

The *linear magnification* M is defined to have a value greater than or equal to one, so for this application

$$M = \frac{U}{V} = \frac{u}{v} \tag{E-3}$$

where the second equality is apparent from similar triangles in Figure E–1(a). Note that

$$u = (M + 1)f \tag{E-4}$$

The sum of u and v is the *track length* ℓ, the total optical path length between object and image, exculsive of the lens length:

$$\ell = u + v \tag{E-5}$$

Track length should be as small as feasible to limit stepper size, but typically it is about 60 to 100 cm. The physical length may be shortened by folding the optical path with mirrors or prisms.

For an object at a finite u as in Figure E–1(a), an *effective* f/#, F_E, can be defined as

$$F_E = \frac{u}{d} = \frac{u}{d} \cdot \frac{f}{f} = \frac{u}{f} F = (M + 1)F \tag{E-6}$$

The effective value $F_E > F$ so the effective diameter $< d$. For a given lens with constant f and d, if the object moves closer to f, $(M + 1)$ decreases as does F_E, so the required exposure is shorter.

The *field size* of a lens is the largest square image, of side s, that it can handle. The diagonal of the square is the diameter d of the circular lens, or

$$d = \sqrt{2}s \tag{E-7}$$

Usually the lens field, or largest chip size, is expressed as a product: 10×10 mm for $s = 10$ mm.

The *numerical aperture* NA of a lens is defined such that $(\text{NA})^2$ is a measure of the len's *light-gathering power*:

$$\text{NA} = \mu \sin\theta \tag{E-8}$$

where μ is the refractive index of the material surrounding the lens [2]. Typically this is air, so $\mu = 1$. As shown in Figure E–1(b) θ is the *angular radius* of the lens as seen from the object location at u on the optical axis. For stepper lenses θ is well under 5°, so we let $\sin\theta = \tan\theta$ and

$$\text{NA} = \tan\theta = \frac{r}{u} = \frac{d}{2u} = \frac{1}{2F_E} = \frac{1}{2(M + 1)F} \tag{E-9}$$

$$= \frac{\sqrt{2}s}{2(M + 1)f} = \frac{s}{\sqrt{2}(M + 1)f} \tag{E-10}$$

For a given stepper lens s and f are fixed. For $\mu = 1$, Eq. (E-8) shows that NA is restricted to values ≤ 1; typical values for stepper lenses are less than 0.45 [3]. Omata gives a detailed discussion of the lenses [4].

The numerical aperture enters into two other important quantities under conditions of *diffraction limitation*, where image formation is limited by the physics of diffraction rather than by lens imperfections [5]. The *depth of focus* δ is the range in v over which the image is in focus, u remaining fixed:

$$\delta = \frac{\lambda}{2(\text{NA})^2} \tag{E-11}$$

and the diffraction limited resolution is approximately

$$\rho \approx \frac{0.61\lambda}{(\mathrm{NA})} \tag{E-12}$$

If the mask is fixed at u and its image is to remain in focus over the topographical variations of the wafer surface located at v, the depth of focus must at least equal the maximum step height on the wafer. Thus, for a maximum step height of, say, 2 μm and $\lambda = 0.4047$ μm (the Hg h-line), the requirement will be $\mathrm{NA} \geq \sqrt{\lambda/2\delta} = \sqrt{0.4047/2(2)} \approx 0.32$ and the corresponding resolution $\rho \approx 0.61\lambda/(\mathrm{NA}) \approx 0.61(0.4047)/0.32 \approx 0.77$ μm.

We desire that δ be as large, and ρ as small, as feasible, or that δ/ρ be maximized. We see from Eqs. (E-11) and (E-12) that

$$\frac{\delta}{\rho} \approx \frac{(\mathrm{NA})}{2(0.61)\lambda} = \frac{(\mathrm{NA})}{1.22\lambda} \tag{E-13}$$

which teaches that NA should be large and λ small. Large NA means values in the range of 0.2 to 0.35. Consider some typical numbers. Given these data for a stepper lens: NA = 0.2, $s = 20$ mm, $d = \sqrt{220} \approx 28$ mm, $M = 5$, $f = 100$ mm. Then $u = 600$ mm, $v = 120$ mm, $\ell = 720$ mm, $F = 3.57$, $F_E = 21.4$. At the Hg h-line, $\lambda = 0.4047$ μm, so $\rho = 1.24$ μm and $\delta = 2.53$ μm. Note that to four significant figures $\tan\theta = 0.02333$, and $\sin\theta = 0.02333$. This justifies substituting $\tan\theta$ for $\sin\theta$.

These relationships can illustrate some of the problems faced in building stepper lenses. Consider wavelength: small resolution and maximum δ/ρ argue for short λ, placing the illumination at least in the UV, preferably in the DUV if possible. On the other hand, normal optical glasses show great changes in refraction index as λ is lowered, and in fact finally lose transparency. Only silica and certain crystalline materials are satisfactory at 365 nm and below, but they are expensive and more difficult to process. Further, lens tolerances are measured in fractions of a wavelength, so lower λ increases lens fabrication problems.

Lenses fail to focus different wavelengths at the same point unless this *chromatic aberration* is compensated for. This is accomplished by making compound lenses for visible light of glasses having different refraction indices. Since these are not available in the DUV, monochromatic illumination must be used, so the standing wave problem increases. Compromise is necessary, in λ and among NA, s, M, and f.

Representative lenses for g-line operation (436 nm) are [6]:

Zeiss 10-78-01: $M = 10$, NA = 0.42, $\rho = 0.8$ μm, $d = 7$ mm

Zeiss 10-78-37: $M = 5$, NA = 0.30, $\rho = 1.1$ μm, $d = 20$ mm

Two lenses are used in a stepper: a condenser between the illumination source and the reticle, and the projection lens that focuses the reticle image on the PR. A parameter encountered in the literature is the *degree of coherence*, S or σ, defined for this lens pair as [3]

$$\sigma = \text{degree of coherence} = \frac{(\text{NA})_{\text{cond}}}{(\text{NA})_{\text{proj}}} \qquad \text{(E-14)}$$

If the σ value ≤ 0.3, the coherence is said to be too high and produces diffraction rings at the corner and edges of pattern features. A typical satisfactory value is 0.7 [7].

References

[1] B. J. Lin, "Optical methods for fine line lithography." In *Fine Line Lithography*, ed. R. Newman. Amsterdam: North Holland, 1980.

[2] E. Hecht and A. Zajac, *Optics*. Reading: Addison Wesley, 1982.

[3] R. K. Watts and J. H. Bruning, "A Review of Fine-Line Lithographic Techniques Present and Future," *Solid State Technol.*, **24**, (5), 95–105, May 1981.

[4] T. Omata, "Reduction Lenses for Submicron Lithography," *Solid State Technol.*, **27**, (9), 173–177, Sept. 1984.

[5] D. A. Doane, "Optical Lithography in the 1-μm Limit," *Solid State Technol.*, **23**, (8), 101–114, Aug. 1980.

[6] P. S. Burggraaf, "Advances Keep Optical Aligners Ahead of Industry Needs," *Semicond. Int.*, **7**, (2), 88–93, Feb. 1984.

[7] M. Lacombat et al., "Laser Projection Printing," *Solid State Technol.*, **23**, (8), 115–121, Aug. 1980.

Appendix F
Graphs and Tables*

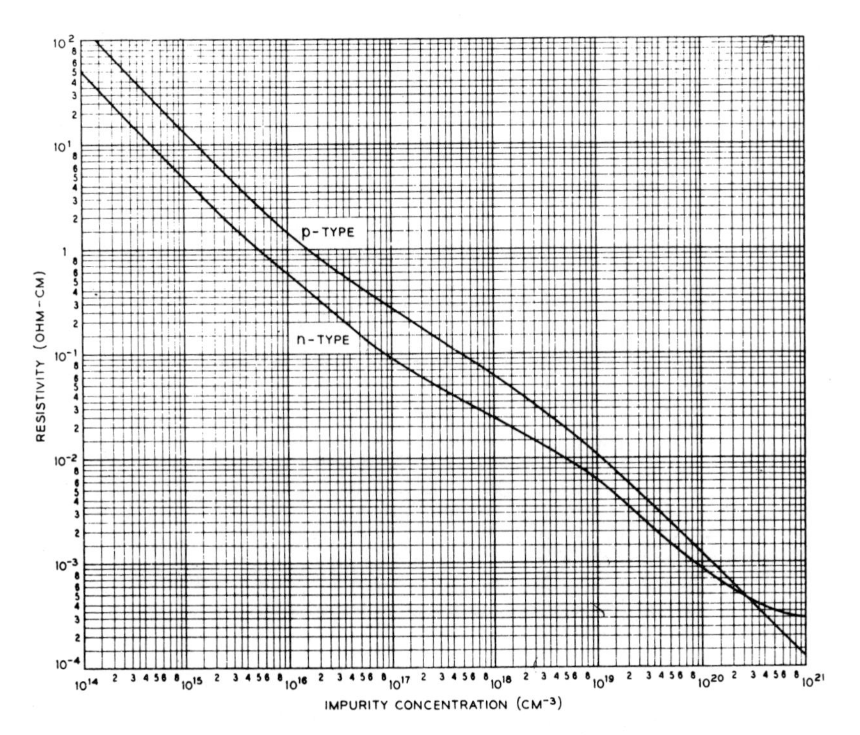

Appendix F.2.1. Irvin Resistivity Curves. Resistivity of silicon at 300 K as a function of acceptor and donor concentration. J. C. Irvin, "Resistivity of Bulk Silicon and of Diffused Layers in Silicon," *Bell Syst. Tech. J.*, **41**, (2), 387–420, March 1962. (Courtesy of AT&T Bell Laboratories.)

* Note that the first digit in an Appendix F number refers to the corresponding chapter.

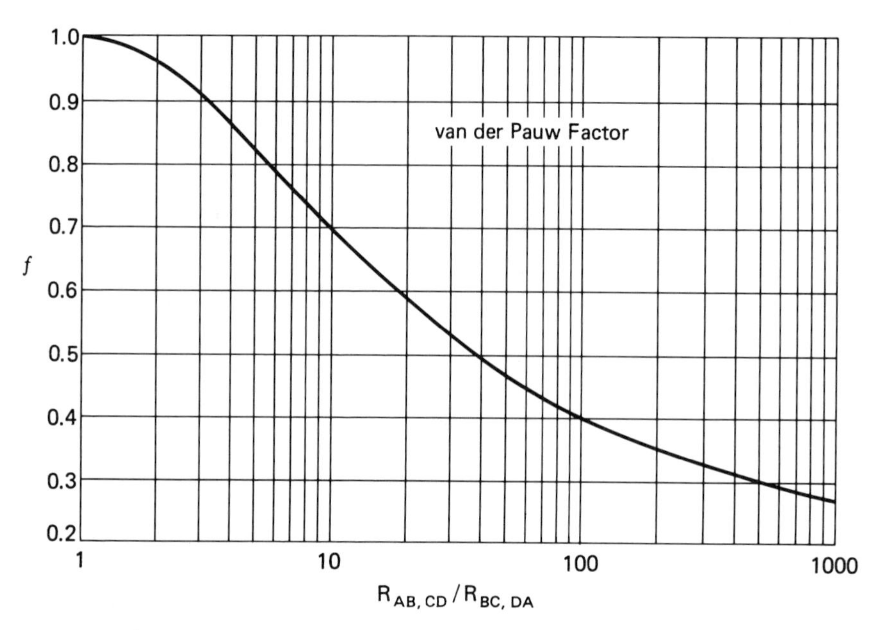

Appendix F.3.1. Van der Pauw correction factor. Moore.

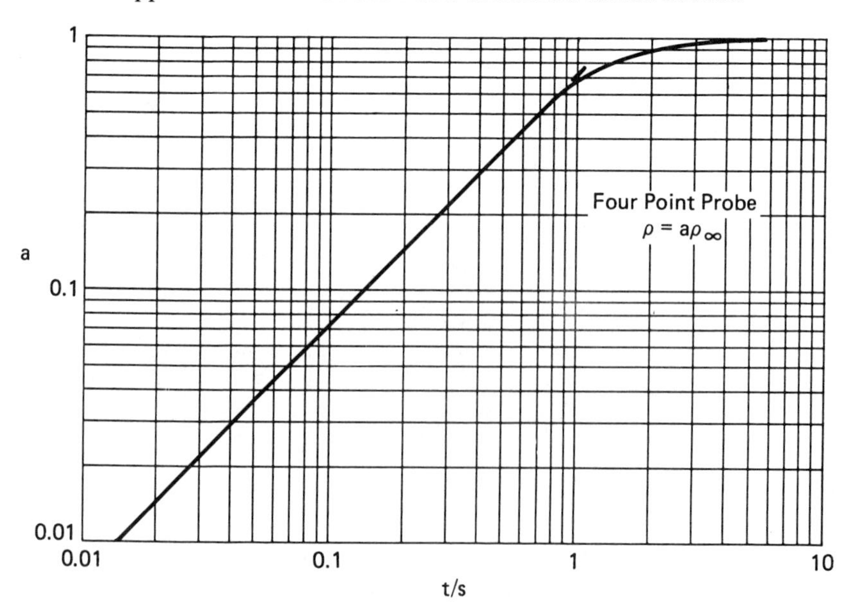

Appendix F.3.2. Thickness correction factor for a four point probe on an infinitely large sample horizontally. Data from L. B. Valdes, "Resistivity Measurements on Germanium for Transistors," *Proc. IRE*, **42**, (2), 420–427, Feb. 1954. © 1954 IEEE.

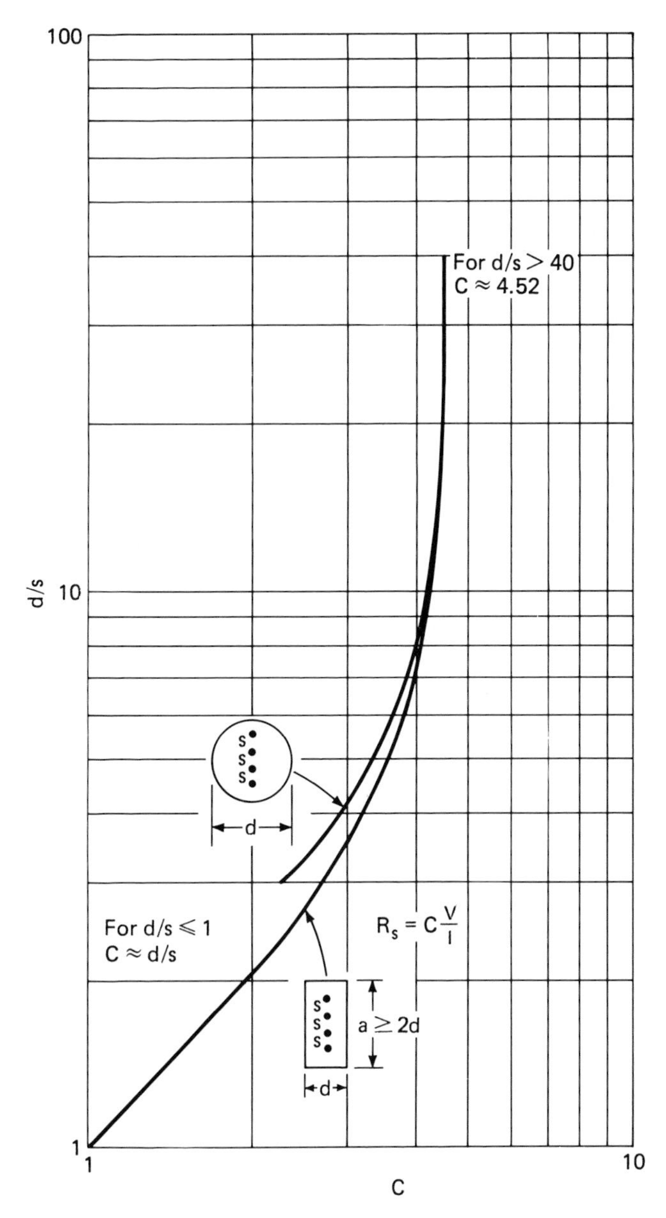

Appendix F.3.3. Sheet resistance correction factors for circular and rectangular samples. F. M. Smits, "Measurements of Sheet Resistivities with Four-Point Probe," *Bell Syst. Tech. J.*, **37**, (3), 711–718, May 1958. (Courtesy of AT&T Bell Laboratories.)

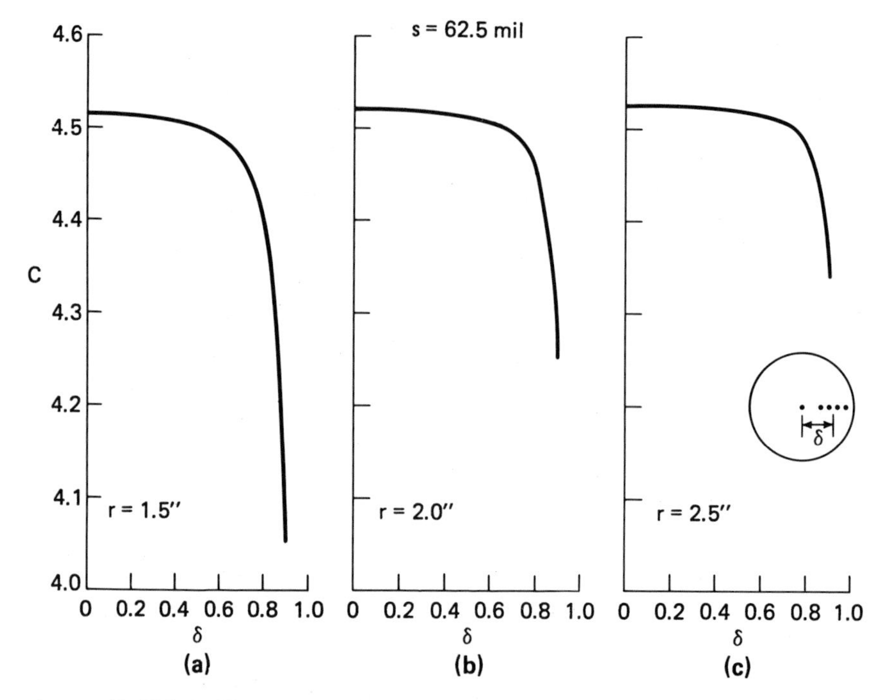

Appendix F.3.4. Sheet resistance correction factors for a collinear four point probe aligned radially on a circular wafer. The distance between centers of the probe array and wafer is δ. Probe tip spacing is 62.5 mils. Calculated from Eq. (3.9-3).

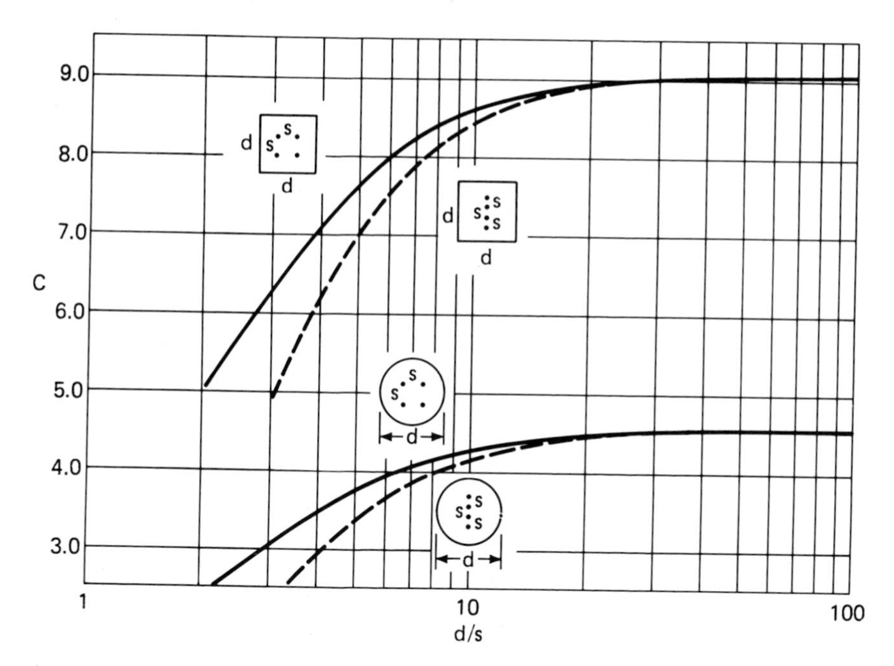

Appendix F.3.5. Sheet resistance correction factors for collinear and square tip probes. Note the relative magnitudes for the two types. After J. S. Glick, M.S. Thesis, Electrical Engineering Department, University of Illinois, 1974.

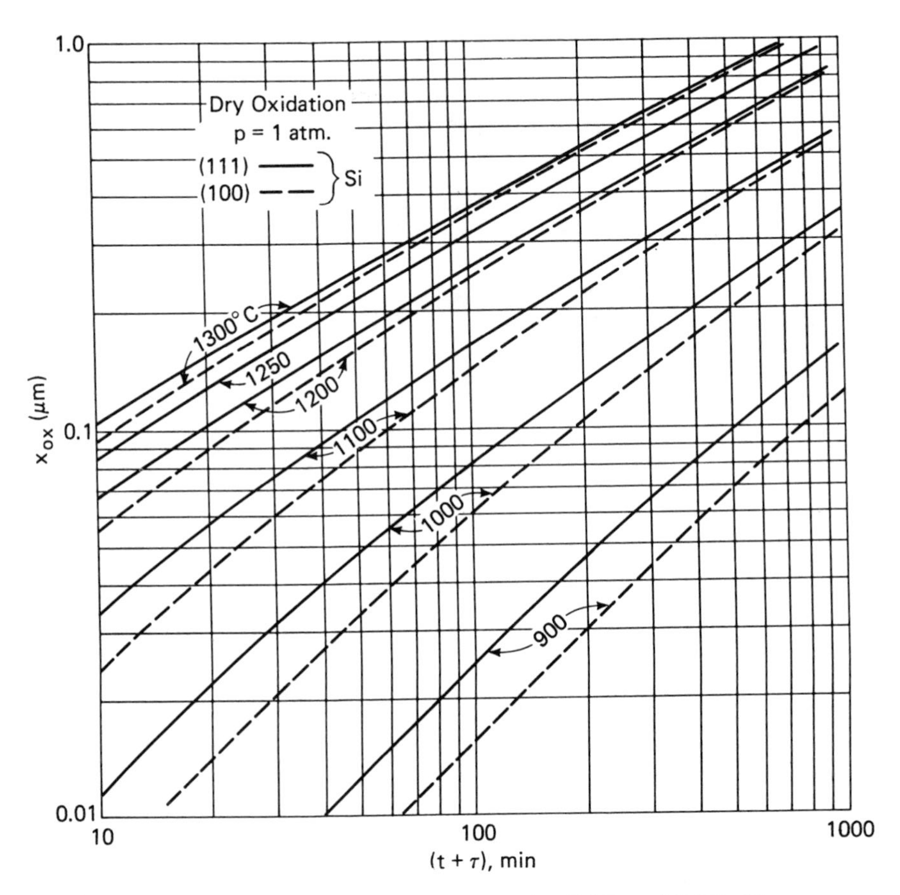

Appendix F.5.1. Dry oxidation curves for (100) and (111) silicon.

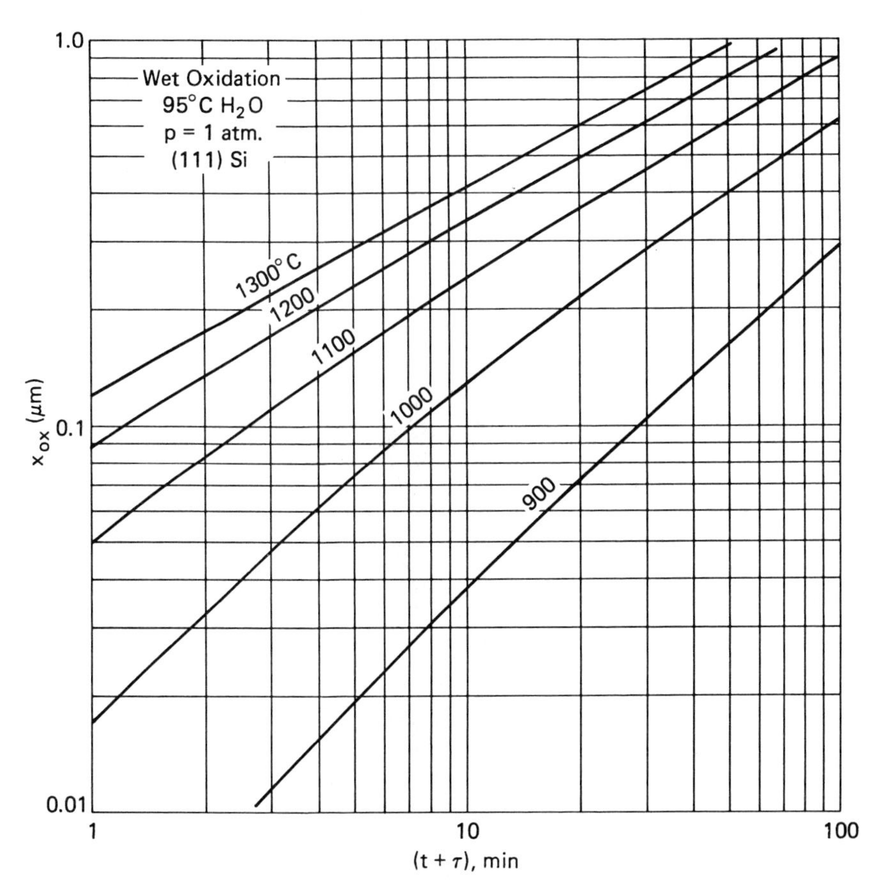

Appendix F.5.2. Wet oxidation curves for (111) silicon.

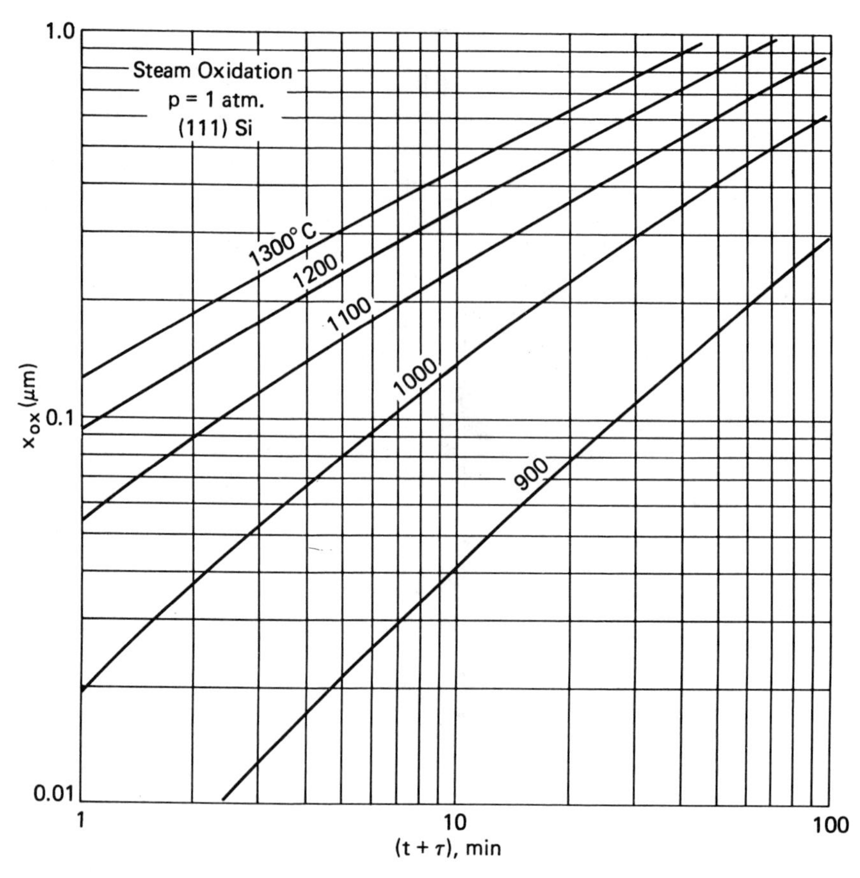

Appendix F.5.3. Steam oxidation curves for (111) silicon.

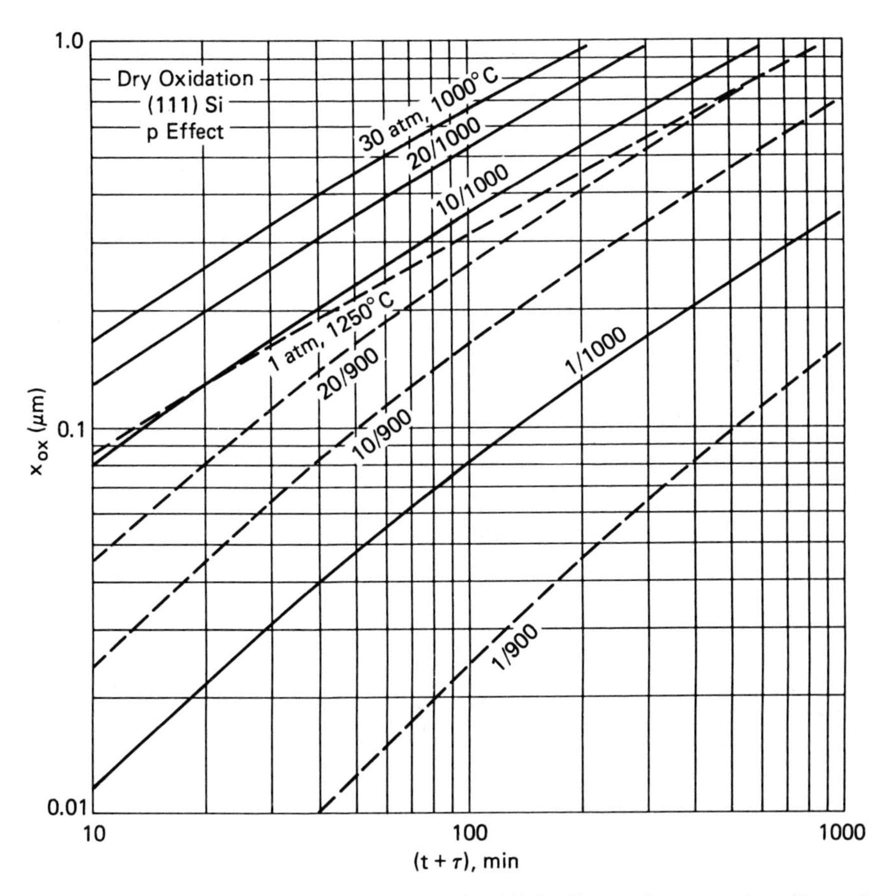

Appendix F.5.4. Dry oxidation curves for (111) silicon showing the effect of oxidant pressure.

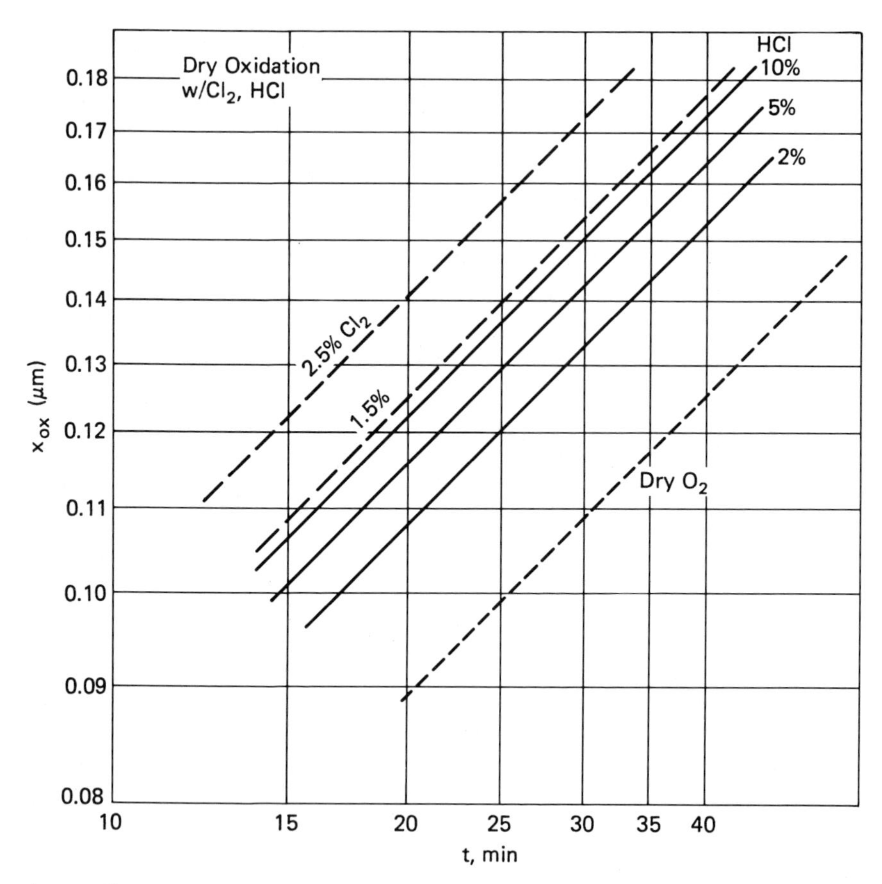

Appendix F.5.5. Dry oxidation curves for (111) silicon with added Cl_2 or HCl. After R. J. Kriegler, Y. C. Cheng, and D. R. Colton, "The Effect of HCl and Cl_2 on the Thermal Oxidation of Silicon" (*J. Electrochem. Soc.*, **119**, (3), 388–392, March 1972. Reprinted by permission of the publisher, The Electrochemical Society, Inc.)

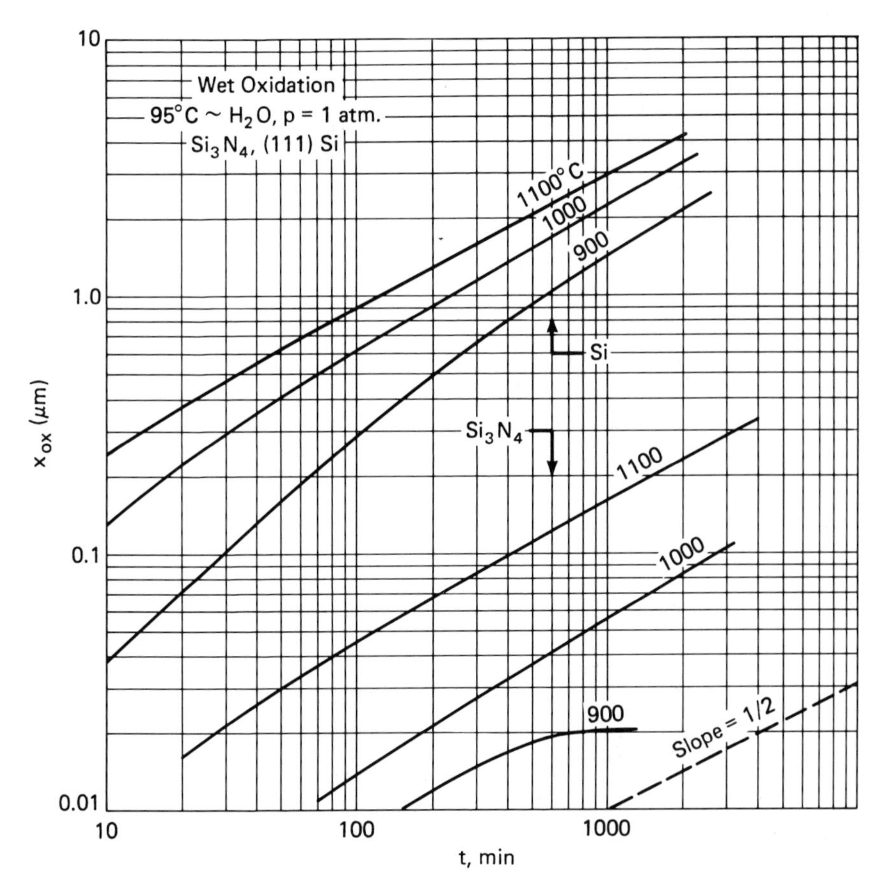

Appendix F.5.6. Wet oxidation curves for Si and Si_3N_4. Nitride curves are after Fränz and Langheinrich. (Reprinted with permission from *Solid State Electronics*, **14**, Fränz and Langheinrich, "Conversion of Silicon Nitride into Silicon Dioxide Through the Influence of Oxygen," Copyright 1971, Pergamon Press PLC.)

```
10 CLEAR
20 PRINT "PROGRAM IS OXIDATN8.BAS"
30 PRINT "P = 1 ATM"
40 'GEA, 1989
50 PRINT "THE OXIDATION TYPE IS: D-DRY, W-WET, S-STEAM"
60 PRINT "WHAT IS THE OXIDATION TYPE?": INPUT TYPE$
70 IF TYPE$ = "D" THEN 150
80 IF TYPE$ = "W" THEN 130
90 IF TYPE$ = "S" THEN 110
100 '111 STEAM PARAMETERS
110 C1 = 1.07E-09: C2 = 4.53: E1 = .78: E2 = 2.05: GOTO 160
120 '111 WET PARAMETERS
130 C1 = 5.94E-10: C2 = 2.49: E1 = .71: E2 = 2!: GOTO 160
140 '111 DRY PARAMETERS
150 C1 = 2.144E-09: C2 = .173: E1 = 1.23: E2 = 2!
160 PRINT "IS THE SILICON 111 OR 100?": INPUT SI
170 IF SI = 100 GOTO 190 ELSE 200
180 '100 PARAMETER CORRECTION
190 C2 = C2/1.68
200 PRINT "WHAT IS OXIDATION TEMPERATURE IN CELSIUS": INPUT TC
210 T = TC + 273
220 PRINT "WHAT IS INITIAL OXIDE THICKNESS IN ANGSTROMS": INPUT XIA
230 XI = XIA*1E-08
240 KT = .0000862*T
250 B = C1*EXP(-E1/KT)
260 D = C2*EXP(-E2/KT)
270 A = B/D
280 TAU =  XI*(XI + A)/B
290 PRINT "CALCULATE: 1-OXIDATION THICKNESS, 2-OXIDATION TIME"
300 INPUT CALL
310 IF CALL = 2 THEN 390
320 IF CALL = 1 THEN 330
330 PRINT "WHAT IS OXIDATION TIME IN MINUTES": INPUT TIMEM
340 TIME = 60*TIMEM
350 XOXCM = (A/2)*SQR(4*B*(TIME + TAU)/A^2 + 1) - 1)
360 XOX = XOXCM*1E+08
370 PRINT "THE OXIDE THICKNESS IN ANGSTROMS IS": XOX
380 END
390 PRINT "THE FINAL OXIDE THICKNESS IN ANGSTROMS IS": INPUT XOXA
400 XOXCM = XOXA*1E-08
410 TIME = XOXCM*(XOXCM + A)/B - TAU
420 TIMEM = TIME/60
430 PRINT "THE OXIDATION TIME IN MINUTES IS": TIMEM
440 GOTO 380
```

Appendix F.5.7. An MBASIC program for oxidation of silicon at atmospheric pressure. Data on lines 100–150, and 190 are from D. T. Antoniadis, S. E. Hansen, and R. W. Dutton, "SUPREM II—A Program for IC Modeling and Simulation," *Technical Report No. 5019-2, Intergrated Circuits Laboratory*, Stanford University, June 1978.

Appendix F.5.8. Color Chart for Thermally Grown SiO_2 Films Observed Perpendicularly under Daylight Fluorescent Lighting

Film Thickness (microns)	Order (5450 Å)	Color and Comments
0.05		Tan
0.07		Brown
0.10		Dark violet to red-violet
0.12		Royal blue
0.15		Light blue to metallic blue
0.17	I	Metallic to very light yellow-green
0.20		Light gold or yellow—slightly metallic
0.22		Gold with slight yellow-orange
0.25		Orange to melon
0.27		Red-violet
0.30		Blue to violet-blue
0.31		Blue
0.32		Blue to blue-green
0.34		Light green
0.35		Green to yellow-green
0.36	II	Yellow-green
0.37		Green-yellow
0.39		Yellow
0.41		Light orange
0.42		Carnation pink
0.44		Violet-red
0.46		Red-violet
0.47		Violet
0.48		Blue-violet
0.49		Blue
0.50		Blue-green
0.52		Green (broad)
0.54		Yellow-green
0.56	III	Green-yellow
0.57		Yellow to "yellowish." (Not yellow but is in the position where yellow is to be expected. At times it appears to be light creamy gray or metallic.)
0.58		Light-orange or yellow to pink borderline
0.60		Carnation pink
0.63		Violet-red
0.68		"Bluish." (Not blue but borderline between violet and blue-green. It appears more like a mixture between violet-red and blue-green and overall looks grayish.)

Appendix F.5.8. Color Chart for Thermally Grown SiO_2 Films Observed Perpendicularly under Daylight Fluorescent Lighting (*Continued*)

Film Thickness (microns)	Order (5450 Å)	Color and Comments
0.72	IV	Blue-green to green (quite broad)
0.77		"Yellowish"
0.80		Orange (rather broad for orange)
0.82		Salmon
0.85		Dull, light red-violet
0.86		Violet
0.87		Blue-violet
0.89		Blue
0.92	V	Blue-green
0.95		Dull yellow-green
0.97		Yellow to "yellowish"
0.99		Orange
1.00		Carnation pink
1.02		Violet-red
1.05		Red-violet
1.06		Violet
1.07		Blue-violet
1.10		Green
1.11		Yellow-green
1.12	VI	Green
1.18		Violet
1.19		Red-violet
1.21		Violet-red
1.24		Carnation pink to salmon
1.25		Orange
1.28		"Yellowish"
1.32	VII	Sky blue to green-blue
1.40		Orange
1.45		Violet
1.46		Blue-violet
1.50	VIII	Blue
1.54		Dull yellow-green

From W. A. Pliskin and E. E. Conrad, "Nondestructive Determination of Thickness and Refractive Index of Transparent Films," *IBM J. Des. Dev.*, **8** (1), 43–51, Jan. 1964.

Appendix F.6.1. Diffusion Data

A. Boron and Phosphorus in silicon

T, °C	B, P D, cm^2s^{-1}	B[a] N_{SL}, cm^{-3}	P[a] N_{SL}, cm^{-3}
900	1.5 E-15	3.7 E20	6.0 E20
950	6.6 E-15	3.9 E20	7.8 E20
1000	2.6 E-14	4.1 E20	1.0 E21
1050	9.3 E-14	4.3 E20	1.2 E21
1100	3.0 E-13	4.5 E20	1.4 E21
1150	9.1 E-13	4.8 E20	1.5 E21
1200	2.5 E-12	5.0 E20	1.5 E21
1250	6.5 E-12	5.2 E20	1.4 E21
1300	1.6 E-11	5.4 E20	1.1 E21
1350	3.7 E-11	5.7 E20	7.1 E20

Note: En = 10^n

B. Data for substitutional dopants in silicon

Dopant	P, B[b]	As[b]	Sb[b]	Al[b]	Ga[b]	In
D_∞, cm^2s^{-1}	10.5	0.058	3.94	1.77	0.573	16.5
E_a, eV	3.69	3.30	3.87	3.26	3.25	3.90
b, 10^{-4}K	4.28	3.83	4.49	3.78	3.77	4.52

C. Data for impurities in SiO_2[c]

Impurity	H_2	O_2	H_2O	B	P
D_∞, cm^2s^{-1}	9.5 E-4	1.5 E-2	1.0 E-6	3.0 E-6	1.0 E-8
E_a, eV	0.69	3.09	0.79	3.50	1.75
b, 10^{-4}K	0.80	3.6	0.92	4.1	2.0

[a] From F. A. Trumbore, "Solid Solubilities of Impurity Elements in Germanium and Silicon," *Bell Syst. Tech. J.*, **19**, (11), 38–43, Nov. 1976. *Courtesy of* AT&T Bell Laboratories.
[b] From S. K. Ghandhi, *The Theory and Practice of Microelectronics*. New York: Wiley, 1968. Used with permission.
[c] From H. Wolf, *Semiconductors*, New York: Wiley, 1971. Used with permission.

Appendix F.6.2. Some Properties of the Error Function

$$\operatorname{erf} u = \frac{2}{\sqrt{\pi}} \int_0^u e^{-z^2}\,dz = \frac{2}{\sqrt{\pi}}\left(u - \frac{u^3}{3 \times 1!} + \frac{u^5}{5 \times 2!} - \cdots\right)$$

Therefore

$$\operatorname{erf}(-u) = -\operatorname{erf} u$$

$$\operatorname{erfc} u = 1 - \operatorname{erf} u = \frac{2}{\sqrt{\pi}} \int_u^\infty e^{-z^2}\,dz$$

Therefore

$$\operatorname{erfc}(-u) = 1 + \operatorname{erf} u$$

If

$$y = \operatorname{erfc} u, \qquad u = \operatorname{erfc}^{-1} y = \operatorname{erf}^{-1}(1 - y)$$

$$\operatorname{erf} u \approx \frac{2u}{\sqrt{\pi}} \quad \text{for} \quad u \ll 1$$

$$\operatorname{erfc} u \approx \frac{1}{\sqrt{\pi}} \frac{e^{-u^2}}{u} \quad \text{for} \quad u \gg 1$$

$$\operatorname{erf} u > 0.995 \quad \text{for} \quad u \geq 2$$

$$\operatorname{erfc} u < 0.005 \quad \text{for} \quad u \geq 2$$

$$\operatorname{erf}(\infty) = 1, \qquad \operatorname{erf}(0) = 0$$

$$\operatorname{erfc}(0) = 1, \qquad \operatorname{erfc}(\infty) = 0$$

$$\frac{d \operatorname{erf} u}{du} = \frac{2}{\sqrt{\pi}} e^{-u^2}$$

$$\int_0^u \operatorname{erfc} z\,dz = u \operatorname{erfc} u + \frac{1}{\sqrt{\pi}}(1 - e^{-u^2})$$

$$\int_0^\infty \operatorname{erfc} z\,dz = \frac{1}{\sqrt{\pi}}$$

$$\int_0^\infty e^{-u^2}\,du = \frac{\sqrt{\pi}}{2}, \qquad \int_0^u e^{-z^2}\,dz = \frac{\sqrt{\pi}}{2} \operatorname{erf} u$$

Appendix F.6.3. Error Function erf(w)

w	erf(w)	w	erf(w)	w	erf(w)	w	erf(w)
0.00	0.000 000	0.43	0.456 887	0.86	0.776 110	1.29	0.931 899
0.01	0.011 283	0.44	0.466 225	0.87	0.781 440	1.30	0.934 008
0.02	0.022 565	0.45	0.475 482	0.88	0.786 687	1.31	0.936 063
0.03	0.033 841	0.46	0.484 655	0.89	0.791 843	1.32	0.938 065
0.04	0.045 111	0.47	0.493 745	0.90	0.796 908	1.33	0.940 015
0.05	0.056 372	0.48	0.502 750	0.91	0.801 883	1.34	0.941 914
0.06	0.067 622	0.49	0.511 668	0.92	0.806 768	1.35	0.943 762
0.07	0.078 858	0.50	0.520 500	0.93	0.811 564	1.36	0.945 561
0.08	0.090 078	0.51	0.529 244	0.94	0.816 271	1.37	0.947 312
0.09	0.101 281	0.52	0.537 899	0.95	0.820 891	1.38	0.949 016
0.10	0.112 463	0.53	0.546 464	0.96	0.825 424	1.39	0.950 673
0.11	0.123 623	0.54	0.554 939	0.97	0.829 870	1.40	0.952 285
0.12	0.134 758	0.55	0.563 323	0.98	0.834 232	1.41	0.953 852
0.13	0.145 867	0.56	0.571 616	0.99	0.838 508	1.42	0.955 376
0.14	0.156 947	0.57	0.579 816	1.00	0.842 701	1.43	0.956 857
0.15	0.167 996	0.58	0.587 923	1.01	0.846 810	1.44	0.958 297
0.16	0.179 012	0.59	0.595 936	1.02	0.850 838	1.45	0.959 695
0.17	0.189 992	0.60	0.603 856	1.03	0.854 784	1.46	0.961 054
0.18	0.200 936	0.61	0.611 681	1.04	0.858 650	1.47	0.962 373
0.19	0.211 840	0.62	0.619 411	1.05	0.862 436	1.48	0.963 654
0.20	0.222 703	0.63	0.627 046	1.06	0.866 144	1.49	0.964 898
0.21	0.233 522	0.64	0.634 586	1.07	0.869 773	1.50	0.966 105
0.22	0.244 296	0.65	0.642 029	1.08	0.873 326	1.51	0.967 277
0.23	0.255 023	0.66	0.649 377	1.09	0.876 803	1.52	0.968 413
0.24	0.265 700	0.67	0.656 628	1.10	0.880 205	1.53	0.969 516
0.25	0.276 326	0.68	0.663 782	1.11	0.883 533	1.54	0.970 586
0.26	0.286 900	0.69	0.670 840	1.12	0.886 788	1.55	0.971 623
0.27	0.297 418	0.70	0.677 801	1.13	0.889 971	1.56	0.972 628
0.28	0.307 880	0.71	0.684 666	1.14	0.893 082	1.57	0.973 603
0.29	0.318 283	0.72	0.691 433	1.15	0.896 124	1.58	0.974 547
0.30	0.328 627	0.73	0.698 104	1.16	0.899 096	1.59	0.975 462
0.31	0.338 908	0.74	0.704 678	1.17	0.902 000	1.60	0.976 348
0.32	0.349 126	0.75	0.711 156	1.18	0.904 837	1.61	0.977 207
0.33	0.359 279	0.76	0.717 537	1.19	0.907 608	1.62	0.978 038
0.34	0.369 365	0.77	0.723 822	1.20	0.910 314	1.63	0.978 843
0.35	0.379 382	0.78	0.730 010	1.21	0.912 956	1.64	0.979 622
0.36	0.389 330	0.79	0.736 103	1.22	0.915 534	1.65	0.980 376
0.37	0.399 206	0.80	0.742 101	1.23	0.918 050	1.66	0.981 105
0.38	0.409 009	0.81	0.748 003	1.24	0.920 505	1.67	0.981 810
0.39	0.418 739	0.82	0.753 811	1.25	0.922 900	1.68	0.982 493
0.40	0.428 392	0.83	0.759 524	1.26	0.925 236	1.69	0.983 153
0.41	0.437 969	0.84	0.765 143	1.27	0.927 514	1.70	0.983 790
0.42	0.447 468	0.85	0.770 668	1.28	0.929 734	1.71	0.984 407

Appendix F.6.3. Error Function erf(w) (*Continued*)

w	erf(w)	w	erf(w)	w	erf(w)	w	erf(w)
1.72	0.985 003	2.15	0.997 639	2.58	0.999 736	3.01	0.999 979 26
1.73	0.985 578	2.16	0.997 747	2.59	0.999 751	3.02	0.999 980 53
1.74	0.986 135	2.17	0.997 851	2.60	0.999 764	3.03	0.999 981 73
1.75	0.986 672	2.18	0.997 951	2.61	0.999 777	3.04	0.999 982 86
1.76	0.987 190	2.19	0.998 046	2.62	0.999 789	3.05	0.999 983 92
1.77	0.987 691	2.20	0.998 137	2.63	0.999 800	3.06	0.999 984 92
1.78	0.988 174	2.21	0.998 224	2.64	0.999 811	3.07	0.999 985 86
1.79	0.988 641	2.22	0.998 308	2.65	0.999 822	3.08	0.999 986 74
1.80	0.989 091	2.23	0.998 388	2.66	0.999 831	3.09	0.999 987 57
1.81	0.989 525	2.24	0.998 464	2.67	0.999 841	3.10	0.999 988 35
1.82	0.989 943	2.25	0.998 537	2.68	0.999 849	3.11	0.999 989 08
1.83	0.990 347	2.26	0.998 607	2.69	0.999 858	3.12	0.999 989 77
1.84	0.990 736	2.27	0.998 674	2.70	0.999 866	3.13	0.999 990 42
1.85	0.991 111	2.28	0.998 738	2.71	0.999 873	3.14	0.999 991 03
1.86	0.991 472	2.29	0.998 799	2.72	0.999 880	3.15	0.999 991 60
1.87	0.991 821	2.30	0.998 857	2.73	0.999 887	3.16	0.999 992 14
1.88	0.992 156	2.31	0.998 912	2.74	0.999 893	3.17	0.999 992 64
1.89	0.992 479	2.32	0.998 966	2.75	0.999 899	3.18	0.999 993 11
1.90	0.992 790	2.33	0.999 016	2.76	0.999 905	3.19	0.999 993 56
1.91	0.993 090	2.34	0.999 065	2.77	0.999 910	3.20	0.999 993 97
1.92	0.993 378	2.35	0.999 111	2.78	0.999 916	3.21	0.999 994 36
1.93	0.993 656	2.36	0.999 155	2.79	0.999 920	3.22	0.999 994 73
1.94	0.993 923	2.37	0.999 197	2.80	0.999 925	3.23	0.999 995 07
1.95	0.994 179	2.38	0.999 237	2.81	0.999 929	3.24	0.999 995 40
1.96	0.994 426	2.39	0.999 275	2.82	0.999 933	3.25	0.999 995 70
1.97	0.994 664	2.40	0.999 311	2.83	0.999 937	3.26	0.999 995 98
1.98	0.994 892	2.41	0.999 346	2.84	0.999 941	3.27	0.999 996 24
1.99	0.995 111	2.42	0.999 379	2.85	0.999 944	3.28	0.999 996 49
2.00	0.995 322	2.43	0.999 411	2.86	0.999 948	3.29	0.999 996 72
2.01	0.995 525	2.44	0.999 441	2.87	0.999 951	3.30	0.999 996 94
2.02	0.995 719	2.45	0.999 469	2.88	0.999 954	3.31	0.999 997 15
2.03	0.995 906	2.46	0.999 497	2.89	0.999 956	3.32	0.999 997 34
2.04	0.996 086	2.47	0.999 523	2.90	0.999 959	3.33	0.999 997 51
2.05	0.996 258	2.48	0.999 547	2.91	0.999 961	3.34	0.999 997 68
2.06	0.996 423	2.49	0.999 571	2.92	0.999 964	3.35	0.999 997 838
2.07	0.996 582	2.50	0.999 593	2.93	0.999 966	3.36	0.999 997 983
2.08	0.996 734	2.51	0.999 614	2.94	0.999 968	3.37	0.999 998 120
2.09	0.996 880	2.52	0.999 634	2.95	0.999 970	3.38	0.999 998 247
2.10	0.997 021	2.53	0.999 654	2.96	0.999 972	3.39	0.999 998 367
2.11	0.997 155	2.54	0.999 672	2.97	0.999 973	3.40	0.999 998 478
2.12	0.997 284	2.55	0.999 689	2.98	0.999 975	3.41	0.999 998 582
2.13	0.997 407	2.56	0.999 706	2.99	0.999 976	3.42	0.999 998 679
2.14	0.997 525	2.57	0.999 722	3.00	0.999 977 91	3.43	0.999 998 770

Appendix F.6.3. Error Function erf(w) (*Continued*)

w	erf(w)	w	erf(w)	w	erf(w)	w	erf(w)
3.44	0.999 998 855	3.58	0.999 999 587	3.72	0.999 999 857	3.86	0.999 999 952
3.45	0.999 998 934	3.59	0.999 999 617	3.73	0.999 999 867	3.87	0.999 999 956
3.46	0.999 999 008	3.60	0.999 999 644	3.74	0.999 999 877	3.88	0.999 999 959
3.47	0.999 999 077	3.61	0.999 999 670	3.75	0.999 999 886	3.89	0.999 999 962
3.48	0.999 999 141	3.62	0.999 999 694	3.76	0.999 999 895	3.90	0.999 999 965
3.49	0.999 999 201	3.63	0.999 999 716	3.77	0.999 999 903	3.91	0.999 999 968
3.50	0.999 999 257	3.64	0.999 999 736	3.78	0.999 999 910	3.92	0.999 999 970
3.51	0.999 999 309	3.65	0.999 999 756	3.79	0.999 999 917	3.93	0.999 999 973
3.52	0.999 999 358	3.66	0.999 999 773	3.80	0.999 999 923	3.94	0.999 999 975
3.53	0.999 999 403	3.67	0.999 999 790	3.81	0.999 999 929	3.95	0.999 999 977
3.54	0.999 999 445	3.68	0.999 999 805	3.82	0.999 999 934	3.96	0.999 999 979
3.55	0.999 999 485	3.69	0.999 999 820	3.83	0.999 999 939	3.97	0.999 999 980
3.56	0.999 999 521	3.70	0.999 999 833	3.84	0.999 999 944	3.98	0.999 999 982
3.57	0.999 999 555	3.71	0.999 999 845	3.85	0.999 999 948	3.99	0.999 999 983

```
10 PRINT "NAME IS ERF4A.BAS"
20 PRINT "SOURCE IS CONTROL DATA CORP. MATH SCIENCE LIBRARY"
30 PRINT "FUNCTION IS ERF(X), DOUBLE PRECISION"
40 PRINT "LIMIT OF X IS 4. ERROR IN ERFC < 1.3% FOR X = 3.65"
50 PRINT "ROUND OFF RESULTS TO FOUR SIGNIFICANT FIGURES"
60 PRINT "SOURCE IS CONTROL DATA CORP. MATH SCIENCE LIBRARY"
70 CLEAR
80 DIM A#(26), B#(28)
90   A#(0) = 3.88730365522904#
100  A#(1) = -1.38163142001979 9#
110  A#(2) = .647316404854584#
120  A#(3) = -.305931024422036#
130  A#(4) = .13867974720203#
140  A#(5) = -.059247456591259#
150  A#(6) = .0236917518249282#
160  A#(7) = -8.847362635240454D-03
170  A#(8) = 3.08566171136092D-03
180  A#(9) = -1.00638635123798D-03
190  A#(10) = 3.07546328843079D-04
200  A#(11) = -8.82619837553631D-05
210  A#(12) = 2.38450961660726D-05
220  A#(13) = -6.07910028505827D-06
230  A#(14) = 1.46597217338083D-06
240  A#(15) = -3.3515993427206D-07
250  A#(16) = 7.280579544232D-08
260  A#(17) = -1.5057911766668D-08
270  A#(18) = 2.970947420555D-09
280  A#(19) = -5.602127393800006D-10
290  A#(20) = 1.011316239D-10
300  A#(21) = -1.7506504485D-11
310  A#(22) = 2.910381390D-12
320  A#(23) = -4.65332645D-13
330  A#(24) = 7.164815D-14
340  A#(25) = -1.063749D-14
350  A#(26) = 1.52467D-15
360  B#(27) = 0
370  B#(28) = 0
380 PRINT "WHAT IS X";
390 INPUT X
400 Y# = ABS(X/4)
410 K# = 4*Y#^2 - 2
420 FOR J = 26 TO 0 STEP - 1
430 B#(J) = K#*B#(J + 1) - B#(J + 2) + A#(J)
440 NEXT J
450 ERF(X) = Y#/2*(B#(0) - B#(2))
460 ERFC(X) = 1 - ERF(X)
470 PRINT "   X,            ERF(X),        ERFC(X)"
480 PRINT   X,           ERF(X),          ERFC(X)
490 INPUT "CALCULATE AGAIN (Y OR N)";R$
500 IF R$ = "Y" THEN 380
510 END
```

Appendix F.6.4. MBASIC listing for calculating erf(x) and erfc(x). (Reprinted by permission of Control Data Corporation.)

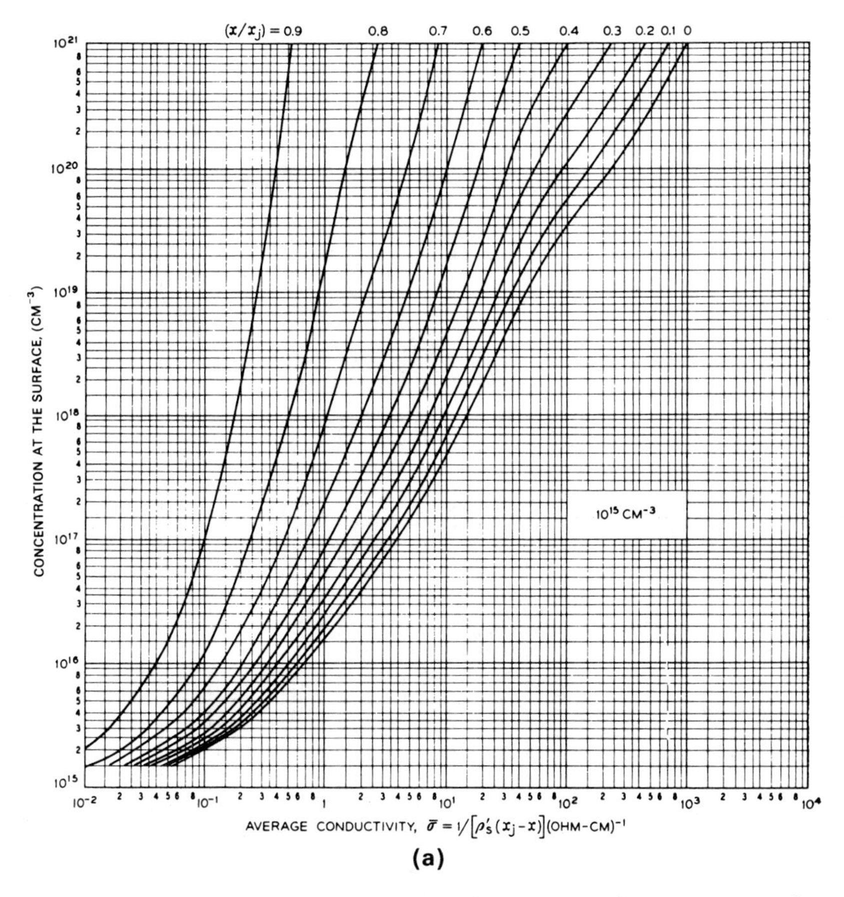

Appendix F.6.5(a). Average conductivity of n-type complementary error function layers in silicon. $N_b = 10^{15}$ cm^{-3}. Parts (a) through (h) of this appendix are from J. C. Irvin, "Resistivity of Bulk Silicon and Diffused Layers in Silicon," *Bell Syst. Tech. J.*, **41**, (2), 387–410, March, 1962. (Courtesy of AT&T Bell Laboratories.)

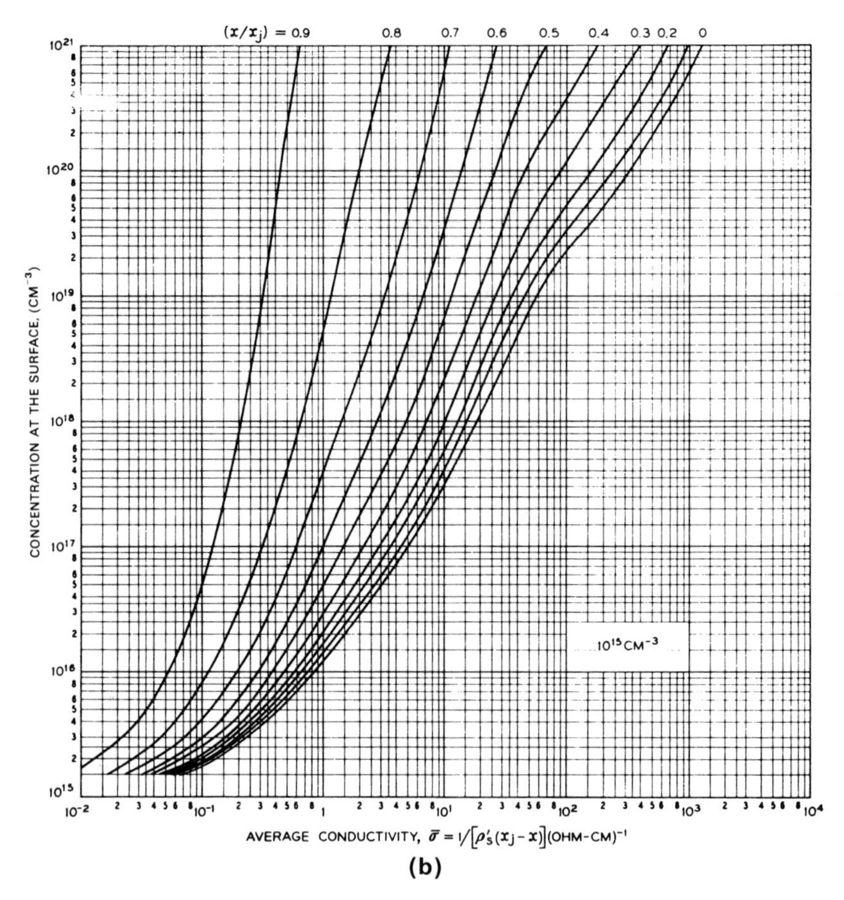

(b)

Appendix F.6.5(b). Average conductivity of n-type Gaussian layers in silicon. $N_b = 10^{15}$ cm^{-3}.

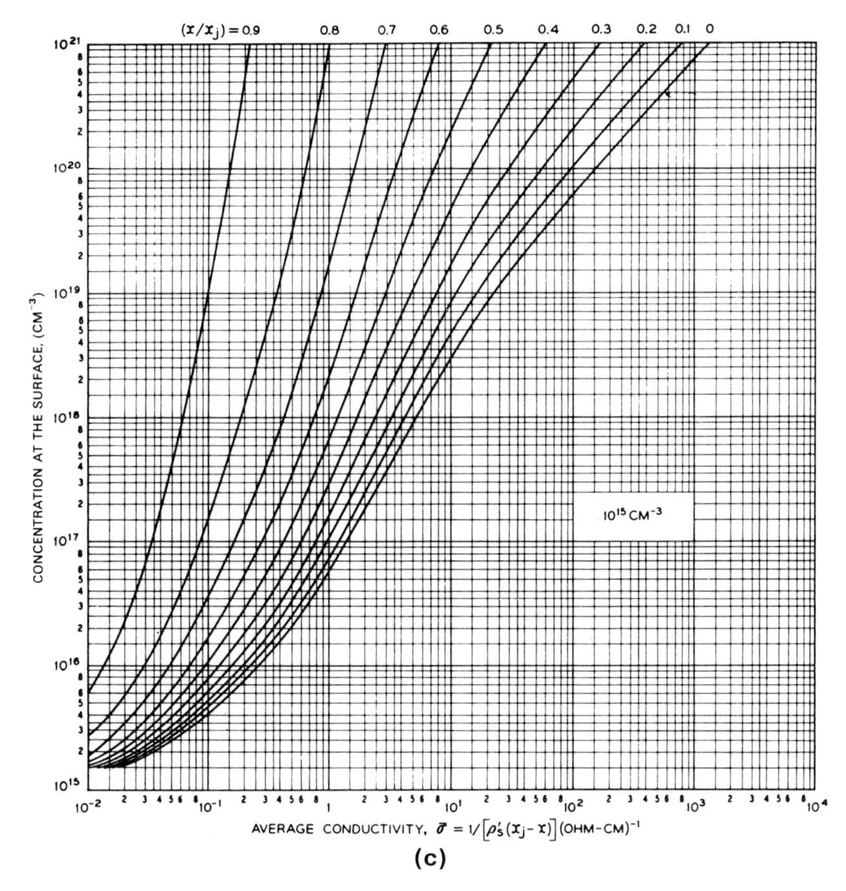

Appendix F.6.5(c). Average conductivity of p-type complementary error function layers in silicon. $N_b = 10^{15}$ cm^{-3}.

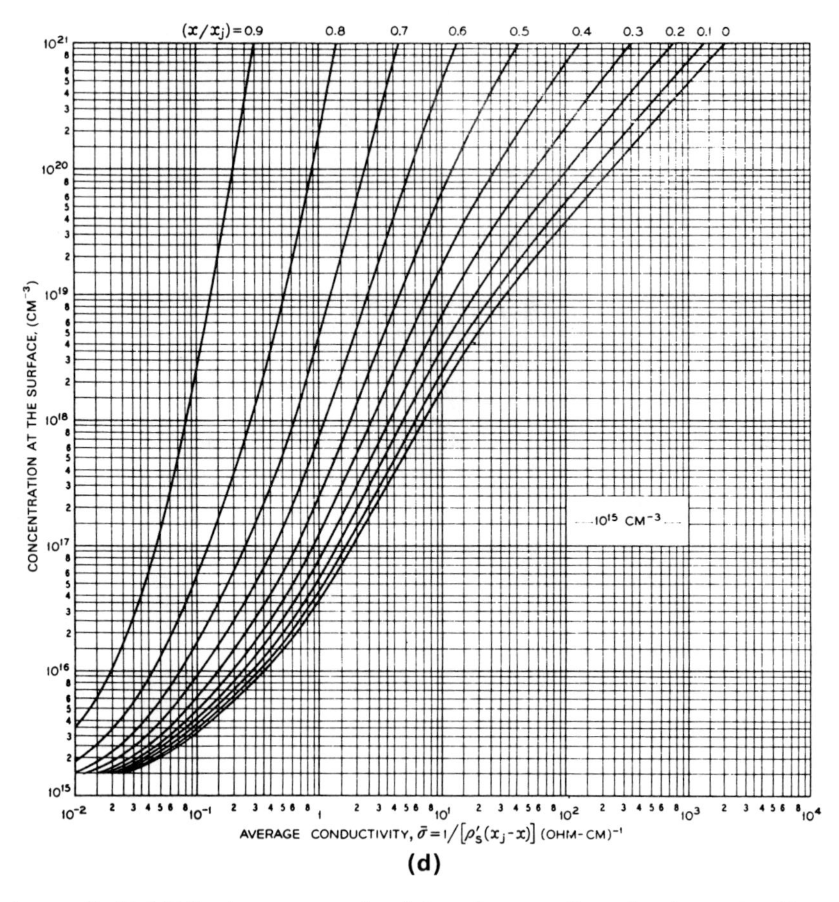

Appendix F.6.5(d). Average conductivity of p-type Gaussian layers in silicon. $N_b = 10^{15}$ cm^{-3}.

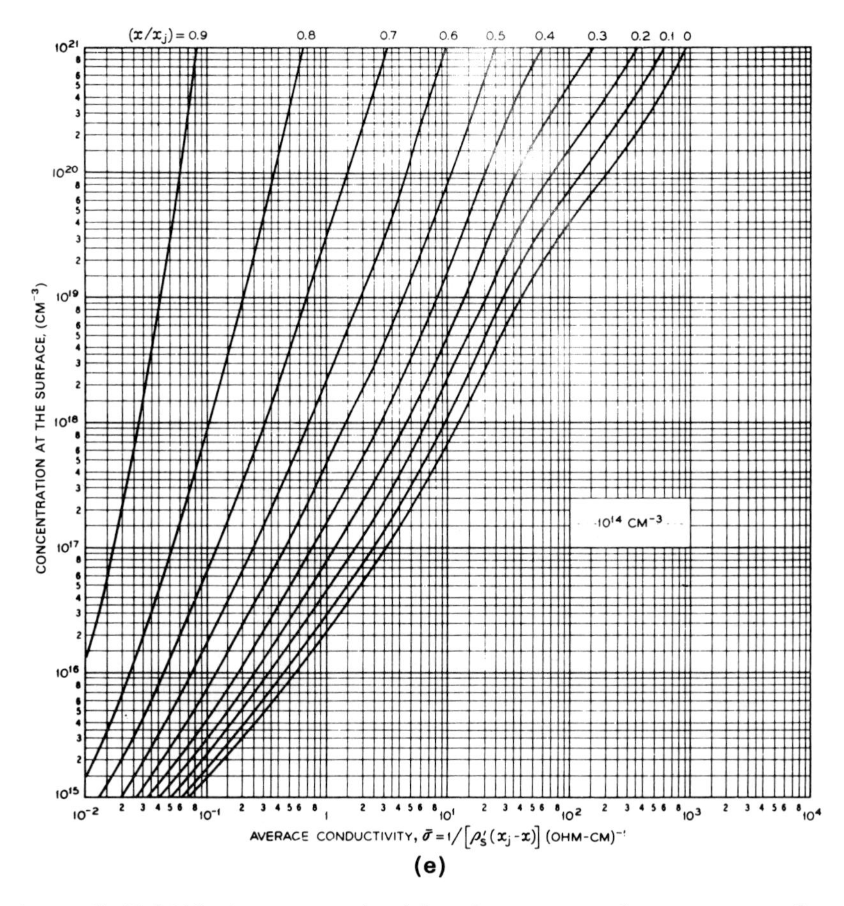

Appendix F.6.5(e). Average conductivity of n-type complementary error function layers in silicon. $N_b = 10^{14}$ cm^{-3}.

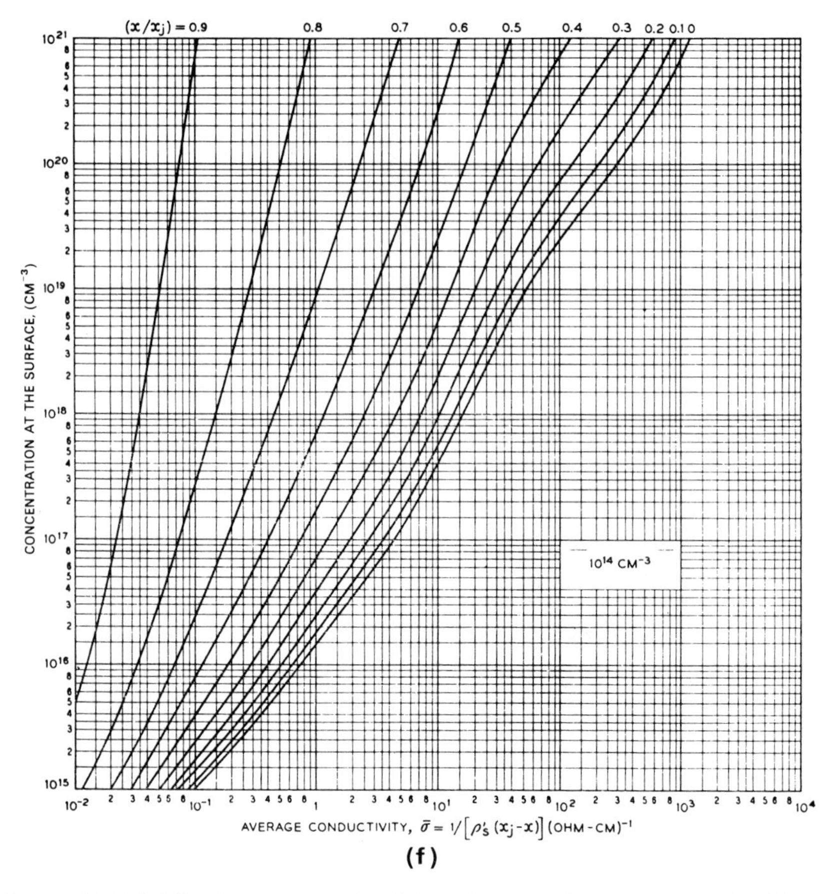

(f)

Appendix F.6.5(f). Average conductivity of n-type Gaussian layers in silicon. $N_b = 10^{14}$ cm^{-3}.

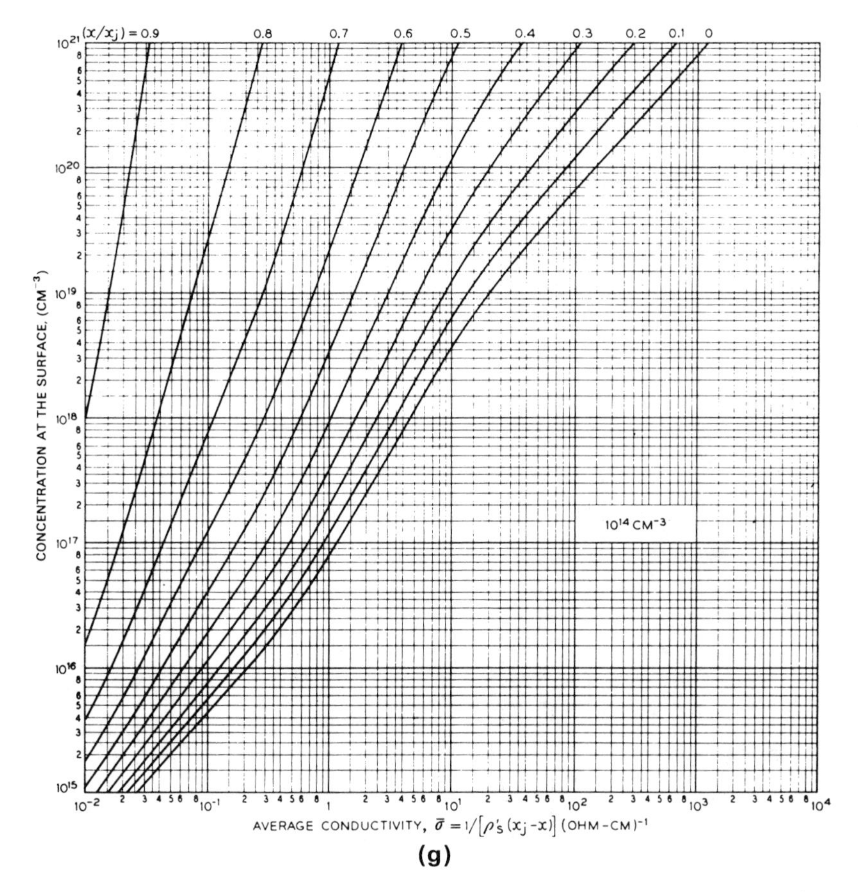

Appendix F.6.5(g). Average conductivity of p-type complementary error function layers in silicon. $N_b = 10^{14}$ cm^{-3}.

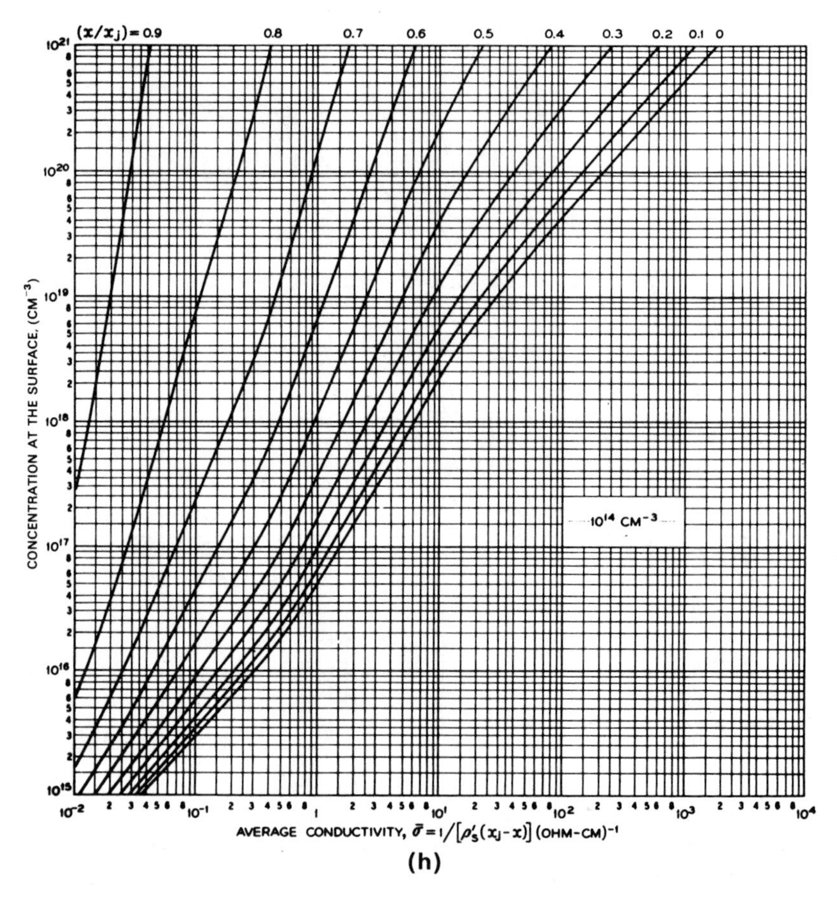

(h)

Appendix F.6.5(h). Average conductivity of p-type Gaussian layers in silicon. $N_b = 10^{14}$ cm^{-3}.

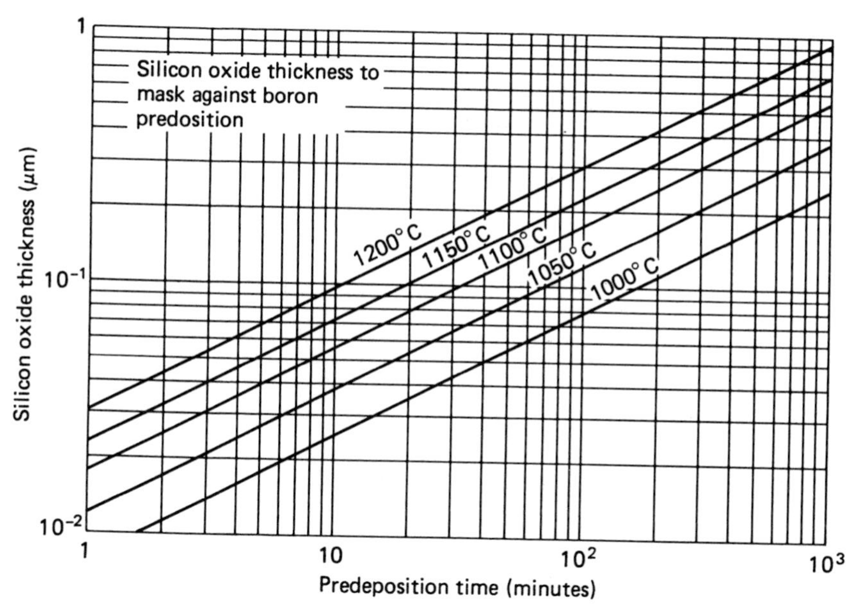

Appendix F.6.6. Oxide masking curves for boron. (From *The Theory and Practice of Microelectronics*, S. K. Ghandhi, © 1968 by John Wiley & Sons, Inc. Used with permission.)

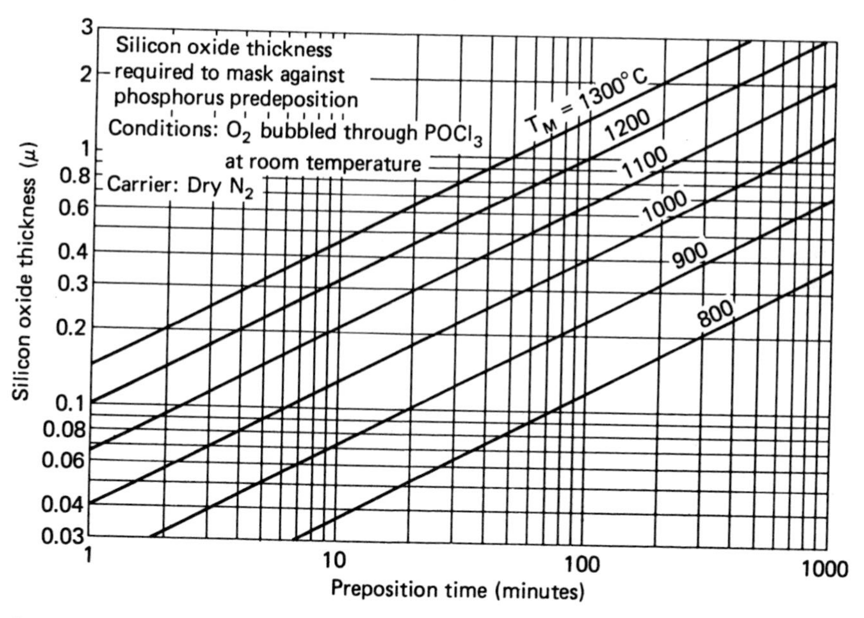

Appendix F.6.7. Oxide masking curves for phosphorus. (From *The Theory and Practice of Microelectronics*, S. K. Ghandhi, © 1968 by John Wiley & Sons, Inc. Used with permission.)

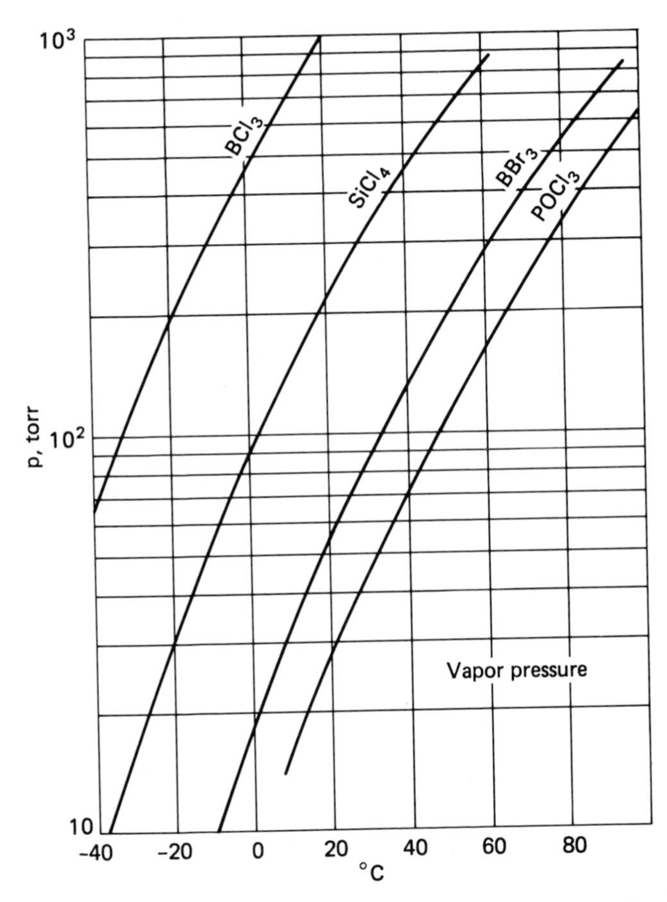

Appendix F.6.8. Vapor pressure curves for four liquid predeposition sources.

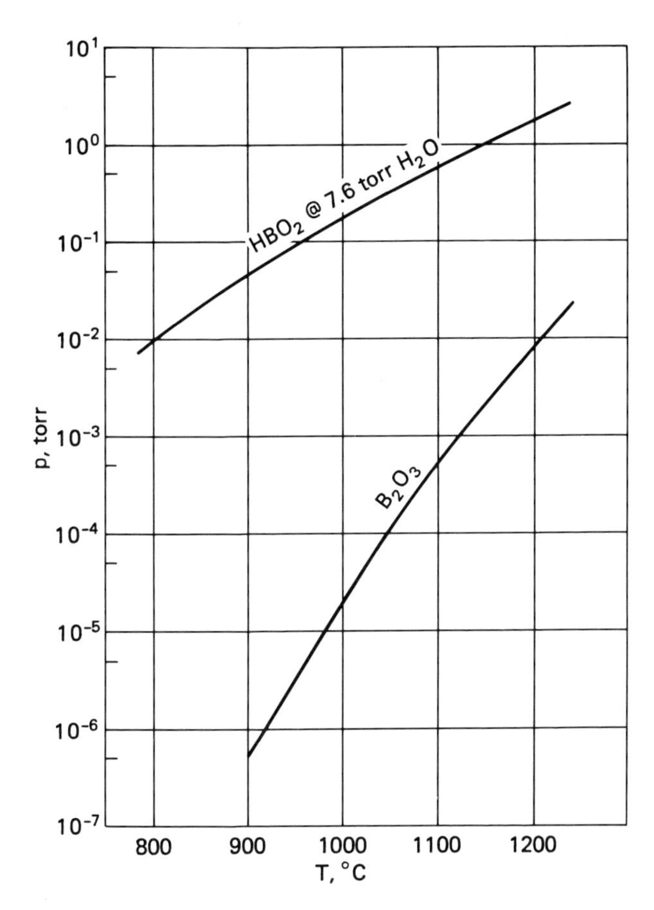

Appendix F.6.9. Vapor pressure curves of HBO_2 and B_2O_3. After *Low Defect Boron Diffusion Process Using Hydrogen Injection, Form A-14,004, 2186.* Niagara Falls: The Carborundum Company, Electronic Ceramics Division, 1985. (Courtesy of The Carborundum Company, Electronic Ceramics Division, © 1985 The Carborundum Company.)

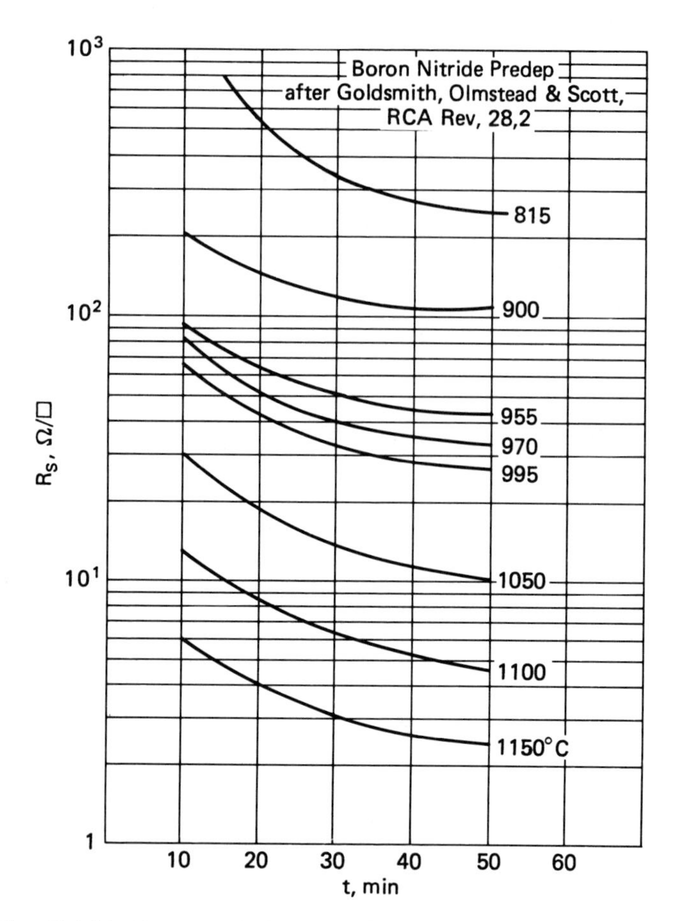

Appendix F.6.10. Boron nitride predeposition curves. N. Goldsmith, J. Olmstead, and J. Scott, "Boron Nitride as a Diffusion Source for Silicon," *RCA Rev.* **28**, (2), 344–350, June 1967. (Courtesy of General Electric Company.)

Appendix F.8.1. Ion Implantation: Effective Range Data

	P in Si		P in SiO_2		B in Si		B in SiO_2	
Energy, keV	R_p	ΔR_p	R_p	ΔR_p	R_p	ΔR_p	R_p	ΔR_p
10	0.0139	0.0069	0.0108	0.0048	0.0333	0.0171	0.0298	0.0143
20	0.0253	0.0119	0.0199	0.0084	0.0662	0.0283	0.0622	0.0252
30	0.0368	0.0166	0.0292	0.0119	0.0987	0.0371	0.0954	0.0342
40	0.0486	0.0212	0.0388	0.0152	0.1302	0.0443	0.1283	0.0418
50	0.0607	0.0256	0.0486	0.0185	0.1608	0.0504	0.1606	0.0483
60	0.0730	0.0298	0.0586	0.0216	0.1903	0.0556	0.1921	0.0540
70	0.0855	0.0340	0.0688	0.0247	0.2188	0.0601	0.2228	0.0590
80	0.0981	0.0380	0.0792	0.0276	0.2465	0.0641	0.2528	0.0634
90	0.1109	0.0418	0.0896	0.0305	0.2733	0.0677	0.2819	0.0674
100	0.1238	0.0456	0.1002	0.0333	0.2994	0.0710	0.3104	0.0710
110	0.1367	0.0492	0.1108	0.0360	0.3248	0.0739	0.3382	0.0743
120	0.1497	0.0528	0.1215	0.0387	0.3496	0.0766	0.3653	0.0774
130	0.1627	0.0562	0.1322	0.0412	0.3737	0.0790	0.3919	0.0801
140	0.1727	0.0595	0.1429	0.0437	0.3974	0.0813	0.4179	0.0827
150	0.1888	0.0628	0.1537	0.0461	0.4205	0.0834	0.4434	0.0851

R_p and ΔR_p in μm

Abstracted from J. F. Gibbons, W. S. Johnson, and S. W. Mylroie, *Projected Range Statistics, Semiconductors and Related Materials, 2nd Ed.* Stroudsburg: Dowden, Hutchinson, and Ross, 1975. Renewed © by John Wiley & Sons. Used with permission.

Appendix G
Numerical Constants

Physical Constants

Boltzmann's constant	$k = 8.62 \times 10^{-5}$ eV/K
	$= 1.38 \times 10^{-23}$ J/K
Planck's constant	$h = 4.14 \times 10^{-15}$ eV-s
electron charge magnitude	$q = 1.6 \times 10^{-19}$ C
permittivity of free space	$\varepsilon_o = 8.85 \times 10^{-14}$ F/cm
voltage equivalent of room temperauture	$\frac{kT}{q} = 2.59 \times 10^{-2}\ \text{V} = \frac{1}{\Lambda}$
Avogadro's number	$n_a = 6.02 \times 10^{23}$ molecules/mole
electron volt	$1\ \text{eV} = 1.6 \times 10^{-19}$ J
unified atomic mass unit	$1\ \text{u} = 1.66 \times 10^{-27}$ kg
Si atomic concentration	$N_{\text{Si}} = 5 \times 10^{22}\ \text{cm}^{-3}$

Semiconductor and Insulator Room-Temperature Parameters

Material	E_g, eV	χ, V	ε_r	ε, pf/cm	n_i, cm^{-3}	N, cm^{-3}
Si	1.1	4.03	11.8	1.04	1.5×10^{10}	5.00×10^{22}
GaAs	1.43	4.07	13.2	1.17	9.0×10^{7}	2.21×10^{22}
$a\text{SiO}_2$	8	1	3.9	0.345		2.30×10^{22}
$a\text{Si}_3\text{N}_4$	5		7.5	0.644		

Metal Work Functions

Metal	Al, Pb	Mo	Cr	Au	Ni	Pt
ϕ, V	4.0	4.3	4.6	4.7	5.1	5.3

Units

1 micrometer = 1 μm = 10^{-4} cm = 10^{3} nm = 10^{4} Å
1 mil = 10^{-3} in. = 10^{3} μin.
1 torr = 1 mm Hg = 133 Pa (Pascal)
1 psi = 6.88×10^{3} Pa = 7.04 kg/cm^{2}

Appendix H
Furnace Construction

The basic construction of a typical single-heating-zone furnace is shown in Figure H–1. Resistance heating is used because, among other things, it provides relative ease of temperature control. The heater is a coil of high-resistance, high-temperature wire or rod surrounding the operating region of the furnace and driven by a controlled current. Relatively inexpensive, high-temperature alloys are available for the heating coils, e.g., Kanthal A (73Fe/22Cr/5Al by weight, MP = 1510°C) [1]. The heater support tube may be made of an SiO_2/Al_2O_3 mix such as Alundum that may contain some alkalis.

Since a furnace operates at hundreds of degrees above ambient temperature, thermal insulation is needed to prevent excessive heat loss. Typical materials are soft ceramic fire brick or Al_2O_3 or SiO_2 *wool*. Usually, a shroud surrounding a cluster of furnaces is cooled with circulating air to prevent excessive warming of the ambient work space. A typical grouping would involve six furnace units, arranged in two side-by-side stacks, each three furnaces high. This arrangment has the added advantage of saving floor space in the clean room.

A fused silica (commonly, but incorrectly, called quartz) tube is located coaxially within the heating coil. This *furnace tube*, shown in Figure H–1, houses the boat of wafers, and appropriate processing gases flow through it. Several less expensive, near-fused silica tube materials are available from several vendors.

The tube serves as a confinement vessel for the wafers and processing gases, including the dopant species, and shields the wafers against contaminants from the heater, its supports, and the ambient atmosphere. A 1-mm-thick silica tube wall provides excellent protection against permeation through the wall by Si, B, P, As, and Sb (the common Si technology materials), but Ca and Na at 1280°C will permeate through in a few minutes and a few ms, respectively [2]. These alkalis cause wafer contamination and hasten devitrification of the

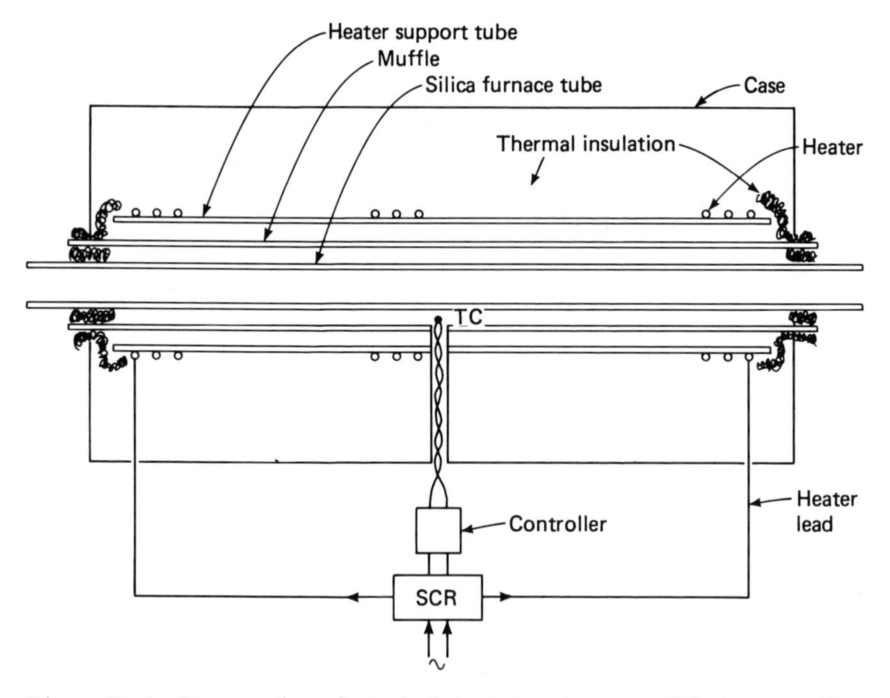

Figure H–1. Cross section of a typical single-heating-zone diffusion or oxidation furnace.

furnace tube, a subject discussed later. Therefore, another alkali barrier is needed.

A *muffle* tube, located between and coaxial to the furnace tube and heater tube, provides the additional protection and also smoothes out small temperature variations within the furnace tube. (These arise because adjacent turns of the heater are some distance apart.) One material of choice for the muffle is Mullite, a very dense, alkali-free SiO_2/Al_2O_3 ceramic.

A feedback loop is used to control the center temperature in the furnace. Figure H–1 shows the loop comprising a thermocouple (TC), used as the temperature sensing element, a controller, and silicon-controlled rectifiers (SCRs) that regulate the flow of power from the supply lines to the furnace heating coil. Ideally, the TC should be located inside the furnace tube, as near to the wafers as possible. If feasible, the TC wires would run into the tube from the open end where the boat is inserted. Where this is not practicable, the TC is located between the silica and muffle tubes as shown in Figure H–1. Such a location requires it to be calibrated against another TC, temporarily placed within the furnace tube, so the control unit may be offset to compensate for the temperature difference between the two locations.

For the range 900 to 1450°C, type R and S TCs, which are formed of platinum/platinum-rhodium wires, are used commonly. For temperatures below 1250°C the less expensive type K, with chromel/alumel wires, is satisfactory. Due to high cost, TC wires are used only in high-temperature regions. They are connected to the controller by matching *extension wires* (designated RX, SX, and KX as the case may be). These are less expensive, and over a limited temperature range around room value, do not form extraneous TC junctions with the R, S, and K types of wires [3].

Because of heat loss from and resulting temperature drop near the furnace tube ends, two methods are used to lengthen the flat zone of temperature. First, the furnace may be made longer wrt the desired flat zone and thermal plugs may be used at the ends. Note also in Figure H–1 that flexible thermal insulation, such as *quartz wool*, is fitted tightly between the muffle and both the furnace case and the furnace tube to further reduce end heat loss.

Secondly, multiple heater zones may be used. For example, in a three-zone furnace, three separate heaters are provided, each with its own TC and control system. The main central unit would be centered on the flat zone and typically might extend over roughly 75% of the furnace tube length. Two other heaters would be located one near each end. Higher power may be supplied to the two end heaters to compensate for the greater heat loss at the ends. Even more zones may be used; five are not unusual in long commercial units.

Recollect from Chapter 5 that vitreous materials gradually soften over a wide temperature range as they are heated. Thus, the furnace tube during oxidation or diffusion will be operating in a range where it really is a supercooled liquid. It will soften to some extent and therefore will deform over a period of time by sagging. Sag takes place along the length of the tube with the center gradually dropping relative to the ends in *beam-bending sag*. The tube also sags along its vertical diameter giving *out of round* sag.

Once a furnace tube has been raised to operating temperatures above about 900°C, sags of both types can be minimized by subsequently keeping the tube temperature from dropping below 600°C between runs. Cycling between operating temperatures and below 600°C also exacerbates *devitrification* of the furnace tube, whereby the material becomes less glasslike and eventually fails.

The failure mechanism has been described this way: Amorphous silica held at high temperatures slowly starts to crystallize into a high-temperature, β-cristobalite phase, usually along the tube's outer surface. At 573°C there is a transition of thermal expansion coefficient. At 272°C another thermal expansion coefficient transition takes place, accompanied by a phase transition from β to α, a low-temperature, cristobalite form that occupies less volume. Repeated cycling between the high and low temperatures causes continued growth of the outer crystalline layer on the tube, and the formation of minute cracks. If temperature cycling is repeated excessively, and if the outer layer gets thick enough, the tube can break [4]. Therefore, furnaces usually are not turned off during long intervals between diffusions, but rather are held above the critical value at about 600°C. The cost of electrical power to sustain this high (wrt room value) temperature when the furnace is on standby is well below the cost of frequent silica tube replacement.

The devitrification process is seeded by contamination on the tube surface, particularly by alkalis. Moisture apparently clusters on these spots and starts the crystallizing process if the temperature is above about 1000°C.

Bare hands can leave alkali deposits; hence, furnace tubes should be handled (cool, of course) with gloves to prevent skin contact with the silica. Periodic cleaning with HF/HNO_3 mixtures, followed by several thorough DI rinses, remove contaminants and thin layers of silica that have been converted to crystalline phases [5]. This can extend tube life. The etching process also can remove predep material that deposits on the inner surface of the furnace tube. This is particularly important in a boron predep furnace: a heavy boron deposit in the tube can cause the boat to stick, making its movement extremely difficult.

References

[1] J. H. Beck, "Metallic Heaters for Electric Furnaces," *Solid State Techol.*, **20**, (11), 59–65, Nov. 1977.

[2] Vendor data on TO-7 silica furnace tubes, Heraeus Amersil, Inc., Sayreville, NJ.

[3] *Complete Temperature Measurement Handbook and Encyclopedia.* Stamford: Omega Engineering, Inc., 1985.

[4] N. G. Grafton and D. J. Rennie, "Larger Quartz Shapes Aid in Lowering Chip Cost," *Electron. Packag. and Prod.*, **14**, (6), 134–136, June 1974.

[5] *Cleaning of Clear Fused Quartz and Fused Silica.* Sayreville, NJ: Heraeus Amersil, Inc.

Index